Modeling and Simulation of Computer Networks and Systems

Methodologies and Applications

Modeling and Simulation of Computer Networks and Systems

Methodologies and Applications

Edited by

Mohammad S. Obaidat

Petros Nicopolitidis

Faouzi Zarai

AMSTERDAM • BOSTON • HEIDELBERG • LONDON
NEW YORK • OXFORD • PARIS • SAN DIEGO
SAN FRANCISCO • SINGAPORE • SYDNEY • TOKYO
Morgan Kaufmann is an imprint of Elsevier

Executive Editor: Steven Elliot
Editorial Project Manager: Benjamin Rearick
Project Manager: Punithavathy Govindaradjane
Designer: Maria Inês Cruz

Morgan Kaufmann is an imprint of Elsevier
225 Wyman Street, Waltham, MA 02451, USA

Notices

Knowledge and best practice in this field are constantly changing. As new research and experience broaden our understanding, changes in research methods, professional practices, or medical treatment may become necessary.

Practitioners and researchers must always rely on their own experience and knowledge in evaluating and using any information, methods, compounds, or experiments described herein. In using such information or methods they should be mindful of their own safety and the safety of others, including parties for whom they have a professional responsibility.

To the fullest extent of the law, neither the Publisher nor the authors, contributors, or editors, assume any liability for any injury and/or damage to persons or property as a matter of products liability, negligence or otherwise, or from any use or operation of any methods, products, instructions, or ideas contained in the material herein.

ISBN: 978-0-12-800887-4

British Library Cataloguing-in-Publication Data

A catalogue record for this book is available from the British Library

Library of Congress Cataloging-in-Publication Data

A catalog record for this book is available from the Library of Congress

For information on all Morgan Kaufmann publications, visit our website at www.mkp.com

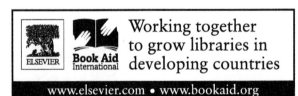

Contents

PART 3 MODELING APPROACHES OF COMPUTER NETWORKS AND SYSTEMS

PART 4 SIMULATION METHODOLOGIES IN COMPUTER NETWORKS AND SYSTEMS

PART 6 MODELING AND SIMULATION FOR SYSTEM SECURITY

List of contributors

Fatih Alagöz
Bogazici University, Istanbul, Turkey

Khitem Ben Ali
University of Sfax, Sfax, Tunisia

M. Amoretti
Università degli Studi di Parma, Parma, Italy

Alagan Anpalagan
Ryerson University, Toronto, ON, Canada

Zemre Arslan
Istanbul Technical University, Ayazaga, Istanbul, Turkey

Hakim Badis
University Paris-Est (UPEM), Cité Descartes, Marne-la-Vallée, France

Şerif Bahtiyar
Provus-A MasterCard Company, Ayazaga, Istanbul, Turkey

Igor Bisio
University of Genova, Genova, Italy

Paolo Bocciarelli
University of Roma "Tor Vergata", Roma, Italy

Meriam Bouzouita
University of Rennes, Rennes, France; Higher School of Communication of Tunis (SUPCOM), Ariana, Tunisia

Berk Canberk
Istanbul Technical University, Ayazaga, Istanbul, Turkey

Lorenzo Carlà
University of Florence, Florence, Italy

Rodrigo Castro
University of Buenos Aires, Buenos Aires, Argentina; National Scientific and Technical Research Council (CONICET), Buenos Aires, Argentina

Soumaya Cherkaoui
Université de Sherbrooke, Sherbrooke, QC, Canada

Francesco Chiti
University of Florence, Florence, Italy

Teong Chee Chuah
Multimedia University, Cyberjaya, Selangor, Malaysia

Andrea D'Ambrogio
University of Roma "Tor Vergata", Roma, Italy

Masoud Daneshtalab
University of Turku, Turku, Finland

Stefano Delucchi
University of Genova, Genova, Italy

Ousmane Diallo
Instituto de Telecomunicações, University of Beira Interior, Covilhã, Portugal; University of Assane Seck of Ziguinchor, Ziguinchor, Senegal

Josep Domènech
Universitat Politècnica de València, València, Spain

Imen El Bouabidi
University of Sfax, Sfax, Tunisia

Muge Erel
Istanbul Technical University, Ayazaga, Istanbul, Turkey

Nicolò Facchi
CNIT/University of Brescia, Brescia, Italy

Romano Fantacci
University of Florence, Florence, Italy

Diogo A.B. Fernandes
University of Beira Interior, Covilhã, Portugal

G. Ferrari
Università degli Studi di Parma, Parma, Italy

Mário M. Freire
University of Beira Interior, Covilhã, Portugal

Daniel F. García
University of Oviedo, Asturias, Spain

José Antonio Gil
Universitat Politècnica de València, València, Spain

Juan C. Granda
University of Oviedo, Asturias, Spain

Francesco Gringoli
CNIT/University of Brescia, Brescia, Italy

Tarik Guelzim
Monmouth University, West Long Branch, NJ, USA

Gürkan Gür
Provus-A MasterCard Company, Ayazaga, Istanbul, Turkey; Bogazici University, Istanbul, Turkey

Yassine Hadjadj-Aoul
University of Rennes, Rennes, France

Sofiane Hamrioui
UHA University, Mulhouse Cedex, France; UMMTO, Tizi-Ouzou, Algeria; USTHB, Bab Ezzouar, Algeria

Pedro R.M. Inácio
University of Beira Interior, Covilhã, Portugal

Vasiliki L. Kakali
Aristotle University, Thessaloniki, Greece

Maryam Kamali
University of Liverpool, Liverpool, UK

Ernesto Kofman
National University of Rosario, Rosario, Argentina; French-Argentine International Center for Information and Systems Sciences (CIFASIS-CONICET), Rosario, Argentina

Constantine A. Kyriakopoulos
Aristotle University, Thessaloniki, Greece

Mustapha Lalam
UMMTO, Tizi-Ouzou, Algeria

Fabio Lavagetto
University of Genova, Genova, Italy

Ying Loong Lee
Multimedia University, Cyberjaya, Selangor, Malaysia

Jaime Lloret
Universidad Politècnica de València, València, Spain

Jonathan Loo
Middlesex University, London, UK

Pascal Lorenz
UHA University, Mulhouse Cedex, France

Valeria Loscri
Inria Lille-Nord Europe, Villeneuve d'Ascq, France

Wahida Mansouri
University of Sfax, Sfax, Tunisia

Mario Marchese
University of Genova, Genova, Italy

Arash Maskooki
Inria Lille-Nord Europe, Villeneuve d'Ascq, France

Nathalie Mitton
Inria Lille-Nord Europe, Villeneuve d'Ascq, France

Lynda Mokdad
Université Paris-Est, Créteil, France

Quentin Monnet
Université Paris-Est, Créteil, France

Eugene David Ngangue Ndih
Université de Sherbrooke, Sherbrooke, QC, Canada

Miguel Neto
University of Beira Interior, Covilhã, Portugal

Pelayo Nuño
University of Oviedo, Asturias, Spain

Mohammad S. Obaidat
Monmouth University, West Long Branch, NJ, USA

Yusuf Ozcevik
Istanbul Technical University, Ayazaga, Istanbul, Turkey

Georgios I. Papadimitriou
Aristotle University, Thessaloniki, Greece

Stefano Paris
Huawei Technologies Co. Ltd., Paris, France; Paris Descartes University, Paris, France

Tommaso Pecorella
University of Florence, Florence, Italy

Raúl Peña-Ortiz
Universitat Politècnica de València, València, Spain

Luigia Petre
Åbo Akademi University, Turku, Finland

M. Picone
Università degli Studi di Parma, Parma, Italy

Ana Pont
Universitat Politècnica de València, València, Spain

Giancarlo Portomauro
University of Genova, Genova, Italy

Abderrezak Rachedi
University Paris-Est (UPEM), Cité Descartes, Marne-la-Vallée, France

Joel J.P.C. Rodrigues
Instituto de Telecomunicações, University of Beira Interior, Covilhã, Portugal

Gabriele Sabatino
Inria Lille-Nord Europe, Villeneuve d'Ascq, France

Julio Sahuquillo
Universitat Politècnica de València, València, Spain

Salima Samaoui
University of Sfax, Sfax, Tunisia

Dharmaraja Selvamuthu
Indian Institute of Technology Delhi, New Delhi, India

Mbaye Sene
Université Cheikh Anta DIOP (UCAD), Dakar, Senegal

Kaisa Sere
Åbo Akademi University, Turku, Finland

Nitin Sharma
BITS, Pilani, RJ, India

Liliana F.B. Soares
University of Beira Interior, Covilhã, Portugal

Francisco J. Suárez
University of Oviedo, Asturias, Spain

Sami Tabbane
Higher School of Communication of Tunis (SUPCOM), Ariana, Tunisia

Kishor S. Trivedi
Duke University, Durham, NC, USA

Kalyanaraman Vaidyanathan
Oracle Corporation, San Diego, CA, USA

Emmanouel (Manos) Varvarigos
Patras University, Patras, Greece

Anna Maria Vegni
Roma Tre University, Rome, Italy

César Viho
University of Rennes, Rennes, France

Nawel Zangar
Higher School of Communication of Tunis (SUPCOM), Ariana, Tunisia

F. Zanichelli
Università degli Studi di Parma, Parma, Italy

Sandro Zappatore
University of Genova, Genova, Italy

Faouzi Zarai
University of Sfax, Sfax, Tunisia

Preface

OVERVIEW AND GOALS

The increased number of network users, services, and applications, along with the many advancements in information technology, make computer networks and systems essential to the survival of all businesses, organizations, and educational institutions. In fact, continued innovation in the communication systems area has a further reckonable impact on the increased demands for mobility and flexibility in our daily lives. For that reason, many researchers, developers, designers, managers, analysts, and professionals are engaged in optimizing network and system performance and satisfying the varied groups that have an interest in network design, operation and implementation by providing new efficient protocols, designs, services or applications. Each time, they have to prove and validate the effectiveness of their design or proposal.

Modeling and simulation techniques are powerful tools for designing and analyzing complex and dynamic systems in order to predict their behavior and performance under different environments, settings, configurations, scenarios and operating conditions on a time scale. It is, thus, the most conventional approach to developing and testing new protocols for wired, wireless and mobile networks, and generally for all kinds of computer systems. Simulation has proven to be a valuable tool in many areas where analytical methods are not applicable and real experimentation and real testing are not feasible. In the context of computer networks, simulators are used for the development, optimization, and validation of new algorithms, architectures, applications or protocols. Improvements of existing algorithms, as well as testing network capacity and efficiency under specific scenarios, can also be a simulator's task. Certainly, simulators model real-world systems using desired granularity and detail.

The aims of this book are to introduce the fundamental principles and concepts in computer networks and systems modeling and simulation, disseminate recent research and development efforts in this fascinating area, and present related case studies and examples as well as trends and challenges.

This book is intended to introduce a broad array of modeling and simulation issues and trends related to computer networks and systems. It focuses on the methodologies, theories, tools, applications and case studies of modeling and simulation of computer systems and networks. It presents recent related efforts in these areas, especially in academia and industry. Moreover, this book describes the new methodologies for modeling and simulation of emerging computer systems and networks, especially the new generations of wireless and mobile networks and cloud and grid computing systems. Because of the increasing concern for the security of these systems, we dedicated a section to cover performance modeling and simulation of the security aspects of these systems.

The chapters of the book have been written by worldwide experts in the field of modeling and simulation of computer networks and systems. Despite the fact that there are different authors of these chapters, we have made sure that the book materials are as coherent and synchronized as possible in order to be easy to follow by all readers.

We hope that the book will be a valuable reference for students, instructors, researchers and industry practitioners, as well as government agency researchers and developers. Every chapter of the book will be accompanied by a set of PowerPoint slides for use by instructors.

FEATURES OF THE BOOK

There are numerous reasons why this book is needed. We summarize below some of the important features of the book, which we believe will make it a valuable resource for its readers:

- There are many published books in this field; however, most of them focus on specific areas/techniques: some analytic modeling, simulation analysis, measurement and testing, simulation tools, or a specific network or system.
- Computer networks and systems are evolving. Hence, researchers and developers and those who work in this area have to review and update the modeling and simulation tools needed to evaluate them. This book aims to make their task easier.
- New topics such as modeling and simulation of next generation wireless networks and wireless sensor networks have attracted a lot of attention. There are not many books available that deal with these.
- We attempt to write a clear and simple reference directed to any reader's level. Indeed, modeling and simulation of computer networks and systems is a complex topic, as it requires knowledge in areas such as mathematical development, queuing theory and modeling, programming, analysis, and statistical inference, among others.
- This book will contribute immensely in improving international research in computer networks and systems because of its coverage of different, new and innovative topics. The authors of the chapters are worldwide technical leaders in the field and will share their expertise and research and development outcomes with the readers.
- The chapters of this book are written by prominent academicians/researchers and practitioners, with solid experience in modeling and simulation, wireless communication, networking, computer systems and pervasive computing areas, who have been working in these areas for many years and have a thorough understanding of the concepts and practical applications of these fields.
- We have chosen more important and timely topics in computer networks and systems. Many critical questions about modeling, simulation, analysis and

security of wireless and mobile networks, especially for next generation wireless networks and new computer systems, are answered in this book.

- This book provides the methodologies, tools, and strategies needed to build computer network and system modeling and simulation from the bottom up, as well as providing solutions to some specific cases.
- This book includes a review and evaluation of many existing simulation tools and methodologies.
- This book takes into account different network performance metrics: mobility, congestion, quality of service, security, among others.
- This book offers a study of modeling methodology examples that can be applied in steps to tackle any computer system and network modeling problem in practice.
- This book provides the reader with the fundamentals of the technologies involved and the related applications.
- This book brings together the most important breakthroughs and recent advances in each of these fields and presents them in a coherent fashion, highlighting the strong interconnections between works in different areas.
- The authors of the chapters of the book are distributed worldwide in a large number of countries and most of them are affiliated with institutions of worldwide reputation. This gives this book an international flavor.
- Each chapter contains a comprehensive bibliography, which should greatly help interested readers to further dig into the topics.
- Each chapter of the book will be accompanied by a set of PowerPoint slides that can aid in the classroom as well as in continuing education and self-study.

ORGANIZATION AND SCANNING OF CHAPTERS

The book is organized into 30 chapters, with each chapter written by topical area expert(s). The chapters are grouped into six parts.

Part I is devoted to topics on modeling and simulation of protocols and services in computer networks and systems. It is composed of six chapters: Chapters 1–5.

Chapter 1 deals with the evolution of wireless and mobile networks, their basic principles of operation, their architectures and modeling and simulation aspects. It presents various network layer protocols and a comprehensive taxonomy of related simulation tools and an analysis and comparison between different simulators. Furthermore, it presents simulation case studies of wireless and mobile networks.

Chapter 2 focuses on simulating the procedure of energy-aware lightpath routing and establishment by means of software tools and concludes that it can lead to decisions about the characteristics, a heuristic method one should use to fulfill one's purpose. This is important as energy efficiency in backbone optical networks can reduce the carbon footprint while preserving performance levels. When embedded in computational logic, core functional traits like lightpath routing will be performed, aiming at better resource utilization from the perspective of energy consumption.

Chapter 3 deals with the interactions between the transport and routing protocols layers. The aim is to provide better planning and design for MANETs (Mobile Ad hoc NETworks). The obtained cross-layer solution is called CL-TCP-OLSR and is evaluated under different network conditions. The obtained performance evaluation results show that CL-TCP-OLSR improves TCP throughput, TCP end-to-end delay, and the energy consumption in the network.

Chapter 4 introduces new extensions to Network Simulator (ns-2) for implementing GATS. It also shows the design of a novel application programming interface (API) that permits seamless integration of new multicast rate adaptation mechanisms by exploiting GATS feedback. This is significant as the widespread utilization of wireless local area networks as leading technology for delivering multimedia streams pushed IEEE to ratify the new 802.11aa amendment that enables reliable multicast transmissions over the air. The new feedback-based Group Addressed Transmission Services (GATS) also permits the development of advanced rate adaptation algorithms for multicast traffic; though not explicitly contemplated by the 802.11aa document, they are essential to improve the overall network throughput. Nonetheless, their performance analysis results on large-scale wireless networks are impractical due to the lack of a unified simulation framework.

Chapter 5 deals mainly with the analytical modeling of communication capacity, energy consumption and congestion, to effectively exploit software defined radio (SDR) and cognitive radio (CR) in wireless sensor networks and body area networks. The latter have numerous civilian and military applications.

Part II addresses the various approaches in performance evaluation of computer networks and systems. It contains Chapters 6–9.

Chapter 6 devises a new testbed that has the ability to reproduce different types of web workloads. After the validation process, the testbed is used to analyze the effect of applying dynamic workloads on the web performance metrics, instead of traditional workloads. The chapter introduces the workload generator used and the testbed design. Moreover, it describes the validation process. Traditional schemes of conducting performance evaluation on such systems have become obsolete as together with the evolution of the WWW, the behavior of the users has changed dramatically, with increased dynamism.

Chapter 7 presents an analysis of the modeling and simulation paradigm applied to assess the performance of computer networks. First, a theoretical background on performance modeling and model validation is introduced. Similarly, well-known fundamentals of event-based simulation are discussed concisely. The set of metrics that can be used at each network layer to assess the performance of a simulated computer network is also described and summarized. Since discrete-event simulation is the most common approach to performing network simulation, the most important network simulators based on such an approach are reviewed and the architecture of a reference network simulator is also depicted. Finally, the chapter presents a case study of performance evaluation using modeling and simulation (M&S) of an overlay network for multimedia interactive communications.

Chapter 8 presents the current state of the art in characterizing and generating workloads for web performance evaluation. First, it reviews a representative subset of the most relevant perspectives to define web workloads, and analyzes the main drawbacks that we have to tackle in order to obtain representative workloads for current web applications. Next, it evaluates and classifies the most commonly used software tools proposed in the open literature according to their main features and ability to generate workloads for the dynamic Web. Finally, the chapter introduces the Dynamic WEB workload model (DWEB), with the aim of characterizing a more realistic workload when evaluating the performance of current web applications.

Chapter 9 proves that the web user's dynamic behavior is a critical issue that must be addressed in web performance studies in order to precisely estimate systems' performance indexes. To this end, it first analyzes and measures the effect of considering different levels of dynamic workload on web performance evaluation, instead of traditional workloads. Then, it analyzes and measures the effect of considering the *user–browser interaction* (UBI) as a part of the user's dynamic behavior on web workload characterization in performance studies.

Part III addresses modeling approaches of computer networks and systems. It consists of four chapters: Chapters 10–13.

Chapter 10 presents a study to assess whether self-similarity is indeed embedded in traffic produced by popular network simulators, namely ns-3 and OMNeT++, and discusses the values for the Hurst parameter obtained using different estimators and for the autocorrelation structure under various network scenarios.

Chapter 11 first defines briefly the fundamental concepts of modeling and performance evaluation of systems with their link to Petri nets, and afterward describes the fundamental concepts of Petri nets, from their applications, ordinary Petri nets (graphical representation and formal definition), stochastic Petri nets, generalized stochastic Petri nets, colored Petri nets, to Stochastic Well-formed Petri Nets, which offer high-level modeling.

Chapter 12 proposes a proving methodology for a communication paradigm for manycore architectures based on the Event-B formal method. The approach is fundamentally reusable, thus addressing the increasing complexity of manycore architectures as well as alleviating the need for proving expertise. These matters are also facilitated by tool support, in the form of the Rodin tool platform.

Chapter 13 proposes the use of Markov chain modeling in preventive maintenance of operational software systems. Its main contribution is the introduction and analysis of Markov modeling techniques for the preventive maintenance of operational software system. This is crucial since the phenomenon of software aging is of great importance, as the state of a software program gradually degrades with time and, if untreated, eventually leads to a crash/hang failure.

Part IV deals with simulations methodologies in computer networks and systems. It is composed of five chapters: Chapters 14–18.

Chapter 14 exploits distributed simulation as a valuable technique to deal with the intrinsic distributed nature of SOA-based systems. Unfortunately, the adoption of distributed simulation requires a significant expertise and a considerable effort due to the complexity of presently available distributed simulation standards and technologies. Hence, the chapter introduces an automated method to reduce the effort and make distributed simulation easier to use. The method takes as input a design model of the SOA-based software system and produces as output the corresponding simulation model, ready to be deployed and transparently executed onto either a local or a distributed simulation engine.

Chapter 15 deals with the most relevant simulation frameworks for wireless systems. The chapter introduces and compares these frameworks. This is of great importance as the standardization activities within the wireless domain require a high degree of interoperability, thus leading to open standards. Even though the benefits are evident, the overall system is usually more complex, with tightly coupled protocol interactions. For this reason, protocol suite optimization is still an open issue. In particular, it is essential to carefully investigate the overall resulting performance before industrial development and real-world deployment of the system. Since a real testbed is often extremely costly, an Open Source network simulator can represent a valid alternative to real device development and testbed deployment for academic and industrial research goals.

Chapter 16 investigates existing approaches on co-simulation of wireless and mobile systems. It then focuses on a recently adopted co-simulation approach, allowing individual components to be simulated by different simulation tools, exchanging information in a collaborative manner.

Widely known discrete event simulation tools, such as ns-2, ns-3, and OMNeT++, are highly specialized for communication networks. As they are not general purpose, they can hardly support the analysis of large-scale distributed applications. Conversely, general-purpose tools like DEUS and CD++ are not provided with all the needed options and scenarios. To fill the gaps of the two families of discrete event simulators, a co-simulation (co-operative simulation) approach has been introduced, which is expected to be very efficient.

Chapter 17 presents some methods, techniques and tools for simulating computer systems, with special emphasis on discrete-event systems such as wired/wireless networks of computer systems. This is important as the success of a simulation method or technique basically depends on the accuracy obtained in modeling the system within the simulation tool, the computational resources required and the time elapsed to complete the simulation.

Chapter 18 presents recently developed theoretical and practical tools for modeling and simulation of hybrid systems with a focus on data networks. An integrative methodology centered on the Discrete EVent Systems specification (DEVS) can enable the M&S of complex systems by choosing the most convenient representation paradigm for each subsystem. This allows discrete time, discrete event and continuous models to coexist. The chapter also focuses on designing Quality of Service (QoS) controllers for data networks at diverse granularity levels, which enables the adoption of control theory.

Part V focuses on next generation wireless network evaluations. It is composed of nine chapters: Chapters 19–27.

Chapter 19 contains the description of a simulative/emulative platform devised by the authors and called "Hybrid Simulative-Emulative Platform (HySEP)." It has been designed and built to simulate long term evolution (LTE) networks by using Network Simulator 3 (ns-3) and to emulate a backhaul network that implements the Differentiated Service (DiffServ) solution to guarantee Quality of Service (QoS). Moreover, this platform enables the formation of a heterogeneous network through the manipulation of personal computers and free software without using any ad-hoc hardware.

Chapter 20 proposes a fluid-based random access model for machine-type communications, used to dynamically determine the value of the Access Class Barring factor that avoids system overload and radio resource underutilization at the same time. The obtained results can help in choosing an appropriate adaptive Access Class Barring factor, which leads to increasing Random Access success probability and hence efficient bandwidth utilization.

Chapter 21 describes the main characteristics and performance requirements that the next generation networks must fulfill. Particularly, the focus is on long term evolution (LTE)/LTE-Advanced technologies where some possible improvements and challenges are explained. Subsequently, the analytical methods and simulation techniques to evaluate the performance of the next generation heterogeneous networks are discussed. Finally, the simulation results for some example scenarios are provided and discussed.

Chapter 22 presents the use of some evolutionary algorithms and associated constrained handling techniques in order to handle the resource allocation problem in emerging OFDMA-based wireless networks. The results obtained though simulation indicate that both the Joint Subcarrier and Power Allocation (JSPA) and Only Subcarrier Allocation (OSA) approaches perform better in terms of sum capacities as compared to linear, modified linear and immune. The sum capacity increases with increased number of users. The sum capacity also increases initially with the increase in number of iterations and population size, but rapidly saturates to a near optimal value. This result suggests that Artificial Bee Colony (ABC) aided resource allocation can provide significant gain in capacity even with a small number of iterations and population size. It can be said that ABC-aided resource allocation is a suitable choice for practical wireless systems like WiMAX (802.16e) where the convergence rate plays a very important role as the wireless channel changes rapidly.

Chapter 23 studies several modeling tools mainly used to evaluate the performance of wireless multi-hop networks. It focuses on Single-Input Single-Output (SISO) and Multi-Input Multi-Output (MIMO) systems. Moreover, it investigates stochastic modeling based on Markov-chain, and conflict graph, particularly graph coloring and asymptotic approaches for large-scale networks. For each tool, it illustrates how it is used in the performance evaluation in terms of throughput and network capacity metrics.

Chapter 24 presents an exploratory treatment of simulation and performance evaluation for resource allocation in LTE femto-cell networks. Two resource allocation schemes, namely centralized dynamic frequency planning (C-DFP) and distributed random access (DRA), are used as reference schemes for simulation modeling and performance evaluation. The chapter outlines the approach used in the two schemes to model and solve the resource allocation problem, and demonstrates the translation of the schemes into network simulation using the open source LTE-Sim simulator.

Chapter 25 deals with the topic of multimedia over wireless networks, focusing on the QoS support for wireless multimedia transmission. After an overview of the wireless networks and the multimedia transmission characteristics, a layered discussion is provided ranging from the application to physical protocol layers. The discussion is extended with a cross-layer perspective entailing two main aspects: cross-layer design and issues inherently spreading multiple network layers. Some of the proposed modifications to these protocols in order to improve multimedia transmission quality in wireless networks are also summarized and investigated. The chapter introduces a number of emerging wireless/mobile networking concepts, including cognitive radio networks, ad hoc and multihop networks and mobile content delivery, providing a discussion of the key challenges and multimedia networking related issues for these systems.

Chapter 26 describes a novel software-defined mobile wireless network management framework that contains topology and admission control mechanisms considering GoS of the overall network and the traffic intensity of the small cells. The described framework observes the Data Plane, and virtualizes the physical infrastructure in the Control Plane with the help of the proposed OpenFlow protocol extension. Performance evaluation shows that the GoS guarantees in the proposed system are better than the ones in conventional wireless system.

Chapter 27 introduces the modeling, analysis, and design of security protocols. It demonstrates why we need security modeling, and then elaborates on two examples: the secure group communication and secure network coding in order to show how to achieve the security goals. Moreover, the chapter introduces the key distribution scheme in group communication, which is based on secret sharing. The proposed protocol achieves key confidentiality due to the security of Shamir's secret sharing, and provides key authentication by broadcasting a single authentication message to all members. Furthermore, the proposed scheme resists against both insider and outsider attacks. The chapter also introduces pollution attacks and entropy attacks, as well as giving a comprehensive description of hemimorphic primitives against these attacks.

Part VI is devoted to modeling and simulation of system security and is composed of three chapters: Chapters 28–30.

Chapter 28 aims at reviewing the applications of wireless sensor networks and investigating the security and energy consumption issues in these systems. By focusing on the network availability, previous studies proposed to protect clustered network against denial of service attacks with the use of traffic monitoring agents on some of the nodes. Those control nodes have to analyze the traffic inside a cluster and to send warnings to the cluster-head whenever an abnormal behavior is detected. However, if the control nodes die out of exhaustion, they leave the network unprotected. Hence, in order to better fight against attacks, this chapter tries to enhance the available solutions by renewing periodically the election process. Furthermore, it proposes two energy-aware and secure methods to designate the cNodes in a hierarchically clustered WSN.

Chapter 29 devises formal schemes for attack modeling and detection. As computer attacks have taken new dimensions with changing complexity, developing formal methods to model them has become crucial. Formal methods allow us to understand an attack vector before it occurs and implement measures to thwart it. Moreover, they permit us to assess the damage caused based on live data analysis if it was successful. The chapter first presents an introduction to computer system attacks and attack modeling. Then it presents formal methods for analysis and prevention.

Chapter 30 sheds some light on the security of computer networks and reviews related concepts and techniques. The security analysis of computer networks has become more essential and more complex due to the proliferation of information and communications systems and applications. This chapter presents a comprehensive review of security analysis of computer networks. Moreover, it discusses key concepts along with methodologies utilized in this field and introduces some emerging topics which will drive future research efforts as well as enable novel approaches and schemes for security analysis of computer networks.

TARGET AUDIENCE

The book is written primarily to target the student community. This includes the students of both undergraduate and graduate levels, as well as students having an intermediate level of knowledge of the topics, and those having extensive knowledge about many of the topics. To achieve this goal, we have attempted to design the overall structure and content of the book in such a manner as to make it useful at all learning levels. The secondary audience for this book is the research community, in academia and industry, as well as practitioners in government agencies. Moreover, we have also taken into consideration the needs of those readers, typically from industry, who need to obtain insight into the practical significance of the topics, expecting to discover how the spectrum of knowledge and ideas presented are relevant for modeling and simulation of computer networks and systems. In order to make the book useful in the classroom, a set of PowerPoint slides has been prepared for each chapter of the book.

ACKNOWLEDGEMENTS

We are enormously thankful to all the authors of the chapters of this book, who have worked very hard to bring forward this unique resource on modeling and simulation of computer networks and systems, to help students, instructors, researchers, and community practitioners. We would like to state that, as the individual chapters of this book were written by different authors, the responsibility for the contents of each chapter lies with the respective authors.

We would like to thank Mr. Steve Elliot, the Elsevier executive acquisitions editor, who worked with us on the project from the beginning, for his suggestions and advice. We also would like to thank Elsevier publishing and marketing staff members, in particular Ms. Kaitlin Herbert, editorial project manager (Elsevier), who left Elsevier before the book was published, but who has helped in the process. Finally, we would like to sincerely thank our respective families, for their continuous support and encouragement during the course of this project.

Mohammad S. Obaidat

Petros Nicopolitidis

Faouzi Zarai

PART 1

Protocols and services in computer networks and systems

CHAPTER

Wireless and mobile technologies and protocols and their performance evaluation

1

Salima Samaoui[1], Imen El Bouabidi[1], Mohammad S. Obaidat[2], Faouzi Zarai[1], and Wahida Mansouri[1]

[1]*University of Sfax, Sfax, Tunisia*

[2]*Monmouth University, West Long Branch, NJ, USA*

1 INTRODUCTION

Currently, wireless and mobile technology has enjoyed rapid growth, an exceptional paradigm shift in design methodology, and considerable improvement in the performance of mobile broadband wireless access technologies, both in professional and general usage. Ensuring communication services anytime, anywhere and even while mobile has been a critical need for connected people. Indeed, the increasing demand for various wireless data and multimedia applications, mobile Internet, and enhanced quality and higher capacity of wireless networks has attracted the attention of the worldwide research community and has become a priority for the standards development organizations.

In parallel with the evolution of the Wireless Local Area Networks (WLAN), we are also seeing an evolution of cellular networks. In fact, the history of cell systems has traditionally been viewed as a sequence of successive generations. The first generation (1G) was entirely analog, whereas the second generation (2G) is a digital system, which appeared in the 1990s with Global System for Mobile Communications (GSM) to improve modulation, voice codecs and security service. This was followed by the third generation (3G), which was envisaged to allow full multimedia data transmission in addition to voice communications. However, 3G systems proved to have limitations with the new multimedia applications, which led to the introduction of the fourth generation (4G) radio access system, referred to as the LTE and LTE-Advanced standards, which are able to offer high bandwidth for the new applications [1].

In order to give background information on this evolution of technologies, this chapter offers a survey of previous, current and emerging wireless and mobile technologies and standards and their performance evaluation using modeling and simulation techniques. We begin by introducing the second generation with its intermediate generations, followed by the 3G system with its associated amendments. Next we describe in more detail the fourth generation (4G) system. We also provide an overview of the Wireless Local Area Network (WLAN) systems, a review of performance metrics and provide case studies. Then we make concluding remarks.

2 WIRELESS AND MOBILE TECHNOLOGIES

2.1 2G AND 2.5G TECHNOLOGIES

While the first generation (1G) mobile network offered a good voice quality, it provided limited spectral efficiency. In fact, the reduced spectral resources, the short battery life and the cost of the terminal were obstacles that have limited the early development of the first generation mobile phones [2]. This is why the evolution towards the second generation (2G) was required in order to overcome the drawbacks of the technology at the beginning of the 1990s. Several digital technologies were developed, including:

- Global System for Mobile Communications (GSM)
- General Packet Radio Service (GPRS)
- Enhanced Data Rates for GSM Evolution (EDGE).

2.1.1 GSM

At first, GSM was designed as a circuit-switched system in a similar way to fixed-line phone networks that establish between two users an exclusive and direct connection on every interface [3]. The GSM chooses a combination of FDMA (Frequency Division Multiple Access) and TDMA (Time Division Multiple Access) to divide the bandwidth in order to exploit the limited radio spectrum resource shared by all users. The GSM architecture consists of four basic components [1]:

- The Mobile Station (MS) consists of the terminal equipment and the Subscriber Identity Module (SIM).
- The Base Station Subsystem (BSS) handles the radio access functions and includes the Base Transceiver System (BTS) and the Base Station Controller (BSC).
- The Network and Switching Subsystem (NSS) is also called the core network, which includes all nodes and functionalities that are necessary for control and switching of calls between different mobile and fixed switching centers and other networks, for subscriber management and mobility management. NSS consists of MSC, which represents a central element responsible for all

processing of voice and data communications, the HLR that contains all subscription details of each subscriber registered in the network, the VLR that contains similar information to the HLR, but on a temporary basis for every active mobile.

- The Operation Sub-System (OSS) contains all the functions necessary for network operation and maintenance. It facilitates the operations of MSCs. The OSS entities are: the authentication center (AuC), which is responsible for the authentication process and security purposes (Sauter, 2011 [3]); and the Operation and Maintenance Center (OMC) which monitors and controls all other GSM network entities (traffic monitoring, status reports of the network entities, subscribers and security management, accounting and billing, among others).

2.1.2 2.5G technology: general packet radio service

The importance of the Internet has shown persistent growth after the transition to the 2G systems. Subsequently, the general packet radio service (GPRS) was developed to enable wireless devices to access to the Internet and transport data in an efficient manner [3]. The principle of GPRS, known also as 2.5G, is to add packet data capabilities by aggregating all time slots together for a single user. This improvement provides data rates up to 140 kb/s [4]. A new class of network stations, named GPRS Support Nodes (GSNs), was introduced. GSNs are in charge of the delivery and routing of data packets between the mobile equipment units and external Packet Data Networks (PDNs). GPRS introduced two new and modified components: the Service GPRS Support Node (SGSN) to enhance the existing GSM infrastructure to facilitate data access at the Radio Interface Technology level, and the Gateway GPRS Support Node (GGSN) to facilitate interconnecting the GPRS network with other data networks, including the Internet [5].

2.1.3 EDGE

Later on, GSM Release '99 introduced a higher data rate than previous generations and provided more enhanced performance in terms of high speed in data service using a higher-level 8-PSK modulation format (up to 473.6 kb/s with uncoded 8-PSK) [4]. This enhancement is called Enhanced Data Rates for GSM Evolution (EDGE), also known as 2.75G; these systems achieved higher bit-rates per radio channel [1,2].

2.2 3G MOBILE TECHNOLOGIES

To meet the ever-rising demand on enhanced data services such as web access and multimedia applications and the convergence of voice and high-speed data services into a single system, the 3G (third generation) wireless communication systems emerged in the late 1990s with greater networking speed and improved multimedia capability. 3G networks are handled under the Third Generation Partnership Project (3GPP), and started with the Universal Mobile Telecommunication System (UMTS). Evolutions of UMTS, such as High Speed Packet Access (HSPA), High

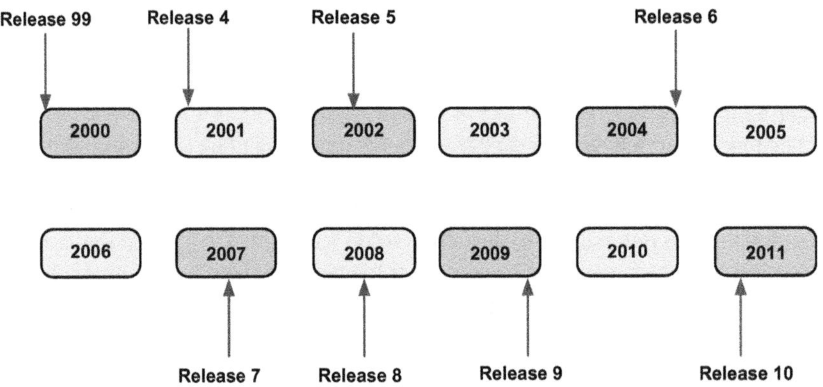

Release 99 :1st UMTS Specifications
Release 4: Originally called Release 2000, Enhancements for the Circuit-Switched Core Network
Release 5: IMS and High-Speed Downlink Packet Access (HSDPA)
Release 6: High-Speed Uplink Packet Access (HSUPA)
Release 7: Even Faster HSPA and Continued Packet Connectivity
Release 8: LTE, Further HSPA and Enhancements and Femtocells
Release 9: Digital Dividend and Dual Cell Improvements
Release 10: LTE advanced fulfilling IMT advanced 4G requirements
Release 11: Advanced IP interconnection of services and non voice emergency services.

FIGURE 1.1

3GPP releases overview.

Speed Packet Access Plus (HSPA+) or 3.5G, have been released as standards providing increased data rates, which enable new mobility for Internet services such as television or high-speed web browsing [1]. (See Figure 1.1.)

2.2.1 UMTS

The main objectives for UMTS systems defined by the International Telecommunications Union are:

- High spectrum efficiency compared to existing systems;
- Interoperability with previous systems such as GSM and GPRS;
- Flexibility to support multiple systems and ensure transparent heterogeneity;
- Supporting packet switched mode to benefit from high-speed transmission rates, diverse applications and allow various subscribers to receive data traffic simultaneously;
- Ensuring fast and seamless handoff by reducing delay and avoiding interruption of service;
- Improving security services, especially in the case of mobility schemes [6].

The first release of UMTS specifications published by 3GPP is referred to as Release 99. UMTS was built on GSM by completely changing the technology used on the air interface from an FDMA/TDMA based system to a Wideband

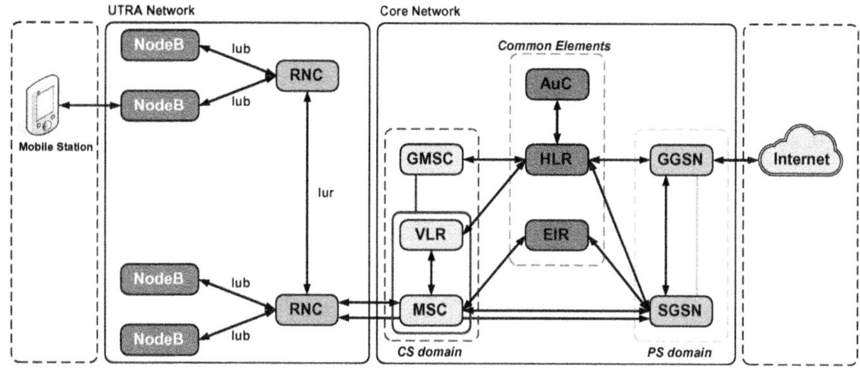

FIGURE 1.2

UMTS architecture.

CDMA (W-CDMA) based air interface, while keeping the core network almost unchanged. With W-CDMA, users are assigned to a single code instead of being separated from each other by timeslots and frequencies. UMTS systems also use a larger bandwidth of 5 MHz for each of the downlinks and uplinks, allowing them to send data with a speed of 384 kbit/s per user in the downlink direction and also a speed of 384 kbit/s in the uplink direction. This amendment was necessary to enable a much faster data transfer than previously possible and to attain a data rate of 2 Mbps connections to be assigned to mobile users [3,7].

Three main components constitute the UMTS system: the UMTS terrestrial radio access network (UTRAN), which handles functions related to radio resources and air interface management; the core network (CN) that performs switching functions and interfaces to external networks such as the Internet; and the user equipment (UE). The UMTS architecture is planned comparably to GSM with some modifications (Figure 1.2). In fact, while its radio access network was completely redesigned, numerous components of the GSM core network that have undergone some modification were reused.

UMTS Radio Network Subsystem: The UTRAN keeps the concept of base stations and controllers from GSM and consists of UMTS base stations called Node B and Radio Network Controller (RNC), respectively [3].

Node B represents the physical entity that contains the transmitter and receiver to communicate with the UEs within the cell. Indeed its principal purpose is to achieve physical layer functions such as modulation, demodulation, coding, interleaving and spreading. In addition, it measures the quality of the connections and transfers the measurement report to the RNC. Moreover, it contributes to power adjustment.

The RNC controls the Node Bs that are connected to it. It is responsible for some mobility management functions, handoff control, handoff decisions, radio resource control, channel allocation, call admission control, encryption, decryption, etc. The RNC communicates not only with the Core Network, but can also communicate with neighboring RNCs to facilitate efficient handover between Node Bs under the control of different RNCs.

UMTS Core Network: The UMTS core network is responsible for transporting user data to the destination. It can operate using both circuit switched mode, which is suitable for voice and limited data communications, and packet switched mode, which is used for other data services. Therefore, the CN may be split into three different areas: circuit switched elements, packet switched elements and shared elements.

- ***Circuit switched elements:*** These entities are principally based on the GSM network entities and transport data using permanent channels throughout the communication. These elements include the mobile switching center (MSC) which is the same as that within GSM; it manages the circuit switched calls underway. The Gateway MSC (GMSC) interfaces to the external networks.
- ***Packet switched elements:*** These entities are designed to transport packet data. They include the Serving GPRS Support Node (SGSN) that provides various functions within the UMTS network architecture such as mobility management, session management, subscriber database management, interaction with other areas of the network and billing; and the Gateway GPRS Support Node (GGSN), which represents the central element within the UMTS packet switched network. It handles inter-working between the UMTS packet switched network and external packet switched networks.
- ***Shared elements:*** The shared elements include the Home Location Register (HLR) that contains all the administrative information about each subscriber along with its last known location; the Equipment Identity Register (EIR), which is the entity that decides whether a given piece of UE equipment may be allowed onto the network; and the Authentication Center (AuC), which is a protected database that contains the secret key also found in the user's USIM card.

2.2.2 High speed downlink packet access

High Speed Downlink Packet Access (HSDPA) enables peak downlink transfer speeds of 14 Mbps (Release 5) and High Speed Uplink Packet Access (HSUPA). The main objectives of HSDPA were to attain a significant increase in network capacity, an increase in peak throughputs and a decrease in latency in the downlink, which were all achieved by introducing new physical and MAC-layer techniques, such as Adaptive Modulation and Coding (AMC), improved scheduling and retransmissions based on Hybrid Automatic Request (HARQ) techniques, and additional channels, all within a new adopted architecture.

2.2.3 High speed uplink packet access

High Speed Uplink Packet Access (HSUPA) was introduced in Release 6 to improve uplink spectral efficiency and further reduce the latency. It has peak data rates of 5.8 Mbps in the uplink [8]. SUPA is the companion technology to HSDPA applied to the uplink transmission directions. HSUPA uses several similar technologies to those applied in HSDPA. Nevertheless, there are some fundamental differences due to the different conditions of links. Although HSUPA provides a considerable growth in the upload speed, it does not provide the same capacity as HSDPA.

2.2.4 HSPA

The 3G system was later developed to improve the performance and provide a higher speed, by introducing the High-Speed Packet Access (HSPA) known as 3.5G technologies (Release 7). HSPA joined the High Speed Downlink Packet Access (HSDPA) and the High Speed Uplink Packet Access (HSUPA). It provides considerable benefits that allow the new service to provide a far better performance for the user. The main benefits include:

- Use of higher order modulation: 16QAM is used in the downlink instead of QPSK to allow data to be transmitted at upper rates.
- Shorter Transmission Time Interval (TTI): Using a shorter TTI decreases the round trip time, enables enhancements in adapting to fast channel variations and reduces latency.
- Use of shared channel transmission: Sharing the resources enables us to achieve high levels of efficiency.
- Use of link adaptation: Adapting the link enables us to maximize the channel usage.
- Fast Node B scheduling: Using fast scheduling with modulation and adaptive coding allows the system to react to the varying radio channel and interference conditions and to receive data traffic which tends to be "bursty" in nature.
- Node B based Hybrid ARQ: HARQ technique reduces retransmission round-trip times and enhances the robustness of the system by permitting soft combining of retransmissions.

2.2.5 HSPA+

Driven by the need for much faster data transfer rates and lower levels of latency, a next evolution of HSPA called HSPA+ (also known as Evolved HSPA) was defined in the 3GPP Release 7 and 8 of the WCDMA specification. Several major new features enable HSPA+ to provide a significant enhancement in performance over that provided by the HSPA systems. These features include the use of Multiple-Input Multiple-Output (MIMO) technologies, higher order modulation schemes (16 QAM (uplink)/64 QAM (downlink)), continuous packet connectivity, enhanced CELL_FACH operation and layer 2 protocol enhancements. HSPA+ provides downlink speeds of 42 Mbps and uplink speeds of 11 Mbps [4].

2.2.6 LTE and LTE-A technologies

To meet the major requirements such as increased data capacity, reduced latency, higher Quality of Service, packet-optimized and more secure service, 3GPP decided to adopt a new approach to the network structure and redesign both the core network and the radio network. The result is referred to as Long Term Evolution (LTE) and LTE-Advanced (LTE-A), which represent a radical step forward for the wireless industry [3]. More details on each of these technologies are given in the following sections.

3 LTE

In fact, LTE was required to offer a peak data rate of 100 Mbps in the downlink and 50 Mbps in the uplink. In addition, it was required to support a spectral efficiency three to four times greater than that of Release 6 W-CDMA in the downlink and two to three times greater in the uplink. Latency represents another central requirement, especially for real-time applications such as voice and interactive games; it should be less than 5 ms. Also, less than 100 ms should be respected when a phone switches from standby to the active state, after an intervention from the user. A fourth requirement related to coverage and mobility is intended. In fact, LTE should be optimized for cell sizes up to 5 km and can work with reduced performance up to 30 km and can even support cell sizes of up to 100 km. Finally, LTE must be designed to work with a diversity of different bandwidths (from 1.4 MHz to a maximum of 20 MHz) [7]. LTE was developed for 4G wireless communication systems adopting the orthogonal frequency-division multiplexing (OFDM) waveform for downlink communications to support wideband transmission and the single-carrier FDM (SC-FDM) waveform for uplink communications, principally to enhance the use for broadband data communications [9]. LTE supports both duplexing modes: the TDD and FDD schemes.

Furthermore, the use of multiple-input multiple-output (MIMO) improves, in a big way, the spectral efficiency. LTE delivers a peak data rate of 100 Mbps in the downlink and 50 Mbps in the uplink [10]. In contrast to 2G and 3G standards, which offer the circuit and packet-switched model, LTE has been designed to support only packet-switched services. It tries to ensure seamless Internet Protocol (IP) connectivity between mobile equipment and the packet data network (PDN), without any interruption to the end users' applications during mobility over the evolved packet core (EPC) [11]. In the LTE architecture, EPC is an IP-based core network that distributes all types of information over IP, including voice.

The EPC's radio communications with mobile equipment is handled by the evolved UMTS terrestrial radio access network (E-UTRAN) which replaces the UTRAN in the previous architecture [7]. EPC principally includes a Mobility Management Entity (MME), a Serving Gateway (S-GW) that interfaces with the E-UTRAN, and a PDN Gateway (P-GW) that interfaces to external packet data networks. This whole new architecture was composed of two 3GPP parts: the System Architecture Evolution (SAE), which covered the core network, and the Long Term Evolution (LTE), which covered the radio access network, air interface and mobile. The entire system is known as the evolved packet system (EPS).

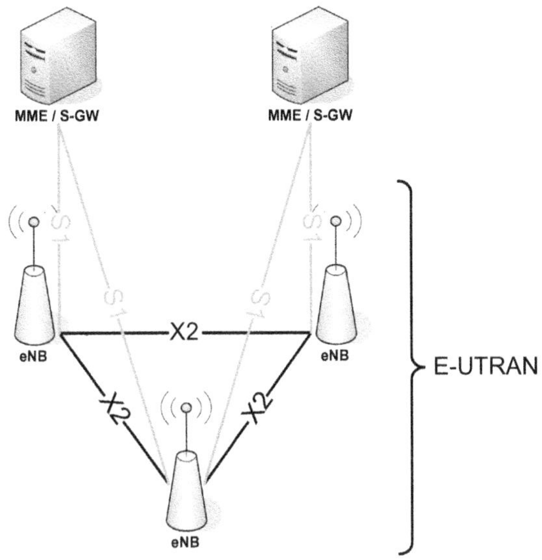

FIGURE 1.3

E-UTRAN architecture.

The rest of this section highlights the design principles of the EPS, principally the radio access network and the core network.

3.1 E-UTRAN

The Evolved UMTS terrestrial radio access network (E-UTRAN) (Release 10) is a new air interface system that offers upper data rates and lower latency. E-UTRAN handles the radio communications between the mobile and the EPC and has just one component, which is the evolved Node B (eNB) [7]. eNB in an LTE network replaces Node B in 3G systems and represents a combination of Node B and the radio network controller (RNC). The eNB interfaces with User Equipment (UE) and can serve one or several cells at one time [9].

Every eNB is connected to the EPC by means of the S1 interface. Precisely, it is connected to the S-GW by means of the S1-U interface and to the MME by means of the S1-MME interface [12]. It can also be optionally connected to another one via the X2 interface, which is mainly used for signaling and packet forwarding during handover [7]. Figure 1.3 shows the E-UTRAN architecture.

3.2 EVOLVED PACKET CORE

The principal component of the SAE architecture is the Evolved Packet Core (EPC), which is realized through the elements described in the following sections.

3.2.1 Serving gateway

The serving gateway (S-GW) acts as a router that is responsible for routing and forwarding packets between UEs and the packet data network (PDN), so it interfaces with the E-UTRAN and a PDN Gateway. Moreover, the S-GW serves as a mobility anchor for handover and interworking with other 3GPP technologies. Therefore, IP data packets are transferred through it. In addition, the S-GW performs several administrative proceedings in the visited network, such as collecting information for charging. Each piece of mobile equipment is assigned to one S-GW, but the S-GW can be switched in case of mobility events [7].

3.2.2 Packet data network gateway

The packet data network gateway (PDN-GW) ensures connectivity to the UE to outer packet data networks, as it represents the EPC's point of contact with the external world. A mobile could have simultaneous connectivity with several PDN-GWs for accessing multiple packet data networks. The PDN-GW is responsible for several functions such as packet filtering, charging support, policy enforcement, packet screening, IP address allocation for the UE, QoS enforcement and lawful interception. The PDN-GW acts also as the anchor for mobility between 3GPP and non-3GPP technologies such as WiMAX [13].

3.2.3 Mobility management entity

The mobility management entity (MME), which presents a key control node for the LTE access network, manages UE access network and mobility, as well as establishing the bearer path for UE's. The MME is also concerned with the bearer activation/deactivation process. In addition, it is responsible for selecting the S-GW for a UE at the initial join and at time of intra-LTE handover. MME is in charge of authenticating the user as well as the generation and allocation of temporary identities to UEs. Furthermore, it verifies the authorization of the UE to camp on the service provider's Public Land Mobile Network (PLMN) and indicates UE roaming restrictions. The Non-Access Stratum (NAS) signaling ends at the MME since it presents the termination point in the network for ciphering/integrity protection for NAS signaling and handles the security key management.

The MME also controls mobility between LTE and 2G/3G access networks. Finally, the MME controls other network elements, by means of signaling messages that are internal to the EPC [13].

3.2.4 Home subscriber server

The Home Subscriber Server (HSS) is a central database that contains information about all the network operator's subscribers as well as any access restrictions for roaming and is shared between the two network domains. It integrates the functions of the Home Location Register (HLR) and Authentication Center (AuC) that generates the vectors for authentication and security keys. In addition, the HSS

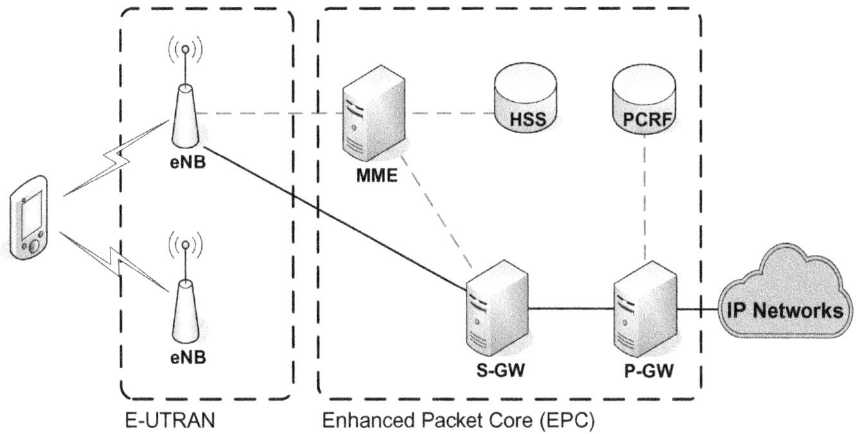

FIGURE 1.4

Overall LTE architecture.

keeps dynamic information such as the identity of the MME to which the user is presently registered or attached [7].

3.2.5 Policy and charging rules function

The Policy and Charging Rules Function entity (PCRF) manages policy and charging rules. It dynamically controls and manages all data sessions and provides appropriate interfaces toward charging and billing systems. In fact, it manages the quality of service required by the network to assign accordingly the appropriate bearer-level QoS parameters that must be in conformity with the user's subscription profile [12].

Figure 1.4 depicts the overall LTE architecture.

3.3 LTE RADIO PROTOCOL ARCHITECTURE

The radio protocol architecture for LTE can be separated into control plane architecture and user plane architecture. The user plane consists of a set of protocols used to transfer the actual user data through the LTE network, whereas the control plane consists of protocols used to control and establish the user connections and bearers within the E-UTRAN.

At the user plane side, the application creates data packets that are processed by protocols such as TCP, UDP and IP, while in the control plane, the radio resource control (RRC) protocol writes the signaling messages that are exchanged between the base station and the mobile. In both cases, the information is processed by the packet data convergence protocol (PDCP), the radio link control (RLC) protocol and the medium access control (MAC) protocol, before being

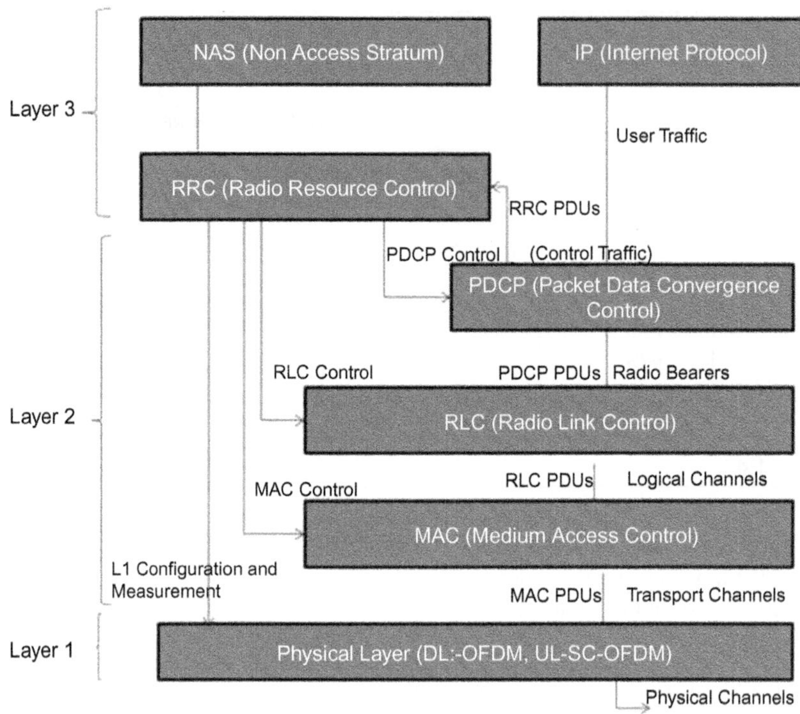

FIGURE 1.5

Diagram of E-UTRAN Protocol stack.

passed to the physical layer for transmission. Figure 1.5 shows the protocol structure of LTE.

3.3.1 Physical layer

The Physical Layer carries all information from the MAC transport channels over the air interface. It is responsible for the radio related issues: e.g., modulation/demodulation, coding/decoding, and MIMO techniques. The Physical Layer uses different channels: Physical Down Link Shared Channel (PDL-SCH), Physical Broadcast Channel (PBCH), Physical Multicast Channel (PMCH), Physical Uplink Shared Channel (PUSCH) and Physical Random Access Channel (PRACH).

3.3.2 Medium access layer (MAC)

The MAC layer is responsible for mapping between logical channels and transport channels, multiplexing and demultiplexing of upper layer PDUs, scheduling air interface resources in both uplink and downlink, error correction through HARQ, priority handling between UEs by means of dynamic scheduling, and priority handling between logical channels of one UE. The MAC layer is located below the RLC layer and it provides services to the RLC by offering logical channels. According to the 3GPP standards [14], the logical channel types are: Broadcast Control Channel (BCCH), Paging Control Channel (PCCH), Common Control Channel (CCCH), Dedicated Control Channel (DCCH), Multi-cast Control Channel (MCCH), Dedicated Traffic Channel (DTCH), and Multicast Traffic Channel (MTCH). The MAC layer uses the services offered by the physical layer in terms of using the transport channels. The LTE transport channels are: Broadcast Channel (BCH), Downlink Shared Channel (DL-SCH), Paging Channel (PCH), Multicast Channel (MCH), Uplink Shared Channel (UL-SCH), and Random Access Channel (RACH).

3.3.3 Radio link control (RLC)

The RLC Layer is responsible for transfer of upper layer PDUs, concatenation, segmentation and reassembly of RLC SDUs. The RLC also performs error corrections using the well-known Automatic Repeat Request (ARQ) methods.

3.3.4 Packet data convergence control (PDCP)

The PDCP Layer is responsible for header compression and decompression of IP data, reducing the overall overhead, which in turn improves the efficiency over the radio interface. This layer also performs additional functionalities such as ciphering and deciphering of user plane data and control plane data, and integrity protection and integrity verification of control plane data. PDCP is used also for SRBs and DRBs mapped on DCCH and DT CH type of logical channels [2].

3.3.5 Radio resource control (RRC)

The RRC sublayer is responsible for the broadcast of system information related to both the non-access stratum (NAS) and the access stratum (AS), the setup and maintenance of the radio bearers, and security functions. In addition, it controls the periodicity of the Channel Quality Indicator (CQI).

3.3.6 Non-access stratum protocols

NAS protocols support the mobility of the UE and the session management procedures to establish and maintain IP connectivity between the UE and a PDN GW.

4 LTE-ADVANCED

Driven by the International Telecommunication Union's (ITU) requirements for IMT-Advanced, 3GPP has been working principally to extend the coverage, improve the system throughput and enhance the capabilities of LTE. In fact, the set of IMT-Advanced high-level requirements represent an elevated degree of commonality of functionality world-wide while retaining the flexibility to support a large range of services and applications effectively—compatibility of services within all other networks as well as compatibility of the Internet working with other radio access systems, suitability of user equipment for worldwide use, high-quality mobile devices, user-friendly services and applications, worldwide roaming capability, very low latency, more efficient interference management and operational cost reduction, in addition to enhanced peak rates to support advanced services and applications up to 1 Gbit/s [15]. The main output from this study was a specification for a system known as LTE-Advanced (Release 10 and Beyond).

4.1 LTE-A KEY TECHNOLOGIES

LTE-Advanced inherits many features from LTE, but improves them to satisfy the IMT-Advanced requirements. There are several key technologies considered for the LTE-Advanced that include enhanced MIMO techniques, ensuring higher efficiency enabled by enhanced uplink multiple access and enhanced multiple antenna transmission, carrier aggregation ensuring wider bandwidths, wireless relays where different levels of wireless multihop relay will be applied, coordinated multipoint transmission and reception (CoMP), enhanced inter-cell interference coordination (eICIC), support for heterogeneous networks, LTE self-optimizing network enhancements, home enhanced-node-B mobility enhancements and fixed wireless customer premises equipment RF [10].

LTE-Advanced was designed to be compatible with LTE; therefore LTE-Advanced mobile can communicate with a base station that is operating LTE and conversely [7]. LTE-Advanced ensures a peak data rate of 1000 Mbps in the downlink, and 500 Mbps in the uplink. Requirements for latencies are achieved by reducing the transition time from 100 ms in LTE to less than 50 ms for idle to connected and from 50 ms in LTE to less than 10 ms for sleeping to be connected. The system supports up to 500 km/h mobility depending on operating band [6].

4.1.1 Carrier aggregation

One of the major goals of LTE-Advanced is to fully utilize the maximum bandwidth of 100 MHz, which represents an extremely large bandwidth. To deal with this problem, a carrier aggregation scheme has been proposed. Carrier aggregation is a technique where multiple carriers of maximum bandwidth of 20 MHz would be aggregated for the same user equipment (UE). Therefore, LTE-Advanced allows a mobile to transmit and receive on a group of five component carriers (CCs) simultaneously, each of which has a maximum bandwidth of 20 MHz [7].

4.1.2 Enhanced MIMO

Among the several important technologies made by Release 10, we mention the advanced antenna techniques, where multiple transmit and receive antennas and multi-cell MIMO techniques will be applied to offer an enhanced downlink MIMO and enhanced uplink MIMO, improving the downlink and the uplink data rate, respectively [5]. Enhanced MIMO is considered to be one of the major aspects of LTE-Advanced that will allow the system to meet the IMT-Advanced rate requirements.

Enhanced MIMO extends the MIMO capabilities of LTE Release 8 to support eight downlink antennas, allowing the possibility in the downlink of 8 × 8 spatial multiplexing and four uplink antennas allowing the possibility of up to 4 × 4 transmission in the uplink when jointed with four eNB receivers [16].

4.1.3 Relays

Relaying presents another method introduced in LTE-Advanced to improve the performance of LTE. In fact, the concept of the relay node (RN) has been introduced to enable traffic/signaling forwarding between UE and eNB to improve the coverage, urban or indoor throughput, group mobility, cell edge coverage, and to extend coverage to heavily shadowed areas in the cell or areas beyond the cell range. It ensures throughput improvement, particularly for the cell edge users by keeping the cell sizes relatively large [17,18]. In fact, relay nodes (RN) are low-cost devices that act like a repeater between users and eNBs. The basic architecture analyzed for LTE-A consists of a single relay node (RN) that is connected to a donor cell of a donor eNodeB. Figure 1.6 represents the basic method of wireless relaying. Relays can be distinguished based on the layers in which their main functionality is realized.

4.1.4 Coordinated multiple point transmission and reception

Coordinated multipoint transmission (CoMP) is one of the new technologies introduced in LTE-Advanced that is based on orthogonalization techniques to achieve higher spectral efficiency and higher peak rates for normal and edge users. The basic idea is to create multi-cells in which information is shared between eNB and the data transmission. With this method the transmission and reception can be processed as in a MIMO scenario, hence inter-cell interference (ICI) can be rejected and spatial diversity can be used to increase signal-to-noise ratio (SNR).

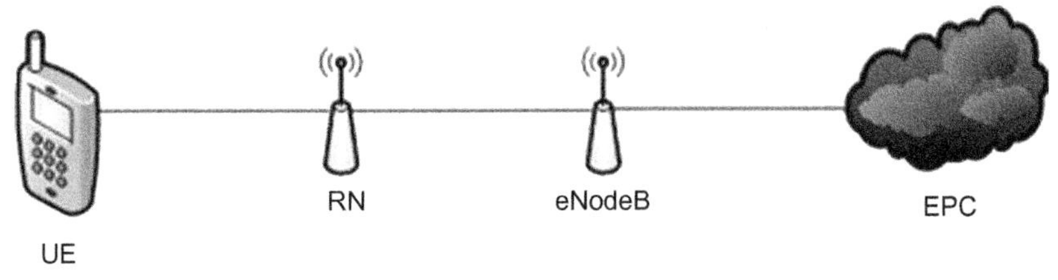

FIGURE 1.6

Wireless relaying.

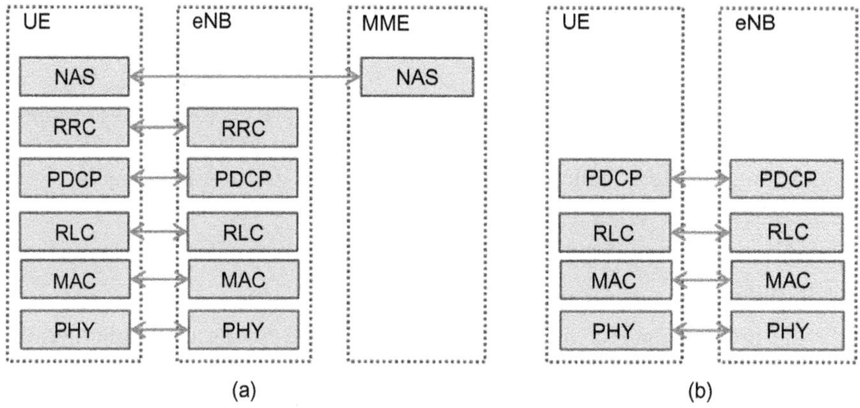

FIGURE 1.7

Protocol stack. (a) Control plane. (b) User plane.

4.2 LTE-A RADIO PROTOCOL ARCHITECTURE

Protocol stacks for the control plane and user plane are illustrated in Figure 1.7 [5]. The protocol for the user plane includes Packet Data Convergence Protocol (PDCP), Radio Link Control (RLC), Medium Access Control (MAC), and the PHY protocol. The control plane stack additionally includes the radio resource control (RRC) and non-access stratum (NAS).

- The Physical Layer (PHY): Like LTE, LTE-A uses OFDM technology to cancel the Inter Symbol Interference (ISI).
- The Medium Access Control (MAC): It performs resource scheduling and Hybrid Automatic Repeat Request (HARQ) for retransmission.
- The Radio Link Control (RLC): It manages the delivery of the data. It is responsible for segmentation of data based on Transport Block Size (TBS).
- The Packet Data Convergence Protocol (PDCP) does ciphering, retransmission and header compression of user data.
- The Radio Resource Control (RRC) manages security functions (authentication, and authorization), handling mobility, roaming, and handovers. The Non-Access Stratum (NAS) is responsible for authentication, registration, connection/session management between UE and the core network.

4.3 COMPARISON OF LTE AND LTE-A

LTE is being enhanced (LTE-Advanced) to support higher peak rates, higher throughput and coverage, and lower latencies, resulting in a better user experience. In fact, there are a lot of similarities in features and there are some differences between the two technologies. Some differences between these two technologies are as follows [9]:

Uplink transmission schemes. LTE uses single-carrier FDMA (SC-FDMA). In LTE-A, carrier aggregation is supported and the single-carrier property of SC-FDMA is no longer preserved.

Channel bandwidth. LTE supports a flexible transmission bandwidth from 1.4 MHz, 3 MHz, 5 MHZ, 10 MHz up to 20 MHz. However, carrier aggregation is used to increase bandwidth up to 100 MHz.

Multi-antenna support. In LTE up to five principal multiple-antenna modes are supported in the downlink, namely transmit diversity, open-loop spatial multiplexing (OLSM), closed-loop spatial multiplexing (CLSM or single-user MIMO (SU-MIMO)), multi-user MIMO (MU-MIMO), and UE-specific reference−symbol-based beam forming. In the uplink, LTE supports one transmit and up to eight receive antennas, and the only multi-antenna scheme supported on the uplink for this technology is MU-MIMO. In LTE-A, the downlink spatial multiplexing scheme was extended to support 8×8 MIMO and enhanced MU-MIMO based on dedicated reference symbols. In LTE-A, uplink spatial multiplexing supporting up to four streams is introduced. Hence the peak data rate supported on the uplink is quadrupled. Also, transmit diversity is supported for the LTE Rel-10 control channel.

Downlink and uplink pilot structure. The LTE supports both common and dedicated reference signals, and the pilot structure is based on frequency-division multiplexing/time-division multiplexing both for the downlink and for the uplink. In LTE-A, the dedicated reference signal is extended to support up to eight streams. A feature comparison between LTE and LTE-A is summarized in Table 1.1 [10].

5 WIRELESS LOCAL AREA NETWORK

The IEEE 802.11 standard committee, whose chief task is producing technical specifications for WLAN implementation, released its first standard "IEEE 802.11" in 1997. The first version of the standard used two radio frequency specifications: direct sequence spread spectrum (DSSS) and frequency hopping spread spectrum (FHSS) operating at 1 Mbps or 2 Mbps over 900 MHz.

The most popular standard is 802.11b, which occupies 83.5 MHz from 2.4 GHz to 2.4835 GHz, and provides 11 channels at 5 MHz intervals. 802.11b is capable of providing peak data rates of 1 to 2 Mbps using DSSS and 5.5 to 11 Mbps in a modified mode called complementary coded keying (CCK). In addition, other 802.11 standards are being developed that add new quality of service features, extend the physical layer options, and provide better interoperability [19].

Table 1.1 Comparison of LTE and LTE-A

Features	LTE	LTE-A
Peak data rate	Downlink: 300 Mbps Uplink: 75 Mbps	Downlink: 1 Gbps Uplink: 500 Mbps
Modulation	Downlink: /UL QPSK,16QAM	64QAM
Peak spectrum efficiency [bps/Hz]	Downlink: 15 Uplink: 3.75	Downlink: 30 Uplink: 15
Network architecture	Very-flat, IP-based eNB + S-GW	Same as for LTE. For heterogeneous network architecture may be different.
Access technology	Downlink: OFDMA Uplink: SC-FDMA	Downlink: Same as for LTE. Uplink: Single-carrier property is not preserved for SC-FDMA uplink.
Channel bandwidth	1.4, 1.6, 3, 5, 10, 15, & 20 MHz	Additionally supports up to 100 MHz for downlink and 40 MHz for uplink with carrier aggregation.
Downlink pilot structure	TDM, common, and dedicated pilots	Dedicated pilot support of up to eight streams, CSI-RS (channel state information reference signal) support for eight antennas
Uplink control channel	FDM, data and control are not transmitted together	Data and control can be transmitted together.
Uplink power control	Fractional OL PC with closed-loop correction, inter-cell interference mitigation using X2 interface	Same as for LTE, modifications to support carrier aggregation
Total overhead	Downlink overhead ~31%–33%	Downlink overhead 25%–28%

Since then, a series of IEEE 802.11 standards have been proposed to improve the performance of WLANs, provide higher data rates and better coverage at very low cost, some of which are listed here:

- **IEEE 802.11a:** It was ratified in 1999 representing a wireless network bearer operating in the 5 GHz ISM band with data rate up to 54 Mbps. IEEE 802.11a provides eight channels using OFDM, with 52 subcarriers spanning over a 20 MHz wide spectrum. Each subcarrier can be modulated with BPSK, QPSK, 16-QAM, or 64-QAM, depending on the wireless environment. 802.11a has less interference than other IEEE 802.11 standards.
- **IEEE 802.11g:** It was ratified in 2003 representing a further higher data rate extension in the 2.4 GHz band with data rate up to 54 Mbps. It is the third modulation standard for WLAN which is compatible with 802.11b. The PHY layer can use either DSSS or OFDM.

- **IEEE 802.11d:** It represents an official amendment to the IEEE 802.11 specification that adds support for additional regulatory domains.
- **IEEE 802.11e:** This is an approved amendment that defines a set of Quality of Service enhancements for the IEEE 802.11 specification by making several modifications to the Media Access Control (MAC) layer. Such enhancements including packet bursting allow the best transmission quality for voice and video applications.
- **IEEE 802.11h:** This is an approved amendment to the IEEE 802.11 specification that adds spectrum and transmit power management extensions.
- **IEEE 802.11i:** This is an approved amendment to the original IEEE 802.11 specification that specifies security mechanisms for wireless networks called Wired Equivalent Privacy (WEP).
- **IEEE 802.11n:** It is an approved amendment to the IEEE 802.11 specification that incorporates multiple-input multiple output (MIMO) technology to improve network throughput. MIMO enhances performance with the use of multiple antennas at both the transmitter and receiver for multiple transmitted data streams, which leads to a considerable increase in data throughput (up to 600 Mbps) without additional cost of bandwidth or transmission power, benefiting from antenna diversity and spatial multiplexing.
- **IEEE 802.11r:** This is an approved amendment to the IEEE 802.11 specification to manage handover in a seamless manner by permitting continuous connectivity aboard wireless devices in motion, with fast and secure handovers from one access point to another.
- **IEEE 802.11u:** This is an approved amendment to the IEEE 802.11 specification to improve interworking with external networks [2].

IEEE 802.11 standards are used both for indoor and outdoor installations. They support two types of networks: Ad-hoc and infrastructure.

In the wireless infrastructure mode, the network consists of a wireless access point (AP) and several wireless clients. The AP coordinates the transmission among stations within its radio coverage area, called Basic Service Set (BSS). It is also responsible for bridging the wireless traffic to the wired local area network as well as being responsible for security management. BSSs are interconnected with each other via a component called a Distribution System (DS). All wireless clients communicate with external networks through the AP. Such a configuration is called an Extended Service Set (ESS).

Ad-hoc networks represent an evolution of infrastructure mode which remove fixed infrastructure. In fact, they consist of self-organized networks, with only wireless clients included which communicate directly with each other without forwarding packets to an access point. This mode is suitable for rapidly setting up a wireless network in a conference room, or anywhere else where enough wired equipment does not exist [2].

6 SIMULATION OF WIRELESS NETWORKS

6.1 WIRELESS NETWORK SIMULATION TOOLS

To simulate wireless networks we can use traditional programming languages such as C++ and Java, simulation languages like SIMSCRIPT III, CSIM, and JAVASIM, and simulation tools like NS2, OPNET, and OMNeT++.

Following is a brief description of examples of these languages and tools.

6.1.1 Network simulator 2

Network Simulator (NS) is simply a discrete event-driven network simulation tool for studying the dynamic nature of communication networks. Network Simulator 2 (NS2) provides substantial support for simulation of different protocols over wired and wireless networks. It provides a highly modular platform for wired and wireless simulations supporting different network elements, protocols, traffic, and routing types [20].

NS2 is a simulation package that supports several network protocols including TCP, UDP, HTTP, and DHCP and these can be modeled using this package [21]. In addition, several kinds of network traffic types such as constant bit rate (CBR), available bit rate (ABR), and variable bit rate (VBR) can be generated easily using this package. It is a very popular simulation package in academic environments.

NS2 has been developed using the C++ programming language and OTcl. OTcl is a relatively new language that uses object-oriented aspects. It was developed at MIT as an object-oriented extension of the Tool command language (Tcl) [21].

6.1.2 Optimized network engineering tool

The Optimized Network Engineering Tool (OPNET) is designed by OPNET Technologies to analyze the performance of communication networks, including wireless systems. The performance is forecasted using discrete event simulation [21]. OPNET Technologies enhances the package continuously and every few years develops a new version of the package. The main features of OPNET are as follows [21]:

1. Simulation and Modeling Cycle: It has powerful means to aid in model building, simulation running and analysis of the simulation outputs.
2. It backs a hierarchical configuration of modeling.
3. It has an efficient set of library modules that support communication protocols and network-related topologies.
4. It has an excellent troubleshooting tool and the model can be easily compiled and run.

OPNET has three sorts of editors for modeling three types of networks [21]:

1. Network topology models are modeled using the Network Editor.
2. Data flow models are designed using the Node Editor.
3. Control flow models are expressed by using the Process Editor.

In OPNET, a number of available tools can be employed to analyze the simulation results. The probe editor is employed for gathering the data. Statistical results are acquired by employing the analysis tool. Processing of data is done using the filter tool. The dynamic actions of the model can be seen using the animation viewer. Entities in the network model are represented as nodes and interaction between the entities is managed with the aid of a link. In order to send data from one entity to all others, we use bus link and radio link for mobile communication [21].

OPNET enables the possibility of simulating entire heterogeneous networks. Simulation in OPNET operates at "packet-level." The main difference with other simulators lies in its power and versatility. This simulator makes it possible to work with OSI models, from layer 7 to the modification of the most essential physical parameters.

6.1.3 OMNeT++

OMNeT++ is an open-source discrete-event simulation tool that is used for simulating computer communication networks, including wireless networks. The programming characteristics in OMNeT++ follow a modular style. It supports three kinds of modules: simple, complex and system modules. Modules in the model communicate with each other by message passing. Modules that are active are named active modules. Complex modules are made by assembling the simple modules. Messages are sent via the gates in the case of simple modules. The input interface and the output interface are called gates and links are used to link input and output gates [21].

Functionalities in the modules can be either co-routine based or event-processing based. In the first scheme, the code in the module runs on its own by generating a thread that is accomplished by the kernel, which bypasses the events. In the event-processing function-based approach, the task is called by the kernel that passes the message as an argument. The latter is handled by the function and is sent back. The network topology can be changed dynamically. Furthermore, there is flexibility to include and remove modules when the simulation is running [21].

One nice feature of OMNeT++ is that it offers a standard library that describes some standard modules which can be used during the modeling process of the system under study. The key modules for troubleshooting, tracing and animation are very efficient in OMNeT++. The library comprises the message classes, container classes, routing classes, random number generator classes and statistical classes. The latter classes are utilized for collecting data when the simulation is on the run so as to assess the performance of the simulated system under study. Message classes are used to provide message packets for various kinds of networks. Container classes offer various storing services including queues and stacks and uphold the general actions on these classes. Routing classes provide the foundation for using a range of routing methods for moving the message packets in the network. Tracing and simple debugging are contained as key features of the OMNeT++ package. To trace the behavior of the system,

OMNeT++ uses three methods: automatic animation, module output windows and object inspectors [21].

Whenever we model, we may generate some data textually as a checkpoint for troubleshooting. This type of data that is employed for debugging is shown in the module output window [21]. The status of the object at any point of time may be shown using object monitors.

In general, OMNET++ is an extensible, modular, component-based C++ simulation library and framework, primarily for building network simulators. It features a generic architecture, so it can be employed in various problem domains including wireless communication and networks.

6.1.4 GloMoSim

GloMoSim is a simulation tool that is mainly used for simulating wireless networks. It was designed using the parallel discrete-event simulation capability provided by PARSEC. PARSEC (Parallel Simulation Environment for Complex Systems) is a parallel simulation language that is written in C. It was developed by the Parallel Computing Laboratory at UCLA, for sequential and parallel execution of discrete-event simulation models. The name GloMoSim originated from the words Global Mobile system Simulator. The GloMoSim library comprises a set of modules in which each module simulates a specific wireless protocol.

There are two options of GloMoSim operation: one for simulating the models in a shared memory setting, and the other one for simulating the models in a distributed memory setting. It can also be used as a parallel programming language. PARSEC can be used to describe the library in GloMoSim [21]. Rather than coding each and every component, a graphical environment called PAVE is provided, which can be used to develop the simulation models. In PARSEC, each node is defined as an entity [21].

Global Mobile system Simulator (GloMoSim) is designed to be extensible and modular. It effectively utilizes parallel execution to reduce the simulation time of detailed high-fidelity models of large communication networks.

6.1.5 LTE-Sim

LTE-Sim encompasses several aspects of LTE networks, including both E-UTRAN and EPS. It supports single and heterogeneous multi-cell environments, QoS management, multi-user environments, user mobility, handover procedures, and frequency reuse techniques. LTE-Sim is presented to provide a complete performance verification of LTE networks. LTE-Sim has been conceived to simulate uplink and downlink scheduling strategies in multi-cell/multi-user environments, taking into account user mobility, radio resource optimization, and frequency reuse techniques, among others [22].

6.2 MOBILITY MODELS

Modeling of user mobility plays an important role in the evaluation of wireless networks. The movement pattern of mobile users and how their location, velocity and acceleration change over time are described by the mobility model. Mobility models are used to simulate and evaluate the performance of mobile wireless systems and the algorithms and protocols [23]. Figure 1.8 illustrates the different categories of mobility models in a mobile ad hoc network based on their specific mobility characteristics.

We present in the following an overview of the commonly used Random Walk Mobility Model.

6.2.1 Random walk mobility model

The Random Walk mobility model is a widely used model to represent purely random movements of the entities of a system in various disciplines from physics to meteorology. The Random Walk with Reflection mobility model is a paradigm in which the domain model is bounded and nonconstrained, while the node model is selfish and erratic. This mobility model was developed to mimic irregular movement in nature. The Random Walk with Wrapping mobility model is similar to the standard Random Walk mobility model, with the difference that it is not bounded.

6.2.2 Random waypoint mobility model

The Random Waypoint mobility model is a simple stochastic model in which a node perpetually chooses destinations (waypoints) and moves towards them. This can be considered as an extension of the Random Walk mobility model, with the addition of pauses between changes in direction or speed.

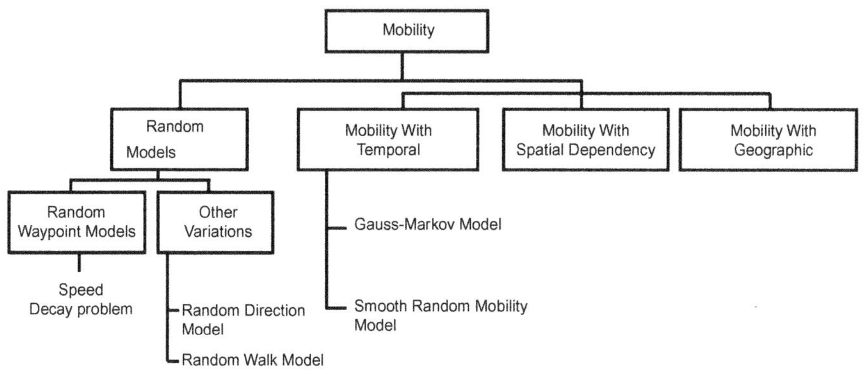

FIGURE 1.8

Categories of mobility models in Mobile Ad hoc Network.

6.2.3 Other mobility models

Starting from the Random Walk and Random Waypoint models, many variations have been proposed. The common characteristic of this class of models is that the movements of the nodes are independent of each other and that the movements are based on random distributions.

- Modified random waypoint model: In this model, the chosen destination points can only be located at the borders of the system area. In this model nodes are located at the border and try to reach from source to destination along with same border with high probability [24].
- Gauss-Markov mobility model: In this model, the velocity of a mobile node is assumed to be correlated over time and modeled as a Gauss-Markov stochastic process. In the Gauss-Markov model, the temporal dependency plays a key role in determining the mobility behavior.
- Smooth Random mobility model: This mobility model considers the temporal dependency of velocity over various time slots. In the Smooth Random mobility model, the frequency of speed change is assumed to be a Poisson process [25].

6.3 METRICS OF MOBILITY MODELS

Mobility models allow mimicking the behavior of mobile nodes when network performance is simulated. The simulation results are strongly correlated with the mobility model. In this subsection, we present some metrics of mobility models which influence the performance of wireless networks, as illustrated in Table 1.2.

Many mobility models can be used in mobile ad hoc networks, and each mobility model has its own mobility patterns that will impact the protocol performance. The parameters considered for direct mobility metrics are as follows: relative velocity, temporal dependence, spatial dependence, and pause time. Bai and Helmy [25] proposed a framework to analyze the impact of mobility on performance of routing protocols for MANET. They proposed two metrics to quantify the spatial and temporal dependence of mobile nodes. In [27], the authors proposed the Improved Degree of Spatial Dependence (IDSD), a spatial mobility metric which is able to capture both movement and pause correlation among mobile nodes. Such mobility metrics that are formulated using some mathematical models based on the direct mobility metrics are well suited to constitute a precise mathematical relationship between network connectivity and node mobility.

6.4 CASE STUDIES AND SIMULATION RESULTS

In this section, we will present a case study of network simulation tools for wireless networks. The authors in [21] used the NS2 package to model the wireless LAN under study.

Table 1.2 Classification of Mobility Metrics and its Characteristics

Metrics	Characteristics
Random based	Without any dependencies and restriction invoked in model
Temporal dependencies	A node actual movement influenced with its past movement
Spatial dependencies	The movement of a node influenced by node around it happens in group mobility
Geographical restrictions	Node movement restricted in certain geographical area
Hybrid structure	All mobility metrics classes are integrated to attain the structure
Pause time	Pause times between changes. A mobile node stays in one location for certain periods of pause time
Relative velocity	Measurement of velocity between two mobile nodes

They present the simulation results of IEEE 802.11 standard/Direct Sequence (DS) with transmission rates of 2, 5 and 11 Mbps. They used an optimized model for the IEEE 802.11 MAC scheme. Obaidat and Boudriga [21] varied the number of nodes using 2, 5, 10, 15 and 20 nodes in the WLAN system. Traffic is assumed to be generated with large packets of size 150 bytes and the network was simulated for different load conditions ranging from 10% to 100% of the channel capacity. The model allows us to determine the maximum channel capacity of the IEEE 802.11 standard. The simulation results are presented in Figures 1.9 to 1.11.

As illustrated in the figures, the normalized channel throughput decreases as the number of nodes increases. Basically, this is a general result of the CSMA protocol. The normalized channel throughput reduces as the data transmission rate increases, a behavior that can be justified due to the fixed overhead in the frames [21].

Another example of simulation is illustrated in [26] which represents a simulation of adaptive ABR voice over ATM networks. The authors analyze the performance of voice quality when sent over the available bit rate (ABR) service in ATM networks using simulation. The simulation model was implemented using C++. The model includes two classes, a Switch class and a Node class. The Switch class implements all the functions performed by an ATM switch, namely receiving cells from sources, scheduling, switching, and running the ERICA congestion avoidance algorithm. The class is driven by two events: (a) arrival of a new connection, and (b) arrival of a new cell [21]. The Node class essentially simulates nodes, whether sources or destinations.

Figure 1.12 illustrates how sources modify their rate as a result of more conflict on the bandwidth. As can be seen, the higher the number of sources, the lower the rate at which sources can send. This can be used by the operator to offer a variety of levels of service based on the customer or application requirements [21].

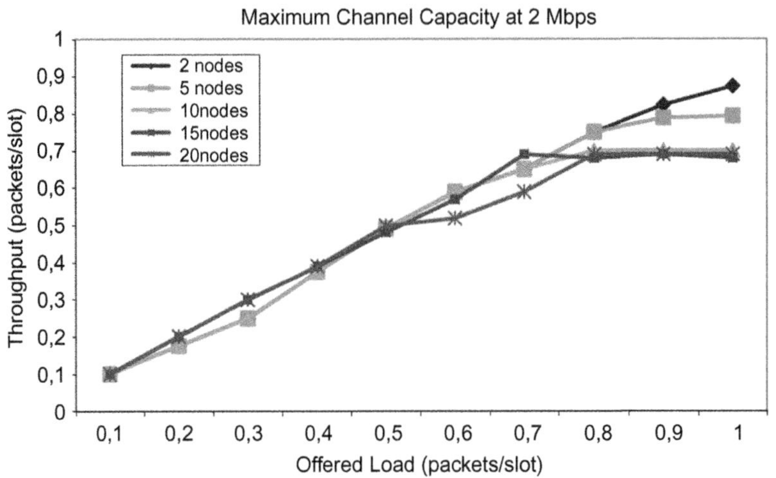

FIGURE 1.9

Throughput vs. offered load for a 2 Mbps WLAN.

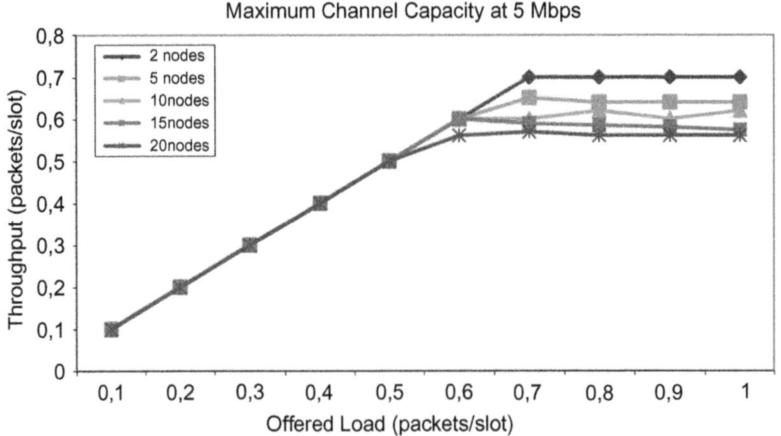

FIGURE 1.10

Throughput vs. offered load for a 5 Mbps WLAN.

Another QoS metric that has been considered is cell delay variation (CDV), which is not a major concern here since ATM has a low CDV, and since this can be taken care of by the playout buffer. Obaidat and Boudriga [21] consider delay-variation bounds of 10 ms and 20 ms, in order to support high quality voice, and thus, two types of voice traffic based on the delay variation bound are defined: one that can afford a 35 ms end-to-end delay, and another that can afford 45 ms.

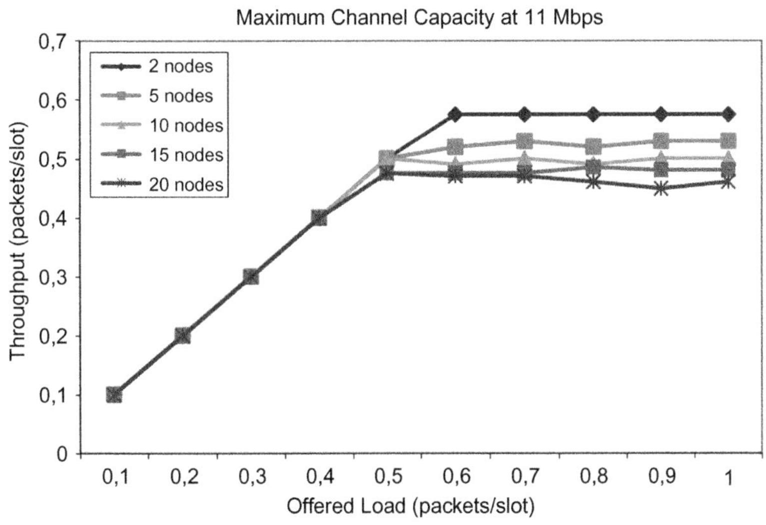

FIGURE 1.11

Throughput vs. offered load for 11 Mbps WLAN.

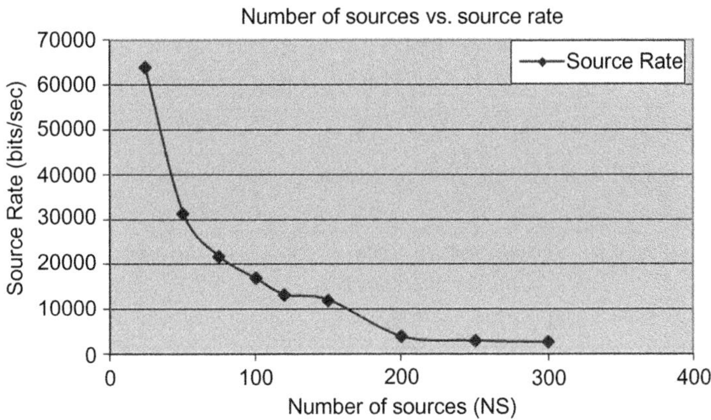

FIGURE 1.12

Number of sources vs. source rate.

Figure 1.13 depicts degradation in voice quality (DVQ) for a delay threshold of 35 ms. In this figure, the higher the number of sources, the worse the quality of voice. This can be explained as follows. The higher the number of sources, the higher the traffic load on the system, and hence the greater the probability of cells being dropped and/or delayed in the switches [21].

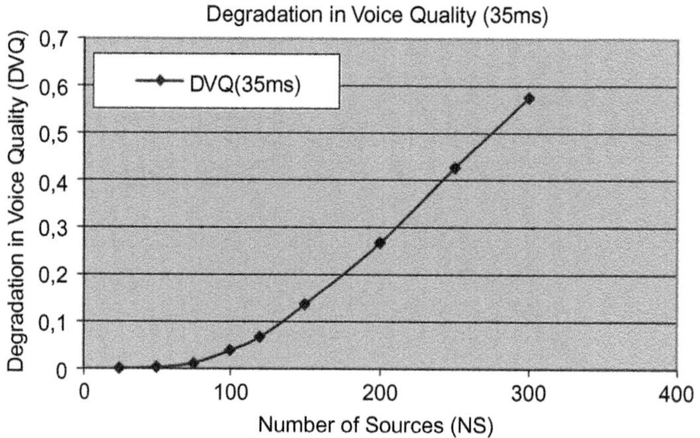

FIGURE 1.13

Number of sources vs. DVQ (35 ms threshold).

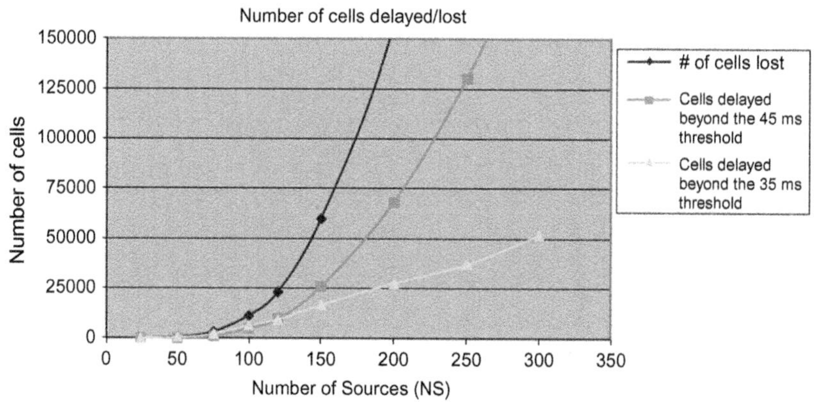

FIGURE 1.14

Number of sources vs. number of cells lost/delayed.

Figure 1.14 shows the consequences of increasing the number of sources on cells lost and cells delayed [21].

7 CONCLUSION

Wireless and mobile technologies have become a very critical part of our everyday life. Their present development is the outcome of various standardizations. In this chapter, we review the various generations of mobile wireless technology, their requirements, portals, performance, architecture, multiplexing methods, advantages and limits of one generation over another. Our review reveals that wireless devices continue to change rapidly. In fact, 5G is the name used in several research papers and specifications to refer to the next major phase of mobile telecommunications standards beyond the 4G standards. The new coming 5G technology will provide a higher peak data rate and much greater reliability than its predecessors.

We also investigate major simulation tools that are used to evaluate the performance of wireless networks, including some examples. We concluded the chapter by presenting two case studies on the use of modeling and simulation to predict the performance of wireless networks: an IEEE 802.11 wireless LAN and adaptive ABR voice over ATM network.

REFERENCES

[1] Nicopolitidis P, Obaidat MS, Papadimitriou GI, Amportsis A. Wireless networks. Hoboken, NJ: John Wiley & Sons; 2013.

[2] Cao J, Zhang C. Seamless and secure communications over heterogeneous wireless networks. New York: Springer; 2014.

[3] Sauter M. From GSM to LTE: an introduction to mobile networks and mobile broadband. Hoboken, NJ: John Wiley & Sons; 2011.

[4] Stuber GL. Principles of mobile communication. New York: Springer; 2011.

[5] Taha AM, Hassanein HS, Abu Ali N. LTE, LTE-advanced and WiMAX Toward IMT-advanced networks. Hoboken, NJ: John Wiley & Sons; 2012.

[6] Stojmenovic I. Handbook of wireless networks and mobile computing. Hoboken, NJ: John Wiley & Sons; 2002.

[7] Cox C. An introduction to LTE: LTE, LTE-Advanced, SAE and 4G mobile communications. Hoboken, NJ: John Wiley & Sons; 2012.

[8] Haloma H, Toskala A. HSDPA/HSUPA for UMTS: high speed radio access for mobile communications. West Sussex, England: John Wiley & Sons; 2006.

[9] Damnjanovic A, Montojo J, Wei Y, Ji T, Luo T, Vajapeyam M, et al. A survey on 3GPP heterogeneous networks. IEEE Wirel Commun 2011;18(3):10–21.

[10] Thien-Toan T, Yoan S, Oh-Soon S. Overview of enabling technologies for 3GPP LTE-advanced. J Wirel Commun Networking 2012;54(1):1–12.

[11] Alcatel-Lucent. The LTE Network architecture: a comprehensive tutorial. A technical overview. Strategic white paper. Available from: <http://www.cse.unt.edu/~rdantu/FALL_2013_WIRELESS_NETWORKS/LTE_Alcatel_White_Paper.pdf>; 2009.

[12] Sassan A. LTE-Advanced: a practical systems approach to understanding the 3GPP LTE releases 10 and 11 radio access technologies. San Diego, CA: Academic Press: An imprint of Elsevier; 2014.

[13] Tara AY. Understanding LTE and its performance. New York, NY: Springer Science & Business Media; 2011.

[14] Goldsmith A. Wireless communications. Cambridge: Cambridge University Press; 2005.

[15] ITU-R. Requirements related to technical performance for IMT-Advanced radio interface(s). Report M.2134. Available from: <http://www.itu.int/pub/R-REP-M.2134-2008/en>; 2008.
[16] Agilent Technologies. Introducing LTE-advanced: application note. Available from: <http://cp.literature.agilent.com/litweb/pdf/5990-6706EN.pdf>; 2009.
[17] Golshan R. LTE advanced: white paper. SAI Technology, Inc. Available from: <http://www.eddywireless.com/yahoo_site_admin/assets/docs/lte_advanced_white_paper_03.9141941.pdf>; 2011.
[18] Yuan Y. LTE-advanced relay technology and standardization. Springer; 2013.
[19] Prakash Agrawal D, Zeng Q. Introduction to wireless and mobile systems. 3rd ed. Cengage Learning; 2011.
[20] Ezreik A, Gheryani A. Design and simulation of wireless networks using NS-2. Second international conference on computer science and information technology (ICCSIT 2012). Singapore; 2012. p. 157–61.
[21] Obaidat MS, Boudriga N. Fundamentals of performance evaluation of computer and telecommunications systems. Hoboken, NJ: John Wiley & Sons; 2010.
[22] Piro G, Grieco LA, Boggia G, Capozzi F, Camarda P. Simulating LTE cellular systems: an open source framework. IEEE Tran Veh Technol 2011;60(2):498–513.
[23] Musolesi M, Mascolo C. Mobility models for systems evaluation: a survey. In: Garbinato B, Miranda H, Rodrigues L, editors. Middleware for network eccentric and mobile applications. Springer; 2009.
[24] Kumar S, Sharma SC, Suman B. Mobility metrics based classification and analysis of mobility model for tactical network. Int J Next-Gener Networks (IJNGN) 2010;2(3):39–51.
[25] Bai F, Helmy A. A survey of mobility models in wireless Ad Hoc networks. USA: University of Southern California. Available from: <http://www.cise.ufl.edu/~helmy/papers/Survey-Mobility-Chapter-1.pdf>; 2004.
[26] Obaidat MS, Obeidat S. Modeling and simulation of adaptive ABR voice over ATM networks, simulation: transactions of the society for modeling and simulation international. SCS 2002;78(3):139–49.
[27] Cavalcanti E, Spohn M. Predicting mobility metrics through regression analysis for random, group, and grid-based mobility models in MANETs. In: Proceedings of the IEEE symposium on computers and communications ISCC. Riccione, Italy; 2010. p. 443–48.

CHAPTER

Network planning and designing

2

Sofiane Hamrioui[1,2,3], Pascal Lorenz[1], Jaime Lloret[4], Joel J.P.C. Rodrigues[5], and Mustapha Lalam[2]

[1]*UHA University, Mulhouse Cedex, France*

[2]*UMMTO, Tizi-Ouzou, Algeria*

[3]*USTHB, Bab Ezzouar, Algeria*

[4]*Universidad Politècnica de València, València, Spain*

[5]*Instituto de Telecomunicações, University of Beira Interior, Covilhã, Portugal*

1 INTRODUCTION

A MANET (Mobile Ad hoc NETwork) [1] is a collection of hosts equipped with antennas which can communicate among each other without any centralized administration, using a wireless communication technology such as WiFi, Bluetooth, etc. In contrast to wired networks where only some nodes, called "routers," are responsible for routing data, in MANET all nodes are both routers and terminals. The choice of nodes that will ensure a communication session in MANET is dynamically dependent on network connectivity; for this reason we use the term "ad-hoc." In such a network, each node can communicate directly (point-to-point) with any other node if it is located in its transmission area, while communication with a node located outside its area of transmission is performed via several intermediate nodes (multi-hop mode).

TCP (Transmission Control Protocol) [2,3] is currently one of the main protocols used in Internet and more than 80% of wired communications use it. It seems logical then to use it for reliable communications involving a wireless link. Unfortunately, TCP was originally developed for fixed networks and it is not suitable to wireless link characteristics. For this reason, TCP suffers from several limitations which tend to grow, especially when the interactions between TCP protocol and the other layers are not taken into account. The study of such interactions and their optimization for better performance of MANET is the aim of this chapter.

Among the current research challenges in MANET is adaptation of the routing algorithms which are designed for static networks. Indeed, these algorithms are unsuitable for this type of network because of the node mobility, which makes the localization of the destination at a given time very difficult. Several routing protocols for MANET have been developed [4]; each protocol tries to maximize network performance by minimizing the delivery time of packets, the use of bandwidth and the energy consumption. Routing algorithms for such networks can be classified into three categories: table-driven protocols, on-demand protocols, and hybrid protocols.

We propose in this chapter to improve MANET performance by combining two of our previous solutions. The first solution is EM-OLSR (Energy Efficient for MANET by improving OLSR) [5], which is an improvement of the OLSR protocol [6], for better energy conservation in MANET. This algorithm is essentially based on the parameters of the communication environment, especially the communication distance of the mobile nodes and their mobility. We added another parameter to the multi-point relay (MPR) technique used by this protocol. This new parameter allows fair energy consumption in the same MPR set. With this adaptation, nodes with low energy are prevented in the routing process in order to maintain similar energy values for all the mobile nodes. The second solution is CL-TCP (Cross Layer TCP) [7]; it proposes an adaptation of the congestion control mechanism of TCP with some information provided from the routing layer and focusing on route length (in terms of number of hops) and the mobility of nodes.

The obtained cross layer solution after combining the CL-TCP and EM-OLSR solutions is called CL-TCP-OLSR. It is simulated and evaluated using the NS-2 simulator according to certain performance parameters: the TCP throughput, the TCP end-to-end delay and the energy consumption in the network. With our cross layer solution, the TCP performance is improved and the energy consumption is fairly used.

The remainder of the chapter is as follows: After a short presentation of the transport and routing protocols in MANET, we give the most significant approaches proposed for improving the TCP performance and the energy consumption in MANET. We focus only on the solutions which are oriented to network and transport levels. Then we turn to the presentation of our CL-TCP-OLSR approach and its implementation in NS-2. We finish the chapter by studying the impacts of CL-TCP-OLSR on MANET performance, particularly on the TCP performance and the energy consumption in the network.

2 TCP IN MANET

Transmission Control Protocol (TCP) [8,9] is a transport protocol that provides a mode connected service to the upper layers (session, presentation and application, and cutting OSI). TCP provides reliable service and is therefore used in the end-to-end packet communication between two computers. TCP controls the transmitter flow in order not to exceed the capacity of the network (congestion control) and assures that this flow will not be too high relative to the receiver flow (flow control). To satisfy all of these properties, TCP is based on several mechanisms: the use acknowledgments (ACK), the mechanisms for the flow control, the mechanism of congestion control based on the slow start, congestion avoidance, fast retransmit and the fast recovery algorithms.

2.1 PROBLEMS OF TCP IN MANET

The performance degradation of the TCP protocol is due to several problems [9–11]. Wireless media, unlike wired media, knows very high BERs (bit error rates). This is mainly due to interference and signal attenuation which corrupt the TCP packet (data and ACK) causing their losses. If the TCP source does not receive an ACK within the retransmission time-out (RTO) time interval following the transmission of a data packet, it concludes that the network is congested, which will result in the retransmission of the lost packet, reducing the congestion window (CWND) to 1, reducing the threshold CWND to half and doubling the RTO. The repeated errors at the transmission channel ensure that the CWND of the TCP source remains too small and then causes a very low transmission rate and a coarse increase of RTO.

Because of the mobility of nodes, the path between the TCP source and destination can be broken at any time. The path breaking between the source and the destination initiates the route discovery mechanism at the source, which takes some time T. In the case of T > RTO, TCP then invokes the congestion control mechanism and the retransmission of the lost packet. Thus, when a new road is discovered, the transmission rate continues to be very low during the Slow-Start phase. It is clear that this behavior is undesirable because, in a highly mobile environment, the TCP connection will never have the opportunity to use the maximum capacity of the channel to transmit. When T ≤ RTO, the TCP source continues to transmit in the new route using the old CWND. However, the old value of CWND may very well not be adapted to the new road, causing the loss of relations between CWND and the transmission rate of data allowed by a road. Indeed, the value of CWND of the old road can be very large for the new, which will cause a sudden congestion of the network.

It is very probable that the network will become partitioned due to moving or stopping of one or more nodes. If the TCP source and destination are in different partitions, the packets from the source will be left by the network, invoking the congestion control mechanism of a time-out. If the network partitioning takes much time, the unnecessary retransmissions of the same data to a disconnected station doubles the RTO of the source until it reaches 64 s. This will cause inactivity of the source for long periods, even if the link between the source and destination TCP is reset.

Some routing protocols maintain multiple routes between the source and the destination (such as TORA) in order to minimize the frequency of recalculation of roads and distribute the load of a TCP connection on several routes. Unfortunately, this can sometimes cause a significant number of packet arrivals out of order (OOO) at the receiver. This can then generate duplicate acknowledgments for each received packet, if the NS does not match the expected NS. Upon receipt of three duplicate ACK packets, the source invokes the Fast Retransmit/Fast Recovery mechanism (reduction of CWND and the threshold to half of CWND). This lowers the transmission rate unnecessarily and degrades network performance.

Many researchers have shown that the MAC layer protocols seriously affect the performance of TCP. Indeed, the nature of the shared wireless medium in an ad hoc environment causes the network stations to always be in competition for access to the channel. If in a certain area the number of nodes is very large, collisions will be more frequent, bringing the node that tries to join another station to conclude, after a number of MAC retransmissions, that the link is broken. This will trigger a useless route discovery mechanism.

Besides this, the schema of random exponential back-off of the IEEE 802.11 standard does not really suit the situation. Indeed, it has serious problems of equity for access to the channel when it favors the node that performed the last successful transmission, which could lead the node to monopolize the channel.

TCP New Reno [13] is an improved version of Reno that avoids multiple reductions of the CWND when several segments from the same window of data are lost. New Reno TCP has been considered because it is the leading Internet congestion control protocol. The initial congestion window is typically one segment. Each time a New Reno source receives an acknowledgment (ACK) packet, it increases the congestion window CWND by one segment. With this approach, the congestion window size expands multiplicatively, doubling every RTT. After reaching the threshold value, New Reno enters into the congestion avoidance phase and, to overcome this fault, it resets its congestion window to half of the ssthresh (Slow Start Threshold) value and then it increases its CWND size by one each RTT, which results in degrading the performance and utilization of Link.

3 MANET ROUTING PROTOCOLS

The routing function in the network [4] is a method of information forwarding to the correct destination through network connection data. It consists of providing a strategy to ensure, at any time, the establishment of correct and effective roads between any pair of nodes belonging to the network. These roads ensure the exchange of messages in a continuous manner. Given the limitations of ad hoc networks, the road construction should be done with a minimum of control and consumption of bandwidth. The problem raised in the context of ad hoc networks is the adaptation of the routing method used to the large number of existing units in an environment characterized by modest computing capacity and backup.

Therefore, it is important that any routing protocol design should consider the following issues:

1. Minimizing network load: the optimization of network resources includes two subproblems—avoiding routing loops, and preventing the concentration of traffic around certain nodes or links.
2. Provide support to perform reliable multipoint communication: the fact that the paths used for routing data packets can evolve should not pose a problem

with the proper routing. The removal of a link, due to failure or due to mobility, should ideally increase the least possible latency.

3. Ensure optimal routing: routing strategy should create optimal paths and take into account different cost metrics (bandwidth, number of links, network delays throughout, resources, etc.. . .). If the construction of optimal paths is a hard problem, maintenance of such roads can become even more complex; the routing strategy must ensure effective maintenance of roads with the lowest possible cost.
4. Latency: quality and latency paths must increase if network connectivity increases.

The MANET routing protocols can be classified as proactive (table-driven), reactive (on-demand) and hybrid routing protocols, depending on how they react to topology changes [12]. Different routing protocols for ad hoc networks can be used to provide consistent services to the mobile node, such as OLSR [6].

The OLSR (Optimization of Link State Routing) protocol [6] was introduced by the IETF MANET working group. It provides a fresh path of destination bases of a table-driven approach. It is an optimization of the pure link state algorithm in an ad hoc network. The routes are always immediately available when needed, due to its proactive nature. It is based on multi-point relays (MPRs) and the MPR set is selected such that it covers all the nodes that are two hop distances. Each node selects a set of its neighbor nodes as MPR. Only nodes selected as MPRs are responsible for generating and forwarding topology information, intended for diffusion into the entire network. The MPR nodes can be selected in the neighbor of the source node. Each node in the network keeps a list of MPR nodes. This MPR selector is obtained from HELLO packets sent between neighbor nodes. These routes are built before any source node intends to send a message to a specified destination. In order to exchange the topological information, the Topology Control (TC) message is broadcasted throughout the network. Nodes in the network send HELLO messages to their neighbors. These messages are sent at a predetermined interval in OLSR to determine the link status.

4 RELATED WORK

In this section we present the various proposals [14–17] which have been made in the literature to improve the performance of TCP and the energy efficiency in MANET. We focus only on the solutions that are oriented to network and transport levels.

TCP-F (TCP-Feedback) [18] offers the possibility to the issuer to distinguish between the route failures and the network congestion. In this scheme, the issuer is forced to stop transmission without reducing the window size on the failure of the route. As soon as the connection is restored, the fast retransmit is permitted. TCP-F is based on the network layer in an intermediate node detecting the route failure due to the mobility of its down neighbor along the route. An issuer may be in an active state or in a snooze state. In the active state, the transport layer is controlled by the normal TCP. Once an intermediate node detects a broken route, it explicitly sends a notification packet of the failed route (RFN) to the sender and logs this event. On receiving the RFN, the sender enters the snooze state in which the issuer ceases completely to send other packets and freezes all timers and values of the state variables such as the RTO and the size of congestion window. Meanwhile, all the intermediate ascending nodes, which receive the RFN, make that particular route invalid to avoid further packet loss. The transmitter remains in the snooze state until it is notified of the restoration of the road by the route recovery notification packet (RRN) of an intermediate node; then it resumes transmission of the frozen state.

Technically based on ELFN [19], this gives a new approach where TCP also interacts with the routing protocol to detect the failure of roads and take appropriate action when it is detected. This is done through explicit link failure notification messages (ELFN) which are returned to the sender node after detecting the failure. Such messages are broadcast by the routing protocol to be adapted for this purpose. In fact, the road failure message of DSR has been modified to carry a payload similar to ICMP (destination unreachable). Mainly, the ELFN messages contain the addresses of the sender and the receiver, the ports and the TCP sequence number. In this way, the modified TCP can distinguish the losses caused by congestion from those due to mobility. When the TCP sender receives an ELFN message, it enters into standby mode, which means that the timers are disabled and probing packets are sent regularly to the destination to detect the restoration of the road. Upon receipt of an ACK packet, the transmitter leaves the standby mode and resumes the transmission normally, using its previous timer values.

ATCP (Ad hoc TCP) [20] uses feedback from the network layer. In addition to the route failures, ATCP tries to address the problem of high bit error rates (BER). The TCP sender can be in a persistent state, congestion control state or in the retransmission state. A layer called ATCP is inserted between the TCP and IP layers of the TCP source nodes. ATCP listens to the network status information provided by ECN messages (explicit congestion notification) and ICMP "destination unreachable" messages, so ATCP puts the TCP agent in the appropriate state. Upon receipt of a message "Destination Unreachable" the TCP agent enters into a persistent state. The TCP agent for this state is blocked and no packet is sent until a new route is found by probing the network. The ECN is used as a mechanism to explicitly inform the sender about network congestion along the route being used. On receipt of ECN, congestion control of TCP is usually called without waiting for a time-out event. To detect packet loss due to channel errors, ATCP monitors the received ACKs. When ATCP sees three double ACKs have been received, it does not send the third duplicate ACK but puts TCP in the persistent state and quickly retransmits the lost TCP packet from the buffer. After receiving the next ACK, ATCP will resume TCP to normal.

TCP-BuS (Buffering capability and Sequence information) is presented in [21]. It uses feedback from the network to detect the failure events of routes and to make the required decision for this event. The new scheme in this proposal is the introduction of buffering opportunities in the mobile nodes. The authors choose the initiated source on demand routing protocol ABR (Associativity-Based Routing). The following improvements are proposed. The first is the explicit Notice with the use of two control messages to inform the source about the route failure and route restoration. The second improvement is the extension of the time-out values: for RCC (Route ReConstruction) phase, the packets along the path from the source to the PN are saved. To avoid time-out events during the RRC phase, the value of the retransmission timer for protected packets is doubled. The last improvement is the selective retransmission request, which consists of not transmitting the lost packets along the path from the source to the PN, while the value of the retransmission timer is doubled, until the expiration of the adjusted retransmission timer. To overcome this, an indication is made to the source to be able to retransmit these lost packets selectively.

In [22] a Split TCP scheme was introduced, to cut the long TCP connections to shorter located segments. The interface node between two located segments is called a proxy. The routing agent decides whether the node is a proxy using the interproxy distance setting. The proxy intercepts the TCP packets, stores them and after that acknowledges their reception to the source (or previous proxy) by sending a local acknowledgment (LACK). Also, the proxy is responsible for delivering the packets, at an appropriate rate, to the next local segment. On receipt of a LACK (of the next proxy or final destination), the proxy will serve the packet's buffer. To ensure the reliability of the source to the destination, an ACK is sent from the destination to the source similarly to the standard TCP. In fact, this scheme also cuts transport-layer functionality to those functions pertaining to congestion control and end-to-end reliability.

Some simulations have been conducted to evaluate the performance of CBR over TCP on MANET using the DSR routing protocol [23]. Although CBR and TCP significantly affect MANET, these differences lead to obtaining significant performance improvements of CBR over TCP, with better throughput and less average maximum end-to-end delay. DSR was able to respond to link failure at low pause time; this led to improving TCP's performance in packet delivery. The authors conclude that TCP traffic models can be used for small networks where frequent topology changes are limited and could be controlled by DSR protocol.

The behavior of TCP Friendly Rate Control (TFRC) and TCP in the presence of DSR and AODV as routing protocols has been studied [24]. The evaluation parameters used are throughput, delay and jitter. This study also allowed the authors to identify which routing protocols have an impact on transport protocols. They showed that the rate of change of throughput between routing protocols is 3.74.

An analysis of TCP performance (including throughput) during break route events is given in [25]. Three routing protocols are employed: AODV, DSR and OLSR, to better understand the different types of TCP behavior during these events, while highlighting the mechanisms that affect each protocol. Their study showed that DSR interacts better with TCP protocol than the others; this happens because of its quick route restoration. AODV showed a good performance by avoiding the RTO by caching TCP packets exiting when there is a link failure. For OLSR, it is maintained by default settings under the optimal value for the studied scenario as if it were used for traffic loads across loaded topologies.

In [26], the authors conducted an investigation of the performance of some routing protocols in the presence of CBR and TCP traffic sources. Two reactive protocols (AODV, DSR) and one proactive (DSDV) are used. Their results showed that reactive protocols have better performance with CBR than TCP in terms of throughput, delay and packet loss. Moreover, reactive protocols are able to respond quickly to broken links, thus avoiding congestion.

In [27], the authors investigate the performance of TCP over DSDV (proactive) and AODV (reactive) protocols using simulations in NS-2 for a range of node mobility with a single traffic source. They found that the proactive protocol consumes more bandwidth, because it transmits routing updates frequently. It reacts slowly in dynamic topologies. Its performance decreases drastically as mobility increases. But the reactive one consumes less bandwidth and has lower overhead of routing information. They concluded that to fight against the performance degradation of TCP under high mobility, it is necessary to have some sort of feedback from the link layer protocol.

The Congestion State Prediction Algorithm (CSPA) and the Group Outlet Directive Algorithm (GODA) are introduced in [28]. They improve the transmission performance by distinguishing between packet loss due to link failure and arbitrary loss of packets. CSPA helps to distinguish between packet loss due to link failure and arbitrary packet loss. Once the congestion contention node is found, GODA attempts to resolve it at the source node and identifies it as a victim of congestion; if congestion is not resolved at the node level, it attempts to handle it at the group level. This process continues with predecessor groups if it fails to control congestion at the current group level.

A new version of AODV for energy efficiency is proposed to reduce energy expenditure due to overhearing [29]. The proposed algorithm controls the level of overhearing. It reduces energy consumption without affecting quality of route information. This algorithm enables the sender to select no overhearing, unconditional overhearing or probability-based overhearing for its neighbors. It is specified in the ATIM frame's subtype field and is made available to its neighbors during the ATIM window. The number of overhearing nodes is controlled by a probability-based overhearing method.

EOLSR [30] is a variant of OLSR, where MPR selection and path calculation is determined by both a node's residual energy level and its number of neighbors. The key insight here is that sending data to a node also forces all its neighbors to consume energy in overhearing the data packet. The simulation results reported in this work show that combining both the new path calculation with the modified MPR selection yields the best performance. EOLSR suggests that a node's residual energy level is propagated by extending the protocol control messages, but does not discuss how accurate this information is.

In [31], the authors propose two novel mechanisms for the OLSR routing protocol, aiming to improve its energy performance in MANETs. They propose a modification in the MPR selection mechanism of the OLSR protocol, based on the Willingness concept, in order to prolong the network lifetime without losses of performance (in terms of throughput, end-to-end delay or overhead). Additionally, we prove that the exclusion of the energy consumption due to overhearing can extend the lifetime of the nodes without compromising the OLSR functioning at all. A comparison of an Energy-Efficient OLSR (EE-OLSR) and the classical OLSR protocol is performed, testing some different well-known energy-aware metrics such as MTPR, CMMBCR and MDR.

EM-OLSR (Energy Efficiency in MANET by improving OLSR protocol) [5] is a new approach to minimizing the energy consumption. EM-OLSR is based on the OLSR routing protocol and adds a new energy fairness parameter to the multi-point relay (MPR) technique. This new parameter is used by our approach and allows fairness energy consumption in the same set of MPRs. In this mechanism, nodes with low power are prevented in the routing process in order to maintain similar power values for all the mobile nodes. The simulation results showed that the proposed EM-OLSR approach allows significant power saving up to 14% and an increase in average lifetime of a mobile node as high as 22%.

An improvement for the interactions between MAC and routing protocols to improve energy consumption in MANETs and the study of its incidences on the performance of the network is presented in [32]. The authors propose a new approach called IMR-EE (Improvement of the Interactions between MAC and Routing protocol for Energy Efficiency), which exploits two communication environment parameters. The first one is the number of nodes; the approach reduces the additional energy used to transmit the lost data by making the size of the back-off interval of the MAC protocol adaptable to the node number in the network. The second parameter is the mobility of nodes; IMR-EE also uses the mobility of nodes to calculate a fairness threshold in order to guarantee the same level of residual energy for each node in the network.

5 PRESENTATION OF CL-TCP-OLSR

The first part of the proposed CL-TCP-OLSR cross layer solution takes into account the strengths of the OLSR protocol. In addition to the MPR technique used by this protocol, we decided to use another parameter that allows fair energy consumption in the same set of multi-point relays. This new parameter get its value according to the parameters of the communication environment used, such as the node number, the distance of the communication, the mobility of nodes, the type of application, the rate of interference, etc. With this new parameter, the energy consumption could be optimized to all nodes of the same multi-point. The modeling of our solution and the pseudo program follow.

For each mobile node i, the most important parameters in the communication environment that can really influence energy consumption the most are the speed of the mobility M_i and the distance of communication D_i. We take into account these two parameters to get the first part of our fairness threshold as follows:

$$F(D_i, M_i) = \frac{M_i + D_i}{\alpha \sum_{J=1}^{N-1} D_j + \beta \sum_{J=1}^{N-1} M_j} \tag{2.1}$$

where:

M_i: the speed mobility of the node i;

D_i: the average distance between the node i and t nodes. It's calculated with the following function:

$$D_i = \frac{\sum_{J=1}^{N-1} d_{i,j}}{N} \tag{2.2}$$

where:

N: the number of nodes in the network;

$d_{i,j}$: the distance between the nodes i and j.

α and β: two coefficients used to make the mobility and the distance parameters more or less important.

Equation (2.1) gives, for the node i, the ratio between the energy dissipated locally and that dissipated in other ways.

In a MANET, we know that not only the distance between the nodes and their mobility lead to higher energy consumption, but other parameters also affect this consumption. For these reasons, our proposed fairness threshold takes into account the maximum of these parameters. Note by:

L: the number of parameters that influence the communication environment;

P_i: parameter number i in the communication environment;

E: the set of the parameters P_i defined as follows:

$$E = \{P_1, \ldots P_i, \ldots P_L\} \tag{2.3}$$

V_i: the numerical value associated with P_i; it is calculated as follows:

$$V_i = \begin{cases} +1 & \text{if } P_i \text{ participates in the rapid dissipation of the energy} \\ 0 & \text{else} \end{cases} \tag{2.4}$$

The average value of the influences of all the P_i auxiliary parameters is noted by G and calculated as follows:

$$G = \frac{\sum_{i=1}^{L} V_i}{L} \tag{2.5}$$

From Equations (2.1) and (2.5), the fairness threshold S_i for each node i is obtained as follows:

$$S_i = F(D_i, M_i) + G \tag{2.6}$$

S_i is used in the MPR set of every node *i* selected by OLSR as MPR. More details about the implementation of this threshold S_i are given in [5].

The second part of our proposed cross layer solution considers the number of nodes in the network and their mobility. CL-TCP-OLSR exploits the information relating to the number of nodes that contain the routing path. As shown in [12,33], slightly delaying acknowledgment of well received TCP packets can improve the performance of the protocol. We take the same principle here, but instead of the number of nodes, we will use the time to go and back for one packet between the source and destination. Using T_r this time, each node *i* after received T_r will run the following treatment:

$$T_r := T_r + T_{i,i+1} \quad (2.7)$$

Here $T_{i,i+1}$ is the time elapsed during the packet transmission from node *i* to the next node in the routing path, the node $i+1$.

The third parameter taken into account by CL-TCP-OLSR is the mobility of nodes. In fact, the mobility of nodes leads often to the breakdown of connectivity between nodes, resulting in data lost. These data losses may (although not always the case) be interpreted by the transport protocol as losses due to congestion. Then, it activates the congestion control mechanism, which will reduce the unnecessary throughput. For this reason, our solution also uses another threshold to guarantee the same level of residual energy for each node in the network.

For this reason, CL-TCP-OLSR takes into account the information about node mobility, which is received from the routing layer. As we know, mobility is generally characterized by its speed and angle of movement. These two factors determine the degree of the impact of mobility on packet loss. In order to model our system, we consider a node *i*, in communication with another node *j*, and then we note:

$\alpha_{i,j}$: the angle between the line (*i*, *j*) and the movement direction of node *i*,

W_i: the speed of mobile node *i*.

We have seen in a previous study [34,35] that for very small speeds, TCP performance parameters are not affected, and larger values of mobility also can help to improve the performance in the case where the communicating nodes do not move far from their neighboring node (with whom it communicates). Based on that, we proceed with the development of the first part of CL-TCP-OLSR.

Let $H(W_i)$ and $G(\alpha_{i,j})$ be two logical functions, whose values are determined as follows:

$$H(W_i) = \begin{cases} \text{True} & \text{if } S_{min} < W_i < S_{max} \\ \text{False} & \text{else} \end{cases} \quad (2.8)$$

Here, S_{min} is the speed of node *i* from which network performance begins to degrade. Its value is determined by the mobility model. In our case, it will be fixed at 10 m/s. S_{max} is the speed of node *i* from which network performance begins to grow again. In our case, it will be fixed at 35 m/s (through our previous results). These two values must be determined according to the mobility model.

G ($\alpha_{i,j}$) is another logic function which relates to the angle $\alpha_{i,j}$ and informs CL-TCP-OLSR if the direction of the movement of node *i* can lead to packet loss due to a broken link. Its value is determined as follows:

$$G(\alpha i, j) = \begin{cases} \text{False} & \text{if} \quad -\Pi/4 \leq \alpha_{i,j} \leq \Pi/4 \\ \text{True} & \text{else} \end{cases} \tag{2.9}$$

CL-TCP-OLSR allows each node *i* to get the value of its current S_i. The easiest way to do this is to deduce it by knowing the time spent between two geographical points. There are many systems for node mobile location such as GPS and power measurement techniques [36,37]. With these systems, each node can know its position at any time, and then it will be able to estimate the distance travelled during an interval of time. With the distance and time we can get the speed of mobility W_i. Moreover, it is possible to determine the direction of movement and the value of the angle $\alpha_{i,j}$. More details about the implementation of this second part of CL-TCP-OLSR are given in [7].

6 EVALUATION OF CL-TCP-OLSR

In this section, we evaluate our proposed solution by showing its impact on the performance of MANET, especially on three parameters previously studied (the TCP throughput, the TCP end-to-end delay and the energy consumption). The obtained results with CL-TCP-OLSR are compared to those obtained with TCP-OLSR.

6.1 EVALUATION ENVIRONMENT

The evaluation of our cross layer solution is carried out through the simulation environment NS-2 (version 2.34) [38]. At the MAC level, the model 802.11 b is used by keeping the default values of the parameters of this model. All nodes communicate through wireless links in half-duplex with an identical bandwidth of 1 Mb/s. Some values, such as the duration of the simulation, the speed of nodes and the number of connections, have been configured properly in order to obtain interpretable results such as those reported in the literature. The simulations are performed for 1000 seconds. Nodes were moving in an area of 1000 m × 1000 m. Each node has a transmission range of 250 m. We chose 1000 seconds in order to analyze the interaction of TCP with the routing alternatives in the simulation environment.

We used OLSR as the routing protocol since this is the protocol used in our previous work [5]. For the transport layer, we opted for TCP New Reno. TCP New Reno is a reactive variant. It is widely deployed, and its performance was evaluated in conditions similar to those conducted here. TCP traffic has been used as the main network traffic.

This study is far from being exhaustive. It presents the results using just one mobility model: Random Waypoint [39]. However, we have chosen it because the network is not designed for particular mobility and this model is widely used in the related literature. In this model, the mobility of the nodes is typically random and all nodes are uniformly distributed in the simulation space. In this model, some mobile nodes are placed in an area where they cannot leave. An initial position, speed and destination are assigned to each mobile node. Whenever the mobile nodes reach their destination within the surface, they leave to another randomly chosen destination after an optional resting period.

6.2 EVALUATION PARAMETERS

In our evaluation of the CL-TCP-OLSR performance, we used three parameters. The first is the throughput, which is given by the received data ratio taking into account all data sent. The second parameter is the end-to-end delay, which is the time a packet takes to travel across the network from a source to a destination. We also evaluate the energy consumption in the network.

6.2.1 Effect of mobility

In this section we consider 20 different mobility scenarios for each outcome. The scenarios consist of 50 nodes moving with a speed between 0 and 40 m/s according to the Random Waypoint mobility model with a time pause of 5 seconds. Ten TCP connections are established between randomly selected pairs.

Figure 2.1 shows that the CL-TCP-OLSR solution improves the TCP throughput. In fact, when mobility increases, the probability of having more broken links becomes important, which leads to having many packet losses in the network. For low speeds the throughput increases but it goes down as soon as the speed

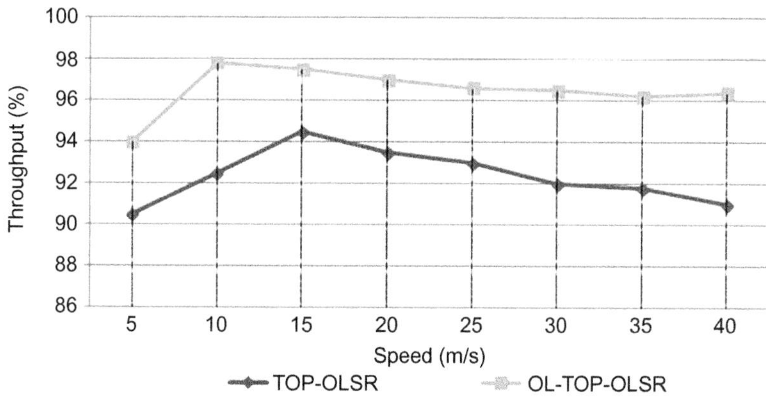

FIGURE 2.1

Effect of mobility on the throughput.

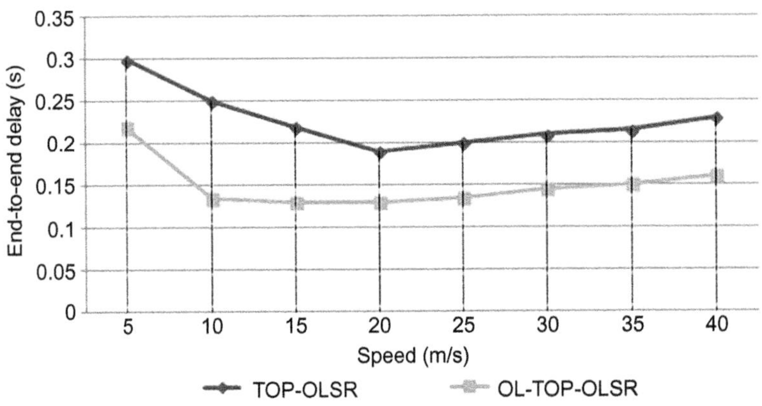

FIGURE 2.2

Effect of mobility on the end-to-end delay.

increases and this happens with both networks (with and without our solution). Initially, the throughput has increased because the number of received packets has increased, but with increasing speed there are more broken links and therefore transmission failures and packet losses, which again degrade the throughput. The problems of route failures in such environments lead to the activation of a congestion control system, which reduces the congestion window and then the throughput is reduced.

In fact, with the increase of mobility, link breaks and node disconnections cause packet losses. With these losses, the congestion control mechanism is activated and reduces the throughput. However, the difference recorded between the two networks is mainly due to the better management of mobility by our proposed cross layer solution. CL-TCP-OLSR informs TCP about the real reason for these losses (which is the mobility, in this case). With this information, TCP will avoid starting or unnecessarily initializing the congestion control mechanism. This explains the maintaining of better throughput with CL-TCP-OLSR, although only a slight decrease was recorded. CL-TCP-OLSR therefore allows both OLSR and TCP protocols to interact and better understand the sources of packet losses in the network.

Figure 2.2 shows that CL-TCP-OLSR also provides an improvement of end-to-end delay. In fact, with our new approach, the congestion control mechanism is not triggered unnecessarily after packet losses due to the mobility. Therefore, the transmission time of packets is minimized, which leads to improving the end-to-end delay. Moreover, with CL-TCP, TCP packets have enough time to go to the receiver and for their acknowledgment (ACK) to come back from receiver to sender. This is because the calculation of a new value of RTO which takes into consideration the new routes formed by the routing protocol. These new routes can be shorter than the previous ones. In this case, the new RTO must be smaller

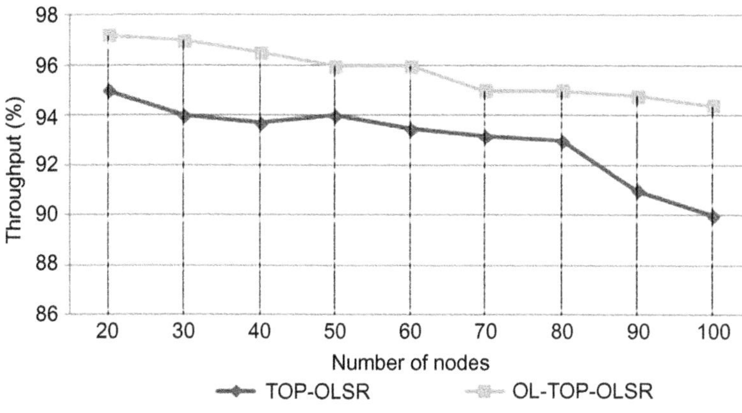

FIGURE 2.3

Effect of the network load on the throughput.

than the previous one. These new routes can also be longer than the previous ones; in this case the new RTO must be taller than the previous one. With this optimization of time, CL-TCP-OLSR improves the end-to-end delay parameter.

6.2.1.1 Effect of network density (number of nodes)

The scenarios considered in this section consist of nodes moving with a speed of 15 m/s according to the Random Waypoint mobility model with a pause time of 5 seconds. Ten TCP connections are established between randomly selected pairs. An average of 20 different scenarios was conducted for each outcome. The number of nodes is varied between 20 and 100 nodes.

Figure 2.3 shows the effect of the number of nodes in the network on the interactions between OLSR and TCP protocols before and after the improvement. For small values of the number of nodes, the two networks recorded better throughput. However, for large values of the number of nodes, there is degradation of TCP throughput, but this degradation is slight with CL-TCP-OLSR. In fact, CL-TCP-OLSR takes into account the information about the number of nodes in the network for the calculation of the new RTO used by the TCP protocol. With this optimization, retransmissions and unnecessary waiting times are avoided, which allows better TCP throughput.

Figure 2.4 shows that the network load also has a significant effect on the end-to-end delay parameter of the TCP protocol. Moreover, as the number of nodes increases, the end-to-end delay increases too. But this increase is lower with our solution. In fact with the increase of the number of nodes, the routes created and used by the routing protocol can become important (in terms of number of nodes contained in the routing path), which makes the time to go and return across the route more important for TCP packets. Our CL-TCP-OLSR cross layer solution calculates the waiting time for TCP acknowledgments by taking into

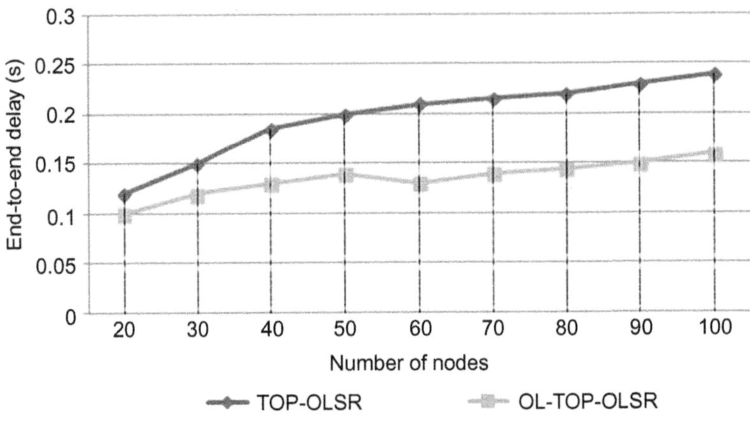

FIGURE 2.4

Evolution of the end-to-end delay with the network load.

account the route lengths in the routing side, which allows TCP to give the required time to receive the TCP acknowledgments before understanding (or assuming) that corresponding data packets for these acknowledgments are lost. Therefore, the additional time for the retransmission of the supposed lost TCP packets is avoided, which improves the TCP end-to-end delay.

6.2.1.2 Energy consumption

The main goal is to evaluate our proposed cross layer solution in terms of energy consumption in MANET. We set the two coefficients α and β to 1, and the set E to empty. This is to validate, for the first time, our energy fairness threshold S with the basic parameters, which are the communication distance and the mobility (number of nodes = 10 nodes, number of CBR sources = 3 sources, maximum speed = 2 m/s, pause time = 50 sec, simulation duration = 1000 sec, initial energy = 200 joules).

Figure 2.5 presents the evolution of the total energy in the network with time. We can see clearly that the total energy of the network using TCP-OLSR decreases faster than the energy of the network using our solution CL-TCP-OLSR and that until the second 600 sec. The difference in the energy consumption during this interval is due to the intelligent management of this energy by our solution. In addition to the MPR principle used par OLSR, TCL-TCP-OLSR uses also a new parameter in every MPR set which helps in an equity consumption of energy. For this reason, the energy level of the network using our solution stays higher during the duration of the simulation. During the interval [600, 1000] sec, we notice that the energy level of the network before improvement stabilizes because the loss of the connectivity between nodes.

This loss of connectivity is due to the fact that the energy of the nodes is used in an "unfair" way, leading to depletion of the total energy of nodes that are the

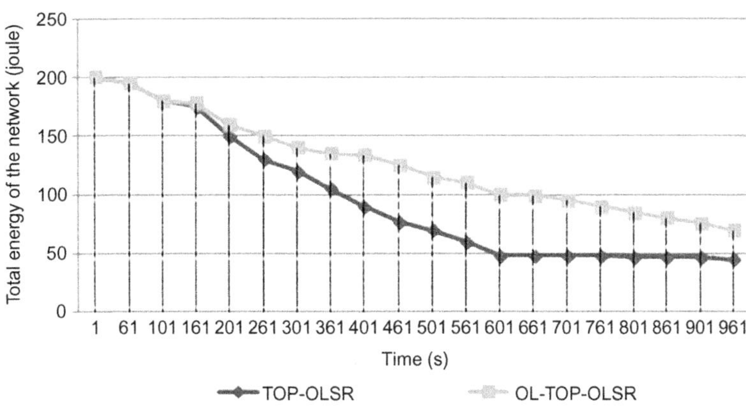

FIGURE 2.5

Evolution of total energy with time.

most solicited by the protocol OLSR. This explains the failure of network connectivity, which stops the activity in the network. Indeed, when t = 600 sec, the number of working nodes in the network using TCP-OLSR is 4, and these nodes probably do not communicate because of their distance. During the same time interval, the energy in the network using our CL-TCP-OLSR improvement continues to decrease, which proves that the nodes are communicating.

We conclude from the graphs shown in Figure 2.5 that our solution, CL-TCP-OLSR, saves an important amount of energy in the network, which improves the QoS by avoiding the loss of connectivity after the unfair using of the energy by the nodes in the network.

Figure 2.6 depicts the lifetime of the network. It shows that the number of working nodes in the network using TCP-OLSR decreases starting from t = 150 sec and stabilizes at t = 650 sec with only 4 nodes because of the loss of the connectivity. This number of nodes starts to decreases again at about t = 950 sec. On the other hand, for our proposed solution, the number of working nodes in the network remains constant until t = 750 sec, and then it starts to decrease rapidly.

Due to the utilization of our fairness threshold, our proposed approach CL-TCP-OLSR favors nodes with a higher residual energy and carries out its equitable use. The results of the simulations showed that our approach allows an important increase in average lifetime of a node, and consequently allows increasing the lifetime of the whole network.

7 CONCLUSION AND FUTURE WORK

In this chapter, we proposed a cross layers solution called CL-TCP-OLSR for better QoS in MANET. CL-TCP-OLSR consists of a combination of two solutions proposed

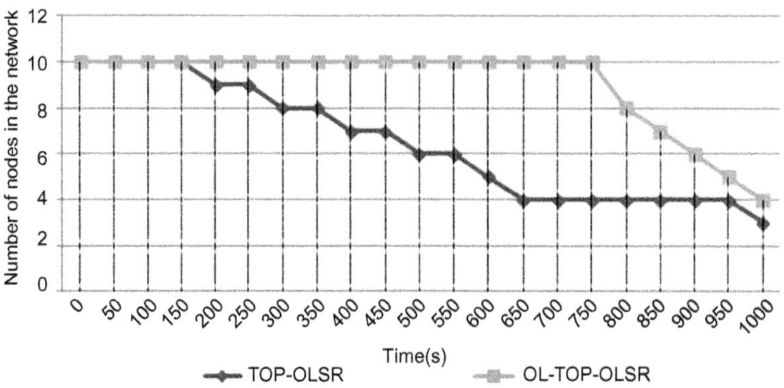

FIGURE 2.6

Evolution of number of nodes with time.

previously and based on the cooperation between two layers which are transport and network layers. CL-TCP-OLSR allows both OLSR and TCP protocols to interact and better understand the sources of packet losses in the network and fairness in energy use by nodes. In the first part, our solution proposes an adaptation of the TCP congestion control mechanism with some information provided from the routing layer and focusing on the route length (in terms of number of hops) and the mobility of nodes. In the second part, CL-TCP-OLSR adds another parameter to the MPR technique used by this protocol. This new parameter allows fair energy consumption in the same set of multi-point relays. With this adaptation, nodes with low power are prevented in the routing process, in order to maintain similar energy values for all the mobile nodes.

After the implementation and simulation of CL-TCP-OLSR, we studied its impact on the QoS performance of the network, most particularly on certain performance parameters: the TCP throughput, the TCP end-to-end delay, and the energy consumption. The results obtained are conclusive and satisfactory. CL-TCP-OLSR helps the transport layer to distinguish between the lost packets due to the congestion and the lost packets due to the nodes load and their mobilities. With this improvement, there is a significant increase of the performance of TCP in terms of throughput and end-to-end delay. Also, we obtained a significant energy saving and an increase in average lifetime of a mobile node.

The achieved results in this chapter are encouraging, justifying further investigation in this direction. In our future work, we will continue to model the maximum number of communication environment parameters. We will try to reflect as much as possible the communication environment. Also, we plan to conduct a study on the communication environment of MANET to better assign meaningful values to the two constants α and β. In fact, these values must be deduced, depending mainly on the complexity of the MANET and the type of application covered by the network. Finally, our CL-TCP-OLSR solution will be tested on a real MANET platform.

ACKNOWLEDGEMENTS

This work has been supported by the University of Haute Alsace, Networks and Telecommunications Research Group of Colmar (GRTC), the University of Mouloud Mammeri of Tizi Ouzou, Computer Sciences Research Laboratory, the *Instituto de Telecomunicações*, Next Generation Networks and Applications Group (NetGNA), Portugal, by the Government of Russian Federation, Grant 074-U01, and by National Funding from the FCT–Fundação para a Ciência e a Tecnologia through the PEst-OE/EEI/LA0008/2013 Project.

REFERENCES

[1] Basagni S, Conti M, Giordano S, Stojmenovic I. Mobile Ad Hoc Networking. USA: Wiley-IEEE Press; 2004.

[2] Holland G, Vaidya N, Analysis of TCP performance not over mobile ad hoc networks. Proc. ACM Mobicom; 1999.

[3] Hanbali A, Altman E, Nain P. A survey of TCP over ad hoc networks. IEEE Comm Sur Tuto 2005;7(3):22–36.

[4] Mohapa P, Kishnamurthy SV. Ad hoc networks: technologies and protocols. New York: Springer Science + Business Media, Inc; 2005.

[5] Hamrioui S, Lalam M, Lorenz P. A new approach for energy efficient in MANET based on the OLSR protocol. Int J Wire Mobile Comp 2012;5(3):292–9.

[6] Clausen T, Jacquet P, Laouiti A, Muhlethaler P, Qayyum A, Viennot L. Optimized link state routing protocol. IEEE INMIC. Pakistan; 2001.

[7] Hamrioui S, Lorenz P, Lloret J, Lalam. M. A cross layer solution for better interactions between routing and transport protocols in MANET. J Comput Inform Technol 2013;CIT 21(3):137–47.

[8] Stevens R. TCP/IP illustrated. The protocols, vol. 1. Addison-Wesley; 1994.

[9] Hamrioui S, Lloret J, Lorenz P, Lalam M. TCP performance in mobile ad hoc networks. Netw Proto Algor J 2013;5(4).

[10] Fu Z, Zerfos P, Luo H, Lu S, Zhang L, Gerla M. The impact of multihop wireless channel on TCP throughput and loss. IEEE INFOCOM '03, San Francisco; 2003.

[11] Fu Z, Meng X, Lu S. How bad TCP can perform in mobile ad-hoc networks. IEEE Symposium on computers and communications, Italy; July 2002.

[12] Hamrioui S, Lalam M. Incidences of the improvement of the MAC-Transport and MAC-Routing interactions on MANET performance. International conference on next generation networks and services, Morocco; 2010.

[13] Floyd S, Henderson T. NewReno modification to TCP's fast recovery. The Internet Society. RFC 2582. Available at: <https://tools.ietf.org/html/rfc2582>; 1999.

[14] Tsaoussidis V, Badr H. TCP-probing: Towards an error control schema with energy and throughput performance gains, Japan, 8th IEEE Conference on network protocols; 2000.

[15] Zhang C, Tsaoussidis V. TCP-probing: Towards an error control schema with energy and throughput performance gains. New York, 11th IEEE/ACM NOSSDAV; June 2001.

[16] Hamrioui S, Lorenz P, LIoret J, Lalam M. A cross layer solution for better interactions between routing and transport protocols in MANET. J Comput Inform Technol 2013;CIT 21(3):137–47.
[17] Hamrioui S, Lorenz P, LIoret J, Lalam M. Incidence of the improvement of the interactions between MAC and Transport protocols on MANET performance. Wireless communications and networking: theory and practice. IGI Global; 2014. p. 275–92. [chapter 10].
[18] Chandran K. A feedback based scheme for improving TCP performance in ad-hoc wireless networks. In: Proceedings of 18th international conference on distributed computing systems. Amsterdam; May 26–29, 1998.
[19] Holland G, Vaidya NH. Analysis of TCP performance over mobile ad hoc networks. Seattle, Annual international conference on mobile computing and networking (Mobicom'99); August 1999.
[20] Liu J, Singh S. ATCP: TCP for mobile ad hoc networks. IEEE JSAC 2001;19 (7):1300–15.
[21] Kim D, Toh C, Choi Y. TCP-BuS: Improving TCP performance in wireless ad hoc networks. J Commun Netw 2001;3(2):175–86.
[22] Kopparty S, Krishnamurthy SV, Faloutsos M, Tripathi SK. Split-TCP for mobile ad hoc networks, University of California Riverside, Technical Report. Available from: <http://64.106.20.52/~michalis/PAPERS/split-glo02.pdf>; 2002.
[23] Bakalis P, Lawal B. Performance evaluation of CBR and TCP traffic models on MANET using DSR routing protocol. CMC'10. Proceedings of the 2010 international conference on communications and mobile computing, vol. 03; 2010. p. 318–22.
[24] Mohd Zaini K, Habbal AMM, Azzali F, Hassan S, Rizal M. An interaction between congestion-control based transport protocols and MANET routing protocols. J Comput Sci 2012;8(4):468–73.
[25] Papanastasiou S, Machenzie LM, Ould-Khaouan M, Charissis V. On the interaction of TCP and routing protocols in MANETs. Advanced international conference on telecommunication and international conference on internet and web applications and services, IEEE, 2006.
[26] Chaudhary MS, Singh V. Simulation and analysis of routing protocol under CBR and TCP traffic source. International conference on communication and network technologies, IEEE, 2012.
[27] Rahman MM, Nashiry MA, Godder TK, Shawkat Ali ABM. TCP performance over proactive and reactive routing protocols for mobile ad hoc network. In: Proceedings of International Conference of Wireless Communication (ICWN), Las Vegas, NV; July 14–17, 2008.
[28] Appaji VV, Sreedhar M. CRT: Crosslayered routing topology for congestion control in mobile ad hoc networks. Int J Sci Eng Res 2012;3(9).
[29] Sumathi N, Thanamani AS. Evaluation of energy efficient reactive routing protocols in QoS enabled routing for MANETS. Int J Comput Appl 2011;14(2).
[30] Mahfoudh S, Minet P. EOLSR: An energy efficient routing protocol in wireless ad hoc sensor networks. J Interconnect Netw 2008;9:389–408.
[31] Rango D, Fotino F, Marano S. EE-OLSR: Energy Efficient OLSR routing protocol for mobile ad-hoc networks. In: IEEE Military communications conference, MILCOM'08. San Diego, CA. November 16–19, 2008.

[32] Hamrioui S, Daoui M, Chamek L, Lalam M, Lorenz P. Incidences of the improvement of the interactions between MAC and routing protocols on MANET performance. J Adv Comput Sci Technol 2012;1(4):250−65.

[33] Hamrioui S, Lalam M. Incidence of the improvement of the transport—MAC protocols interactions on MANET performance. In: 8th Annual international conference on new technologies of distributed systems (NOTERE'08). Lyon, France; 2008.

[34] Hamrioui S, Lalam M, Lorenz P. Effets de la Mobilité sur les Protocoles de Transport et de Routage dans les MANET, 12émes Journées Doctorales en Informatique et Réseaux (JDIR'11), UTBM, Belfort, France; Novembre 2011. p. 24−5.

[35] Hamrioui S, Bouamra S, Lalam Mustapha M. Les Effets de la Mobilité des nœuds sur la QoS dans un MANET. The Maghrebian Conference on Software Engineering and Artificial Intelligence (MCSEAI'08), Oran, Algérie; April 2008. p. 28−30.

[36] Elliott DK. Understanding GPS: principles and applications. 2nd ed. Artech House Publishers; 2005.

[37] Doherty L, Pister KSJ, El Ghaoui L, Convex position estimation in wireless sensor networks. In: Proc. of the IEEE INFOCOM, Alaska, vol. 3; 2001. p. 1655−63.

[38] NSNAM Webpages. The network simulator ns2. Information Sciences Institute (ISI). Available from: <http://www.isi.edu/nsnam>.

[39] Hyytiä E, Virtamo J. Random waypoint model in n-dimensional space. Oper Res Lett 2005;33:567−71.

CHAPTER 3

Rate adaptation algorithms for reliable multicast transmissions in wireless LANs

Stefano Paris[1,2], Nicolò Facchi[3], and Francesco Gringoli[3,*]
[1]*Huawei Technologies Co. Ltd., Paris, France*
[2]*Paris Descartes University, Paris, France*
[3]*CNIT/University of Brescia, Brescia, Italy*

1 INTRODUCTION

IEEE 802.11 is becoming the most common technology for providing broadband access to user equipment over unlicensed bands: while all new smartphones and laptops integrate Wi-Fi connectivity, manufacturers are increasingly dropping the adoption of Ethernet connectors because of space and budget constraints. This wireless technology is hence superseding wired networking in many contexts, including hotspot classic applications, but also high-speed home and corporate access, relegating Ethernet to specific scenarios that require 10 Gb/s or more. Thanks to its general low cost and ease of deployment, validated Medium Access Control (MAC) algorithms like the *distributed coordination function* (DCF) and the possibility of adapting the transmission rate to channel conditions, Wi-Fi technology has proven to be almost unbeatable for the low latency and reliable delivery of unicast data. Unfortunately, IEEE did not provide adequate maintenance of multicast access, which was downgraded to a supporting facility (e.g., for nonconfigured DNS resolutions [1]) instead of being promoted to a core technology for multimedia delivery; though there are many proposals for handling reception failures with both retransmission and forward error correction [2–8], no proper mechanisms have yet been included in the standard. This has also led to a few attempts [9–12] to adapt the rate to the environment conditions.

Recently, however, IEEE tried to counter this trend with the ratification of the 802.11aa document that introduces new feedback-based mechanisms for reliable delivery of multicast traffic [13]. Although it has raised research interest [14], no chipset provides support for it so far. Apart from a preliminary and experimental

*Corresponding author: francesco.gringoli@unibs.it

implementation working on off-the-shelf equipment [15], many contributions have focused only on modelling the maximum throughput it can achieve under different operating conditions [16,17]. We hence provide in this chapter a thorough and detailed framework for assisting experimentation with this new standard inside Network Simulator v2 [18].

A key contribution of this work is an easily extensible application programming interface for enabling a multicast transmitter to dynamically adapt the *modulation and coding scheme* (MCS) to the environment conditions, using the feedback made available by the 802.11aa protocol itself. As with unicast transmission, in fact, the correct choice of MCS allows increasing the throughput while keeping losses negligible, by maximizing the reception probability according to the quality of channel. Rate adaptation algorithms have received significant attention, as the problem they are called to solve is complicated by the proliferation of flexible PHY supporting the different transmission modes (single/multi-antennas, short/long preambles, bonding of multiple channels, etc.) and coding schemes dictated by the standard, with a specific behavior characteristic of the chipset. For ease of implementation and deployment over CPU limited equipment such as inexpensive off-the-shelf Access Points (APs), such algorithms have to be simple; for this reason, apart from minor changes, they usually build on straightforward feedback mechanisms for assessing the channel quality towards each receiver and opportunistically selecting the PHY mode and the corresponding data rate as a trade-off between modulation robustness and channel occupancy time. In the case of unicast frame delivery service, based on positive acknowledgments from the receivers, feedback information is easily available. Previously to the introduction of the 802.11aa amendment the multicast delivery service lacked such information (frames had to be transmitted once and with no acknowledgment) and multicast transmitters always had to stick to the basic rate. The new 802.11aa amendment is pushing the multicast service into a rapid evolution: first, it defines a new family of Group Address Transmission Services (GATS), based on feedback from the receiver, that improve the reliability of multicast transmission; second, it enables the transmitter to select the best MCS, according to the (multiple) feedback received by the stations belonging to the same Group Address. It is worth mentioning that proprietary mechanisms are now appearing in closed source drivers from major manufacturers, such as the *wl* configuration tool from Broadcom [19]. Motivated by the lack of an "almost definitive" rate-control algorithm for the multicast traffic equivalent to the Minstrel-HT for the unicast case [20], we also provide NS-2 with APIs for using the reports collected from stations by the 802.11aa protocol and setting up the rate.

The main contributions of the work presented in this chapter are hence: (i) an exhaustive implementation in NS-2 of 802.11aa with Block-Acknowledgment (whose performance has been validated comparing simulations against experimental results) that can be used to study the dynamics of this new protocol with its overhead for providing reliability to the multicast access; (ii) an extensible API for using the stations' reports for rate-adaptation purposes, allowing developers to use our extensions and study new algorithms for adapting the frame rate to the environment; (iii) three rate-adaptation mechanisms that demonstrate the flexibility of our framework and the possibility offered by the 802.11aa protocol itself.

The rest of the chapter is organized as follows: we discuss following this section the general reasons behind rate adaptation while we report in Section 2 known rate-adaptation approaches for unicast and multicast access. We introduce in Section 3 the 802.11aa mechanisms known as Group Addressed Transmission Services (GATS), describing also a real implementation that we tried to accurately model in our NS-2 extension that we present in Section 4. Section 5 describes the rate-adaptation mechanisms that we add to validate the MCS selection APIs. Finally, we show in Section 6 the overall performance and we conclude the chapter in Section 7.

1.1 MODULATION SCHEMES AND THEORETICAL THROUGHPUT

The reliability and performance of wireless communications are highly affected by several time-varying effects, like signal attenuation, channel fading due to multipath propagation, and interference caused by other transmissions occurring on overlapping frequencies [37]. Node mobility and the presence of moving obstacles, such as people, make the modelling of the communication channel even more complex so that it is basically not possible to know in advance the overall node-to-node signal-to-noise ratio (SNR). To counter this issue, wireless transmitters can choose among different types of modulations of increasing reliability and decreasing line-rate. For every specific modulation, the transmitter transforms the data stream (i.e., the frame) into a sequence of symbols, which are later encoded as changes of physical properties of a waveform (i.e., amplitude and phase). The amount of information that can be transmitted depends therefore on the number of possible values carried by each symbol (i.e., the constellation size) and this reflects on the line-rate if symbol duration is fixed. However, the higher the constellation density, the higher the SNR needed to correctly decode the symbols as they get closer and closer in the signal-space diagram, or conversely, the higher is the symbol-error-rate (SER) for a fixed SNR. For this reason, if the SNR is below a modulation specific threshold, the receiver may incorrectly decode some symbols and recovering the original data (the frame) might be impossible, even when some *forward error correction* (FEC) technique is used [38]. To help the receiver understand whether a frame is correct or not, a wireless transmission should embed a *frame check sequence* (FCS) somewhere in the data payload. The receiver could then compute the same value and compare it with that received, discarding the frame if they do not match, otherwise acknowledging it; failed frames may be retransmitted later by the source. Apart from the transmitted data, a wireless transmission should also include a preamble, for specifying the modulation type used in the data part and for helping the receiver to synchronize; it is worth it to say that this preamble should be transmitted using a fixed modulation so that all receivers may easily sense when a transmission is occurring. In any case, for a given modulation and SNR it is possible to compute the SER, that together with the frame length and the channel coding (i.e., number of symbols composing the frame and FEC style) set the packet-error-probability (PER). Finally, knowing the preamble duration, it is possible to compute the theoretic throughput as we did in Figure 3.1 for a wireless transmission employing a 802.11-like receiver. There we show the theoretical throughput as a function of the SNR measured at the receiver for different modulations, assuming 1 MSymbol/s, additive Gaussian white noise (AGWN), no FEC, frames of 1500 bytes and an ideal zero-length preamble.

More specifically, the IEEE 802.11 standard [21] defines many different modulations with rates ranging from a minimum of 1 Mb/s (802.11b, encoded with binary-phase-shift-keying, no FEC, in a 20-MHz bandwidth) up to a maximum of 866.7 Mb/s (802.11ac, encoded with 256-QAM, 5/6 convolutional encoding FEC, short guard interval, in a 160-MHz bandwidth) for every single spatial stream: when multiple spatial streams are used in MIMO-enabled networks, the rate scales up with their number. Unicast data must be acknowledged with ACK frames and transmitted up to a maximum number of times that each manufacturer may fix to its own optimal value. For accessing the channel, the standard employs *carrier sense multiple access with collision avoidance* (CSMA/CA) as the medium access control (MAC) protocol. It is implemented by running on all nodes a distributed coordination function (DCF) algorithm based on a binary exponential backoff (BEB) mechanism that defers user transmissions for a random number of empty slots in the attempt of reducing the probability of collisions. Given all these premises, choosing the right modulation can be a tough task: e.g, considering only the PER can be misleading, as a modulation can exhibit a higher PER than another but still allow a higher throughput because of the more transmitted packets per unit of time. For this reason, apart from fairness issues that we do not want to take into account in this chapter, the modulation that guarantees the highest throughput in a given environment can be the optimal

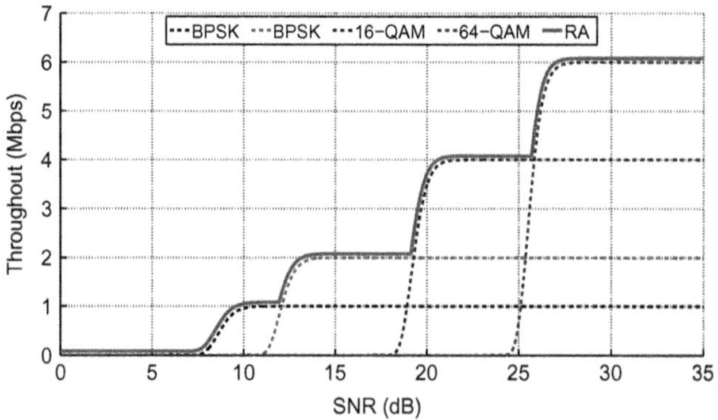

FIGURE 3.1

Theoretical throughput in Mb/s as a function of the SNR for different modulations using a 802.11-like receiver, assuming AWGN, 1 MSymbol/s, frames of 1500 bytes and no preamble.

candidate: however it could change both over time, because of varying channel conditions, and over space, for instance when the node moves as we report in Figure 3.2 with the variation of the received power (Power, top) and the PER (bottom) for a fixed modulation set to 48 Mbps. At the beginning the node is not moving (and still the Power changes), until when in $t = 20$ sec, the node leaves its original position and moves far away from the receiver; at some point in $t = 32$ sec the SNR is so low that the PER hits its maximum and all packets are lost. As soon as the node again gets close to the Access Point, the received power is enough to allow good recovery of the modulated frames.

This variability calls for algorithms, which we call *rate controllers*, able to select the best modulation according to channel conditions, choosing the best setting for achieving the maximum theoretical transmission rate. In other words, rate adaptation algorithms try to approximate the envelope of the theoretical throughput (e.g., the grey curve "*RA*" of Figure 3.1). However, such a task requires the design of a feedback mechanism to feed the transmitter with the channel quality measured at the receiver, since the transmitter cannot know such information due to the asymmetry of the wireless channel. For 802.11 unicast traffic, feedback is provided by the ACK frames, whereas for multicast traffic there has been no feedback till the introduction of 802.11aa, which we will describe later on.

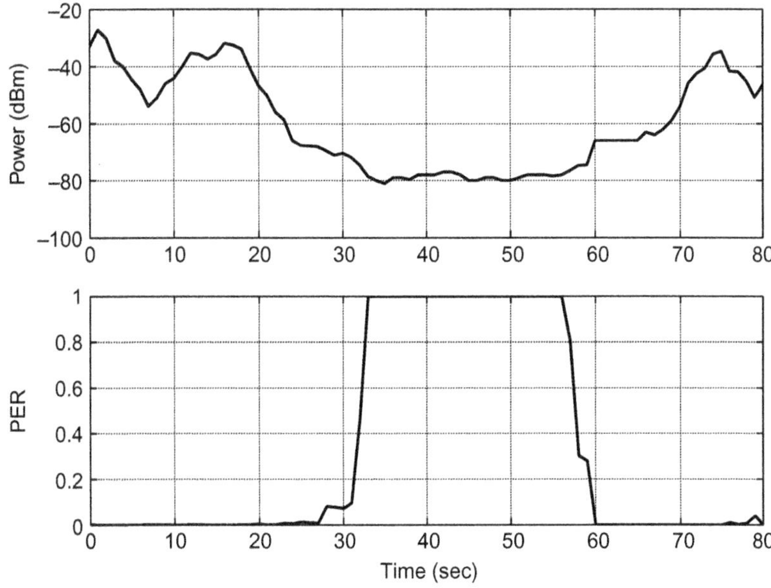

FIGURE 3.2

Received Power and Packet Error Rate when the transmitter (mobile node) moves away from the receiver (AP) using a fixed transmission rate of 48 Mbps.

To conclude, the design of any rate adaptation mechanism requires the specification of the information on the channel quality used to select the best modulation and its sampling rate, as well as the timescale at which the rate selection is performed. Note that the sampling and rate selection timescales may differ. Indeed, the rate adaptation algorithm should be synchronized with the variation speed of the channel quality to avoid unnecessary changes of the transmission rate, during short variations of the SNR due, for example, to temporary signal attenuation or interference (e.g., the presence of a mobile obstacle or the interference caused by a beacon transmission from another WLAN operating on the same frequency).

The type of information used to estimate the channel quality permits us to broadly classify rate adaptation techniques into three main categories: frame reception/loss rate, BER, and SNR. Following this chapter, we will describe the main rate adaptation schemes proposed by the research community and industry.

2 RELATED WORK

Following the evolution of the IEEE 802.11 documents [21], all radio manufacturers have introduced in the last few years extremely flexible PHYs that support different modulations: e.g., legacy *direct sequence spread spectrum* (DSSS, 2.4 GHz only), and Extended Rate PHY-Orthogonal Frequency Division Multiplexing (ERP-OFDM, originally defined by 802.11a for the 5-GHz band and backported to 2.4.GHz). While DSSS supports only four different rates (1, 2, 5.5 and 11 Mb/s), ERP-OFDM supports several Modulation and Coding Schemes (MCSs) that can be obtained by combining different basic parameters, including: guard interval between consecutive symbols (long/short); bandwidth (20 MHz, 40 MHz for 802.11n, 80 MHz, 160 MHz for 802.11ac); total number of carriers; and, finally, the number of spatial streams that can be transmitted together thanks to the MIMO approach; this leads to a data rate of the MAC payload ranging from 6.5 Mb/s up to approximately 6.933 Gb/s. (802.11ac, 8 spatial streams, Short Guard Interval, 160 MHz) Although the standard document does not mandate any particular strategy for selecting the MCS, it requires that receivers not be bound to any in particular and must synchronize and decode all MCSs they support. For this reason, the variety of MCSs pushed the development of rate adaptation algorithms (RAAs) that can properly select the most convenient rate according to channel conditions—e.g., the one that maximizes the throughput, or, considering more sophisticated cost-functions, the one that maximizes the throughput and minimizes losses (or, equivalently, the number of retransmissions that can also be performed with different MCSs). The key issue to solve is the design of an effective feedback mechanism for assessing the channel quality towards each receiver and opportunistically select the MCS (and the corresponding data rate) as a trade-off between modulation robustness and air–time. Multicast scenarios increase the overall complexity, as the transmitter must handle multiple feedback simultaneously.

Minstrel-HT [20] is today a well-known RAA that can be used with many drivers in the Linux kernel: it keeps evaluating the delivery probability performance of every MCS by using look-around frames during normal operations, and selects on average the MCS with the best performance. Except for Minstrel-HT, published as open-source software, many of the other available RAAs are proprietary and usually integrated with state-of-the-art closed-source drivers that power common off-the-shelf Access Points from Linksys, D-Link, and Netgear (which in turn use radios from Broadcom [22], or Qualcomm Atheros [23], the last a follow-up of the MadWifi technology [24]).

Apart from these implementations, solutions explored in the scientific literature consider both open-loop and closed-loop approaches based on different physical parameters, such as signal-to-noise ratio (SNR), or bit error rate (BER), related to channel quality. In many cases these schemes require changing the frame formats (for coding BER measures performed at the receiver side), the channel access operations (using RTS/CTS or basic access according to losses) or the frame handshake sequences (using a variable CTS transmission rate as a parameter robust to collisions for coding the channel quality). As no commercial card supports such customizations, most of these schemes have been only simulated or validated with simplified implementations over open-source drivers and SDR platforms. In particular, the Collision Aware Rate Adaptation scheme (CARA) proposed in [25] combines information obtained both from the RTS/CTS frame exchange and the CCA-based back-off to evaluate the channel status and selects the transmission rate accordingly. In [26] the authors improve the stability of frame loss based schemes through the design of the Robust Rate Adaptation Algorithm (RRAA) that uses a transmission rate as long as rate switching does not lead to goodput enhancements. Reference [27] further extends RRAA considering past transmission outcomes to avoid the selection of those data rates that offered low performance.

SNR-triggered schemes select the highest transmission rate that permits obtaining a target frame delivery rate using SNR–BER relationships. To improve the accuracy of the prediction mechanism in the presence of external interference, which affects the SNR–BER relationship, [28] proposes the SGRA (SNR-Guided Rate Adaptation).

BER-based solutions like SoftRate [29] and AccuRate [30] estimate respectively the bit error rate (BER) and the error vectors of every PHY symbol to predict the expected frame delivery rate achievable using a given transmission rate.

Slightly more complicated, H-RCA [31] aims at maximizing the total throughput by minimizing the average time spent on the medium by each packet for being delivered either after the first attempt or with retries. To this end authors developed a technique for determining when a failure is due to a collision or to channel noise, so that they can base the MCS decision only on the latter, which proved by simulations to be a robust choice with respect to different environments.

Conversely, multicast transmissions have received less attention due to the complexity of implementing an efficient feedback mechanism to collect the information used by the rate selection routine. For example, multicast rate adaptation schemes like [9,10] adjust the data rate according to the lowest SNR among those perceived by the stations of the multicast group. However, the rate control based on the worst SNR does not necessarily guarantee the same delivery probability to all stations of the multicast group, since the joint reception probability may differ. In [11] the data rate is instead adjusted by measuring the quality-of-experience at the receivers where a Pseudo-Subjective Quality Assessment tool is run for computing mean opinion score in real-time. In [12] authors detect collisions by taking into account both the *received signal strength indication* (RSSI) and the *packet error rate* (PER) that they use together to understand when losses are due to collision and decide the best rate when there are no collisions. Unlike previous solutions, rateless mechanisms like Strider [7,8] exploit forward error correction schemes to improve the throughput of wireless communications.

Finally, [32] and [33] have recently introduced some simple feedback-based methods for the evaluation of the joint delivery probability measured by receivers: one uses a *Linear increase/Multiplicative decrease* approach for choosing the MCS, while the other evaluates continuously all the MCSs by transmitting some look-around frames while keeping on average the best-performing one.

3 THE 802.11AA GROUP ADDRESS TRANSMISSION SERVICE

The multicast delivery service was introduced with the first publication of the 802.11 standard in 1997. To keep the complexity of the hardware within acceptable limits and avoid a dramatic increase of the production cost, the standard provided reliability only to the unicast delivery service; a protocol for collecting positive feedbacks from multiple destinations requires, in fact, a complex state machine that was not compatible with the hardware designed for mass production of cheap wireless Network Interface Cards (NICs). For this same reason, the standard mandated also a fixed modulation and coding scheme (MCS) for multicast frames, corresponding to the Basic Rate of the Basic Service Set, so that no feedback had to be collected for estimating the channel quality jointly perceived by the intended destinations. Surprisingly, given both the evolution of the hardware capabilities and the impressive improvement of the available MCS introduced by 802.11n and later by 802.11ac (up to gigabit/s with multiple spatial streams), the multicast service did not undergo major changes till the ratification of the 802.11aa amendment in 2012 [13]. This document finally enabled both reliability and improved performance in the multicast service through the definition of the Group Address Transmission Services (GATS), which we introduce in the following subsections.

3.1 THE 802.11AA GROUP ADDRESSED TRANSMISSION SERVICE

Before introducing GATS, the key novelty of 802.11aa, we quickly review the default multicast scheme, which we call "legacy" in Figure 3.3. The multicast

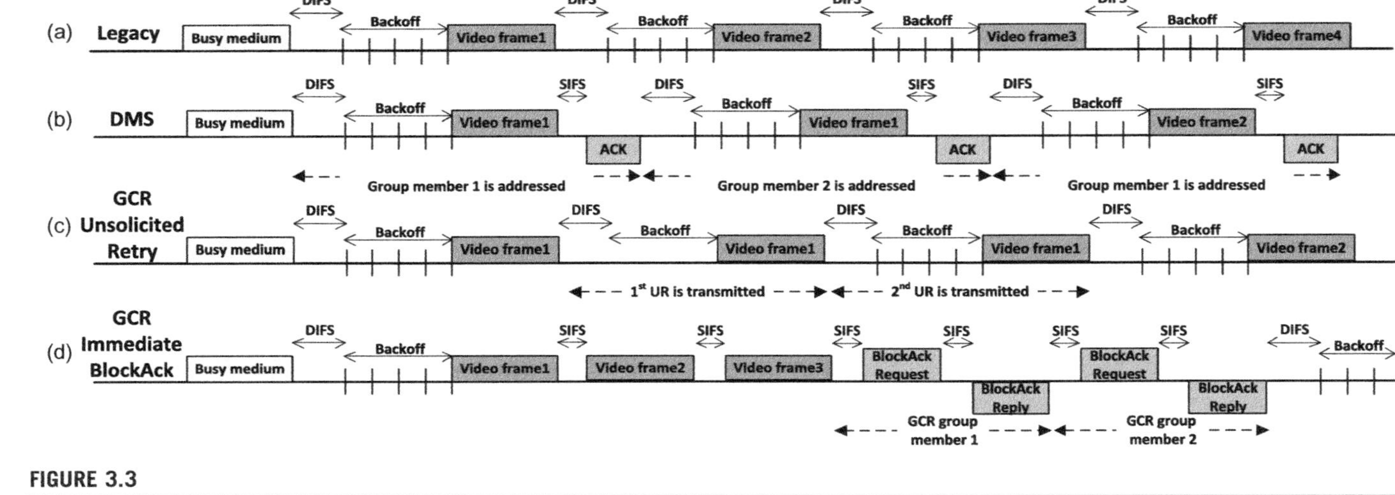

FIGURE 3.3

Channel access times for different multicast mechanisms.

source transmits each frame only once at the Basic Rate: since for every frame it is always the first channel access attempt, the minimum Contention Window (CW) is used (in case 802.11e extensions are used, then the value of CW and the AIFS specific to the multicast queue are used), although backoff is always performed and the multicast traffic has to compete with traffic coming from other stations, and may eventually collide. Unfortunately, no acknowledgments are collected from the intended destinations and some or even all receivers may lose the data. For this reason, multicast applications had to take care of retransmissions by themselves when running over a legacy interface. Besides reliability issues, legacy service may affect the achievable performance of the whole network. Indeed, forcing a low MCS triggers the well-known performance anomaly problem [34].

GATS addresses all these problems with three different approaches of increasing complexity, all building on the concept of transmitting every frame multiple times, either *pre-emptively* or *on demand*. All approaches share also a new addressing scheme that groups a set of destination stations under the same Groupcast Concealment Address (GCA). The election of the GCA and the procedures for adding new stations to the group upon request lie outside the definition of the three GATS mechanisms and can be based on existing protocols such as IGMP; for this reason we do not add these features to our proposed simulation framework, which is instead focused on the evaluation of pure performance. Programmers have to manage group formation, either statically, as done in the released example, or by implementing a signaling protocol.

In the design of the GATS mechanisms, IEEE exploits MAC techniques that have been proposed in the past for different purposes (e.g., Block Acknowledgment) and have been tested by users for a long time: for this reason the implementation of the standard on real equipment should be pretty straightforward. However, as recently reported in [15], a port to existing and not explicitly 802.11aa designed NICs could face some performance issues: we report some information on this implementation at the end of this section as it inspired our NS-2 extensions.

GATS consist of one scheme inherited from 802.11v, and two new schemes based on the GroupCast with Retries (GCR) service, as illustrated in Figure 3.3. We introduce a detailed description of all these schemes in the following paragraphs.

3.1.1 Directed multicast service

Directed Multicast Service (DMS) was specified by 802.11v to add reliability for specific types of management traffic. DMS achieves this by performing as many unicast transmissions as the number of original receivers. Such a mechanism requires at least as many transmissions as the number of stations that belong to the multicast group. This represents the easiest service to include in the NS-2 network simulator, since we only have to intercept the multicast frame at the application layer and replace it with N unicast frames, copying the destination addresses from the static list we are using to represent the GCA. With respect to the performance, transforming multicast into unicast frames is revealed to be very resource-consuming and inefficient as the number of channel accesses (each with its own backoff as depicted in Figure 3.3) can waste a lot of airtime. However, given that the rate is chosen by the rate controller for the specific destination, as in the case of unicast traffic, DMS can achieve higher throughput than legacy service. In order to benefit from retransmissions but without such extreme inefficiency, 802.11aa specifies two GCR mechanisms, one relatively simple but not very efficient, and another that ends up more complex but that improves the usage of wireless resources.

3.1.2 GCR unsolicited retries

Unsolicited Retries (UR) retransmits each multicast frame up to R times, with each transmission being an independent channel access so that the backoff is always executed with the same CW parameter. GCR-UR obviously reduces the impact of channel errors but at the cost of a huge waste of resources as the retransmission is done in a preemptive fashion without considering any feedback from receivers; in Figure 3.3 we reported the case $R=3$. Given no feedback is deployed at the MAC layer, this mechanism should not require a great implementation effort, as it is only needed to duplicate each frame at the transmitter output queue. Unfortunately, it does not provide any feedback to the transmitter and for this reason we left it out of our evaluation.

3.1.3 GCR with immediate block acknowledgment

As the name states, this mechanism uses the Block Acknowledgment (BA) feature introduced by the 802.11e amendment. Even though in 802.11e BA was defined to acknowledge multiple unicast frames addressed to the same receiver with a single ACK, in 802.11aa it is extended to the multicast case by simply sending the BA request to all GCR stations in a round-robin fashion as we pointed out in Figure 3.3 (bottom). Each destination stores the success/failure history of the most recently received frames with their sequence numbers (SNs); when asked by the transmitter, it builds the BA including the SN of the first successfully received frame and a bitmap where bit k indicates whether frame SN + k has been received or not. After collecting all the BAs, the transmitter can reschedule the transmission of only those frames that were not acknowledged by all receivers. Since the document standard does not mandate a specific scheduling policy, we implemented our own: the multicast source first waits for M frames in the upper layer queue; it then schedules the first transmission of an M-sized burst, with packets transmitted back-to-back spaced only an SIFS; polls all the stations by sending BA Request (BAR), and collects all the BA responses. In case a BA is lost, it retransmits the BAR to the corresponding station up to N_1 times. Note that the definition of the best behavior after N_1 consecutive BA losses is still an open question, since a station that cannot receive several consecutive management frames transmitted at the basic rate without interfering transmissions will be unable to correctly decode data frames. Given its underlying complexity, this scheme required major changes at the simulator. With respect to efficiency, it is worth noting that in the case of perfect channel conditions, a wrong setting of the parameters may result in GCR-BA performing worse than GCR-UR (with $R = 0$) or even legacy, as the BAR-BA procedure is performed anyway. Also, in this case the standard does not specify the transmission rate but, differently from GCR-UR, we can profitably use the underlying MAC layer feedbacks to collect some indication of the channel state. As from [15], this scheme turned out to be the best performing one under many operating conditions. Therefore, in the numerical evaluation we illustrate only the performance of GATS with BA using our extensions of NS-2.

3.2 IMPLEMENTATION OF GATS: A REAL CASE

Before moving to the NS-2 implementation, we want to emphasize that the current generation of NIC is simply an incremental evolution of the legacy 802.11a/b/g architectures, which have been designed to support single-packet-single-acknowledgment transmissions within the time constraint of the DCF access strategy. When they are used to transmit burst of packets separated by SIFS interval in a TXOP (e.g., with HT-PHY in 802.11n), it's the kernel that handles the transmission of the BAR, processes the BA and retransmits the missing frames. When state-of-the-art VHT-PHY NIC (i.e., 802.11ac compliant) transmits an Aggregated Mac Protocol Data Unit (A-MPDU), the retransmission of the MPDUs that have not been correctly received is handled by the NIC, but as can be verified by sniffing the channel, all MPDUs between the first and the last not acknowledged (included) are retransmitted. To understand what the consequences are of the possible implementation of 802.11aa on top of these NICs we report in Figure 3.4 the high-level architecture of the transmission path of a 802.11 NIC where direct memory access (DMA) transfers data between the memory of the main host running the operating system (OS) and the NIC circuitry. A kernel thread writes frames into a consecutive set of memory pages which are arranged in a circular ring style: it then programs the DMA controller on the NIC to produce incremental addresses for fetching contiguous rows of memory cells and storing them into the internal NIC buffer. This technique emulates a simple FIFO queue for pushing frames to the NIC; if more FIFO queues are needed, such as for supporting Quality-of-Service (QoS) with the more recent Enhanced Distributed Channel Access (EDCA) function, multiple DMA controllers are programmed to present different memory windows to the NIC. The benefit of the DMA-based approach is twofold: i) it guarantees that the NIC is always fed with frames without having to ask for new ones to the main host; ii) it relieves the main host CPU while the NIC is transmitting frames. Once the NIC has handled a frame, it acknowledges the kernel raising an IRQ, the corresponding DMA controller starts fetching the next frame from memory and the kernel reclaims the DMA slot in the ring for new frames that will eventually come.

On the NIC side, the internal CPU can focus on re/transmission operations, as host communication is offloaded by the DMA subsystem. This internal CPU, that

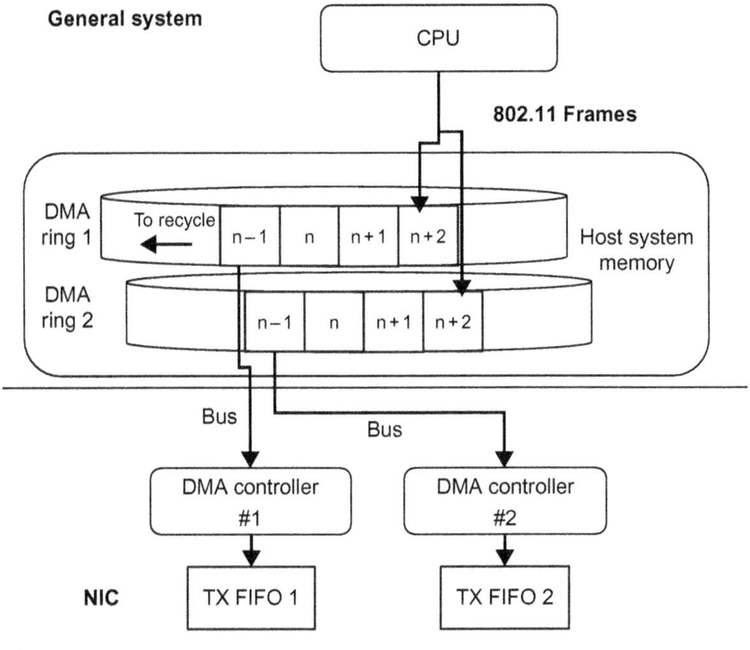

FIGURE 3.4

Overview of a WiFi NIC: every packet is available only once. After (re)transmission the FIFO advances to the next packet and the previous one cannot be accessed anymore.

has direct access to each HOL frame in every queue, executes an endless loop and continuously checks for received frames from the PHY. It also schedules frame transmission of the queue with the smaller backoff counter at the correct slot. Thanks to this approach, the NIC can easily satisfy the strict time requirements imposed by the standards without introducing the significant and unpredictable delays that would appear if the main host were involved.

It is worth noting that this approach is perfectly sound for those mechanisms (e.g., the regular operation of DCF or EDCA) where each Head-of-Line frame is transmitted up to a maximum number of attempts before passing to the next one in queue. It also matches the requirements of burst transmission with BA, as the packet to be retransmitted will be regenerated by the kernel. And finally it meets those of AMPDU as the aggregated data that includes all the missing frames will be retransmitted from the queue by simply skipping the initial MPDUs, successfully delivered in previous attempts, at the cost of transmitting also those frames in the middle that have already been received and acknowledged.

Unfortunately, as it turned out from [15], this architecture cannot easily handle GCR with BA in its fastest variant, where missing frames are retransmitted immediately after the BAR-BA procedure. Although the results reported in the paper hold for a specific chipset produced by Broadcom (the 4318KFBG, supporting

Table 3.1 Time Required to Flush Retransmission Queues is Not Negligible and Leads to Inefficiency (η as Percentage of the Used Air-Time)

M	Theoretical R_i (Mb/s)	Experimental R_e (Mb/s)	Flushing (ms)	η
8	36.76	22.75	1.25	38.11%
16	40.13	26.82	1.65	33.16%
32	42.06	27.95	2.25	33.55%

802.11bg), authors analyzed the behavior of a very recent one (the 43460, supporting 802.11abgn/ac) and they found out that the architecture at the base is exactly the same, with the same limitations. The main problem is that the NIC cannot cache internally frames in a burst as there is not a specific memory for storing temporal data: frames can be simply drawn from the queues, following the FIFO access style. Once a frame is transmitted and the NIC switches to the next one in the queue, the previous one is lost. Another problem is that, according to 802.11aa, the BAR-BA procedure might start immediately after the transmission of the last frame in the burst, and retransmission of the first lost frame could be scheduled immediately after the reception of the last BA. For these reasons, the kernel cannot handle the collection of the BAs: the poll must be driven by the NIC itself. The only way for implementing 802.11aa on top of the current generation of NIC was hence to use the available FIFO queues, which are normally used for running EDCA channel function, to store the same burst of packets multiple times and guarantee to the NIC the availability of the same frame in case it must be retransmitted. So the burst in the first queue is completely transmitted as an *original* transmission, while the burst in the second queue is used to retrieve lost frames. However, as reported in Table 3.1, this approach has a drawback, as the time required to flush unneeded frames from the additional queues is not negligible and depending on the particular combination of lost frames efficiency does not hit 100% if compared to theoretical time.

We stressed this because if people plan to model 802.11aa with our NS-2 extensions, which we describe in the next section, they must keep in mind that an experimental setup with real devices could show less efficiency than the simulation if the DMA subsystem is not taken into account. For the sake of completeness we added in our framework an initial support for simulating this problem by adding configurable idle time to every transmission round. We plan to better simulate the DMA subsystem in a future work.

4 NS-2: RATE ADAPTATION LIBRARY AND GATS EXTENSIONS

In this section we introduce the Rate Adaptation Library that was added to Network Simulator 2 (NS-2) [18]: we considered it as a starting point for our multicast rate adaptation framework that we detail in the next section. We also introduce the GATS Extensions that we developed to evaluate the performance of the multicast rate adaption framework using the feedback information provided by the 802.11aa protocol.

4.1 RATE ADAPTATION LIBRARY

Network Simulator does not support natively any rate adaptation mechanism for the IEEE 802.11 standard. The dei80211mr library proposed in [35] represents the first attempt to provide a revised and more realistic implementation of the IEEE 802.11 protocol for Network Simulator Version 2. The library improves several functionalities of the CMU implementation of the 802.11 standard [36], providing a more realistic behavior of the Physical and MAC layers. More specifically, the dei80211mr library provides all the Modulation and Coding Schemes defined in the IEEE 802.11b/g standards to simulate the different transmission rates that can be used for unicast transmissions in a real wireless LAN. Moreover, the library introduces a packet error model based on the *signal to interference plus noise ratio* (SINR) to decide whether the 802.11 frames can be decoded and received correctly. The PER, which permits the decision whether to receive or drop a frame, is computed using predetermined PER-SINR curves and the packet size. These curves, along with the noise power, can be dynamically modified by the user at simulation time to accurately simulate the WLAN environment. The SINR-based reception model represents an important enhancement with respect to the CMU implementation of the 802.11 standard (the basic implementation integrated in NS-2), which simply uses a threshold on the received signal to decide whether to receive or drop a frame. Indeed, such an approach does not take into account the interference caused by simultaneous transmissions that highly affect the received signal and the PER. Furthermore, to provide a more realistic interference model, the interference is computed using a Gaussian model to account for all conflicting transmissions that occur simultaneously with the one performed by the source node. The dei80211 library replaces also the capture threshold with a model to simulate the capture effect, namely the ability of wireless cards to decode correctly the received frame even if there are simultaneous interfering transmissions that overlap the transmitted signal at the receiver.

Finally, the Carrier Sense (CS) threshold that was used to identify nodes affected by an interfering data transmission is replaced using a fixed value in meters for the distance representing the interference range, which can be set at the beginning of the simulation. Indeed, with the utilization of the Gaussian interference model, a node transmission below the CS threshold often provides a non-negligible contribution to the overall interference. The default value for the interference range is set so that all nodes are considered for the interference computation (i.e., all nodes are in the reciprocal interference ranges). While this value yields accurate simulations, the computational time to obtain the solution may be too large. Therefore, the dei80211mr library permits modification of this parameter at the beginning of the simulation to trade off between computational load and simulation accuracy.

4.2 ARCHITECTURE OF THE RATE ADAPTATION LIBRARY

The library defines a set of classes for the MAC and PHY layers of the IEEE 802.11 as well as for the accurate simulation of the signal propagation, the interference computation and the rate adaptation. We first describe the architecture and the main functions of the MAC layer. Then, we briefly describe the rate adaptation classes implemented by the library.

4.2.1 MAC layer

The most important class, namely Mac80211_mr, which represents the main entry point for data transmission, implements the Medium Access Control scheme defined by the IEEE 802.11a/b/g protocols. Specifically, the class implements the CSMA/CA (Carrier Sense Multiple Access with Collision Avoidance), the exponential backoff and the ARQ (Automatic Repeat Request) mechanisms to coordinate the access to the channel in a distributed fashion. For an accurate modeling of the 802.11 standard, the class also provides the RTS/CTS (Request to Send and Clear to Send) message exchange to protect data transmissions from hidden nodes. Similarly to real wireless cards, the threshold on the packet size that triggers the RTS/CTS exchange can be adjusted by the user during the simulation. The processing of the events that are generated at the MAC layer, like packet transmission, packet reception, acknowledgement, RTS/CTS exchange and transmission deferral, are handled by specific timers implemented as friend classes of Mac80211_mr.

The packet transmission path (i.e., the set of main steps performed to simulate the packet transmission on the wireless channel) can be split broadly into two phases: the preparation and the transmission. The preparation phase is implemented by the send(), sendDATA() and sendRTS() functions that prepare data and RTS frames by setting the header's fields with the corresponding values. The functions sendACK() and sendCTS() prepare the transmission of the acknowledgment and the CTS, but differently from sendDATA() and sendRTS(), are invoked from the procedures that manage the reception of the data and CTS frames, respectively.

The transmission phase starts when the backoff timer has expired; the backoff handler, which is the function executed when the timer expires, invokes the check_pktTx() and transmit() functions that, in turn, compute the duration of the data transmission and pass the data frame down to the radio interface for the transmission over the wireless channel. Note that the computation of the interference is performed at the receiver side, when the reception event is triggered.

Similarly, the packet reception path can be split into three main phases: the preverification, the postverification and reception. The main function that implements the reception path is recv() that first verifies if the frame can be decoded and received correctly. Specifically, the function checks if the interface on which the frame should be received is not in transmission mode. If the interface is idle, the function schedules the reception timer, whose handler (i.e., the procedure invoked when the timer expires) performs the remaining postverification and reception phases.

The postverification phase, which is implemented within the function recv_timer(), computes the noise and the interference caused by all transmissions of nodes within the interference range that overlap temporally the transmission of the received frame. If the SINR is below the reception sensitivity threshold the frame is discarded; otherwise the frame can be correctly decoded. However, before sending the packet to the upper layer, the MAC checks the destination address in order to filter those frames whose destination address differs from the node address, and updates the NAV (Network Allocation Vector) information for the virtual carrier sense. If the frame passes both the preverification and postverification phases, the MAC layer can correctly receive it. For each type of frame, different functions are called according to the frame subtype to simulate different types of behavior that the MAC needs to implement. In particular, four functions have been defined: recvDATA(), recvACK(), recvRTS(), and recvCTS(). The function recvDATA() sends the packet to the upper layer and schedules the transmission of the acknowledgment frame to inform the transmitter of the correct reception. The function recvACK() stops the timer for the retransmission of the data frame and schedules the backoff timer for the transmission of a new frame. The function recvRTS() updates the NAV information and sends back a CTS frame, whereas recvCTS() stops the transmission and backoff timer to defer the transmission of the current frame.

4.2.2 Rate adaptation framework

The rate adaptation has been implemented as standalone class whose objects can be plugged into the Mac80211_mr class at the beginning of the simulation to modify and add functionalities that enhance its basic behavior. In particular, the dei80211mr library defines an abstract class called RateAdapter that adds the main functions and parameters to dynamically adjust the transmission rate of data frames. Every rate adaptation algorithm for unicast transmissions extends the RateAdapter class by defining further functions and members that implement the specific rules to adjust the transmission rate. More specifically, each class, which implements a specific rate adaptation algorithm, modifies the member dataMode_ of the class Mac80211_mr that stores the Modulation and Coding Scheme that will be used for the transmission at the PHY layer of the current data frame. The update of the MCS (i.e., the dataMode_ parameter) is implemented within a function that extends the basic behavior of sendDATA() to perform the rate adjustment on a per-packet basis or through a timer to collect statistics over a set of consecutive frame transmissions.

The dei80211 library defines natively three rate adaptation algorithms and corresponding classes: the Auto Rate Feedback (ARF), the Receiver Based Auto Rate (RBAR), and the SNR-based Rate Adaptation (RA-SNR).

When the ARF mechanism is set as rate adaptation, each transmitter increases the transmission rate after a fixed number of successful transmissions at a given rate and switches back to a lower rate after few consecutive transmission failures (usually one or two).

The ARF algorithm decreases the current transmission rate and starts a timer when two consecutive transmissions fail in a row. If the transmitter successfully receives ten acknowledgments or the timer expires, the ARF algorithm increases the transmission rate and resets the timer. In ARF, the first transmission after the rate increase is used as a probing transmission, since its failure causes the instantaneous decrease of the rate and the restart of the aforementioned timer.

In RBAR, each transmitter computes the MCS for the transmission of the current data frame by comparing the SNR (Signal to Noise Ratio) of previous transmissions for the same destination against a set of SNR thresholds computed using a wireless channel model. RBAR selects the transmission rate such that the frame's SNR estimated from previous statistics is higher than the SNR threshold of the corresponding transmission rate. In other words, RBAR attempts to use the highest transmission rate that satisfies the estimated SNR. Similarly, RA-SNR uses the highest transmission rate that guarantees a maximum frame error probability. Such parameter is provided by the user at the beginning of the simulation in order to limit the number of lost data frames. The relationships between the transmission rate and the frame error probability are estimated using the SINR and a wireless channel model. Furthermore, these relationships are improved using the information collected during the simulation.

Both RBAR and RA-SNR implement their rate adaptation rules by defining a function sendDATA() that is invoked by the corresponding function of the class MAC80211_mr. Conversely, the ARF mechanism implements its adaptation algorithm by defining a timer and two further functions that are invoked by the MAC80211_mr class each time a frame is successfully received or lost, respectively.

4.3 THE GATS EXTENSIONS TO THE MAC LAYER

The implementation of the IEEE 802.11aa amendment and the framework that permits exploitation of the information collected by the GATS polling mechanism to develop new rate adaptation mechanisms for multicast traffic required a deep revision of the MAC layer of the dei80211mr library. In particular, the class Mac80211_mr, which implements the MAC layer, has been extended to accommodate the information defined in the amendment, like the list of multicast stations, and the functions that permit simulation of the GATS polling and retransmission of lost multicast frames.

Regarding the GATS polling, we follow the same philosophy implemented by the library for management and control frames (RST, CTS, and ACK frames), dividing the transmission and reception paths into two and three phases, respectively.

Specifically, the transmission path of any GATS frame is divided into preparation and transmission phases, which are liable respectively for creating the GATS frame and transferring it to the radio interface for the following transmission on the channel. To this end, the library defines sendGATSPoll() and sendGATSAck () functions for the preparation of the GATS Poll sent by the AP and the successive GATS Acknowledgment sent by a multicast STA. Conversely, the function check_pktGATS(), which computes the duration of the GATS frame and passes it to the radio interface for simulating its transmission, is called each time the backoff timer expires. Note that during the GATS polling the backoff expires after a SIFS to prevent the transmissions of other stations during the polling period.

The GATS polling is activated when the AP has transmitted a preconfigured number of multicast frames, or equivalently when the transmission opportunity for the transmission of multicast frames expires. To store the information about stations that belong to the multicast group and the outcome of the GATS polling (i.e., the bitmap of each station and the success or failure of the GATS polling), the MAC has been extended with a list of multicast stations, whose entries are added whenever a new station joins a multicast group; it is worth saying that 802.11aa leaves the task of group formation to a different protocol (e.g., IGMP), which can be directly implemented in the *tcl* script that defines the simulation, since the list of multicast stations is mapped to the corresponding Otcl object for direct access in *tcl*.

Differently from the transmission path, the GATS reception path shares the preprocessing and postprocessing phases of other frames, and it simply defines a different reception function for each control frame, in order to implement different operations for the AP and multicast STA according to the received control frame. The reception of a GATS Poll, which is managed by the function recvGATSPoll() and executed only by multicast STAs, triggers the transmission of the GATS Ack that contains the information about the multicast frames that have been lost in the last transmission opportunity. The reception of a GATS Ack, whose operations are implemented by recvGATSAck() and executed only by the AP, activates either the transmission of a GATS Poll to a new multicast station or, in case all STAs have replied, the computation of the joint reception probability. At the end of the polling period, the MAC returns in the state before the execution of the GATS polling to restore the old behavior.

In addition to the extension of the dei80211mr library, the implementation of the IEEE 80211aa amendment required the addition of the control frames defined by the GATS service to the MAC layer of NS-2. To this end, we extended the MAC to add the necessary structures that define GATS control frames.

5 RATE ADAPTATION SYSTEMS FOR MULTICAST TRANSMISSIONS

In this section we describe the rate adaptation systems we have developed using the extended dei80211 library and that we have tested in real network scenarios

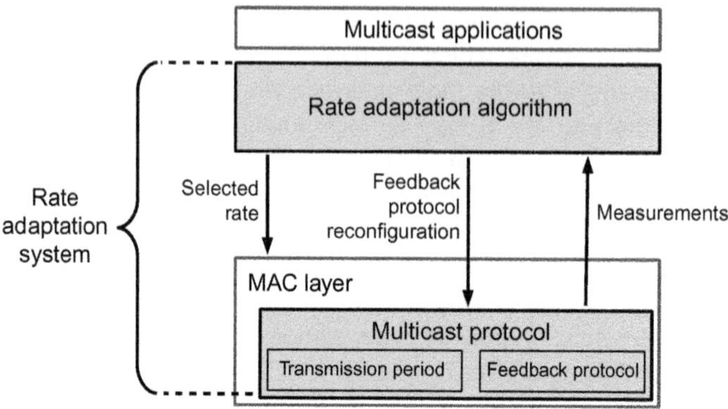

FIGURE 3.5

Rate Adaptation System Architecture.

[32,33]. We begin introducing the general architecture of a Rate Adaptation System (RAS) with a brief description of its main parts. Then, we dig into the details of the Rate Adaptation Algorithms (RAAs) that we designed. We explain how an RAA integrates with the presented RAS architecture and which parameters and information are used to select the best transmission rate.

5.1 RATE ADAPTATION SYSTEMS: ARCHITECTURE

The general architecture of a rate adaptation system can be thought of as composed of two levels, as illustrated in Figure 3.5. At the lower level there is the multicast protocol that runs inside the Medium Access Control (MAC) Layer. The multicast protocol provides functionalities both for multicast data frames transmission (transmission period) and for feedback collection (feedback protocol). In particular, the feedback protocol has the responsibility to periodically assess the status of the wireless channel between the Access Point (AP) and the stations (STAs) of the multicast group. Considering the GCR-BA scheme proposed by the 802.11aa standard that we described in Section 3, the transmission period corresponds to the transmission of the M-sized multicast data frames burst and the feedback protocol corresponds to the BA request/response exchange. When the feedback protocol terminates, it reports to the RAA how many multicast data frames transmitted during the last transmission period have been lost by each STA of the multicast group. A key aspect to take into account is that, in a real system, the feedback protocol must be part of the MAC layer. This is really important for two reasons. The first one is that if it is part of the MAC layer it is easier to make it a standard improving interoperability between vendors (see 802.11aa description in Section 3). The second reason, more technical and related to performance, is that the feedback protocol must be executed periodically in order to have constantly updated information about the status of the wireless channel. Moreover, in order to keep the collected information coherent with the real status of the wireless channel, the time that elapses between two consecutive runs of the feedback protocol must not be too long. This means that during a communication that can last for minutes or hours the feedback protocol can be executed hundreds or even thousands of times so, to reduce its impact on the multicast transmission performance, it must be executed as fast as possible. The only way to satisfy this requirement is to make it part of the MAC layer or, to be more precise, to make it part of the lower MAC layer, that is the part of the MAC layer that is usually executed inside the wireless NIC. In this way it is possible to satisfy the high timing requirements of the feedback protocol and give higher priority to frames of the feedback protocol with respect to data frames.

RAA lies on top of the multicast protocol: it takes as input the measurements collected by the feedback protocol and, using a heuristic logic, selects the best rate that will be used to transmit the multicast data frames during the next transmission period. The chosen transmission rate is then communicated to the MAC layer that has the responsibility to set the Modulation and Coding Scheme (MCS) for every transmitted frame. If the multicast transmission protocol provides retransmission functionality, the most sophisticated RAAs can provide the MAC layer with a list of transmission rates: the primary rate for the first transmission and the fallback rates for each retransmission. Finally, the RAA can also reconfigure dynamically the behavior of the multicast protocol. For example, it can set how often the feedback protocol must be executed (burst size) and which transmission rate to use for the feedback protocol frames (that are treated separately from the multicast data frames). A comprehensive discussion of the rate adaptation algorithms can be found in Sections 5.2 and 5.3.

To simplify the description, we set some limitations on the RAS that we present in the next sections. First of all, the RAA that we are going to describe does not dynamically reconfigure the multicast protocol. We suppose that the number of multicast data frames transmitted during a transmission period is fixed and never changes during the multicast session and that the frames used by the feedback protocol are transmitted with a fixed rate. Second, the RAAs that we are going to describe use a heuristic approach for selecting the best rate, considering only the single transmission case (if the multicast protocol provides retransmissions, the fallback rates will be equal to the primary rate).

5.2 RATE ADAPTATION SYSTEMS: OVERVIEW

In this section we describe in a more detailed way the blocks that compose a RAS, how they interact with each other and the parameters they use to achieve their goals. We will use the activity diagram illustrated in Figure 3.6 as the reference point for our description. Table 3.2 and Table 3.3 summarize the nomenclature and the parameters used by the RAAs we are going to describe.

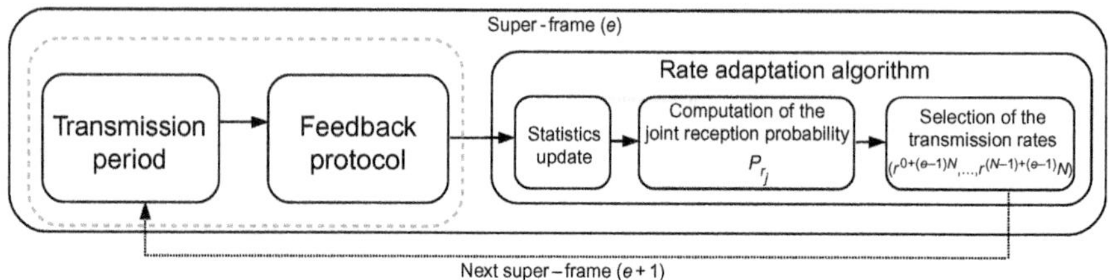

FIGURE 3.6

Activity diagram of a Rate Adaptation System.

Table 3.2 Nomenclature

Name	Description
f_i	Refers to the multicast data frame whose sequence number is i
r_j	Refers to the transmission rate of index j. We suppose the rate list contains the transmission rate in a nondecreasing order (where the transmission rates are compared to each other considering the bit rate in Mb/s). In case of 802.11g and 802.11n standards the index j corresponds to the MCS value provided by the standard. For example, considering 802.11g, $j=0$ corresponds to MCS 0 that is equivalent to a bit rate of 6 Mb/s; instead, $j=7$ corresponds to MCS 7 that is equivalent to a bit rate of 54 Mb/s.
$r_j(e)$	Refers to the transmission rate of index j used during super-frame number e.
r_b	Used to denote the best transmission rate.
r_l	Used to denote a look-around rate, that is the rate used for look-around frames.
r^i	Used to denote the transmission rate used for the transmission of the multicast data frame whose sequence number is i.
$b_s(e)$	It denotes the bitmap received from station s during the execution of the feedback protocol of super-frame e.
$b_{s,n}(e)$	It denotes the n-th bit of bitmap received from station s during the execution of the feedback protocol of super-frame e.
$b_{AND}(e)$	Bitmap resulting from the logical AND of all the bitmap $b_s(e)$ received from the STAs of the multicast group during the execution of the feedback protocol of super-frame e.
$b_{AND,n}(e)$	It denotes the n-th bit of the bitmap resulting from the logical AND of all the bitmap $b_s(e)$ received from the STAs of the multicast group during the execution of the feedback protocol of super-frame e.
$P_{r_j}(e)$	Joint Reception Probability of super-frame e for multicast data frames transmitted with rate r_j. It represents the probability that a multicast data frame, transmitted with the rate r_j during the e-th super-frame, is jointly received by all the STAs of the multicast group.
P_{r_j}	Estimation of the Joint Reception Probability for future super-frames. It represents an estimation of the probability that a multicast data frame, transmitted with the rate r_j, will be received by all the STAs of the multicast group. The value is updated every time $np_{r_j} \geq \beta$ (see Table 3.3 and Table 3.4).
S_{r_j}	Measure of successfulness of multicast data frame transmitted with rate r_j. Recomputed with P_{r_j}.
p_{r_l}	Probability that the look-around rate r_l will be chosen to transmit the next look-around frame.

Table 3.3 Parameters Used by the Rate Adaptation Algorithms

Name	Description
e	Super-frames counter. It is initialized to 1 and increased at the end of every super-frame.
N	Denotes the number of multicast data frames transmitted during each transmission period. Its value is fixed before the multicast communication starts and it never changes during the multicast session. If its value is small (e.g., 8–16) the feedback protocol will be executed more often introducing more overhead but, at the same time, the channel status information will be maintained updated becoming more reliable. Instead, if its value is big (64–128) the overhead due to the feedback protocol will be reduced, but the channel status information will be updated less frequently with the risk that they will not represent the real status of the wireless channel any more.
γ	$\gamma \in (0, 1)$ is the percentage of look-around frames transmitted during a transmission period. Look-around frames are used to maintain the channel status information up to date for every possible transmission rate. For this reason γ is an important parameter. As we will see, increasing the value of γ can help the RAA to find the best transmission rate faster. At the same time, it must be kept in mind that look-around frames are transmitted with a suboptimal rate. The look-around rate r_l can be lower than r_b resulting in an underutilization of the wireless channel or, the look-around rate r_l can be higher than r_b resulting in an increase of frame losses. Both cases affect the throughput. Typical values of γ are in the range [0.1, 0.2]
β	P_{r_l} is recomputed only if $np_{r_l} \geq \beta \geq 1$ (see Table 3.4). High values may lead to better estimation of the Joint Reception Probability, reducing oscillation in the choice of the optimal rate. Nevertheless, high values can decrease the algorithm's convergence time.
α	Threshold value that is used to assure that every look-around rate r_l has a probability greater than zero to be chosen to transmit the next look-around frame.
λ	EWMA parameter $\in [0, 1]$. It is used to adapt convergence to new channel conditions.
i	Sequence number of the multicast data frames. It starts from 0 and is incremented after every transmission.
x	Loss threshold value for the limited losses version of the look-around algorithm (see Section 5.3). Used to try to limit frame losses probability to $(1-x)$
$\sigma_1, \sigma_2, \sigma_3$	Weight terms used to compute p_{r_l} (see Table 3.2).

An RAS can be thought of as an infinite loop where every cycle, which we called *super-frame*, is split into three ordered steps:

- *Transmission period*: during which the AP transmits the multicast data frames;
- *Feedback protocol*: during which the AP runs the feedback protocol;
- *Rate Adaptation Algorithm*: during which the AP applies the heuristic to select the best transmission rates for the subsequent multicast data frames.

More specifically, during the transmission period the AP transmits N multicast data frames using the list of transmission rates selected by the RAA at the end of the previous super-frame. More formally, we use the parameter $e \in [1, \infty)$ as a super-frames counter. At the end of the $(e-1)$-th super-frame the RAA provides the MAC layer with a list of N transmission rates $(r^{(e-1)N}, r^{1+(e-1)N}, ..., r^{(N-1)+(e-1)N})$. Then, during the e-th super-frame, the AP transmits the N multicast data frames f_i, whose sequence number i is such that $(e-1)N \leq i < eN$, with the corresponding transmission rate r^i. The multicast data frames f_i are classified into two groups that affect the choice of the transmission rate r^i performed by the RAA. If the sequence number i satisfies $(i \bmod \lfloor \gamma N \rfloor) \neq 0$, where $\gamma \in (0, 1)$, then $r^i = r_b$, with r_b being the best transmission rate selected according to the RAA's rules. For the remaining $\lfloor \gamma N \rfloor$ multicast data frames, that we call *look-around* frames, the algorithm sets $r^i = r_l \neq r_b$. We will describe later how r_b and r_l are selected, but the basic idea behind these two groups of frames is that if the parameter γ is set carefully (typical values are in the range 0.1–0.2) then most of the multicast data frames are transmitted with the best transmission rate but, at the same time, thanks to the look-around frames, it is possible to continuously evaluate the channel quality experienced with data rates other than the optimal one. In this way the RAA is able to keep updated statistics for every possible transmission rate. Special attention is required to the case $e = 1$ that identifies the first super-frame. In this case r_b is not chosen by the RAA but is fixed before the multicast transmission starts. There are alternative strategies to select the initial value of r_b. For example, it is possible to set it to a low and robust transmission rate in order to avoid frame losses with the drawback of reducing the throughput in the first part of the multicast communication, or it is possible to be more aggressive and set it to a higher transmission rate. In the latter case, if the channel conditions are good, it is possible to obtain a performance boost at the beginning of the communication. Obviously, if the chosen transmission rate is too high with respect to the actual wireless channel conditions, there is the risk of losing most of the transmitted data frames, limiting the initial throughput. At any rate, the RAA usually converges quickly and from a long run transmission point of view the impact of the initial choice of r_b on the average throughput is negligible.

The transmission period is followed by the feedback protocol. We assume the use of the Block-Ack request/response exchange of the GCR-BA scheme provided by the 802.11aa standard and that, at the end of its execution, the RAA receives as input one bitmap $b_s(e)$ for each STA s of the multicast group; as we use the 802.11aa standard, every bitmap refers to the same set of multicast data frames. Finally, we assume that the STAs not responding to the Block-Ack request left the multicast group; in this way we can ignore the missing feedbacks. Figure 3.7 illustrates an example of feedback received from station s during the feedback protocol of the e-th super-frame. It is made of: 1) the sequence number *Seq* of the first multicast data frame transmitted during the transmission period of the e-th super-frame; 2) an N bits bitmap $b_s(e)$, whose n-th bit represents the

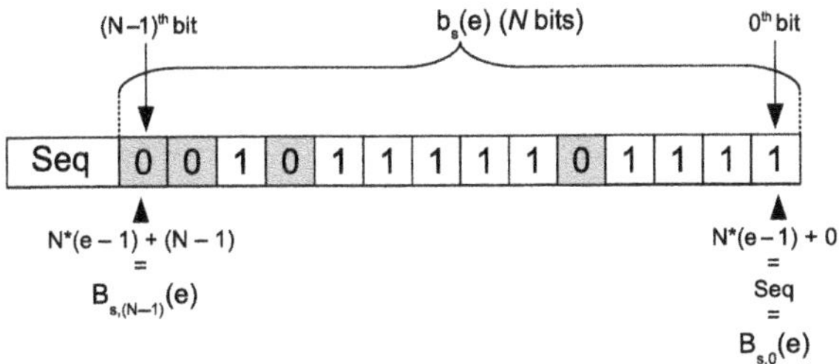

FIGURE 3.7

Bitmap received from station *s* during super-frame *e*.

reception ($b_{s,n}(e) = 1$) or the loss ($b_{s,n}(e) = 0$) of the multicast data frame identified by sequence number $(i = Seq + n) \in [(e-1)N, eN)$.

Once the feedback protocol collected the bitmaps from every STA of the multicast group, the RAA can update the transmission statistics. Given that the bitmaps are aligned, the RAA computes their logical AND ($b_{AND}(e)$) and then counts how many bits in the result are set to 1: this represents the number of frames received jointly by all the STAs of the multicast group. We now introduce the concept of Joint Reception Probability (JRP) of the *e*-th super-frame $P_{r_j}(e)$ as the probability that a multicast data frame, transmitted with rate r_j, is received by all the STAs of the multicast group. $P_{r_j}(e)$ is computed as:

$$P_{r_j}(e) = \frac{\sum_{n=0}^{(N-1)} b_{AND,n}(e)}{N} \tag{3.1}$$

This formula is valid only under the assumption that all the N frames transmitted during the transmission period of the *e*-th super-frame were transmitted using the same rate r_j (this can be obtained setting $\gamma = 0$). As we will see, this fact is true in one of the algorithms that we propose. Moreover, we will see that the concept of JRP can be extended also to the case where $\gamma > 0$. Indeed, all the algorithms we propose rely on the value of the JRP to choose the best transmission rate. In order to compute the JRP also when $\gamma > 0$, the RAS keeps a lot of information at the AP during every super-frame. Thanks to this information it is possible to compute how many multicast data frames have been transmitted with the rate r_j during the last transmission period ($np_{r_j}(e)$) and how many of these frames have been jointly received by all the associated STAs ($nj_{r_j}(e)$). We then compute the JRP as:

$$P_{r_j}(e) = \frac{nj_{r_j}(e)}{np_{r_j}(e)} \tag{3.2}$$

It is easy to understand that, by taking into account $P_{r_j}(i), i \leq e$, it is possible to estimate the JRP for the rate r_j for subsequent super-frames (P_{r_j}), which is an estimation of the probability that a multicast data frame, transmitted with the rate r_j, will be received correctly by all the STAs of the multicast group. Table 3.4

Table 3.4 Information Maintained at the AP for Each Rate r_j

Name	Description
np_{r_j}	Number of multicast data frames transmitted with rate r_j since the last time P_{r_j} has been computed. This value is reset to 0 every time P_{r_j} is updated.
$np_{r_j}(e)$	Number of multicast data frames transmitted with rate r_j during super-frame number e. This value is reset to 0 at the beginning of the super-frame $(e+1)$
nj_{r_j}	Number of multicast data frames transmitted with rate r_j and jointly received by all the STAs of the multicast group since the last time P_{r_j} was updated. This value is reset to 0 every time P_{r_j} is updated.
$nj_{r_j}(e)$	Number of multicast data frames transmitted with rate r_j and jointly received by all the STAs of the multicast group during super-frame number e. This value is reset to 0 at the beginning of the super-frame $(e+1)$
ls_{r_j}	Sequence number of the last multicast data frame transmitted with rate r_j

summarizes the information maintained at the AP for every transmission rate r_j. As we will see, not all the RAAs that we describe use all the information listed in the table.

After the analysis of the feedback received during the execution of the feedback protocol and the computation of the statistics representing the channel quality, the RAA can select the data rate for the transmission of the successive multicast burst. As we will see in the next section we propose two different algorithms:

- *Linear increase/Multiplicative decrease algorithm*: this algorithm does not use look-around frames (during the transmission period of the e-th super-frame all multicast frames are transmitted with the same rate) and, in order to choose the best transmission rate for super-frame $(e+1)$ ($r_b(e+1)$), it compares the average time that would have been necessary to deliver reliably a multicast data frame to every STA of the multicast group during the two previous super-frames (e and $(e-1)$);
- *Look-around algorithm*: this algorithm uses look-around frames to continuously estimate the JRP P_{r_j}, $\forall j$. Moreover, P_{r_j} is updated using an EWMA (Exponentially Weighted Moving Average) technique. There are two versions of the algorithm: the first one, which we call *best throughput*, aims to maximize the throughput; the second one, which we call *limited losses*, aims to maintain losses under a threshold value.

5.3 RATE ADAPTATION ALGORITHMS

We start this section with the description of the *Linear increase/Multiplicative decrease* RAA. Given that this algorithm does not use look-around frames, it must only compute the best transmission rate $r_b(e+1)$ to be used for all the multicast data frames that will be transmitted during the super-frame $(e+1)$. To achieve these goals it compares the average time of the super-frame e and $(e-1)$ that would have been necessary to deliver reliably a multicast data frame to all members of the multicast group. The average number of transmissions that should be performed by the AP to successfully deliver a multicast data frame to all members of the multicast group during the super-frame e can be computed as follows:

$$N(P_{r_b}(e), r_b(e)) = \sum_{c=1}^{\infty} i \cdot P_{r_b}(e) \cdot (1 - P_{r_b}(e))^{(c-1)} = \frac{1}{P_{r_b}(e)} \quad (3.3)$$

Now, assuming the same size of L bytes for all the multicast data frames, the average time needed to deliver reliably such a data frame is equal to:

$$T(e) = T(P_{r_b}(e), r_b(e)) = N(P_{r_b}(e), r_b(e)) \cdot \frac{L}{r_b(e)} = \frac{L}{r_b(e) \cdot P_{r_b}(e)} \quad (3.4)$$

Similarly, the average time necessary needed to reliably deliver a multicast data frame to all members of the multicast group during the super-frame $(e-1)$ is $T(e-1) = T(P_{r_b}(e-1), r_b(e-1)) = \frac{L}{r_b(e-1) \cdot P_{r_b}(e-1)}$. At this point the selection of $r_b(e+1)$ is performed according to the following rule:

$$j = index\_of(R, r_b(e))$$

$$r_b(e+1) = \begin{cases} R[j+1], & \frac{T(e)}{T(e-1)} \leq 1 \\ R[j-2], & \frac{T(e)}{T(e-1)} > 1 \end{cases} \quad (3.5)$$

where R is the list of available transmission rates, while the function $index\_of(R, r_b(e))$ returns the index of rate $r_b(e)$ within the list R. We assume that when $j \leq 2$, $R[j-2]$ is simply equal to $R[0]$. This is a *Linear increase/Multiplicative decrease* adaptive rule. When the transmission time between two consecutive transmission periods decreases or remains constant, the RAA attempts to further reduce the transmission time by increasing the data rate that will be used in the successive super-frame. Indeed, assuming that the JRP does not vary sharply between two consecutive transmission rates, the rate increase would reduce the overall transmission time, thus increasing the goodput perceived by the STAs of the multicast group. On the contrary, when the transmission time increases, the RAA halves the data rate to recover quickly from an unreliable operating state. In this case, the increase of the transmission time is caused by the degradation of the JRP, which would require more transmissions to successfully deliver a single data frame to all stations.

The main drawback of the *Linear increase/Multiplicative decrease* RAA is that the choice of $r_b(e+1)$ depends only on the transmission rates $r_b(e)$ and $r_b(e-1)$ and on the feedback collected during super-frames e and $(e-1)$ (that, in turn, are used to compute $P_{r_b}(e)$ and $P_{r_b}(e-1)$). This means that the algorithm is affected by short-term variations of the wireless channel status that can lead to oscillation in the selection of the best transmission rate. Moreover, the selection of $r_b(e+1)$ is performed in accordance with a static rule that does not take into consideration any statistics about $r_b(e+1)$.

These problems are successfully addressed by the *look-around* RAA that uses look-around frames to continuously evaluate the channel quality experienced with all the available transmission rates, uses an EWMA technique to update the estimation of the JRP in order to take into account the long-term variation of the wireless channel status and bases the choice of the rate $r_b(e+1)$ on its measure of successfulness S_{r_b}. More specifically, in the *look-around* RAA, at the end of the feedback protocol, the value of the JRP P_{r_j} is updated for every rate r_j such that $np_{r_j} \geq \beta$ using the following rule:

$$P_{r_j} \leftarrow (1-\lambda)P_{r_j} + \lambda\left(\frac{nj_{r_j}}{np_{r_j}}\right) \tag{3.6}$$

where $\lambda \in (1,0)$ is the EWMA (Exponentially Weighted Moving Average) weight and for every rate r_j the corresponding P_{r_j} is initially set to 0. For every rate r_j, whose corresponding JRP P_{r_j} has been recomputed after the execution of the feedback protocol, the algorithm updates the measure of successfulness S_{r_j} that is a function $f(^*)$ of the JRP P_{r_j}. $f(^*)$ differs in the two versions of the algorithm. For the *best throughput* version, S_{r_j} is given by P_{r_j} multiplied by the bit rate of the rate r_j and can be expressed as:

$$S_{r_j} = f(P_{r_j}) = P_{r_j} \cdot r_j \tag{3.7}$$

In this case the algorithm sets the transmission rate $r_b(e+1)$ to be used in the next transmission period to the highest rate r_j whose measure of successfulness is maximum among all the available transmission rates:

$$r_b(e+1) \leftarrow \max_j \left\{ r_j \middle| S_{r_j} = \max_k \{S_{r_k}\} \right\} \tag{3.8}$$

Note that with this rule, the algorithm tries to maximize the throughput. In the case of the *limited losses* version, the measure of successfulness corresponds to the JRP:

$$S_{r_j} = P_{r_j} \tag{3.9}$$

and the algorithm sets the transmission rate $r_b(e+1)$ to be used in the next transmission period to the highest rate r_j such that $S_{r_j} \geq x$, where $x \in (0,1)$ is a fixed threshold value:

$$r_b(e+1) \leftarrow \max_j \left\{ r_j \middle| S_{r_j} \geq x, 0 < x < 1 \right\} \tag{3.10}$$

Note that in this case the algorithm tries to limit the probability of losing a frame to $(1-x)$. An important aspect of the look-around algorithm is the selection of the look-around rate to be used for the next look-around frames. A rate $r_l \neq r_b$ is chosen with a probability p_{r_l} that is influenced by the current JRP P_{r_l}, by the number of multicast frames that have been sent since the last multicast data frame transmitted with rate r_l $(1 - sl_{r_l})$ and by the total number of multicast frames transmitted with the rate r_l since the last time P_{r_l} has been updated (np_{r_l}). On the one end, each rate r_l must have a positive probability p_{r_l} to be chosen for the next look-around frame (we want to continuously evaluate channel quality for all available rates). On the other hand, we do not want to transmit too many multicast frames with a rate that has shown bad performance to limit frame losses. The rule used to compute p_{r_l} for the next look-around frame is expressed by the following five equations:

$$p_{rl} = \frac{W_{r_l}}{\sum_k W_{r_k}} \tag{3.11}$$

$$W_{r_t} = \sigma_1 A_t + \sigma_2 B_t + \sigma_3 C_t \tag{3.12}$$

$$A_t = \begin{cases} \frac{\beta - np_{r_t}}{\beta} & \beta - np_{r_t} \geq 0 \\ 0 & otherwise \end{cases} \tag{3.13}$$

$$B_t = \frac{i - ls_{r_t}}{\max\limits_k (i - ls_{r_k})} \tag{3.14}$$

$$B_t = \frac{i - ls_{r_t}}{\max\limits_k (i - ls_{r_k})} \tag{3.15}$$

Term A_t states that the probability of choosing rate r_t for the next look-around frame must decrease when the total number of multicast frames transmitted at that rate since P_{r_t} has been computed is close or equal to β, where the latter is the minimum value that must be reached by np_{r_t} before P_{r_t} can be recomputed. Term B_t expresses the fact that the probability to choose rate r_t must be higher for rates that have not been tested for a long time. Term C_t states that the probability to choose rate r_t must be higher for rates that have shown good performance until now. Parameter α is used as a threshold value so that all rates have $C_t > 0$ and σ_1, σ_2 and σ_3 are used as weights for the three terms A_t, B_t and C_t.

6 VALIDATION AND NUMERICAL RESULTS

In the following, we first present the validation of the GATS implementation comparing the performance of the fixed transmission rate approach of the NS-2 solution against those obtained on a real testbed. Then, we illustrate a comparative evaluation of our rate adaptation algorithm implemented in NS-2 [32,33] to show the validity of the proposed NS-2 extensions as a means of testing innovative solutions for the multicast delivery in wireless networks. Finally, we show the flexibility of our architecture by illustrating the results that we obtain in several network topologies in heterogeneous network conditions, considering different types of data traffic.

6.1 VALIDATION OF THE GATS IMPLEMENTATION

In order to validate our library extensions, we compare the throughput of a multicast data traffic stream in saturated conditions obtained respectively using the

Table 3.5 Comparative Evaluation of the Throughput Obtained Using Our NS-2 Library Extension

Burst Size (Frames)	Theoretical Rate (Mb/s)	Simulation Rate (Mb/s)	Experiments		Simulation Rate* (Mb/s)
			Exp. Rate (Mb/s)	Flushing (ms)	
8	36.76	36.01	22.75	1.25	20.11
16	40.13	39.87	26.82	1.65	26.31
32	42.06	42.29	27.95	2.25	30.05

**Additional overhead introduced to simulate the time wasted to flush the queue storing replica of multicast packets.*

theoretical model presented in [15], the real-life implementation discussed in [33] and the simulations using our NS-2 modified library. To this end, we simulate a network scenario composed of a single AP and four static STAs placed at 10 m from the AP. On the AP we configure a CBR (constant bit rate) source that broadcasts data frames at a rate large enough to saturate the wireless channel capacity. The CBR packet size is fixed to 1470 bytes to fit the 802.11 frame payload (1500 bytes), considering the additional overhead of the IP and UDP headers.

All network nodes use the IEEE 802.11g MAC protocol with the GATS extensions defined in the IEEE 802.11aa amendment (we choose GCR-BA) and use the same wireless channel. The physical layer models accurately the cumulative SINR computation, the preamble and PLCP header processing, and the frame capture. Regarding the radio propagation, we use the Nakagami model, since it permits consideration of several effects that cause signal attenuation like fading and multipath. To evaluate the performance degradation caused by the GATS frame exchange, we fix the transmission rate to 54 Mb/s and we vary the size of the multicast burst (and the corresponding bitmap of 802.11aa control frames) considering 8, 16, and 32 frames.

Table 3.5 shows the throughput predicted by the theoretical model and measured using the NS-2 implementation of the 802.11aa protocol. As can be observed the results obtained from the simulations well approach those estimated using the theoretical model, thus validating our implementation of the 802.11aa amendment within the NS-2 simulator. Furthermore, as expected, the burst size affects the raw downlink throughput, since after the transmission of the multicast burst the AP polls all stations to obtain the bitmaps corresponding to the received/lost frames. Therefore, the smaller the burst, the higher the signaling overhead and the airtime dedicated to the polling phase. Note, however, that a small burst size permits retransmitting the lost frames quickly, thus increasing the transmission reliability and the goodput experienced by the application layer, which is more important than the raw throughput for video and audio streaming applications. Similarly, the performance of the multicast rate adaptation algorithm depends on the burst size, since a large burst permits an accurate estimate of the joint reception probability, but at the same time may result in several lost frames transmitted with the wrong data rate.

Table 3.5 shows also the comparative evaluation of the throughput measured using the real-life (column identified by Experimental Rate) and NS-2 implementations of the 802.11aa protocol (column identified by Simulation Rate*). The implementation on the wireless NIC keeps a replica of all frames of a multicast burst in several queues to quickly retransmit the lost frames according to the specifications of the 802.11aa standard. However, such an approach introduces a computational overhead due to the management of the queues. Indeed, the NIC CPU must remove from the queues the data frames that have been correctly received. In other words, it must flush the queues to collect the frames of the new multicast burst sent from the upper layers. The simulator takes into account the computational overhead by waiting the additional time due to the flushing operation before starting the transmission of a new multicast burst.

In the experimental activity, the computational overhead caused by the queues management has been quantified computing the time wasted to perform the flushing operation (during this activity the MAC protocol cannot transmit any data frames). The flushing time corresponding to different multicast burst sizes is also illustrated in Table 3.5. It can be observed from the table that such an additional contribution introduced in the simulation allows a good approximation of the achievable throughput obtained on the real-life testbed (the performance gap between the experiments and the simulations is always inferior to 10%).

6.2 COMPARATIVE EVALUATION

The first comparative experiment between the NS-2 versions of the proposed rate adaptation algorithms considers a network scenario composed of one Access Point (AP) and four Stations (STAs), as illustrated in Figure 3.8; the AP is placed

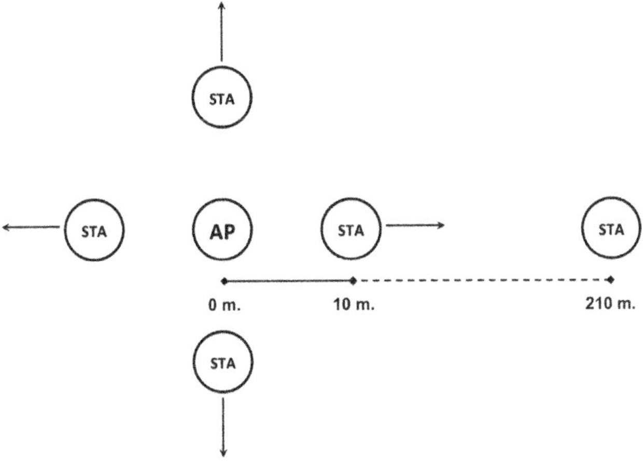

FIGURE 3.8

Simulated scenario with one AP and four mobile stations.

in a fixed position, whereas the four stations are free to move. The parameters of the radio interfaces of each NS-2 node (e.g., reception sensitivity, SINR thresholds of data rate) are configured to match those of the real testbed devices used in [32,33], which are equipped with a Broadcom 4318 wireless cards. As can be seen from Figure 3.8, at the beginning of the experiment, all stations are radially placed at 10 meters from the AP. We moved the mobile nodes farther away from the AP towards the directions indicated by the black arrows, with steps of 20 m, for testing the behavior of the proposed techniques with increasing attenuation and fading effects over the received signal.

In the simulation, we increase the noise affecting the mobile station to obtain the same effects that decrease the SINR at the receiver in the real scenario. The multicast burst size has been fixed to 32, since this allows the highest throughput as illustrated in Table 3.5. As for the two transmission phases of the rate adaptation algorithms, namely transmission at the optimal rate and transmission of look-around frames, we evenly divided the multicast burst. Therefore, 16 frames are transmitted with the best Modulation and Coding Scheme selected according to the rules of the rate adaptation scheme, whereas the remaining 16 frames are used to test all data rates according to the probabilistic rule described in the previous section.

For each position at which we placed the mobile nodes, the AP performs five tests with fixed transmission rate and five tests for each technique proposed in the previous section. During each test the AP works as a greedy CBR source generating multicast packets that saturate the channel for 60 seconds. Note that the channel is interference free since there are no surrounding WLANs and the GATS mechanism is scheduled at the end of the multicast burst. The packet payload size is fixed to 1470 bytes, which is equal to the value used by several video-streaming applications based on the Real Time Protocol (RTP).

We consider as performance metrics the *average goodput*, namely the average bandwidth actually used for successful transmission of useful data (the application-layer payload), and the *average delay* of the multicast transmissions measured by the four STAs. Figure 3.9 shows the *average goodput* measured by all four STAs, as a function of the mobile node position. The performance of the *Linear increase/Multiplicative decrease*, *best throughput*, and *limited-losses* versions of the proposed rate adaptation algorithms is indicated by the labels LIMD, BR, and LL, respectively. The curve identified by the label *Basic* depicts the results of the IEEE 802.11 standard, which uses a fixed rate equal to 6 Mb/s. It can be observed that the *best-throughout* algorithm outperforms the other schemes at distances lower than 190 meters. In particular, we obtain a performance increase ranging from 60% to 300% with respect to the standard *fixed rate* solution, from 10% to 140% with respect to the *Linear increase/Multiplicative decrease*, and from 5% to 110% if compared with the *limited-losses* version of the algorithm. Even though the *limited-losses* algorithm performs slightly worse than the *Linear increase/Multiplicative decrease*, it provides better performance in terms of losses. Indeed, the *Linear increase/Multiplicative decrease* represents

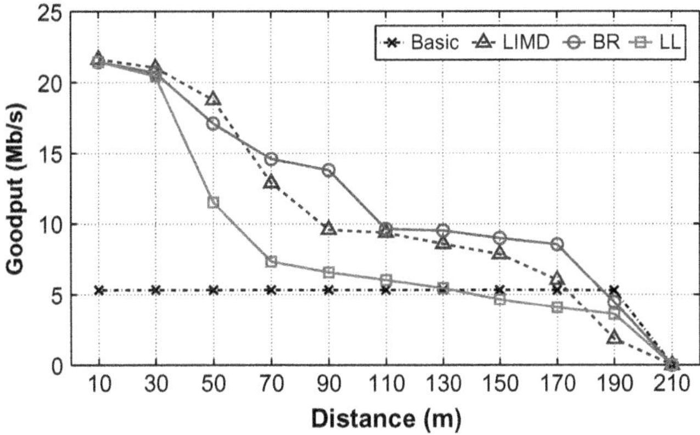

FIGURE 3.9

Average goodput measured in the scenario with one AP and four mobile STAs.

the solution with the worst performance in terms of frame losses, since it greedily uses the transmission rates for all frames of the multicast burst.

We further observe that the *limited-losses* algorithm experiences worse performance than the *fixed rate* approach for distances ranging from the 130 to 200 meters. This is mainly due to the signaling overhead of the GATS polling mechanism and the test performed by the rate adaptation algorithm on nonoptimal rates necessary to collect statistics on the channel quality at different Modulation and Coding Schemes (look-around frames are used to test other data rates than the optimal one). However, the performance degradation remains always inferior to 20%. To reduce the performance degradation, the rate adaptation algorithm can be disabled when the highest SNR among the stations of the multicast group falls below a threshold.

Figure 3.10 shows the average *one-way transmission delay* measured by the four STAs in the downlink direction. Results are specular with respect to the goodput: the standard *fixed rate* solution always exhibits higher delays than the *Linear increase/Multiplicative decrease* and the *limited-losses* solutions, while the *best-throughput* algorithm achieves the lowest delay. Quantitatively the last three algorithms reduce the *one-way delay* up to approximately four times with respect to the fixed rate solution.

To provide a more in-depth comparison, we also measured the average *goodput* and *delay* incurred by broadcast data transmissions in a network scenario composed of an AP and eight STAs, which receive the broadcast data transmissions of a CBR application operating on the AP. The corresponding results, which are illustrated in Figures 3.11 and 3.12, confirm the trends obtained in the scenario with four mobile STAs. In particular, the *best-throughput* rate adaptation technique always achieves the best performance for distances between the AP and the STAs inferior to 190 meters, whereas the *limited-losses* algorithm performs worse even than the fixed

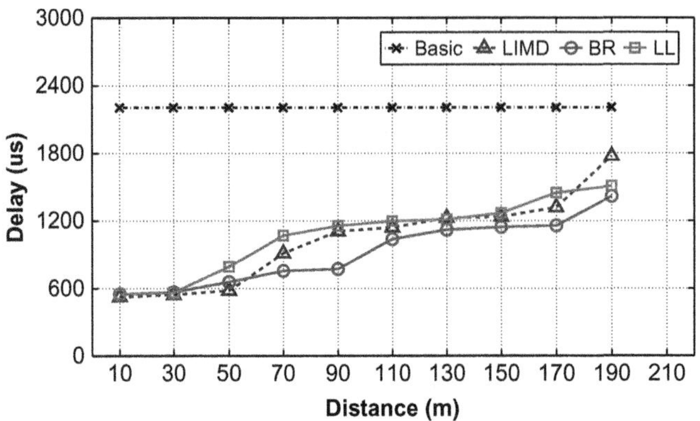

FIGURE 3.10

Average one-way delay measured in the scenario with one AP and four mobile STAs.

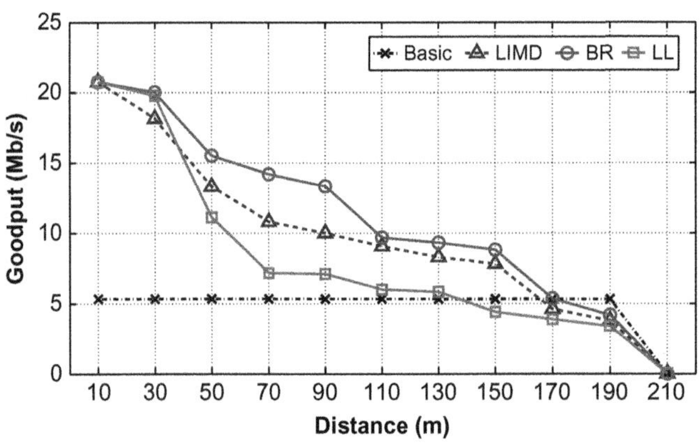

FIGURE 3.11

Average goodput measured in the scenario with one AP and eight mobile STAs.

rate approach for distances higher than 130 meters, due to the signaling overhead and the look-around frames transmitted at nonoptimal rates.

6.3 PERFORMANCE EVALUATION IN HETEROGENEOUS CONDITIONS

In order to analyze our framework in heterogeneous network settings, we further evaluate the performance of our algorithms considering a network scenario where multicast and unicast traffic is transmitted simultaneously by applications operating on different devices. Specifically, we consider a network topology composed of one AP and eight STAs placed radially at 10 meters around the AP as

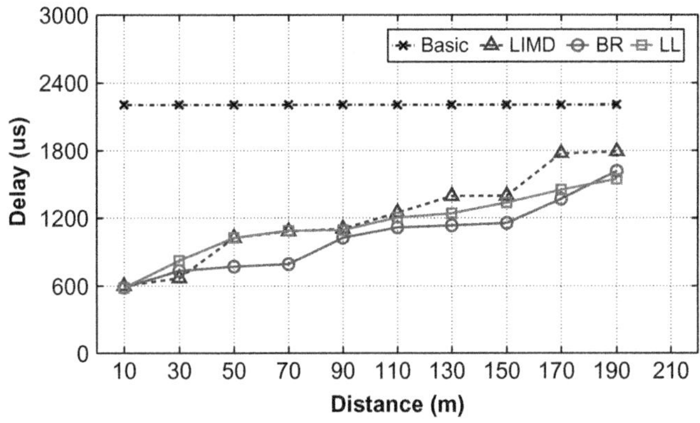

FIGURE 3.12

Average one-way delay measured in the scenario with one AP and eight mobile STAs.

illustrated in Figure 3.13. As in previous scenarios, the parameters of all radio interfaces are configured to match those of real wireless cards.

On the AP we installed a CBR source that generates multicast traffic for a subset of four STAs, which are indicated by the label M1, M2, M3, and M4 in the figure. The packet size of the CBR application is set to 1470 bytes to fill the payload of an 802.11 data frame. The remaining four out of eight STAs (U1, U2, U3, and U4) transmit UDP unicast traffic towards the AP using the same parameters (i.e., packet size and bitrate) of the CBR source installed at the AP. Therefore, there are five traffic sources in the network that compete for the same channel.

We varied the bitrate of all CBR sources in the range [2; 42] Mb/s to evaluate the performance of the overall system using the 802.11aa protocol coupled with our rate adaptation techniques and the fixed rate approach. In all simulations, the multicast burst size has been fixed to 8 data frames ($\gamma = 0.5$).

Figures 3.14, 3.15, 3.16, and 3.17 illustrate the average goodput of the multicast and unicast traffic as a function of the bitrate of the CBR sources (recall that all sources have the same bitrate) using respectively the *fixed rate* approach (54 Mb/s), the *Linear increase/Multiplicative decrease*, *best throughput*, and *limited-losses* versions of the proposed rate adaptation algorithms. The large performance gap, which can be observed in the figures when the system approaches saturated conditions, confirms that the multicast transmission always captures the channel, thus obtaining a higher share of the channel resource. This is due to the reservation mechanism of the 802.11aa protocol, which permits the transmission of the entire burst of multicast packets without contention. Therefore, the unicast transmissions of other stations can compete for the channel access only at the end of the multicast transmission and polling phase (GCR-BA). Indeed, only at the end of these two phases the MAC of the AP contends the channel using the classical CSMA/CA parameters (i.e., it waits for a DIFS plus an additional backoff time before transmitting the data frame). We can further observe that the *best throughput* and *limited-losses* algorithms

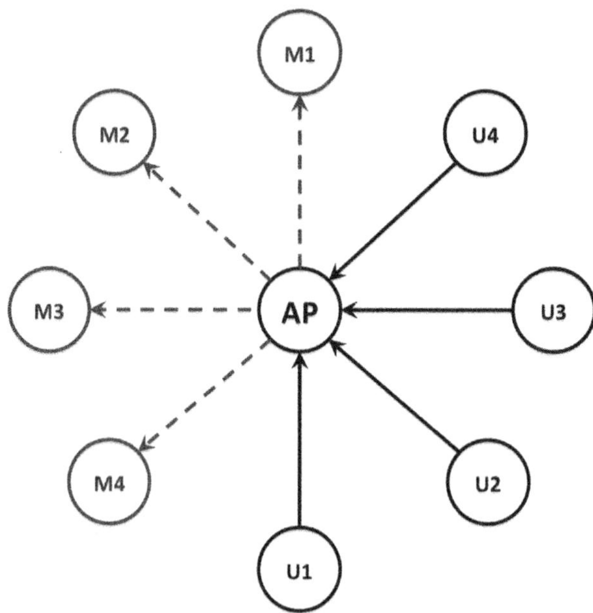

FIGURE 3.13

Average goodput measured in the scenario with five traffic sources operating on the AP and 4 out of 8 STAs. Solid black arrows depict unicast transmissions, whereas dashed grey arrows represent the multicast transmission and the corresponding stations of the multicast group.

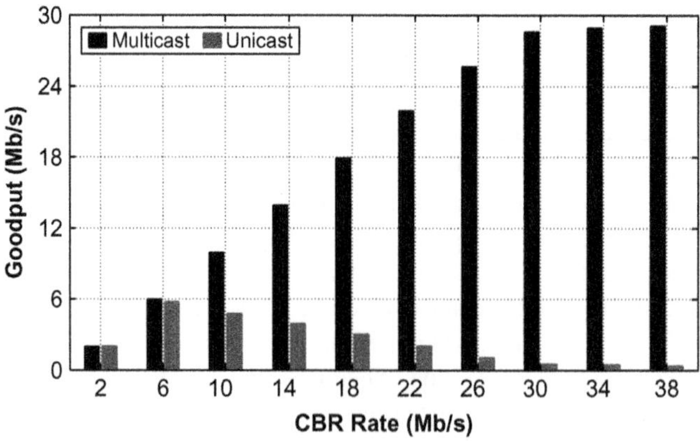

FIGURE 3.14

Average goodput of multicast and unicast transmissions obtained using the 802.11aa GATS with fixed data rate (54 Mb/s).

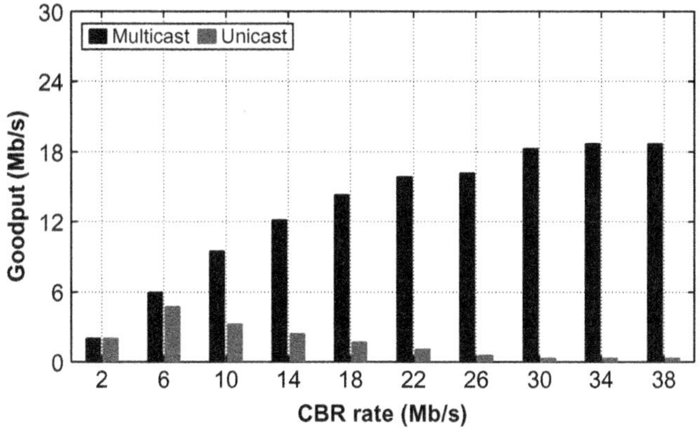

FIGURE 3.15

Average goodput of multicast and unicast transmissions obtained using the 802.11aa GATS with the *Linear increase/Multiplicative decrease* Rate Adaptation algorithm.

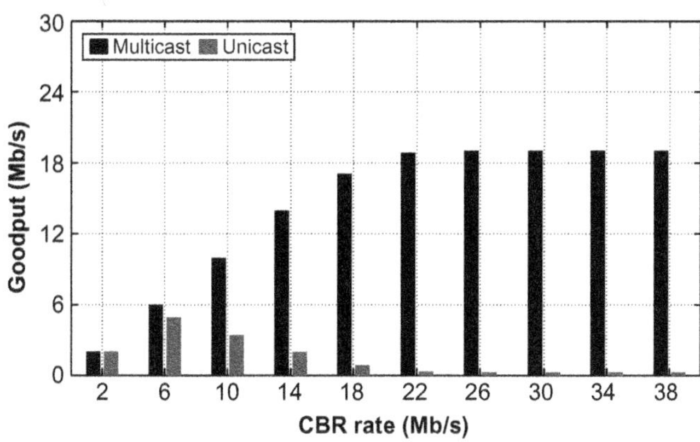

FIGURE 3.16

Average goodput of multicast and unicast transmissions obtained using the 802.11aa GATS with the Best Throughput Rate Adaptation algorithm.

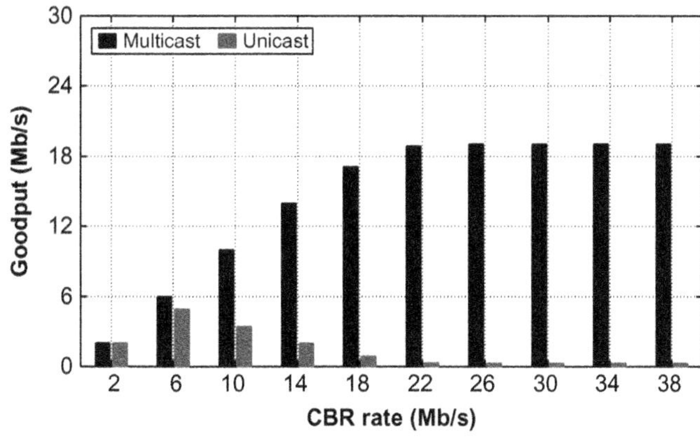

FIGURE 3.17

Average goodput of multicast and unicast transmissions obtained using the 802.11aa GATS with the Limited Losses Rate Adaptation algorithm.

achieve higher performance for multicast transmissions than the *Linear increase/Multiplicative decrease* technique, since they exploit the historical information on the reliability of all transmission rates. For example, when the bitrate of the CBR source is equal to 14 Mb/s, the STAs of the multicast group observe a goodput of approximately 14 Mb/s with the *best throughput* and *limited-losses* algorithms and 12 Mb/s with the *Linear increase/Multiplicative decrease* approach. Finally, the performance gap that can be observed between the *fixed rate* approach and the rate adaptation algorithms is simply due to the tests that they perform continuously with the look-around frames. Since these frames are transmitted using non-optimal rates to collect statistics on their reliability and performance, a portion of the available airtime is wasted. We underline however that the *fixed rate* approach with transmission rate at 54 Mb/s represents only an upper bound to the achievable performance, since this rate can be used only with high SINR (e.g., for Line-of-Sight distances lower than 35 m).

7 CONCLUSIONS

We introduced in this chapter a complete framework for the simulation of the recently introduced IEEE 802.11aa protocol inside the NS-2 simulator. The extensions that we developed allow users to study the dynamics of the mechanisms that support reliable multicast delivery services in wireless networks. Furthermore, the availability of feedback collected from receivers also permits the design of a new generation of algorithms for selecting the best Modulation and Coding Scheme for multicast transmissions according to the link quality. To this end, we added a set of APIs for customizing these operations and we presented some working examples based on a *Linear increase/Multiplicative decrease* approach and on look-around frames. These approaches can be specialized to increase the overall throughput or to limit the amount of frame losses. Apart from the two techniques that we present as test cases, we believe that our framework could be a valid instrument for studying new multicast rate adaptation techniques based on the IEEE 802.11aa protocol, for two main reasons. First, the lack of commercial NICs compliant with the new protocol makes experiments extremely complex, since the feedback protocol must be executed with tight timing requirements and cannot be emulated from the user space. The availability of a simulation framework enables a preliminary analysis of rate adaptation techniques, focusing on the most important parameters that can affect the achievable performance. Second, as we reported in this Chapter, we validated our techniques by comparing results with those obtained from preliminary experiments on a real-life testbed where prototype firmware was used, showing that the feedback protocol we introduced in the simulator respects the correct timings.

ACKNOWLEDGMENT

This work has been partially supported by the European Community through the WISHFUL project, H2020-645274.

REFERENCES

[1] Stuart C. Multicast DNS, Available from: <http://www.multicastdns.org>; 2015.

[2] Srinivas V, Ruan L. An efficient reliable multicast protocol for 802.11-based wireless LANs. In: Proceedings of world of wireless, mobile and multimedia networks & workshops, Kos, Greece; 15–19 Jun, 2009.

[3] Sun M, Huang L, Arora A, Lai T. Reliable MAC layer multicast in IEEE 802.11 wireless networks. Proceedings of the international conference on parallel processing, Vancouver, Canada; 18–21 Aug, 2002.

[4] DirCast: A practical and efficient Wi-Fi multicast system. In: Proceedings of the IEEE International conference on network protocols, Princeton, NJ; 13–16 Oct, 2009.

[5] Kuri J, Kasera S. Reliable multicast in multi-access wireless LANs. Wirel Netw 2001;7(4):359–69.

[6] Rubenstein D, Kasera S, Towsley D, Kurose J. Improving reliable multicast using active parity encoding services. Comput Netw. 2004;44(1):63–78.

[7] Choi M, Sun W, Koo J, Choi S, Shin KG. Reliable video multicast over Wi-Fi networks with coordinated multiple APs. In: Proceedings of IEEE INFOCOM conference on computer communications, Toronto, Canada; 27 Apr.–2 May, 2014.

[8] Gudipati A, Katti S. Strider: automatic rate adaptation and collision handling. ACM SIGCOMM 2011;41(4):158.

[9] Villalón J, Cuenca P, Orozco-Barbosa L, Seok Y, Turletti T. Cross-layer architecture for adaptive video multicast streaming over multirate wireless LANs. IEEE J Sel Areas Commun 2007;25(4):699–711.

[10] Basalamah A, Sugimoto H, Sato T. Rate adaptive reliable multicast MAC protocol for WLANs. In: IEEE vehicular technology conference, vol. 3. p. 1216–20; 2006.

[11] Piamrat K, Ksentini A, Bonnin J, Viho C. Rate adaptation mechanism for multimedia multicasting in wireless networks. In: Proceedings of the international conference on broadband communications, networks, and systems. Madrid, Spain; 14–16 Sept. 2009.

[12] Zhou C, Zhang X, Lu L, Gou Z. Collision-detection based rate-adaptation for video multicasting over IEEE 802.11 wireless networks. In: Proceedings of the international conference on image processing; 26–29 Sept. 2009.

[13] IEEE Standards Association. IEEE Get Program. IEEE 802.11aa document standard. Available from: <http://standards.ieee.org/getieee802/download/802.11aa-2012.pdf>; 2015.

[14] Kosek-Szott K, Krasilov A, Lyakhov A, Natkaniec M, et al. What's new for QoS in IEEE 802.11? IEEE Netw 2013;27(6):95–104.

[15] Salvador P, Cominardi L, Gringoli F, Serrano P. A first implementation and evaluation of the IEEE 802.11aa group addressed transmission service. ACM Comput Commun Rev 2014;44(1).

[16] Kosek-Szott K, Throughput A. Model of IEEE 802.11aa Intra-access category prioritization. Wirel Pers Commun 2013;71(2):1075–83.

[17] Ángeles Santos M, Villalón J, Orozco-Barbosa L. Evaluation of the IEEE 802.11aa group addressed service for robust audio-video streaming. In: Proceedings of the IEEE international conference on communications, Ottawa, Canada; 10–15 June, 2012.

[18] The network simulator – ns-2 website. Available from: <http://www.isi.edu/nsnam/ns/>; 2014.

[19] Debian/Wiki website. Broadcom wl Proprietary Driver for Debian distribution, Available from: <https://wiki.debian.org/wl>; 2014.

[20] Linux Wireless website. Minstrel HT rate control algorithm. Available from: <http://wireless.kernel.org/en/developers/Documentation/mac80211/RateControl/minstrel>; 2015.

[21] IEEE Standards Association. IEEE Get Program. IEEE 802.11 document standard, Available from: <http://standards.ieee.org/getieee802/download/802.11-2012.pdf>; 2015.

[22] Broadcom 802.11ac solutions. Available from: <http://www.broadcom.com/products/Wireless-LAN/802.11-Wireless-LAN-Solutions/BCM4360>; 2015.

[23] Qualcom Atheros 802.11ac solutions. Available from: <http://www.qca.qualcomm.com/mobile-connectivity/wi-fi/qualcomm-vive/>; 2015.

[24] The MadWifi project. Available from: <https://madwifi-project.org/wiki/UserDocs/RateControl>; 2015.

[25] Kim J, Kim S, Choi S, Qiao D. CARA: Collision-aware Rate Adaptation for IEEE 802.11 WLANs. In: Proceedings of IEEE INFOCOM Conference on computer communications, Barcelona: Spain; 23–29 Apr. 2006.

[26] Wong SHY, Yang H, Lu S, Bharghavan V. Robust Rate Adaptation for 802.11 Wireless Networks. In: Proceedings of the International Conference on Mobile Computing and Networking, Los Angeles, CA; 24–29 Sept. 2006.

[27] Pefkianakis I, Wong SHY, Yang H, Lee SB, Lu S. Towards history-aware robust 802.11 rate adaptation. IEEE Trans Mobile Comput 2012;12(3):502–15.

[28] Zhang J, Tan K, Zhao J, Wu H, Zhang Y. A Practical SNR-guided Rate Adaptation. Proceedings of IEEE INFOCOM conference on computer communications, Phoenix, AZ; 13–18 Apr. 2008.

[29] Vutukuru M, Balakrishnan H, Jamieson K. Cross-layer wireless bit rate adaptation. ACM Comput Commun Rev 2009;39(4):3–14.

[30] Sen S, Santhapuri N, Choudhury RR, Nelakuditi S. AccuRate: Constellation Based Rate Estimation in Wireless Networks. In: Proceedings of the USENIX NSDI Conference on Networked Systems Design and Implementation, San Jose, CA; 28–30 Apr. 2010.

[31] Huang KD, Duffy KR, Malone D. H-RCA: 802.11 collision-aware rate control. IEEE Trans Netw 2013;21(4):1021–34.

[32] Facchi N, Gallo P, Gringoli F, Paris S, Tinnirello I, Capone A. Flexible and Modular Support for Multicast Rate Adaptation in WLANs. In: Proceedings of the future network and mobile summit, Lisbon, Portugal; 3–5 July, 2013.

[33] Paris S, Facchi N, Gringoli F, Capone A. An innovative rate adaptation algorithm for multicast transmissions in wireless LANs. In: Proceedings of the vehicular technology conference, VTC 2013 Spring, Dresden, Germany; 2–5 June, 2013.

[34] Heusse M, Rousseau F, Berger-Sabbatel G, Duda A. Performance Anomaly of 802.11b. In: Proceedings of IEEE INFOCOM Conference on Computer Communications, San Francisco, CA; 1–3 April, 2003.

[35] GitHub. The dei80211mr library. Available from: <https://github.com/paultsr/ns-allinone-2.35/tree/master/dei80211mr-1.1.4>; 2015.

[36] The Rice University Monarch Research Project. The CMU Monarch project's Wireless and mobility extensions to ns, Available from: <http://www.monarch.cs.cmu.edu>; 2015.

[37] Gibson JD. Mobile communications handbook. Boca Raton, FL: CRC Press; 2013.

[38] Lin S, Costello DJ. Error control coding: fundamentals and applications. Englewood Cliffs, NJ: Prentice-Hall; 2004.

CHAPTER

4 Simulation techniques for evaluating energy-efficient heuristics for backbone optical networks

Constantine A. Kyriakopoulos[1], Georgios I. Papadimitriou[1], Vasiliki L. Kakali[1], and Emmanouel (Manos) Varvarigos[2]

[1]*Aristotle University, Thessaloniki, Greece*

[2]*Patras University, Patras, Greece*

1 ENERGY EFFICIENCY IN OPTICAL BACKBONE NETWORKS

1.1 INTRODUCTION

Power consumption of information and communication technology (ICT) [1] is becoming an increasingly important problem for the research community, Telecom Equipment companies and internet service providers (ISPs). Internet will continue to increase in size [2], with the introduction of new multimedia, cloud, and other bandwidth-hungry services, leading to growth of up to 50 times within the next 10–15 years. Energy consumption of telecom networks has recently become a major concern, to the extent that it is conjectured that internet growth may ultimately be constrained by energy consumption rather than bandwidth. Ensuring low energy consumption per transferred bit and controlling the energy density of large switching centers is regarded as a key economic, environmental, social and political issue [3].

Energy-aware lightpath establishment—a procedure for interconnecting distant network vital components—in optical backbone networks can reduce energy consumption while preserving performance levels. In this chapter, the whole procedure will be presented in detail and the main internal parts of a software simulator suitable for energy-aware lightpath routing and establishment will be exposed. At the same time, a simple yet efficient energy-aware heuristic[1] will be designed and simulated.

[1]Method for routing and establishing virtual links

The whole procedure of energy-aware lightpath establishment is complicated and resource-sensitive (when the network operates), so strong motivation for utilizing simulated methods emerges. That way, low level system internals can be modeled and scrutinized from the network's energy perspective and, at the same time, parameters that affect the outcome (concerning the network's operation) can be set according to a heuristic's functionality. Comparison between different heuristics can be performed through simulation that leads to evaluation of energy efficiency or the network's performance, according to predefined criteria these heuristics are based on. Conclusions from a heuristic simulation can be drawn concerning its energy footprint, propagation time, path length, required physical processing time and bandwidth distribution on the topology's edges, among others.

The main task a simulator must accomplish is to construct the network's virtual topology[2] based on heuristic principles and then calculate the amount of consumed energy. This virtual topology can represent the network under its full load as a snapshot that is invariable through time, or can evolve through simulation time with connection requests between node-pairs starting at different timestamps. The former case is not based on discrete event simulation, so requires fewer computing resources to be fulfilled. The latter case can be more realistic, though resource-heavy.

The simulating procedure follows three main phases: initialization with traffic matrix generation, virtual topology construction and, finally, power consumption calculation. First, the traffic needs between all node-pairs will be produced according to simple mathematical models. Second, these requests will be lined up in time (in case of discrete event simulation) and an energy-efficient heuristic method will try to route them upon a virtual topology. Last, power consumption will be calculated according to the produced topology with its required and occupied resources.

1.2 PLANNING PHASE

One of two phases where design decisions are made concerning connectivity between nodes at a low level is the *planning phase*. Regarding network planning,[3] an operator wishing to reduce energy consumption, carbon footprint and operational cost of his network must perform an optimization over the physical and the network layers to yield a cross-layer solution. The cross-layer Routing and Wavelength Assignment (RWA) solution will result in an energy-efficient translucent (when regenerators are used to eliminate physical impairments) or transparent (when impairments are negligible and regenerators are not used) optical network, according to its current size and physical distances. There is also efficient planning of Mixed

[2]logical connections between nodes

[3]The off-line design of network's virtual topology, i.e., connections between node pairs above the physical layer

Line Rate (MLR) WDM systems that support a variety of channel rates. In addition, the problem of energy efficient RWA can be viewed during network operation,[4] which will be discussed next.

1.3 OPERATION PHASE

In addition to planning, there is another phase where lightpath design decisions can be made. In the *operation phase*, traffic grooming[5] issues can also be examined. Algorithms can be designed to identify how to efficiently utilize the network capacity, and decide upon the routing of new or the rerouting of existing connections and the placement of regenerators (when needed) with the objective of minimizing the overall energy consumption of the network for a given blocking performance. A key issue in minimizing energy consumption is the capability of turning off or putting into sleep mode some network components. Cognitive solutions pertaining to (a) traffic prediction and (b) energy consumption estimation can also be examined to further improve the performance of the operation algorithms.

Progress beyond the state of the art, in most research projects trying to minimize power consumption in optical networks, has been achieved through network planning [4,5], traffic grooming [6–9], wavelength routing [10] and by setting network components into sleep mode [11]. Power consumption in Optical Burst Switching (OBS) is also studied by researchers [12,13], concerning lower layers [14]. Mixed Line Rate (MLR) systems that are power-aware are examined in [15]. In [11], energy efficiency is achieved not only through a proper utilization of resources for primary and protection lightpaths, but also through reducing the network power consumption of redundant resources. Since those resources are idle until a failure occurs, they can be set in sleep mode. In [4], to reduce energy consumption in the IP over a WDM network, lightpath bypass in the optical layer is applied due to its ability to reduce the number of required IP router ports. IP routers play a major role in the total energy consumption in an IP over a WDM transport network. In [6], the minimization of power consumption is achieved through traffic grooming.

1.4 TRAFFIC GROOMING

Traffic engineering when performed and exploited properly can lead to energy efficiency. So, *traffic grooming* addresses the gap between the capacity of wavelength channels and bandwidth requirements. Total power consumption of the network is calculated by considering power consumption of individual lightpaths. In [16], some heuristic algorithms are proposed that promise to reduce the number of used links and nodes up to 30% and 50% respectively during off-peak hours, while offering the same

[4]The on-line design of network's virtual topology

[5]The aggregation of many low bandwidth requests into a single one

service quality. In [17], Energy-Aware Traffic engineering (EATe) is presented, which is a technique that takes energy consumption into account while achieving the same traffic rates as the energy-oblivious approaches. EATe uses a scalable, on-line technique to spread the load among multiple paths so as to increase energy savings. Finally, in [18] a simple algorithm based on Integer Linear Programming (ILP) methodology is proposed. By selectively turning off spare devices whose capacity is not required to transport off-peak traffic, it is shown that it is possible to easily achieve more than 23% of energy savings per year. Research in this field tackles all mentioned areas and new and innovative methods are being developed, trying to achieve better results concerning power consumption and efficiency than those reported in the literature so far.

2 SOFTWARE SIMULATOR FOR ENERGY-EFFICIENT OPTICAL NETWORKS

2.1 INTRODUCTION

The main task an optical network's (energy-efficient) simulator has to perform is power consumption calculation. This procedure can take place when the network is represented under its full load, i.e., there is a higher layer[6] connection between each node-pair. In such cases, there can be no dynamic network development under a running simulation time, just a representation of a snapshot under full load.[7] When this snapshot is created, power consumption will be calculated according to a power model that relates to a specific moment and not to a wide time span.

Another type of simulator software would be the role of creating a network state that develops when the simulation time moves forward. This can be a discrete event or slot-based. Power models in this case have to take into account consumption in correlation to running time, e.g., in kWh. A simulator that belongs to the former case will be extensively analyzed in the next section.

There are three main phases a simulator has to go through to fulfill its purpose: initialization with traffic matrix generation, lightpath routing for virtual topology construction, and power consumption calculation. Details are in Figure 4.1.

2.2 PHASE 1: INITIALIZATION AND TRAFFIC GENERATION

The first step for a simulator is to parse all important settings, either from an external file or from console input. These settings include the network's physical

[6]IP layer

[7]There is a light connection between each node-pair

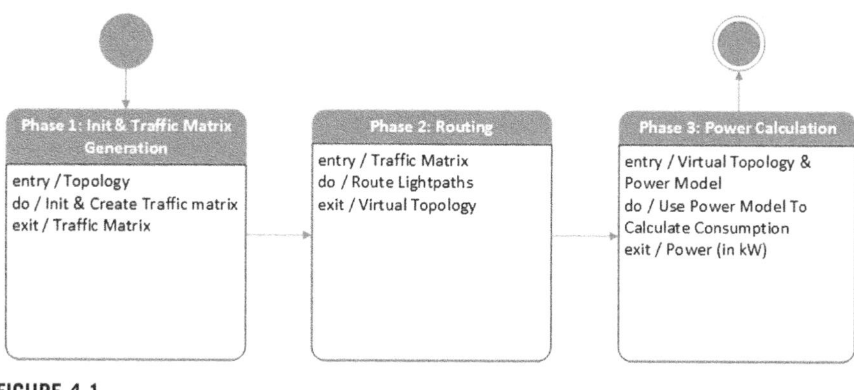

FIGURE 4.1

Simulator's UML state diagram.

topology, the heuristic to be utilized, power model and average traffic demand, among others. Topology has to be represented internally as a special data structure that can be used anytime for fetching every required shortest path between node-pairs, using an algorithm such as Dijkstra's. This topology differs from the virtual one that will be constructed piece by piece during the next phase. The heuristic has to be initialized now so it can be employed as a tool for creating the required virtual lightpaths later on. The power model will be needed during the last phase for calculating power consumption.

Average traffic demand value is needed for the traffic matrix's initialization. Every node-pair will communicate with a specific bandwidth requirement that has to be stored in advance, so it can be used later on. This value (not the average one) will be produced by using a mathematical distribution with predefined average value and other important parameters when needed. If there is only one bandwidth requirement between node-pairs and no explicit correlation between traffic needs of neighboring nodes, this utilized distribution can be a uniform one. A candidate data structure is the two-dimensional array that can hold floating point values. Every value can be replaced by a linked list of values, in case more transmissions between node-pairs have to be carried out.

2.3 PHASE 2: VIRTUAL TOPOLOGY CONSTRUCTION

Requests that reside in the matrix will be fetched one by one and the employed heuristic will try to route them, creating in that way the virtual topology. The amount of energy efficiency that will be achieved depends upon the heuristic's design and functionality. Two types of heuristics can be employed: on-line and off-line. The first type doesn't use history concerning the incoming requests, trying to fulfil the next one independently of the previously established lightpaths.

The latter frequently revisits the virtual topology, trying to achieve better resource utilization, for example by performing elaborate combinations.

Virtual topology (VT) consists of lightpaths that can use use bypass technology for node interconnection as well. This topology is constructed piece by piece. If there are no available resources for a new request to be fulfilled, new lightpaths are introduced, raising in that way the resources used, and consequently more energy is consumed. There are two ways to search VT for existing resources (preestablished lightpaths): by using a brute force technique or utilizing adaptive functionality such as Ant Colony Optimization (ACO) or Learning Automata (LA).

Brute force guarantees the best results but suffers from high complexity since it is being applied on an NP-Hard problem. A recursive method is needed to be applied upon nodes stemmed from VT just to find all possible paths (or the first one when the First-Fit strategy is used) that connect the new request's end nodes. If paths are found successfully, resource reuse increases, which eventually leads to a more energy-efficient virtual topology. When the brute force strategy finds all available paths per node-pair, they can be scrutinized in order to choose the most suitable ones for the needs and design decisions of the employed energy-efficient heuristic.

Results from adaptive techniques usually diverge slightly (proven to be below 1% on exposed experimental results) from the best possible ones, but these techniques compensate due to the polynomial complexity they offer. The advantage is prominent; they can be applied on very large virtual topologies running in feasible physical time. For every new traffic request, the current virtual topology is provided to the adaptive mechanism as input. Its purpose is to find the most suitable paths from source to destination with specified properties, e.g., the highest available bandwidth.

2.4 PHASE 3: POWER CALCULATION

The simulator's last phase is the power consumption calculation. The previous phase's output is the virtual topology that was designed based on principles stemming from the employed heuristic. The purpose of this phase's procedure is to analyze the virtual topology's energy-consuming individual components according to their individual power consumption and calculate the summary (energy footprint). A modern pluggable simulator design mandates the use of independent software units as power models. That way, different calculating mechanisms can be used on a virtual topology for producing results according to different principles, simultaneously.

The main components that are considered from most power models' perspective when calculating energy are router ports, line amplifiers and transponders.

There are also models taking into account accumulated traffic from low-end routers along with traffic coming from backbone routers. This phase can be extended to include, instead of power metrics, other data of interest (statistics) like hop-count. To summarize, the VT output of the previous phase feeds this phase as input and the new output are values representing power consumption.

3 OBJECT-ORIENTED PRINCIPLES FOR DESIGN

3.1 TOPOLOGY REPRESENTATION

Well-known and correctly used object-oriented principles and design patterns can offer coding flexibility and safety and can ease the maintenance procedure. There are two types of topologies, those that are well known in research cycles and are static[8] and the dynamic ones concerning their properties.[9] There are also two main ways of representing topologies in software simulators, i.e., hard-coded or parsed dynamically from an external representation,[10] independent from hardware details.

Hard-coded static topologies can take advantage of inheritance. Common parts or functionality to all of them, like a shortest path algorithm, can as well reside on a base class, being accessible to all concrete implementations. Unique topologies can differentiate by employing (implementing) different edge weights, connectivity and nodes. An example of a hard-coded exible topology relation can be found in UML, Figure 4.2.

Dynamic topologies can be constructed on the fly from external input[11] for efficiency. A data description language like XML can be utilized, having the advantage of carrying a standardized form, independent from hardware technicalities and easily being parsed.

3.2 HEURISTICS

Energy-efficient heuristics should extensively use inheritance to enable exibility and avoid code duplication. An abstract class can represent the base (API) functionality and concrete implementations of the heuristics themselves. Since there are two types[12] of them, the inheritance tree should include middle classes (semi-implemented abstract classes) that carry functionality needed by (common to) all operation and all planning heuristics, respectively.

When exploiting parallelism, polymorphism comes in handy for the need of dispatching at the same time different heuristics on multiple processing units that will run simultaneously, with each one producing a unique virtual topology, using the same or different traffic matrix on the same network configuration. Specific code can manipulate (with a simple dependency) only the base class and send different concrete implementations to available processing resources. The UML heuristic diagram can be found in Figure 4.3, where the ease of adding or removing heuristics, code duplication avoidance and other object-oriented principles are implied.

[8]e.g., NSFNet and USNet

[9]Edge weights and connectivity can vary through time

[10]XML is a candidate solution

[11]A settings file or stream

[12]On-line and off-line, i.e., Operation or Planning

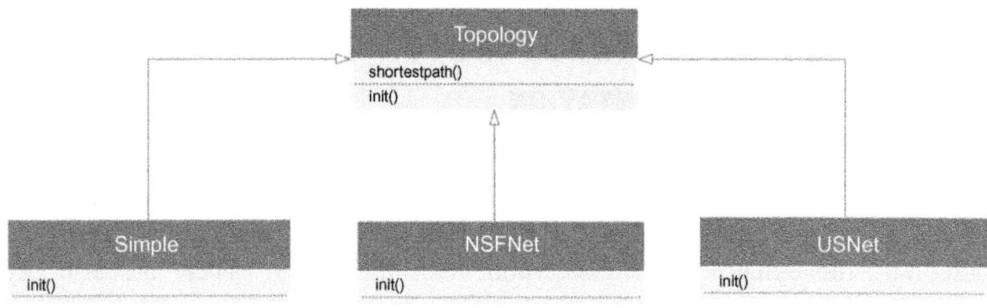

FIGURE 4.2

Flexible topology diagram.

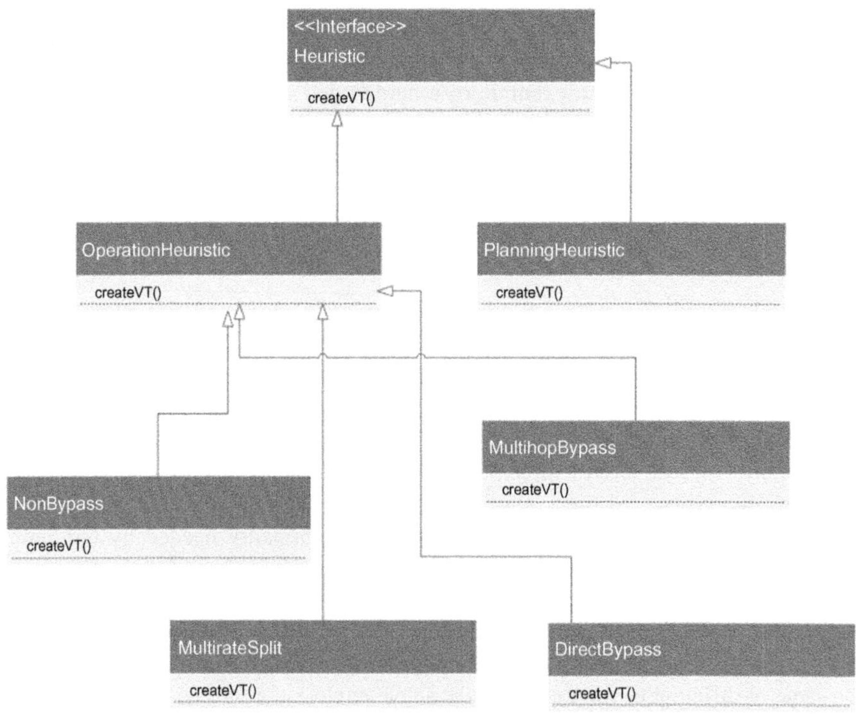

FIGURE 4.3

Heuristic inheritance.

3.3 POWER MODELS

Sometimes there is a need for calculating power consumption with different methods on the same virtual topology, sequentially or in parallel. In some other cases, multiple power models can be executed in parallel on the same or different virtual topologies, or be chosen upon user request. Polymorphism is invaluable in

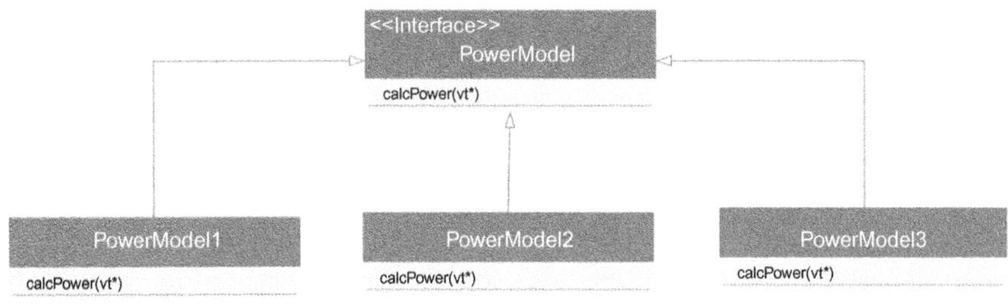

FIGURE 4.4

Power model relation.

execution environments such as this. An API provided by a base class can be utilized with its concrete implementation to be chosen or changed during runtime. Power models can inherit API methods but provide different implementations that can be changed at runtime by using specific design patterns.

Another design decision is to implement power models as autonomous components that can be loaded at runtime as external plug-ins. In this case, there is a need for a standardized communication protocol that can give implementation freedom even on a different programming language. A typical power model inheritance is shown in Figure 4.4.

3.4 ASSISTING TOOLS

A modeling language like UML can come in handy for an abstract representation of a simulator's design. High-level hierarchies, dependencies, transitions and other properties can be depicted in a standardized manner, so efficient communication between programmers can take place, among other things. Diagrams that can be useful are Class, Sequence, Activity and Collaboration-related.

External tools can be useful during implementation. There is a need for finding the shortest paths on a typical graph structure, so a shortest-path algorithm such as Dijkstra's is required. It can be used from a graph library like Boost.[13] When heuristics fail to create efficient lightpaths by reusing preexisting resources, the preference would be to establish new ones along the shortest path—a viable solution.

Containers are a concrete part of a high-level object-oriented language. Temporary memory storage needs can be fulfilled by using them. Since they provide type safety, a lot of obscure bugs can be avoided, in contrast to using traditional low-level language arrays which are prone to errors. Maps, Sets, Lists and Vectors (even Array class simulators) are important for writing efficient code.

[13]www.boost.org

The right choice is correlated to current coding needs; for example, if random access is important, Vector can be a choice. If not, a Linked List provides fast traversal. If ordering is important, a Set is needed. If not, an Unordered Set would provide more performance.

Lambda functions can make a standard library's algorithm use safer (less obscure bugs). They can be used, for example, when searching for properties on existing lightpaths.

4 NETWORK ENVIRONMENT

4.1 NETWORK COMPONENTS

The network environment that is under examination from an energy perspective consists of interconnected backbone routers that are also capable of accumulating traffic of other low-end routers. The latter intervene between end users and the backbone network. This type of network environment is suitable for the needs of midsize to large ISPs that cover the area of small nations to large continents. Since most Internet traffic is being routed upon such networks, it is really important for routing and resource utilization, in general, to be aware of energy issues. That way, the energy factor gets embedded as a main parameter into a network's functionality.

The main network components that consume high amounts of energy according to Cisco white papers [2,19] are router ports, transponders and line amplifiers (Table 4.1). So, a heuristic method that tries to create the VT (virtual topology, a layer above physical) has to consider minimizing their presence, without performance penalty.[14] This can be achieved by smart reuse of existing lightpaths that carry enough available bandwidth, when trying to route new connection requests.

4.2 TESTBED TOPOLOGIES

In order for research to take place in this field, static topologies are needed that represent realistic optical networks, to be used for performance evaluation of new resource utilization methods. Three backbone network topologies (Figures 4.5, 4.6 and 4.7) can be used for such testing purposes: i.e., a simple 6-node topology, NSFNet and the large USNet. They can be used as part of, or as the whole, backbone network on a national or continental scale. Creating a network operation snapshot of these topologies under full load doesn't require heavy use of computer resources during the lightpath establishment procedure, as long as the search domain of a brute force method is minimized or a smart/ adaptive heuristic is utilized.

[14]That can be high propagation time or Quality of Service

Table 4.1 Simulation Parameters

Parameter	Value
Wavelength max bandwidth	40 Gbps
Distance between two line ampliers	80 km
Available wavelengths per fiber	16
Router port power consumption	1000 W
Transponder power consumption	73 W
Line amplifier power consumption	8 W

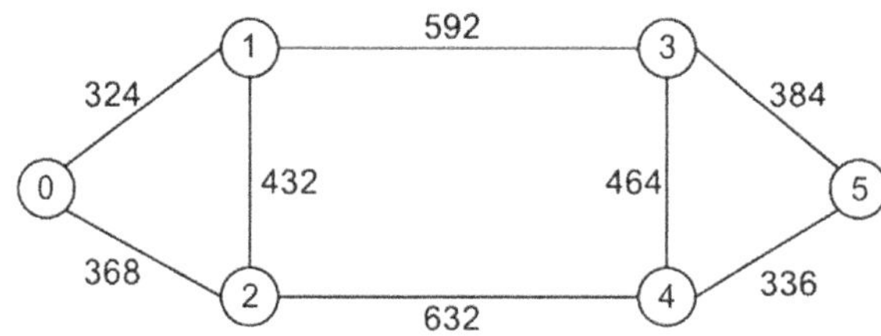

FIGURE 4.5

Simple 6-node test topology.

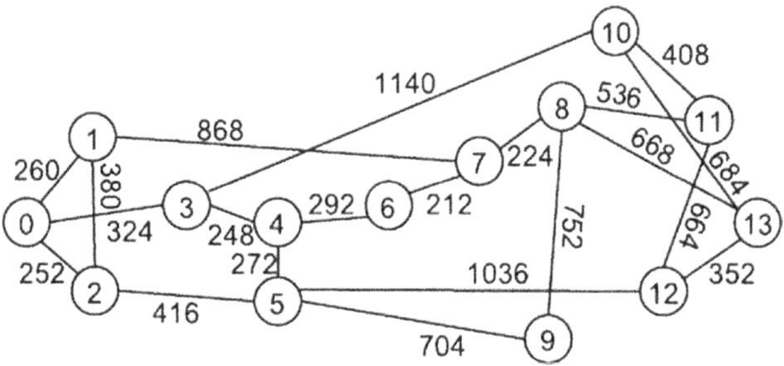

FIGURE 4.6

NSFNet test topology.

The first topology is simple enough for practical deployment, yet very useful for testing the correctness of advanced techniques concerning lightpath routing. Execution time in this topology for every setup and utilized heuristic is extremely low on a typical modern CPU, due to the small number of nodes. Even NP-Hard problems can be executed upon it, so extensive testing can be realized. The second one can represent the backbone network of a typical European country, so energy minimization of this topology becomes a key factor of ongoing research. Even brute force algorithms can be executed upon it, though they are resource

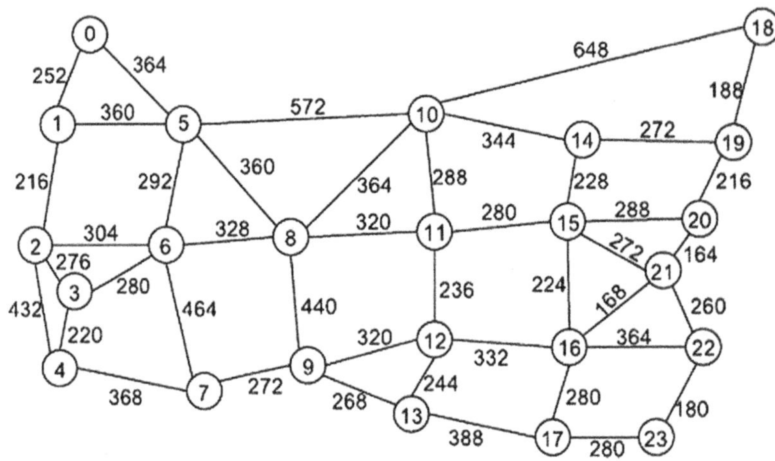

FIGURE 4.7

USNet test topology.

hungry. The last one can cover the network of big countries and continents. It is important that every new heuristic is able to be executed upon all three topologies in a predictable amount of time.

4.3 TRAFFIC MATRIX

The traffic matrix usually contains one request per node-pair. That is the case when a network's snapshot under full load has to be captured and analyzed to determine its power consumption and other performance parameters. When simulation runs through virtual time, more connection requests per node-pair can be simulated and fulfilled.[15] A data structure that resembles a two-dimensional array is sufficient for the former case. As for the latter (multiple requests per node-pair), the values of this table can also be carried by linked lists for efficiency.

The upper bandwidth limit of a lightpath can be set at 40 Gbps for current state optical networks. Thus, if the matrix carries higher values, multiple lightpaths have to be fulfilled per request. A mathematical distribution can be the tool for producing these bandwidth values. A simple uniform distribution is adequate for the single request per node-pair case. For other paradigms, Pareto distribution can be utilized.

4.4 POWER MODELS

For energy consumption calculations, special models should be considered. Every one of them can implement a different strategy, which tries to stay close to

[15]Routed, for creating the virtual topology

real-world scenarios by predicting consumption as accurately as possible. Energy-consuming components [2,19] (Table 4.1) are considered by some models to be the line amplifiers E_e (8 W per unit), transponders E_t (73 W per unit) and router ports E_r (1000 W per unit). For the majority of cases, every physical link can include two (pre- and post-) amplifiers at its end points and one every 80 km in between. The maximum per-wavelength bandwidth is usually 40 Gbps and every fiber can include up to 16 of these. Every neighboring node-pair can be interconnected with an unlimited number of fibers, the actual number being dependent on the heuristic being applied. Total energy consumption can be computed with the following formula [4] in watts. This represents a simple power model, though more elaborate models can be employed as well.

$$\sum_{i\in N} E_r \times \left(D_i + \sum_{j\in N:i\neq j} C_{ij}\right) + \sum_{m\in N}\sum_{n\in Nm} E_t \times W_{mn} + \sum_{m\in N}\sum_{n\in Nm} E_e \times A_{mn} \times f_{mn} \quad (4.1)$$

D_i is the number of ports that are used to aggregate data traffic from low end routers and equals to $\lceil(\sum d_{\in}N\, l^{id})/B\rceil$; l^{id} is the traffic demand between node-pair (i,d) and B is the maximum wavelength bandwidth. C_{ij} is the number of wavelength channels (wavelengths that start from i and end at j with an uninterrupted physical single-hop light connection) upon the lightpath (i,j) and the summary denotes how many wavelength units are starting from node i and reach to all other topology nodes. Next, w_{mn} is the number of used wavelengths on the physical link (m,n). Finally, f_{mn} is the number of deployed bers on the same physical link and A_{mn} is the number of amplifiers which equals to $\lceil L_{mn}/S-1\rceil+2$. L_{mn} is the distance measured in kilometers and S is constant and equals to 80 km (the distance between two consecutive intermediate line amplifiers except the last one). The addition of 2 to this formula covers both amplifiers at the beginning and end of the physical link which perform pre- and post-light amplification. N_m is the set of m's neighbors and N the set of the network's nodes.

5 EXPLOITING TRAFFIC GROOMING AND RELATED HEURISTICS

5.1 TRAFFIC GROOMING

When lightpath reuse is absent from the establishment procedure, some wavelengths (network resources) will be underutilized. For example, if a traffic request between a node-pair consists of 17 Gbps, the other 23 Gbps will not be available for utilization by another low-traffic request (maximum 40 Gbps per wavelength), resulting in higher demand for spare wavelengths and fibers, leading to higher energy consumption. Avoiding such bandwidth fragmentation is the route to energy efficiency. The traffic grooming problem was extensively studied and a mathematical formulation was presented [20], along with several typical heuristics. The assumptions that took place in this study were: (a) there is at most one fiber link between each node-pair, (b) there is no wavelength conversion capability, (c) node transceivers are tunable to any wavelength, (d) a connection request cannot be divided into several lower-speed connections which can be routed separately and (e) there is unlimited (de)multiplexing and timeslot interchange capability in each node. Two main heuristics were proposed with the first one being Maximizing Single-hop Traffic (MST) and the other Maximizing Resource Utilization (MRU). The former attempts to establish lightpaths between each source and destination node with the higher demand traffic values. The connection will be carried on a new established lightpath as much as possible. If there is enough spare bandwidth in the network, only single-hop lightpaths will be created. If not, the currently available spare capacity will be used. The latter heuristic defines a resource utilization parameter and tries to establish lightpaths between node-pairs with the maximum resource utilization values. If it fails, the spare capacity will be used.

Dynamic Traffic Grooming was also studied [20] and an analytical model was developed allowing heterogeneous data rates for sub-wavelength connections, arbitrary routing in both logical and physical topologies and arbitrary wavelength connections. This model can be used to evaluate the performance of a specific network or new grooming algorithm and to study optical network design with consideration of traffic grooming. It can also be used to analyze the blocking performance of dynamic calls under mixed, static and dynamic traffic.

5.2 NONBYPASS

A standard nonbypass method (reusing or not available preexisting lightpaths) can be used as the low anchor during the measurements throughout a simulation. This method is considered resource hungry, so its utilization for practical deployment is not recommended. Every node along the path of this typical heuristic is a point of energy dissipation, i.e., O-E-O signal conversion is required and takes place. So, it's necessary for a high number of router ports to be powered and ready to be utilized, leading to consumption of high amounts of energy. Dijkstra's shortest path algorithm is applied for finding the route for every traffic request. So, every edge of the graph's structure (representing the network topology) requires a high number of lightpaths passing over it, since their distribution to topology cannot be uniform. This leads to the need for more fibers per physical connection, which can be interpreted to mean more transponders and amplifiers, consuming considerable amounts of energy. The average hop-count for requests is higher when compared to most of the other heuristics that can use bypass as a strategy for establishing new lightpaths.

5.3 DIRECT BYPASS

Direct bypass is simple concerning its implementation, yet powerful, achieving low energy consumption. A new direct lightpath is established per traffic request, bypassing intermediate nodes. Data remain in the optical domain throughout the shortest path connecting request's end nodes. This means that no O-E-O conversion is required or performed at intermediate nodes. The side effect of this method is that there is no lightpath reuse when there is only one traffic request per node-pair, leading to lightpaths with spare bandwidth that gets unexploited. So, more fibers and router ports are needed for deployment, leading to consumption of high amounts of energy.

5.4 MULTIHOP BYPASS

Multihop bypass [4], the third method, was the one targeted for performance improvement from the proposed heuristic's (Multirate-Split Bypass) perspective, which will be presented in the last sections of this chapter. Multihop bypass creates lightpaths consisting of multiple hops—each one of them can contain multiple nodes within the physical node sequence—but it manages to save energy due to lightpath reuse from previously established connection requests. So, when the whole path of a new traffic request is covered by preexisting lightpaths with adequate free capacity, all of these will be reused to create the new groomed lightpath (first-fit strategy to curb complexity). The basic idea is that no new router port utilization will be required, which would consume high amounts of energy, in contrast to the other two energy-consuming components (line amplifiers and transponders) that are also considered when power consumption is taken into account. However, there is a slight penalty concerning light signal propagation delay due to the existence of multiple hops along path's node sequence, where O-E-O conversions are required and performed. Also, there is elongation of some paths due to searching (utilizing brute force techniques) of the whole topology for finding pre-established lightpaths with spare bandwidth, not just within the shortest path connecting the two end nodes.

ALGORITHM 4.1 BASIC ACO OUTLINE

```
set ACO parameters
  initialize pheromone levels
  while stopping criteria not met
    for each ant k
      select source initial node
      repeat
        select next node based on decision policy
      until destination node is reached
    end for
    update pheromone levels
  end while
```

6 ADAPTIVE METHODS TO REDUCE COMPLEXITY

6.1 COMPLEXITY

Finding routes by reusing lightpaths that carry adequate bandwidth to fulfill a request is an NP-Hard problem. Complexity becomes a hurdle to similar research and Multihop Bypass suffers from it. It is performing, for every new traffic request, a brute force attack to find a path consisting of lightpaths carrying enough spare bandwidth. This problem becomes difficult to solve for large network topologies with low average traffic requests (leading to large virtual topology), wasting CPU and memory resources. Multihop Bypass has the advantage of shortening the search domain by choosing lightpaths for participating in it that only carry enough bandwidth to fulfil the request, but it still suffers from low traffic requests.

On the other hand, using heuristics based on Swarm Intelligence or Learning Automata, complexity becomes polynomial, i.e., $O(N^3)$, so development and deployment become feasible for the majority of large network topologies. For example, that can be confronted by looking at the basic ACO outline [21], which shows that complexity is not a hurdle for implementation and deployment. The outer loop (Algorithm 4.1) represents the number of iterations and the inner depends on the number of ants. Every ant uses a recursive method that cannot be deeper than the number of topology nodes. So, the total algorithm's complexity is described by notation $O(N^3)$. The ant overhead is minor and linearly related to the topology's size. At the same time, divergence from optimal path finding—due to ACO being a heuristic method—is extremely low [21].

6.2 ANT COLONY OPTIMIZATION

ACO along with its extensions has been applied successfully [21,22] to the Traveling Salesman Problem (TSP), converting complexity to a polynomial form. The purpose is to visit all cities sequentially with the minimum cost—represented by edge weights. The virtual mapping of lightpaths to physical connections for creating survivable topologies [23] can also be performed using ACO. It has also been applied to other hard combinatorial optimization problems such as quadratic assignment, vehicle routing, job-shop scheduling and graph coloring.

The implementation of energy-efficient heuristics in optical backbone networks implies that the best paths with specific properties have to be obtained to be exploited. This is an NP-Hard problem, so ACO serves the purpose of reducing complexity to a polynomial scale. An Ant System which is based on ACO principles (a set of main rules that describe the behavior of a class of heuristics) can be transformed as well to solve the shortest path problem. Thus, restrictions have been applied, e.g., visiting only physically connected neighboring nodes and all ants starting their traversal from a specific source node to reach a specific destination node. It was demonstrated that the best paths can be easily obtained using low computational resources.

Network routing can also be confronted with ACO. The basic network characteristics such as traffic load and topology can vary stochastically in a time-varying way. Considering physical disturbances on the lower network layer, ACO becomes a promising tool to make routing more efficient. AntNet [21] is an ACO implementation that confronts these challenges efficiently. It has been extensively tested in practice and it was proved that it serves its promises efficiently.

6.3 LEARNING AUTOMATA

The main problem that has been tackled by using ACO can also be confronted with the use of Learning Automata (LA), i.e., to nd the shortest path [24] on a virtual topology. The edge weight can be swapped with another property of importance, in the case of energy-efficient optical networks, which is the available bandwidth a lightpath carries. That way, paths with the highest available bandwidth can be obtained under polynomial complexity and be utilized during routing upon virtual topology. An explicit consequence is the advanced resource utilization that eventually leads to energy efficiency. If the network's performance is more important than energy efficiency, short paths with fewer hops can be obtained, instead of those with high available bandwidth.

Every virtual topology's node can contain an LA that holds transition probabilities to its neighboring nodes. When a destination node is to be reached by starting from a specific source, a recursive method is executed (for a predefined number of iterations) that chooses the next node according to current probability numbers. If the destination node is reached, a feedback value is applied to all LAs along the path at the end of each iteration, which raises the corresponding probabilities rendering that path more preferable during subsequent iterations. That way, the best paths that comply with predefined properties can be obtained. Being heuristics, these methods are unable to find the optimal paths in every case, but the divergence from corresponding brute force results can be less than 1%, which renders them invaluable.

7 A SIMPLE YET EFFICIENT HEURISTIC: MULTIRATE-SPLIT BYPASS

7.1 MAIN OUTLINE

The Multihop Bypass method is more efficient than Direct Bypass according to simulation results, but leaves space for further performance improvement.

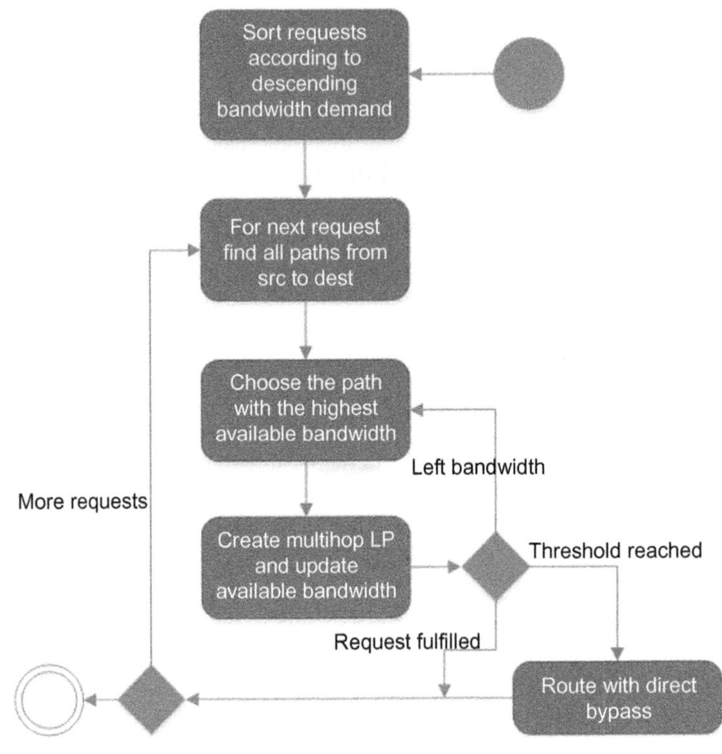

FIGURE 4.8

Multirate-Split activity diagram (asymmetric).

As long as the node sequence of a new lightpath request is covered with preexisting lightpaths with adequate spare bandwidth, this method is suitable for reducing energy consumption. If this is not the case, a new heuristic is required that is capable of taking advantage of asymmetric bandwidth split through multiple streams, routed to the destination through different node sequences—introducing Multirate-Split Bypass.

The proposed Multirate-Split Bypass heuristic (Figure 4.8) improves performance when trying to create a new lightpath, by asymmetrically splitting bandwidth along different paths that offer less available bandwidth against the one of the initial request. If a route is found that is capable of carrying the initial request's bandwidth, it is used in the same way that Multihop Bypass would take advantage of. If not, asymmetric split takes place into several streams (until a threshold value, i.e., 7 is reached, used for the purpose of reducing complexity with no actual benefit to power consumption for higher values) and are routed to the same destination using different node paths, using Multihop Bypass. The first stream uses the path with the highest available bandwidth. The second stream uses the path with the next available one and the rest accordingly, until a threshold value (dependent on current processing

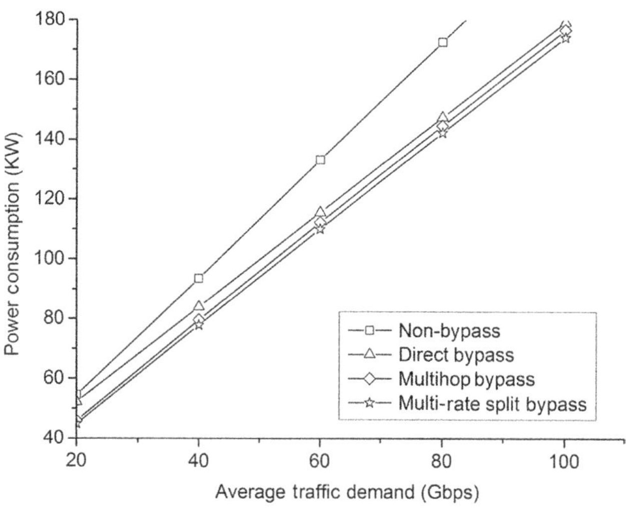

FIGURE 4.9

Power consumption of 6-node topology.

resources) is reached. If this is not enough for the whole bandwidth to be routed, a simple Direct Bypass lightpath is established using the shortest path between end nodes, carrying the initial request's bandwidth. When the Multirate-Split Bypass heuristic is applied, every new request's bandwidth adapts to the current topology's spare resources (available bandwidth) and fragmentation is minimized, i.e., lightpaths with little available bandwidth that cannot be easily reused by the following requests. Simulation results showed that higher threshold values won't reduce energy consumption significantly (Figures 4.9, 4.10), and at the same time, routing complexity will increase to a point that makes the heuristic practically undeployable.

Finding the streams with enough available bandwidth to fulfill the initial request is an NP-Hard problem. If a lightpath is going to be routed with a single stream, all pre-established lightpaths with higher spare bandwidth from that of the request's are needed to participate in a recursion that leads to the destination node. This minimizes the recursion domain and subsequently, complexity. When there is a need to find a number of paths with the highest available bandwidth, all available lightpaths should participate. That way, complexity is maximum. To overcome this hurdle, a method that uses a decreasing step was employed. First, an attempt is being made to find all paths with higher available bandwidth than the initial request's. Inside this recursion, only those lightpaths participate that have higher available bandwidth. If there is no success, there is a decrease by 1 Gb and the set of lightpaths with higher available bandwidth is fetched again, repeating the procedure. This step decreases by the same amount until all initial bandwidth is routed through different paths. Lower values increase complexity since almost all lightpaths participate in the search domain.

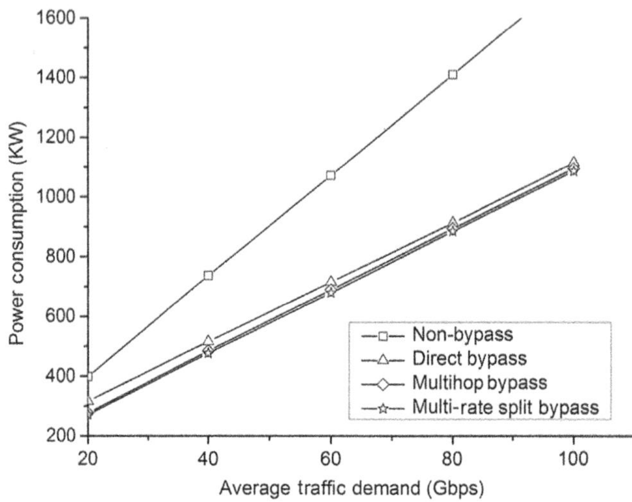

FIGURE 4.10

Power consumption of NSFNet topology.

7.2 HEURISTIC TYPE

The Multirate-Split method is an online algorithm that won't preprocess traffic requests before initiating lightpath establishment, except from an initial sorting that takes place in descending order, according to every request's bandwidth requirement. Then, for every new request, an attempt is made for creating a lightpath based on full reuse of previously established lightpaths. If there is success, energy consumption is minimized due to the absence of new router ports along the path. If not, a Direct Bypass lightpath will be established through the shortest path between the source and destination node-pair. That way, the required transponders and amplifiers along the path are minimized and also there is a chance for higher reuse rate of the next lightpath establishments. The internals of this heuristic are depicted in the UML activity diagram of Figure 4.8.

Asymmetric split is more efficient concerning the energy footprint when compared to Multihop Bypass, but requires more computer resources when it is implemented. This is due to the lack of knowledge of the actual required bandwidth that will be needed by every candidate route. For symmetric split this is already known in advance, so the graph structure carries only the adequate connections that can fulfill the next request, minimizing the recursion domain. This won't apply to the asymmetric form.

Summing up, the main idea embedded in Multirate-Split Bypass is the asymmetric division of initial bandwidth into smaller fragments that will be routed independently using the preexisting Multihop Bypass heuristic. When all these fragments reach the destination, assembly takes place in the electrical domain.

That way, spare bandwidth of preexisting lightpaths is being exploited in an advanced way, i.e., more demand for smaller bandwidth requests takes place that eventually leads to energy efficiency.

8 MULTIRATE-SPLIT BYPASS: SIMULATION RESULTS

8.1 INTRODUCTION

The simple topology (Figure 4.5) along with NSFNet (Figure 4.6) were extensively tested with various values as average bandwidth demand, calculating power consumption at the end of each network snapshot creation. The traffic matrix carries bandwidth requirements between each node-pair (one request) and is initialized in the beginning of the simulation. For the traffic matrix's initialization, the formula of [4] can be used, i.e., $X \in \{20,40,\ldots,100\}$ Gbps (that is the average value used for executing simulation instances) and the actual node-pair demand as a real number is produced by using a uniform distribution in the range of [10, 2X − 10] Gbps. Since the average traffic value can fluctuate considerably between different simulations, the average output value of many executions should be recorded.

8.2 CONSUMED POWER

The next step in the experiments was to calculate the percentage of saved power from benchmarked heuristics, using the Non-bypass method as low anchor, i.e., power saving in comparison to this method. All parameters that define the simulation environment and were used in this study can be found in Table 4.1. Since each simulation run requires low physical computation time, the average output power value of 100 runs of each configuration is depicted as resulting waveforms.

In Figures 4.9 and 4.10, the power consumption in kW is depicted on the vertical axis for ascending values of the average traffic demand between node-pairs (horizontal axis). These figures relate to both network topologies. The most power-consuming heuristic is Non-bypass which is used as the low anchor in every test case. After a large margin, the other three heuristics prove their efficiency against it. According to their specific values of power consumption, Direct Bypass is less efficient, while Multirate-Split Bypass outperforms all three of them.

The second ranking heuristic is Multihop Bypass, which takes its advantage from the third Direct Bypass due to its ability to reuse previously created lightpaths when carrying available bandwidth. The differences among them are clearly verified in Figures 4.11 and 4.12, which are produced under the same network configuration.

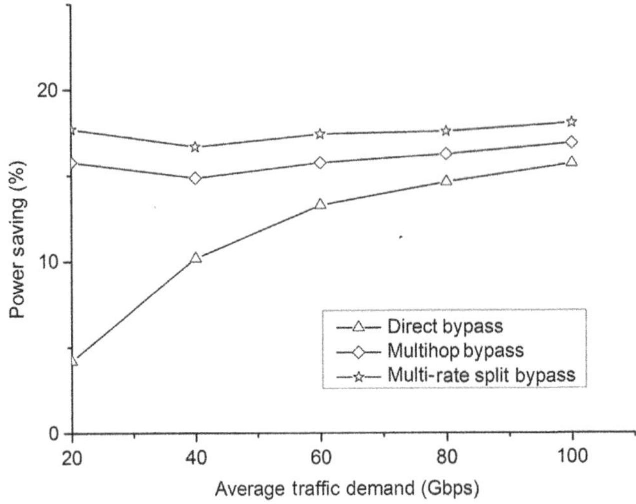

FIGURE 4.11

Power saving in 6-node topology.

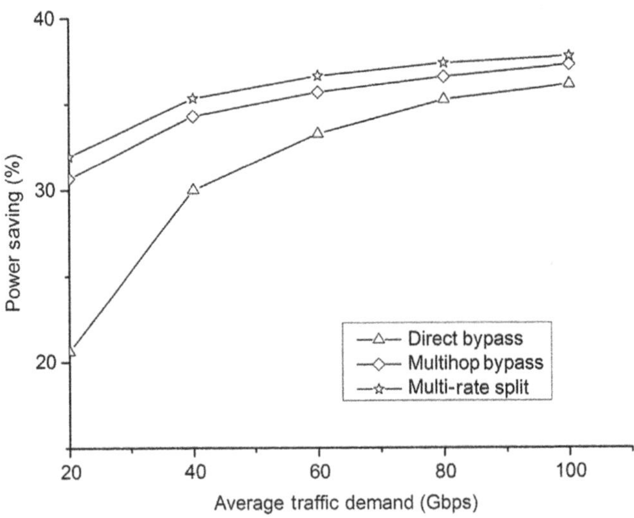

FIGURE 4.12

Power saving in NSFNet topology.

8.3 POWER SAVING PERCENTAGE

In Figures 4.11 and 4.12, power savings of Multirate-Split Bypass methods outperform those of competing heuristics. When the values of average traffic demand get higher, differences between heuristics tend to persist. The upper bandwidth limit of lightpaths, i.e., 40 Gbps, is quite low when compared to the increasing average traffic demand, which eventually reaches 100 Gbps. So, there are fewer chances for a lightpath with spare bandwidth to be found and be reused. Energy efficiency is therefore less prevalent for all participating heuristics within the high range. Energy efficiency is more important when there are low (in this scale) bandwidth requirements that define current consumer networks.

ACKNOWLEDGEMENTS

This work has been funded by the NSRF (2007–2013) Synergasia 2011/EPAN-II Program titled "Energy Efficient Optical Network Planning and Operation," General Secretariat for Research and Technology, Ministry of Education, Religious Affairs, Culture and Sports (contract no. 11SYN_6_1942).

REFERENCES

[1] Zhang Y, Chowdhury P, Tornatore M, Mukherjee B. Energy efficiency in telecom optical networks. IEEE Commun Surv Tutorials 2010;12(4): Fourth Quarter.

[2] Cisco Systems. In: Cisco visual networking index: forecast and methodology, 2009–2014, Cisco White paper.

[3] Stichting Green Touch. GreenTouch website. Available from: <http://www.greentouch.org/>; 2015.

[4] Shen G, Tucker RS. Energy-minimized design for IP over WDM networks. IEEE J Sel Areas Commun 2009;27(3).

[5] Ricciardi S, Careglio D, Palmieri F, Fiore U, Santos-Boada G, SolePareta J. Energy-oriented models for WDM networks. Lecture notes of the institute for computer sciences, social informatics and telecommunications engineering, vol. 66. 2012, p. 534–48.

[6] Yetginer E, Rouskas GN. Tubitak. Power efficient traffic grooming in optical WDM Networks. UEKAE, Ankara, Turkey, Global Telecommunications Conference. GLOBECOM '09. IEEE; 2009.

[7] Hasan MM, Farahmand F, Jue JP. Energy-awareness in dynamic traffic grooming, optical fiber communication conference. San Diego, CA; March 21–25, 2010.

[8] Huang S, Seshadri D, Dutta R. Traffic grooming: a changing role in green optical networks, global telecommunications conference. GLOBECOM '09. IEEE; 2009.

[9] Hasan MM, Farahmand F, Patel AN, Jue JP. Traffic grooming in green optical networks, communications (ICC). IEEE International Conference; 2010.

[10] Wong SW, et al. Sleep mode for energy saving PONs: advantages and drawbacks, GLOBECOM Workshops; 2009.

[11] Muhammad A, et al. Energy-Efficient WDM network planning with dedicated protection resources in sleep mode, Univ., Linkop ing, Sweden, Global telecommunications conference (GLOBECOM), IEEE; 2010.

[12] Bathula BG, Elmirghani JMH. Energy efficient optical burst switched (OBS) Networks, GLOBECOM Workshops, IEEE; 2009.

[13] Kang D-K, et al. Impact of traffic patterns and burst assembly on energy consumption in OBS networks computer modelling and simulation (UKSim). 12th International Conference; 2010.

[14] Dorize C, Morea A, Rival O, Berde B. An energy-efficient node interface for optical core networks, transparent optical networks (ICTON). 12th international conference; 2010.

[15] Chowdhury P, Tornatore M, Mukherjee B. On the energy efficiency of mixed-line-rate networks, optical fiber communication conference San Diego, California, United States. March 21–25, 2010.

[16] Chiaraviglio L, Mellia M, Neri F. Reducing power consumption in backbone networks, Communications. ICC '09. IEEE International Conference; 2009.

[17] Vasic N, Kostic D. Energy-aware traffic engineering. In: Proceedings of the first international conference on energy-efficient computing and networking, p. 169–78, 2010.

[18] Chiaraviglio L, Mellia M, Neri F. Energy-aware backbone networks: a case study, communications workshops. ICC workshops '09. IEEE international conference; 2009.

[19] Cisco Systems Inc. Cost Comparison between IP-over-DWDM and Other Transport Architectures, 1992–2007, Cisco white paper.

[20] Xin C. Blocking analysis of dynamic traffic grooming in Mesh WDM optical networks. IEEE/ACM Trans Netw 2007;15(3).

[21] Dorigo M, Stutzle T. Ant colony optimization. Cambridge, Massachusetts: The MIT Press; 2002.

[22] Gaabowski M, Musznicki B, Nowak P, Zwierzykowski P. Shortest path problem solving based on ant colony optimization metaheuristic. Image Process Commun 2012;17(1–2).

[23] Kaldrm E, Ergin FC, Uyar S, Yayml A. Ant colony optimization for survivable virtual topology mapping in optical WDM networks, September 14–16, 2009 METU Northern Cyprus Campus 344.

[24] Misra S, Oommen BJ. Dynamic algorithms for the shortest path routing problem: learning automata-based solutions. IEEE Trans Syst Man Cybern B Cybern 2005;35(6).

CHAPTER

5 Wireless cognitive network technologies and protocols

Valeria Loscri[1], Arash Maskooki[1], Nathalie Mitton[1], and Anna Maria Vegni[2]

[1]*Inria Lille-Nord Europe, Villeneuve d'Ascq, France*

[2]*Roma Tre University, Rome, Italy*

1 INTRODUCTION

Software-defined radio (SDR) can be considered as a technology that allows the implementation of some modulation or demodulation schemes through modifiable software or firmware.

The concept of *cognitive radio* (CR) [1,2] can be summarized as devices equipped with the capability to observe and learn from the operating environment, and adapt their wireless communication parameters in order to optimize network performance. The key attributes of cognitive radio are "learn," "sense," and "adapt." The learning mechanisms are necessary to acquire information about communication parameters and to capture the underutilized spectrum by sensing the surroundings. Adaptive and dynamic adjustment of the transmission parameters (i.e., transmission power) allows the achievement of better utilization of the spectrum.

Future wireless communications will require an increasingly opportunistic use of the licensed radio frequency spectrum, and the CR paradigm provides a suitable framework for this aim. The main objective of the CR paradigm is to improve spectrum usage efficiency, while minimizing the problem of spectrum overcrowding [3]. Indeed, a CR system is based on the fact that the different parts of the channel are not only allocated to fixed and pre-assigned users, but idle segments of the spectrum are made available to other users.

Recent advances in cognitive radio networks (CRNs) have dealt with multi-hop networks, representing a promising design to leverage the full potential of CRNs. One of the main features in multi-hop networks is the routing metric used to select the best route to forward packets. In [4], Youssef et al. survey the state-of-the-art routing metrics for CRNs. The problem of reliability of a multi-hop, and multichannel CRN, is investigated also by Pal et al. in [5]. In such a scenario, to support multilink operations and networking functions, traditional spectrum sensing is not enough, and there is a need for developing a CR tomography to meet the general needs of networking operations. This topic has also been investigated by Kai-Yu in [6]. Well-designed multi-hop CRNs can provide high bandwidth efficiency by using dynamic spectrum access technologies, as well as provide extended coverage and ubiquitous connectivity for end users. Prior to cognitive radio approaches, a better usage of the underused spectrum bandwidth was obtained by focusing on the scheduler at Medium Access Level (MAC), as shown in [7] and [8], that were supposed to improve the assignment of the time slots for each user. The advent of cognitive radio represents a new open research direction that presents specific challenges and issues.

In [9], Sengupta and Subbalakshmi survey unique challenges and open research issues in the design of multi-hop CRNs; for example, they focus on the MAC and network layers of the multi-hop CR protocol stack. They investigate the issues related to efficient spectrum sharing, optimal relay node selection, interference mitigation, end-to-end delay, and many others. As an instance, the concept of spectrum sharing represents an effective method to fix the spectrum scarcity problem. Indeed, spectrum-sharing solutions allow unlicensed users to coexist with licensed users, under the condition of protecting the latter from interference. In [6], Stotas and Nallanathan investigate the throughput maximization in spectrum sharing CRNs by introducing a novel receiver and frame structure, and then deriving the optimal power allocation strategy that maximizes the capacity of the proposed CR system.

Another survey on recent advances in CR is presented by Wang et al. in [10], while Naeem et al. [11] present the issue and suitable solutions for efficient resource allocation in cooperative CRNs, in order to meet the challenges of future wireless networks. The authors in [11] also highlight the use of power control, cooperation types, network configurations, and decision types used in cooperative CRNs.

The integration of the SDR capabilities and the CR paradigm in wireless sensor networks (WSNs) is an active research field that is attracting attention from various directions. This increasing interest of integration of the SDR and WSNs mainly stems from the potential advantages that can be derived. First of all, by integrating the SDR into WSN, and considering the same hardware, it is possible to have more standards. Also reprogramming the sensor nodes rather than the hardware and circuits will drastically reduce the design time. The same reasoning can be applied to the integration of SDR capabilities and the cognitive radio paradigm in wireless body area networks (WBANs).

This chapter is mainly devoted to the description of the main contributions of the integration of SDR and CR into WSNs and WBANs, respectively. After a general description of the key features deriving from the integration, we provide some insights into the modeling aspects and the simulation tools that are mostly considered and exploited in both contexts of WSNs and WBANs. The last contribution considers the routing approaches by highlighting the main differences among traditional approaches applied to both wireless sensor networks and wireless body area networks, and routing techniques based on SDR and CR paradigms.

Table 5.1 List of Acronyms Used in the Book Chapter

Acronym	Definition
ACQUIRE	Active Query Forwarding in Sensor Networks
AMC	Automatic Modulation Schemes
AWGN	Additive White Gaussian Noise
BAN	Body Area Networks
BNC	Body Network Controller
CR	Cognitive Radio
CRCN	Cognitive Radio Cognitive Network
CRN	Cognitive Radio Network
CRSN	CR Sensor Node
CWBAN	Cognitive Wireless Body Area Network
CWSN	Cognitive Wireless Sensor Network
MAC	Medium Access Control
MBAN	Medical Body Area Networks
MEMS	Micro-Electro-Mechanical Systems
MICS	Medical Implant Communications Service
PU	Primary User
QoS	Quality-of-Service
RF	Radio Frequency
SDR	Software-Defined Radio
SPIN	Sensor Protocol for Information via Negotiation
SU	Secondary User
WBAN	Wireless Body Area Network
WMTS	Wireless Medical Telemetry Service
WPAN	Wireless Personal Area Network
WSN	Wireless Sensor Network
WBSN	Wireless Body Sensor Network

For the sake of clarity, in Table 5.1 we provide a list of the main acronyms used in the chapter.

2 COGNITIVE WIRELESS SENSOR NETWORKS AND COGNITIVE WIRELESS BODY AREA NETWORKS

In this section we introduce the basic concepts of Cognitive Wireless Sensor Networks (CWSN), and Cognitive Wireless Body Area Networks (CWBAN), from traditional wireless sensor and body area networks, where the concept of cognition is applied to the definition of routing protocols for such specific networks. We observe how adding cognition to the existing wireless sensor networks brings many benefits, and we highlight the main differences between traditional WSN/BAN and CWSN/CWBAN.

The following subsections present an overview of CWSNs and CWBANs, respectively, and then discuss the emerging topics and recent challenges in such areas.

2.1 COGNITIVE WIRELESS SENSOR NETWORKS

Nowadays, the increasing demand for wireless communications represents a challenge for efficient spectrum utilization. To address this challenge, CR has emerged as a key technology, which enables opportunistic access to the spectrum. A CR is an intelligent wireless communication system that is aware of its surrounding environment and accordingly adapts its internal parameters to achieve reliable and efficient communications [1]. Following an opportunistic manner, CR enables unlicensed (*secondary*) users to exploit the spectrum allocated to licensed (*primary*) users. As a consequence, innovative and energy efficient MAC and routing protocols must be implemented to improve the coexistence between different users, as well as managing the scarce resources in an efficient way without degrading primary user communication performances.

Moreover, the next generation of CR networks will be supplied by renewable energy from natural resources, such as solar, wind and radio frequency (RF) energy. This energy could be used overnight to increase the battery charge, or to prevent power leakage. In a hazardous situation, if a battery or a solar-collector/battery package completely fails, harvested energy from radio waves can enable the system to transmit a wireless distress signal, while potentially maintaining critical functionalities.

Finally, in future applications [3] such as upcoming smart-dust networks, an opportunistic use of spectrum would be inevitable. Smart dust is a system of many tiny micro-electro-mechanical systems (MEMS), such as sensors, robots, or other devices, that can detect light, temperature, vibration, and so on. They are usually distributed over some area to perform tasks, usually sensing through RF identification. In such a scenario, the growing number of such short-range nodes distributed in a region that use different radio technologies at unknown locations cause the spectral environment to be sharply variable, even over short distances and brief time periods. Hence, time-limited measurements carried out at a distance from the local area under study cannot characterize the local conditions [4].

In general, the main aspects of the cognitive network consist of a *behavior-oriented architecture* with agents that have a sensor-based robust behavior with slow rate of processing, distributed control, small size, and inexpensive low power consumption hardware [5]. Since a WSN is comprised of low power consumption devices with limited processor capabilities, cognition should be implemented in such an infrastructure. Today, WSNs are one of the areas with the highest demand for cognitive networking. Traditionally, WSNs have been exploited as enabling technology for *ambient intelligence* [6], and adding cognition capabilities to the existing WSN infrastructure is expected to provide many benefits and advantages such as (*i*) higher transmission range, (*ii*) a lower number of nodes required to cover a specific area, and (*iii*) lower energy consumption per bit. Indeed, the higher communication range provides CWSNs with a smaller number of hops per route. This also provides a smaller average end-to-end delay.

The term "cognition" applied to networks refers to the process of making decisions and action based on the network conditions in order to achieve end-to-end goals. Specifically, when applied to WSNs [12], cognition can help the network to achieve a better performance by the means of awareness and information sharing in the network. In more detail, when incorporated into sensor networks, cognition will enable achieving two main objectives: (*i*) to make the network aware of, and dynamically adapt to, application requirements and the environment in which it is deployed, and (*ii*) to provide a holistic approach to enable the sensor network to achieve its end-to-end goals, i.e. gather information about the network status from network and MAC layers, application requirements from the application layer and achieve the objectives of the network [9].

For the first objective, cognitive WSN nodes should change their transmission and reception parameters according to the radio environment. Cognitive capabilities are based in four technical components: (*i*) sensing spectrum monitoring; (*ii*) analysis and environment characterization; (*iii*) optimization for the best communication strategy based on different constraints such as reliability, power consumption, security, etc.; (*iv*) adaptation, and collaboration strategy. Indeed, the cognitive technology will not only provide access to new spectrum but also provide better propagation characteristics. By adaptively changing system parameters like modulation schemes, transmit power, carrier frequency and constellation size, a wide variety of data rates can be achieved. This can improve power consumption, network life and reliability in a WSN.

The ultimate goal is to design WSNs that are more aware of the concurrent conditions of the network, but above all can make decisions based on the information, and take actions. In [13] Baumgarten and Mulvenna present the potentials and challenges of future sensor networks, and provide the foundations for sensor nodes with self-adapting ambient intelligence. As an instance, CWSN could provide access not only to new spectrum (rather than the worldwide available 2.4 GHz band), but also to the spectrum with better propagation characteristics. A channel decision of lower frequency leads to more advantages in a CWSN such as higher transmission range, fewer sensor nodes required to cover a specific area, and lower energy consumption.

In WSNs, nodes are constrained mainly in terms of battery and computation power, but also in terms of spectrum availability. With cognitive capabilities, WSN could find a free channel in the unlicensed band to transmit or in the licensed band to communicate. A cognitive WSN should be aware of the amount of sensory data being communicated, and know when and where to forward it. The energy available at each node is fed back to determine a maximum average power [5]. In addition, because cognitive sensor networks should have such a high level of knowledge about the environment and the types of information exchanged, they must be application specific.

Cognitive nodes are then designed such that they use the same infrastructure as sensor nodes but are able to handle cognition processes and manage decisions and actions by commanding other nodes. By using cognitive nodes in the network, the performance of the network would be improved, as well as the cost, like the energy consumption and the delay [10]; in particular, the cost added due to cognition will be as low as possible, and can benefit from previous developments in the design of sensor nodes.

Leveraging on all the previous features, we can distinguish some basic differences between WSN and CWSN. In a WSN, each node either sends/receives data, or it is in idle state. Similarly, a CWSN consists of many tiny and inexpensive sensors where each node operates on limited battery energy; moreover, in CWSN there is another state, called the *sensing state*, where the sensor nodes sense the spectrum to find spectrum opportunities or spectrum holes. However, among various tasks for each node, the transmission and reception of data are the most energy-consuming tasks.

The spectrum sensing task in the CWSN sensing state can be performed either by a distributed or centralized scheme. In a distributed scheme, each sensor competes with other sensors to access the available spectrum [11], and then it must have the ability to sense the whole channel, and determine an optimal scheme to maximize its benefits, such as the number of transmissions over time. However, due to the fact that CWSN sensor nodes are mostly low-powered with limited capabilities, it may not be feasible to deploy the full functionalities of a distributed scheme in these networks. Thus, in many applications a centralized scheme is preferred, where spectrum opportunities are detected by a single entity called network coordinator [14,15]. The network coordinator broadcasts a channel switch command to indicate an alternate available channel (i.e., the alternate channel could be another licensed channel or an unlicensed channel in the ISM band). The broadcast message could be retransmitted by multiple nodes, in order to reliably deliver the message. Typically, there exist two traffic load configurations in a CWSN: (*i*) the regular status report, where each sensor sends regular status update to the coordinator, and (*ii*) the control commands, where control messages are sent by the coordinator (e.g., in a heat control application, the coordinator sends commands to switch on/off the heaters, thus providing an automatic domotic system).

The presence of a coordinator node is typical of a distributed network architecture. Figure 5.1 compares a distributed sensing scheme with a centralized method; the main difference is the presence of the spectrum coordinator, as well as the task of spectrum hole information gathering, in the distributed scheme.

Sensor nodes in a CWSN can measure and provide accurate information at various locations within the network. The higher communication range provides CWSNs with the smaller number of hops needed per route. Thus, the average end-to-end delays will be likely to be smaller. In addition, a CWSN could provide access not only to new spectrum (rather than the worldwide available 2.4 GHz band), but also to the spectrum with better propagation characteristics [11]. As an

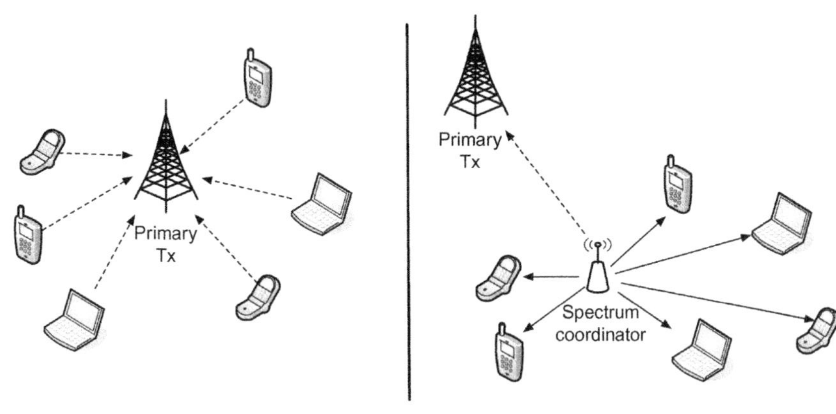

FIGURE 5.1

Schematic of (left) centralized, and (right) distributed architecture for Cognitive Wireless Sensor Networks. Nodes perform different tasks, i.e., channel sensing (dotted lines), and spectrum hole information gathering (solid lines).

Table 5.2 Main Features and Differences of Wireless Sensor Networks and Cognitive Wireless Sensor Networks (Gray Cells Represent an Advantage for a Given Network)

Parameter	WSNs	CWSNs
Transmission range	Lower	Higher
Sensor nodes required	Higher	Lower
Energy consumption	Higher	Lower
End-to-end delay	Higher	Lower
Accuracy of sensing algorithm	Lower	Higher
Protocol complexity	Lower	Higher
Network control overhead	Lower	Higher

instance, if the transmission power of the secondary user remains the same, its transmission range increases at lower frequencies [16].

Based on all previous highlights, it is expected that CWSNs will be widely used in the future. The performance gains can be obtained at the cost of a slight increase in the protocol complexity, and network control overhead. Table 5.2 compares WSNs and CWSNs, by means of main features. It follows that CWSN is envisaged as a new concept [4] with many advantages, such as higher transmission range, fewer sensor nodes required to cover a specific area, a better use of the spectrum, lower energy consumption and delays, better communication quality and data reliability, and also a better use of sensing frequency based on the changes in the channel environment. In Table 5.2 we highlight the features that represent a strength point for a given technology (e.g., high transmission range is an advantage for CWSNs, while low power complexity is an advantage for WSNs).

2.2 COGNITIVE WIRELESS BODY AREA NETWORKS

In this subsection, we investigate how CR technology is applied to future body area networks (BAN). Indeed, CR features are expected to affect BAN, such as to supply all the nodes of a BAN without the need of replacement of the primary source of energy (i.e., batteries). At the same time, new cross-layer algorithms are envisaged to adapt to the changes in the transmission link, based on the quality of the received signal, radio interference, radio node density, network topology or traffic demand.

Although many papers [1,4,11,16] have discussed the different facets of CR, just a few works [17–19] address CR application to medical environments. The most representative scheme of a typical BAN scenario with CR is depicted in Figure 5.2. A medical BAN (i.e., a WBAN for medical applications) comprises multiple sensor nodes, each of them capable of sampling, processing, and communicating one or more vital signals—i.e., electroencephalography (EEG), electromyography (EMG), electrocardiography (ECG), etc. This biomedical information is transmitted to a Body Network Controller (BNC) [17], and on the basis of the spectrum availability, sensor nodes transmit their results from the sensing to the next hops, and eventually to the sink in an opportunistic manner [18].

The recent trend in telemedicine is towards the use of wearable health monitoring devices [17]. This has motivated the standardization of the so-called Wireless Body Area Networks (WBANs) through the IEEE 802.15.6 working group [20]. The IEEE 802.15.6 standardization working group has produced the first draft of a document specifying the physical (PHY) and MAC layer characteristics of the radio interfaces for WBAN applications [19].

The requirements for a wireless communication system to be used in healthcare have been identified in [17] and a CR system has been proposed for a

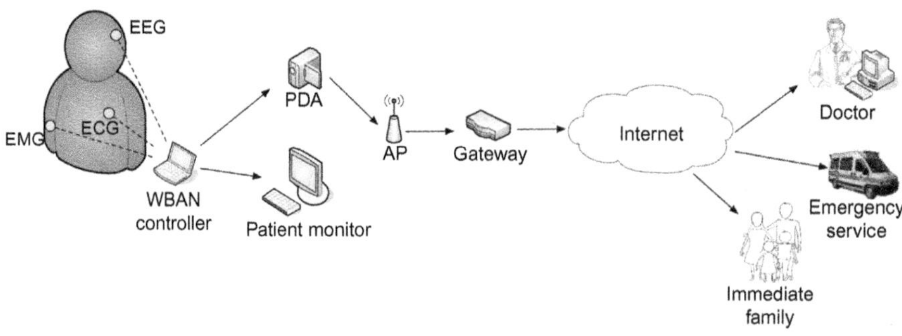

FIGURE 5.2

Basic architecture of a telemedicine system.

hospital scenario. The proposed system is designed to avoid electromagnetic interference to electronic medical devices, which are considered to be protected users. At the same time, the system guarantees Quality-of-Service (QoS) for different wireless applications.

Some medical applications, namely the Wireless Medical Telemetry Service (WMTS) and the Medical Implant Communications Service (MICS), have been licensed to operate in exclusive parts of the spectrum. Specifically, the 608–614 MHz, 1395–1400 MHz, and 1427–1432 MHz bands are assigned to WMTS, whereas the 402–405 MHz band is for MICS. However, spectrum regulatory policies affect the ability to deploy MBAN technologies on a global scale; nevertheless, they also offer unique opportunities for a CR-based solution.

Regardless of the different spectrum regulatory constraints across the world, several bands designated for medical applications are mostly available on a license-exempt secondary basis. Specifically, UWB technology shows interesting applicability for MBANs. Both the United States and Europe have regulated the parts of the spectrum that can be used by UWB on a license-exempt basis. The band made available in the United States corresponds to 3.1–10.6 GHz, whereas in Europe two spectrum segments have been defined, 3.4–4.8 GHz and 6–8.5 GHz.

3 ANALYTICAL MODELS AND SIMULATION TOOLS FOR WIRELESS SENSOR NETWORKS BASED ON SOFTWARE-DEFINED RADIO

The combination of the CR paradigm and WSNs is a promising field of research, since it allows the coexistence of overlapping wireless networks in ISM bands by minimizing the interference [21]. As outlined in [22], the advantages of opportunistic usage of the lower frequency in the spectrum in terms of better propagation characteristics, and the fact that this type of frequency allows a better penetration to obstacles, makes CWSNs a very interesting and promising solution to a better management of the spectrum. In Figure 5.3, we show a general model for a CWSN, where nodes send their readings to the sink via multiple hops in an ad hoc manner. This topology imposes less communication overhead (i.e., control data), but due to the hidden terminal problem, spectrum sensing may be inaccurate.

3.1 ANALYTICAL MODELS

In this section we present the main analytical models adopted for wireless sensor networks by applying the CR paradigm. Fundamentally, we can notice two main classes of models: (*i*) Queuing Traffic Modeling, and (*ii*) Topology Modeling.

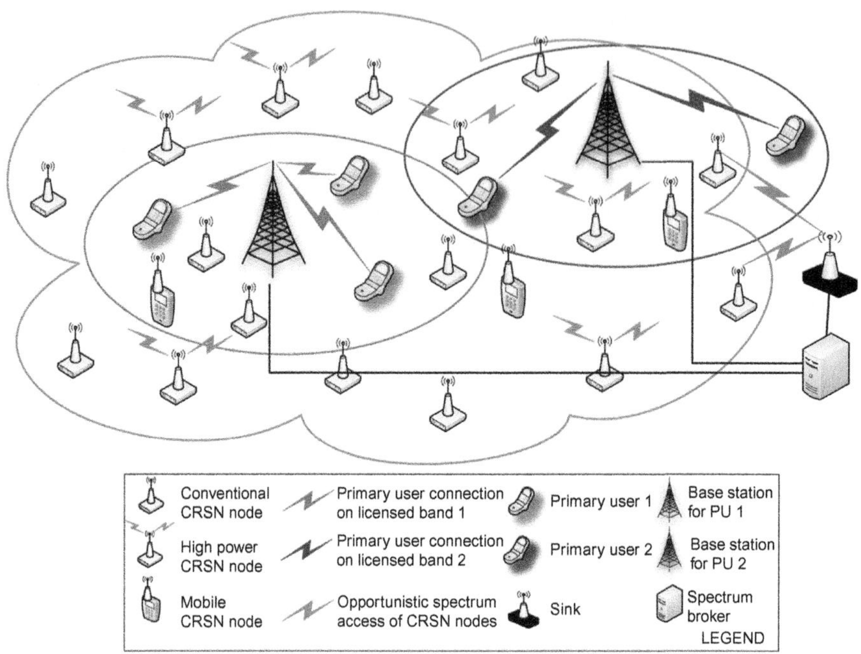

FIGURE 5.3

A Cognitive Radio Wireless Sensor Network model [23].

3.1.1 Queuing traffic modeling

Generally, the modeling approaches related to the CR-based wireless sensor networks regard primary user (PUs) and secondary user (SUs) modeling traffic. In [24], Zhang et al. model the hierarchical structures, by considering the secondary user as a relaying terminal for assisting primary communication. They present a system model of the cooperative CRNs, and in order to evaluate the effectiveness of the cooperation on their system, they also consider a noncooperative model based on a priority queuing system. The M/G/1 priority queuing system proposed for the cognitive radio network allows evaluation of the performance in terms of delay and throughput for each user. At this modeling they add the cooperative diversity and show the effectiveness of the cooperation on secondary user throughput. Another user model has been presented in [25], where the authors focus on the waiting time analysis in a time slotted system. CR is modeled by considering a priority queuing and arrival of the packets is modeled as a modified M/D/1 system. Gao and Jiang in [26] also take account of the errors deriving from a nonideal spectrum sensing and model these "imperfections" through a stochastic approach, namely stochastic network calculus. The authors also consider the retransmission of the packets in their model by considering three retransmission schemes: (*i*) without retransmissions, (*ii*) retransmission until success, and (*iii*) maximum-N-time retransmissions. By using stochastic network calculus, they show expressions for backlog and delay bounds and their results could be exploited to design efficient retransmission schemes in CR-based WSNs.

3.1.2 Topology modeling

In Figure 5.4 we show two possible topologies that can be figured out for CR-based WSNs. How to model the deployment in this type of network is perhaps more important than in "traditional" CRNs, since CWSNs are prone to change more frequently than CRNs. Just to provide an example, the main difference between the hardware structure of a classical sensor and a CR Sensor Node (CRSN) is represented by the cognitive radio transceiver. Indeed, a CRSN is able to dynamically adapt the communication parameters, such as carrier frequency, transmission power, and modulation.

In [23], Akan et al. propose four different types of topologies, namely (*i*) Ad Hoc CWSNs, (*ii*) Clustered CWSNs, (*iii*) Heterogeneous and Hierarchical CWSNs, and (*iv*) Mobile CWSNs. They outline that, despite the increased challenges and complexity, the last two types of topologies, due to the heterogeneity and the mobility aspects, can be very effective and beneficial for increasing the potentiality of the network. In fact, either the presence of more powerful devices, or the possibility of moving some devices towards a specific and useful position, can be exploited to increase the lifetime of the whole network, by decreasing the energy consumption.

Figure 5.4(a) describes a clustered CWSN topology, where there is a common channel to exchange various control data, such as spectrum sensing results, spectrum allocation data, neighbor discovery, and maintenance information. This type of topology is an appropriate choice for effective dynamic spectrum management, with a local common control channel approach. On the other hand, Figure 5.4(b) depicts the architecture for heterogeneous and hierarchical CWSN, which incorporates special nodes (i.e., actor nodes) equipped with renewable power sources, and acting in additional tasks like local spectrum bargaining.

3.2 SIMULATION TOOLS

In this section, we present the main simulation tools that have been used to implement the SDR paradigm in the context of wireless sensor networks.

3.2.1 Software-defined radio as simulation tool

First, in this section we consider the matter of link quality estimation and control in a wireless sensor network, and the possibility of exploiting the SDR paradigm to make that happen in a very effective and efficient way, as shown in [27]. In this case, the authors have built an IEEE 802.15.4 physical layer communication connection based on SDR, and then used it as a kind of spectrum analyzer. The authors simulated three different types of link quality metrics and performed

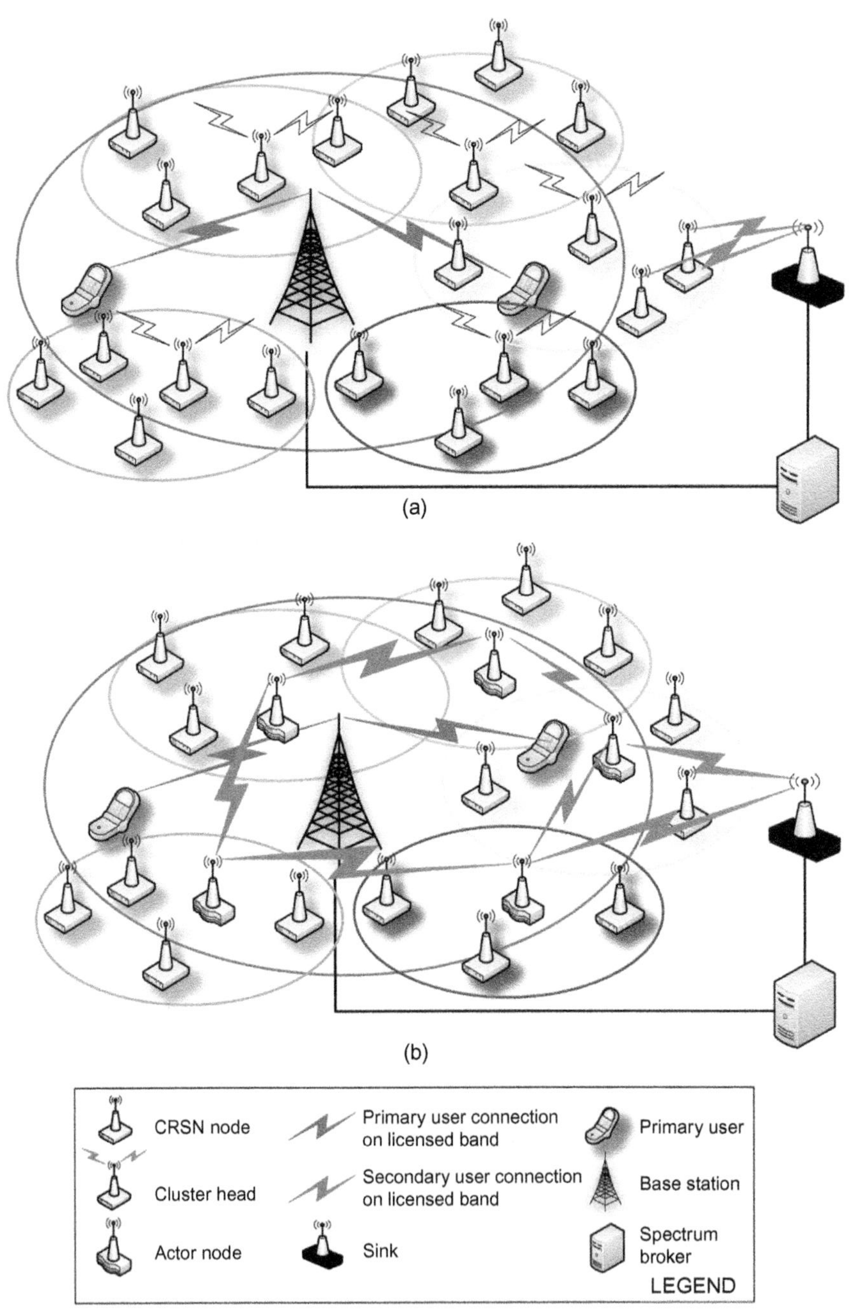

FIGURE 5.4

Two possible topologies for CR based WSNs: (a) clustered, and (b) heterogeneous and hierarchical CWSNs [23].

simulations by considering four of the most common radio environments: (*i*) clean channel, with no interference and a low level of thermal noise, (*ii*) additive white Gaussian noise (AWGN), that represents the most common channel model considered in wireless systems, (*iii*) in-band noise system, where a noise with a spectrum similar to the spectrum of the signal is considered as additional noise, and (*iv*) adjacent channel interference, where two sensor nodes transmitting simultaneously on adjacent channels can interfere with each other. The authors envisage very interesting potential future research based on the novel link quality estimation method they obtained, that consists of modeling various types of channels.

3.2.2 Monte carlo simulation

In [25], Suliman and Lehtomaki introduce two types of errors, namely false alarms and missed detections, and evaluate their impact on the system performance by considering Monte Carlo simulations. Through the simulation tool, the authors were able to show a match with the analytical results. In [28], Marković and Dukić consider Monte Carlo simulations to study and evaluate Automatic Modulation Schemes (AMC) in terms of average probability of correct classification. They conclude that the cooperative based schemes are able to reach larger gains than classical AMC schemes. In [29] Ho et al. propose a Sensor Network Controlled Indoor Cognitive Radio system. This system is conceived to make the secondary users (SUs) able to access the licensed spectrum inside a building, located in a space where SUs are excluded. The authors considered Monte Carlo simulation to show the effectiveness of the system in terms of control of the outdoor interference caused by unlicensed access of indoors users.

3.2.3 OPNET

Many researchers make use of OPNET as a simulation tool to test their proposals in the context of CR-based WSNs in which the standard ZigBee/802.15.4 model is available. Cavalcanti et al. [30] leverage on this model and built an enhanced CR mode on the ZigBee/802.15.4 protocol stack. The authors aim to analyze and study the performance of a CWSN considered for specific applications, and compare their performance with a standard WSN based on the ZigBee/802.15.4 paradigm. In their work they show, by means of simulation results, that the application of CR to WSNs can reduce the number of hops required to route packets, by allowing a decrease in terms of energy expenditure and improving the network lifetime. These results derive from the fact that the authors obtained that by using the same transmission power; the transmission range of 680 MHz UHF TV band frequency doubles that of 2.4 GHz.

3.2.4 Cognitive wireless sensor network simulator

The Cognitive Wireless Sensor Network simulator [31] is based on Castalia [32], and OMNET++ [33]. The main advantage of this kind of simulator is that the physical and MAC layers are realistic. The authors added explicitly the difference between primary users (PUs) and SUs. In Figure 5.5 the Castalia cognitive radio

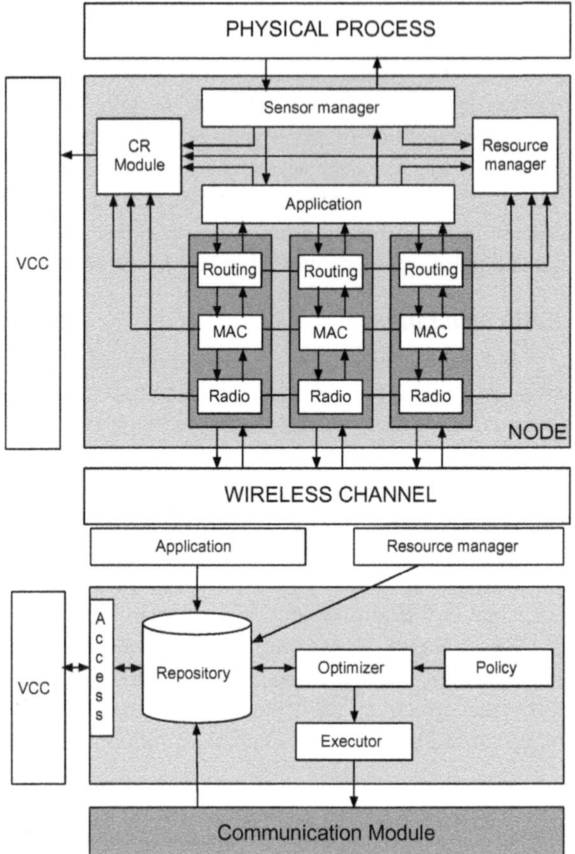

FIGURE 5.5

Castalia cognitive radio module.

module is shown. This module is composed from four main elements: a repository, an optimizer, a policy, and an executor. The Virtual Control Channel (VCC) element represents the access interface. The authors showed the effectiveness of this new simulation tool to validate and test new schemes and optimization mechanism.

3.2.5 MATLAB and network simulator 2 simulation

Another very common simulation tool in the wireless scientific community is the Network Simulator 2 (NS-2). The version 2.31 of NS-2 incorporated the Cognitive Radio Cognitive Network (CRCN) [34]. In [22] Oey et al. used MATLAB to analyze the metrics when they had to evaluate their analytical model. In the same contribution, the authors propose a routing approach based on cognitive radio for WSNs to make an efficient usage of the energy, named Energy and Cognitive radio aware Routing (ECR). The simulation results show that they are able to obtain better results in terms of lifetime and throughput compared to AODV. In this work the authors envisage the main issues related to the integration of the CR paradigm in WSNs, namely: (*i*) enabling dynamic spectrum access implies a joint node-channel assignment, and (*ii*) energy consumption in hardware constrained networks.

Finally, a recent work by Al-Ali and Chowdhury [35] proposes a framework for CRs in Network Simulator 3 (NS-3). The authors introduce several CR capabilities, such as spectrum sensing, primary user detection, and spectrum hand-off. Compared to NS-2, the simulator in [30] demonstrates improvements in execution time and memory usage.

3.2.6 FREVO

FREVO [35] is an open source simulation tool. In [36], the authors propose to compute the best modulation schemes for fixed/mobile devices equipped with SDR capabilities by implementing a multi-objective and distributed neural/genetic algorithm. The objectives that nodes have to achieve are multiple and in some ways opposite, namely a better coverage of a specific sensing field and connectivity with a central node that plays the role of sink. In [36] the authors show the potential of mobile SDR communication sensors both from operation flexibility and dynamical reconfiguration. The use of the FREVO framework allowed the developing of a simulator for evolutionary design, and the strategy considered has been validated in different scenarios, by varying the number of nodes supporting the SDR capabilities and the number of nodes equipped with mobility capabilities. The use of SDR capabilities is envisaged as providing the possibility of forming self-evolving wireless networks able to support high data rate and connectivity in various communication scenarios. In Figure 5.6, we show a mobile node supporting SDR capabilities.

In Figure 5.7 we show that several modulation/demodulation software blocks can be developed within the generic SDR architecture for both transmitter and receiver. In this way, more powerful devices, able to support the dynamic modulation changing, can be considered.

4 ANALYTICAL MODELS AND SIMULATION TOOLS FOR WIRELESS BODY AREA NETWORKS BASED ON SOFTWARE-DEFINED RADIO

Wireless body area networks are a new and intriguing concept that is gaining more and more interest. The possibility of including CR approaches in WBAN architecture represents a very recent and hot topic research field. In [38] Wang et al. specify the application of cognitive radio to Medical Body Area Networks (MBANS) as a new and emerging cognitive radio application.

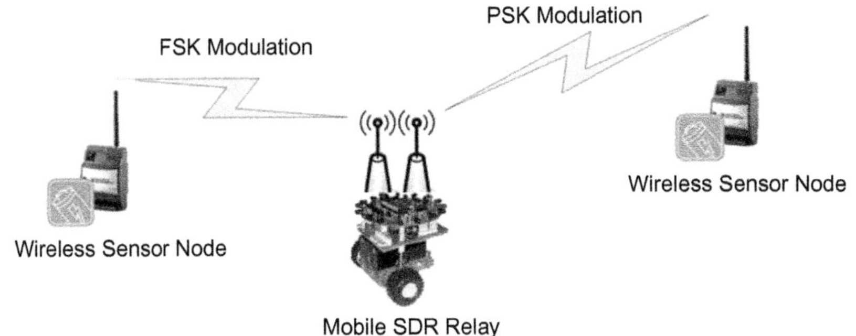

FIGURE 5.6

Mobile nodes supporting software-defined radio capabilities.

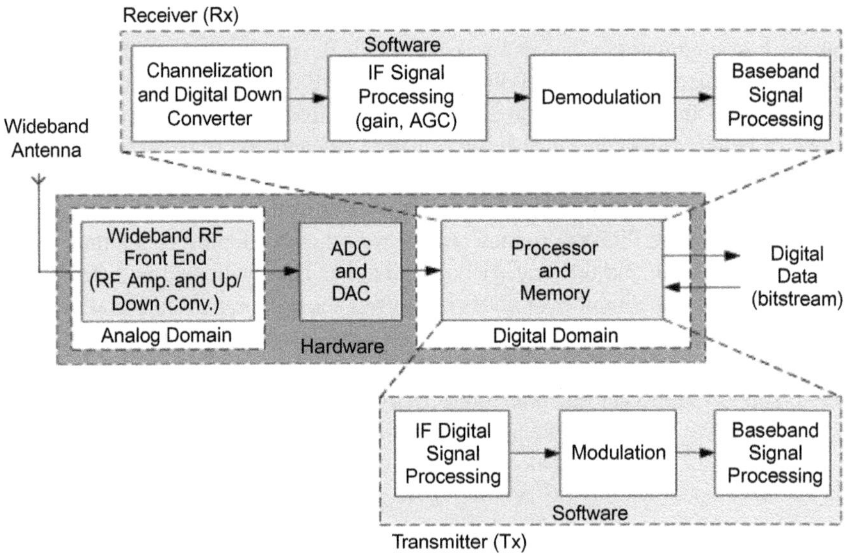

FIGURE 5.7

Software-defined architecture for a sensor node [37].

In the following subsections, we present the contributions that can be found in literature by distinguishing them from a modeling point of view and from a simulation point of view.

4.1 ANALYTICAL MODELS

The possibility of exploiting the CR paradigm in an effective way when applied to WBANs has been a fervent research topic for some time, but in the last few

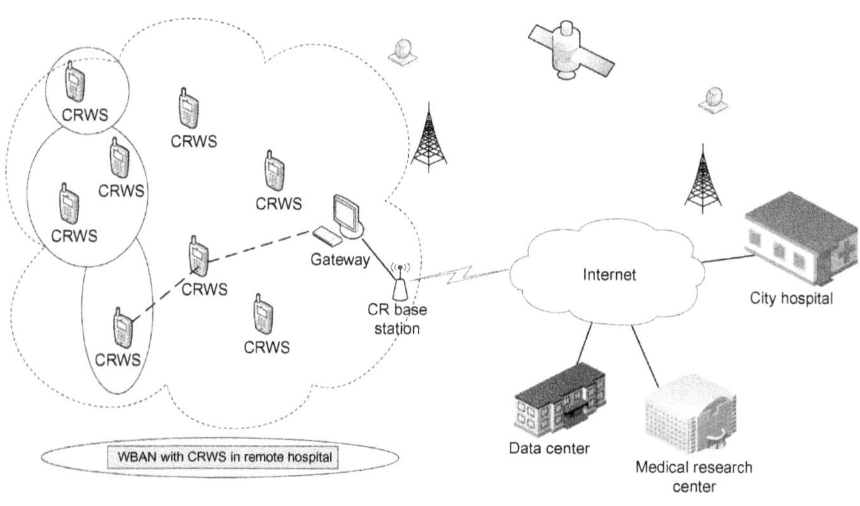

FIGURE 5.8

A model of wireless body area networks with cognitive wireless sensor networks.

years, it is acquiring more and more interest. In [39] Joshi et al. show a model of WBAN with the CWSNs, well depicted in Figure 5.8.

4.1.1 Channel modeling

One of the most important and crucial aspects of the CR-based WBANs is the statistical characterization and modeling of the channel. Indeed, in April 2009, the channel modeling subcommittee of the IEEE 802.15 task group 6 (TG6) worked on channel models of wireless body area networks, by focusing on in-body and on-body communications [40]. In order to exploit the cognitive radio paradigm in the context of Wireless Body Area Networks, a correct channel modeling is a fundamental prerequisite. A very useful contribution is represented by [41], where the authors present various channel models developed at the Centre for Wireless Communications (CWC) in Finland. Specifically, the authors focus on two WBAN channel models. Another interesting contribution in terms of channel modeling for UWB WBAN with Finite Integration Technique (FIT) is given in [42]. The interesting conclusion achieved in [42] is that FIT is individuated as a promising technique to model a realistic on-off body communications channel, in the case where no analytical model is available.

In [43], a different perspective of the channel modeling problem is presented. Cotton and Scanlon propose a modeling of an indoor radio channel for Wireless Personal Area Networks (WPANs). From the measurements performed, they established that at the specific frequency 868 MHz, the Nakagami-*m* distribution presents the best propagation behavior in all the environments they considered.

4.1.2 Configuration modeling

Fundamentally we can consider two main configuration models in the context of CR-based WBANs: centralized and distributed. Both models present their advantages. In the centralized configuration, sensors are physically tethered to the sensor interface and this allows the relief of some of the concerns with communication and power. In this case, scheduling is not required and this makes the delivery of data packets easier and more effective. In a distributed model, the communication and power components are present on each distributed node. For sure, this type of configuration is more "open" to the integration of additional devices/sensors, by presenting a higher level of scalability. This type of configuration model is useful for meeting patient needs and risk level.

4.2 SIMULATION TOOLS

4.2.1 OPNET and MATLAB

In [44] Çalhan and Atmaca propose a new Network Coordinator Node design for the selection of the most suitable technologies among those available, both in in-door and out-door applications. The performance evaluation of the selection mechanism is realized through OPNET Modeler, and the implementation of the fuzzy logic based algorithm incorporated into the NCN has been realized in MATLAB. The authors compared their approach with a more "traditional" approach based on the received signal strength indicator (RSSI), and showed how in an RSSI-based AP selection algorithm, where the value of RSSI is the only parameter used to make the selection, the results can be sometimes misleading. In [41] the authors use MATLAB to develop a simulation tool to evaluate the performances of two single-band ultra wide band (UWB) systems: Direct-Sequence UWB (DS-UWB) and UWB Frequency-Modulation (UWB-FM). Their work is relevant since they were able to show that the choice of the channel model plays a crucial role in the performance of the system.

5 ROUTING APPROACHES FOR WIRELESS COGNITIVE SENSOR NETWORKS

In this section, we address the routing techniques that have been specifically studied for CWSNs, and we focus on the most common simulation tools that have been considered to evaluate their performance. We start with a rapid overview of the routing protocols that have been proposed for the "traditional" WSNs, by outlining the main simulator used for this type of network. Then, we introduce the relatively new concept of WBANs and the routing approaches related to them and the simulation tools mostly used for their performance evaluation. Finally, we attempt to outline the need to define and study new effective solutions in terms of routing for the CR-based WSNs and we present the main simulators that have been considered also in this case. We conclude this part of the chapter with a dissertation of the CR paradigm applied to Wireless Body Area Networks.

5.1 TRADITIONAL ROUTING TECHNIQUES AND SIMULATORS ADOPTED

Networking is an important functionality of WSNs. Networking enables WSN to transmit the sensed information to a central device for monitoring, and decision making. Classical IP-based routing techniques could not be utilized in most of the sensor networks due to the sheer number of the nodes. Hence, a decentralized routing system must be exploited to efficiently route to the designated sink.

Flooding is a simple classical solution to disseminate information through the network [45,46]. In flooding every node will retransmit every newly received packet to all its neighbors until the packet reaches the destination or the maximum number of hops allowed. In this way, the information is spread through the network without any networking overhead. However, flooding causes significant energy waste in large networks due to unnecessary packet forwarding and duplicate receptions [47]. In gossiping, nodes avoid this problem by randomly choosing a neighbor to disseminate the information. Although gossiping solves the duplicate transmission problem, it increases the average delay in data delivery in large networks significantly.

In general, routing solutions for wireless sensor networks are application dependent—that is, there is no unique protocol that could be used for any WSN application. Here we discuss three main classes—data centric, heretical and geographical—briefly, and describe modeling and simulation approaches used for each protocol.

5.1.1 Data centric

The first class is the data centric protocols. Data centric protocols are query based and generally use data labeling to address data. Neighbor nodes aggregate the received data with their own and transmit to the next hop [46]. Routing can be advertisement based as in Sensor Protocol for Information via Negotiation (SPIN) [47], where nodes advertise for their information by transmitting a description of their data. Nodes that are interested will request and receive the data. The authors have compared SPIN with gossiping and flooding by extending the functionality of NS-2 software package. They have shown that SPIN can perform better than gossiping in terms of throughput while it has equal performance with flooding but with less energy consumption. Reportedly, SPIN can save more energy than flooding by avoiding unnecessary transmissions by a factor of 3.5 and halves the redundant data. However, it does not guarantee data delivery in larger networks as the middle nodes may not be interested in the data of the faraway nodes and hence do not route it to the sink. Directed diffusion [48] is a query-based protocol where a query is flooded in the network by the sink where multiple routes are established between the sink and source. The sink reinforces one of the paths and receives data in a shorter interval through this reinforced path. Authors in [48] have altered the NS-2 software radio energy model, which was originally designed for 802.11 radio, to analyze the energy consumption of the sensor network using directed diffusion protocol. It is shown that directed diffusion could save energy by choosing the good path. Nevertheless, directed diffusion cannot be applied to the applications that require constant data delivery such as monitoring applications. Energy-aware routing [49] argues that using the same minimum energy path will deplete the nodes in this route of energy and hence probabilistically choose between different existing paths. Simulation results using OPNET show that the energy dissipation is spread over different paths and results in increasing the lifetime of the network by 44% comparing to directed diffusion. In *rumor routing* [50], instead of flooding the network with queries, the source nodes will create long-lived data packets corresponding to certain events called agents and inject them to the network. Agents travel through the network and inform the nodes about the event. If any node is interested in the data it will request it from a node that knows the route. Monte Carlo simulations show that rumor routing can result in significant energy savings compared to flooding. However, rumor routing is only applicable when the number of events is small. The energy cost of maintaining the agent when the number of events is large is significant. *Gradient-based routing* [51] defines the height of each node as the number of hops to the sink and the difference between the heights of the nodes as gradients. The next hop is chosen as the node with the highest gradient. Numerical simulations show that this scheme can reduce the overall energy compared to directed diffusion. Active Query Forwarding in Sensor Networks (ACQUIRE) [52] sees the network as a distributed database. The query is generated and transmitted by the sink node. Nodes that receive the query respond by sending their pre-cached information. If the pre-cached information is not up to date, the nodes will request the information from their neighbors. A mathematical model for the energy consumption is provided and the performance of the protocol is evaluated and compared with flooding-based queries. Results show that ACQUIRE can outperform its counterpart.

5.1.2 Hierarchical routing

Hierarchical protocols create clusters of nodes where a cluster head has the task of aggregating data, removing redundancy and avoiding overloading the gateway. Low-Energy Adaptive Clustering Hierarchy (LEACH) [53] forms clusters of nodes in the network and uses the cluster heads to route data to the sink. The cluster heads collect and aggregate the data from the nodes in their cluster before transmitting to the sink. The nodes in a cluster become the cluster head in turn to distribute the energy consumption in the cluster. MATLAB is used to show that LEACH can reduce the energy dissipation by a factor of 8 compared to conventional routing protocols. Power-Efficient Gathering in Sensor Information Systems (PEGASIS) [54] avoids the dynamic clustering overhead of forming clusters as in LEACH by forming a chain of nodes where each node transmits the data to its neighbor and one node will transmit the aggregated data to the sink. Numerical analysis shows that PEGASIS can achieve up to 300% improvement in energy efficiency compared to LEACH. Hierarchical-PEGASIS [55] decreases the delay incurred by the chain transmission of the information by using concurrent transmission separated by CDMA or other signal-processing techniques. Threshold sensitive Energy Efficient sensor Network protocol (TEEN) [56] is designed for the applications where the sensed attribute changes suddenly. Closer nodes form a cluster with a cluster head. The cluster head in turn forms a second level cluster and the process goes on until the sink is reached. Nodes are activated when the sensed attribute passes a hard threshold and report the value when the value changes at least equal to a soft threshold. The hard and soft thresholds are defined by the user and transmitted to the sensor by the cluster heads. NS-2 network simulator with LEACH extension is used to show that TEEN can achieve better energy efficiency than LEACH.

5.1.2 Geographical routing

Geographical routing protocols [46] exploit the location information of the nodes for a more efficient routing of data. Minimum Energy Communication Network (MECN) [57] exploits the fact that transmission through a relay will dissipate much less energy than transmitting directly for a distant node. Each node has the location information of its neighbors and hence can find the most energy-efficient path. Mathematical models for the energy consumption and numerical analysis of the network confirm the energy efficiency advantage of the protocol. Geographic Adaptive Fidelity (GAF) [58] forms a virtual grid for the covered area and activates only one node in every point of the grid. The active node is randomly changed to preserve the energy of the nodes. NS-2 simulations show that GAF can substantially conserve energy. Geographic and Energy Aware Routing (GEAR) [59] saves energy by limiting the region in which a certain query is disseminated, which is confirmed by the NS-2 simulator. In this merit, it could be considered as a location aware directed diffusion.

5.2 COGNITIVE RADIO APPROACHES FOR ROUTING AND MEDIUM ACCESS IN GENERIC WIRELESS SENSOR NETWORKS

Wireless sensor networks often use the Industrial, Scientific and Medical (ISM) radio bands for communication since these bands are license free and available globally. However, this comes at a cost. As the band is license free, many other applications use the same band, including microwave ovens, cordless phones, RFID devices, Bluetooth technology, etc. This can have catastrophic effects on the performance of the WSN if it shares the spectrum with one or more of these applications due to high levels of interference.

An interesting solution to this problem is to use licensed bands for WSN communications as the secondary user of the band using CR techniques. Studies [60] show that the licensed spectrum is largely underutilized temporally and spatially. Due to the large number of nodes in WSN, using a central scheduling system is not the best option. Hence, studies have been mostly focused on decentralized medium access and routing techniques for cognitive radio WSN.

In [61], a cognitive spectrum access method is proposed to optimize the performance of the individual nodes without a centralized scheduling system. The proposed MAC protocol is modeled and analyzed mathematically by the Partially Observable Markov Decision Process (POMDP). It is shown that the best strategy for the secondary user to utilize the spectrum is a policy of POMDP that maximizes the reward function. Finding the optimum policy for POMDP can be computationally prohibitive when the number of accessible channels is large. This is because the number of POMDP states increase exponentially by the number of available channels. Hence, a suboptimal greedy approach is proposed where the statistics of the channels are assumed independent which leads to less complexity of the POMDP. The greedy approach tries to maximize the reward function at each individual time slot. The performance of the suboptimal approach is compared with the optimal protocol using numerical simulations. The results show that the greedy approach has comparable performance to the optimal approach and both have significant improvements over random channel selection. Subsequently, the effect of error in sensing the spectrum is investigated. It is shown that as the percentage of collision with the primary network is relaxed, the performance of the greedy policy in the presence of channel sensing error approaches the optimal policy with no sensing errors.

In [62], an energy-efficient opportunistic spectrum access strategy is proposed for energy harvesting secondary users of the spectrum. The optimum strategy is developed as the solution to a Partially Observable Markov Decision Process (POMDP) as the mathematical model of the spectrum access for the secondary user. In this model, statistics of the primary user's transmission is assumed known. Based on this information and the knowledge of the available energy reserve, the secondary user decides on sensing and transmitting according to the optimum policy of the POMDP model, which aims to maximize the immediate throughput of the secondary user. Numerical simulations are used to show that the optimum policy outperforms the random channel selection in most scenarios.

CR-based routing is studied in [63]. The Spectrum Aware Routing Protocol for Cognitive ad-Hoc networks (SEARCH) is proposed, which jointly selects path and channel to minimize end-to-end delivery time. NS-2 network simulator is used to compare SEARCH with existing routing protocols. Results show that SEARCH has advantage over other schemes in minimizing the end-to-end delivery time.

5.3 ROUTING TECHNIQUES FOR WIRELESS BODY SENSOR NETWORKS

One of the most promising applications of WSNs is the Wireless Body Sensor Network (WBSN). A wireless body sensor network consists of a network of heterogeneous wireless devices reporting certain physiological data to a Network Coordinator (*NC*). *NC* is usually more powerful in terms of computational and energy resources [64–66]. *NC* aggregates data received from sensors and transmits the data to a central processing and logging server over an external link.

WBSN shares some of the challenges of the generic sensor network. However, unique features of WBSN such as small number of nodes, heterogeneity of the devices connected to the network, and dynamic nature of the human body as the propagation medium, makes it necessary to develop medium access and routing approaches specifically designed for WBSN.

Routing data in a body area network has been investigated in the literature. Design of a personal gateway for communication with outside networks is investigated in [67]. Narrowband radio is compared with FM-UWB for connecting the sensor nodes in an on-body network in [68]. Numerical results implemented in OMNET++ discrete event simulator show that FM-UWB outperforms the narrowband radio in terms of energy consumption. In [69], communication between an implanted sensor node and an outside base station is investigated. A MAC layer protocol is provided for a star topology of WBSN, which uses a flexible bandwidth allocation to reduce the energy consumption of the nodes and is simulated in [70] by using Open-ZB toolset, and open source implementation of IEEE 802.15.4.

In [71], cooperation between nodes in a multi-BAN network to route data to a sink node belonging to one BAN is studied. A stochastic route selection mechanism based on the maximum perceived outage probability, maximum queue utilization factor, and the remaining battery power information, is provided. The performance of the proposed scheme is assessed through system level simulation and has been shown to outperform random and best route selection schemes. In general, existing routing protocols for WBSN could be sorted into three categories. Some protocols use a star topology. Star topology is the simplest network topology for BAN. In this configuration, communication between sensor nodes and *NC* is direct or single hop. Direct communication between nodes and network coordinator simplifies the network protocol and reduces packet delivery delay when the link quality is high enough to satisfy the quality of service merit. However, as mentioned previously, the human body is a dynamic environment. As a result, the topology of the BAN can change drastically over time.

Such variations in the network topology can cause high path loss between the source node and the network coordinator even though the nodes might be very close to each other. Thus, to compensate for the severe attenuation of the signal, transmission power has to be increased proportionally so that the signal reaches the destination with the desired SNR. Nevertheless, increasing transmit power severely affects the energy resources of the nodes and reduces the lifetime of the network. The reason for this is that, unlike a generic sensor network where nodes are usually similar and can undertake each other's task when one of them runs out of energy or fails, in a body area network, each node has a specific task that is different from other nodes. Hence, if a node fails, the whole network might malfunction.

To avoid increasing transmission power proportional to the path loss, cooperative communication is suggested for WBSNs. In this type of protocol, each node is assigned to a higher-level node. The network coordinator has the highest level. To reach NC, the source node transmits its packets to higher-level nodes in a multi-hop scheme until it reaches NC. Based on the number of nodes and network topology, the number of hops can vary between two to several hops. Multi-hop communication is investigated for routing data in BAN [72–75]. In [72], using extra nodes and cooperation between nodes to relay information to the base is compared with the single hop scheme. The simulation results show that the relay network can greatly improve the lifetime of the network. Network lifetime is defined as the time interval between the time the network starts working to the time in which the first node dies in the network. Numerical analysis shows that in the single hop scheme, the nodes with highest distance from the sink will be drained out of energy and die faster while in the multi-hop case the relay nodes closer to the sink have much higher energy consumption due to the traffic they carry.

In [75], performance of noncooperative communication is compared with cooperative communication in the body area network environment. Numerical simulations show that cooperation can reduce energy consumption in a higher path loss environment while direct transmission is preferred in a lower path loss environment. In addition, it is suggested that prior knowledge of the body posture can improve the energy efficiency. While such protocols provide higher packet delivery success rates and reduce outage probability, they severely increase energy usage of the nodes due to excessive and unnecessary packet retransmissions.

Opportunistic routing is suggested and analyzed for BANs in [76–79]. In opportunistic routing, each node is assigned to a predefined node. During the packet transmission, the relay node assigned to the source node tries to pick up the packet. If the relay receives the packet successfully it would forward it to the destination node (NC) at another session. Numerical analysis has shown that opportunistic routing can improve packet delivery success rate while reducing the outage probability in comparison with direct and multi-hop routing protocols.

Sensor nodes in a body area network must be small and ergonomic. The small dimensions means small antenna size, which in turn translates to higher frequency band usage. On the other hand, the human body is considered a hostile environment in terms of electromagnetic propagation for higher frequencies (i.e., microwave and above). This is because higher frequencies are greatly attenuated inside and in the close vicinity of the human body. This means that two on-body nodes may experience very weak connectivity even though they are located in very close proximity. To make everything worse, the human body is a dynamic environment. This means that the on-body network topology is changing drastically by time. Such changes could be quite abrupt and can affect the channel severely even between two nearby nodes. Hence, there is a great need for cognitive approaches that can adapt to the changes of the parameters of the environment.

5.4 COGNITIVE RADIO BASED WIRELESS BODY SENSOR NETWORKS

A cognitive radio can intelligently use the information about the transmission medium to adapt its parameters and enhance its performance. An example of such an adaptive approach for WBSNs is proposed in [80] to improve the energy efficiency of the nodes in the network. The key idea is to adaptively change the routing strategy based on the quality of the channel.

Assume node *A* is an on-body device which has information to transmit to the NC, which is another on-body device with more computational and energy resources. It has the choice to transmit directly or through an on-body relay *R*. Due to the dynamic nature of the human body, channel quality between the nodes changes by time. In the proposed adaptive routing, *A* obtains information on the channel quality through an RTS-CTS scheme; *A* transmits a Request-To-Send frame (RTS) to indicate that it has data to transmit and reserve the channel. *NC* will reply by CTS indicating that it is ready to receive the data. Node *A* obtains the channel information through CTS preamble. If the channel quality is above a certain threshold, *A* transmits directly to NC. Otherwise, it will request a relay path by transmitting a Request-To-Relay (RTR). The relay will reply by CTR, and reserve the channel for node *A*, which will transmit its data, and both *R* and *NC* try to receive the packet. Upon successful reception, both *NC* and *R* will transmit an acknowledgement packet (ACK) to indicate the reception of the packet. If relay receives the packet correctly but does not hear any ACK from *NC* it will retransmit the packet to *NC*. If *A* receives no ACK from either *R* or *NC* it will retry transmission in another session.

A two-dimensional Markov chain model is proposed to model the adaptive protocol and the protocol is mathematically analyzed based on the medium access procedures of the IEEE 802.15.6 standard. Based on the standard, a node can access the medium when the channel is idle for a certain amount of time. To access the medium, the node will initiate a random counter between zero and a maximum window size (W_i) after the idle period, where i is the stage of the counter. The node will transmit when the counter reaches zero. If the channel becomes busy during the countdown, the counter freezes. When two or more nodes transmit at the same time, collision occurs, which results in increasing the counter stage. The maximum window size doubles every even stage number. The Markov model of the adaptive protocol is depicted in Figure 5.9, where rows indicate the stage of the counter and columns indicate the state of the counter, m is the maximum number of retries before the packet is discarded, p_b and p_f indicate the probability that the channel is busy, and that of a failed transmission due to collision or unfavorable channel conditions, respectively. The energy consumption model, i.e., the energy consumed per bit [J/bit], for node *A* is derived as follows:

$$E_b^A = \frac{E_{bo}^A + E_o^A + E_r^A + E_s^A}{L} \tag{5.1}$$

where E_{bo}^A [J] is the average energy consumed during the back-off procedure, E_o^A [J] is the average energy consumed to overhear the ongoing communications to assess the transmission opportunity, E_r^A [J] is the average energy consumed for retransmission of a failed data frame, E_s^A [J] is the average energy of transmitting a data frame successfully, and L [bit] is the payload size.

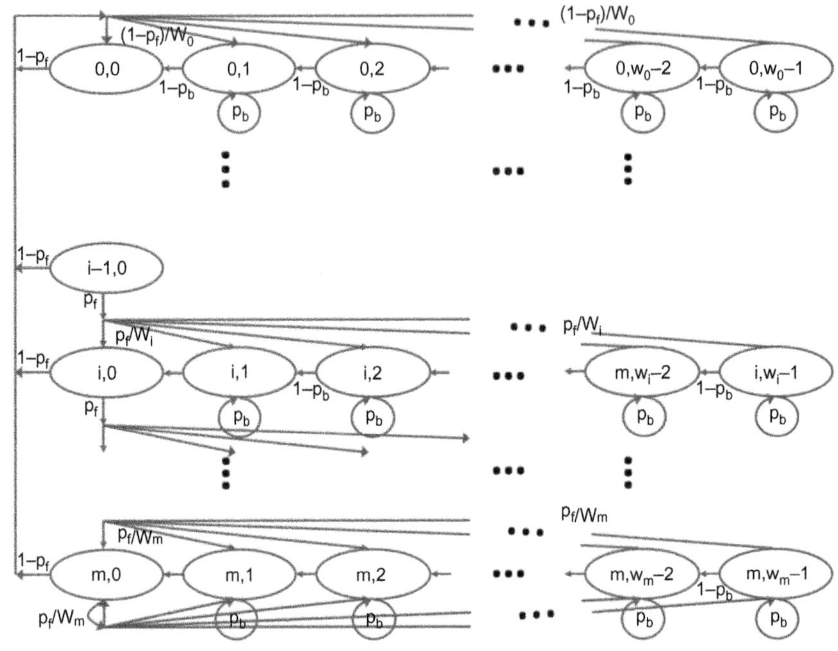

FIGURE 5.9

Markov model of the back-off procedure. Rows indicate the stage of the back off, while columns represent the state of the back-off counter.

The average energy consumption could be derived mathematically based on the Markov chain steady state equations. More details on the derivations are given in [80]. Again, the energy consumption model, i.e., the energy consumed per bit [J/bit], for the relay node is as follows,

$$E_b^R = \frac{E_{RX}^R + (1 - p_d)(E_{bo}^R + E_o^R + E_r^R + E_s^R)}{L} \tag{5.2}$$

where E_{RX}^R [J] is the average energy consumed to receive the packet, and p_d is the probability that the packet is received with errors. Hence, from (1) and (2), the total energy consumed per bit in adaptive routing derives as:

$$E_b^T = E_b^A + p_R E_b^R \tag{5.3}$$

where p_R is the probability of using relay.

The analytical model is then validated through MATLAB simulations. Medium access functions and routing procedures are implemented in MATLAB and the results are compared with predictions from the mathematical model. Simulation results show that the mathematical model can closely predict the energy consumption of the nodes. In addition, the adaptive routing scheme is compared with fixed direct and relayed transmission schemes provisioned in the IEEE 802.15.6 standard. Results show that adaptive routing can reduce the transmission energy cost per bit by 54% compared to the other schemes.

6 FUTURE DIRECTIONS

This section is devoted to summarizing the main and key features of the technologies described in the chapter. Moreover, we will try to sketch out some useful directions in terms of research, without pretending to be exhaustive.

Table 5.3 characterizes and summarizes main aspects of WSNs, CWSNs, BANs, and CBANs.

Cognitive radio technologies seem promising for the exponential rise in the demand for bandwidth. However, as remarked in Table 5.3, there are still open problems that need to be addressed before such applications could be utilized in practice.

In the following subsections we detail some of the most significant open problems: spectrum decision, sharing, mobility and handover, and energy efficiency.

6.1 SPECTRUM DECISION

Spectrum decision strategy is still an open problem in cognitive radio networks. An optimal strategy aims to minimize the delay and energy consumption by selecting the best channel at the right time, in order to access the medium, and transmit data. A promising solution to this problem is to create an accurate model of channel availability rate, and update the model based on the observed data in real time. This model could be used to make decisions on the optimum channel to access, and transmit data.

6.2 SPECTRUM SHARING

Conventional spectrum sharing protocols cannot be applied in cognitive radio setting. This is because the operating frequency is constantly changing according to the observations and the model predictions. Hence, cognitive MAC protocol has to be able to assign transmission time fairly to the nodes operating in the same band. This approach is still broadly under investigation by both academic and industrial communities.

6.3 SPECTRUM MOBILITY AND HANDOVER

The spectrum mobility and handover mechanism is another open problem in the cognitive radio paradigm. To guarantee seamless communication, hopping to a new frequency band should be handled efficiently by the MAC and PHY layer. In

Table 5.3 Key Features of WSNs, CWSNs, BANs, and CBANs

	WSNs	**CWSNs**	**BANs**	**CBANs**
Standards	IEEE 802.15.4, ZigBee, ISA 100, IEEE 1451	Not yet defined	IEEE 802.15.6	Not yet defined
Hardware Features	Small devices, low processing capacity, low memory capacity	Small intelligent devices, cognition capabilities, moderate processing capacity, moderate memory capacity	Low power devices, small devices, low memory capacity	Small intelligent devices, cognition capabilities
Inter-communication	Normally operates as an autonomous system	Normally operates as an autonomous system	Seldom works alone	Seldom works alone
Interaction with Human Being	Focus on the interaction with the environment	Focus on the interaction with the environment	Very close to the humans	Very close to the humans
x/centric	Generally data/centric	Generally data/centric	Data/user/centric	Data/user/centric
Application Specific	Yes	Yes	Yes	Yes
Routing Type	Broadcast, from/to sink	Broadcast, from/to sink	Broadcast, from/to sink	Broadcast, from/to sink
QoS Parameters	• Energy • Scalability • Throughput • Approximation Accuracy	• Energy • Scalability • Approximation Accuracy • Spectrum Efficiency Utilization • Interference to PUs	• Guaranteed Bandwidth • Delay • Jitter • Error Rates • Energy	• Guaranteed Bandwidth • Delay • Jitter • Error Rates • Energy • Spectrum Efficiency Utilization • Interference to PUs

Research Directions	Many areas related to WSNs have been very deeply explored. Among the most critical topics we individuate: • Trust and Security • Management of big data	Research is still in infancy. Several areas are still to be explored, such as: • Spectrum Decision • Spectrum Sharing • Spectrum Mobility and Handover • Energy Efficiency	QoS handling is a challenging job in BANs. Interesting research directions are concerning MAC protocol definition and routing approaches.	QoS handling is a challenging job in BANs. Interesting research directions are concerning MAC protocol definition and routing approaches. Cross-layer techniques to include: • QoS • Spectrum Decision • Spectrum Sharing • Spectrum Mobility and Handover • Energy Efficiency

other words, the frequency hopping should be invisible to higher layers especially for delay intolerant applications, such as audio and video transmissions.

6.4 ENERGY EFFICIENCY

Energy efficiency in cognitive radio is of great importance, as some of the main "users" of cognitive radio are energy constraint networks such as wireless sensor networks. Particularly, cognitive radio systems must be able to sense the spectrum, transmit data and predict the spectrum availability in an energy efficient way. Channel sensing and prioritizing the channels to access should be efficiently implemented in CR protocols, in order to avoid unnecessary energy overhead on CR devices. As an instance, in [81] Maleki et al. present a scheme for minimizing the energy consumed in distributed sensing in a CRN, subject to constraints on the detection performance. Specifically, they consider the availability of prior knowledge about the probability of primary user presence.

7 CONCLUSIONS

This chapter focuses on the relatively new notion of software defined and cognitive radio for wireless sensor networks. Specifically, analytical modeling of capacity, energy consumption and congestion for Cognitive Wireless Sensor Networks and Cognitive Wireless Body Area Networks is discussed and evaluated. Moreover, routing approaches and modeling techniques to evaluate the performance of routing for both generic wireless sensor networks and wireless body sensor networks are discussed in detail.

Several studies show that cognitive approaches are among the most promising approaches to target higher capacity demand and lower energy consumption for the abovementioned wireless networks. As an example of a cognitive solution for future body area networks, an adaptive routing scheme for wireless body sensor networks is discussed. The key idea in this approach is to adaptively change the routing strategy based on the quality of the channel.

Mathematical models for the MAC protocol and the energy cost per bit are provided and analyzed based on the medium access procedures of IEEE 802.15.6. The adaptive scheme is then compared with the existing methods, and it is shown that the adaptive routing can reduce the energy cost of information per bit by 54%. However, there is still a need for more efficient cognitive techniques to sense the spectrum, and exploit this information to increase the quality-of-service in the future wireless sensor networks. Modeling and analysis of these techniques not only provide more insight into the limitations of the system but also lead to more practical and efficient solutions.

ACKNOWLEDGEMENTS

This work is partially supported by CPER NPdC/FEDER CIA.

REFERENCES

[1] Mitola J, Maguire Jr. GQ. Cognitive radio: making software radios more personal. IEEE Pers Commun 1999;6(4):13–18.

[2] Pace P, Loscri V. OpenBTS: a step forward in the cognitive direction, In: Proceedings of 21st IEEE international conference on computers, Communications and Networks (ICCCN), 2012.

[3] Tawk Y, Costantine J, Christodoulou CG. Cognitive-radio and antenna functionalities: a tutorial [Wireless Corner]. IEEE Antennas Propagat Mag 2014;56(1):231–43.

[4] Youssef M, Ibrahim M, Abdelatif M, Lin C, Vasilakos AV. Routing metrics of cognitive radio networks: a survey. IEEE Commun Surv Tutorials 2014;16(1):92–109 First Quarter.

[5] Pal R, Idris D, Pasari K, Prasad N. Characterizing reliability in cognitive radio networks, In: Proceedings of first international symposium on applied sciences on biomedical and communication technologies, (ISABEL), p. 1–6, 25–28 Oct. 2008.

[6] Stotas S, Nallanathan A. Enhancing the capacity of spectrum sharing cognitive radio networks. IEEE Trans Veh Technol 2011;60(8):3768–79.

[7] Loscri V. A queue based dynamic approach for the Coordinated Distributed scheduler of the IEEE 802.16. In: Proceedings of IEEE symposium on computers and communications (ISCC), 2008.

[8] Loscri V. A new distributed scheduling scheme for wireless mesh networks, In: Proceedings of IEEE 18th international symposium on personal, indoor and mobile radio communication (PIMRC); 2007.

[9] Sengupta S, Subbalakshmi KP. Open research issues in multi-hop cognitive radio networks. IEEE Commun Mag 2013;51(4):168–76.

[10] Wang B, Liu KJR. Advances in cognitive radio networks: a survey. IEEE J Sel Top Signal Processing 2011;5(1):5–23.

[11] Naeem M, Anpalagan A, Jaseemuddin M, Lee DC. Resource allocation techniques in cooperative cognitive radio networks. IEEE Commun Surv Tutorials 2014;16 (2):729–44. Second Quarter.

[12] Yu C-K, Chen K-C, Cheng S-M. Cognitive radio network tomography. IEEE Trans Veh Technol 2010;59(4):1980–97.

[13] Baumgarten M, Mulvenna M. Cognitive sensor networks: towards self-adapting ambient intelligence for pervasive healthcare, In: Proceedings of IEEE 5th international conference on pervasive computing technologies for healthcare (PervasiveHealth), 2011, p. 366–9, 23–26 May 2011.

[14] Mitola J. Cognitive radio: an integrated agent architecture for software defined radio, Ph.D. dissertation, Stockholm, Sweden; Royal Institute of Technology; 2000.

[15] Zahmati AS, Hussain S, Fernando X, Grami A. Cognitive wireless sensor networks: emerging topics and recent challenges, In: Proceedings of IEEE International conference on science and technology for humanity (TIC-STH), 2009 Toronto; p. 593–6, September 26–27, 2009.

[16] Shankar NS, Cordeiro C, Challapali K. Spectrum agile radios: utilization and sensing architectures, In: Proceedings of 1st IEEE International symposium on new frontiers in dynamic spectrum access networks, DySPAN. p. 160–9, November 2005.

[17] Bidr E, Ibnkahla M. Performance modeling of cognitive wireless sensor networks applied to environmental protection, In: Proceedings of IEEE global telecommunications conference (Globecom), Honolulu, December 30–January 4, 2009, Honolulu, Hawaii, USA.

[18] Vijay G, Ben Ali Bdira E, Ibnkahla M. Cognition in WSNs: a perspective. IEEE Sens J 2011;11(3).

[19] Reznik L, Von Pless G. Neural networks for cognitive sensor networks, In: Proceedings of IEEE International joint conference on neural network, IJCNN. 2008. p. 1235–41.

[20] Aalamifar F, Vijay G, Khozani PA, Ibnkahla M. Cognitive Wireless Sensor Networks for Highway Safety, In: Proceedings of ACM DIVANET; 2011.

[21] Otto C, Milenkovic A, Sanders C, Jovanov E. System architecture of a wireless body area sensor network for ubiquitous health monitoring. J Mob Multimedia 2006;1 (4):307–26.

[22] IEEE 802.15. 2015. IEEE 802.15 WPAN. Available at: <http://www.ieee802.org/15/pub/TG6.html>.

[23] Akyildiz IF, Lee W, Vuran MC, Mohanty S. NeXt generation/dynamic spectrum access/cognitive radio wireless networks: a survey. Comput Netw 2006;50:2127–59.

[24] Cavalcanti D, Das S, Jianfeng W, Challapali K. Cognitive radio based wireless sensor networks, In: Proceedings of 17th International conference on computer communications and networks (ICCCN); August 2008. p. 1–6.

[25] Gao S, Qian L, Vaman DR, Qu Q. Energy Efficient adaptive modulation in wireless cognitive radio sensor networks, In: Proceedings of IEEE International conference on communications 2007 (ICC); June 2007. p. 3980–6.

[26] Byun S, Balasingham I, Liang X. Dynamic spectrum allocation in wireless cognitive sensor networks: improving fairness and energy Efficiency, In: Proceedings of 68th IEEE Vehicular technology conference (VTC); September 2008. p. 1–5.

[27] Romer K, Mattern F. The design space of wireless sensor networks. IEEE Wirel Commun 2004;11(6):54–61.

[28] Punchongharn P, Hossain E, Niyato D, Camorlinga S. A Cognitive radio system for E-health applications in a hospital environment. IEEE Wirel Commun 2010;17 (no. 1):20–8.

[29] Kwak KS, Ullah S, Ullah N. An overview of the IEEE 802.15.6 standard, In: Proceedings of 3rd International symposium of applied science and biomedical and communication technologies. Rome, Italy. November 2010.

[30] Qin Y, He Z. A communication monitor for wireless sensor networks based on software defined radio, SICS Technical Report T2011:03, ISRN: SICS-T-2011/03-SE, January 26, 2011.

[31] Suliman I, Lehtomaki J. Queueing analysis of opportunistic access in cognitive radios, In: Proceedings of CogART'09; May 2009. p. 153–7.

[32] Kailas A. Power allocation strategies to minimize energy consumption in wireless body area networks, In: Proceedings of annual international conference of the IEEE In: Engineering in Medicine and Biology Society (EMBC); p. 2204–7. IEEE, 2011.

[33] Ferrand P, Maman M, Goursaud C, Gorce JM, Ouvry L. Performance evaluation of direct and cooperative transmissions in body area networks. Ann Telecommun 2011;66(3):213–28.

[34] Gao Y, Jiang Y. Performance analysis of a cognitive radio network with imperfect spectrum sensing, In: Proceedings of IEEE INFOCOM'10, March 2010, p. 1–6.

[35] Al-Ali A, Chowdhury KR, Simulating dynamic spectrum access using NS-3 for wireless networks in smart environments, IEEE SECON workshop on self-organizing wireless access networks for smart city. Singapore, June 2014.

[36] Oey CHW, Christian I, Moh S. Energy- and cognitive-radio-aware routing in cognitive radio sensor networks, Hindawi Publishing Corporation, Int J Distri Sen Netw 2012. Article ID 636723.

[37] Quwaider M, Biswas S. On-body packet routing algorithms for body sensor networks, In: Proceedings of the IEEE 1st international conference on networks and communications (NETCOM). p. 171–7, 2009.

[38] Akan OB, Karli OB, Ergul O. Cognitive radio sensor networks, IEEE Netw, 2009; 23. p. 34–40.

[39] Marković GB, Dukić ML. Cooperative AMC schemes using cumulants with hard and soft decision fusion, In: Proceedings of 20th Telecommunications Forum TELFOR 2012. Serbia, Belgrade, November 20–22, 2012.

[40] Ghaboosi K, Pahlavan K, Pomalaza-Raez CA. A cooperative medical traffic delivery mechanism for multi-hop body area networks, In: Proceedings of the IEEE 22nd international symposium on personal indoor and mobile radio communications (PIMRC); 2011; p. 2239–43.

[41] Braem B, Latre B, Moerman I, Blondia C, Reusens E, Joseph W, et al. The need for cooperation and relaying in shortrange high path loss sensor networks, In: Proceedings of the IEEE international conference on sensor technologies and applications, SensorComm; 2007. p. 566–71.

[42] Braem B, Latre B, Blondia C, Moerman I, Demeester P. Improving reliability in multi-hop body sensor networks, In: Proceedings of the IEEE 2nd international conference on sensor technologies and applications (SENSORCOMM); 2008. p. 342–7.

[43] Ho M-J, Berber SM, Sowerby KW. Sensor network controlled indoor cognitive radio systems, In: Proceedings of telecommunication networks and applications conference (ATNAC) Australasian, p. 1–6. November 7–9, 2012.

[44] Yau KLA, Komisarczuk P, Teal PD. Cognitive radio-based wireless sensor networks: conceptual design and open issues, In: Proceedings of the IEEE 34th Conference on local computer networks (LCN '09); October 2009. p. 955–62.

[45] Cavalcanti D, Das S, Wang J, Challapali K. Cognitive radio based wireless sensor networks, In: Proceedings of the 17th International conference on computer communications and networks (ICCCN); August 2008. p. 491–6.

[46] OPNET Technologies. Academic Research and Teaching with OPNET Software. 2008. Available from: <http://netlab.boun.edu.tr/opnet.html>. accessed on-line on 06-March-2014.

[47] Araujo A, Romero E, Blesa J, Taldriz ON. Cognitive wireless sensor networks framework for green communications design. In: Proceedings of the 2nd International conference on advances in cognitive radio, COCORA; 2012.

[48] Zhong J, Li J. Cognitive radio cognitive network simulator. Michigan Tech University; Available from: <http://stuweb.ee.mtu.edu/~ljialian/>; 2009.

[49] Wang J, Ghosh M, Challapali K. Emerging cognitive radio applications: a survey. IEEE Commun Mag 2011;49(3):74–81.

[50] Joshi GP, Nam SY, Kim SW. Cognitive radio wireless sensor networks: applications, challenges and research trends. Sensors 2013;13:11196–228.

[51] Cotton SL, Scanlon WG. Characterization and modeling of the indoor radio channel at 868 MHz for a mobile bodyworn wireless personal area network. IEEE Antennas Wirel Propagat Lett 2007;6.

[52] Çalhan A, Atmaca S. A new network coordinator node design selecting the optimum wireless technology for wireless body area networks. KSII Trans Internet Inf Syst 2013;7(5).

[53] Loscrì V, Pace P, Surace R. Multi-objective evolving neural network supporting SDR modulations management, In: Proceedings of the IEEE 24th International symposium on personal, indoor and mobile radio communications: mobile and wireless networks, PIMRC, 2013.

[54] SourceForge. Frevo. Available from: <http://sourceforge.net/p/frevo/wiki/Tutorials/>; 2015.

[55] Hedetniemi S, Liestman A. A survey of gossiping and broadcasting in communication networks. Networks 1988;18(4):319–49.

[56] Akkaya K, Younis M. A survey on routing protocols for wireless sensor networks. Ad Hoc Netw 2005;3(3):325–49.

[57] Heinzelman W, Kulik J, Balakrishnan H. Adaptive protocols for information dissemination in wireless sensor networks, In: Proceedings of the 5th Annual ACM/IEEE international conference on mobile computing and networking (MobiCom'99). Seattle, WA: August 1999.

[58] Intanagonwiwat C, Govindan R, Estrin D. Directed diffusion: a scalable and robust communication paradigm for sensor networks. In: Proceedings of the 6th annual ACM/IEEE international conference on mobile computing and networking (MobiCom'00). Boston, MA; August 2000.

[59] Shah R, Rabaey J. Energy aware routing for low energy ad hoc sensor networks, In: Proceedings of the IEEE wireless communications and networking conference (WCNC). Orlando, FL; March 2002.

[60] Braginsky D, Estrin D. Rumor routing algorithm for sensor networks, In: Proceedings of the 1st workshop on sensor networks and applications (WSNA). Atlanta, GA; October 2002.

[61] Schurgers C, Srivastava MB. Energy efficient routing in wireless sensor networks, In: Proceedings of MILCOM communications for network-centric operations: creating the information force. McLean, VA; 2001.

[62] Sadagopan N, Krishnamachari B, Helmy A. The acquire mechanism for efficient querying in sensor networks, In: Proceedings of the 1st international workshop on sensor network protocol and applications, Anchorage, Alaska; May 2003.

[63] Heinzelman W, Chandrakasan A, Balakrishnan H. Energy-efficient communication protocol for wireless sensor networks, In: Proceedings of the hawaii international conference system sciences. Hawaii; January 2000.

[64] Lindsey S, Raghavendra CS. PEGASIS: Power efficient gathering in sensor information systems, In: Proceedings of the IEEE aerospace conference. Big Sky, Montana; March 2002.

[65] Lindsey S, Raghavendra CS, Sivalingam K, Data gathering in sensor networks using the energy delay metric, In: Proceedings of the IPDPS workshop on issues in wireless networks and mobile computing. San Francisco, CA; April 2001.

[66] Manjeshwar A, Agrawal DP. TEEN: a protocol for enhanced efficiency in wireless sensor networks, In: Proceedings of the 1st International workshop on parallel and distributed computing issues in wireless networks and mobile computing. San Francisco, CA; April 2001.

[67] Rodoplu V, Ming TH. Minimum energy mobile wireless networks. IEEE J Sel Areas Commun 1999;17(8):1333–44.

[68] Xu Y, Heidemann J, Estrin D. Geography-informed energy conservation for ad hoc Routing, In: Proceedings of the 7th annual ACM/IEEE international conference on mobile computing and networking (MobiCom'01). Rome, Italy; July 2001.

[69] Yu Y, Estrin D, Govindan R. Geographical and energy-aware routing: a recursive data dissemination protocol for wireless sensor networks, UCLA computer science department technical report, UCLA-CSD TR-01-0023, May 2001.

[70] FCC, ET Docket No. 03-222 Notice of proposed rule making and order, December 2003. Available from, <http://web.cs.ucdavis.edu/ ~ liu/289I/Material/FCC-03-322A1.pdf>.

[71] Zhao Q, Tong L, Swami A, Chen Y. Decentralized cognitive MAC for opportunistic spectrum access in ad hoc networks: a POMDP framework. IEEE J Sel Areas Commun 2007;25(3):589–600.

[72] Park S, Lee S, Kim B, Hong D, Lee J. Energy-efficient opportunistic spectrum access in cognitive radio networks with energy harvesting, In: Proceedings of the ACM 4th international conference on cognitive radio and advanced spectrum management, 2011.

[73] Chowdhury K, Di Felice M. Search: a routing protocol for mobile cognitive radio ad-hoc networks, In: Proceedings of IEEE sarnoff symposium; March 30–April 1, 2009.

[74] Pantelopoulos A, Bourbakis NG. A survey on wearable sensor-based systems for health monitoring and prognosis. IEEE Trans Syst Man Cybern Part C Appl Rev 2010;40(1):1–12.

[75] Chen M, Gonzalez S, Vasilakos A, Cao H, Leung VCM. Body area networks: a survey. Mob Netw Appl 2011;16(2):171–93.

[76] Latre B, Braem B, Moerman I, Blondia C, Demeester P. A survey on wireless body area networks. Wirel Netw 2011;17(1):1–18.

[77] Wang CC, Yang CY, Huang CC. Personal gateway design for portable medical devices used in body area networks, IEEE Digest of technical papers international conference on consumer electronics (ICCE); 2010. p. 189–90.

[78] Rousselot J, Decotignie JD. Wireless communication systems for continuous multiparameter health monitoring, in Proceedings of IEEE International Conference on Ultra-Wideband, ICUWB'09; 2009. p. 480–4.

[79] Forouzandeh FF, Mohamed OA, Sawan M, Awwad F. TBCD-TDM: Novel ultra-low energy protocol for implantable wireless body sensor networks, In: Proceedings of IEEE global telecommunications conference (GLOBECOM); 2009. p. 1–6.

[80] Fang G, Dutkiewicz E. BodyMAC: energy efficient TDMA-based MAC protocol for wireless body area networks, In: Proceedings of the IEEE 9th International symposium on communications and information technology, ISCIT'09; 2009. p. 1455–9.

[81] Maleki S, Pandharipande A, Leus G. Energy-efficient distributed spectrum sensing for cognitive sensor networks. IEEE Sens J 2011;11(3):565–73.

PART

2

Approaches in performance evaluation

CHAPTER

Generating realistic workload for web performance studies

6

Raúl Peña-Ortiz, José Antonio Gil, Julio Sahuquillo, Ana Pont, and Josep Domènech

Universitat Politècnica de València, València, Spain

1 INTRODUCTION

There are few technological success stories as dramatic as that of the Web. Originally designed to share static contents among a small group of researchers, the Web is being used today by many millions of people as a part of their daily routines and social lives. This incessant evolution has been possible thanks to the continuous changes in technology that have introduced new features in the current and incoming Web, both in its applications, users, and infrastructure [1].

With the emergence of current Web applications and services, users are no longer passive consumers, but they become participative contributors to the dynamic content accessible on the Web [2,3]. Therefore, a new user's dynamic behavior can be distinguished and is being more relevant and meaningful in the incoming Web, also referred to as Web 3.0 [4] or Future Internet [5].

As a system that is continuously changing, both in the offered applications and infrastructure, performance evaluation studies are necessary in order to provide sound proposals when designing new Web-related systems [6] such as Web services, Web servers, proxies or content distribution policies.

As in any performance evaluation process, accurate and representative workload models must be used in order to guarantee the validity of the results. Regarding Web systems, the user's dynamic behavior makes the design of accurate Web workload representing realistic users' navigations difficult. In general, there are three main challenges that must be addressed when modeling the user's dynamic behavior on representative workloads:

- **Challenge I:** The dynamism in user's behavior when surfing the Web must be taken into account [6]. That is, users' behaviors as they interact with Web contents and services have to be characterized, modeling the different aspects that determine users' navigation decisions. For instance, personal preferences, navigation goals, visited resources or connectivity conditions.

- **Challenge II:** The different user's roles when navigating a website must be identified and defined as user's behaviors [7]. For instance, searcher and surfer roles refer to users who start navigations with a query in a given searcher engine, or navigate the Web by following direct hyperlinks, respectively [8].
- **Challenge III:** Continuous changes in these user's roles during the same navigation session must be modeled and considered [9]. That is, changes in users' behaviors over time have to be characterized.

2 WORKLOAD MODELS AND THE CURRENT WEB

Web workload characterization studies that help us to model and reproduce users' behaviors grow in importance with the massive use of Web applications and services. Moreover, both types of applications are developed using new technologies that have a strong impact on the system performance. Some previous attempts have been published to reflect this fact. For instance, Cecchet et al. [10] investigate the effect of different J2EE application deployments on the performance scalability of application servers. Schneider et al. [11] point out that the use of AJAX and mashups generates more aggressive and bursty network usage compared to the overall HTTP traffic. Similar conclusions but also considering server performance are presented in [12]. Unfortunately, these studies only consider specific Web paradigms; thus the workload used is not well representative of current users' navigations. In a more recent work, Kosir et al. [13] propose a new method to build user profiles by using the user's history of visited websites using large datasets.

Web workload models are abstractions of the real workload that reproduce users' behaviors and ensure that a particular Web application performs as it would do when working with real users. To this end, the model represents a set of users of the Web application and avoids those characteristics that are superfluous for a particular study. Workload models can be classified into two main groups: trace-based and analytical models. Traces log the sequence of HTTP requests and commands received by a Web application during a certain period of time under given conditions. Traces are obtained for a particular environment; that is, specific server speed, network bandwidth, browser cache capacity, etc. This means that if any system parameter varied, the obtained trace would be different. Therefore, the main challenge of trace-based models is to achieve a good representativeness, especially when requests received by different servers exhibit a large variability. Consequently, trace-based models are not appropriate to model changes in the user's behavior.

The analytical approach uses mathematical models to simulate the user's behavior or the characteristics of specific workloads. These models allow us to consider different scenarios by setting some input parameters that specify the main

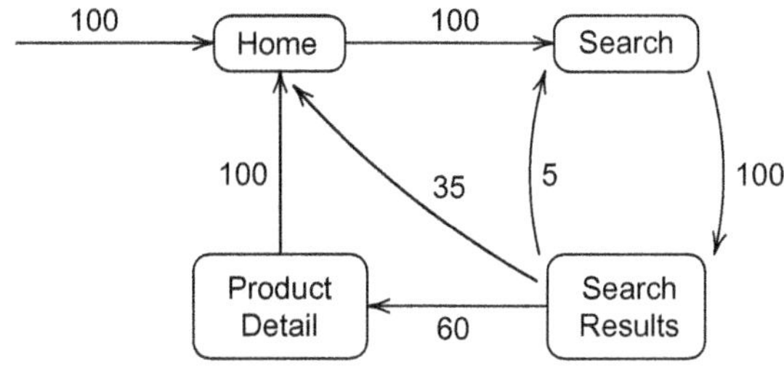

FIGURE 6.1

Example of a simplified CBMG model.

characteristics of the workload to be characterized. Analytical models are a flexible approach for modeling changes in the user's behavior. Some studies [14,15] confirm how difficult it is to model and generate representative Web requests, especially when trying to model the characteristics of the dynamic Web and its users.

There have been few but interesting efforts to model users' behavior in order to obtain more representative user workloads for specific Web applications. Menascé and Almeida [16] introduced the Customer Behavior Model Graph (CBMG) that describes patterns of users' behavior in the workloads of e-commerce websites by using an approach based on finite state machines (FSMs). The CBMG model consists of all pages of an on-line bookstore and the associated transition probability. For illustrative purposes, Figure 6.1 depicts an example of CBMG for a search process, showing that users may visit several pages and move among these pages according to the arc's weight. Numbers in the arcs indicate the probability of taking that transition. For example, the probability of going to the Product Detail page from the Search Results page is 60%. This value means that after a search, regardless of whether the search returns a list of books or a void list, the Product Detail page will be visited 60% of the time.

The Visitor Behavior Model Graph (VBMG) for workload definition of the blogspace was introduced by Duarte et al. [17], extending the CBMG model. Blog visitors can be grouped into different categories according to their visiting patterns. These categories are characterized by different VBMGs in terms of the state transition probabilities. For example, Figure 6.2 shows the typical behavior of blurkers, who tend to read a lot of blogs but never post any comments. This behavior only considers that a blurker can start reading a new blog or can continue reading the same blog. Notice that if a *blurker* reads the same blog at least twice, he can also leave the blog with a probability of 43% (exit transition).

An application modeling methodology to handle interrequest and data dependencies was proposed in [18]. The methodology relies on extended finite state machines (EFSMs) that can model applications with higher-order request dependencies without encountering the state explosion problem [19] typical in FSM-based approaches. Consequently, EFSM is better suited for modeling Web applications than CBMG and VBMG. Figure 6.3 depicts an example of EFSM for an

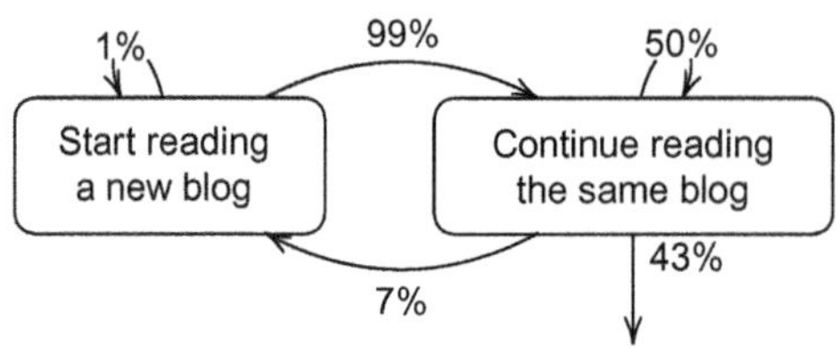

FIGURE 6.2

Example of a VBMG model for blurkers.

Source: [17].

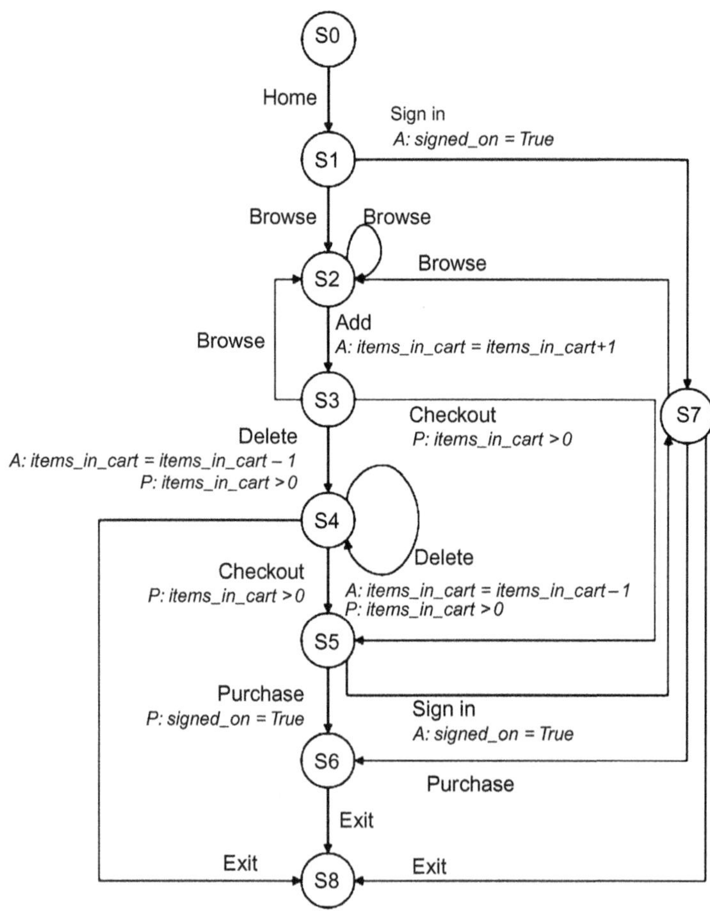

FIGURE 6.3

Example of a simplified EFSM model for an e-commerce system.

Source: [18].

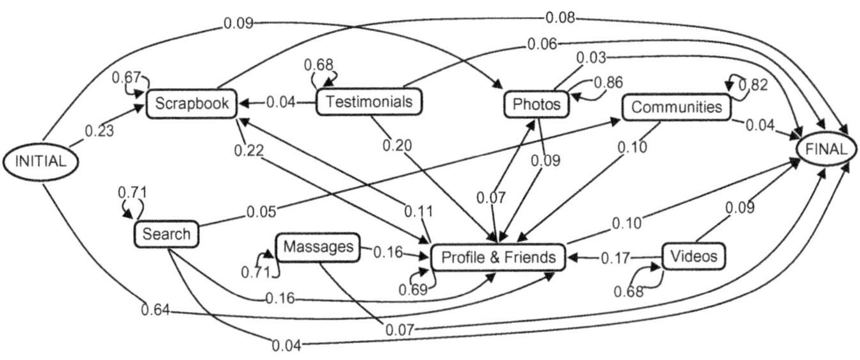

FIGURE 6.4

Transition probability in the Clickstream Model for an OSN.

Source: [20].

e-commerce system where nodes are states in the user's navigation and arcs are requests to the Web application. Two request dependency state variables are used to enforce interrequest dependencies. The *items_in_cart* is an integer variable that indicates the number of items in the shopping cart and the *signed_on* is a Boolean variable that states whether a user has signed on or not. For example, the *items_in_cart* variable is incremented by 1 when the user executes the *Add* request type S_2 to S_3 and it is decremented by 1 when the user executes the *Delete* request type (transition from S_3 to S_4). Therefore, the *Checkout* request type (transition from S_4 to S_5) is only allowed when the previous sequences of requests have resulted in at least one item in the shopping cart ($items\_in\_cart > 0$).

Benevenuto et al. [20] introduced the Clickstream Model to characterize user's behavior in On-line Social Network (OSN). This approach identifies and describes representative users' behaviors in OSNs by characterizing the type, frequency, and sequence of their activities. The modeling of the system implies two steps: i) to identify dominant user's activities in clickstreams, and ii) to compute the transition rates between activities. For illustrative purposes, Figure 6.4 shows the transition probability in the Clickstream Model for an OSN.

In a more recent work [21], they provide an in-depth workload characterization of a SN, obtaining models that describe these user access patterns within sessions when they log into these systems.

These four models only characterize Web workload for specific paradigms or applications, but they either do not model users' dynamic behavior for a general context and in an appropriate and accurate way *Challenge I* or do not consider users' dynamic roles *Challenges II* and *III*. In a recent work [22] Abramson introduces temporal information as context in this model for a more accurate user representation.

On the other hand, there is an evidence of an important change of user interaction with the Web. For instance, a recent study showed that 57.4% of Web sessions involve parallel browsing behavior [23]. This behavior was originally found in the experienced users, who surf the Web by using multiple browser tabs or windows to support backtracking or multitasking with the aim of enhancing their navigation [24,25]. Moreover, the history-back button, included in any current Web browser, is still one of the world's most heavily used user interface components in the Web context, and accounts for up to 31% of all revisits [26]. This important change has been considered in several studies and tools to improve the website usability [27] to test Web applications [28] or learning user preferences [29]. However, to the best of our knowledge, UBI have not been taken into account yet when modeling user dynamism on workload characterization in Web performance studies.

3 WEB WORKLOAD GENERATORS OVERVIEW

Workload generators are software products based on workload models to generate HTTP request sequences similar to real requests. They are designed and implemented as versatile software tools for performing tuning or capacity planning studies.

Comparing Web workload generators is a laborious and difficult task since they offer a large amount and diversity of features. In this section we contrast generators according to a wide set of features and capabilities, focusing on their ability to reproduce user dynamism in performance studies for current Web.

To this end, this section analyzes a representative subset of state-of-the-art workload generators as a first step, highlighting their main features as Web performance evaluation software and their main disadvantages when reproducing accurate workload for the current Web. After that, we evaluate and classify these generators, concentrating on those that consider user dynamic behavior.

3.1 SOFTWARE TOOLS STUDY

3.1.1 WebStone

WebStone [30] was designed by Silicon Graphics in 1996 to measure the performance of Web server software and hardware products. Nowadays, both executable and source actualized code for WebStone are available for free.

The benchmark generates a Web server load by simulating multiple Web clients navigating a website as shown in Figure 6.5. These clients can be considered as users, Web browsers, or other software that makes requests to the website files, which can be classified in different categories according to their size. The simulation is carried out using multiple clients running on one or more computers to generate large loads on a Web server. All the testing done by the benchmark is controlled by a Webmaster, which is a program that can be run on one of the client computers or on a different one. The Webmaster distributes the Web client software and test configuration files to the client computers. After

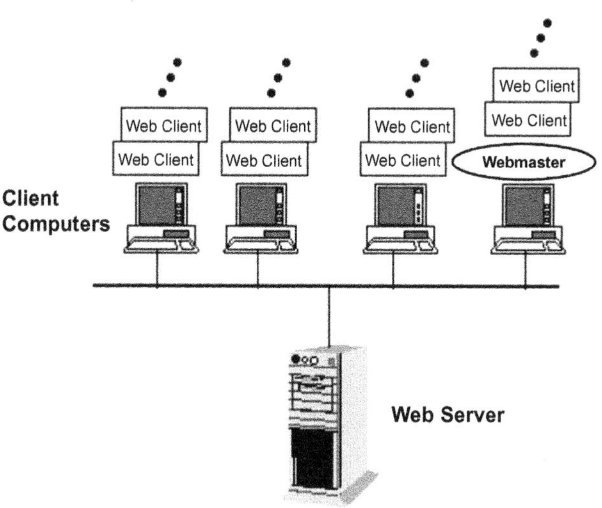

FIGURE 6.5

WebStone architecture.

Source: [30].

BOX 6.1 MAIN FEATURES AND DISADVANTAGES OF WEBSTONE

Main Features

- Parameterized workload.
- Distributed model for workload generation.
- Open performance reports.
- Open source solution.

Disadvantages

- Basic HTTP protocol only.
- No users' navigation characterization.
- No facilities to consider user's dynamism.

that, it starts the execution and waits for the clients to report the performance they measured. Finally, the Webmaster combines the performance results from all the clients into a single report.

To sum up, WebStone is one of the first software products proposed to measure the performance of Web systems but it seems obsolete for the current Web. Box 6.1 summarizes its main features and disadvantages.

3.1.2 SPEC's benchmarks for web servers

The Standard Performance Evaluation Corporation [31] has commercialized benchmarks for Web Servers from 1996 to early 2012. This benchmark family is designed to measure the performance of systems offering services in the Web.

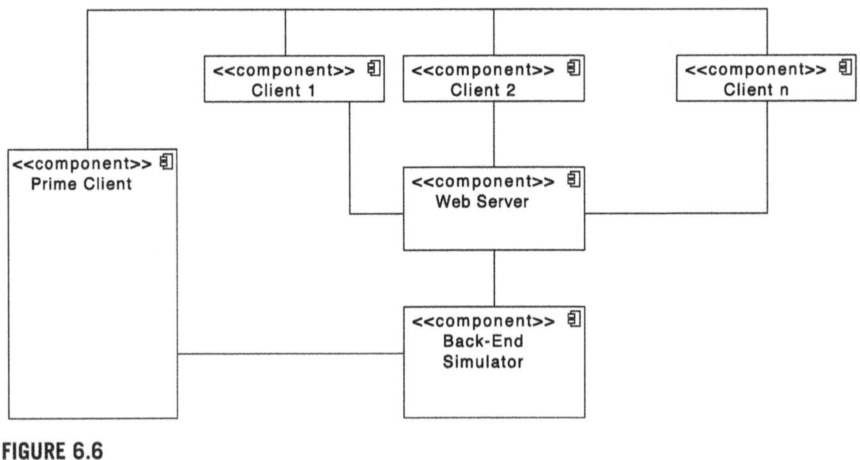

FIGURE 6.6

Logical components of SPECWeb2009.

Source: [31].

The last member of the family, named SPECWeb2009, includes many sophisticated and state-of-the-art enhancements to meet the modern demands of the current Web, such as requests to static and dynamic content (ASP, JSP, and PHP), simultaneous user sessions, parallel HTTP connections to request page images or simulate browser caching effects.

Figure 6.6 shows the logical components of SPECWeb2009. The prime client initializes and manages the other clients, sets up the Web server and the back-end simulator, and stores the results of the benchmark tests. The Web server handles the requests issued by the clients by itself or by communicating with the back-end simulator in order to retrieve specific information needed to complete HTTP responses. This simulator emulates the communication between a Web server and a back-end application server. Each benchmark client generates HTTP requests according to certain workloads that are defined by studying three representative types of Web applications (banking, e-commerce, and support).

Box 6.2 presents the main features and disadvantages of SPECWeb2009. As observed, SPEC software is a mature benchmark that has evolved with the Web; nonetheless it has not achieved the ability to reproduce realistic workload in the performance studies for the current Web because it does not consider user dynamism on workload characterization.

3.1.3 Surge

The Scalable URL Reference Generator (SURGE) was developed by Barford in 1998 [32] with the goal of measuring server behavior while varying the user load. The need to develop SURGE appeared with the difficulty of generating representative traces for the Web because workloads generated by Web users have a number of unusual features, such as the highly variable demands experienced by

BOX 6.2 MAIN FEATURES AND DISADVANTAGES OF SPECWEB2009

Main Features

- Parameterized workload.
- Different types of workloads according to the kind of Web application.
- Distributed model for workload generation.
- Full HTTP protocol (cookies, HTTPS, dynamic con- tent, etc.).
- Performance reports.
- Proprietary software.

Disadvantages

- No users' navigation characterization.
- No facilities to consider user's dynamism.

the Web servers or the self-similarity shown by the network traffic. To tackle these drawbacks, SURGE performs an analytical characterization of the user load and a set of mathematical models that generate the HTTP requests in the server [33]. These models characterize:

- The distribution of sizes of unique files requested from Web servers.
- The distribution of sizes of all files transferred from Web servers.
- The popularity of all requested files.
- The temporal locality of requested files.
- The active (ON) and inactive (OFF) periods of time for the emulated users.
- The number of documents transferred during an active period.

SURGE was designed as a scalable software framework where the previous models are combined according to the various components of the Web [34]. The software resides on a set of clients that are connected to a Web server as depicted in Figure 6.7. Each client executes a set of threads that request sets of documents, which are then transferred by the server (ON time). After receiving a set, the thread sleeps for some amount of time (OFF time), simulating the user's think time.

In summary, SURGE was a step forward on modeling accurate workload for evaluating the performance of Web 1.0. Specifically, it was able to produce self-similar network traffic under conditions of both high and low workload intensity. However, it also seems to be obsolete for the current Web because its generation process is based on analytical models that do not consider user dynamism, and it cannot model 3-tier architectures for dynamic content generation. Box 6.3 summarizes the main features and disadvantages of SURGE.

3.1.4 S-Clients

A new improved methodology for HTTP request generation was proposed by Banga & Druschel [35]. In this context, S-Clients was designed with the aim of reproducing bursty traffic with peak loads exceeding the capacity of the server as well as the modeling delay and loss characteristics of Wide Area Networks

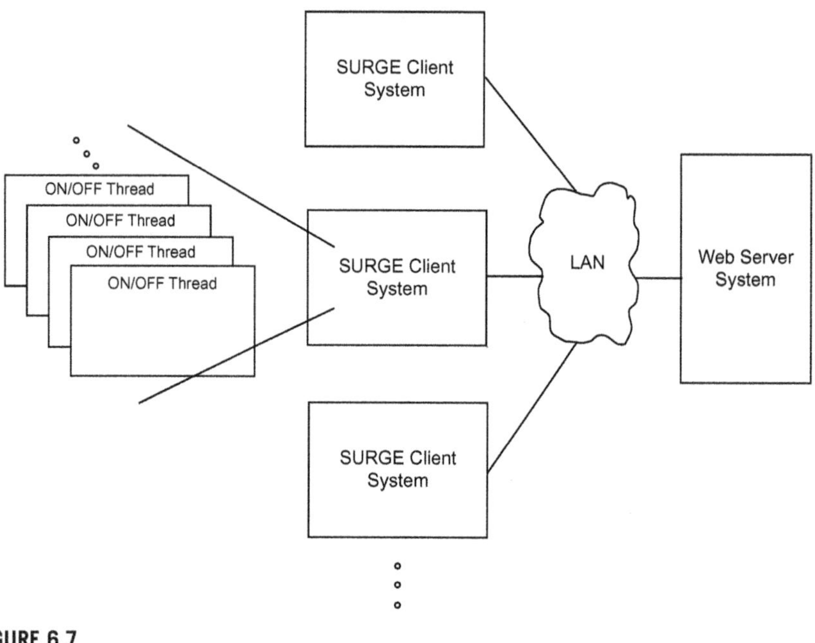

FIGURE 6.7

SURGE Architecture.

Source: [32].

BOX 6.3 MAIN FEATURES AND DISADVANTAGES OF SURGE

Main Features

- Workload generation architecture based on analytical models.
- Distributed model for workload generation.
- Open source solution.

Disadvantages

- Basic HTTP protocol only.
- No users' navigation characterization.
- No facilities to consider user's dynamism.

(WANs). Figure 6.8 shows the S-Clients design. It defines an architecture (Figure 6.8a) where a set of client machines are connected to the server machine being tested through a router, which has sufficient capacity to support the maximum client traffic specification. The purpose of the router is to simulate WAN effects by introducing an artificial delay and/or dropping packets at a controlled rate. Each client machine runs a number of scalable client processes. S-Clients splits the process of generating traced HTTP requests into two subprocesses: one for obtaining the connection and the other for recovering the content (Figure 6.8b), thus enabling a relative parallelism.

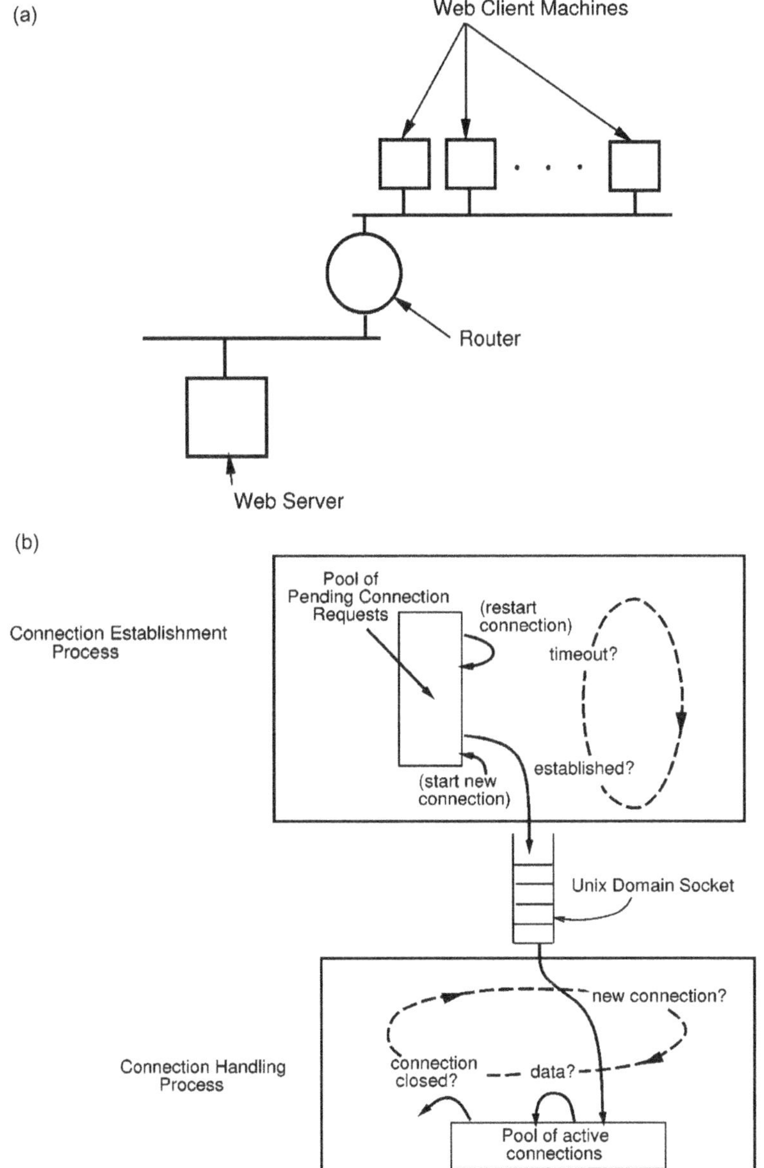

FIGURE 6.8

S-Clients design. (a) Basic structure; (b) A scalable client.

Source: [35].

BOX 6.4 MAIN FEATURES AND DISADVANTAGES OF S-CLIENTS

Main Features

- Parameterized workload.
- A router to simulate WAN effects.
- A split generation process to avoid limits on HTTP requests.
- Open source solution.

Disadvantages

- An architecture only for workload generation.
- Basic HTTP protocol only.
- No user navigation characterization.
- No facilities to consider user's dynamism.

BOX 6.5 MAIN FEATURES AND DISADVANTAGES OF WEBJAMMA

Main Features

- Easy to use as a baseline of other software generators.
- Open source solution.

Disadvantages

- Basic stressing functionalities.
- No users' navigation characterization.
- No facilities to consider user's dynamism.

To sum up, S-Clients was an architecture devised to improve workload generators for Web 1.0 that is still interesting to be considered in Web 2.0. Box 6.4 presents the main features and disadvantages of S-Clients.

3.1.5 WebJamma

WebJamma was a library to generate HTTP traffic written by the Network Research Group at Virginia Tech [36]. It is aimed at serving as baseline for developing a full Web workload generator. This library works in a simple way by taking a URL file that provides the source of the HTTP requests to be generated, so it cannot represent users' dynamism. It uses a multiprocessing architecture based on distributed generation nodes to test the performance of Web caching subsystems.

In summary, WebJamma was an interesting open source library to generate HTTP requests in an easy way. Box 6.5 shows its main features and disadvantages.

3.1.6 TPC benchmark™ W

TPC Benchmark™ W (TPC-W) is a transactional Web benchmark defined by the Transaction Processing Performance Council [37]. It models a representative e-commerce evaluating the architecture performance on a generic profile. To this

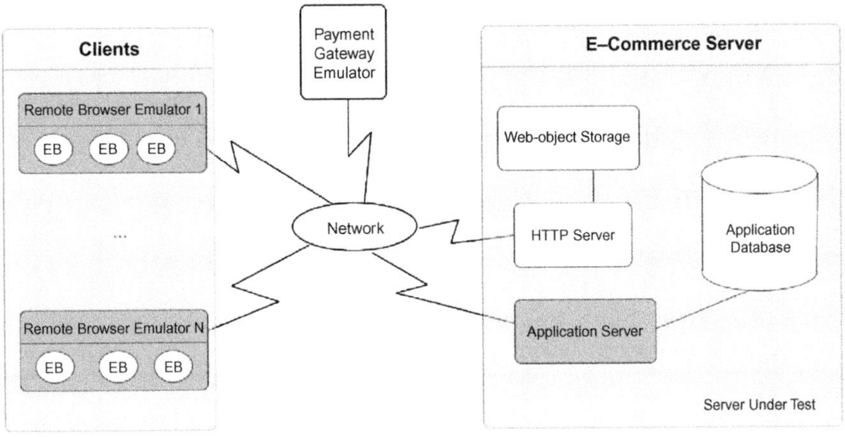

FIGURE 6.9

TPC-W architecture.

Source: [37].

end, the benchmark provides both models of business-client and business-business and examines real features of e-commerce applications, such as: catalog, searcher, security, etc.

As shown in Figure 6.9, TPC-W presents client-server architecture. The remote browser emulators are located in the client side and generate workload towards the e-commerce Web application, which is located in the server side (*e-commerce server*). With the aim of reproducing a representative workload, the emulators simulate real users' behaviors when they surf the Website by using the CBMG model, which is composed of all pages of the online bookstore and the associated transition probability. The server hosts the *system under test*, which consists of a Web server and its storage of static contents, and an application server with a database system to generate dynamic content. The *payment gateway emulator* represents an entity to authorize user payments. These three main architecture components are interconnected through a dedicated network.

To sum up, TPC-W was the first benchmark for e-commerce considering the users' behaviors on workload generation. To this end, TPC-W adopts the CBMG model to define Web workload in spite of this model only characterizing user dynamic behavior partially, as mentioned in the previous section. The benchmark has been commonly accepted by the scientific community in many research projects [38–40], even recent ones despite its evident shortcomings and age, as in [41].

The main features and disadvantages of this benchmark are shown in Box 6.6.

3.1.7 Web polygraph

Web Polygraph is a performance testing tool for caching proxies, origin server accelerators, L4/7 switches, content filters, and other Web intermediaries. It was

BOX 6.6 MAIN FEATURES AND DISADVANTAGES OF TPC-W

Main Features

- Parameterized workload.
- Different types of workloads according to the type of scenario.
- Distributed model for workload generation.
- Full HTTP protocol (cookies, HTTPS, dynamic content, etc.).
- Basic facilities to consider user's behavior.
- Performance reports.
- Open source solution.

Disadvantages

- No users' navigation characterization.
- No advanced facilities to consider user's dynamism.

originally developed at the University of California by Rousskov & Wessels [42] in the context of the IRCache project. Nowadays, it is copyrighted by The Measurement Factory [43], which authorizes the use of Polygraph under the Apache License.

The benchmark consists of virtual clients and servers glued together with an experiment configuration file [44]. Clients, named robots, generate HTTP requests for the simulated objects. These requests may be sent directly to the servers (e.g., Web servers), or through an intermediary (e.g., proxy cache or load balancer) using a configurable mix of HTTP/1.0 and HTTP/1.1 protocols, optionally encrypted with SSL or TLS. The benchmark can be configured to produce a variety of realistic and unrealistic workloads based on a synthetic workload characterization. As Polygraph runs, measurements and statistics are gathered for a detailed postmortem analysis.

In summary, Web Polygraph is a versatile tool for generating Web traffic and measuring proxy performance that has been chosen for several industry-wide benchmarking events. Box 6.7 shows its main features and disadvantages focusing on workload generation.

3.1.8 LoadRunner

LoadRunner is one of the most popular industry-standard software products for functional and performance testing. It was originally developed by Mercury Interactive, but nowadays it is commercialized by Hewlett-Packard [45].

Figure 6.10 shows how LoadRunner works. As observed, it tests a Web application by emulating an environment where multiple users work concurrently. Moreover, it accurately measures, monitors, and analyzes performance and functionality of the application while it is working under load. The testing process is controlled by a central console.

LoadRunner supports the definition of user navigations, which are represented using a scripting language, to characterize users' families. Figure 6.11 depicts the sequential approach to scripting a Web 2.0 application using LoadRunner. First

BOX 6.7 MAIN FEATURES AND DISADVANTAGES OF WEB POLYGRAPH

Main Features

- Synthetic workload characterization.
- Distributed model for workload generation.
- Full HTTP protocol.
- Performance reports.
- Successful industrial solution.
- Apache License.

Disadvantages

- No users' navigation characterization.
- No facilities to consider user's dynamism.

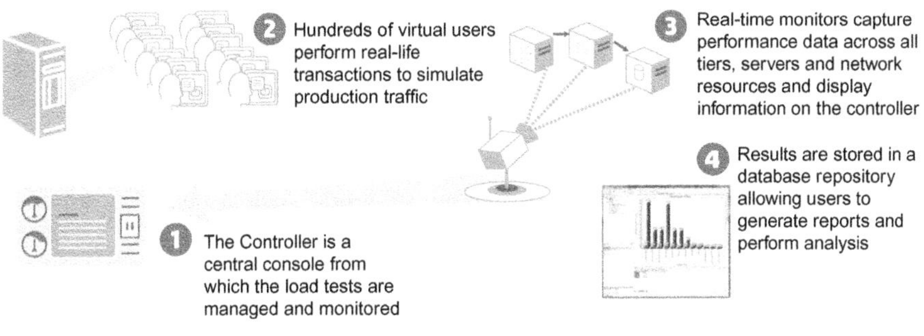

FIGURE 6.10

LoadRunner working example.

Source: [45].

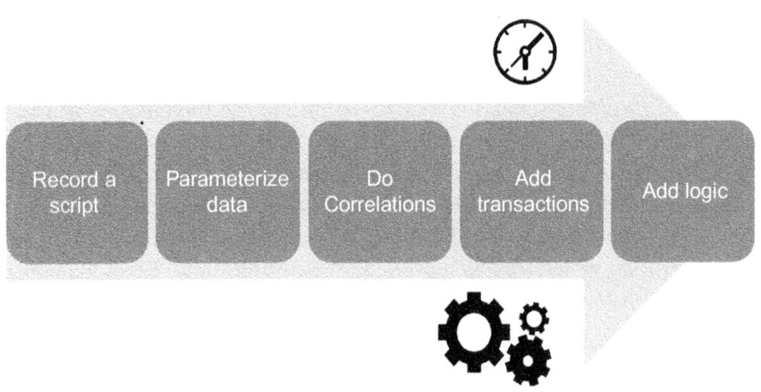

FIGURE 6.11

LoadRunner scripting for Web 2.0 applications.

Source: [45].

BOX 6.8 MAIN FEATURES AND DISADVANTAGES OF LOADRUNNER

Main Features

- Parameterized workload.
- Different types of workloads according to the type of users' families.
- Distributed model for workload generation.
- Full HTTP protocol (cookies, HTTPS, dynamic content, etc.).
- Basic facilities to consider user's behavior.
- Advanced reports during performance evaluation studies.
- Multiplatform.

Disadvantages

- No users' dynamic navigation characterization.
- No advanced facilities to consider user's dynamism.

the basic steps are recorded, creating a shell script. Next, this script is then taken off-line, and undergoes further manual steps such as data parameterization and correlations. Finally, the desired performance scripts are obtained after adding transactions and any other required logic. LoadRunner scripting only permits partial reproduction of user dynamism when generating Web workload, because it cannot define either advanced interactions of users, such as parallel browsing behavior, or continuous changes in user's behaviors.

To sum up, LoadRunner is one of the most important software products to test the functionality and performance of a Web application. It presents some facilities to consider user's dynamism on the workload generation process, but only partially. Box 6.8 summarizes its main features and disadvantages.

3.1.9 WebLOAD

WebLOAD [46] is a software tool for Web performance commercialized by RadView. It is oriented to explore the performance of critical Web applications by quantifying the utilization of the main server resources.

Figure 6.12 depicts the WebLOAD architecture. The *authoring environment* is a software tool to create scenarios that try to mimic the navigations of real users. To this end, it provides facilities to record, edit and debug test scripts, which are used to define the scenarios on workload characterization. The *execution environment* is a console to manage test execution, whose results are analyzed in the *Analytics* application. Since WebLOAD is a distributed system, it is possible to deploy several load generators to reproduce the desired load. Load generators can also be used as probing clients where a single virtual user is simulated to evaluate specific statistics of a single user. These probing clients resemble the experience of a real user using the system while it is under load. In summary, WebLOAD is a commercial software application that presents some capability to generate user's dynamic behavior, but only in a partial way, when evaluating performance of a given Web application.

Box 6.9 shows the main features and disadvantages of WebLOAD.

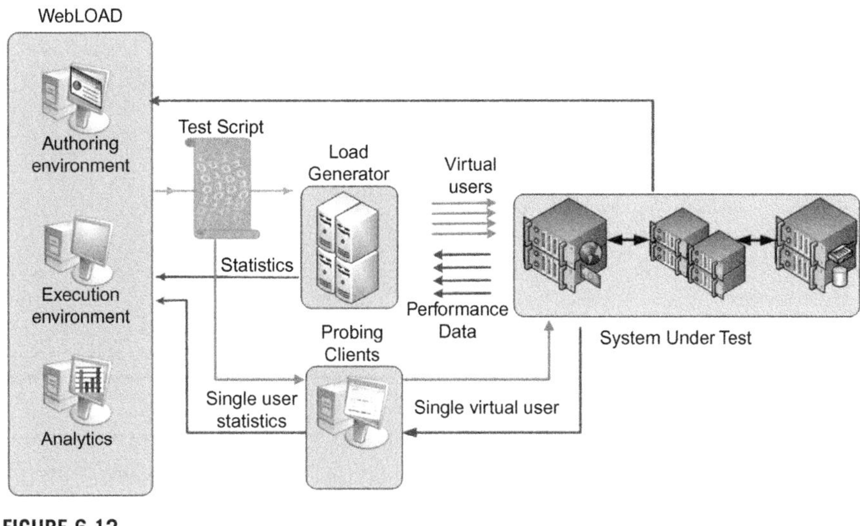

FIGURE 6.12

WebLOAD architecture.

Source: [46].

> **BOX 6.9 MAIN FEATURES AND DISADVANTAGES OF WEBLOAD**
>
> **Main Features**
> - Parameterized workload.
> - Different types of workloads according to the type of scenario.
> - Distributed model for workload generation.
> - Full HTTP protocol (cookies, HTTPS, dynamic content, etc.).
> - Basic facilities to consider user's behavior.
> - Advanced reports during performance evaluation studies.
> - Multiplatform.
>
> **Disadvantages**
> - No users' dynamic navigation characterization.
> - No advanced facilities to consider user's dynamism.

3.1.10 JMeter

JMeter [47] is an open source solution presented by the Apache Software Foundation and designed to generate Web workload with the aim of testing client/server software, such as Web applications and services.

The generator is written entirely in Java and provides an easily configurable and visual API to define, execute and analyze Web performance tests from the client side. It presents partial capability to generate dynamic user's workload by defining a navigation test based on patterns (e.g., regular expressions). Additionally, JMeter

BOX 6.10 MAIN FEATURES AND DISADVANTAGES OF JMETER

Main Features

- Parameterized workload.
- Different types of workloads according to the type of scenario.
- Distributed model for workload generation.
- Full HTTP protocol (cookies, HTTPS, dynamic content, etc.).
- Basic facilities to consider user's behavior.
- Advanced reports during performance evaluation studies.
- Multiplatform.

Disadvantages

- No users' dynamic navigation characterization.
- No advanced facilities to consider user's dynamism.

presents some facilities to check the functionality of a Web application, such as test scripts which use assertions to validate that the application returns the expected results. Box 6.10 summarizes its main features and disadvantages.

3.1.11 *Testing scripts and tools*

With the increasing popularity of Web applications, some software and testing factories or Web developers have created a number of scripts and tools, which are usually basic and open source approaches. These tools capture HTTP requests and reproduce them for the purpose of stressing applications and testing their functionalities.

For instance, HTTPERF [48] and Deluge [49] were developed as tools for measuring Web server performance in Hewlett-Packard and Thrown Clear Productions, respectively. HTTPERF is not focused on implementing one particular benchmark but on providing a robust high-performance tool that facilitates the construction of both micro- and macro-level benchmarks. In contrast, Deluge is a final stressing tool that includes three main components: i) `dlg-proxy`, which records HTTP requests, ii) `dlg-attack`, which generates workload by reproducing recorded users' requests, and iii) `dlg-eval` that elaborates statistics from the generated results.

In addition, HAMMERHEAD 2 [50], PTester [51], Siege [52] and Autobench [53] are examples of scripts and utilities deployed by the open source community to evaluate the quality of its developments.

Box 6.11 summarizes common features and disadvantages of these tools.

3.2 A SURVEY ON REPRODUCING USERS' DYNAMISM

In this section we classify the studied tools according to a wide set of features and capabilities. Below, the 12 features and capabilities used are defined to ease understanding of the comparison study.

1. Distributed architecture. This refers to the ability to distribute the generation process among different nodes. The distribution of the workload generation significantly helps us to improve the workload accuracy.

BOX 6.11 MAIN FEATURES AND DISADVANTAGES OF TESTING SCRIPTS AND TOOLS

Main Features

- Easy to use and introduce in both development and testing processes.
- Simple reports for functional and performance tests.
- Open source solutions.

Disadvantages

- Basic stressing functionalities.
- No users' navigation characterization.
- No facilities to consider user's dynamism.

2. Analytical-based architecture. This feature represents the capability to use analytical and mathematical models to define the workload. These models allow improving the workload quality by using them as workload parameters (e.g., user's behavior models or simulation architectures).
3. Business-based architecture. When defining a testing environment, the simulator architecture should implement the same features as the real environment (e.g., e-commerce architectures typically include a catalog, a product searcher or a payment gateway), so it is quite important to model the business logic deployed by the Web application under test.
4. Client parameterization. This is the ability to parameterize generator nodes (e.g., number of users, allowed navigation set, or changes between navigations). In general, Web dynamism highlights the need for a workload characterization based on parameters, and specially related to user behavior.
5. Workload types. Some generators organize the workload in categories or types, each one modeling a given user profile (e.g., searcher or buyer user profiles).
6. Testing the Web application functionality (functional testing). This capability permits definition of functional tests related to a real Web application. These tests allow a guarantee of the application correctness; that is, the application provides the defined functionality, which fulfills the quality and assurance requirements.
7. Multiplatform refers to a software package that is implemented in multiple types of computer platforms, interoperating among them.
8. Differences between LAN and WAN. Simulations usually run in local area network (LAN) environments. Most of the current simulators cannot model differences between LAN and WAN, where applications are usually located.
9. Ease of use. The generator should be a friendly application carrying out usability guidelines, mainly in commercial products.
10. Performance reports. The elaborated results by the generation process are usually presented by using both on-line and off-line graphical plots.
11. Open source. This feature allows the open source community to develop extensions or different generation alternatives over the generator architecture.

12. Users' dynamism. This is the main feature we are interested in, because the dynamism in contents and users is the most relevant characteristic in the current Web that makes workload generation difficult.

Table 6.1 summarizes the studied software packages used to generate Web workload as well as the grade (full or partial) in which they fulfill the features described above. These software packages can be classified into three groups according to their main application contexts:

- Group I: Benchmarks that model the client and server paradigm in Web context. In this case, among the five studied benchmarks, only TPC-W provides a workload generation process that considers user dynamism, but only partially. The others do not model user dynamism because: i) they are simulation approaches that do not reproduce real workload (WebStone and Web Polygraph), or ii) they are based on analytical models that do not consider user dynamism as a parameter (SPECWeb and SURGE).
- Group II: Software products to evaluate performance and functionality of a given Web application, such as LoadRunner, WebLOAD and JMeter. All of them provide abilities to generate Web workload taking into account user dynamism in a partial way.
- Group III: Testing tools and other approaches for traffic generation that cannot reproduce user dynamic behavior due to the fact that they are based on HTTP traces.

As observed, only four of the studied approaches (i.e., TPC-W, LoadRunner, WebLOAD and JMeter) present some capability to reproduce users' dynamism. Table 6.2 deals with the ability of considering users' dynamism in depth, and explores how each approach takes into account the three challenges. Notice that the four generators provide some capability to partially reproduce the dynamism of users when they surf a website (Challenge I) but in a different way. For instance, TPC-W only considers a probabilistic approach to define users' navigations by using the CBMG model. On the other hand, LoadRunner, WebLOAD and JMeter provide scripting languages that permit definition of users' navigations considering conditional transitions between their pages. Among these three generators, only the commercial products (LoadRunner and WebLOAD) offer software artifacts to represent the different behaviors of users (Challenge II), but they do not mind continuous changes in these behaviors.

4 DWEB: MODELING USER DYNAMISM ON WEB WORKLOAD CHARACTERIZATION

This section presents the Dynamic WEB workload model (DWEB) [54] with the aim of characterizing a more realistic workload when evaluating the performance of current Web applications.

Table 6.1 Web Workload Generators and Grade in Which Main Features Are Fulfilled

	Group I					Group II			Group III							
Generator	**WebStone**	**SPECWeb**	**SURGE**	**Web Polygraph**	**TPC-W**	**LoadRunner**	**WebLOAD**	**JMeter**	**S-Clients**	**WebJamma**	**Deluge**	**HAMMERHEAD 2**	**PTester**	**Siege**	**HTTPERF**	**Autobench**
Feature/Capability																
Analytical-Based Architecture	◆	◆	◆	◆	◆	❖	❖	❖								
Distributed Architecture	◆	◆	◆	◆		◆	◆									
Business-Based Architecture		❖		❖	◆	◆	◆	◆								
Client Parameterization	◆	◆		❖	◆	◆	◆	◆		◆		❖			❖	
Workload Types		◆			◆	◆	◆	❖								
Functional Testing						◆	◆	❖				❖	❖	❖		❖
LAN and WAN									◆							
Multi-platform	◆	◆	◆	◆	◆	◆	◆	◆	◆	◆	◆	❖	◆	❖	◆	◆
Ease of Use						◆	◆	❖								
Performance Reports	❖	◆		◆	◆	◆	◆	◆			❖	❖	❖	❖	❖	❖
Open Source	◆		◆	❖	◆			❖	◆	◆	◆	◆	◆	◆	◆	◆
User's Dynamism					❖	❖	❖	❖								

◆ *Full support* ❖ *Partial support.*

Table 6.2 Web Workload Generators and How Challenges of User's Dynamism Are Fulfilled

	TPC-W	LoadRunner	WebLOAD	JMeter
Challenge I	*Analytical Approach*	*Scripting*	*Scripting*	*Scripting*
Challenge II	*Not covered*	*Software Artifact*	*Software Artifact*	*Not covered*
Challenge III	*Not covered*	*Not covered*	*Not covered*	*Not covered*

DWEB tackles in a progressive way the three previously mentioned challenges when modeling the user's behavior on representative workloads. To this end, it defines a couple of new concepts: user navigations and user roles. These concepts characterize different levels of dynamism in the workload definition by means of modeling Web users' behaviors.

4.1 USER NAVIGATION

The concept of user navigation defines a first level of user dynamism and satisfies Challenge I by modeling users' dynamic behavior when interacting with the contents and services offered by the Web.

For instance, a typical navigation of a user searching for specific information usually begins with a query on a Web finder. Queries are frequently cancelled when the response time surpasses a certain value, which is characteristic for each user and his current navigation conditions. In the case of obtaining results, users usually visit the first site on the list or refine the search when receiving too much information. Analyzing this simple example, one can see that each user request depends not only on the response itself but also on other issues related to the quality of service (e.g., response time length or content amount), and the users' states. That is, users make their navigation decisions according to their personal preferences, navigation goals, visited resources, network and connectivity conditions, etc.

The navigation concept is not just limited to reproducing human behavior, since it can be further applied to any Web client, such as software automatons, that are easier to model than users because they follow a given navigation pattern. Nevertheless, the strong point of the concept lies in the flexibility to represent dynamism in user's behavior when interacting with the Web.

Formally, a user's navigation N is defined as a sequence of n URLs of HTTP requests where each visited URL depends on the previously visited ones, as defined in Equation (6.1).

$$N = \{url_1, url_2, \ldots url_n\} / \forall i = 2 \ldots n : url_i \text{ depends on } url_k \text{ for } k < i \text{ and user's-state}_{i-1} \quad (6.1)$$

where url_i refers both to the content related to the resource i and its associated characteristics, and user's-state$_{i-1}$ denotes the user's state resulting from the interaction with the previously visited resource.

A graph where nodes were pages of websites and arcs were transitions between pages is not enough as a visual representation of a user's navigation, because we need a visual modeling language that allows us to define a user's state. Moreover, this language has to provide mechanisms to easily model the user's dynamism when navigating.

There are several successful extensions of Unified Modeling Language (UML) applied to Web engineering. For instance, User eXperience diagrams [55] are introduced to model the storyboards and the dynamic information of pages in building model-driven Web applications. Moreover, these diagrams were extended to rapidly develop and deploy public administration portals considering usability factors [56].

Due to these reasons, we decided to represent a navigation using a state machine view of UML. In general, a state machine is a graph of states and transitions that describes the response of an object to the events that it receives [57]. We simplify this model by considering only a reduced set of graphical elements where states are Web pages, as shown in Table 6.3. The result has been adopted for visual representation of the navigation concept.

For illustrative purposes, Figure 6.13 shows the visual navigation corresponding to a Google search where some dynamic issues in the user's behavior (e.g., dynamic think time or conditional and parallel requests) are introduced. Two main parts can be distinguished in this navigation:

1. The upper part of the diagram (before reaching branch b1) shows the two ways in which the search can be initiated:
   a. On the left side, the user makes use of a search toolbar of a Web browser (e.g., Google toolbar for Mozilla Firefox) to make the query directly.
   b. On the right side of the figure, the user reaches branch b1 after the Google. HOME node, where the user requests the main page to the Websearcher engine (www.google.com). After that, he waits for a while (time referred to as the *user's think time*) and then the user makes the query.
2. In the bottom of the diagram, the user analyzes the results for a dynamic think time, which depends on the number of results, and makes a decision (conditional request):
   a. If the Web search engine provides results (path from *b1* to *b2*) the user analyzes them. After that, he can refine the results by making a new query, *refined query* in the figure (path from *b2* to *b1*), or access the top 10 sites provided by using multiple browsers tabs (one for each result). Finally, he finishes the navigation (path from *b2* to black dot through *X TH RESULTS.HOME* node).
   b. When no results are provided (path from *b1* to *b3*), the user can make a new query, other query in the figure (path from *b3* to *b1*), or finish the process (path from *b3* to black dot).

Table 6.3 User's Navigation Notation

Notation	Name	Description
●	Beginning	The initial state in a state machine represents the beginning of a navigation.
◉	End	The final state in a state machine defines the end of a navigation; that is, the user leaves the Website.
Page internal action[guard condition]	Page	Pages are states where a user can execute actions (e.g., user's think time) when their guard conditions are true. These conditions can also consider probabilities in the same way as CBMG.
action [guard condition] →	HTTP request	Transitions are HTTP requests to pages that are executed when their guard conditions are true. These conditions can also consider probabilities in the same way as CBMG.
[Guard condition] true false	If/else	A simple condition is introduced to ease understanding of branches in user's way.
action [guard condition]	Parallel HTTP request	A parallel request starts parallelism on navigating to execute *n* HTTP requests using *n* different threads
	End of parallelism	It transforms the parallel navigation in a sequential navigation again by killing the threading.
	Extension	It introduces a new extension point in the navigation (e.g., call to external constraints or functions), which is used to highlight issues of dynamism.
H	Call other navigation	It calls another navigation that is defined outside the model.
	States set/get	It allows users to store (set) and recover (get) some information at their states (e.g., cookies or visited contents).

4.2 USER ROLES

The DWEB model proposes the concept of user's roles to fulfill Challenges II and III by introducing a second level of dynamism in user behavior. This level is related to the roles that users play when navigating the Web. Continuous changes of these roles define users' behaviors.

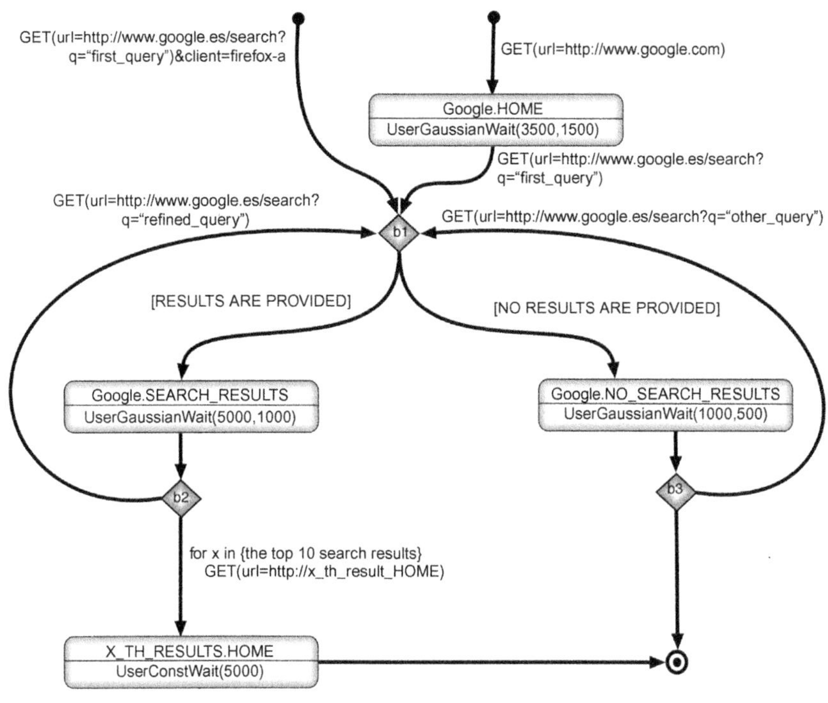

FIGURE 6.13

Google Search navigation pattern.

For instance, let's consider people at the Import and Export Department in a typical multinational company who usually access the Web during their working time. Assume also that the company has an intranet (Web ERP) that allows Web access. Most of the time, workers use Internet for professional purposes (e.g., intranet navigations, supplier site navigations, or professional Web searches), but sometimes they use the Web for leisure purposes (e.g., reading the news with Google Reader, performing personal searches, or checking mail). Therefore, we can distinguish between two roles when a department member navigates the Web: working behavior (professional navigations), and leisure behavior (personal navigations). Figure 6.14 defines the working and leisure behaviors of the example, and the likelihood to change between behaviors by using balanced arcs (the arc weight is the probability to change from the source behavior to the destination behavior). These behaviors are defined as automatons, where their nodes represent navigations, and their balanced arcs indicate the transitions between navigations (the arc weight indicates the probability to take that arc).

Formally, we define user's roles $R(C,\phi:CxC \rightarrow N)$, where:

- C is the set of navigations; that is, $C = \{n_1,n_2 \ldots ,n_k\}$ with $n_i \in N$.
- $\phi:CxC \rightarrow N$. ϕ is the function that provides the next navigation to be executed in terms of probabilities.

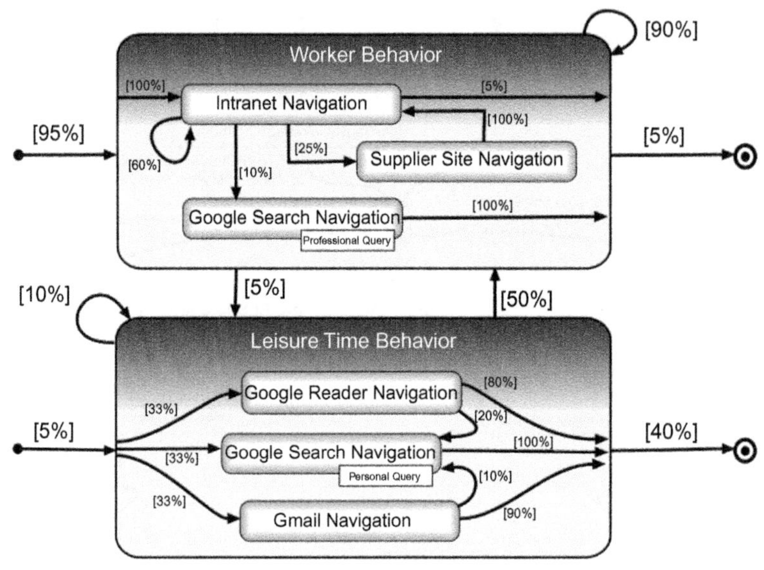

FIGURE 6.14

User role examples: working and leisure behaviors.

5 SUMMARY

This chapter has analyzed state-of-the-art workload models and generators for Web performance evaluation, focusing on the capability of modeling the user's dynamic behavior when navigating the current Web.

With the aim of improving workload models, few approaches (CBMG, VBMG, EFSM and Clickstream Model) provided some capabilities to represent a user's behavior on Web workload characterization, but they do not offer an accurate solution to model users' dynamism.

Furthermore, among the studied software tools, only one benchmark for e-commerce (TPC-W) and three software solutions to evaluate functionality and performance of a given Web application (LoadRunner, WebLOAD and JMeter) provide mechanisms to reproduce user navigations in the current Web. However, these mechanisms do not consider all challenges when reproducing users' dynamic behavior, so they are not sufficient to mimic real patterns of HTTP requests.

These deficiencies in models and software motivate us to propose a more accurate workload model in order to develop a new workload generator with the aim of analyzing the effect of using dynamic workloads on Web performance evaluation, instead of traditional workloads.

To this end, DWEB tackles in a progressive way the main challenges when modeling the user's behavior on representative workloads by using a couple of new concepts: user's navigation and user's roles. The user's navigation defines a first level of user dynamism by modeling a user's dynamic behavior when interacting with the current Web; also, the concept of user's roles permit the modeling of a second level of dynamism related to the roles that users play when navigating the Web, and the continuous changes of these roles. Each role is explained in the terms defined by the navigation concept.

For illustrative purposes, some examples of the use of DWEB concepts have been presented, and the main visual notation has been described in this chapter.

ACKNOWLEDGMENTS

This work has been partially supported by Spanish Ministry of Economy and Competitiveness under grant TIN2013-43913.

REFERENCES

[1] Rodriguez P. Web infrastructure for the 21st Century. In: Proceedings of 18th International conference on World Wide Web. April 20–24, Madrid, Spain. Available from: <http://www2009.eprints.org/213/>; 2009.

[2] Cormode G, Krishnamurthy B. Key differences between Web 1.0 and Web 2.0. First Monday J 2008;13(6) [Online]. Available from: <http://firstmonday.org/ojs/index.php/fm/article/view/2125>.

[3] Radinsky K, Svore K, Dumais S, Teevan J, Bocharov A, Horvitz E. Modeling and predicting behavioral dynamics on the Web. In: Proceedings of international conference on the World Wide Web; 2012. p. 599–608. ACM.

[4] Hendler J. Web 3.0 emerging. Computer 2009;42(1):111–13.

[5] Tselentis G, Domingue J, Galis A, Gavras A, Hausheer D, Krco A, et al. Towards the future internet: a european research perspective. IOS Press; 2009 [Ebook].

[6] Barford P, Crovella M. Generating representative Web work-loads for network and server performance evaluation. In: SIGMETRICS '98/PERFORMANCE '98, Joint international conference on measurement and modeling of computer systems; 1998. p. 151–60.

[7] Weinreich H, Obendorf H, Herder E, Mayer M. Off the beaten tracks: exploring three aspects of Web navigation. In: Proceedings of international conference on the World Wide Web; 2006. p. 133–42.

[8] Pirolli P, Pitkow JE. Distributions of surfers' paths through the World Wide Web: empirical characterizations. In: Proceedings of international conference on the World Wide Web, vol. 2(1/2); 1999. p. 29–45.

[9] Goel S, Broder A, Gabrilovich E, Pang B. Anatomy of the long tail: ordinary people with extraordinary tastes. In: ACM international conference on web search and data Mining; 2010. p. 201–10. Yahoo Research.

[10] Cecchet E, Marguerite J, Zwaenepoel W. Performance and scalability of EJB applications. In: Conference on object-oriented programming, systems, Languages, and Applications; 2002. p. 246–61. ACM.

[11] Schneider F, Agarwal S, Alpcan T, Feldmann A. The new Web: characterizing AJAX traffic. In: International conference on passive and active network measurement; 2008. p. 31–40.

[12] Ohara M, Nagpurkar P, Ueda Y, Ishizaki K. The Data-centricity of Web 2.0 workloads and its impact on server performance. In IEEE International symposium on workload characterization; 2009. p. 133–42.

[13] Kosir D, Kononenko I, Bosnić Z. Web user profiles with time-decay and prototyping. New York: Springer Science + Business Media; 2014.

[14] Floyd S, Paxson V. Difficulties in simulating the internet. IEEE/ACM Trans Netw 2001;9(4):392–403.

[15] Song Shi W, Collins E, Karamcheti V. Modeling object characteristics of dynamic Web content. J Parallel Distrib Comput 2003;63(10):963–80.

[16] Menascé D, Almeida V. Scaling for E-Business: technologies, models, performance, and capacity planning. Upper Saddle River, NJ: Prentice Hall; 2000.

[17] Duarte F, Mattos B, Almeida J, Almeida V, Curiel M, Bestavros A. Hierarchical characterization and generation of blogosphere workloads. Technical report. Computer Science Department, Boston University; 2008.

[18] Shams M, Krishnamurthy D, Far B. A model-based approach for testing the performance of Web applications. In: International workshop on software quality assurance; 2006. p. 54–61. ACM.

[19] Lee D, Yannakakis M. Principles and methods of testing finite state machines—a survey. Proc. IEEE 1996;84(8):1090–123.

[20] Benevenuto F, Rodrigues de Magalhanes T, Cha M, Almeida V. Characterizing user behavior in online social networks. In: Internet measurement conference; 2009. p. 49–62.

[21] Lins T, Pereira A, Benevenuto F. Workload characterization of a location-based social network. Social network analysis and mining. Berlin: Springer Verlag; 2014.

[22] Abramson M. Learning temporal user profiles of web browsing behavior. In: ASE BIGDATA/SOCIALCOM/CYBERSECURITY Conference, Stanford University; May 27–31, 2014.

[23] Huang J, White RW. Parallel browsing behavior on the Web. In: Conference on hypertext and hypermedia; 2010. p. 13–18. ACM.

[24] Aula A, Jhaveri N, Käki M. Information search and re-access strategies of experienced web users. In: Proceedings of international conference on World Wide Web; 2005. p. 583–92.

[25] Thatcher A. Web search strategies: the influence of Web experience and task type. Inform Processing Manage 2008;44(3).

[26] Obendorf H, Weinreich H, Herder E, Mayer M. Web page revisitation revisited: implications of a long-term click-stream study of browser usage. In: Conference on human factors in computing systems; 2007. p. 597–06.

[27] Arroyo E, Selker T, Wei W. Usability tool for analysis of Web designs using mouse tracks. In: Conference on human factors in computing systems; 2006. p. 484–89. ACM.

[28] Di Lucca G, Di Penta M. Considering browser interaction in Web application testing. In: International workshop on web site evolution; 2003. p. 74–81. IEEE.

[29] Seo YW, Zhang BT. Learning user's preferences by analyzing Web-browsing behaviors. In International conference on autonomous agents; 2000. p. 381–87. ACM.

[30] Mindcraft. WebStone: The benchmark for web servers. [online]. Available from: <http://www.mindcraft.com/Webstone/>; 2002.

[31] Standard Performance Evaluation Corporation (SPEC). Benchmarks: web servers. Available online from: <https://www.spec.org/benchmarks.html#web>; 2009.

[32] Barford P. The SURGE Web workloaad generator. Available from: <http://pages.cs.wisc.edu/~pb/software_data.html>; 1998.

[33] Barford P, Bestavros A, Bradley A, Crovella M. Changes in Web client access patterns: characteristics and caching implications. World Wide Web Internet and Web Information Systems 1999;2(1):15–28.

[34] Barford P, Crovella M. 1997. An architecture for a WWW workload generator. In: World Wide Web Consortium Workshop on Workload Characterization.

[35] Banga G, Drusche P. Measuring the capacity of a Web server under realistic loads. World Wide Web Internet and Web Information Systems 1999;2(1/2):69–83.

[36] Chen H, Abrams M, Johnson T, Mathur A, Anwar I, Stevenson E. Wormhole caching with HTTP PUSH method for a satellite-based Web content multicast and replication system. In: Web Caching Workshop; 1999.

[37] TPC: Transaction Processing Performance Council. TPC Benchmark™ W Specification. Version 1.8. Technical report; 2002.

[38] Dodge R, Menascé D, Barbara D. Testing e-commerce site scalability with TPC-W. In: Computer Measurement Group Conference; 2001. p. 457–66.

[39] Amza C, Chanda A, Cox AL, et al. Specification and implementation of dynamic web site benchmarks. In: proceedings of the IEEE international workshop on workload characterization; 2002. p. 3–13.

[40] García D, García J. TPC-W e-commerce benchmark evaluation. Computer 2003;36 (2):42–8.

[41] Hashemian R, et al. Characterizing the scalability of a Web application on a multicore server. Concurr Comput Pract Exp 2014;2017–50 Wiley Online Library.

[42] Rousskov A, Wessels D, Chisholm G, The First IRCache Web Cache Bake-off—the Official Report. In: Web Caching Workshop; April 1999.

[43] MF: The Measurement Factory. Web Polygraph [online]. Available from: <http://www.Web-polygraph.org>; 2012.

[44] Rousskov A, Wessels D. High performance benchmarking with Web Polygraph. Softw Pract Exp 2003;1(1–10).

[45] Hewlett-Packard. HP LoadRunner. Available from: <http://www8.hp.com/us/en/software-solutions/software.html? compURI = 1175451>; 2012.

[46] RadView Software. WebLOAD [online]. Available from: http://www.radview.com/product/Product.aspx; 2012.

[47] ASF Apache Software Foundation. Apache JMeterTM [online]. Available from: <http://jmeter.apache.org/>; 2012.

[48] Mosberger D, Jin T. HTTPERF. Available from: <http://www.hpl.hp.com/research/linux/httperf/>; 2008.

[49] Blakeley M. Deluge: website stress test tool. [online]. Available from: <http://deluge.sourceforge.net/>; 2010.

[50] Wong G, Dwyer M, Gifford J. Hammerhead 2.0. Available from: <http://sourceforge.net/projects/ hammerhead/>; 2011.

[51] Eriksson P. PTester [online]. Available from: <http://www.lysator.liu.se/~pen/ptester/>; 1999.

[52] Fulmer J. SIEGE. [online]. Available from: <http://www.joedog.org/siege-home/>; 2012.

[53] Midgley J. AUTOBENCH. Available from: <http://www.xenoclast.org/autobench>; 2004.

[54] Peña-Ortiz R, Sahuquillo J, Pont A, Gil JA. DWEB model: representing Web 2.0 dynamism. Comput Commun J 2009;32(6):1118–28. ISSN 0140-3664.
[55] Conallen J. Building web applications with UML. Addison-Wesley Professional; 2003.
[56] Foglia P, Prete CA, Zanda M. Modelling public administration portals. Encyclopedia of portal technologies and applications; 2007. p. 606–14.
[57] Rumbaugh J, Jacobson I, Booch G. The unified modeling language reference manual. 2nd ed. Addison-Wesley Professional; 2004.

CHAPTER 7

Computer networks performance modeling and simulation

Francisco J. Suárez, Pelayo Nuño, Juan C. Granda, and Daniel F. García
University of Oviedo, Asturias, Spain

1 INTRODUCTION

Computer networks are an inherent substrate in many daily tasks of business, e-commerce, e-government, education or leisure. The complexity and the deployment of computer networks continue to grow, since new protocols, architectures and applications constantly emerge, and computer networks are in a continuous process of adaptation. Furthermore, new network systems and applications with the purpose of serving massive real-time user demands appear periodically. Coexistence with the former architectures and systems may hinder the evaluation of these new developments. Thus, studying and testing the performance of a network may pose a difficult task to achieve, due to several reasons. First, studying the network may be excessively costly. Second, studying some features may penalize, or even interrupt, the performance of the network and the quality of service perceived by users would be affected, which is unacceptable. Third, interacting directly with the network may be physically impossible either because of its size or its state (the network could be in a development phase, still nonoperative). Therefore, modeling and simulation (M&S) may be the only alternative for examining the network behavior and performance under different scenarios.

Modeling a system involves the abstraction of its features and properties, focusing exclusively on those that are of interest to the study [1]. As a result, a model can be understood as the logical representation of a system with different levels of complexity (normally less complex than the real system). Simulation is the imitation of a real-world system through a computational representation of its behavior according to the rules described previously in a model. When a system is simulated, it is mandatory to consider a limited number of characteristics, properties or behaviors of interest, so as to make the model tractable; otherwise, it will be infinitely more complex and detailed [2].

1.1 COMPUTER-BASED MODELS

Computer-based models are usually classified as follows:

- **Deterministic vs. Stochastic:** A deterministic model predicts a specific output from a given set of inputs with neither randomness nor probabilistic components. A given input will always produce the same output given the same initial conditions. In contrast, a stochastic model has some inputs with randomness, so the model predicts a set of possible outputs weighted by their likelihoods or probabilities [3].
- **Steady-state vs. Dynamic:** A steady-state model tries to establish the outputs according to the given set of inputs when the system has reached steady-state equilibrium. In contrast, a dynamic model provides the system reactions facing variable inputs. Steady-state approaches are often used to provide a simplified preliminary model [4].
- **Discrete vs. Continuous values:** A discrete model is represented by a finite co-domain; hence, the state variables take their values from a countable set of values. In contrast, a continuous model corresponds to an infinite co-domain. Therefore, the state variables can take any value within the range of two values [5]. However, there are some systems that need to be modeled showing aspects of both approaches which bring about combined discrete-continuous models [6].
- **Discrete vs. Continuous time:** In a discrete model the state changes can only occur at a specific instant in time. These instants correspond to significant events that impact the output or internal state of the system. In contrast, in a continuous model the state variables change in a continuous way and not abruptly from one state to another [7]. Therefore, continuous models encompass an infinite number of states. Discrete models are the most commonly used for network M&S.

Simulations can be carried out following two approaches: local and distributed. Distributed simulation is such that multiple systems are interconnected to work together, interacting with each other, to conduct the simulation. In contrast, a local simulation is carried out on a single computer. Historically, the latter approach has been the most widely used to simulate computer networks, but the increasing complexity of simulations is fostering the importance of the former approach [4].

Figure 7.1 summarizes the M&S process. Behavioral information extracted from a real system is used jointly with system specifications and requirements. Based on these inputs, a system model is built and subsequently validated through simulation. Usually, several performance metrics are determined during simulations, which can be compared with results extracted from experimentations with the real system. If both are similar, the model is considered to be valid, while if they are not, the system model must be corrected. Similarly, performance metrics are used to determine if the system model fulfills the requirements, so the designed system can be refined and extended in a controlled way.

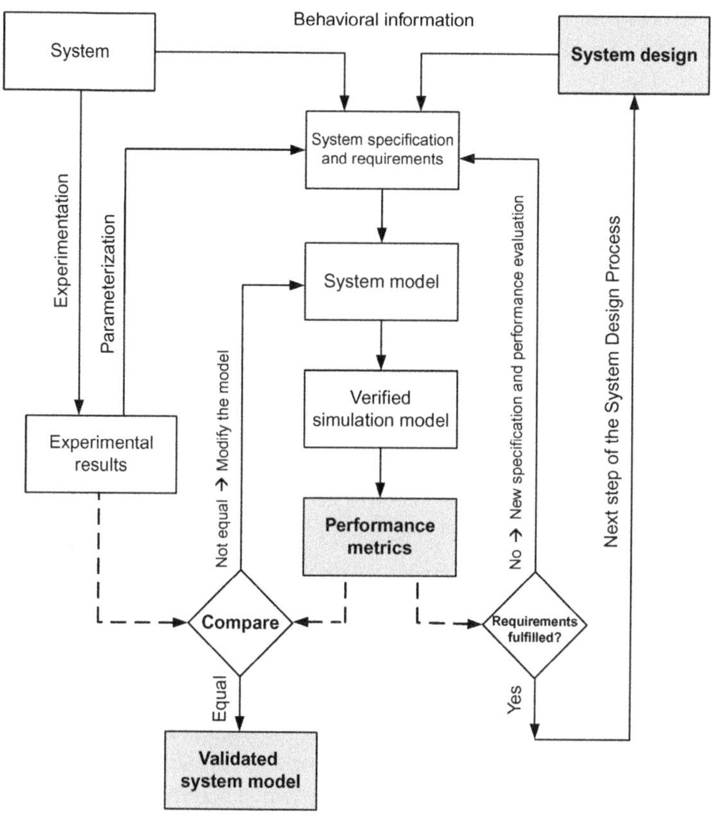

FIGURE 7.1

System modeling and simulation scheme.

1.2 ADVANTAGES, DISADVANTAGES AND COMMON PITFALLS OF M&S

The use of M&S has some advantages and disadvantages when compared with empirical or real testing [8]. First is the ability to make more reliable decisions based on the simulation of a wide range of operational scenarios, allowing for testing every aspect of a system, a proposed change or a specific circumstance. Another advantage is flexibility, since the time and pace of the simulation can be adapted to allow the user to speed up or slow down transitions or events, facilitating precise analyses. M&S fosters and improves learning. Users can evaluate the occurrence or behavior of a specific scenario by reconstructing and examining it closely, so they develop understanding by observing the operation of the system. M&S is also very useful with regard to design time, allowing for specifying the requirements of a system in detail, or even diagnosing problems derived from the interaction between system components. The reduced cost is another important aspect of M&S. Simulations can be carried out less expensively than with empirical testing. Finally, M&S avoids disruptions in the real system, enabling the assessment of new processes, operating procedures or methods.

However, there are also disadvantages to using M&S. First, and most notorious, M&S is only a representation of the system, rather than the real system itself, so this may lead to inappropriate use of M&S when an analytical solution can be obtained. Second, when M&S implies high randomness, assumptions, simplifications or complex interrelationships between system members, the interpretation of the results may be difficult or lead to misleading conclusions. Third, there may be issues with the statistical significance of the results, even though they are correct, depending on the number of replicas of the simulation. Finally, special training is needed for building models and conducting simulations properly, so timing and monetary costs may be high.

Related work on M&S suggests several common mistakes that can originate inaccuracy and unreliability. Most of the pitfalls that lead to misleading models or simulation results are discussed in [2], [4] and [6]:

1. **Using M&S when not necessary:** Although there are many scenarios where M&S is very useful, there are also scenarios where it is unnecessary. It is mandatory to evaluate and decide if M&S is the right choice for the case study.
2. **Modeling a misunderstood system:** It is almost impossible to obtain significant results if the knowledge about the system is incomplete. Aspects such as the functionality, the environment, the composition and the properties of the system must be well known. Therefore, it is mandatory to have a full understanding of the system to be modeled.
3. **Loose understanding of the model:** A model, although correct, is not trustworthy if the developer or the researcher is not able to understand either the behavior or the results of the simulation. Therefore, understanding what has been modeled is as important as understanding the model.
4. **Unverified models:** Simulation software developments can contain several programming and logical errors. Verification processes are mandatory to prove correctness, and hence to avoid incorrect results and misleading conclusions.
5. **Unrealistic models:** A model may be correctly implemented, but if it does not represent properly the behavior of the real system, it is of no use. Therefore, model validation is also extremely important.
6. **Inappropriate level of detail:** It is mandatory to model what is needed and no more. A high level of detail implies much more time needed to develop and run the simulations. Furthermore, and even more important, too many details introduce a large number of interdependent variables influencing the system, so it becomes difficult to determine its performance reliably.

7. **Excessive simplifications:** Sometimes it is necessary to simplify complex aspects of the system to be modeled. However, excessive assumptions or simplifications can lead to misleading results or statistically nonsignificant results, masking key variables affecting the system performance. Therefore, simplifications must be assumed with caution.
8. **Accuracy of the results:** A model encompasses a high number of variables affecting the results. These variables may be a source of error inducing bias and variance which leads to inaccurate models. A tradeoff to minimize both bias and variance must be established to improve the degree of confidence of the results.
9. **Inadequate simulation runtimes:** Short runtimes lead to no significant results, preventing the model from reflecting the long-term system performance properly. In contrast, excessively long runtimes may lead to wasting computational resources, which may be unacceptable in distributed simulations. Therefore, the duration of the simulations must be established carefully.
10. **Weak random number generation:** Random number generators and seeds have considerable impact on simulation results. Thus, it is mandatory to use trustworthy and widely known random number generators and independent seeds.
11. **Unachievable targets:** A simulation model may be useless if it defines a set of goals that are neither realistic nor achievable.
12. **Results not independently repeatable:** A simulation can be considered reliable when its results are repeatable.
13. **Lack of essential skills:** Some skills are crucial for the proper development of a simulation project. People with a background in mathematics, statistics and modeling, software developers highly experienced in programming and debugging, and even some members with extensive know-how on team leading are desirable project members.
14. **Incorrect programming language selection:** The programming language selected to implement the model has an impact on the software development process. Therefore, it is recommended to seek a compromise between the learning curve, the time required for the implementation, and the efficiency of the programming language at runtime.
15. **Inadequate project scheduling:** A simulation project requires an adequate estimation of the time needed for the development, implementation and testing of the model.
16. **Unmanaged simulation projects:** Simulation projects are complex, making it necessary to employ tools, such as code repositories and project management tools, to control the progress of the software implementation.

1.3 COMPUTER NETWORK SIMULATION

Nowadays, computer network simulation is a fundamental element in network researching, development and teaching. Network simulators are widely used not

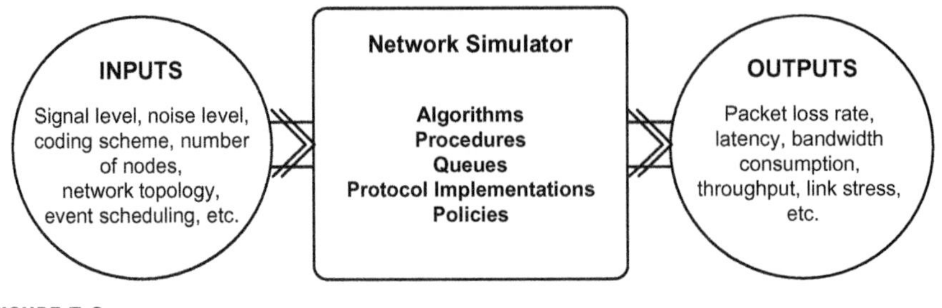

FIGURE 7.2

Network simulator abstraction.

only by practitioners and researchers, but also by students to improve their understanding of computer networks. Network simulation can be extremely useful when applied to scenarios such as protocol analysis, complex network deployment, evaluation of new services, prototypes or architectures, and so on. Broadly speaking, a network simulator can be understood as a black box, as illustrated in Figure 7.2.

A network simulator implements a computer network model through algorithms, procedures and structures according to a given programming language. The implemented network model receives a set of parameters as inputs, which impose the rules and constraints of the simulation. In other words, these parameters set the course of the simulation. When the simulation process ends, the network simulator returns a set of results as outputs that can be used with a dual purpose: validating and verifying the correctness of the implemented model for a particular domain of system states, and analyzing the behavior of the network modeled when the former has been successful.

Computer networks are most often simulated with discrete-event simulation. The main reason behind this widespread use is that discrete-event simulation adapts better to represent the behavior of computer networks, since computer network protocols can be modeled as finite state machines. In a computer network there is a steady state between two consecutive events, and discrete-event simulation allows for jumping from one steady state to another, leading to faster simulations. Other interesting aspects of discrete-event simulation are flexibility and lower computational overhead.

2 PERFORMANCE MODELING

As pointed out previously, modeling is the first step in the evaluation of the system performance, where the system is abstracted through a properly detailed model that reproduces the system behavior under the environment influence (workload). In the most specific performance evaluation studies, the main objective is to obtain a set of performance metrics such as throughput and response

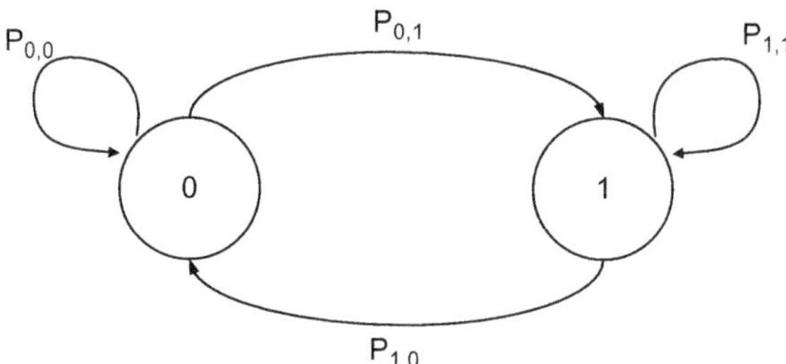

FIGURE 7.3

Example of a Markov Chain model.

times in the steady-state of the system. First, modeling techniques for performance evaluation are introduced in this section. Then, the key aspects to consider when modeling protocol layers and workloads are outlined. Finally, some possibilities to model the network topology are presented.

2.1 MODELING TECHNIQUES

The most common techniques for performance modeling of discrete-event dynamic systems in general and computer networks in particular are Markov Chains, Queuing Networks and Petri Nets [5]. Markov Chains provide a general framework to model discrete-event dynamic systems, while Queuing Networks and Petri Nets allow for building high-level models that can also be translated to Markov Chains. These high-level models are closer to the structure of the real systems and so it is easy to match system and model components, whereas Markov Chains put the emphasis on describing the behavior of the system at the state space level. All these techniques use stochastic processes for representing times as the way to properly analyze the performance of the modeled systems.

Markov Chains are stochastic models described by graphs, where nodes correspond to system states and transition between states are represented by links. The probability associated to each transition is also annotated in the graphs. Figure 7.3 shows an example of a Markov Chain model with only two states and the corresponding probabilities of changing the state or staying in the same. An important property of Markov Chains is that the future states only depend on the present state, but not on the previous states—that is, they are memoryless. Thanks to this property, Markov Chains have the great advantage of low analysis complexity. *Continuous Time Markov Chains* (CTMC) are the Markov Chains commonly used to model discrete-event dynamic systems.

When the systems are complex, the number of possible states is high, and modeling them directly with CTMC becomes difficult. In these cases, more abstract or high-level models like Queuing Networks and Petri Nets are better alternatives.

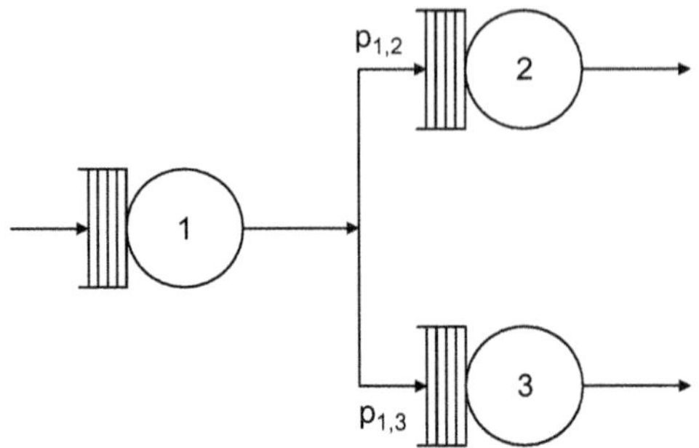

FIGURE 7.4

Example of a Queuing Network model.

Queuing Networks (QN) are models where customers (service requests) arrive at service stations (servers) to be served. When customers arrive at a busy service station, they are queued for a waiting time until the service station is free. Both the arrival and service times are described as stochastic processes. Important parameters of Queuing Networks are the number of customers and servers, the size of the waiting queues and the queuing discipline (priorities, preemption, FIFO, etc.). Figure 7.4 shows a model composed of three queues with the probabilities of the customers arriving at an alternative second queue after being served in the first. Queuing Networks are very suitable to detect performance bottlenecks due to shared resources on distributed systems. An important advantage of simple Queuing Network models is the low complexity of the steady-state solutions (polynomial in number of queues and customers), because they can be obtained as a product of the steady-state solution for each of the individual queues in the network [9].

Queuing Networks do not properly model the common synchronization mechanisms of distributed systems. Thus, Petri Nets support more powerful models.

Petri Nets (PN) are models of concurrent systems also described by graphs. There are two types of nodes in the graph, *places* (circles) and *transitions* (bars). Places represent system states and transitions represent system evolution. Finally, links can only connect places with transitions and vice versa. The graph only gives a static view of the system. To study the dynamics of the system, the Petri Net must be executed. This is possible thanks to the special marks (tokens) that are assigned to the places and commonly associated to resources. There is no limit for the number of tokens per place.

Petri Nets evolve by executing (firing) the enabled transitions, that is, the transitions where all the input places have at least one token. The firing is an atomic operation that implies taking one token from each input place and placing one token on every output place. Finally, if more than one transition is enabled, they are fired in a nondeterministic fashion. Figure 7.5 shows the initial states of a simple Petri Net with a transition enabled and the final states after the transition firing.

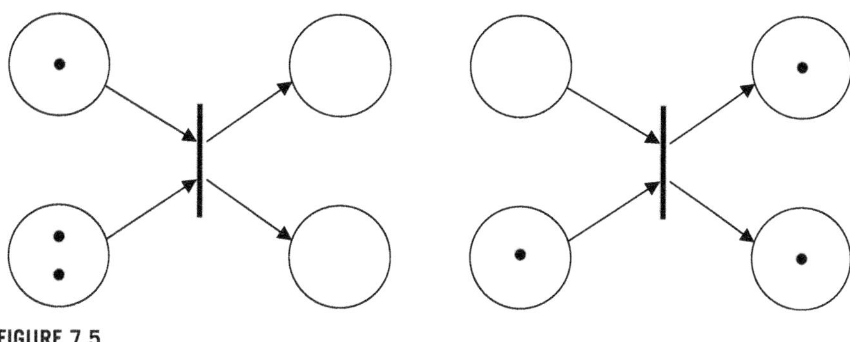

FIGURE 7.5

Example of a Petri Net model before and after transition firing.

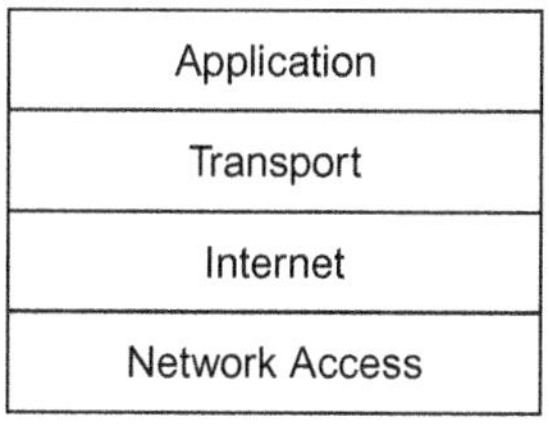

Application
Transport
Internet
Network Access

FIGURE 7.6

Layers of the TCP/IP protocol stack.

The use of Petri Nets for performance assessment requires the addition of temporal specifications to the basic untimed Petri nets, resulting in the so-called *Timed Petri Nets* (TPN). Time is commonly associated to the transitions as delays using stochastic processes, defining two more specific kinds of PN: *Stochastic Petri Nets* (SPN), where all the transitions have non-null delays, and *Generalized Stochastic Petri Nets* (GSPN), where non-null delay and null delay transitions are considered [9].

Performance metrics (throughput, response times, etc.) can be obtained from CTMCs, QNs and SPNs either by computing an analytic solution (low complexity models), approximating solutions by applying numerical algorithms or using performance analysis tools (medium complexity models). Discrete-event simulation can also be applied (high complexity models), executing the model and collecting enough measures to derive the performance metrics [10].

2.2 PROTOCOL MODELING

In this subsection, some important modeling aspects for the layers of the TCP/IP protocol stack (Figure 7.6) are introduced. The focus is only on the lower layers, that is, *Network Access*, *Internet* and *Transport* layers. The modeling of the *Physical* layer is currently a challenge only for wireless networks. The higher *Application* layer is addressed in the next subsection.

2.2.1 Network access layer

Models for this layer have often been less valued than higher layer models. As a result, the Ethernet links are simply modeled as single point-to-point links with fixed propagation delay. Analytical solutions for models of these protocols become intractable when they include all the behavior details of CSMA/CD or CSMA/CA mechanisms, and simulation is the only way to characterize these protocols in many cases. The main components of a detailed CSMA/CD model are propagation delay on the LAN, exponential back-off, carrier sensing and the jamming signal mechanism.

Several studies conclude that differences in the results from detailed versus more abstracted models become important with average network loads above 40%, reaching differences of up to 50% for average loads above 70% [4]. In conclusion, the level of detail necessary for the Network Access layer depends on the traffic load scenario of the performance study.

2.2.2 Internet layer

The network can be modeled at this level as a graph with nodes connected by weighted links (with properties such as *bandwidth*, *delay*, etc.). Taking advantage of this abstraction, routing problems can be translated to find paths in the graph that satisfy constraints such as the shortest path between two nodes. Algorithms from *Dijkstra* and others can then be applied to the simulation model in order to evaluate the quality of the routes discovered by the routing protocols [11].

The models related to path selection can be extended by considering traffic patterns (workload) and resources available in the nodes, in particular memory resources. These extensions can be modeled with queuing networks, minimizing the average packet delay under network cost constraints or, alternatively, minimizing the cost under a given maximum tolerable delay. Models for the Internet layer are closely related with topology models of the computer network that will be discussed later.

2.2.3 Transport layer

There are two approaches in transport layer modeling, *direct models* and *performance models* [12].

Direct models are oriented to simulation and aim to reproduce the protocol interactions. In fact, even the full protocols are sometimes implemented under simulation. These models are useful for studying the protocol behavior itself or supporting application layer protocols. Typical parameters at this layer are *throughput* between two hosts, *end-to-end delay* (mainly constituted by the aggregation of individual channel delays of the network links modeled in lower layers) and end-to-end delay variation or *jitter*.

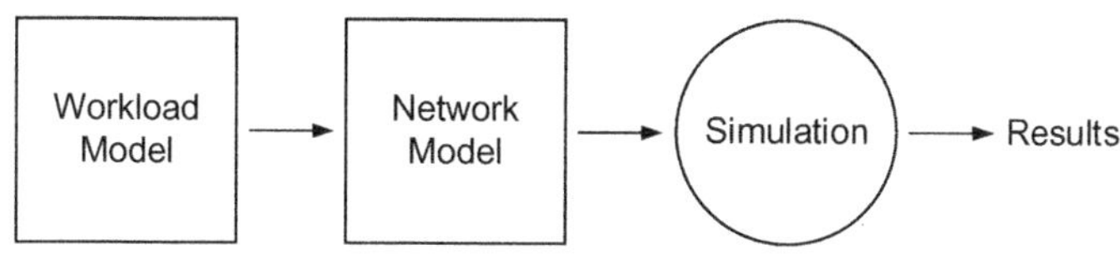

FIGURE 7.7

Integration of the workload model.

Performance models are often analytical models based on stochastic processes and mainly focused on the influence of TCP congestion mechanisms on throughput. Hybrid solutions combining analytical and simulation techniques have also been used for large networks. In this scenario, the transport layer properties of the network are characterized by a fluid model (direct graph composed of *routers* and *links*) described with a set of differential equations. Moreover, the behavior of the smaller networks connected to the backbone network is simulated at packet level.

2.3 WORKLOAD MODELING

The traffic generated by protocols at the application layer represents the influence of the environment over the network behavior, so the traffic model, also known as the workload model, is a critical part of the modeling process in order to obtain accurate and credible results [13]. Figure 7.7 shows the integration of the workload model in an M&S scenario.

Measurements of the real network may be important before modeling and necessary to obtain the relevant information about the application traffic. These can be obtained using three approaches: server logs, client logs and packet traces. Some crucial aspects to consider when modeling three of the most common forms of application traffic in networks (web, voice and video) are presented here.

2.3.1 Web traffic

Two important kinds of web traffic models are *page-oriented models* and *ON/OFF models*, derived from packet traces and client logs, respectively.

Page-oriented models are hierarchical models with four levels: *session* level, *page* level, *connection* level and *packet* level. Examples of modeling parameters at session level are the *web sessions per period*, the *session inter-arrival time*, the *viewing time* and the *number of pages per session*. Examples of modeling parameters at page level are the *time between two consecutive pages* and the *size of page objects* (text, images, sound, etc.) and the *parsing time of page objects*. Connection level relates to the TCP connections per page necessary to serve the web page objects. The modeling parameters at this level are the *number of connections per page*, the *time between two consecutive connections within the same page* and the *duration of the connections*. Almost none of the traffic models consider the last level of granularity, the packet level, with parameters such as *packet size distribution* and the *packet inter-arrival time.*

ON/OFF models simply model ON phases, corresponding to the download period of pages, and OFF phases, corresponding to the time between the end and beginning of the downloading of consecutive pages.

2.3.2 Voice traffic

There is increasingly more voice traffic in computer networks as a result of the availability of low-cost voice communication services on the market. Voice traffic models should consider components both at user and packet levels. Components at user level are the *speaker model*, that describes user behavior during the conversation, and the *session model*, based on SIP or H.323 protocols. The components at packet level are the *codec model* for digital voice encoding and the *transport model* based on RTP *(Real Time Protocol).* The most important parameters of the speaker model are the *session duration* and the *duration of the speech events* (*talkspurts* and *pauses*). Moreover, the codec model determines when the packets are sent for a specific bit rate, and there are two main parameters to consider: *packet size* and *packet inter-arrival time*.

2.3.3 Video traffic

Video conferencing and streaming services are very popular in modern networks and demand a huge amount of resources, so accurate video traffic models are needed. Like voice traffic models, video traffic models consider components at user and packet levels. Components at user level are the *user model*, that describes the requests from the client side to the server side, and the *session model*, based on protocols like RTSP *(Real Time Streaming Protocol).* Regarding packet level, the components are the *codec model* and the *transport model* based on RTP. The main parameters of the session model are *session duration* (between client and server) and *session inter-arrival time* (between client requests). Finally, the most important parameters of the codec model are the same as for voice codec models, that is, *packet size* and *packet inter-arrival time*.

2.4 NETWORK TOPOLOGY MODELING

The network topology defines how the nodes of the computer network are interconnected and so how information may flow [11]. Both physical and logical level topologies can be considered. Logical or overlay topologies are built on top of the physical network and allow data exchange arranged via network and application layers. The network topology has a strong influence on protocol performance and routing behavior, so network topology modeling is an important issue.

Network topologies are modeled through graphs, where vertices correspond to nodes (hosts, switches, routers, etc.) and edges correspond to links. Properties from graph models like the *average node degree* (number of connections), the *mean path length* between nodes and the *betweenness* (node load) allow extracting performance information about these topologies.

There are three main approaches to constructing a graph that satisfies specific properties: random graphs, geometric random graphs (accounting for the distance between nodes) and hierarchical topologies.

3 PERFORMANCE METRICS IN COMPUTER NETWORK SIMULATION

After being modeled using some of the techniques previously presented, the performance of a computer network can be evaluated at different layers. To extract accurate information from a simulation, it is mandatory to select the proper metrics according to the studied layer. In this section, the most widely used metrics to evaluate network performance are summarized. These metrics are related to the TCP/IP stack. They are also classified according to their usefulness to wired and wireless networks.

3.1 LINK LAYER

A metric commonly used to determine throughput at this layer is the ratio of bits received to the duration of the transmission. Another important throughput metric is the *nominal channel capacity* (NCC), which is the maximum number of bits that can be transmitted per time unit. NCC is subsequently used to calculate the *effective channel capacity* (ECC) by not considering the overhead introduced by the protocols used in the communication. Throughput can also be estimated using the *channel utilization* (CU), which is the ratio between NCC and the number of bits received per transmission time [14].

Transmission errors are also important at this layer. There are two main metrics to evaluate transmission errors: *bit-error rate* (BER) and *packet-error rate* (PER). The former represents the ratio of the number of received bits that have been altered while traversing the communication channel to the number of sent bits, while the latter indicates the ratio of the number of incorrectly received data packets to the number of received packets. In both cases, the higher the metric value, the worse the network performance.

The aforementioned metrics can be applied both to wired and wireless networks. However, there are metrics to determine performance exclusively related to the latter. For example, the *spectral efficiency*, which is the number of received bits per unit time per unit bandwidth and per unit area (b/s)/(Hz $\times$ m^2) [15].

There are other metrics that can be used to estimate other network performance aspects at link layer in wireless networks. The *received signal strength indication* (RSSI) represents the signal strength observed at the receiver's antenna during packet reception [16]. The *access point transition time* (APTT) evaluates the duration of the handover [17]. The *Jain's index* evaluates fairness when multiple nodes are attempting to access the wireless medium [18].

Furthermore, there are metrics to evaluate the quality of the signal both in wired and wireless networks. The *signal-to-noise ratio* (SNR) metric is used in wired networks to calculate the ratio of the signal power to the background noise. In wireless networks the analogous metric is the *signal-to-interference-plus-noise ratio* (SINR), where the interference power of other signals is also considered.

3.2 INTERNET LAYER

This subsection focuses on describing the metrics used to estimate the performance of two main routing tasks: path selection and network topology management. First, the path selection process of a routing protocol tries to determine the best path to a destination node, using several metrics to evaluate the cost of the path between source and destination. In wired networks, the most widely used metric is *hop count*, which helps find the minimum hop-count routing between source and destination. Other metrics related to path selection in wired networks are the available bandwidth of the link and the *round trip time* (RTT), which measures the round trip delay of unicast probes between neighboring nodes [19].

On the other hand, although the aforementioned metrics can also be used in wireless networks, other metrics are specifically applied in practice. For example, path selection metrics exclusively used in wireless networks are the *expected transmission count* (ETX) and the *expected transmission time* (ETT). The former predicts the number of retransmissions needed to send a data packet over a link [20], while the latter determines the time a data packet needs to be correctly transmitted over a link. The *expected transmission time* was extended considering channel diversity to obtain a tradeoff between delay and throughput in a path selection metric called *weighted cumulative expected transmission time* (WCETT) [21]. Other metrics related to path selection in wireless networks are the *metric of interference and channel-switching* (MIC), the *exclusive expected transmission time* (EETT), the *interference aware routing metric* (iAWARE) and the *WCETT-load balancing* (WCETT-LB), surveyed in [22].

A network topology imposes how network entities are interconnected with each other and hence establishes how data is forwarded through the network. Communications developed on higher layers depend on the organization and management of the underlying network, so a specific network topology affects the routing scheme, the scalability and the complexity of the network. Metrics commonly used to estimate the correctness of a network topology are *betweenness* and *node degree*. The former is the number of shortest paths between any two nodes that go through a particular node, quantifying the importance of such a node in the information exchange [23]. The latter determines the number of nodes that depend on a specific node.

3.3 TRANSPORT AND APPLICATION LAYER

The most widely used performance measure at the transport layer is throughput. Although such a measure was previously presented in the link layer subsection, in practice it can also be calculated at every layer to analyze the behavior of the protocols and to estimate their impact on performance. Thus, TCP and UDP throughput are frequently used at transport layer, so the performance of the network can be analyzed when specific parameters of each protocol are modified, such as the TCP window size or the packet size [24]. In addition, there also exist metrics at the transport layer depending on time such as the *end-to-end delay* and the *jitter*. The *end-to-end delay* is the time required to transmit a packet along the path between source and destination, and the *jitter* is the packet delay variation. Although both metrics can be analyzed at the network layer, they are usually analyzed at the transport layer to include several features of the protocols of this layer like additional delays when performing checksum verification or buffering.

A subtle reinterpretation of throughput is *goodput*, which refers to the total number of bits received at the application layer of the receiver divided by the simulation time [14]. However, there are many other specific metrics to evaluate the performance of each application protocol, and even aspects related to human behavior. An interesting group of applications at this layer is overlay networks. They are used for content distribution, P2P data sharing and real-time interactive communications. Overlay networks are complex data distribution networks, deployed over existing networks providing services unavailable in the latter, where management and routing tasks are performed by the nodes of the overlay at application level. Some of the metrics exclusively used to evaluate the performance of overlay networks are the *delay stretch*, *bandwidth* and the *link stress*. The *delay stretch* determines the ratio between the application-level delay and the IP-level delay between two nodes [25]. The *bandwidth* measures the overall bandwidth consumption in the overlay network, which depends on packet replication. In the same way, the *link stress* quantifies the number of packet replicas traversing the same physical link. There are also other specific metrics that can be applied to evaluate the performance of P2P overlays such as the *cutoff delay clustering*, *spatial growth* and *proximity* [26].

4 DISCRETE-EVENT SIMULATION

Discrete-event simulation deals with system models in which changes happen at discrete instants in time (events), rather than continuously, and state variables do not change in the intervals between these discrete instants [27].

Computer networks are most often simulated with discrete-event models, and there are two basic types of discrete-event simulation: *trace-driven* simulation and *stochastic* simulation. In *trace-driven* simulation, the simulation inputs come from data captured on the real system (traces), so the accuracy is maximized, but is not applicable to all types of systems. In *stochastic* simulation, the system workload is characterized by probability distributions, using random values as inputs to the simulation model. This section first introduces random value generation. Then, the three simulation approaches for a stochastic simulation program (event-driven, process-oriented and parallel) are presented.

4.1 RANDOM VALUE GENERATION

To develop simulations properly, it is mandatory to generate random values with predefined statistical distributions [6,28]. Generally, this is accomplished in two steps, which are briefly introduced in the following paragraphs.

- First, a step called random-number generation, in which a sequence of random numbers between 0 and 1 is generated with a uniform distribution.
- Second, a step called random-variate generation, in which the previous sequence is used to generate a new sequence with the desired distribution.

4.1.1 Random number generation

A common technique to generate a sequence of random numbers is using a recursive function (x_n), where the last number generated is a function of a set of the previous numbers generated. As an example:

$$x_n = f(x_{n-1}, x_{n-2}, \ldots) \tag{7.1}$$

The sequence generated is not fully random, but pseudorandom, which is better for simulation because a sequence of pseudorandom numbers can be precisely repeated in several simulations. The type and properties of random number generators mainly depend on the generation function. Common types are Linear Congruential Generators (LGCs) [29,30] and Tausworthe Generators [31]. Currently, Mersenne Twister [32] is the common generator used in many software packages. The length of the period of the pseudorandom sequence corresponds to a prime number of Mersenne. The most used version of this generator is based on the Mersenne prime $2^{19937} - 1$. This generator is based on a matrix linear recurrence, which is an evolution of the Generalized Feedback Shift Register pseudorandom number generators. There is a standard implementation in portable C-code that uses 32-bit word length called MT19937 and another variant using 64-bit word length called MT19937-6, which generates a different sequence.

Before using a pseudorandom sequence in a simulation, a thorough analysis of the randomness of the sequence should be carried out. The first test involves plotting the distribution of the sequence to check if it seems uniformly distributed. Additionally, several statistical tests can be applied to the sequence to check if it matches a uniform distribution. The most common are the Chi-square [33] and the Kolmogorov-Smirnov [34] goodness of fit tests.

4.1.2 Random-variate generation

A sequence of numbers with a uniform distribution can be transformed into another sequence with a specific distribution using several methods. The most common is the inverse transformation method. It is based on the following observation: for any random variable x, with cumulated distribution function (CDF) $F(x)$, variable $u = F(x)$ is distributed uniformly between 0 and 1. Therefore, x can be generated from u using the inverse of $F(x)$:

$$x = F^{-1}(u) \tag{7.2}$$

This inverse transformation can be carried out analytically for many distributions, like exponential, geometric, logistic, Pareto or Weibull. The typical example is for the exponential distribution:

$$\begin{aligned} f(x) &= \lambda e^{-\lambda x} \\ F(x) &= 1 - e^{-\lambda x} = u \\ x &= -\left(\frac{1}{\lambda}\right)\mathrm{Ln}(1-u) = -\left(\frac{1}{\lambda}\right)\mathrm{Ln}(u) \end{aligned} \tag{7.3}$$

But the inverse transformation can also be done numerically, generating random uniform values u^* and finding the correspondent values x^* that satisfy the equation $u^* = F(x^*)$, typically using bisection or Newton methods. Figure 7.8 illustrates this technique which is a good approach when an analytic inversion of $F(x)$ is not possible.

There are additional techniques that can be applied in special cases. The composition technique can be used when the desired $F(x)$ can be expressed as a weighted sum of other F(x)'s, and the convolution technique can be used when the random variable x can be expressed as a sum of other random variables.

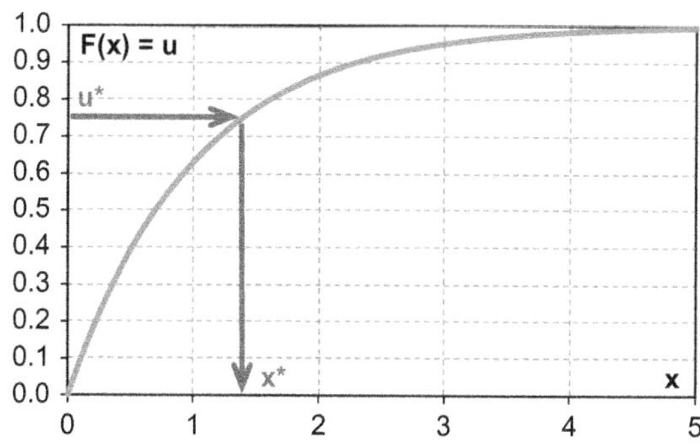

FIGURE 7.8

Numeric inverse transformation.

4.2 EVENT-DRIVEN SIMULATION

In event-driven simulation, system state changes are caused by events, and the changes are considered instantaneous. Two facts are necessary for the simulation to evolve: when the events take place and how they change the system state. At the heart of the event-driven simulation is the so-called event list, a time-sorted list of the current planned events for the system. The first event on the list (at the head) is the closest in the future. As shown in Figure 7.9, each entry on the list is composed of four fields: event identity, event time, event type and event info. The simulation also maintains a clock with the time corresponding to the event being currently processed. The type field is used to select the event service routine to be executed, which also uses the info field during execution. As a result of the execution, the system state variables are changed and possible new events are added to the event list.

Figure 7.10 describes the basic execution behavior of an event-driven simulation, where all the events of the list are processed in time order. There are three possible conditions for the simulation to finish: emptiness of the event list, reaching the planned maximum simulation time or stability of the computed performance metrics in a steady state.

4.3 PROCESS-ORIENTED SIMULATION

All discrete-event simulations are event-driven simulations, but process-oriented is a high-level approach that makes implementation and debugging of the

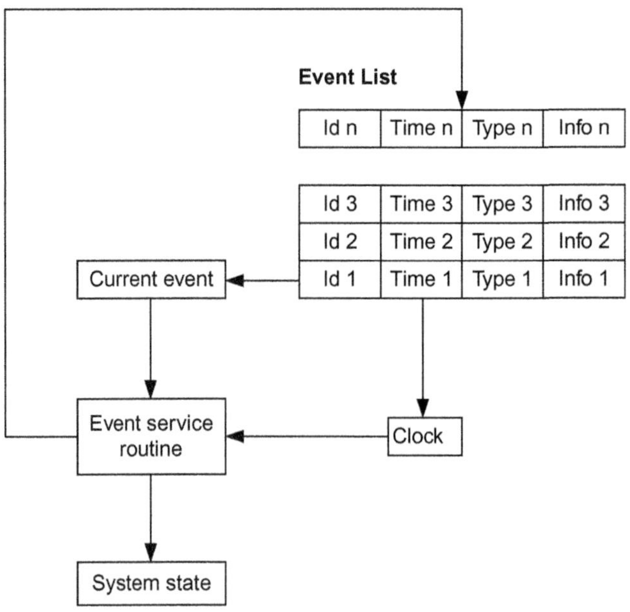

FIGURE 7.9

Structure of an event-driven simulation.

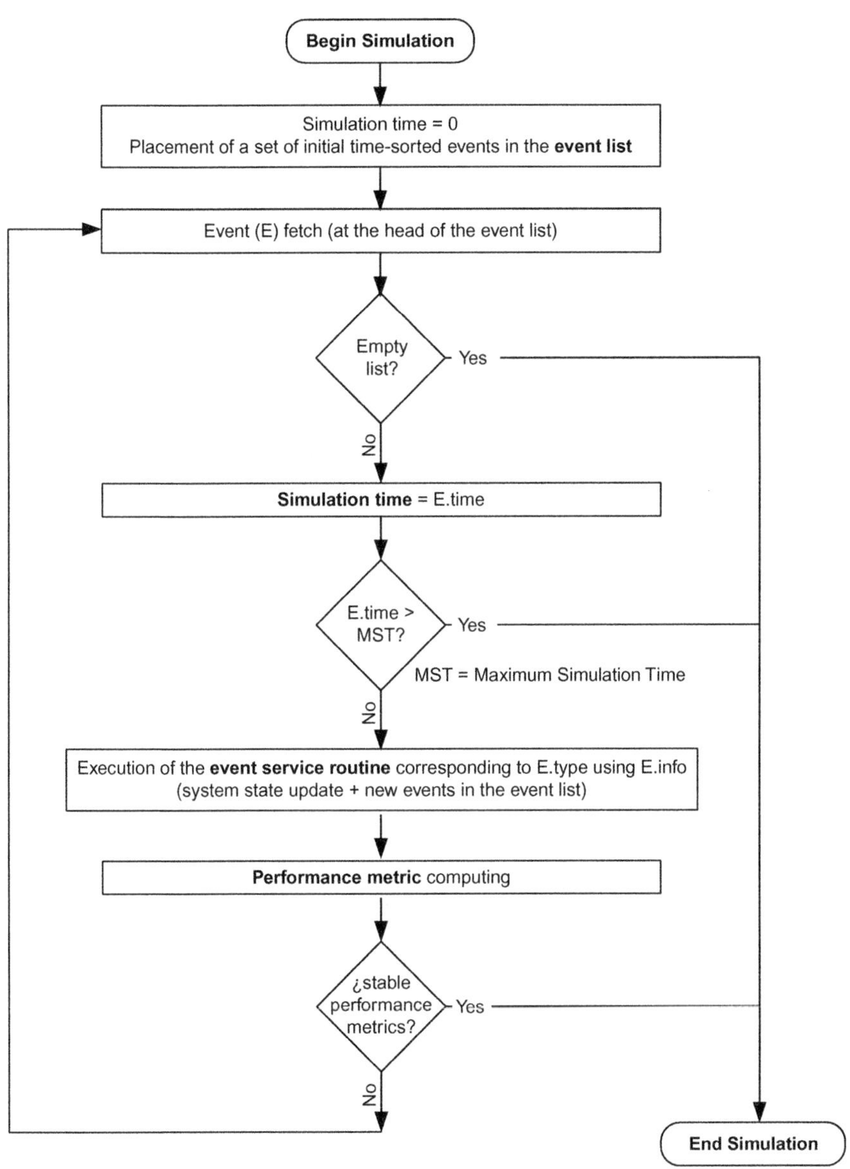

FIGURE 7.10

Execution of an event-driven simulation.

simulation model easier. This approach aggregates related events in processes that are active during time intervals and interact with each other. Processes are composed of code, resources (memory mainly) and state (current execution point and values of variables). As stated in Figure 7.11, the event list is now a process list with identities and activation times. A process can be active or suspended

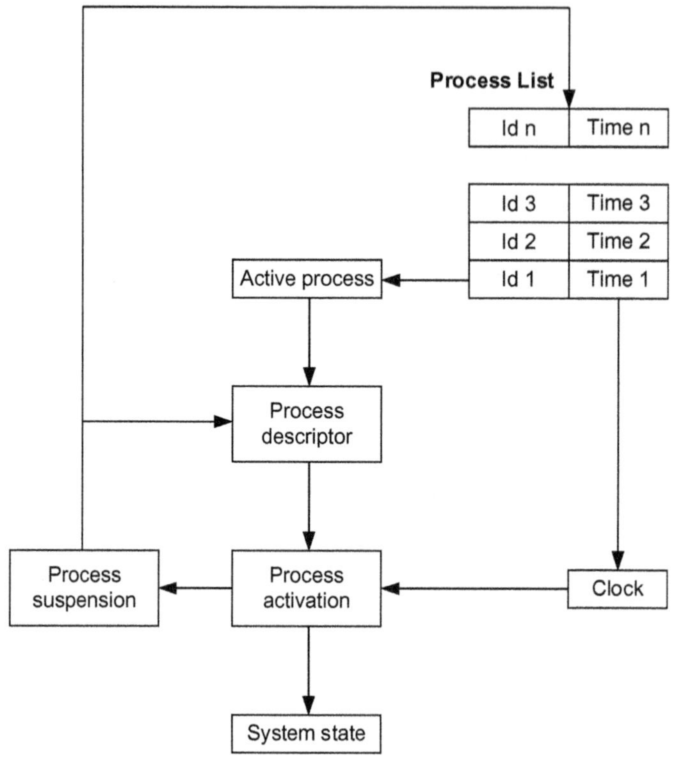

FIGURE 7.11

Structure of a process-oriented simulation.

(waiting for a delay to expire or on a semaphore closed by another process). The process descriptor allows for finding and restoring the execution context of the process.

4.4 PARALLEL SIMULATION

Discrete-event simulation of computer networks is limited both by the amount of available memory, which decreases with the number of nodes in the network, at least linearly, and the execution time, which increases with the packet traffic in the network and the model complexity.

In *Parallel Discrete-Event Simulation* (PDES), the simulation is executed in parallel on several processors. It implies the availability of more computation and memory resources, and so the opportunity to increase *simulation performance* (execution speed) and *simulation scalability* (the size of networks that can be modeled). As a result, more complex networks with higher packet traffic can be simulated in reasonable execution time.

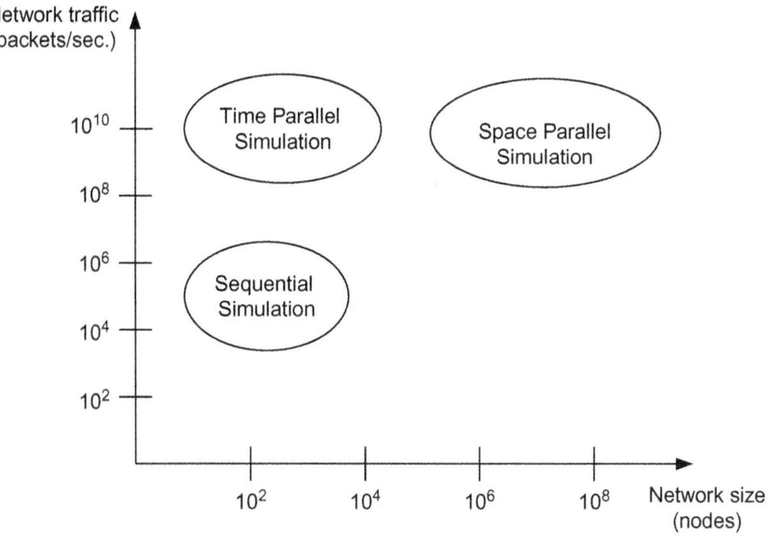

FIGURE 7.12

Real-time simulation alternatives.

4.4.1 Simulation performance and scalability

PDES can improve the simulation both in terms of execution speed (performance) and network size (scalability) depending on the adopted parallel strategy [35]. Figure 7.12 shows the approximate limits of the two main parallel strategies and the sequential baseline for conducting real-time simulation, that is, simulate x seconds of network operation in no more than x seconds.

Time Parallel Simulation only achieves performance improvement, whereas Space Parallel Simulation can improve both performance and scalability. In some specific cases, both Time and Space Parallel Simulation can be used, further improving the execution speed. These strategies will be detailed below. An additional nonparallel and trivial strategy to reduce the execution time of a set of simulation experiments consists of concurrently executing different simulations on different machines.

The performance of a packet-level network simulator, where simulation involves packet processing and transmission from source to destination nodes through routers, can be characterized by the *Simulated packet transmissions per second* (PTS). This metric allows for estimating simulation times when the packet traffic and the average number of intermediate nodes (routers) are known:

$$T = N_{pt}/\text{PTS} \tag{7.4}$$

$$N_{pt} = N_f{}^*P_f{}^*H_f. \tag{7.5}$$

In the equations above, T is the execution time of the simulation, N_{pt} the number of packet transmissions, N_f the number of flows, P_f the average number of

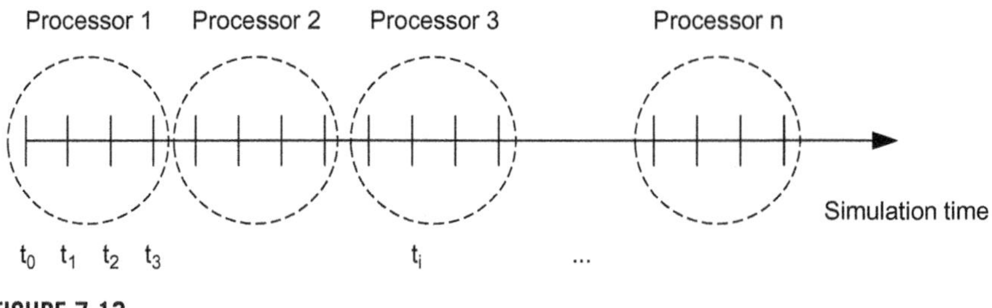

FIGURE 7.13

Time parallel simulation.

packet transmissions per flow and H_f the average number of hops from source to destination nodes.

4.4.2 Time parallel simulation

Time parallel simulation, as indicated in Figure 7.13, is a parallel strategy that divides the simulation time axis in equal-sized intervals, and assigns them to different processors [35]. The requirement for the technique to be useful is that the simulation of each interval can be considered independent from the others, and hence the initial state of the simulation for each interval must be known in advance. This is a strict requirement because the state of the simulation at the beginning of an interval must be known without completing the simulation of all the previous intervals.

Time parallel simulation has potential to do massive parallel processing and drastically reduce simulation time. However, it depends strongly on the model characteristics and is only applicable to specific networks, for instance those modeled as queuing networks or Petri nets with relatively simple behavior.

4.4.3 Space parallel simulation

Space Parallel Simulation is a more flexible and widely applicable strategy for reducing the execution time, but also for increasing the simulation scalability. It partitions the network to simulate in subnetworks and assigns logical processes to those that can be concurrently executed on different processors. Each logical process maintains state variables, a time-stamped list of events and a local clock, and communicates with the other processes during the concurrent simulation through message passing mechanisms. Figure 7.14 shows a scheme representing the assignment of four sub-networks to different processors.

Important issues that arise in space parallel simulations are partitioning (of the network in sub-networks), load balancing (mapping of processes on available processors, trying to equilibrate the load of processors and minimize the communications between them) and synchronization (ensuring that the events are processed in a correct order, as in the sequential simulation) [35,36].

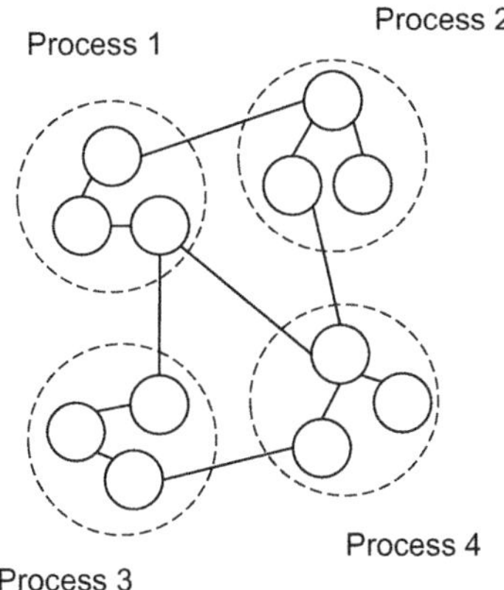

FIGURE 7.14

Space parallel simulation.

Synchronization is the most relevant characteristic of Space Parallel Simulation. The goal of each logical process is to ensure that events are processed in timestamp order *(local causality constraint)*, considering both the events scheduled by the process itself and the events scheduled for it by other processes.

5 VALIDATION AND VERIFICATION

Simulation models must be validated and verified before starting the analysis of simulation results and taking decisions based on the analysis. The validation of a simulation model is the process to assure that the design and assumptions taken are reasonable, and therefore, the model will provide results consistent with those of the real system. The verification of a simulation model is the process to check the correctness of the implementation of the design and the assumptions in a simulation program [28]. A simulation model can be in four possible states:

- invalid and unverified
- invalid and verified
- valid and unverified
- valid and verified

When the modeling of a system and the programming of the model are developed by different teams, the modeling team is responsible for validation and the programming team is responsible for verification.

5.1 MODEL VALIDATION

Validation depends on the system being modeled and the simplifications assumed to relax the inherent complexity of the real system. There are three main aspects of a model that must be validated:

1. Design and assumptions
2. Input parameters and distributions
3. Outputs

Validity tests compare any of the previous aspects with the information obtained from one or several of the following sources:

1. **Expert opinions:** Expert opinion is the most common approach to validation of a simulation model. Meetings should be held with people involved in the development, maintenance and utilization of the real system under modeling. As the development of the model progresses, at least three successive meetings are required. The first one is for validating the preliminary design of the model, the second to validate the model, and the third to validate the initial results provided by the model. Later, more meetings should be carried out to validate successive refinements and improvements of the model. A quick validation test consists in presenting a mixture of outputs from the real system and the simulation model to experts in order to see if they are able to distinguish between them.
2. **Real system measurements:** One of the preferred techniques to validate a simulation model is the comparison with measurements of the real system. Although obtaining the measurements may be very expensive, they make the validity of the model noticeably trustworthy. The measurements required include input and output values, the design and configuration of the real system and the workloads supported.
3. **Theoretical results:** For some systems, a very simple model can be built, which can be solved analytically. There are also well-known elemental models which can be configured to reproduce the behavior of any particular system with a moderately low level of detail. These simple models can be used to define the expected theoretical behavior of the system. Thus, the validation can be carried out by checking if the simulation results are similar to the theoretical results.

Although an analyst could use the three sources of information to validate each of the aspects (nine possibilities), generally very few sources will be available, if any, for each aspect.

Normally the validation of the model is only carried out for several scenarios. The selected scenarios should include the most common utilization cases of the real system. A complete validation of the model to guarantee that it reproduces the behavior of the real system under any possible condition would require excessive time and resources. However, the validation process could also include extreme utilization cases in order to check the range of conditions within which the simulation model represents the behavior of the real system properly. This will increase the confidence in the results provided by the model.

5.2 MODEL VERIFICATION

Many techniques can be used to verify the correctness of the computer program implementing the model. The general techniques used to verify programs can be used with simulation programs, but there are also validation techniques that are specifically for simulation programs.

The technique of *anti-bugging* consists of including checks in the program to detect bugs. For example, in a computer network simulation the difference between the number of packets sent by the source nodes and the number of packets received by the destination nodes must be equal to the number of lost packets. The unfulfillment of this equality indicates a programming error in the simulation program. Another good practice consists of tracking the reference count of the entities (packets, nodes, connections, etc.) created during the simulation. At the end of the simulation all the references must be eliminated.

One of the main issues to face when debugging simulation programs is that many of their variables contain data generated randomly, typically using statistical distributions. A common verification technique consists of changing the statistical distributions for constant distributions that allow for determining the expected values in the variables.

The execution of simplified cases can also be of great help to verify a simulation program: for example, using only one source node and one destination node, sending and receiving a single packet respectively. Then, intermediate nodes can be included progressively. The simplified cases must be analyzed by hand prior to the execution of the simulation program in order to compare the results of both approaches. However, the correct operation of the simulation program for simplified cases does not guarantee its correct operation for more complex cases, but the simplified cases provide the starting steps for the verification process.

6 NETWORK SIMULATORS

As previously commented in Section 4, the discrete-event simulation is the most widely used approach to study the behavior of a network. The main reason for using this approach is that it allows for obviating all those system states between two discrete points in time. Discrete-event simulation makes it possible to simplify a complex real process into a set of primitive actions. Thus, each primitive corresponds to a state change, and is represented by an event in the simulation. For example, the process of sending a data packet from a source to a destination is composed of several complicated tasks related to encoding and decoding data,

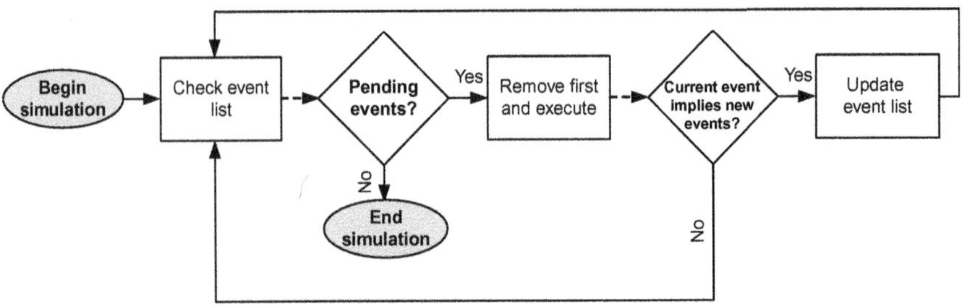

FIGURE 7.15

Simulation life cycle.

signal transmission and data processing and verification. In contrast, under the discrete-event approach the aforementioned process could be depicted by only two events: 1) when the source has sent the packet, and 2) when the destination has received the packet, but a higher number of events might be defined for a deeper and more exhaustive analysis of the process.

A network simulator based on the discrete-event approach has two mandatory structures: a simulation time variable and a list of pending future events. The former represents the time at which the current state of the system is known and represented in the simulation environment. The latter is a list containing state changes that have been scheduled to occur in the future, which rule the course of the simulation [35]. Each network simulator has an API and a programming language that encompass these mandatory structures. Additionally, more specialized structures can be implemented such as a scheduler entity to manage the pending list by adding and removing events.

The life cycle of the simulation is implemented over the two structures presented above, as illustrated in Figure 7.15. The simulator keeps track of a list of pending future events that have been scheduled to be executed at a specific simulation time. The simulator executes the events in sequential increasing time order. Specifically, the simulation time immediately jumps forward from the scheduled execution time of an executed event, to the execution time of the next event. Furthermore, once the execution of an event ends, the simulator checks the list of pending events. Then, the simulator will move to the next event, or will terminate the simulation if there are no more events in the pending list. It should be noted that an event execution may imply the emergence and the scheduling of one or more additional future events.

Before a simulation takes place, several processes may be developed in the network simulator such as a network topology definition, and a link and node configuration. When the simulation ends, the performance can be determined through the analysis of the data available on time-stamped traces returned by the simulator.

Table 7.1 summarizes the most important network simulators based on the event-driven simulation paradigm. This summary includes diverse types of

Table 7.1 Network Simulators

No	Simulator	Developer	URL	Commercial	Language
1	Cnet	Univ. Western of Australia	http://www.csse.uwa.edu.au/cnet	No	C
2	EstiNet	EstiNet Technologies	http://www.estinet.com	Yes	C++
3	GloMoSim	UCLA	http://pcl.cs.ucla.edu/projects/glomosim	No	C, C++
4	GTNetS	Georgia Tech Univ.	http://www.ece.gatech.edu/research/labs/MANIACS/GTNetS	No	C++
5	IKR SimLib	Univ. of Stuttgart	http://www.ikr.uni-stuttgart.de/IKRSimLib	No	C++, Java
6	J-Sim	UIUC	http://j-sim.cs.uiuc.edu	No	Java, Perl, Tcl, Python
7	NCTUns	Nat. Chiao Tung Univ.	http://nsl10.csie.nctu.edu.tw	No	C++
8	NetSim	Tetcos	http://www.tetcos.com/netsim_gen.html	Yes	C++
9	NS-2	USC ISI	http://www.isi.edu/nsnam/ns	No	C++, OTcl
10	NS-3	ns-3 project	http://www.nsnam.org	No	C++, Python
11	OMNeT++	Technical Univ. of Budapest	http://www.omnetpp.org	No	C++, C#, Java
12	OMNEST	Simulcraft	http://www.omnest.com	Yes	C++
13	OPNET	Riverbed Technology	http://www.opnet.com	Yes	C, C++, Proto-C
14	QualNet	SCALABLE Network Technologies	http://web.scalable-networks.com/content/qualnet	Yes	C
15	SENSE	Rensselaer Polytechnic Institute	http://www.ita.cs.rpi.edu	No	C++
16	SimPy	MIT	http://simpy.readthedocs.org	No	Python
17	SSFNet	Renesys	http://www.ssfnet.org	No	C++, Java
18	SWANS	Cornell Univ.	http://jist.ece.cornell.edu	No	Java
19	SWANS++	AquaLab	http://www.aqualab.cs.northwestern.edu/legacy/swans++	No	Java
20	YANS	INRIA	http://sourceforge.net/projects/yans-netsim	No	C, C++

event-based simulators such as discrete-event and parallel discrete-event simulators. Furthermore, it also encompasses both commercial and research network simulators.

A node is the key entity of the network simulator architecture, representing any computing device that is connected to a network. It is an abstraction that encapsulates all the representations of computing devices in a simulation. Thus, nodes can refer to elements such as routers, switches and hubs at the backbone of the network, and personal computers or servers as endpoints of the network. The main characteristic of a node is packet transmission. An endpoint can be the source or destination of data packets while backbone elements perform forwarding tasks. A node may be represented by several state variables referring to CPU and memory consumption, battery power and physical location.

A node is composed of other entities such as network interface cards (NICs). An interface abstracts a node from transmitting, receiving and processing packets. An interface has state variables representing whether the interface is installed or not, and if it is idle or busy. Similarly to a node, an interface includes other entities, like a queue and a link, in order to represent a realistic operative scenario. A queue is an entity that represents the buffers used in the outgoing and incoming packet processing. It encompasses a packet list of a limited size. A link is an entity that represents the connectivity between two nodes, so it allows for delivering packets between interfaces, managing the aspects related to the communication medium. Usually a link is modeled by establishing parameters such as the available bandwidth, the propagation delay and the jitter, and even a pre-established packet loss rate.

Queues and links process packets created by a source node. A packet is an entity that contains the data being conveyed between nodes through the network. The size of the packet depends on the protocol being used. Therefore, each node also includes entities representing each protocol stack implementation. A protocol entity manages the outgoing and incoming packets by adding and removing packet headers. A special protocol entity is the application entity, which is responsible for generating the data packets sent by the source node.

Finally, there are other entities that can be found in a network simulator. These entities do not represent elements of a real network, but they are needed to evaluate the performance of the simulations or to facilitate the model implementation. For example, a logging entity collects information about each packet retransmitted or the link usage. In other cases, helper entities are provided to help build network topologies.

7 CASE STUDY: PERFORMANCE EVALUATION OF AN OVERLAY NETWORK

In this section, a simulation to assess the performance of an overlay network for supporting synchronous e-training activities is described. In general terms, an

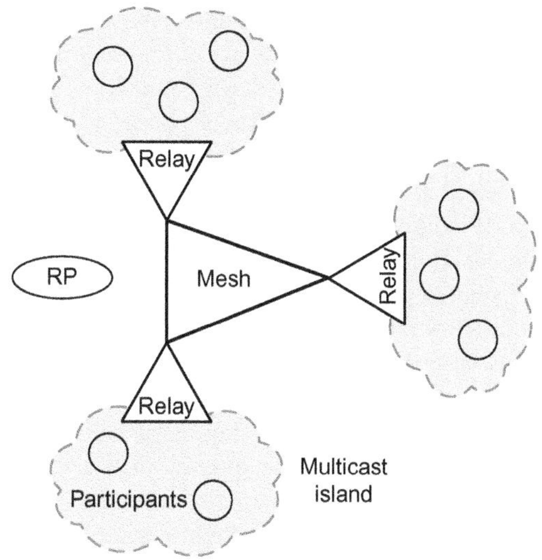

FIGURE 7.16

Architecture of the overlay network.

overlay network is a virtual network deployed on top of an existing network to provide additional services. In this case, the overlay is used to provide a multicast delivery service between the participants of an e-training activity, so the multimedia data streams generated by a participant are delivered efficiently to the rest of the participants. This simulation is a simplified version of that carried out in [37].

7.1 OVERLAY ARCHITECTURE

The overlay network is organized in three planes. A simplified illustration of the overlay is depicted in Figure 7.16. These planes are managed by a central entity called the Rendezvous Point (RP).

First, the relay mesh is composed of several relays that are responsible for forwarding data to participants. Each of these relays is located in a multicast island where IP multicast is available and they are all interconnected through the Internet forming a full-mesh topology to reduce latency. Multimedia data streams are delivered using RTP, so the relays must forward RTP streams between participants. Various media types are used in e-training sessions, with audio, video, instant messaging and shared whiteboard annotations among them [38], and an RTP session is used for each media type.

The participants use IP multicast to send their data streams to the relay located in the same multicast island, which forwards them to the rest of the relays in the mesh. The streams are subsequently forwarded to the participants in the multicast islands of the relays. The efficiency of the relay mesh relies heavily on the availability of native multicast in the underlying network and the clustering of participants (number of multicast islands).

Second, a signaling plane is established between the participants of an e-training activity and the RP. This link allows participants to join the activity and facilitates the negotiation of the multimedia configuration (media types, encoding parameters, IP addresses and ports, etc.). The Session Initiation Protocol (SIP) is used as the signaling protocol. The RP plays the role of an SIP focus to implement a tightly coupled conferencing service, so every participant establishes a SIP dialog with the RP while the activity is running. The configuration of the multimedia sessions is described using the Session Description Protocol (SDP).

Finally, a mesh control plane is used to communicate the RP with the relays. The RP uses these links to reorganize the relay mesh according to the joining and leaving of participants and failures in the overlay in order to keep the real-time data delivery service efficient. Relays register with the RP as soon as possible, so the RP can establish TCP connections with them during activities. These connections are used as a keep-alive mechanism to regularly gather link status information from the relays. It is not necessary that all the relays participate in the ongoing activity. In fact, a relay will not be part of the relay mesh if there is no participant in its multicast island.

The overlay network includes a self-optimization technique that ensures a minimum number of streams interchanged among multicast islands. As previously mentioned, the RP maintains a list of registered relays, associating them with the identifier of the multicast island in which they are located. A relay is active if at least one participant joins the activity from the multicast island where the relay is located. The RP is responsible for including and excluding relays from the mesh as participants join and leave, so the relay mesh is only composed of active relays.

7.2 SIMULATION

One of the main reasons for using simulation to assess the performance of the overlay is the impossibility of requesting a high number of users to participate in synchronous e-training activities for testing. This would be extremely expensive and disruptive for users. Besides, the network resources required to evaluate the performance of the overlay thoroughly might not be available or may be too expensive.

The ns-3 simulator is used to simulate the operation of the network of an organization geographically dispersed in several sites, the entities of the overlay and the participants within a synchronous e-training activity. The ns-3 simulator includes a vast number of models of wired and wireless channels and many of the elements that can be found in modern networks. It is written in C++ and uses the combined multiple-recursive random number generator MRG32K3a proposed in [39]. The whole TCP/IP stack is implemented on top of the channel models, so programmers can reuse the algorithms (even the code) of their implementations to build their ns-3 models. Thus, existing RTP and SIP libraries are integrated in the simulator by using the socket library of the simulator instead of the socket library of the operating system in order to simulate the overlay.

The model assumes that there exist various sites where IP multicast is available and there is an RTP relay in each site. An ideal wide area network (WAN) connects every site through 20 Mbps network links with a delay of 20 ms. The data rate of the network link connecting the RP to the WAN is 10 Mbps and introduces no delay in communications. These parameters are obtained empirically from the corporate network of a large corporation [38]. A CSMA channel is used to simulate the WAN between the sites. This channel is modeled as a simplistic Ethernet-like network in which the state of the medium is instantaneously shared among all the devices, so there is no need for collision detection. A queuing approach is used in this model. The local area network (LAN) of each site is also modelled as a CSMA channel. A point-to-point channel, modeled as a simplistic point-to-point serial line link, interconnects each LAN with the ideal WAN. Both the CSMA and point-to-point channels model both the physical and link layers of the protocol stack.

The range of possible overlay deployments is extremely high. Many parameters such as the number of relays in the mesh, the number of participants and the balance of participants between relays can be varied. However, in order to evaluate the impact of each parameter individually, only one is varied at a time during the simulations. For simplicity, only balanced deployments are considered—that is, all the relays serve the same number of participants.

A participant has to be modeled to simulate the traffic generated during a synchronous e-training activity. The participant can use several kinds of media types to communicate with other participants in an activity. In this case, the participant is represented as an entity that generates one audio stream, one video stream, annotations on the shared whiteboard and telepointer movements, resembling the data generated by a synchronous e-training tool [38].

The key decision is how to model the traffic generated by a participant. Traces reporting the behavior of users can be collected from real e-training activities, so the model of a participant can be derived. For example, a participant may join the activity after a *waiting time*, simulated using an exponential random variable Exp(1/45) s. Then, the participant establishes an SIP dialog with the RP to join the activity and negotiate the media configuration. The audio and video streams of a participant are activated regularly. A video stream is a simulated 160×120 variable bitrate H.264 video stream at 10 frames per second (the size of the video packets sent to the network also has to be modeled). An audio stream is a simulated constant bitrate iLBC audio stream with a packetization time of 20 ms. Both the interval between activations and the duration of the streams are assumed to be normal random variables. Thus, a participant generates an audio stream of duration N(15, 2.6) s after an elapsed time N(25, 3.3) min, and generates a video stream of duration N(30, 9.5) s after an elapsed time N(40, 8.2) min.

The formats of the annotations on the shared whiteboard and the telepointers used by participants are described in [38]. A participant generates an RTP stream containing their annotations on the shared whiteboard, while another RTP stream is used to convey the information of the telepointer. Similarly to audio and video, the activation time and the duration of these streams are assumed to be normal random variables. A participant generates an annotation stream of duration N(60, 10) s after an elapsed time N(16.6, 1.3) min, and generates a telepointer stream of duration N(15, 2.5) s after an elapsed time N(20, 1.6) min.

Real e-training activities are moderated in order to manage the interactions between participants and avoid excessive network resource consumption due to many multimedia streams. Floor control protocols can be used to share the data channels between the participants of the activity. The maximum number of data streams that can be activated simultaneously in the simulations are: 4 audio streams, 4 video streams, 2 annotation streams and 2 telepointer streams.

Moreover, two roles are usually identified in the participants of a synchronous e-training activity: one instructor and many regular participants. The instructor is modeled as a participant who uses all the media types continuously during the activity. The rest of the participants issue floor requests to use the data channels. This situation closely resembles the behavior of users in synchronous e-training activities where participants usually interrupt the activity to ask questions and the instructor can be seen and heard throughout the duration of the activity. In all cases, the requests by participants to use the data channels are granted in a first-in first-out (FIFO) order.

Various simulations have to be conducted to assess the performance of the overlay when varying the number of participants and the number of sites where they are located. The duration of a typical e-training activity is one hour.

7.3 ANALYSIS OF RESULTS

Different performance metrics can be used to analyze the overall performance of the overlay. The workload at the relays when forwarding streams can be estimated using the average bandwidth consumption. The user experience when participating in synchronous e-training activities supported by the platform can be estimated using the average packet loss and the interarrival jitter of the audio and video packets. As an example, Figure 7.17 shows the cumulative average network bandwidth consumed in the links of the sites to the Internet when varying the number of participants in the activity and the number of relays in the mesh. This proves that the network bandwidth consumed in the links of the sites grows asymptotically due to the floor control policy limiting the maximum number of active streams in the overlay.

Since stochastic processes modeling different entities of the overlay are involved in the simulations, confidence intervals are very useful to describe the accuracy of the results. As can be seen in Figure 7.18, these intervals are quite small due to the huge number of replicas in the tests.

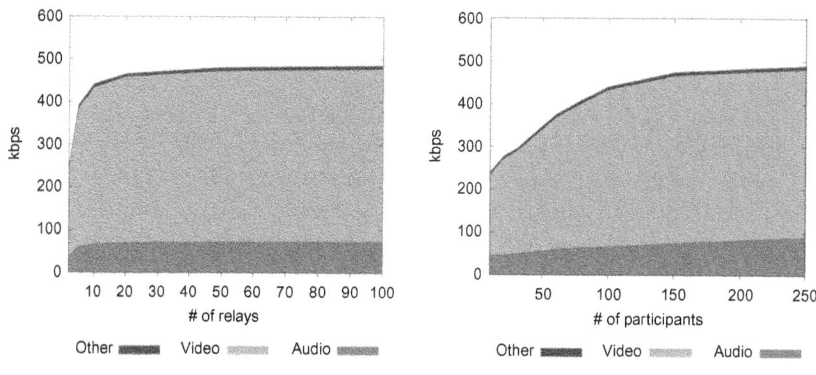

FIGURE 7.17

Cumulative average network bandwidth consumption in the network links of the sites.

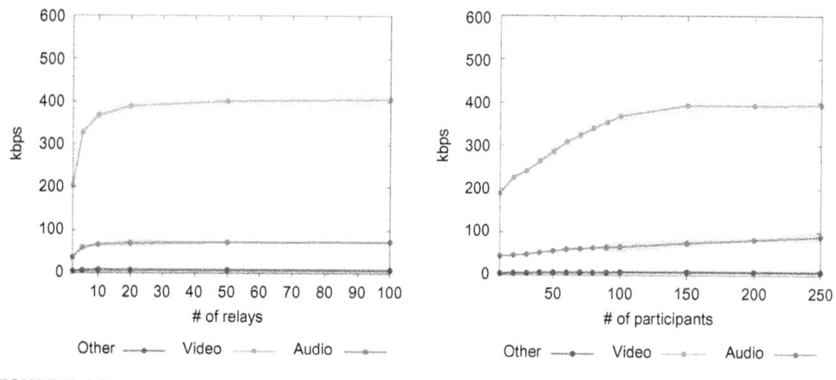

FIGURE 7.18

Average network bandwidth consumption in the network links of the sites for each media type.

8 SUMMARY

Modeling and simulation (M&S) are attractive and widely used techniques for the study of the performance of computer networks. They provide detailed results without disturbing network operation or even without the need of network availability. This chapter summarizes the whole topic of performance M&S applied to computer networks. After an introduction, the main modeling techniques and performance metrics used in computer networks have been presented. Computer networks are discrete-event systems and so discrete-event simulation has been highlighted as the most common simulation technique for executing computer network models. Random number generation, event-driven or process-based simulation, and parallel discrete-event simulation are important techniques related to discrete-event simulation covered in the chapter. Confidence in simulation results is only possible if the models are validated against the real systems and the simulation algorithms are verified against the models, so conditions for both have been detailed. The architecture of a simulator and the characteristics of the most used simulators complete this high-level panoramic overview of the performance M&S of computer networks. The chapter finishes with a case study of performance M&S for an application layer overlay network.

GLOSSARY

APTT Access Point Transition Time
BER Bit Error Rate
CDF Cumulated Distribution Function
CPU Central Processing Unit
CSMA/CA Carrier Sense Multiple Access with Collision Avoidance
CSMA/CD Carrier Sense Multiple Access with Collision Detection
CTMC Continuous Time Markov Chains
CU Channel Utilization
ECC Effective Channel Capacity
EETT Exclusive Expected Transmission Time
ETT Expected Transmission Time
ETX Expected Transmission Count
FIFO First In First Out
GSPN Generalized Stochastic Petri Nets
H.264 Recommendation by the ITU Telecommunication Standardization Sector (ITU-T)
H.323 Recommendation by the ITU Telecommunication Standardization Sector (ITU-T)
iAWARE Interference Aware Routing Metric
iLBC Internet Low Bitrate Codec
IP Internet Protocol
LAN Local Area Network
LGCs Linear Congruential Generators
MAC Medium Access layer
MIC Metric of Interference and Channel-şwitching
M&S Modeling and Simulation
NCC Nominal Channel Capacity
NIC Network Interface Card
NS-3 Network Simulator 3
P2P Peer to Peer
PDES Parallel Discrete-Event Simulation
PER Packet Error Rate
PN Petri Nets
PTS Simulated packet transmissions per second
QN Queuing Networks
RP Rendezvous Point

RSSI Received Signal Strength Indication
RTP Real Time Protocol
RTSP Real Time Streaming Protocol
RTT Round Trip Time
SDP Session Description Protocol
SINR Signal-to-Interference-Plus-Noise Ratio
SIP Session Initiation Protocol
SNE Signal-to-Noise
SPN Stochastic Petri Nets
TCP Transmission Control Protocol
TCP/IP Transmission Control Protocol/Internet Protocol
TPN Timed Petri Nets
UDP User Datagram Protocol
WAN Wide Area Network
WCETT Weighted Cumulative Expected Transmission Time
WECTT-LB WCETT-load balancing

REFERENCES

[1] Garzia RF, Garzia MR, editors. Network modeling, simulation, and analysis. Electrical and computer engineering. Taylor & Francis; 1990.

[2] Guizani M, Rayes A, Khan B, Al-Fuqaha A. Network modeling and simulation: a practical perspective. Hoboken, NJ: John Wiley & Sons; 2010.

[3] Taylor HE, Karlin S. An introduction to stochastic modeling. 3rd ed. San Diego, CA: Academic Press; 1998.

[4] Burbank J, Kasch W, Ward J. An introduction to network modeling and simulation for the practicing engineer. The ComSoc guides to communications technologies. Hoboken, NJ: John Wiley & Sons; 2011.

[5] Wainer GA. Discrete-event modeling and simulation: a practitioner's approach. Boca Raton, FL, USA: CRC Press, Inc.; 2009.

[6] Law AM, Kelton DM. Simulation modeling and analysis. New York, NY: McGraw-Hill Higher Education; 1999.

[7] Pidd M. Computer simulation in management science. Hoboken, NJ: John Wiley & Sons; 1988.

[8] Sokolowski J, Banks C. Principles of modeling and simulation: a multidisciplinary approach. Hoboken, NJ: John Wiley & Sons; 2009.

[9] Marsan MA, Bobbio A, Donatelli S. Petri nets in performance analysis: an introduction. In: Reisig W, Rozenberg G, editors. Petri nets. Vol. 1491 of lecture notes in computer science. Springer; 1996. p. 211–56.

[10] Bolch G, Greiner S, de Meer H, Trivedi KS. Queueing networks and Markov chains—modeling and performance evaluation with computer science applications. Hoboken, NJ: John Wiley & Sons; 2006.

[11] Wählisch M. Modeling the network topology. Springer; 2010. p. 471–86 [chapter 22].

[12] Sasnauskas R, Weingärtner E. Modeling transport layer protocols. Springer; 2010. p. 389–99 [chapter 17].

[13] Aktas I, King T, Mengi C. Modeling application traffic. Springer; 2010. p. 397–26 [chapter 18].

[14] Mühleisen M, Jennen R, Kirsche M. Wireless networking use cases. Springer; 2010. p. 305–25 [chapter 13].

[15] Alouini MS, Goldsmith AJ. Area spectral efficiency of cellular mobile radio systems. IEEE Trans Veh Technol 1999;48(4):1047–66.

[16] Vlavianos A, Law LK, Broustis I, Krishnamurthy SV, Faloutsos M. Assessing link quality in IEEE 802.11 wireless networks: which is the right metric? In: Personal indoor and mobile radio communications (PIMRC). IEEE 19th International Symposium, 2008. p. 1–6.

[17] Bangolae S, Wright C, Trecker C, Emmelmann M, Mlinarsky F. November, 14–18, 2005. Test methodology proposal for measuring fast BSS/BSS transition time. doc. 11-05/537, IEEE 802.11 TGt Wireless Performance Prediction Task Group, Vancouver, Canada, substantive Standard Draft Text. Accepted into the IEEE P802.11.2 Draft Recommended Practice.

[18] Jain R, Chiu D-M, Hawe W. A quantitative measure of fairness and discrimination for resource allocation in shared computer systems. DEC Research Report TR-301. 1984.

[19] Draves R, Padhye J, Zill B. Comparison of routing metrics for static multi-hop wireless networks. In: Proceedings of the 19th conference on applications, technologies, architectures, and protocols for computer communications (SIGCOMM '04). Portland, OR, USA; 2004a. p. 133–44.

[20] De Couto DSJ, Aguayo D, Bicket J, Morris R. A high-throughput path metric for multi-hop wireless routing. In: Proceedings of the ninth annual international conference on mobile computing and networking (MobiCom '03). San Diego, California, USA; 2003. p. 134–46.

[21] Draves R, Padhye J, Zill B. Routing in multi-radio, multi-hop wireless mesh networks. In: Proceedings of the tenth annual international conference on mobile computing and networking (MobiCom '04). Philadelphia, PA, USA; 2004b. p. 114–28.

[22] Di P, Wählisch M, Wittenburg G. Modeling the network layer and routing protocols. Springer; 2010. p. 359–84 [chapter 16].

[23] Freeman LC. A set of measures of centrality based on betweenness. Sociometry 1977;40(1):35–41.

[24] Xylomenos G, Polyzos GC. TCP and UDP performance over a wireless LAN. In: Proceedings of the IEEE 18th conference on computer communications (INFOCOM 1999). New York, NY, USA; 1999. p. 439–46.

[25] Yang L, Guo M. High-performance computing: paradigm and infrastructure. Wiley series on parallel and distributed computing. Wiley-Interscience; 2006.

[26] Kaune S, Wählisch M, Pussep K. Modeling the internet delay space and its application in large scale P2P simulations. Springer; 2010. p. 427–46 [chapter 19].

[27] Sinclair JB. Simulation of computer systems and network: a process oriented approach. Cambridge, England: Cambridge University Press; 2004.

[28] Jain R. The art of computer systems performance analysis—techniques for experimental design, measurement, simulation, and modeling. Wiley professional computing. New York, NY: Wiley; 1991.

[29] Knuth DE. Art of computer programming, volume 2: seminumerical algorithms. Addison-Wesley Professional, Section 3.2.1: The linear congruential method; 1997. p. 10–26.

[30] Park SK, Miller KW. Random number generators: good ones are hard to find. Commun ACM 1988;31(10):1192–201.

[31] Tausworthe RC. Random numbers generated by linear recurrence modulo two. Math Comput 1965;19(90):201–9.

[32] Matsumoto M, Nishimura T. Mersenne twister: a 623-dimensionally equidistributed uniform pseudo-random number generator. ACM Trans Model Comput Simul 1998;8 (1):3–30.

[33] Greenwood PE, Nikulin MS. A guide to Chi-squared testing. Wiley-Interscience; 1996.

[34] Massey FJ. The Kolmogorov-Smirnov test for goodness of fit. J Am Stat Assoc 1951;46(253):68–78.

[35] Fujimoto RM, Perumalla KS, Riley GF. Network simulation. Synthesis lectures on communication networks. Morgan & Claypool Publishers; 2006.

[36] Kunz G. Parallel discrete event simulation. Springer; 2010 [chapter 8], p. 121–31.

[37] Granda J, Nuño P, García D, Suárez F. Autonomic platform for synchronous e-training in dispersed organizations. J Netw Syst Manage 2013;1–27.

[38] Granda J, Nuño P, Suárez F, Pérez M. E-psylon: a synchronous e-learning platform for staff training in large corporations. Multimedia Tools Appl 2013;66(3):431–63.

[39] L'Ecuyer P. Good parameters and implementations for combined multiple recursive random number generators. Oper Res 1999;47(1):159–64.

CHAPTER

A new testbed for web performance evaluation

8

Raúl Peña-Ortiz, José Antonio Gil, Julio Sahuquillo, Ana Pont, and Josep Domènech

Universitat Politècnica de València, València, Spain

In this chapter we devise a new testbed that has the ability to reproduce different types of web workloads. This new testbed must accomplish three main goals. First, it must define and reproduce traditional web workloads by using a parameterized and extensible architecture that allows us to integrate a new workload generation process based on a previously developed generator. Second, it must be able to provide client and server metrics with the aim of being used for web performance evaluation studies. Finally, it should be representative of web transactional systems that have been established in recent years, such as e-commerce websites, blogs or OSNs.

The well-known benchmark TPC-W is the best candidate to provide an appropriate testbed for our purposes, because it satisfies the previous goals and also considers the user's behavior on workload generation, although in a partial way. Consequently, with the aim of contrasting a dynamic web workload against the traditional approach to workload generation of TPC-W, we deploy a new testbed for web performance evaluation by integrating our generator into the benchmark.

The remainder of this chapter is organized as follows. Section 1 introduces the workload generator GUERNICA, its architecture and how it can be applied to devise a new testbed. This workload generator allows performance studies to be carried out based on five phases, which are illustrated in Section 2 with an example of web performance evaluation using the generator. Section 3 validates the testbed by integrating it into a well-known benchmark. Here, the experimental setup and the main measured performance metrics in the validation process are described. Finally, we present some concluding remarks in Section 4.

1 GUERNICA

This section introduces and describes the Universal Generator of Dynamic Workload under WWW Platforms (GUERNICA), which is a web workload generator and testing tool to evaluate performance and functionality of web applications. GUERNICA was developed as a result of cooperation among the *Web Architecture Research Group*

(*Universitat Politècnica de València*), *iSOCO S.L*, and the Institute of Computer Technology, thereby bridging the gap between academia and industry.

The main aim of GUERNICA is the workload generation process. This process is based on the DWEB model that characterizes the web workload modeling user's behavior.

1.1 THE APPLICATION SUITE

GUERNICA is software made up of three main applications (*workload generator, performance evaluator* and *performance tests planner*) as shown in Figure 8.1. Each application, described in the following, permits an autonomous distribution among different machines of the main activities in the evaluation of performance and functional specifications of a web application.

- The *performance tests planner* is the application in charge of managing the evaluation procedure. Its main functionalities are: (i) to define both performance and functional test cases, (ii) to plan and schedule their execution, (iii) to provide us with on-line monitoring features, in order to check the results generated by the distributed clients, and (iv) to elaborate final reports combining the obtained results.
- *Workload generators* reproduce web workload for stressing purposes. These generators are not required to be executed in the same machine as the planner,

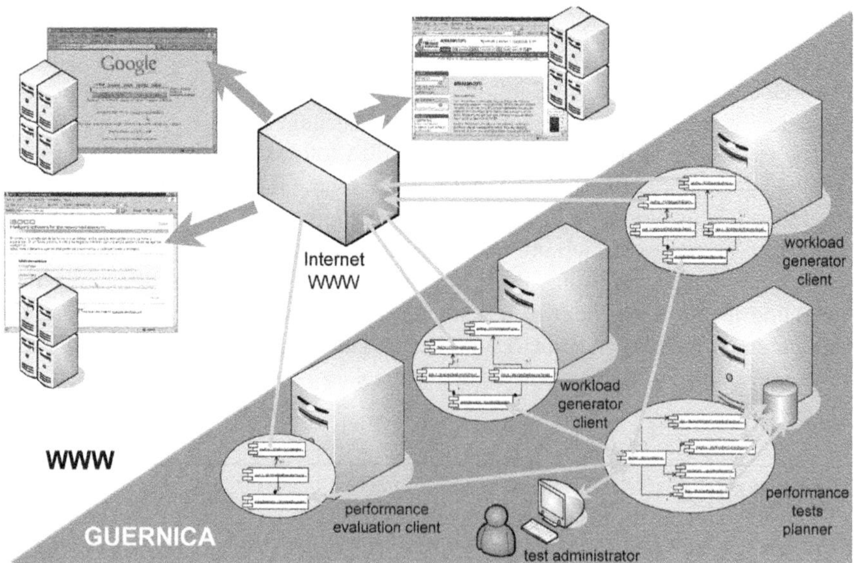

FIGURE 8.1

Main applications of GUERNICA.

which configures generators and controls the executions. Each generator mimics the user's behavior and notifies navigation statistics and results to the planner, which implements the corresponding graphical representation. Additionally, a generator can be stopped by the planner when it is required to do so.

- The *performance evaluator*, also known as *probe client*, is aimed at evaluating from the user's point of view the major functionalities of a given web application while it is stressed by the workload generators. A probe client also can execute functional tests by notifying their results to the planner.

In addition to these main applications, the software extension *SemViz* [1] assists the suite. *SemViz* is an application to visualize knowledge based on ontologies by using 3D technology. GUERNICA transforms results of functional tests in an external format based on an ontology that *SemViz* can read and display graphically in a 3D representation.

1.2 TESTING PHASES

GUERNICA allows a website evaluation of performance and functional specifications to be carried out by following the next five phases (see Figure 8.2):

1. The *navigation definition* phase characterizes the first level of the user's dynamism by modeling his/her interactions with the dynamic web contents,

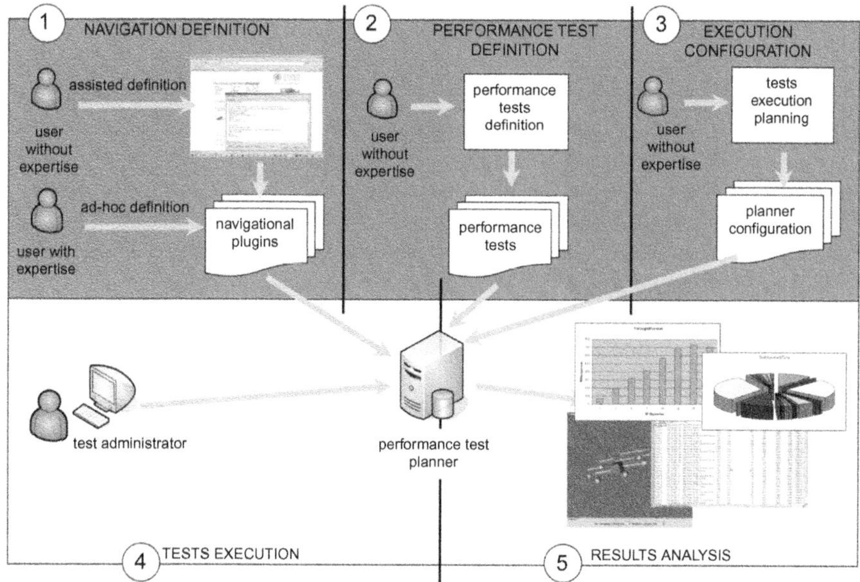

FIGURE 8.2

Testing phases in GUERNICA.

which is a common characteristic of the current Web. To this end, GUERNICA adopts the user's navigation concept of DWEB, and implements it as a *navigational plugin*. To assist the process of defining the user's navigations in GUERNICA, it could be useful to capture and analyze real HTTP sequences that could be converted, in a second step, to navigational plugins.

In this context, the definition of navigation should include how web users employ the features of current browsers, as they might severely affect the server performance [2]. For instance, using tabs when browsing increases the burstiness of requests and makes it difficult to track the user navigation [3].

2. The *performance test* definition phase specifies the behaviors that users play in a website using *workload tests* that implement the user's roles concept of DWEB. A workload test specifies a set of navigational plugins that define user's behaviors by considering the capability of changing them with time.
3. The *execution configuration* phase sets other execution parameters, such as the number and type of users, the types of reports or the *workload distribution*. The distribution allows GUERNICA to improve the workload accuracy by distributing its generation among different machines. Figure 8.3 shows the three main workload generation approaches that can be used by the *performance test planner* to coordinate the set of secondary workload generation processes.

   In the *ideal distributed model*, the planner, the generators, and the probe client are located in different machines. This is the best way to evaluate the performance of web applications because the probe is in an individual machine; therefore its measurements are not influenced by other generators or by the planner.

   The *advanced distributed model* stipulates only three machines to host the planner, the generators and the probe. The main drawback of this approach is that all generators are in the same machine. In this model, the generated workload is not as real as in the ideal model because in a real environment different users are usually in different machines. Even so, this model can be used when there are not enough machines to perform the evaluation.

   In the *basic model*, the generators, the probe client, and the planner are located in the same machine. Thus, if the performance of this machine is not powerful enough, the approach will introduce *noise* in the tests. This model can be used as a first attempt in the workload generation to identify performance bottlenecks in web applications. Nevertheless, ideal or advanced approaches are required for a more accurate performance evaluation.
4. *Tests execution*. This phase executes workload tests gathering performance and functional statistics. Results collected by generators and probe clients are given to the planner, which groups, classifies and reaches a consensus among them in order to obtain a uniform set of results.
5. *Results analysis*. Finally, GUERNICA analyzes the performance of the system under test and represents performance indexes in different formats (e.g., graphical plots or tabular text). As introduced, *SemViz* visualizes results by using 3D technology.

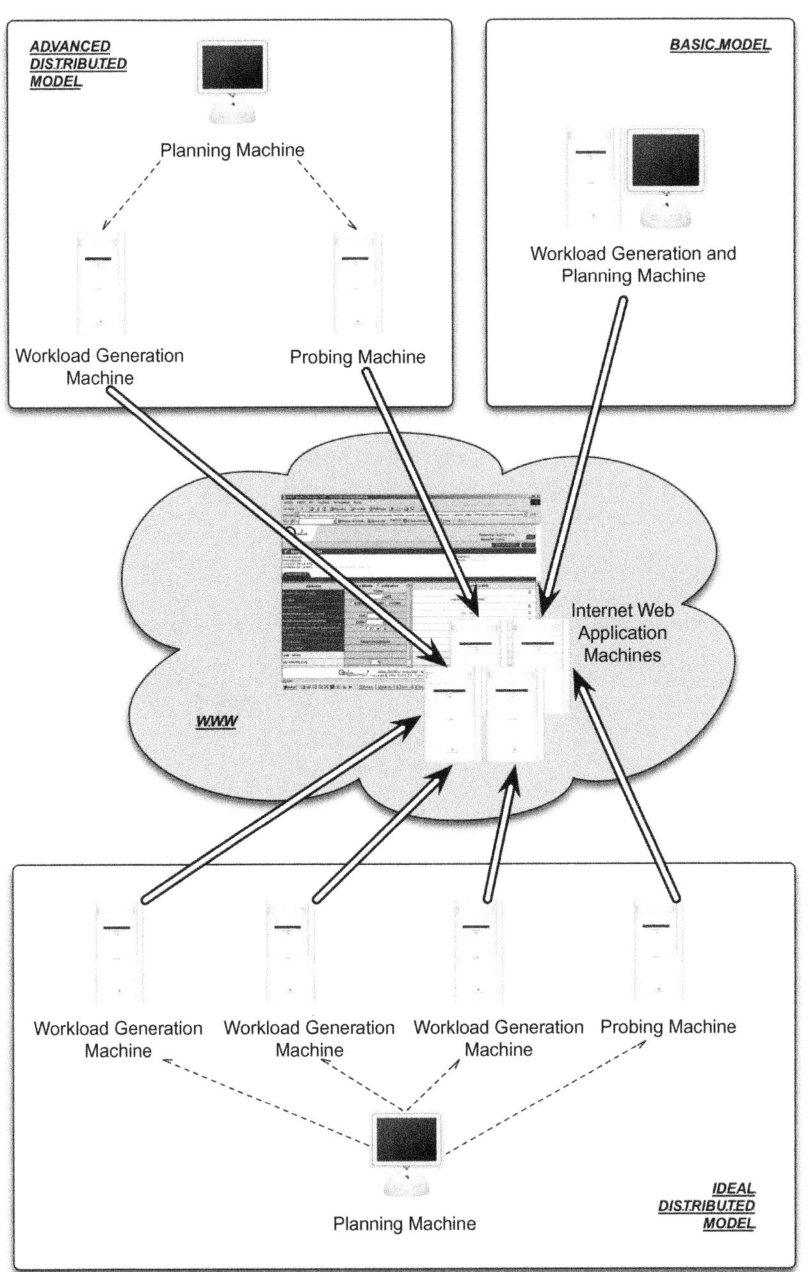

FIGURE 8.3

Distribution of workload generation.

1.3 ARCHITECTURE

The GUERNICA suite presents a distributed software architecture as depicted in the deployment diagram (UML 2.1) of Figure 8.4.

The three main applications (WorkloadGeneratorApp, PlannerApp and ProbeApp) have been programmed in Java using web services technology, so they run independently of the execution platform. Their business logics are provided by three interfaces of the core library: WorkloadPlanner, Probe and WorkloadGenerator.

The core is the main component of the architecture and carries out the workload generation process by using DWEB. Navigational plugins and workload tests are implemented by WorkloadNavigation and WorkloadTest interfaces, respectively. The NavigationEngine defines an API to reproduce the user's behavior; its configuration is described in terms of DWEB, and it is stored in a repository called WorkloadTestRespository. The engine can operate with any technology fulfilling its API, but it currently supports Personal Content Aggregator (PCA) technology to implement the navigational plugin execution. To this end, the PCANavigation and the PCANavigationEngine classes have been provided.

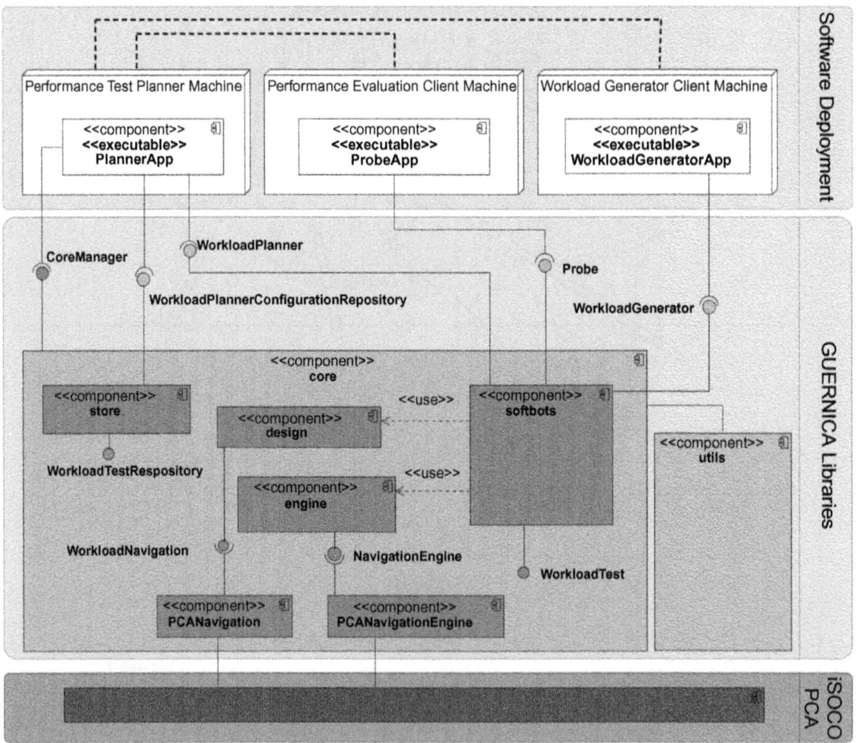

FIGURE 8.4

Architecture of GUERNICA.

The PCA technology, developed by iSOCO S.L., offers: (i) a scripting language that allows GUERNICA to easily define dynamic user's navigations for the current Web, and ii) an engine to carry out automatic executions for the navigational plugins.

Finally, the `util` package helps the main library, which can be accessed by using the `CoreManager` in a centralized way.

1.4 MAIN FEATURES

GUERNICA supports, totally or partially, the main features available in those well-known workload generators proposed in the open literature, as well as the capability to represent the user's dynamism, as shown in Table 8.1.

Regarding the completely supported features, the generator presents a distributed architecture based on web services. That is, GUERNICA allows distributing the generation process among different nodes which emulate users working on different workstations. Furthermore, the generator implements the DWEB model that offers important capabilities. For instance, it provides an analytical approach (user's roles) to characterize users' dynamic behaviors as well as their continuous changes between them. That is, GUERNICA can organize the workload in dynamic categories or types, each of them modeling a given user profile by using a workload test. The model also introduces some client variables to parameterize the user behavior (e.g., the user's think time by means of a Gaussian distribution), which are provided by GUERNICA as a part of the scripting language. This language allows the generator to reproduce users' dynamic navigations.

Table 8.1 GUERNICA Features

Feature/Capability	GUERNICA
Analytical-Based Architecture	◒
Distributed Architecture	●
Business-Based Architecture	◒
Client Parameterization	●
Workload Types	●
Functional Testing	◒
LAN and WAN	◒
Multiplatform	●
Ease of Use	●
Performance Reports	●
Open Source	◒
User's Dynamism	●

Legend: ● Full support; ◒ Partial support.

Among other features supported by GUERNICA, one can observe its ease of use, the ability to generate performance reports (e.g., 3D graphical representation), and the capability to model differences between LAN (where generators are usually run) and WAN (where applications are usually located). Moreover, the workload tests can be used as web application functional tests, and the generation process can be easily applied to different business architectures. Finally, it should be noted that a significant part of the code (the applications and the core packages that do not use PCA technology) has been written under an open source license.

2 CASE STUDY

The objective of this case study is not to present a detailed performance evaluation but to show how GUERNICA can be used in web performance studies.

The case study considers that users can behave as *searchers* or *surfers*, as described by [4] for electronic commerce sites, but also by [5] for general-purpose web browsing. Searchers are users who start their navigations with a query in a search engine like Google. However, surfers prefer to navigate through the Web using its direct hyperlinks. With the aim of illustrating these two behaviors, different navigations looking for some information related to the *US president election in 2012* are chosen.

Figure 8.5 presents an automaton that defines the surfer and searcher users' behaviors, and the continuous changes of behavior when they are looking for

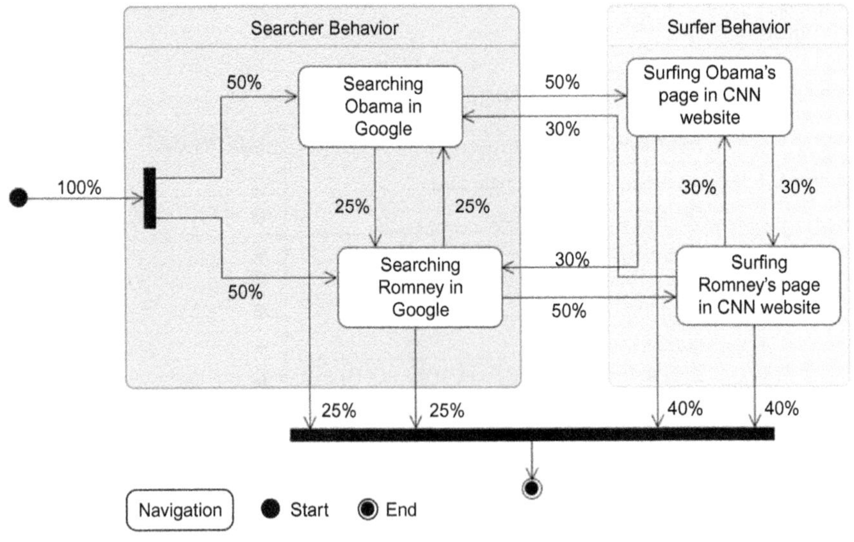

FIGURE 8.5

Web searcher and surfer user's behaviors.

FIGURE 8.6

A simple search in Google.

some information about the candidates. The automaton shows how a typical user can proceed: (i) to search the name of a candidate in the search engine, or ii) to navigate through one of the most important news sites (e.g., the CNN website). The automaton assumes the same probability (50%) to start a user's navigation by searching in Google for the Democratic candidate name (Obama) or the Republican (Romney). Then, the user can search the other candidate in Google (25%), can browse the CNN website to look for some information about the same candidate (50%), or can finish the navigation (25% probability). When the user has found the information in the CNN website, he can navigate to the other candidate page in the site (30% probability), or he can change his behavior by executing a new search looking for some information about the other candidate (30%). The third alternative is to finish the navigation (40%).

For illustrative purposes, Figure 8.6 specifies the searcher behavior by modeling a simple search in Google. As observed, once the main page of Google (www.google.es) is accessed, the user spends some time thinking (think time: 1000 ms)

```
<? xml version ="1.0" encoding ="UTF-8" ?>
< WorkloadTest id="test_searcher_vs_surfer" >
< UsersNumber >2</ UsersNumber >
< NavigationGraph >
< InitialNavigations >
< InitialNavigation id="google_obama" prob ="0.50" />
< InitialNavigation id="google_romney" prob ="0.50" />
</ InitialNavigations >
< NavigationTransitions >
< NavigationTransition from ="google_obama" to="cnn_obama" prob ="0.50" />
< NavigationTransition from ="google_obama" to="google_romney" prob ="0.25" />
< NavigationTransition from ="google_romney" to="cnn_romney" prob ="0.50" />
< NavigationTransition from ="google_romney" to="google_obama" prob ="0.25" />
< NavigationTransition from ="cnn_obama" to="google_romney" prob ="0.30" />
< NavigationTransition from ="cnn_obama" to="cnn_romney" prob ="0.30" />
< NavigationTransition from ="cnn_romney" to="google_obama" prob ="0.30" />
< NavigationTransition from ="cnn_romney" to="cnn_obama" prob ="0.30" />
</ NavigationTransitions >
</ NavigationGraph >
</ WorkloadTest >
```

LISTING 8.1

Workload Test.

until he asks Google for some information (www.google.es/search). After that, Google returns the results page (SEARCH_RESULTS) and a new user's think time (given by a Gaussian distribution as 3500 ms as average with 1500 ms standard deviation) is provided. Then, the user has two options represented by two branches in the graph: if the search engine provides results for the *candidate name* (left branch) the user will access the candidate site (first site provided); otherwise, the user will finish the process (right branch). If the user accesses the candidate home page (www.first_result.home), he spends time thinking about the returned contents before finishing the navigation (black dot).

GUERNICA was implemented by using a model language based on XML labels to define *workload tests*, *navigational plugins* and *workload distribution*. Listing 8.1 shows the XML file that implements as a workload test the automaton of Figure 8.5. On the other hand, Listing 8.2 presents the navigational plugin for a simple search in Google (Figure 8.6). The `PCA-Plugin` is the XML file that defines, by using the PCA technology, the user's navigation.

Finally, Listing 8.3 shows the XML file that distributes the workload generation process. This file configures `GUERNICA` in the basic generation approach with only a workload generator process (`generator-1`) in a single machine.

Once the workload test has been run, we obtain the plan of conducted navigations and their HTTP requests, each one having a different outcome. Listing 8.4 illustrates an example showing the results of an execution for two users. As observed, a set of global statistics is obtained, such as the total

```
<? xml version ="1.0" encoding ="UTF-8" ?>
< Navigation id="google_obama" >
< InputData >
< Param name ="phrase" value ="Obama" >
</ InputData >
< ExecutionCode >
< PCA-Plugin name ="google.pca.xml" >
</ ExecutionCode >
< StatisticsConfiguration >
< StatisticAttribute name ="NavigationTime" >
< StatisticAttribute name ="ExecutionTime" >
< StatisticAttribute name ="HttpRoute" >
< StatisticAttribute name ="URL" >
< StatisticAttribute name ="HttpMethod" >
< StatisticAttribute name ="Stablished" >
< StatisticAttribute name ="StablishmentTime" >
< StatisticAttribute name ="TransferTime" >
< StatisticAttribute name ="ThinkUserTime" >
< StatisticAttribute name ="ContentSize" >
</ StatisticAttribute >
</ StatisticsConfiguration >
</ Navigation >
```

LISTING 8.2

Searching Obama in Google for the US President Election 2012.

```
<?xml version ="1.0" encoding ="UTF-8"?>
< WorkloadPlannerConfiguration >
< PlannerIdentifier >election-2004</ PlannerIdentifier >
< WorkloadGenerators >
< WorkloadGenerator >
< Id >generator-1< /Id >
</ WorkloadGenerator >
</ WorkloadGenerators >
< WorkloadTestAssignations >
< WorkloadtestAssignation testId ="test_searcher_vs_surfer" generatorId ="generator-1"/>
</ WorkloadTestAssignations >
</ WorkloadPlannerConfiguration >
```

LISTING 8.3

Workload Distribution for the US President Election 2012.

execution time of the experiment or the navigation time. For each HTTP request, the test execution produces different statistics, such as the method (`GET` or `POST`), the time for establishing the connection, the transfer time, the user think time, the content size, accessed URL, or the successfulness when making the connection (`Established`).

```
<? xml version ="1.0" encoding ="UTF-8" ?>
< WorkloadTestStatistics id="..." testId ="test_searcher_vs_surfer" >
< UserNavigations >
< NavigationStatistic navigationId ="google_obama" date ="..." >
< NavigationTime >17065 </ NavigationTime >
< ExecutionTime >18453 </ ExecutionTime >
< HttpRoute >
< HttpRouteElement >
< URL >http://www.google.es</ URL >
< HttpMethod >GET </ HttpMethod >
...
</ HttpRouteElement >
< HttpRouteElement >
< URL >http://www.google.es/search?q=Obama</ URL >
< HttpMethod >GET </ HttpMethod >
< Established >true </ Established >
< EstablishmentTime >116 </ EstablishmentTime >
< TransferTime >1084 </ TransferTime >
< ThinkTime >5622 </ ThinkTime >
< ContentSize >19896 </ ContentSize >
</ HttpRouteElement >
< HttpRouteElement >
...
< TransferTime >257 </ TransferTime >
< ThinkTime >5000 </ ThinkTime >
< ContentSize >15545 </ ContentSize >
</ HttpRouteElement >
</ HttpRoute >
</ NavigationStatistic >
< NavigationStatistic navigationId ="cnn_obama" date ="..." >
...
</ NavigationStatistic >
< NavigationStatistic navigationId ="google_romney" date ="..." >
...
</ NavigationStatistic >
< NavigationStatistic navigationId ="cnn_romney" date ="..." >
...
</ NavigationStatistic >
</ UserNavigations >
< UserNavigations >
< NavigationStatistic navigationId ="google_obama" date ="..." >
...
</ NavigationStatistic >
</ UserNavigations >
</ WorkloadTestStatistics >
```


LISTING 8.4

Testing Results for the US President Election 2012.

3 VALIDATION

To ensure that GUERNICA is valid for conducting web performance studies, it must be validated against a traditional approach to workload generation. To this end, we devise a new testbed with the ability of reproducing different types of workloads. After the validation process, the testbed could be safely used to analyze the effect of applying dynamic workloads on the web performance metrics, instead of traditional workloads. This section describes the validation process.

Among the benchmarks available in the open literature, TPC-W is the best candidate to provide an appropriate testbed for our purposes, because it satisfies the previous goals and also considers user's behavior on workload generation, although in a partial way. Consequently, with the aim of validating GUERNICA against the traditional approach to workload generation of TPC-W, we deploy a new testbed for web performance evaluation by integrating our generator into the benchmark.

3.1 THE TPC-W FRAMEWORK

The TPC-W is a transactional web benchmark that models an on-line bookstore environment. The benchmark specification [6] defines a full website map for the on-line bookstore that consists of 14 unique pages and their navigation transitions. Figure 8.7 depicts a reduced TPC-W website map, where pages with related functionality are included in the same group: *ordering*, *shopping*, *browsing*, *admin* and *search*. Navigation hyperlinks among them are also indicated.

The *search* group provides a book searcher by using the Search page to request the query and the Search Results page to show a list of results. The *browsing* group embraces the Bestsellers and the New Products pages, which arrange the bookstore catalog according to the sales and the publication date, respectively. The *shopping* group is the largest set of pages and provides (i) selling functionality by managing the shopping cart (ShoppingCart page), (ii) the purchase request and its confirmation (Buy Request page and Buy Confirm page, respectively), and (iii) the payment through a secured navigation (Customer Registration page). The *ordering* group includes a set of pages that allows users to check the status of an order (Order Inquiry page and Order Display page). The *admin* group manages the catalog of books (using Admin Request and Product Updated pages). Finally, the most referred pages (Home page and Product Detail page) are also included. *Search* and *shopping* groups implement the most interactive and personalized functionality in the website, so they are potentially interesting as a dynamic user workload.

The benchmark provides a standard environment that is independent of the underlying technology, designed architecture and deployed infrastructure. A TPC-W Java implementation developed by the UW-Madison Computer Architecture Group [7] was selected as framework of our testbed. As shown in Figure 8.8, the client side of the architecture is a Java console application that provides two interfaces for generating workload: an Emulated Browser (EB) and a factory (EBFactory) to

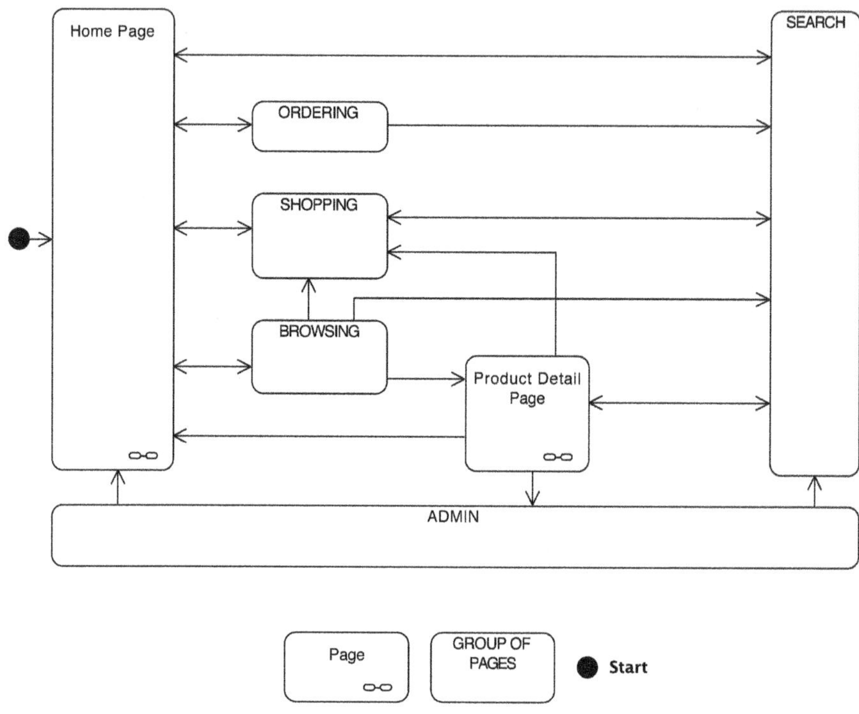

FIGURE 8.7

TPC-W reduced website map.

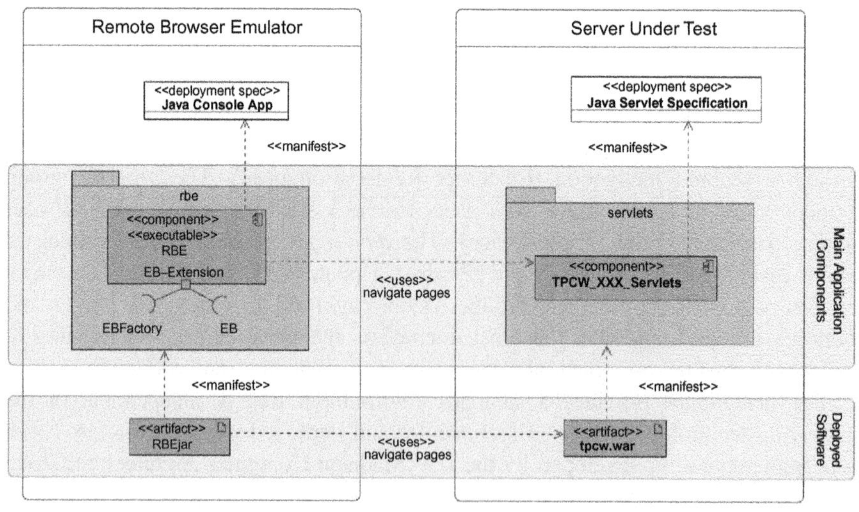

FIGURE 8.8

Main software components of TPC-W Java implementation.

create, configure and manage it. These interfaces allow us to define new processes for workload generation. The server side was developed as a `Java` web application made of a set of `Servlets`. Each `Servlet` resolves client requests by looking in the database information.

3.2 TESTBED ARCHITECTURE

The architecture of integrating `GUERNICA` into `TPC-W` is organized in three main layers as depicted in Figure 8.9 and detailed in the following:

- The top layer is defined at the client side of `TPC-W` and supplies the two interfaces related to the workload generation process (`EB`, `EBFactory`), as introduced previously.

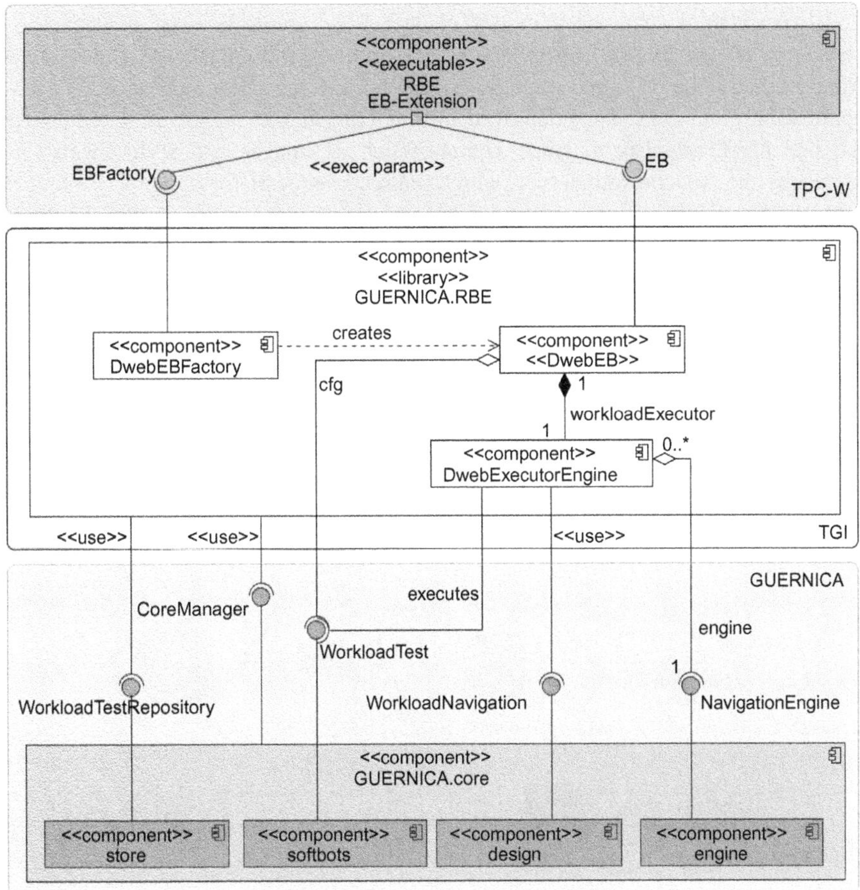

FIGURE 8.9

Testbed architecture.

- The bottom layer is related to the process of workload generation in GUERNICA, detailed in Section 1.1.
- Finally, the intermediate layer defines the integration between GUERNICA and TPC-W. This integration is provided by an independent Java library named TGI. This library implements a new type of EB (DwebEB) that uses the GUERNICA core to reproduce the user's dynamic behavior in the workload generation process. In order to simplify the new EB, a workload generation engine (DwebExecutorEngine) is implemented to carry out the generation process. A browser factory (DwebEBFactory) is also developed to manage the creation and configuration of the new EB.

3.3 EXPERIMENTAL SETUP

The experimental setup used in this chapter is a typical two-tier configuration consisting of an Ubuntu Linux Server back-end tier and an Ubuntu Linux client front-end tier. The back-end runs the on-line bookstore, whose core is a Java web application (TPC-W web app) deployed on the Tomcat web application server. On the one hand, requests to static content, such as images and style sheets, are served by the Apache web server, which redirects requests for dynamic content to Tomcat. TPC-W web app generates the dynamic content by fetching data from the MySQL database. On the other hand, the front-end tier is able to generate the workload either using conventional or dynamic models. Both web application and workload generators are run on the SUN Java Runtime Environment 5.0 (JRE 5.0). Figure 8.10 illustrates the hardware/software platform of the experimental setup used in the validation process.

Given the multi-tier configuration of this environment, system parameters (both in the server and in the workload generators) have been properly tuned to

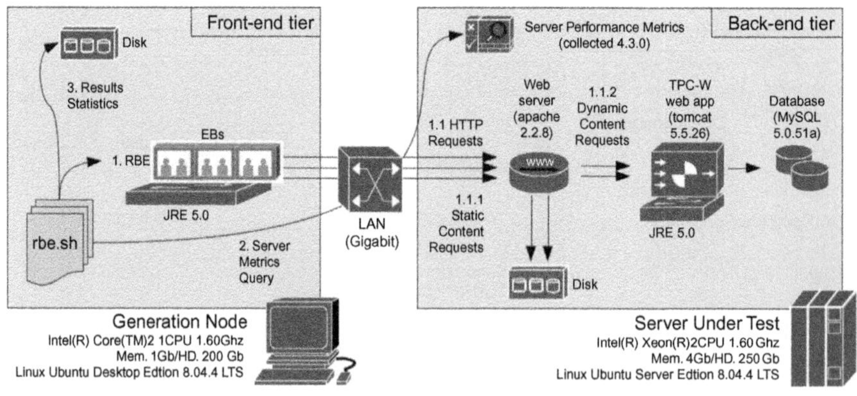

FIGURE 8.10

Experimental setup.

avoid middleware and infrastructure bottlenecks interfering in the results. The on-line bookstore has been configured with 300 EBs and a large number of items (100,000 books) that forced us to balance accesses in the database (e.g., the pool connection size), static content service by Apache (e.g., the number of processes to attend HTTP requests), or dynamic content service by Tomcat (e.g., the number of threads providing dynamic contents). For each experiment, the measurements were performed for several runs with a 20-minute collecting-data phase after a 15-minute warm-up phase.

3.4 PERFORMANCE METRICS

Table 8.2 summarizes the performance metrics available in the experimental setup. The main metrics measured on the client side are the *total number of requests per page* and the *response time*, which is expressed as *Web Interaction Response Time* (WIRT). On the server side, the platform collects the server performance statistics required by the TPC-W specification (*CPU* and *memory utilization*, *database I/O activity*, *system I/O activity*, and *web server statistics*) as well as other optional statistics. These metrics allow a better understanding of the system behavior under test and permit to check the techniques used to improve performance when applying a dynamic workload. The collected metrics can be classified in two main groups: metrics related to the usage of main hardware resources, and performance metrics for the software components of the back-end. For evaluation purposes, we used a middleware named collectd [8] that collects system performance statistics periodically.

3.5 GUERNICA VALIDATION

This section validates GUERNICA by using the devised testbed to compare our workload generation approach against the TPC-W approach. According to the TPC-W specification three scenarios are defined when characterizing the web workload: shopping, browsing, and ordering. The shopping scenario is characterized by intensive browsing and ordering activities while the browsing and ordering scenarios have a reduced ordering and browsing activities, respectively. TPC-W describes these scenarios as three different full CBMGs.

Regarding the validation test, we contrast both workload characterization approximations (i.e., CBMG and DWEB) for each scenario. Figure 8.11 depicts the shopping scenario workload as an illustrative example. Note that we are able to model the same workload by using only the navigation concept of DWEB, and disabling all the parameters used to include user dynamism in the workload characterization. The validation test considers 50 EBs because the Java implementation of the TPC-W generator presents some limitations in the workload generation process. The measurements were performed for 50 runs and obtaining confidence intervals with a 99% confidence level.

Table 8.2 Performance Metrics Classification According to the Evaluated Resource

Resource			Metric	Description/Formula
Client Side			Response Time (*WIRT*)	WIRT is defined by TPC-W as t2 − t1, where t1 is the time measured at the Emulated Browsers when the first byte of the first HTTP request of the web interaction is sent by the browser to the server, and t2 is the time when the last byte of the last HTTP response that completes the web interaction is received.
			Average Response Time ($\overline{WIRT}$)	$\overline{WIRT} = \frac{\sum_{i \in Pages} WIRT_i Req_i}{\sum_{i \in Pages} Req_i}$
			Req_{page}	Requests per Page (Req_{page}) are the number of connections for a page requested by Emulated Browsers and accepted by the server.
Server Side	Hardware	CPU Memory Disk Network	U_{CPU} U_{mem} U_{disk}, X_{disk} U_{net}, X_{net}	Metrics for hardware resources include utilization for all of them, and throughput for the disk and the network
	Software	Apache Tomcat MySQL	X_{apache}, U_{apache}, Mem_{apache} X_{tomcat}, U_{tomcat}, Mem_{tomcat} X_{MySQL}, U_{MySQL}, Mem_{MySQL}	Performance metrics for software components of server include: their throughput, the CPU and memory consumption, the number of processes or threads, etc.

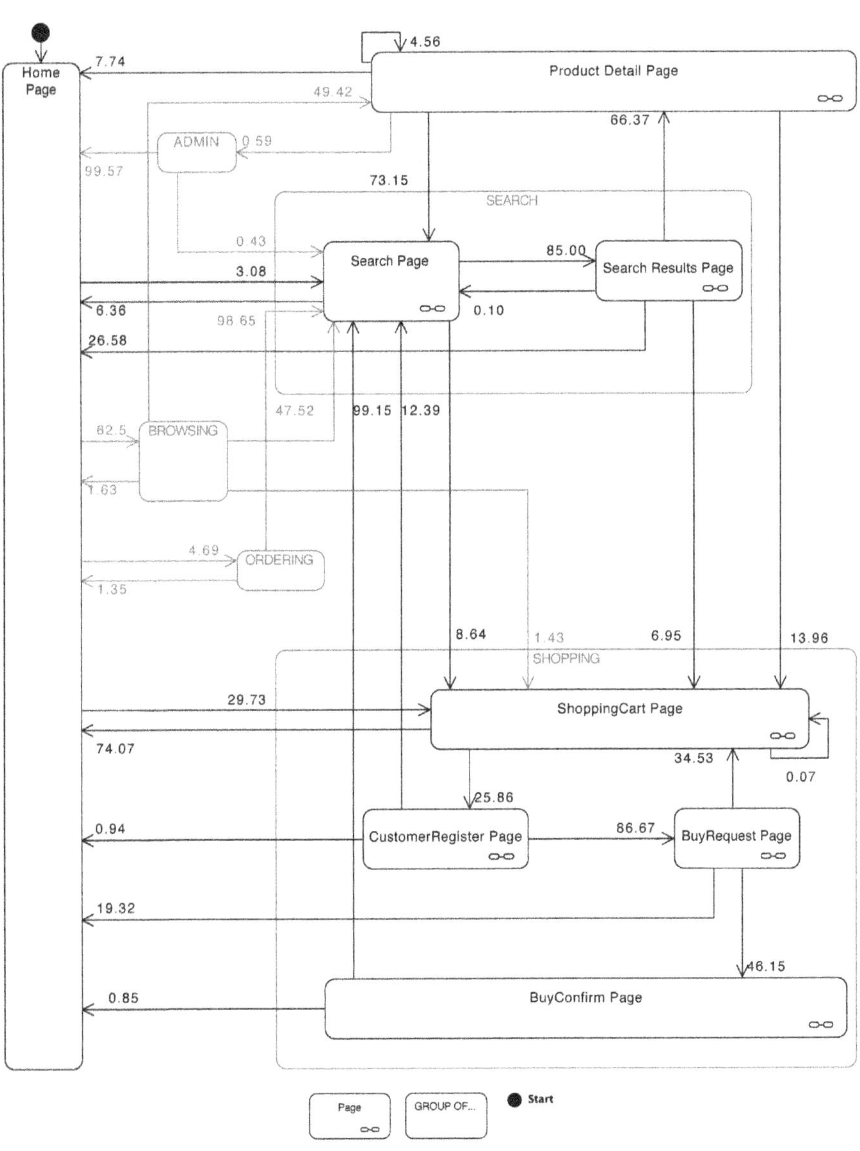

FIGURE 8.11

CBMG model for shopping scenario in GUERNICA validation.

For illustrative purposes, this section presents results of a subset of the most significant metrics when running TPC-W for the three scenarios defined by CBMG and DWEB.

Figure 8.12 and Figure 8.13 depict client and server performance metrics for the shopping scenario, respectively.

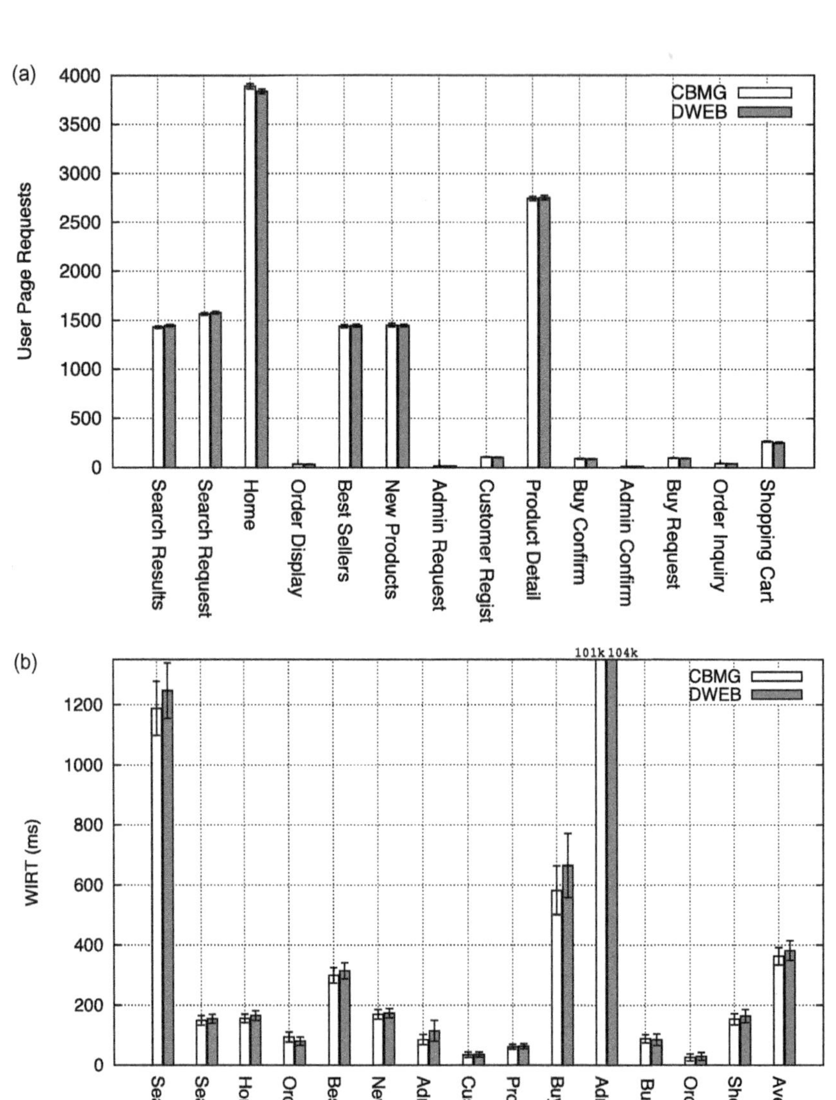

FIGURE 8.12

Client metrics obtained for the shopping scenario in GUERNICA validation: (a) User page requests; (b) WIRT.

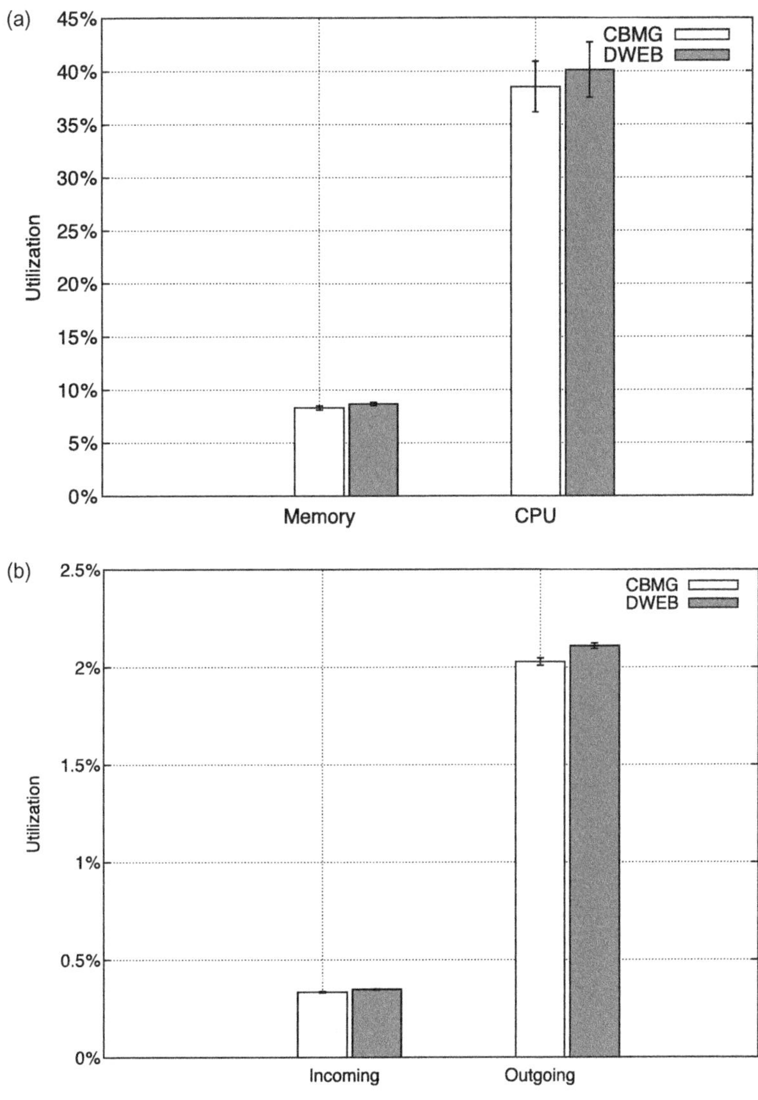

FIGURE 8.13

Server metrics obtained for the shopping scenario in GUERNICA validation: (a) Server memory and CPU utilization; (b) Server network utilization.

As shown in Figure 8.12a, both approximations generate a similar number of page requests. Figure 8.12b exhibits that the DWEB response time is, on average, 5% higher than that of CBMG, because some pages (e.g., Search Results or Buy Confirm) present very wide confidence intervals in this scenario. However, this difference does not affect the server performance metrics since, as observed in

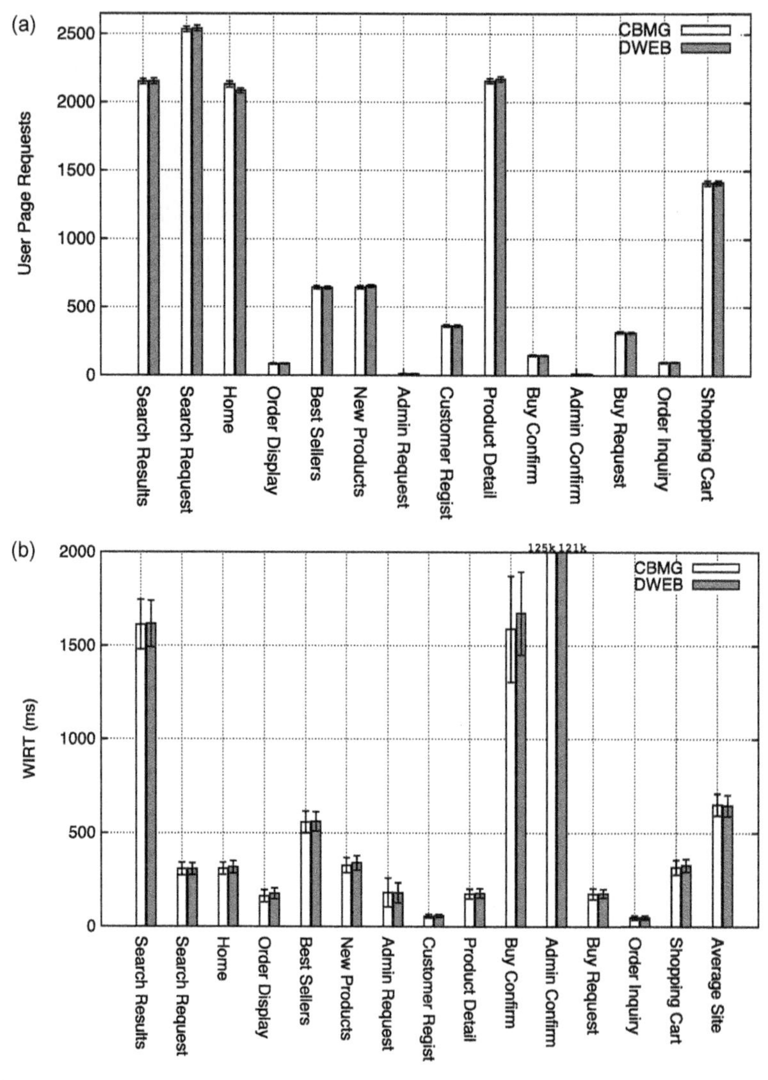

FIGURE 8.14

Client metrics obtained for the browsing scenario in GUERNICA validation: (a) User page requests; (b) WIRT.

Figure 8.13, the highest utilization is below 40% in both cases. The utilization for CPU and memory is rather low and similar in both cases (see Figure 8.13a). Incoming and outgoing traffic does not increase network utilization more than 2% in the studied workloads (see Figure 8.13b). Finally, the disk utilization is lower than 0.2% in both workloads (not shown in the figures). Browsing scenario results are illustrated in Figure 8.14 and Figure 8.15. Both workloads generate a similar

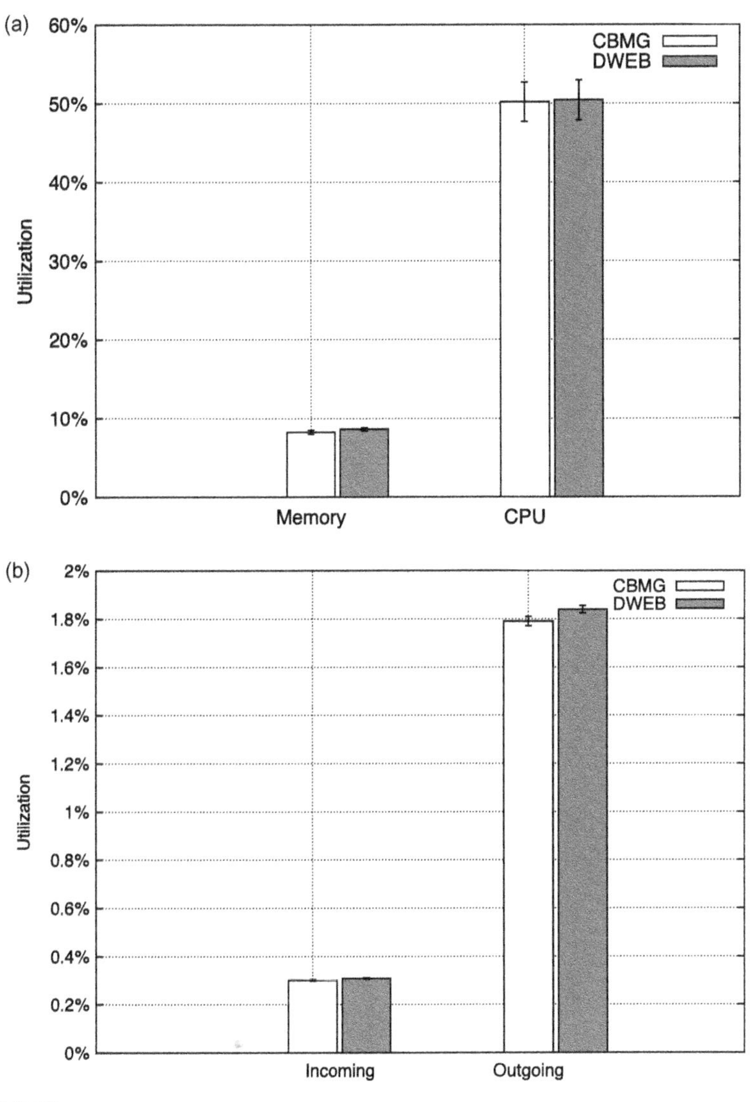

FIGURE 8.15

Server metrics obtained for the browsing scenario in GUERNICA validation: (a) Server memory and CPU utilization; (b) Server network utilization.

number of page requests and response time as shown in Figure 8.14a and Figure 8.14b, respectively. On the other hand, the server is characterized by a moderate level of stress in both cases. CPU utilization is about 50%, while the memory, network and disk utilizations are low, as observed in Figures 8.15a and Figure 8.15b, respectively.

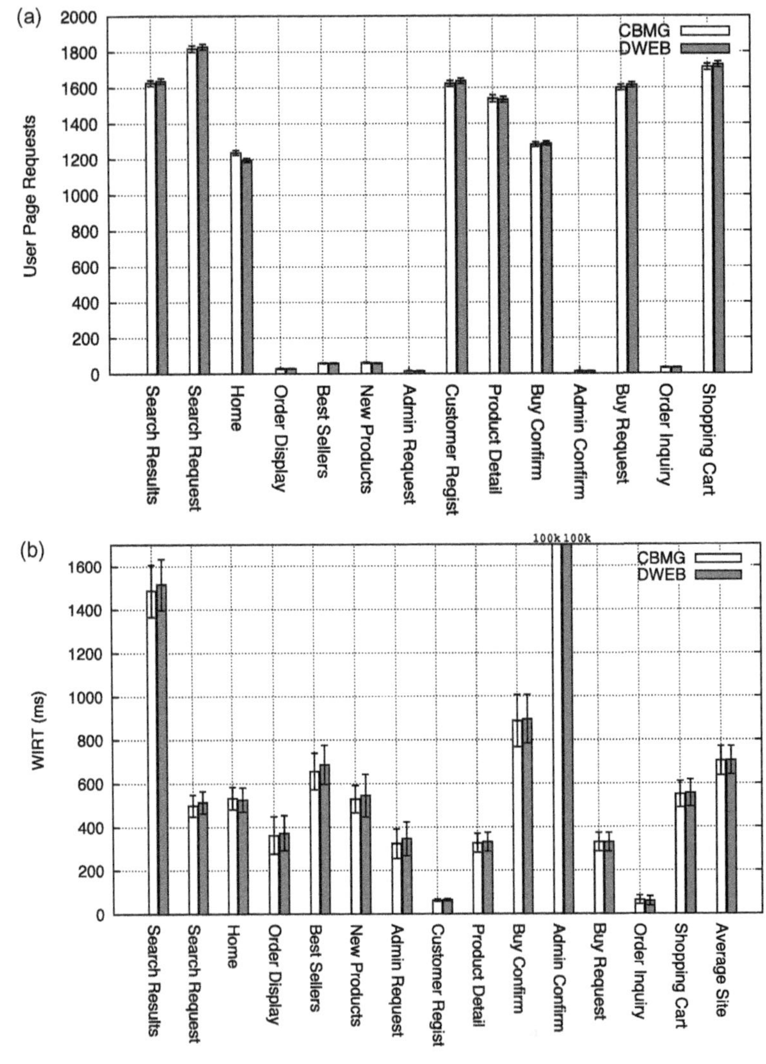

FIGURE 8.16

Client metrics obtained for the ordering scenario in GUERNICA validation: (a) User page requests; (b) WIRT.

Figure 8.16 and Figure 8.17 depict the results for the ordering scenario. The former shows that both workloads exhibit similar levels in the considered client metrics. The latter demonstrates that the highest server utilization is lower than 40% in both cases.

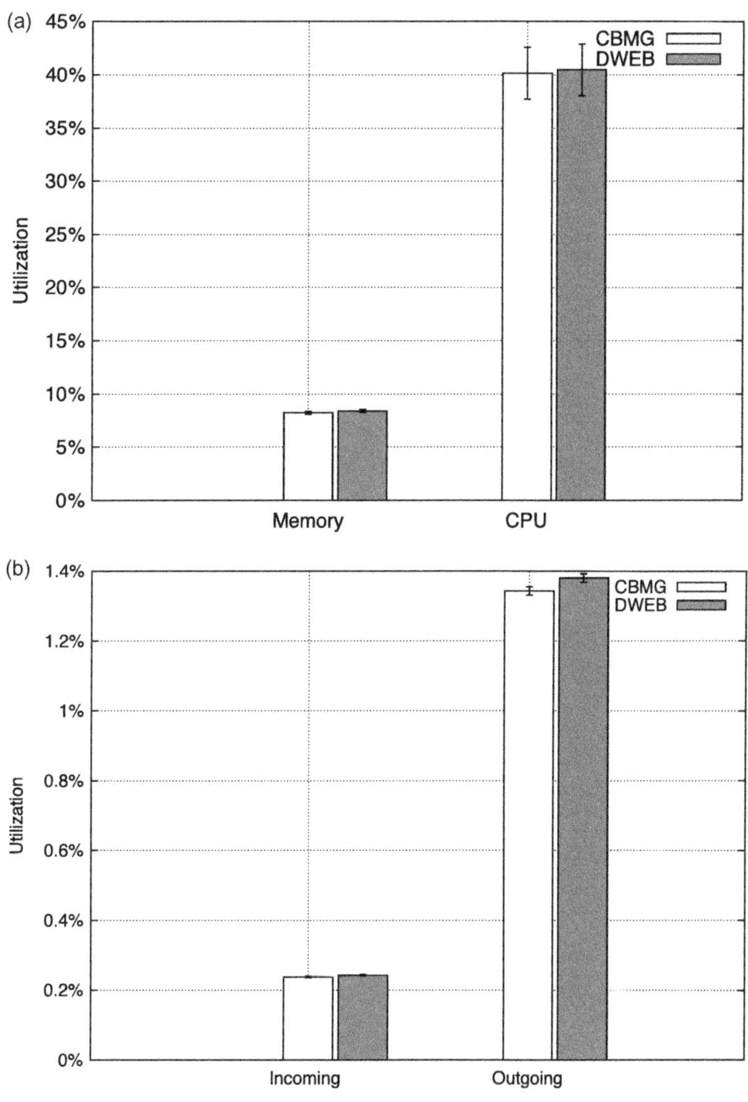

FIGURE 8.17

Server metrics obtained for the browsing scenario in GUERNICA validation: (a) Server memory and CPU utilization; (b) Server network utilization.

To sum up, we can conclude that the DWEB model and GUERNICA can generate accurate traditional workloads for web performance studies according to TPC-W. Moreover, due to their designs, our new testbed can be used to generate web workloads with a user's dynamic behavior.

4 SUMMARY

This chapter has presented GUERNICA, a new web workload generator based on a DWEB model, which aims to reproduce a dynamic user's workload in a more accurate and appropriate way than traditional approaches to workload generation. In addition, GUERNICA has been validated against a traditional approach to workload generation. In this way, we have developed a new testbed for performance evaluation with the ability of generating dynamic user workload by integrating GUERNICA into TPC-W.

GUERNICA implements the main DWEB concepts in order to adopt the model in several ways: (i) the user's navigation is incorporated with a scripting approach named *navigational plugin*, and the *workload test* carries out the concept of the user's roles from an analytical perspective. Furthermore, the generator represents the physical distribution of users in the Web by providing a distributed architecture, which permits one to set up different approaches (basic, advanced, and ideal) when generating Web workload. Additionally, GUERNICA supports, totally or partially, the main features introduced in well-known workload generators proposed in the open literature. These features work together under a five-phase methodology allowing GUERNICA to carry out performance and functional evaluation of web application in the current Web.

We have validated our approach by contrasting the new testbed main functionalities and behavior against TPC-W, and found that both implementations present similar behavior in traditional web workloads. Moreover, our approach represents a more valuable alternative because DWEB and GUERNICA are able to model and reproduce the user's dynamism on workload characterization in an accurate and appropriate way, respectively.

An excerpt of the main results introduced in this chapter was reported in [9].

ACKNOWLEDGMENTS

This work has been partially supported by Spanish Ministry of Economy and Competitiveness under grant TIN2013-43913.

REFERENCES

[1] Blazquez-Civico M, Contreras-Cino J, Peña-Ortiz R, Benjamins VR. Trends on legal knowledge, the semantic web and the regulation of electronic social systems. Chapter: visualization of semantic content. European Press Academic Publishing; 2007.

[2] Peña-Ortiz R, Gil JA, Sahuquillo J, Pont A. The impact of user-browser interaction on web performance. In: Proceedings of the 28th annual ACM symposium on applied computing, Coimbra, Portugal, March 18–22, 2013.

[3] Torres LM, Magaña E, Izal M, Morato D. Characterizing webpage load from the perspective of TCP connections. In: Proceedings of the computer science and information systems (FedCSIS), Warsaw, Poland; September 7–10, 2014.

[4] Lin W-S, Cassaigneb N, Huanc T-C. A framework of online shopping support for information recommendations. Expert Syst Appl 2010;37(10):6874–84.

[5] Chi EH, Pirolli P, Chen K, Pitkow J. Using information scent to model user information needs and actions and the Web. In: Conference on human factors in computing systems. Seattle, WA; 31 Mar–5Apr, 2001. p. 496–7.

[6] Transaction Processing Performance Council. TPC Benchmark™ W Specification. Version 1.8. Technical report, February 2002. Available from: <http://www.tpc.org/tpcw/spec/tpcw_v1.8.pdf>.

[7] Cain HW, Rajwar R, Marden M, Lipasti MH. An architectural evaluation of Java TPC-W. In: International symposium on high-performance computer architecture. Barcelona, Spain; January 2001. p. 229.

[8] Forster F. Collectd—The system statistics collection daemon [online]. 2014. Available from: <http://collectd.org/> [cited April 2014].

[9] Peña-Ortiz R, Sahuquillo J, Pont A, Gil JA. Dweb model: representing Web 2.0 dynamism. Comput Commun 2009;32(6):1118–28.

CHAPTER

The impact of dynamic user workloads on web performance: the e-commerce case study

9

Raúl Peña-Ortiz, José Antonio Gil, Julio Sahuquillo, Ana Pont, and Josep Domènech
Universitat Politècnica de València, València, Spain

1 INTRODUCTION

As in any performance evaluation process, accurate and representative workload models must be used in order to guarantee the validity of the results. Regarding web systems, the implicit user's dynamism hinders the design of accurate web workload representing users' navigations. Due to this fact, many research studies are still currently using nonrepresentative workloads of web navigations, thus affecting the credibility of results.

To deal with this shortcoming, in a previous work [1] we focused on modeling this dynamism, introducing the new web workload model DWEB (detailed in Chapter 6). This model is able to represent dynamic changes of user's behavior during a navigation session by adopting different roles (e.g., roles of browsing or ordering in an e-commerce environment). Another major concern when characterizing web workloads is the effect of considering the User-Browser Interaction (UBI) as a part of the intrinsic user's dynamism.

In this chapter we analyze and measure the effect of using dynamic workloads instead of traditional workloads, when evaluating web performance.

To this end, we evaluate a typical e-commerce scenario and compare the obtained results for different behaviors that take the user interaction into account, such as different user's reactions to the dynamic contents and services, the use of the back button and the parallel browsing originated by using browser tabs or opening new windows when surfing a website. Using the testbed introduced in Chapter 8, where the web workload generator GUERNICA was presented, we performed this evaluation.

The aim of this chapter is to explore how typical web performance metrics are affected by introducing different degrees of dynamism, rather than presenting a detailed dynamic workload based on actual users' behavior.

2 CONSIDERING DYNAMISM ON USERS' NAVIGATIONS

In a first step, a dynamic workload (DWEB workload 1, DW1) is defined with the aim of introducing users' dynamic navigations on workload characterization. For this purpose, we assume a common scenario of e-commerce where the main objective is to avoid the defection of customers. A large percentage of new customers—more than 60% in some sectors—defects before their third anniversary with an e-commerce website [2]. Consequently, these websites care deeply about customer retention and consider loyalty vital to the success of their on-line operations.

For the studied scenario, regarding customer retention in the on-line bookstore, we define a loyalty promotion consisting of a general discount only for those customers who buy at least once a month. The promotion introduces a new behavior with four cases of dynamism as summarized in Table 9.1.

The DWEB model allows us to model such cases of dynamism, which cannot be represented with this level of accuracy using traditional approaches such as the Customer Behavior Model Graph (CBMG) proposed by [3].

Figure 9.1 depicts the workload of a shopping scenario for an on-line bookstore modeled using CBMG. Alternatively, Figure 9.2 shows the workload characterization using DWEB (DW1 workload). Both graphs focus on searching and shopping groups of the TPC-W website, and highlight their main pages and transitions.

The CBMG workload considers the think time based on the TPC-W specification [4]. This benchmark defines the user think time (TT) as TT = T2 − T1, where T1 is the time measured at the emulated browser when the last byte of the last web interaction is received from the server, and T2 is the time measured when the first byte of the first HTTP request of the next web interaction is sent from

Table 9.1 Cases of Dynamism in Loyalty Promotion Behavior

Case	Description
1	If customers do not remember their last order status, they will check them by navigating into the ordering group of pages.
2	Because the customer has to buy at least once a month to keep the discount, a buying session must finish with a payment when he has not bought anything during that month.
3	An experienced customer only buys a book when its cost is 25% cheaper than in other markets.
4	The higher the number of provided search results, the longer the time that a user takes to read and think about them.

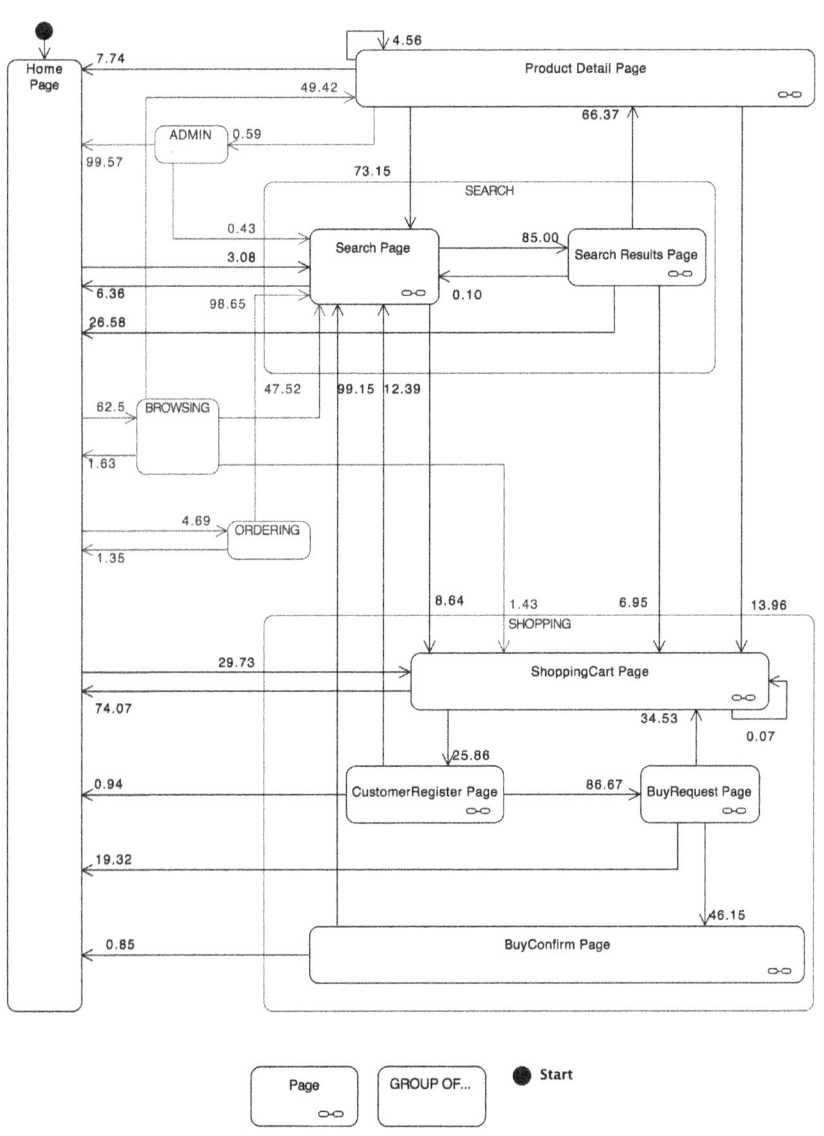

FIGURE 9.1

CBMG workload.

the emulated browser to the server. TPC-W considers that each think time must be taken independently from a negative exponential distribution, with the restriction that the average value must be greater than 7 seconds and less than 8 seconds, as shown in Equation (9.1).

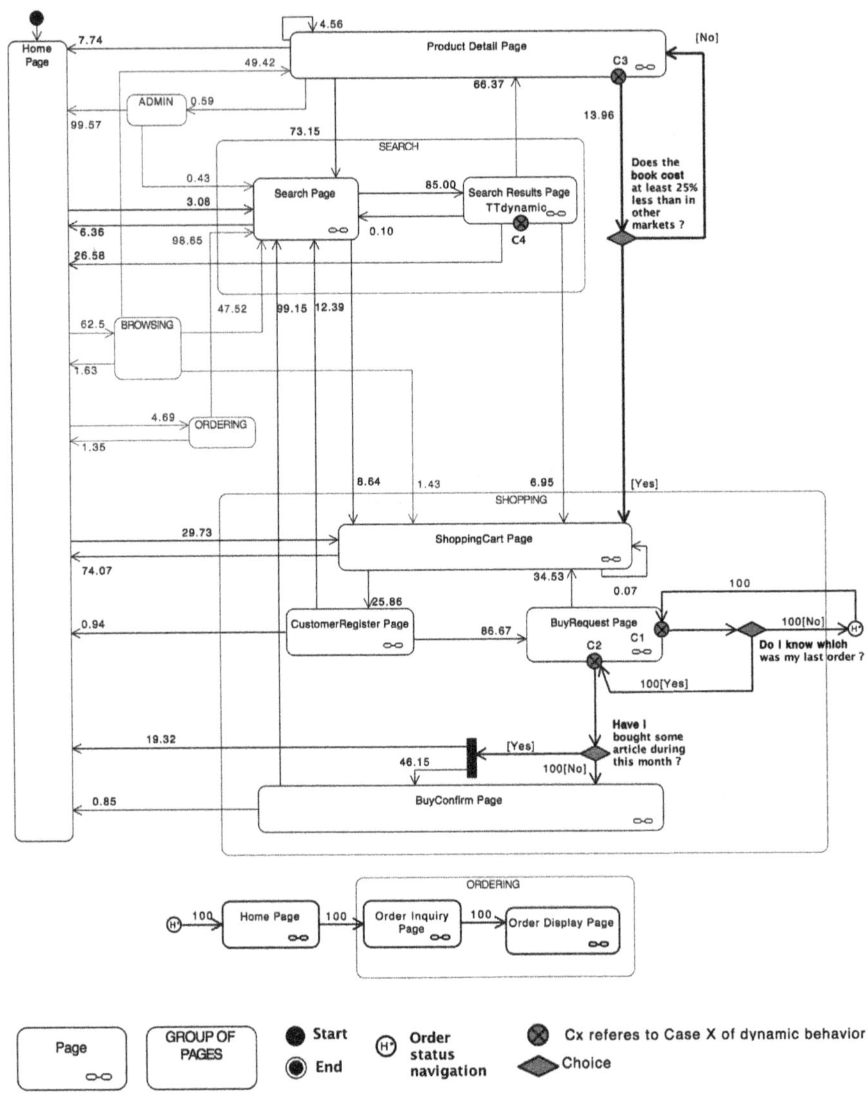

FIGURE 9.2

DWEB workload I DW1: navigation for loyalty promotion behavior.

An important drawback of this approach is that it does not consider the influence of current search results (web contents) on the user's think time, which exists in real navigations.

$$TT_{TPC\text{-}W} = -\ln(r)^{*}p, \text{ where } 0 < r < 1 \text{ and } (7 \le p \le 8) \tag{9.1}$$

In contrast, DWEB allows us to define a dynamic think time ($TT_{dynamic}$) according to the number of items returned by the search as shown in Equation (9.2), which is closer to real web activities like the one defined in Case 4. DW1 workload is still assuming TT for the less dynamic pages in the website (e.g., the Home page or the Product Detail page), but it uses $TT_{dynamic}$ in the Search Results page.

$$TT_{dynamic} = -\ln(r)^{*}p, \text{ where } 0 < r < 1 \text{ and } \left(p = 7 + \frac{\text{Number Of Search Results}}{\text{Max. Search Results}}\right) \quad (9.2)$$

The remaining cases of dynamism have been characterized using conditional transitions with DWEB. The transition from the Buy Request to the Buy Confirm pages depends on the last customer's purchase. If the customer did not buy any book during a given month, he has to commit the buying process; otherwise, he may finish the purchase or navigate to the Home page according to estimated probabilities of arcs as defined in Case 2. Notice that, when a customer does not remember the date of his last purchase, he must visit the ordering group in order to find it, as defined in Case 1. Finally, Case 3 has been implemented in DW1 workload with a conditional transition between the Product Detail and the Shopping Cart pages. This transition allows users to add a book to the shopping cart only in the case when its cost is 25% cheaper than in other markets.

3 ONE STEP AHEAD: EVOLVING USER'S PROFILE USING DYNAMIC ROLES

A common behavior of web users is the dynamism in the user's roles: i.e., the different roles a given user can adopt and the characteristics of the switching among roles. For instance, Chang et al. [5] reported three phases of marketing in an e-commerce website that can induce the mentioned dynamic evolution of web users. These phases are: presales, on-line, and after sales. The presales phase includes company efforts to attract customers by advertising, public relations, new products or service announcements, and other related activities such as discounts in some products or freebies (e.g., Apple promotions: Back to school and 12 Days of Christmas). Customers' electronic purchasing activities take place in the on-line sales where orders and charges are done through web facilities. The after-sales phase includes customer service, problem resolution, etc. User behavior evolution in e-commerce implies different navigation patterns that lead to different correlations between browsing and purchasing related metrics, such as reading and writing operations in a database [6]. In a second scenario, we define a new DWEB workload (DWEB workload II, DW2) that reproduces how user's behavior evolves from presales to on-line user profiles. We adopt the loyalty promotion behavior presented above as an example of on-line user profile, and define a new behavior based on a presales promotion on the studied on-line bookstore. The pre-sales promotion consists of 1000 bonus points to acquire common books

Table 9.2 Cases of Dynamism in the New Pre-Sales Promotion Behavior

Case	Description
1	Books that are bestsellers or new products cannot be added to the shopping cart because the bonus is only valid for common books.
2	The customer can buy a book only if the cost of the resulting shopping cart is lower than the bonus value.
3	A buying session (navigation session) finishes with a payment when the shopping cart cost is at least 75% of the bonus value.
4	A customer leaves the website when the buying session finishes.
5	The higher the number of provided search results, the longer the time that a user takes to read and think about them.

in a buying session. This promotion should present a different user behavior with five cases of dynamism as summarized in Table 9.2.

Figure 9.3 depicts the DW2 workload. This characterization defines the presales promotion behavior using the DWEB navigation concept (Figure 9.3a). Case 1 and Case 2 have been implemented with a conditional transition between the Product Detail and the Shopping Cart pages. This transition depends on the user's state (the available user's bonus) and the user's navigation path. That is, it only allows users to add a book to the shopping cart in the case when they arrive at the Product Detail page from other pages than the Bestsellers or New Products ones. The transition from the Shopping Cart to the Customer Register page depends on the content of the first page (Case 3), and it implies the end of the user navigation when the buying process is committed (Case 4). We also use the dynamic think time based on the number of items returned by the search (Case 5).

Finally, we combine both promotion behaviors (behaviors for loyalty and presales promotions) as dynamic roles by using the user's role concept of DWEB (Figure 9.3b). The user's role automaton considers that the average response of the presales promotion is 25% of users, based on the suggestions made in [7]. The transition between the presales promotion and the loyalty promotion behaviors models the evolution from a new user to a loyal customer, according to the average percentage of customer retention (40%) reported in [2], while the arcs arriving at a final state represent the users' defection (60%).

The *object requests per second* generated by CBMG workload are 45% to 60% higher than what is generated by dynamic workloads (see Figure 9.4a). However, the response time for DWEB workloads presents exponential curves with more pronounced slopes than the CBMG curve (see Figure 9.4b); for instance, DW2 workload increases the difference with respect to CBMG workload by 10%. On the other hand, the *total number of page requests* shows different values depending on the stressing server conditions (see Figure 9.4a). When the server is characterized by a low stress level (e.g., less than 40 browsers), the number of page requests generated by the dynamic workloads is 3% higher than that generated by CBMG, because $TT_{dynamic}$ reduces idle times when there are simultaneous requests on a search process.

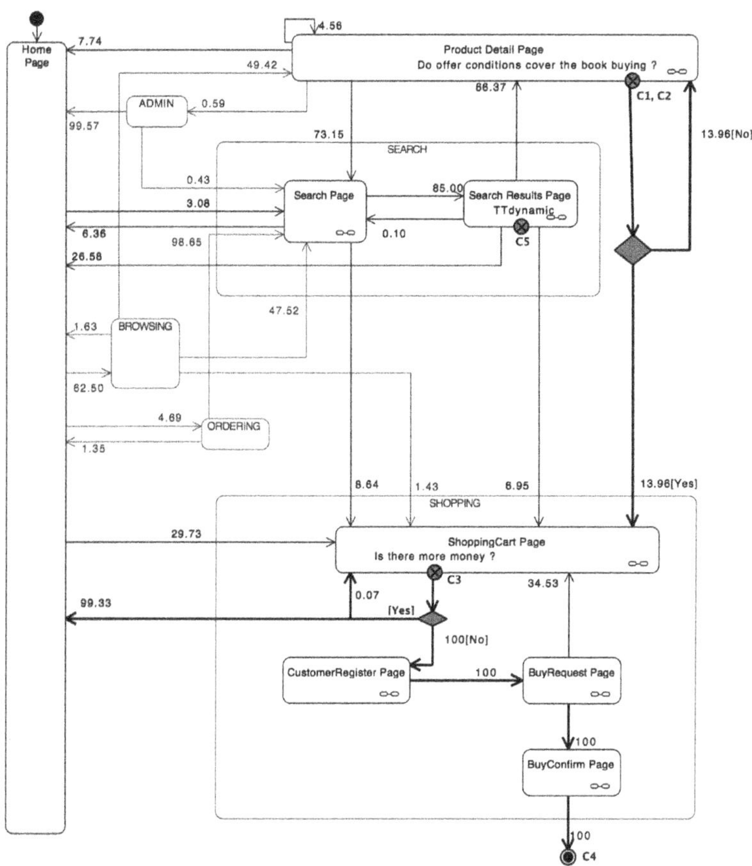

(a) Navigation for pre-sales promotion behavior

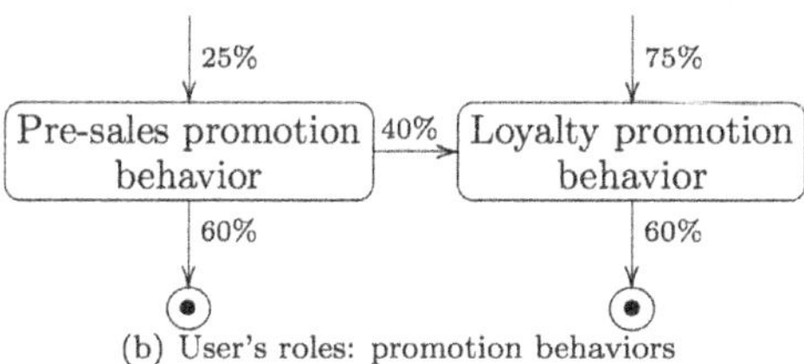

(b) User's roles: promotion behaviors

FIGURE 9.3

DWEB workload II DW2: characterization based on user's dynamic roles.

However, the CBMG workload requests a higher number of pages than DWEB workloads for a significant level of stress (e.g., more than 40 browsers), because the dynamism produces, in general, more complex requests that require extra service time, thus reducing service rate.

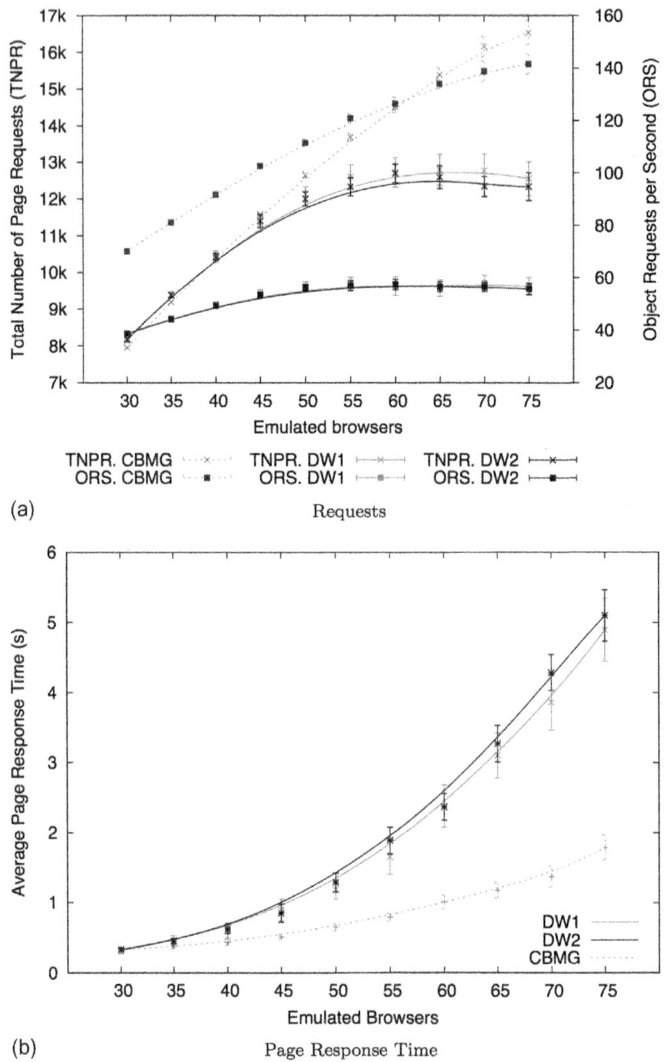

FIGURE 9.4

Main client performance metrics.

Regarding the utilization of the main hardware resources (CPU, memory, network and disk), only the processor presents significant differences. As expected, the CPU utilization increases with the number of emulated browsers as shown in Figure 9.5a. However, although the workloads present similar CPU utilization for a low number of browsers, differences between dynamic and traditional workloads can be as large as 30% when considering dynamism. Notice that a high CPU utilization means that the processor acts as the main performance bottleneck.

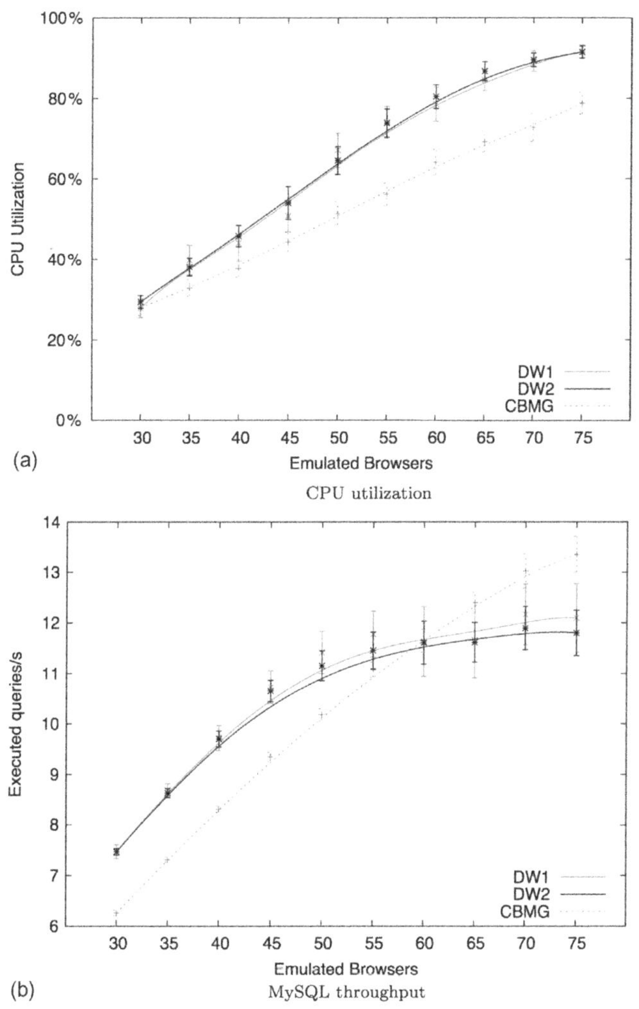

FIGURE 9.5

Main server performance metrics.

To better understand why the CPU utilization is so high, we studied how the main software components use the processor (Apache, Tomcat and MySQL). As observed in Table 9.3, MySQL almost monopolizes the processor since its execution time is more than two orders of magnitude higher than the time devoted to Tomcat, especially with dynamic workloads. Figure 9.5b shows the rate of queries executed by MySQL for each workload. As observed, for a relatively low number of emulated browsers (e.g., 45), this rate is 15% higher when considering dynamism. Nevertheless, more than 60 browsers cause the number of executed queries

Table 9.3 CPU Consumption (in jiffies[1]) for Each Application

EBs	Workload	Apache	Tomcat	MySQL
30	CBMG	7.71	4.55	886.48
	DW1	2.64	4.65	972.87
	DW2	2.61	4.93	1044.66
35	CBMG	9.36	5.41	1103.24
	DW1	3.44	6.47	1390.73
	DW2	3.25	6.34	1462.59
40	CBMG	11.01	6.45	1301.23
	DW1	3.87	7.51	1612.87
	DW2	3.86	7.92	1693.67
45	CBMG	13.16	8.03	1573.35
	DW1	4.52	9.05	1862.63
	DW2	4.67	9.59	2003.51
50	CBMG	14.95	9.64	1760.76
	DW1	5.45	11.52	2486.49
	DW2	5.32	11.45	2398.61
55	CBMG	17.13	11.59	1979.87
	DW1	6.00	12.95	2703.26
	DW2	6.04	13.03	2732.06
60	CBMG	19.04	12.98	2171.47
	DW1	6.45	13.82	2868.51
	DW2	6.55	14.11	2948.50
65	CBMG	21.97	15.42	2489.40
	DW1	6.81	14.97	3043.73
	DW2	7.00	15.19	3152.04
70	CBMG	23.51	16.81	2575.40
	DW1	6.97	14.99	3190.34
	DW2	7.28	15.72	3222.43
75	CBMG	25.68	18.08	2748.36
	DW1	7.23	15.33	3246.03
	DW2	7.32	15.80	3264.57

[1]A jiffy is a CPU consumption metric in Linux that indicates elapsed ticks since the system was started.

to become almost constant with dynamic workloads. This is due to a higher CPU utilization (greater than 80%) as depicted in Figure 9.5a. Consequently, the MySQL database is the major candidate to become a software bottleneck.

With the aim of evaluating the impact of dynamism on server stress peaks, we analyze the database usage associated with dynamic workloads. Before this study, we need to understand how MySQL database works. This database includes qcache as a cache of executed queries, where queries result in hit or miss. We also distinguish not-cached queries as a part of misses that cannot be cached for

```
SELECT  ol_i_id FROM orders , order_line
WHERE
        orders.o_id = order_line.ol_o_id
        AND NOT (order_line.ol_i_id=93234)
        AND orders.o_c_id  IN  (
                SELECT  o_c_id FROM orders , order_line
                WHERE orders.o_id= order_line.ol_o_id AND
                        Orders.o_id >(SELECT MAX(o_id)  ...)
)   ...
```

LISTING 9.1

Not cached queries example at TPC-W website.

their dynamic nature or their complexity [8]. Listing 9.1 shows a not-cached query example at TPC-W website.

For a deeper study, we compared the cache status of the executed queries versus CPU utilization for several stress levels. Figure 9.6 shows the values for 35, 55 and 75 emulated browsers as examples of low stress, significant stress, and overloaded situations at the server side, respectively. DW2 workload (the right column) presents a higher number of misses than DW1 and CBMG workloads (center and left columns, respectively). CPU overload peaks become larger and larger with the increase of not-cached queries, which is caused by the dynamic query nature, especially for the DW2 workload that presents higher increase than the DW1 workload. Note that hits are not represented because their execution time is around 1 millisecond, while the time taken by misses might be more than 8 seconds.

Finally, we present an example illustrating how the system level of stressing caused by user's dynamism might induce a higher probability of user abandonment. According to a Jupiter Research report [9] commissioned by Akamai, web page rendering should be no longer than 4 seconds to avoid user abandonment. Figure 9.7 depicts how dynamic workloads increase the probability that a response takes over 4 seconds, especially when considering changes of user behaviors in an overloaded system. Consequently, if the server is not properly tuned, the extra workload induced by the user's dynamic behavior can increase the probability of user abandonment (up to 40%).

In summary, results show that considering user's dynamism when characterizing web workload affects system performance. Dynamism on workload characterization introduces new patterns of HTTP requests. These patterns, in general, reduce the number of requests to objects but increase the number of requests to dynamic web content, thus incurring differences in the system performance metrics, especially on the server's processor usage and the database throughput. As a result, the server performance degradation affects service conditions, increasing the average response time that induces a higher number of user abandonments. Results also proved that considering a user's dynamic navigations on workload characterization produces a higher impact on the system performance metrics than modeling changes in user roles, because users' decisions when navigating the Web are more dynamic than changes in their roles.

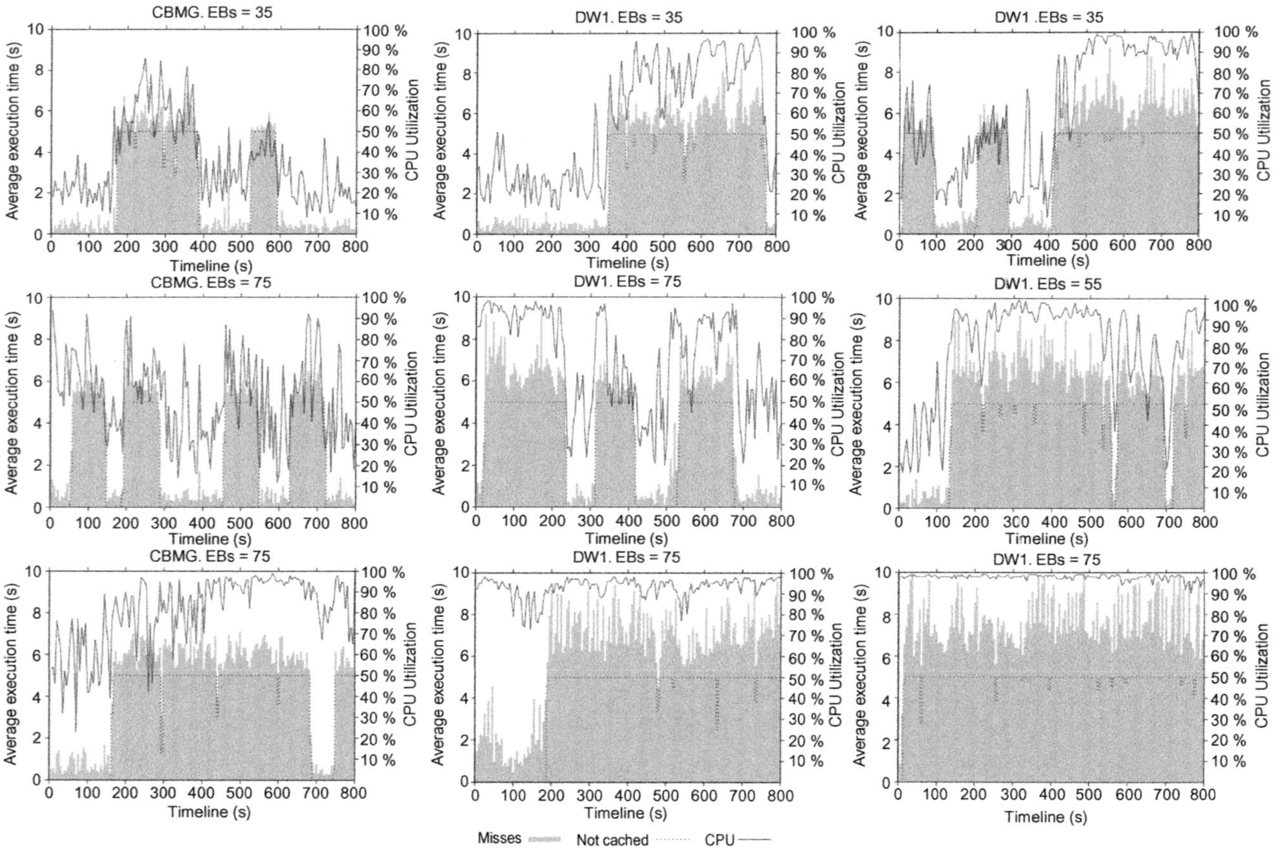

FIGURE 9.6

CPU utilization by query cache status.

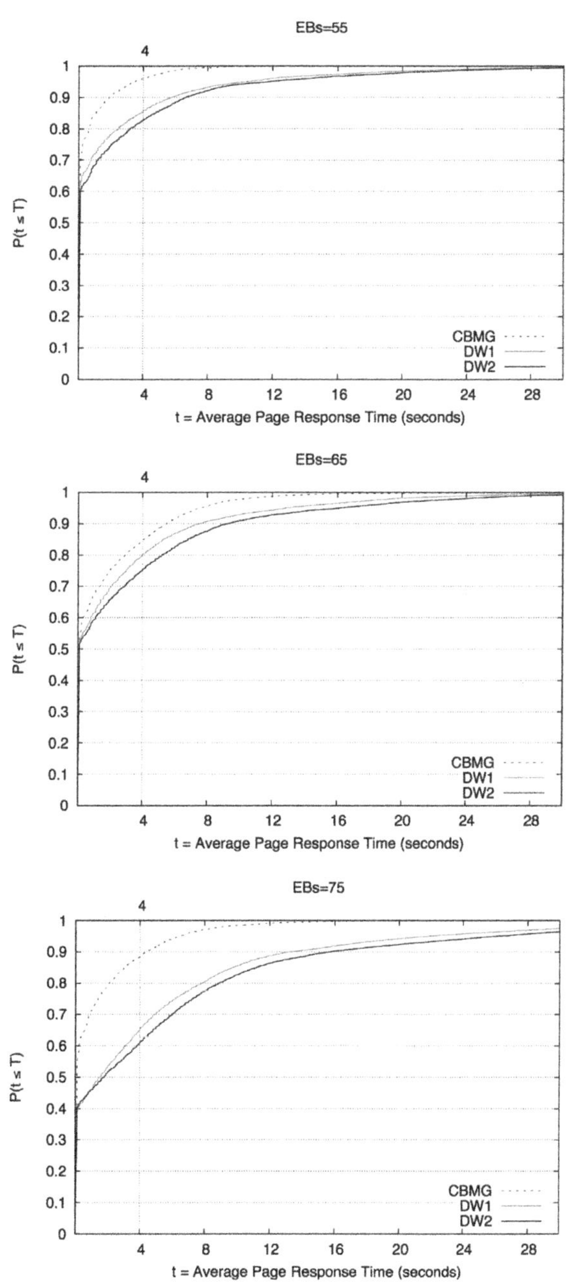

FIGURE 9.7

Cumulative distribution for page response time.

4 MODELING THE USER-BROWSER INTERACTION

Using DWEB we introduced different levels of user dynamism on workload characterization, specifically dynamism related to the interaction with the web browser. In Section 4.1 we focus on user's goals when surfing a website and define user's navigations according to this end. After that, a more realistic dynamic workload is defined considering rapid return to recently visited pages by using the history-back button on web browsers. Finally, Section 4.2 introduces parallel tab browsing behavior on workload characterization.

4.1 THE BACK BUTTON: RAPID RETURN TO RECENTLY VISITED PAGES

Web browsers include a stack-based model for web page navigation [10]. In this model, there are two traditional ways of displaying pages in the browser: load and reload. Pages are loaded when the user clicks on a link, types a URL or selects a favorite page. The effect of load is to add the page to the top of the visited page's stack. Pages are reloaded with the back and forward buttons, and the effect is to alter the position within the stack. Each back click shows the next page down the stack until the stack bottom is reached. Forward clicks show pages up the stack until the top is reached.

An important factor of the back button success is the chance of rapid return to recently visited pages [11], which can avoid new HTTP requests and consequently changes the causes of the stressing conditions of a given website. However, only certain visited pages can be cached by a web browser with the aim of going back using the back button. That is, some web contents such as audio or video streaming, dynamic contents or web forms are not cached according to their HTTP headers. For these types of contents the back button does not have any effect because pages are completely reloaded.

With the purpose of measuring the effect of UBI in web performance evaluation we compare different dynamic workloads. In a first step, a workload (LOY) conducted by user's goals is defined. To this end, we extend the loyalty promotion behavior presented in Section 3 by adding a new case of dynamism to establish that a user leaves the website when his goals are satisfied. The new behavior is characterized by five cases of dynamism as summarized in Table 9.4. Secondly, a new DWEB workload (LOYB) is defined by extending the LOY workload with an extra case of dynamism that characterizes the use of the back button on the web.

Figure 9.8 and Figure 9.9 show the workloads generated using DWEB for the loyalty promotion behavior conducted by goals (LOY) and its extended version (LOYB) considering the back button, respectively. Both workloads assume *TT* (see Equation (9.1)) as *think time* for the less dynamic pages in the

Table 9.4 Cases of Dynamism in the Loyalty Promotion Behaviors Conducted by Goals

Case	Description
1	If customers do not remember their last order status, they will check them by navigating into the ordering group of pages.
2	Because the customer has to buy at least once a month to keep the discount, a buying session must finish with a payment when he has not bought anything during that month.
3	An experienced customer only buys a book when its cost is 25% cheaper than in other markets.
4	The higher the number of provided search results, the longer the time that a user takes to read and think about them.
5	A customer leaves the website when the buying session finishes because his goals have been satisfied.
Extra case in the extended behavior	
6	A customer can return to recently visited listings of books (browsing listings or search results) without repeating a request to the web server through the back button.

website (e.g., the Home page or the Product Detail page), and $TT_{dynamic}$ (see Equation (9.2)) in the Search Result page, that is, closer to real web activities like that defined in *case 4* listed in Table 9.5.

The remaining cases of dynamism have been characterized using conditional transitions with DWEB. The transition from the Buy Request page to the Buy Confirm page depends on the last customer's purchase. If the customer does not buy any book during a given month, he has to commit the buying process, which means the end of the navigation session (*case 5*). Otherwise, he may finish the purchase or navigate to the Home page according to estimated probabilities of arcs as defined in *case 2* and shown in both figures. Notice that when a customer does not remember the date of his last purchase, he must visit the ordering group in order to find it out, as defined in *case 1*. *Case 3* has been implemented with a conditional transition between the Product Detail and the Shopping Cart pages. This transition represents users adding a book to the shopping cart because its cost is 25% cheaper than in other markets. Finally, the back button has been characterized in LOYB workload as a cache that only stores the immediately previous listing page (*case 6*).

4.2 OPTIMIZING USER PRODUCTIVITY: THE PARALLEL TAB BROWSING BEHAVIOR

Parallel browsing describes a behavior where users visit web pages in multiple concurrent threads by using web browsers' tabs or windows. To help the understanding of how parallel browsing works, Figure 9.10 illustrates a parallel tab

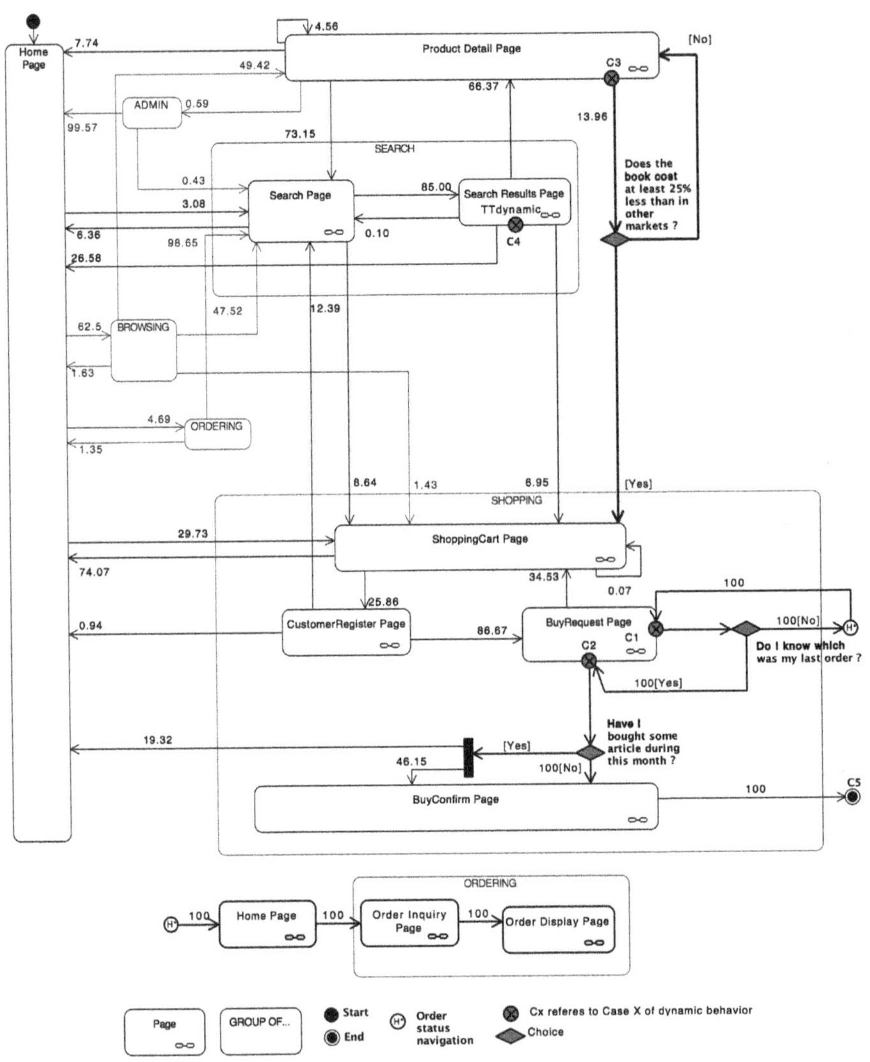

FIGURE 9.8

LOY workload: loyalty promotion behaviors conducted by goals.

browsing session applied to the on-line bookstore. It shows how a web user uses parallel browsing based on three web browser tabs to improve his navigation time by avoiding searches of books. The user begins a navigation session in a window with the aim of buying a book that fulfills several requirements as soon as possible. After he visits some pages, a search result is provided. At this point of time, he starts a parallel tab browsing behavior by opening three new tabs with the detail of three different books selected from the results. Then, he takes some time

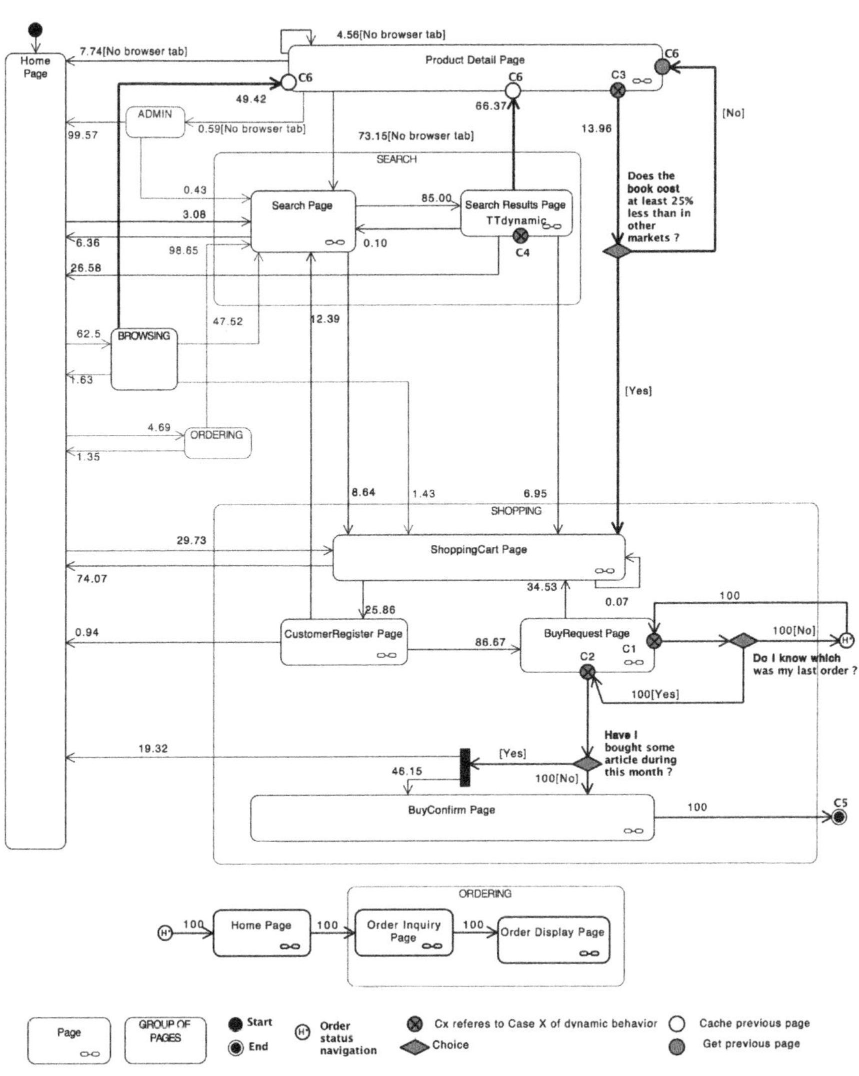

FIGURE 9.9

LOYB workload: LOY workload considering the use of the back button.

calculated according to Equation (9.1), such as Think Time 4 or Think Time 5 (see Figure 9.10), to evaluate the content of each tab until he finds a book satisfying his requirements. When a book does not fulfill the requirements, the user closes its associated tab and switches to the next one. Otherwise he discards the rest of the tabs and adds the book to the shopping cart, finishing the parallel browsing session. Notice that the user does not take any think time on a book detail page when he discards its tab.

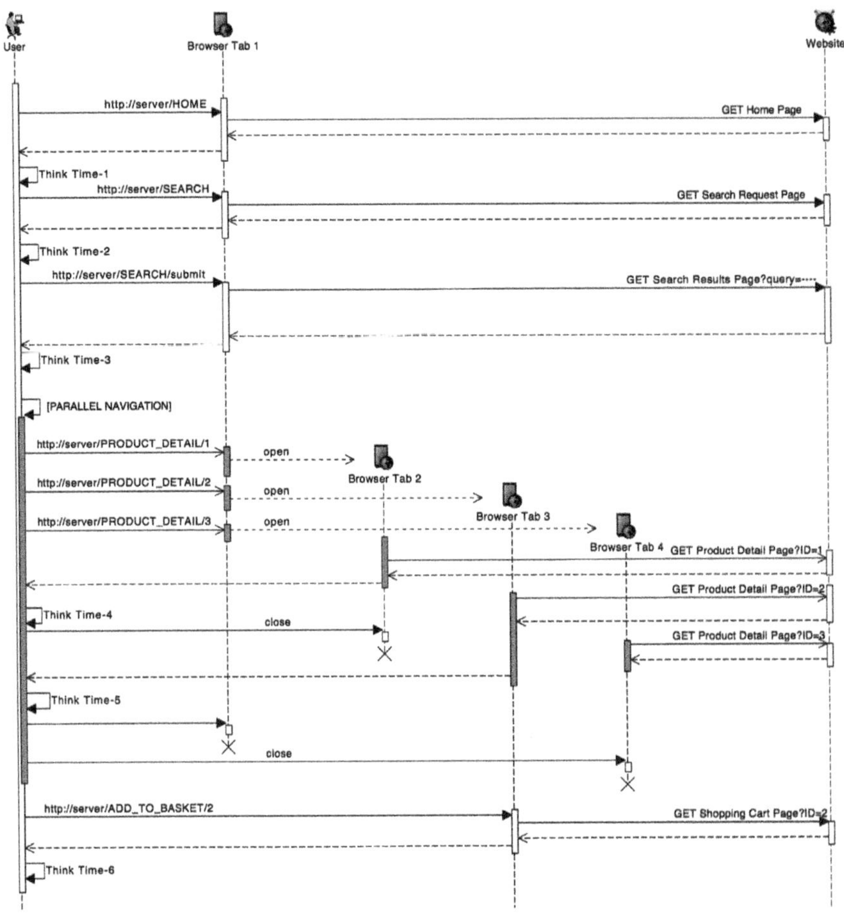

FIGURE 9.10

Example of parallel tab browsing session.

In the third scenario, we define a new DWEB workload (LOYT) that reproduces parallel tab browsing behavior in the loyalty promotion. With this aim, we extended the loyalty promotion behavior presented above by characterizing the navigation on web browser tabs as summarized in Table 9.5.

Figure 9.11 depicts the LOYT workload. This characterization defines the same behavior as the LOY workload (Figure 9.8), except for the navigations related to parallel tab browsing (*cases 6* and *7*). *Case 6* has been implemented with a pool of three parallel navigation threads, one for each tab. A navigation thread is killed when its book does not fulfill the user's requirements, or when the user finds the required book in another thread, as defined in *case 7*. The

Table 9.5 Extra Cases of Dynamism in the Loyalty Promotion Behavior Conducted by Goals to Represent Parallel Tab Browsing

Case	Description
6	When a user has to review a listing of books, such as the result of a search or a browsing request, he begins a parallel tab browsing session with three tabs.
7	A user closes a tab when its book does not fulfill his buying requirements or when he has found the desired book in another tab.

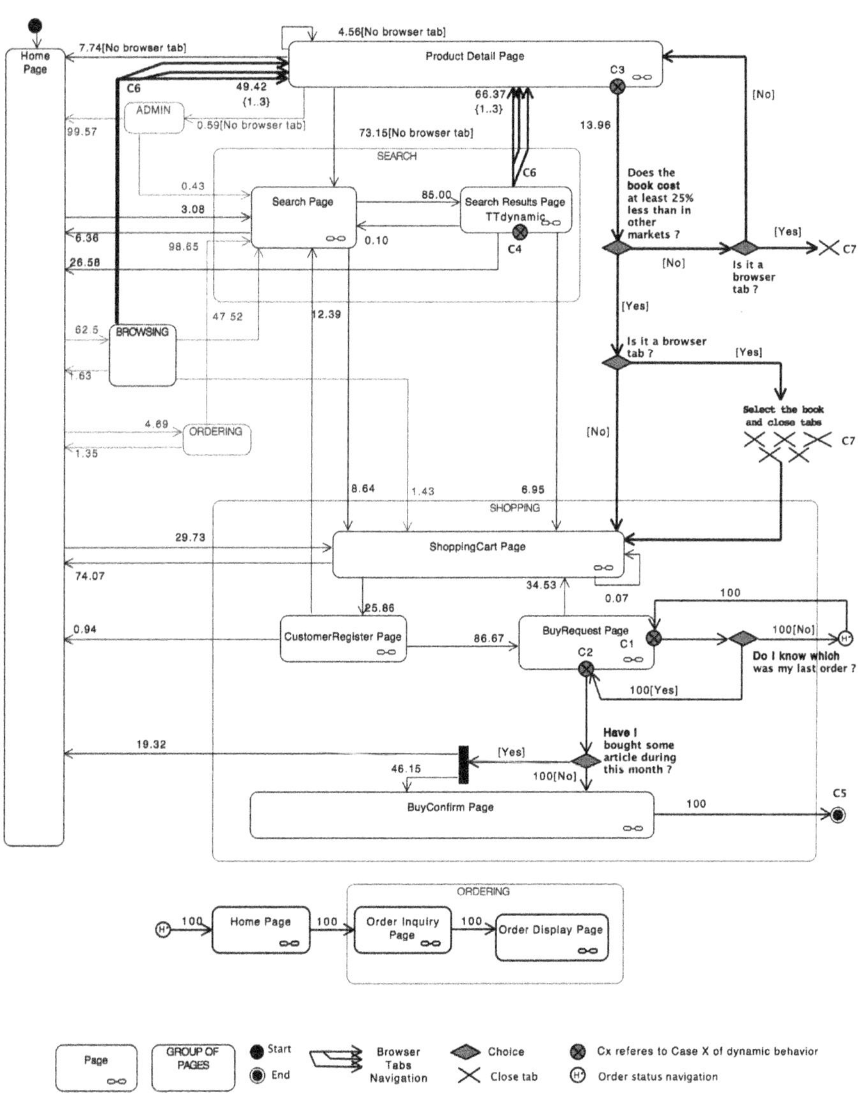

FIGURE 9.11

LOYT workload: parallel tab browsing behavior in LOY workload.

navigation becomes sequential again when only a thread is kept alive. Notice that, despite the fact that the LOYT workload does not consider the possibility of opening new tabs from another existing tab in this study, DWEB and GUERNICA provide mechanisms to model and execute multilevels in tab branching, respectively, if required.

4.3 IMPACT OF UBI ON WEB PERFORMANCE

Experimental tests have been carried out to compare the performance achieved with the loyalty promotion workload (LOY) against those of the extended versions (LOYB and LOYT), which consider UBI when modeling the user's behavior. With the aim of finding out the stress borderline of the server for each workload, we varied the number of users ranging from 50 to 250 in 50-user steps.

Additional metrics measured at the client side have been defined in order to quantify the user's productivity, such as number of finished sessions, session length and number of visited pages per session. Notice that a navigation session finishes when a user leaves the website after his goals are achieved; that is, after he buys some books in the case study. Thus, the higher the number of finished sessions, the higher the user's productivity.

Although we measured all the performance metrics listed in Table 9.6, only those showing significant differences in the studied workloads are discussed in the following paragraphs.

Figure 9.12 shows how the user's productivity (measured in number of finished sessions) increases when considering UBI on workload characterization, especially when a parallel browsing behavior is introduced, because both the session length (in seconds) and the number of visited pages per session are lower in these workloads (see Table 9.7 considering 100 simultaneous users as example). Therefore, the user's productivity is improved when the user changes his navigation patterns as a result of parallel browsing behavior (up to 200%) or using the back button (up to 100%).

In general, the extended workloads also increase the server throughput, despite the LOY workload degrading the service conditions more than considering the interactive behaviors as depicted in Figure 9.13. Considering a significant number of simultaneous users (e.g., from 100 users on), the total served pages is on average 25% and 90% higher for the extended workloads (Figure 9.13). To identify the causes of this high increase, served pages are classified into three main types: search results page, product detail page and others. As observed in Figure 9.14, the interactive behaviors generate fewer requests to the search engine (Figure 9.14a) and increase the number of requests to the product detail page (Figure 9.14b) and to the others (Figure 9.14c), especially for the LOYT workload. This new pattern of HTTP requests reduces the complexity of database queries when decreasing searches, so the throughput of the web server increases as shown in Figure 9.15. Specifically, the Apache HTTP requests per second

Table 9.6 Performance Metrics Classification According to the Evaluated Resource

Resource			Metric	Description/Formula
Client Side			Response Time (WIRT)	WIRT is defined by TPC-W as t2 – t1, where t1 is the time measured at the Emulated Browsers when the first byte of the first HTTP request of the web interaction is sent by the browser to the server, and t2 is the time when the last byte of the last HTTP response that completes the web interaction is received.
			Average Response Time ($\overline{WIRT}$)	$\overline{WIRT} = \frac{\sum_{i \in Pages} WIRT_i * Req_i}{\sum_{i \in Pages} Req_i}$
			Req_{page}	Requests per Page ($Reqpage$) are the number of connections for a page requested by Emulated Browsers and accepted by the server.
Server Side	Hardware	CPU Memory Disk Network	U_{CPU} U_{memory} U_{disk}, X_{disk} U_{net}, X_{net}	Metrics for hardware resources include utilization for all of them, and throughput for the disk and the network.
	Software	Apache Tomcat MySQL	$X_{apache}, CPU_{apache}, MEM_{apache}$ $X_{tomcat}, CPU_{tomcat}, MEM_{tomcat}$ $X_{mysql}, CPU_{mysql}, MEM_{mysql}$	Performance metrics for software components of server include: their throughput, the CPU and memory consumption, the number of processes or threads, etc.

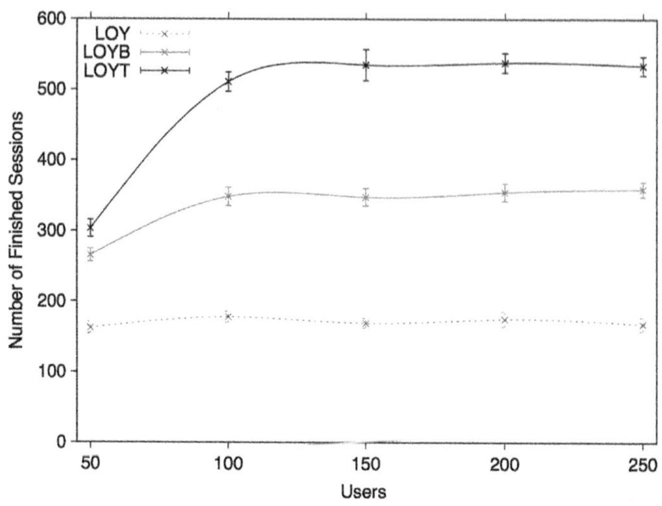

FIGURE 9.12

User's productivity evolution.

Table 9.7 Mean User Productivity Considering 100 Simultaneous Users in the System

Metric	LOY	LOYB	LOYT
Number of finished sessions	177.530	348.180	526.380
Session length (sec)	598.024	230.032	132.839
Number of visited pages per session	78.909	26.146	18.814

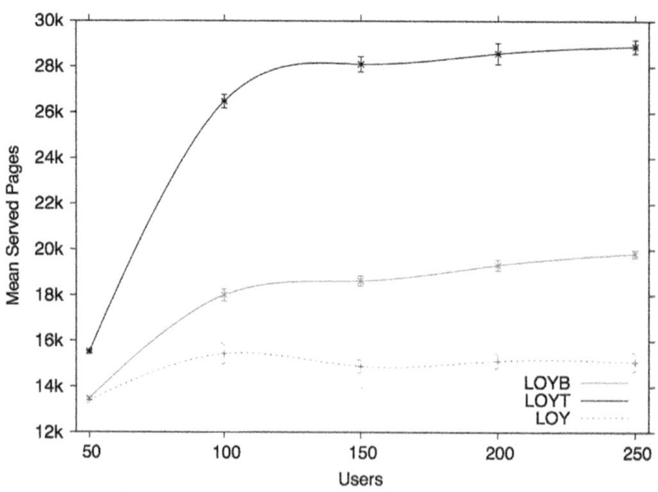

FIGURE 9.13

Total served pages.

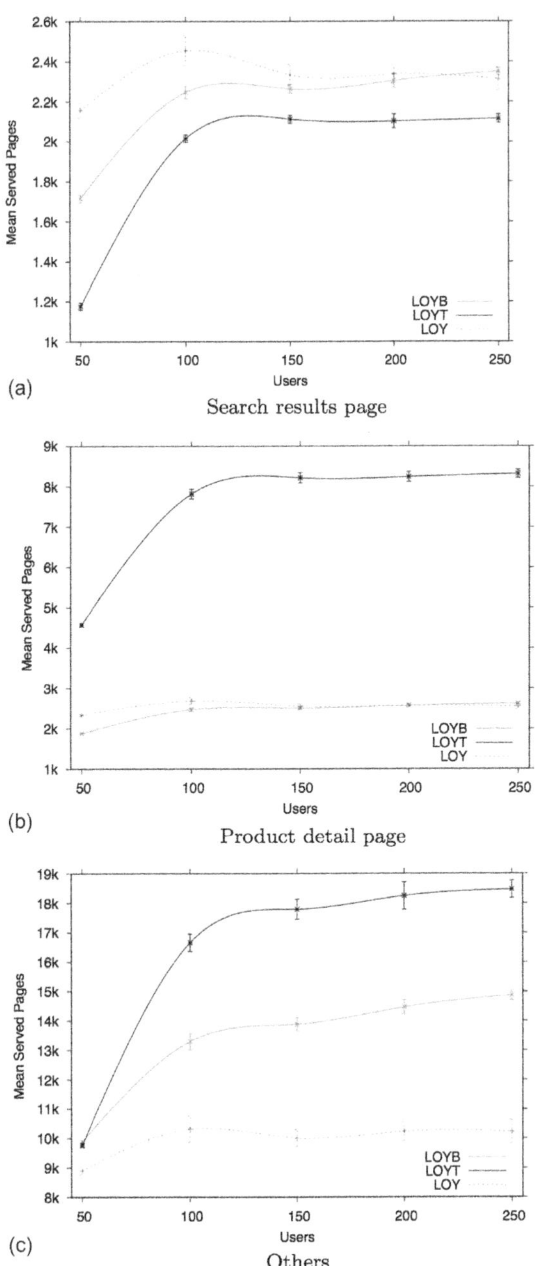

FIGURE 9.14

Mean served pages by type.

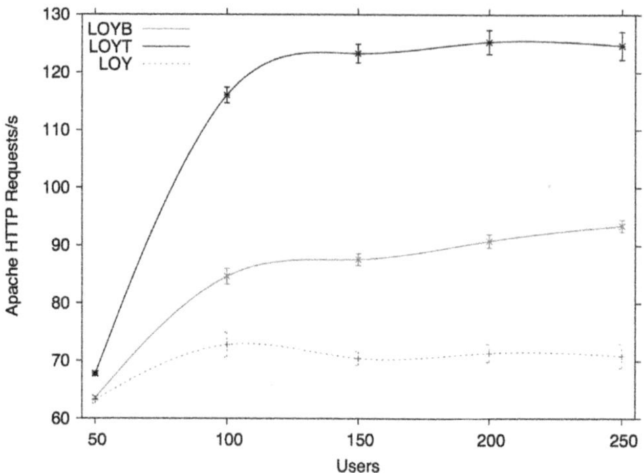

FIGURE 9.15

Apache throughput.

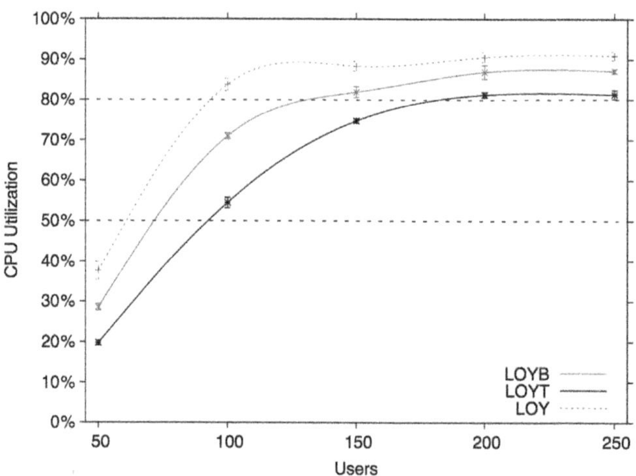

FIGURE 9.16

Server CPU utilization.

generated by the LOY workload is 25% and 75% lower than those generated by the LOYB and the LOYT workloads, respectively.

Figure 9.16 shows the *CPU utilization*, which is the only element that presents significant differences among the main hardware resources. Stress conditions have been classified into three levels according to the CPU utilization values: low stress ($U_{UCP} < 50\%$), significant stress ($50\% \leq U_{UCP} \leq 80\%$), and

Table 9.8 CPU Consumption (in jiffies) for Each Application

(a) LOY **Users**	**50**	**100**	**150**	**200**	**250**
Soft.					
MySQL	1315	2883	2807	2768	2897
Tomcat	10	22	22	23	23
Apache	4	9	10	10	10
(b) LOYB **Users**	**50**	**100**	**150**	**200**	**250**
Soft.					
MySQL	931	2439	2606	2540	2624
Tomcat	9	24	26	27	28
Apache	3	9	11	12	13
(c) LOYT **Users**	**50**	**100**	**150**	**200**	**250**
Soft.					
MySQL	567	1808	2381	2434	2424
Tomcat	8	27	33	34	34
Apache	3	10	15	17	18

high stress—here by overload at the server—($U_{UCP} > 80\%$). As can be seen, the processor utilization for the extended workloads is always lower than for the LOY workload. For instance, considering 100 users, the utilization decreases by 15% and 45% for the LOYB and the LOYT workloads, respectively. Notice that the high CPU utilization value shows that the processor acts as the main performance bottleneck but the stress borderline moves from 100 users for the LOY workload to 150 and 200 users for the LOYB and the LOYT workloads, respectively. This means that the web server allows the system to serve about 50% and 100% more users for the LOYB and the LOYT workloads, respectively.

To better understand why the CPU utilization decreases, we studied how the main software components (Apache, Tomcat and MySQL) make use of the processor. As observed in Table 9.8, MySQL almost monopolizes the processor for the different workloads since its execution time is more than two orders of magnitude higher than the time devoted to Tomcat and Apache. Consequently, MySQL database is the major candidate to be a software bottleneck. However, despite the fact that the executed queries rate for the extended workloads can be as much as 30% and 93% higher than for the LOY workload (Figure 9.17), the CPU time consumed by MySQL for the extended workloads

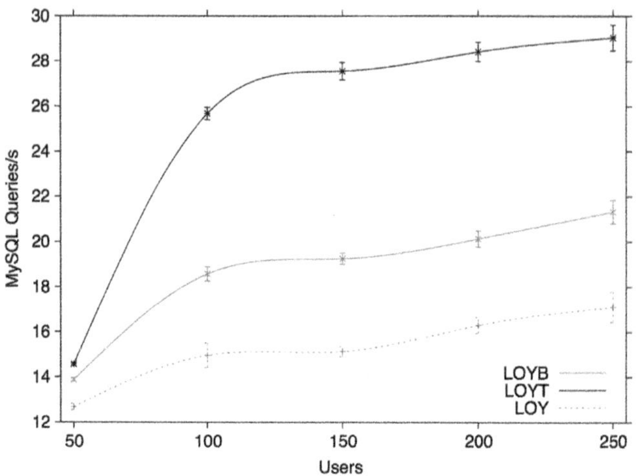

FIGURE 9.17

MySQL throughput.

is 30% and 55% lower than for the LOY workload. Moreover, the CPU consumption of Tomcat and Apache increases for the extended workloads. That is, there is a change in the patterns of HTTP requests and consequently in the type of executed queries at MySQL.

We performed a deeper study to provide a sound understanding of how the database is used by these workloads. As explained before, the MySQL database includes `qcache` [7] as a cache of executed queries, where a query results in a hit or miss. Figure 9.18 shows the mean execution time for hits, misses and total queries considering 50, 100 and 250 simultaneous users in the system as examples of the different stress conditions: low stress (Figure 9.18a), significant stress (Figure 9.18b) and overload at server (Figure 9.18c). As observed, the mean execution time for a query when considering the extended workloads is always lower than for the LOY workload load because decreasing the number of searches reduces the complexity, on average, of misses (the execution time decreases for the extended workloads) and increases the number of hits.

In summary, results show that considering user interaction resorting to web browser features positively affects the system performance. This is because considering UBI when modeling the user's behavior introduces new patterns of HTTP requests. These patterns, in general, increase the number of requests to objects but reduce requests to the search engine, allowing users to achieve a higher productivity. This fact causes noticeable differences in the performance metrics, especially in the processor utilization and the server throughput. As a result, the stress borderline of the server is affected.

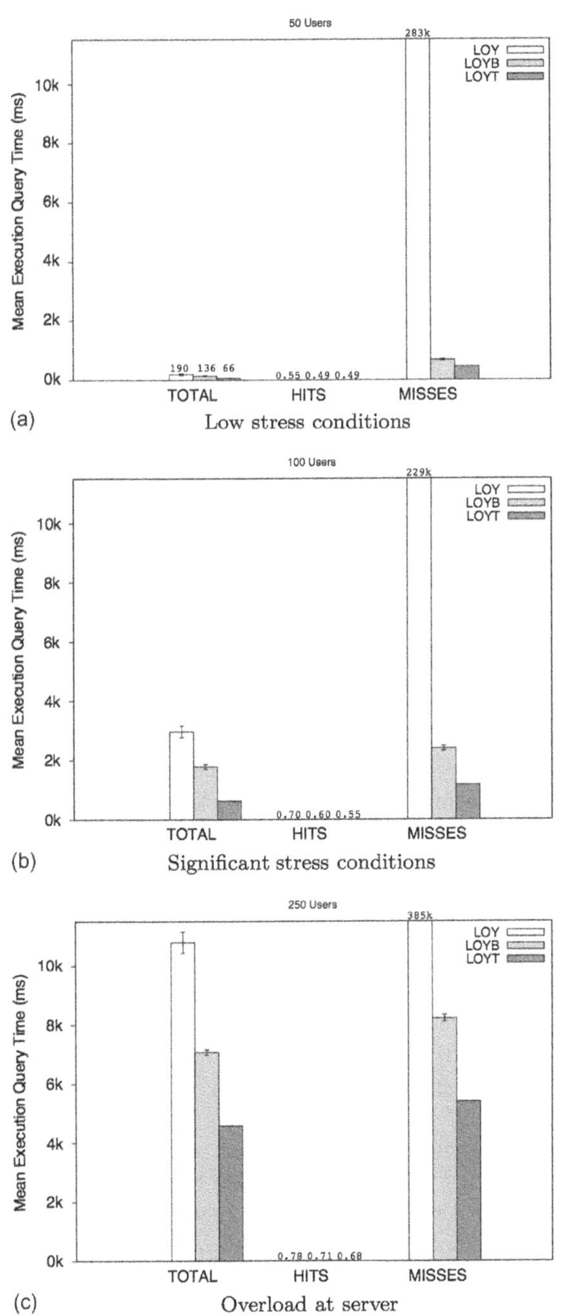

FIGURE 9.18

Execution time per query type.

5 SUMMARY AND CONCLUSIONS

This chapter has explored the effects of using dynamic user workload on web performance evaluation studies. The obtained results have been compared to those generated with traditional workloads. To this end, a scenario based on a typical e-commerce website has been recreated and different user dynamic behaviors have been reproduced. We have used the DWEB model to define dynamic workloads, as it simplifies the modeling of such behaviors, which cannot be represented at this level of accuracy with traditional approaches, such as the CBMG type of model. Furthermore, DWEB allowed us to define the workloads considering not only the dynamism in user interactions but also the dynamic changes in the roles when navigating a website. Moreover, DWEB also allows reproduction of a dynamic workload driven by the user's goals.

With the aim of evaluating the effect of these levels of dynamism on the system performance, a large and representative set of the most commonly used performance metrics has been measured during experimental tests, e.g., client metrics (total number of page requests or response time), server hardware metrics (CPU utilization), or metrics for the main software components at the server (executed queries rate).

Experimental results have shown that dynamic workloads affect the stress borderline on the server, which is degraded further than when using traditional workloads in the studied cases. The object requests rate decreased between 45% and 60%, but dynamic workloads have also changed the request nature, which became more dynamic. This change implied an important growth by 15% in the number of executed queries by the database with a significant increase in their dynamic nature that led the database to be an overloaded application. Consequently, the CPU utilization increased by 30%, consolidating the CPU as the main performance bottleneck. The server performance degradation affected the service conditions, increasing the average response time exponentially, which leads to higher probability of user abandonment (up to 40%). Furthermore, we observed that considering users' dynamic navigations on workload characterization has a stronger impact on the system performance metrics than modeling changes in the role, because dynamism is more present in navigations than in user roles.

Additionally, with the aim of improving the user's productivity, we included the use of the back button and the parallel tab browsing in the navigation patterns, as examples of more realistic user interactions with web browser features.

In addition to traditional performance metrics, new indexes have been defined in order to quantify productivity in web navigations from the user point of view, such as number of finished sessions, number of visited pages per session or session length.

This study proved that navigations using the back button or opening new tabs, which result from considering dynamic user interaction with the offered contents, clearly allow users to achieve their goals in less time, thus increasing their navigation productivity. These browsing patterns also affect the utilization and the throughput of the main system resources, and consequently the stress borderline on the server. Experimental results have shown that parallel browsing with tabs noticeably increases the user's productivity (measured in number of finished sessions) up to 200% with respect to browsing the website in a serialized way. The new navigation patterns also increase the number of served pages (up to 90%). Nevertheless, they generate fewer requests to the search engine, thus reducing the complexity of the executed database queries and decreasing their execution time. This implied an important drop in the processor utilization (up to 45%) and a noticeable rise in the web server throughput (up to 75%). As a consequence, the stress borderline is relaxed, permitting the system to either support more applications or serve more users.

A summary of the study introduced in this chapter was presented in [12–15].

ACKNOWLEDGEMENTS

This work has been partially supported by Spanish Ministry of Economy and Competitiveness under grant TIN2013-43913.

REFERENCES

[1] Peña-Ortiz R, Sahuquillo J, Pont A, Gil JA. DWEB model: representing Web 2.0 dynamism. Comput Commun J 2009;32(6):1118–28. ISSN 0140-3664.

[2] Reichheld F, Schefter P. E-Loyalty: your secret weapon on the web. Harv Bus Rev Mag 2000;78:105.

[3] Menascé D, Almeida V. Scaling for e-business: technologies, models, performance, and capacity planning. Saddle Hill, NJ: Prentice Hall; 2000.

[4] TPC: Transaction processing performance council. TPC Benchmark™ W Specification. Version 1.8. Technical report; 2002.

[5] Chang L, Arnett K, Capella L, Beatty R. Web sites of the fortune 500 companies: facing customers through home pages. Inform Manage J 1997;31(6):335–45.

[6] Wang T, Wei J, Zhang W, Zhong H, Huang T. Workload-aware anomaly detection for web applications. J Syst Softw 2013;89(2014):19–32.

[7] Srinivasan S, Anderson R, Ponnavolu K. Customer loyalty in e-commerce: an exploration of its antecedents and consequences. J Retailing 2002;78(1):41–50.

[8] Oracle. MySQL 5.1 Reference manual: how the query cache operates. Available from: <http://dev.mysql.com/doc/refman/5.1/en/query-cache-operation.html>; 2012.

[9] Jupiter Research. Retail web site performance: consumer reaction to a poor online shopping experience. Technical report, Akamai, 2006.

[10] Cockburn A, McKenzie B, JasonSmith M. Pushing back: evaluating a new behaviour for the back and forward buttons in web browsers. Int J Hum Comput Stud 2002;57 (5):397–414.

[11] Cockburn A, Greenberg S. Issues of page representation and organisation in web browser's revisitation tools. Australas J Inf Syst 2000;7:2.

[12] Peña-Ortiz R, Gil JA, Sahuquillo J, Pont A. The impact of user's dynamic behavior on web performance. 11th IEEE international symposium on network computing and applications. Massachusetts, USA: Cambridge; 2012. p. 143–50.

[13] Peña-Ortiz R, Gil JA, Sahuquillo J, Pont A. The impact of user-browser interaction on web performance. In: 28th Symposium on Applied Computing (SAC); 2013a.

[14] Peña-Ortiz R, Gil JA, Sahuquillo J, Pont A. Analyzing web server performance under dynamic user workloads. Comput Commun J 2013;6(4):386–95. ISSN 0140-3664.

[15] Peña-Ortiz R, Gil JA, Sahuquillo J, Pont A. Surfing the web using browser interface facilities: a performance evaluation approach. J Web Eng 2015;14(1–2):003–21. ISSN: 1540-9589.

PART

3

Modeling approaches of computer networks and systems

CHAPTER

On the self-similarity of traffic generated by network traffic simulators

10

Diogo A.B. Fernandes, Miguel Neto, Liliana F.B. Soares, Mário M. Freire, and Pedro R.M. Inácio

University of Beira Interior, Covilhã, Portugal

1 INTRODUCTION

Network traffic simulators aim to imitate as faithfully as possible the many different properties of real network traffic. This enables modeling network traffic accurately and, therefore, makes it possible to study simulated network traffic that could be otherwise impossible to obtain in real network environments. In addition, it might also be useful to first design and study networks on a simulator in order to check if the setup is correct and works well under the requirements, before setting up the actual physical infrastructure. However, simulating networks and network traffic is a complicated task because it requires modeling network components and equipment, as well as links connecting them. Additionally, modeling networked applications is a highly complex task, to say the least, since the user and protocols influence the behavior of the applications. For example, studies [1] have focused on matching the best distributions and inherent parameters to model the packets lengths, the bit count per time unit, and the interarrival times of network traffic at both sources and aggregation points for web browsing, streaming, instant messaging, Voice over IP (VoIP), and Peer-to-Peer (P2P) traffic profiles, for instance. The generation of network traffic is directly dependent on the user behavior and interaction with the computer and installed applications. On the other hand, simulating network protocols may be a less arduous task since it is only required to implement them according to the specifications.

Network traffic simulation is useful for both researchers and industry practitioners. Several tools provide the capability of simulating network environments together with live interactions within the simulation between network nodes and services or applications installed atop. Such tools allow defining the duration of simulations and feature monitoring functionalities to watch the simulation workflow evolve through time according to the defined parameters. This is helpful to solve optimization and simulation problems seeking suitable inputs to some desired outputs and searching appropriate outputs given known inputs, respectively. However, how realistic the simulated network traffic is may be somewhat questionable, because modeling all the network details cannot be done either completely or perfectly. In this respect, making simulation results more reassuring requires assessing whether the properties embedded within simulated network traffic flows are comparable and compliant with the ones observed on real computer networks.

Self-similarity is known to be a statistical property of the bit count per time unit of network traffic in network aggregation points of local area network (LAN) and wide area network (WAN) environments. Self-similarity implies the network traffic is characterized by a *fractality character* and by the well-known *burstiness* phenomenon. The former means that an object appears the same regardless of the scale, while the latter means that network traffic volume activity may be composed of lengthy periods of data transmission followed by periods of weak activity. These two network traits imply network traffic spikes to ride on bigger waves that, in turn, ride on even larger swells [2]. The knowledge of self-similarity is crucial in order to efficiently design routers in terms of both hardware and software, notably with respect to the lengths of packet queues. Many methods available in the literature allow generation of sequences of values with the self-similar property embedded by default, the aggregation of network traffic mentioned above being one of those methods. In turn, the intensity of the self-similar effect is usually measured by means of the well-known Hurst parameter, for which several estimators also exist in the literature.

In the light of what was discussed above, it is therefore important to determine if the self-similar property is present at aggregate traffic produced by network simulators. As it will be best detailed in a subsequent section, the tools this chapter studies are the Network Simulator 3 (NS3) and OMNeT++. Herein, these tools are utilized to design network topologies and simulate aggregate traffic for posterior analyses. Such analyses focus on estimating the Hurst parameter by means of the Rescaled Range Statistics (R/S) and Variance Time (VT) methods, and on computing the autocorrelation of the resulting datasets. Therefore, the contributions of this chapter are twofold. First, the self-similarity property is explained with emphasis on its influence and impact on network traffic aggregation points and on network traffic analysis in general. Second, the mentioned network simulators are reviewed and are studied for their compliance with self-similarity in simulated network traffic.

The remainder of this chapter is structured as follows. Section 2 explains the theoretical background of self-similarity and of the Hurst parameter. Section 3 then describes the self-similar phenomena observed in network traffic and discusses this property thoroughly. Section 4 demonstrates by means of empirical analyses whether the self-similar effect is noticed on traffic simulated by popular tools. Finally, Section 5 concludes the chapter.

2 SELF-SIMILARITY AND THE HURST PARAMETER

This section explains the theory of self-similarity and the relation of this property with the Hurst parameter. The focus then shifts to the generation of self-similar sequences and finally to the estimation of the Hurst parameter.

2.1 THE SELF-SIMILARITY PROPERTY

For its many relations with artificial and natural processes, the property of self-similarity has been the focus of many studies since its discovery in the fifties. Its origin is due to the early studies of the hydrologist Harold E. Hurst who found the flows of the Nile River to be self-similar [3]. Subsequently, researchers have been motivated to study self-similarity in several areas of knowledge, mostly because its statistical impact can help predict the behavior and future state of a given process, provided observations about the past of the process are available to analysis. Besides hydrology, self-similarity has been studied in fields like biomedicine [4,5], network traffic analysis [2,6,7], economics [8], seismology [9], and human motion [10], receiving scrutiny throughout research academia. Economics is a particular area where the knowledge of this property may have some financial impact. Studies have pointed out the feasibility of foreseeing stock prices through the analysis of self-similarity, which can influence stock pricing and purchasing decisions. Discerning the self-similar nature is therefore crucial in order not only to understand, but also to model natural phenomena and artificial processes. Ultimately, one can make more realistic and authentic representations of nature and of real-world applications by imprinting the self-similarity property into the data. The latter case covers network traffic modeling.

The formal description of self-similarity is typically done for *stationary stochastic processes*, among which *time series* are of importance. The term itself is due to fractal theory that says that an object appears the same regardless of the scale, and further states that the moments of the distributions of a given process remain constant as time goes by. Although defining self-similarity should be done for *continuous-time stochastic processes*, it is more useful to do so for the case of *discrete-time stochastic processes* with stationary increments, because the analysis of self-similarity is mostly done for countable (finite) sets of values. A discrete-time stochastic process $X = \{X(t)\}_{t \in N}$ is said to be self-similar if the statistical description of its finite-dimensional distribution is equal to that of the process obtained by scaling the amplitude by a^{-H} and its time axis by a, where H $(0 < H < 1, H \in \mathbb{R})$ denotes the Hurst parameter. Formally, this is defined as

$$X(t) \underset{=}{d} a^{-H} X(at), \text{ with } a \in \mathrm{N} \tag{10.1}$$

where the symbol $\underset{=}{d}$ denotes equality in all finite-dimensional distributions.

Self-similarity can be further described in terms of the *first order differences process* and of the *aggregate processes*. In this case, X is said to be self-similar if the finite-dimensional distribution of its first-order differences process $\{Y(t)\}_{t\in N}$ (Equation (10.3)) equals the finite-dimensional distribution of its *aggregate processes* $\{Y^{(m)}(i)\}_{i\in N}$ (Equation (10.4)) when scaling their amplitude by m^{1-H}, such that

$$Y(t) \underline{\underline{d}}\, m^{1-H} Y^{(m)}(i), \text{ with } m \in N \tag{10.2}$$

where the processes $\{Y(t)\}_{t\in N}$ and $\{Y^{(m)}(i)\}_{i\in N}$ are each defined by the following respective expressions:

$$Y(t) = X(t+1) - X(t) \tag{10.3}$$

and

$$Y^{(m)}(i) = \frac{Y(mi) + (mi+1) + \cdots + Y((m+1)i - 1)}{m} \tag{10.4}$$

Note that m represents the size of each aggregation scale or block of the resulting aggregate series $\{Y^{(m)}(i)\}_{i\in N}$, which by design are *non-overlapping* and *consecutive*. From here onward, interpret m_k as a particular instance of m, and also take that aggregate processes or scaling series are used interchangeably to refer to $\{Y^{(m)}(i)\}_{i\in N}$.

The aggregation and scaling procedure is important due to its inherent exploration of the fractality character of self-similar processes, for which several estimators of the Hurst parameter require to be executed prior to their main calculations. Because of this, self-similar processes are sometimes called *scale-invariant*, since the self-similarity property continues to hold for every m_k. If the aforementioned condition holds for any $m_k \in N$, then the process is said to be *exactly second order self-similar*. If the self-similarity holds as only $m \to \infty$, then the process is said to be *asymptotically second order self-similar*. Self-similarity is commonly assimilated through the classes of fractional Brownian motion (fBm) and of its first-order differences process called fractional Gaussian noise (fGn). Both are a type of stochastic Gaussian process, as other less popular models are, like the Autoregressive Integrated Moving Average (ARIMA) and the Autoregressive Fractional Integrated Moving Average (FARIMA).

2.2 THE HURST PARAMETER

The Hurst parameter is a dimensionless factor used to estimate the presence and magnitude of the self-similarity property. It is a measure not only for self-similarity, but also for the statistical properties that self-similarity entails (see below). Estimations of the Hurst parameter are useful to understand the autocorrelation structure and the evolution of a process, and to thus attain the aforementioned goals which the study of self-similarity is based on. The Hurst parameter corresponds to a fractal dimension d by the relation $d = 2 - H$, where $1 < d < 2$ so that H takes values bigger than 0 (exclusive) and smaller than 1 (exclusive).

A given process is said to have *memory* when current observations of the process exhibit dependencies from past observations of the process, and when the actual status will influence its future state some instant ahead in time. Such processes are characterized by values of the Hurst parameter between 0 and 1 (both exclusive), but different from 0.5. When the values of the Hurst parameter are bigger than 0 (exclusive) and smaller than 0.5 (exclusive), the process is said to be *anti-persistent*. Anti-persistent processes are best known as *mean-reverting* processes in financial analysis for the high probability of the process to converge to its mean that characterizes them. A similar and widely used concept of anti-persistence is *short-range dependence*. Short-range dependence complements the notion of anti-persistence by emphasizing the fact that the aforementioned statistical dependence has a more limited (local) scope in the time-domain. On the other hand, processes exhibiting *persistent* behavior take values of the Hurst parameter bigger than 0.5 (exclusively) and smaller than 1 (exclusive). Contrarily to the mean-reverting phenomenon, persistent processes are trendy in the sense that there is a higher likelihood of maintaining their current trend (ascending or descending) rather than falling back to the mean. As in anti-persistence, persistence processes are complemented by the concept of *long-range dependence*, which enlarges the mentioned scope of the statistical dependence in the time domain. In other words, observations further back in time still have an impact in the actual status of a process. Additionally, it is commonly accepted that long-range dependence is the slow power-law decrease of the autocorrelation function. Graphically speaking, the lower H is, the noisier or more volatile the process is, while the higher H is, the smoother it is. A process absent of trends is said to be *memoryless* when values of the Hurst parameter are equal to 0.5. Random walks and white noises are examples of memoryless processes.

2.3 SIMULATING SELF-SIMILARITY

Because of the interest simulating self-similarity has gained along the years among the scientific community, several algorithms have been proposed to synthesize series with such a property. From the statistical point of view, the synthesis of artificial sequences with the self-similarity property embedded in the whole data is a complex engineering task in computer science. In cases where the property of long-range dependence is desired, the statistical bond each point may be subjected to implies the recall of the entire past of the series to generate the next point, so as to comply with the typical autocorrelation structure this type of process has. Because of that, the computational complexity of some procedures (the exact methods such as the Hosking method [11]) available to generate self-similar series is usually higher than $O(n)$, though some (but fewer) methods (e.g., the fractional Brownian motion Sequential Generation Algorithm (fBm-SGA) [12]) present a linear computational complexity at the cost of being approximate, as some mechanisms to generate Gaussian variables are.

The literature on the subject mostly focuses on devising generators whose outputs are self-similar sequences that are long-range dependent, notably fBm (e.g., see [13]). Some generators produce sequences that are exactly self-similar while others provide approximate solutions, which often counterbalance performance with the quality of the resulting series. Modeling of aggregate network traffic falls within the category of approximate generators. Some of the generators [12] are even directly engineered with a basis on the way most methods for estimating the Hurst parameter work, as is best described in the following subsection. Equalities (10.1) and (10.2) give origin to several ramifications that are susceptible to elaborate generators of self-similar sequences while having an undefined Hurst parameter variable H somewhere among the theoretical background. This is useful, as algorithms are thus written by allowing a target value of the Hurst parameter to be provided as input prior to their execution in order to synthesize quality sequences with a predefined Hurst parameter value. Other means explore the spectral and wavelet domains [14–16] to achieve the purpose discussed in this subsection. As perceived, simulating the self-similarity is an easy task, and so modeling the aggregation effect of network aggregation points must take some conditions into consideration, as subsection 3.2 details.

2.4 ESTIMATING THE HURST PARAMETER

It was previously stated that the interest in self-similarity has given rise to many generators of self-similar sequences. Such effort would be worthless without the procedures capable of producing estimates of the degree of self-similarity embedded in the data. For this particular purpose, several algorithms have been developed to produce estimates of the Hurst parameter of a given process. Most methods are built upon the equality drawn in Equation (10.2), from which is possible to create various mechanisms that address a particular statistical law of the process under observation. It is important to explore the compliance of the process under analysis with the multiple statistical properties ruling series that exhibit self-similarity, and as such estimating the Hurst parameter is often best done while resorting to more than one estimator.

Some estimators of the Hurst parameter elaborate on sound signal theory (e.g., the wavelets-based estimator [17]), while others analyze the process by scanning for signs of long-range dependence in the time-domain (e.g., the Embedded Branching Process (EBP) estimator [18]). Other estimators are based on a graphical analysis between all scaling series m_k considered just to produce a single estimate of the Hurst parameter. As a product of the analysis of each aggregate process by the procedure implemented by each estimator, a specific statistic is taken to produce a point in a chart plotted against the associated scale m_k. After all aggregate processes have been computed, linear regression analysis is typically used to retrieve the Hurst parameter estimate. The R/S and the VT methods, which are described below, follow this approach.

2.4.1 Rescaled range statistics

Mandelbrot and Wallis [19] formalized the R/S method after Harold E. Hurst studied the Nile River flows. Summarily, this estimator explores the time-domain structure of a given process such that it rescales the observations within each aggregation block with basis on the underlying respective mean, so as to compute a series of the maximum and the minimum of series of readjusted sums to produce a series of ranges, which are then used to compute the final R/S statistic. In the following calculation of the readjusted sums $\{S_k^i(j)\}_{j\in N}$, note that $Y^{(m_k)}(i)$ represents a particular aggregation block of the series concerning a particular scale m_k:

$$S_k^i(j) = \sum_{l=im_k}^{im_k+j} (Y(l) - Y^{(m_k)}(i)) \tag{10.5}$$

At this point, the series of the maximum and minimum are computed as the following respective expressions denote:

$$Max_k(i) = max(0, S_k^i(1), S_k^i(2), \ldots, S_k^i(m_k - 1)) \tag{10.6}$$

$$Min_k(i) = min(0, S_k^i(1), S_k^i(2), \ldots, S_k^i(m_k - 1)) \tag{10.7}$$

Based on Equations (10.6) and (10.7), the aforementioned range series are determined as follows:

$$Range_k(i) = Max_k(i) - Min_k(i) \tag{10.8}$$

Finally, dividing each value of the range series by the standard deviation of the respective process of the readjusted sums yields the rescaled range series of the R/S statistic:

$$RS_k(i) = \frac{Range_k(i)}{\sigma(i)} \tag{10.9}$$

By applying linear regression to the points whose coordinates are the logarithm of each scale m_k plotted with the average of each respective rescaled range series $\{RS_k(i)\}_{i\in N}$, the equation of the following curve is obtained:

$$\log(E(RS_k)) = \log(C) + H \times \log(m_k) \tag{10.10}$$

where E denotes the expectation operator and C is a positive number that is irrelevant to estimate the Hurst parameter. The slope of the line can be directly interpreted as an estimate for the Hurst parameter.

2.4.2 Variance time

The VT method is formulated upon the equality in the moments of the distributions drawn in Equation (10.2), namely the variance, herein denoted by V, for which the following expression can be derived:

$$V(Y) = V(m^{1-H} Y^{(m)}) \tag{10.11}$$

The properties of the variance and of the logarithm are used in the deduction to derive Equation (10.12):

$$\log(V(Y^{(m)})) = log(V(Y)) + (2H - 2) \times log(m) \tag{10.12}$$

If the same reasoning behind the R/S is applied here, where the logarithm of each scale m_k is plotted with the respective logarithm of the variance $V(Y^{(m_k)})$, the slope of the line that best fits the points in Equation (10.12) can be directly used to retrieve Hurst parameter estimates. By applying linear regression to obtain the mentioned slope, herein represented by β, H can be obtained from:

$$H = 1 + \frac{\beta}{2} \tag{10.13}$$

3 SELF-SIMILARITY IN NETWORK TRAFFIC

This section focuses on bridging self-similarity and network traffic at aggregation points. It first discusses seminal works on the subject, followed by the formalization of aggregated network traffic in terms of self-similarity. Its impact is summarized at the end.

3.1 NETWORK TRAFFIC MODELING AND ANALYSIS

In the initial moldings of modeling and analyzing network traffic, it was thought that the aggregation of network traffic coming from several sources, in terms of the information per time unit, could be represented by memoryless compound Poisson or Markovian arrival processes. Such processes entailed the network traffic to be smoother as the aggregation scale increased, i.e., as $m \rightarrow \infty$. In this case, averaging the network traffic over long time scales would smooth its burstiness. However, later breakthroughs by Leland et al. [2,6] provided empirical evidence of the self-similarity in aggregate network traffic that can be accurately modeled by the fGn and FARIMA models, which depict statistical laws typical of fractals that are actually different from the ones of memoryless Poisson and Markovian processes. Leland et al. discovered the self-similar property to be embedded in aggregate Ethernet traffic composed of several network streams sourced from distinct networked applications and computers. The authors studied the aggregate network traffic with different network topologies at various contexts in the network hierarchy, following a bottom-up approach. They started by looking at the aggregate traffic of a simple LAN, then moved on to an interdomain routing point, and finally ended up on the backbone of the Bellcore facility network. These topologies are important in the work depicted in this chapter, since they were on the basis of the topologies studied and of the results presented below. It was later concluded [14,20] that WAN traffic is also not accurately modeled through Poisson processes, and that some of its aspects show the properties of asymptotically second-order self-similar processes.

The discovery of the self-similar nature of network traffic has produced several ramifications in research on network traffic monitoring and analysis during subsequent years, up until present day. In order to make more rigorous and realistic models of everyday stochastic processes, it is required to discern the statistical laws of the inherent self-similarity property and make both analytical expressions and simulations comply with the properties therein observed. In the networking field, this can help in carrying out general assessments of the network (e.g., performance and optimization), stress tests, or troubleshooting experiments prior to mounting a physical network infrastructure, either by means of mathematical constructions or simulation runs, both being relatively cheap. Now, consider network traffic sampling techniques that do not account for the statistical properties of the network traffic. In such cases, important information about the network traffic may not be captured at high-speed aggregation points because most packet sampling mechanisms are stochastic or adaptive (based on linear prediction or fuzzy logic). Research on this subject has thus started estimating the network traffic by including its most salient aspects, such as traffic distributions [21], the spectral density (periodicity) [22] and self-similarity [7], so as to devise more precise techniques in terms of estimation while considering the sampling performance.

3.2 MODELING SELF-SIMILAR AGGREGATE NETWORK TRAFFIC

The theory behind the modeling of self-similar aggregate network traffic states that the superposition of many renewal processes exhibiting the *Noah effect* (processes following a heavy-tailed distribution) converges to a process showing the *Joseph effect* (in other words, self-similarity). In the networking analogy, this means that the bit rate per time unit process at network traffic aggregation points exhibits the property of self-similarity when such aggregation points handle network traffic originating from various network sources. Because network traffic generated at the source level depends directly on user interactions and on applications behavior, it is a complex affair to faithfully model its exact characteristics. However, it is commonly known that network sources produce traffic as *bursts*, in which a certain number of bits are transferred in succession during a certain period of time. Each burst transfer is immediately followed by a *period of silence*, during which no transfer occurs. In turn, each period of silence is succeeded by a burst, and each burst by silence, and so on. Suppose there are S independent and identically distributed (i.i.d.) renewal sources, where each renewal process s is modeled by a stationary binary process $\{W^{(S)}(t)\}_{t\in \mathrm{N}}$. Each process s is at an *on* state if $W^{(s)}(t)=1$, otherwise $W^{(s)}(t)=0$. Rescaling time by a factor T, let the aggregate cumulative *on* count be defined as follows in the interval $[0,Tt]$:

$$W_S(Tt)=\int_0^{Tt}\left(\sum_0^S W^{(s)}(u)\right)du \tag{10.14}$$

For $\{W_S(Tt)\}_{t \in \mathbb{R}^+}$, the following limit holds [21]:

$$\text{dlim}_{T \to \infty} \text{dlim}_{S \to \infty} T^{-H} S^{-\frac{1}{2}} \left(W_S(Tt) - \frac{TSt}{2} \right) = \sigma B_H(t) \qquad (10.15)$$

where B_H denotes fBm with Hurst parameter H and dlim denotes *limit in distribution*.

A generator of self-similar sequences whose operation is based on this method can model the lengths of the *on* and *off* periods by a heavy-tailed distribution, namely Pareto, for which the Hurst parameter is directly linked in the light of the relation $H = (3 - \alpha)/2$, where α is the shape parameter of the Pareto distribution. For the realization of a persistent series where $0.5 < H < 1$, $1 < \alpha < 2$, and the closer to one α is, the *burstier* the traffic is. For a certain time t, the aggregation procedure can then be formalized as

$$W_S(t) = \sum_{s=0}^{S} W^{(s)} \qquad (10.16)$$

Despite the simplicity of the modeling of renewal processes depicted here, the aggregation effect hides many complex modeling details.

Another commonly adopted aggregation model for simulating long-range dependence with asymptotic self-similarity uses the $M/G/\infty$ queuing model, in which customers arrive according to a Poisson process and are served from a distribution with infinite variance. Though a trade-off between the degree of self-similarity and the computational complexity is also present in this model [13], the method of aggregation of renewal processes is well regarded in network traffic modeling. When the lengths of the *on* and *off* periods are determined stochastically and independently, each renewal process is uncorrelated and short-range dependent. In this regard, however, more recent empirical studies [1] of several network traffic traces containing various application profiles, namely Voice over Internet Protocol (VoIP) and file sharing, have shown the persistent properties of self-similar traffic to be present in some combinations of flows at network sources and destinations. So, in the worst case it is safe to assume each source contributes with short-range dependent increments, for which the aggregation effect intensifies the memory of the aggregate process. Self-similarity is further enhanced when considering the overlap of several traffic streams, among which some may be other network aggregation points. Notice that the formalizations in Equations (10.14) and (10.15) abstract the sources in the sense that they do not have to necessarily be end-computers. Furthermore, other empirical studies [22] showed that both the processes of the interarrival times and of the packet sizes of aggregate traffic appear to be long-range dependent also, in addition to the packet and byte count.

3.3 IMPACT OF SELF-SIMILARITY IN NETWORK TRAFFIC

The role of self-similarity in network traffic is mostly discussed when the topology of the network contains points where network traffic originating from several sources is aggregated for processing. Packet data units transverse several network nodes when traveling from one system to another. The Internet Protocol (IP), the main actor in the *forwarding* and *routing* of network traffic flows in the Internet, together with other standards and protocols, provide the necessary information to hop packets between the intermediate nodes so as to reach the final destination. Several network utilities, namely `traceroute`, allow analysis of the network path and troubleshooting. On the other hand, *switching* is mainly ruled by the Ethernet standard. In either case, network traffic aggregation points are unavoidable when considering interdomain routing and the core and backbone of large LANs or WANs. When processing the packet flows arriving at the possibly many network interface cards, routers are required to put packets in buffers to decide on their outcome with basis on routing tables, Quality of Service (QoS) requirements, and resource availability. Although the self-similarity property is a consequence of the aggregation, its origin lies in the sources of information, as duly formalized in the previous subsection.

If network traffic were not self-similar, the effect of the *law of large numbers* would make the process of information per time unit converge as the aggregation scale increases, as Poisson or Markovian models suggest. However, self-similarity causes the network traffic to be resistant to change as m_k increases, and so poor assumptions about the network design or the conceiving of network devices may be made. The main outcome of the self-similarity property in network traffic is so-called *burstiness*. Instead of packets arriving at devices randomly spaced in time, the flows they are part of form bursts that make the network traffic unbalanced in the sense that spikes may appear suddenly and unexpectedly. The aggregation point is statistically likely to handle amounts of data above the average longer than expected, followed by similarly statistical likelihoods of periods where the amount of data is actually lower than the expected value (long-range dependence). In this regard, estimating the Hurst parameter in networking also measures the degree of burstiness of the network traffic, though other less-precise metrics like the index of dispersion or the coefficient of variation exist for assessing the burstiness. Understanding self-similarity can thus help define suitable packet buffers at routers that account for the mentioned bursts during both congested and noncongested periods.

4 ANALYSIS OF SELF-SIMILARITY IN SIMULATED NETWORK TRAFFIC

This section describes the means used to study self-similarity in simulated network traffic. It starts by describing the tools and traces under analysis and then briefly describes the library used for testing self-similarity. The simulated network topologies and the respective results are discussed at the end.

4.1 NETWORK SIMULATION TOOLS

The network simulators analyzed in this chapter are the widely known NS3 [23] and OMNeT++ [24]. Although other popular tools, notably the Cisco Packet Tracer, provide many other simulation features and a useful graphical interface, it was opted to study these tools because they are especially popular for research. The NS3 and OMNeT++ are open-source, discrete-event simulators that provide application programming interfaces (APIs) and are written in ANSI C and in C++, respectively, which allow schematizing several types of wired and wireless networks along with network nodes and applications programmatically. OMNeT++ provides the simulation engine, but it is its INET framework that gives the package the ability of implementation for several networking protocols and applications. Although these tools require coding, OMNeT++ provides a graphical view of the simulation as it runs, facilitating the pictorial analysis of the produced network traffic. NS3 provides no such feature. NS3 is a modular tool and is the successor of NS2 (now deprecated), whose core was rewritten. As such, NS3 is not backward compatible. On the other hand, OMNeT++ is the academic version of OMNEST.

These tools take into account the layers two, three and four of the Open Systems Interconnection (OSI) model and populate the headers of network packets accordingly, but the payload may be left empty. This is not an issue in this work because the size of each packet is set, and so the processes of the bit count per time unit are duly constructed. Note that it was not attempted to link real networks with the simulation environment, as NS3 allows, since the objective here is to keep the work aimed toward full simulations.

4.2 THE TESTING LIBRARY AND RELATED CONFIGURATIONS

To compute Hurst parameter estimates for the several traces that were collected in the scope of this chapter, a statistical library was used whose main purpose is to study self-similar stochastic processes. The library is called *TestH*, whose authorship is of the authors of this chapter. It is written in ANSI C and provides a user-friendly API to build customized programs. Though many other parts of the library support its operation behind the scenes, the library is essentially divided into a set of estimators of the Hurst parameter and into another set of generators of self-similar sequences. Because *TestH* can handle generic processes, it has applicability in a wide array of applications. *TestH* is the main subject of separate publications that explain and describe its inner workings together with the implemented methods. For more information about the library, contact the authors of this chapter.

Various programs were developed to carry out the necessary tests. The modeled networks under observation were set up and various simulations were executed. The resulting traces were preprocessed to filter out the timestamps of the packet arrivals and the packet sizes, so as to study the bit rate per time unit process in terms of self-similarity. The scaling procedure discussed in subsection 2.1 is implemented in *TestH* to compute aggregate processes according to a time interval (in seconds), rather than a fixed m_k for each aggregation block. In the analyses included in the subsection 4.5, for each of the datasets obtained from the simulations, the original process was built with the precision of a tenth of a second, which means that all packet arrivals within a given second were split into ten intervals, each a tenth of a second, and were afterwards aggregated. Both of the estimators of the Hurst parameter formalized in subsection 2.4 are implemented in *TestH* and were used to retrieve Hurst parameter estimates. Preliminary analysis revealed that the R/S often output Hurst parameter estimates bigger than 1, which means that the slope of the line in the respective pox-plots was accentuated. To overcome this obstacle and to check for asymptotical self-similarity, it was opted to compute the scaling procedure for scales of type $m_k = 2^x$ (in *TestH*, this is called *power scaling*), where $x = \{6, 7, 8, 9, 10, 11, 12, 13, 14\}$. For demonstration purposes, the following C code depicts how programs using *TestH* are usually developed for reading data stored within files and for subsequent estimation of the Hurst parameter:

```
#include “process.h”
#include “generators.h”
#include “estimators.h”
void main (void) {
  proc_Process *proc = NULL;
  proc_ScalesConfig *conf = NULL;
  proc = gen_ReadFileTime ("data.txt", "Name", TestH_fGn, 1);
  conf = proc_CreateScalesConfig (TestH_POW, 6, 14, 2);
  proc_CreateScales (proc, conf);
  est_VarianceTime (proc);
}
```

4.3 MODELED NETWORK TOPOLOGIES

The main objective of the network topologies setup in the considered network simulators was to generate traffic under several conditions from various sources and analyze the resulting aggregate network traffic. To that end, the architecture of the topologies was composed by at least one aggregation point of network traffic, functioning like a router. In fact, the topologies modeled in this work imitate the architecture of some of the real Ethernet networks Leland et al. analyzed in their seminal works [2,6].

Two topologies were modeled in both NS3 and OMNeT++ and various simulations were respectively run to obtain the traces containing the timestamps of the packet arrivals and the packet lengths in bytes. One of the topologies is a LAN environment composed of over 70 network nodes simulating workstations (clients), servers, and interconnecting equipment. The LAN was split across two segments, each having a reasonable number and type of nodes. The network traffic passing by the interconnecting equipment was sniffed and saved to several traces. The illustration shown in the left part of Figure 10.1 concerns the first topology just discussed. The second topology imitates a backbone network point where the network traffic originates from multiple internal LANs with over 280 network nodes. Both directions of the communication were captured and no attempt to communicate with the Internet was done. This topology is illustrated in the right-hand side of Figure 10.1. In order to give the simulations some degree of realism, some of the traces contain traffic concerning specific combinations of protocols, while others have mixed combinations, as one would expect in real aggregation points where routers handle several types of communications. No wireless connections were considered in both topologies.

4.4 THE SELF-SIMILARITY OF THE DATASETS OF LELAND ET AL.

The traces (a total of four) publicly available online at [25] were utilized as a baseline comparison for the results obtained with the simulation tools. Those traces are a subset of the datasets used by Leland et al. [2], and each one contains the timestamp and the length in bytes of the packets of one million packet arrivals. The traces were captured in 1989 at the Bellcore Laboratory. Two of them, which are named BC-pAug89 and BC-pOct89, are mostly composed of LAN traffic, while the other two, which are named BC-pAug89Ext and BC-pAug89Ext4, contain only bidirectional WAN traffic flowing between the backbone of the Bellcore network and the Internet. At the time, the nodes connected to that laboratory network which participated in the generation of traffic were dataless workstations, minicomputers, personal computers, fileservers, and printers, many of which were operated by humans.

For demonstration purposes, Figure 10.2 contains the bit count per time unit process of the aggregate (summed) network traffic at different time scales for the trace BC-pAug89, which was aggregated with *TestH*. The figure illustrates the self-similarity property in network traffic by observing its burstiness. As is perceivable, the network traffic is burstier when the time scale is smaller, though the burstiness is also visible at bigger time scales. This is expected because the variances of the sample means actually decrease, though more slowly than the reciprocal sample size and not exponentially fast as other models imply. The same behavior happens with the autocorrelation, which decreases hyperbolically. The figure further illustrates the fact that network traffic spikes ride on long-term ripples which, in turn, ride on even longer term swells, as mentioned in the introduction, thereby demonstrating the fractal-like behavior of network traffic.

All the datasets are now analyzed in terms of self-similarity with *TestH* also. Figure 10.3 contains a summary of the estimations obtained for the Hurst

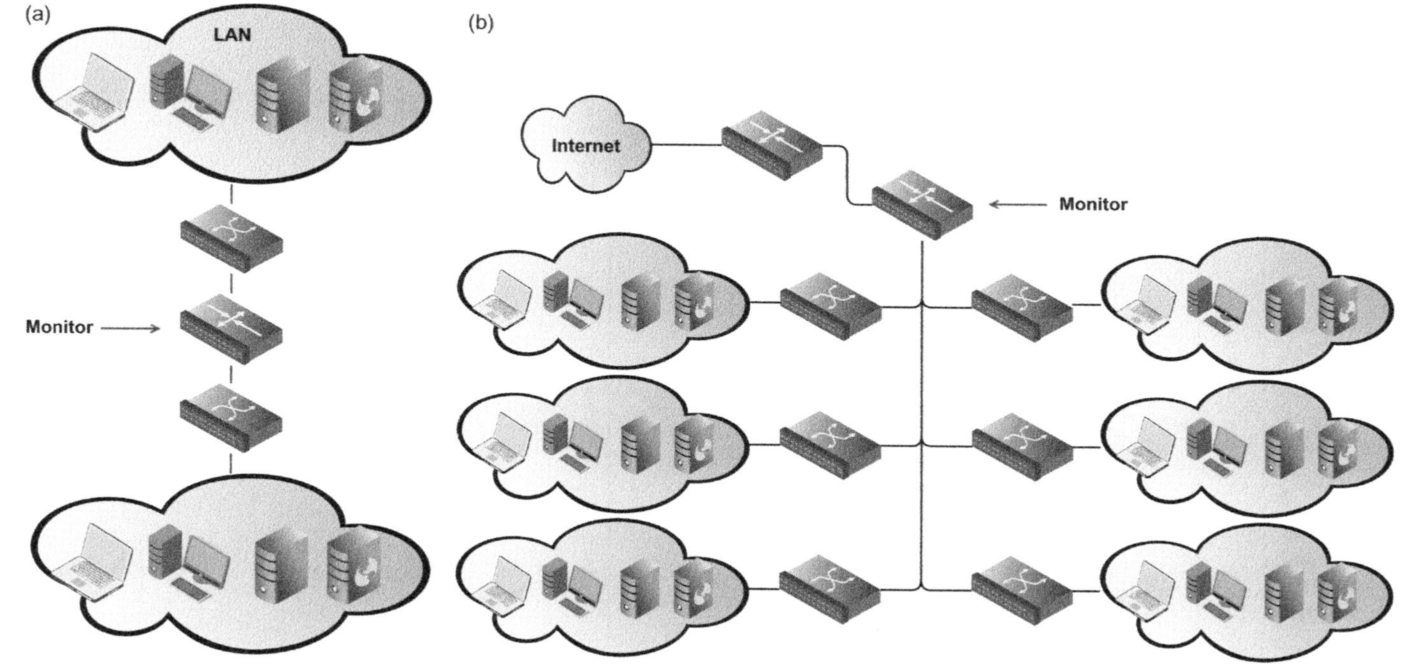

FIGURE 10.1

Modeled network topologies with the inherent interconnecting devices and end nodes. The topology on the left is named T1 and the one on the right is called T2.

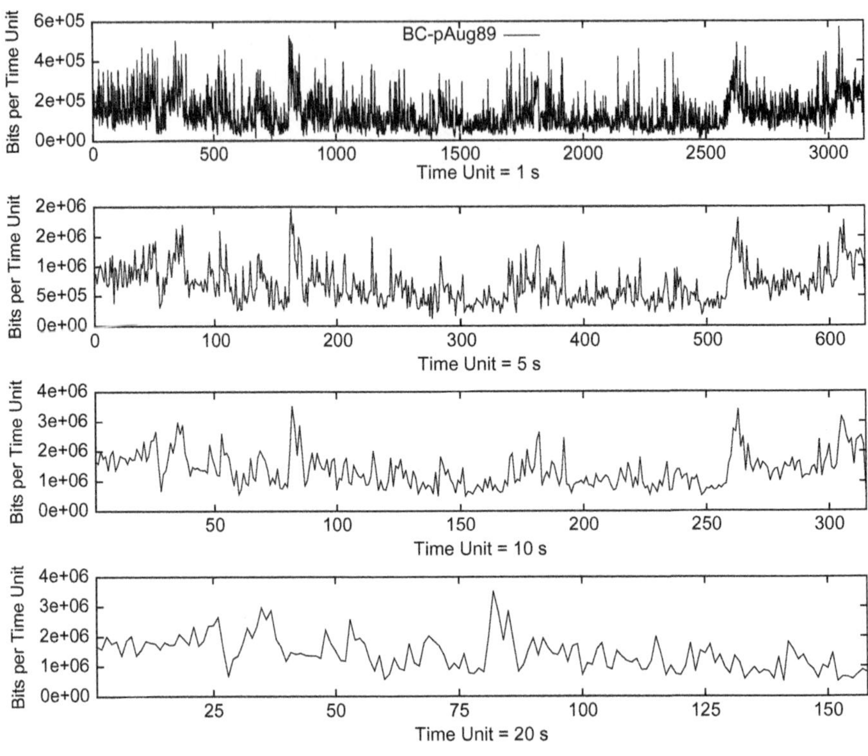

FIGURE 10.2

Pictorial demonstration of self-similarity on different time scales for the dataset named BC-pAug89.

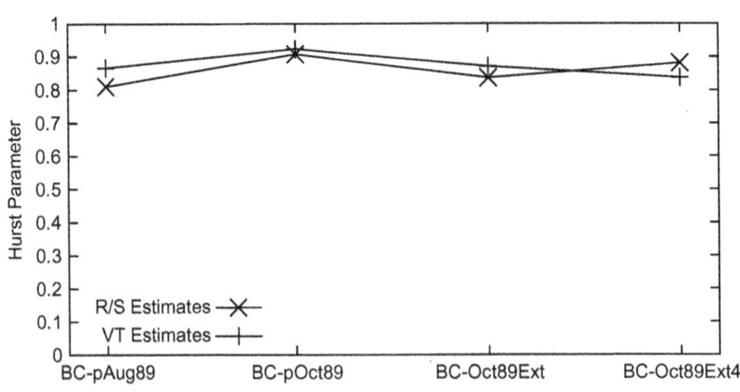

FIGURE 10.3

Hurst parameter estimates for all of the datasets available at [25].

parameter embedded in the series of each dataset. The scaling procedure applied to each original dataset was configured to produce a total of 51 scale series, with the first scale size m_k equal to 12 and the last one equal to 512, with an incremental step equal to 10. The results pictured in Figure 10.3 are corroborated by the conclusions reached by Leland et al. [2], where the network traffic is self-similar with Hurst parameter estimates around 0.8 and 0.9 for all the network environments studied therein. The network traffic is thus long-range dependent where the bursts created by the flows of data may last longer than expected, as discussed in subsection 3.3.

4.5 THE SELF-SIMILARITY OF SIMULATED NETWORK TRAFFIC

This subsection first describes all the traces analyzed in the scope of the simulation tools. Then, the self-similarity of the network traffic produced by NS3 is analyzed, followed by the analysis of the network traffic generated by OMNeT++.

4.5.1 Traces under analysis

To obtain a good baseline for comparison, the same kind of experiment was carried out in both NS3 and OMNeT++, meaning that the same protocols and applications were set up to generate the same types of traces. However, each simulator has a few particularities in terms of programming and simulation construction that made it impossible to exactly replicate each experiment. A variety of protocols and applications were used, namely the Hypertext Transfer Protocol (HTTP), the User Datagram Protocol (UDP) carrying data of video streaming, and the Transmission Control Protocol (TCP), used by the File Transport Protocol (FTP) and the Telnet remote control utility, the Internet Control Message Protocol (ICMP) used by the ping utility, and finally the Secure Shell (SSH) remote control application. The simulation of such protocols and applications in both NS3 and OMNeT++ is handled by the functions of the simulators themselves, only requiring configuration of the source and destination of the network traffic and simulation-specific options.

A total of 17 traces were obtained, which are all summarized in Table 10.1. As can be seen in that table, HTTP traffic is present in all traces, regardless of the simulator and topology, and each trace has a number of packet arrivals superior to the number of packet arrivals available in the public datasets of Leland et al. [25]. Moreover, the several simulations ran for different amounts of time with similar protocol and application configurations until a satisfactory dataset size was obtained. During runtime, however, the simulator engine can define a different behavior for a given application, depending on the simulation state, which can result in distinct traffic profiles for different traces. This happens due to the inherent randomness of the simulations and available resources like link

Table 10.1 Description of the Synthetic Traces Obtained from NS3 and OMNeT++

Simulator	Topology	Trace	Duration (Seconds)	Number of Packet Arrivals	Protocols/ Applications
NS3	T1	NT1_HI	9999	2263928	HTTP, ICMP
		NT1_HU	100000	1850429	HTTP, UDP
		NT1_HUS	100000	1850409	HTTP, UDP, SSH
	T2	NT2_HF	2153	4191176	HTTP, FTP
		NT2_HU	1018	1985897	HTTP, UDP
		NT2_HUI	2003	3902333	HTTP, UDP, ICMP
		NT2_HUF	2033	3957606	HTTP, UDP, FTP
OMNeT++	T1	OT1_H	1796329	1492798	HTTP
		OT1_HF	837215	2322162	HTTP, FTP
		OT1_HT	10022	19166766	HTTP, Telnet
		OT1_HU	1986119	1642182	HTTP, UDP
		OT1_HUT	1715	3259875	HTTP, UDP, Telnet
		OT1_HUIF	5720	1057934	HTTP, UDP, ICMP, FTP
	T2	OT2_HF	10828	1905826	HTTP, FTP
		OT2_HU	24457	1930003	HTTP, UDP
		OT2_HUT	17568	1736263	HTTP, UDP, Telnet
		OT2_HUFT	45742	1401660	HTTP, UDP, FTP, Telnet

bandwidth, and because of the different distributions (e.g., exponential, normal and uniform) used to define the start time of a web server, the packet sizes, response times, client requests, and so on. In fact, the available distributions in the simulators may not be the most suitable to define packet sizes; it has been demonstrated in [1] that the packet sizes of several types of traffic have bimodal or trimodal distributions.

4.5.2 Analysis of the network traffic of NS3

For demonstration purposes, Figure 10.4 contains a pictorial view of the bits per time unit for the datasets NT1-HU and NT2-HUF. In Figure 10.4(a), it can be seen that the self-similarity property may be embedded in the data for the dataset NT1-HU because the process varies (visually speaking) similarly to the

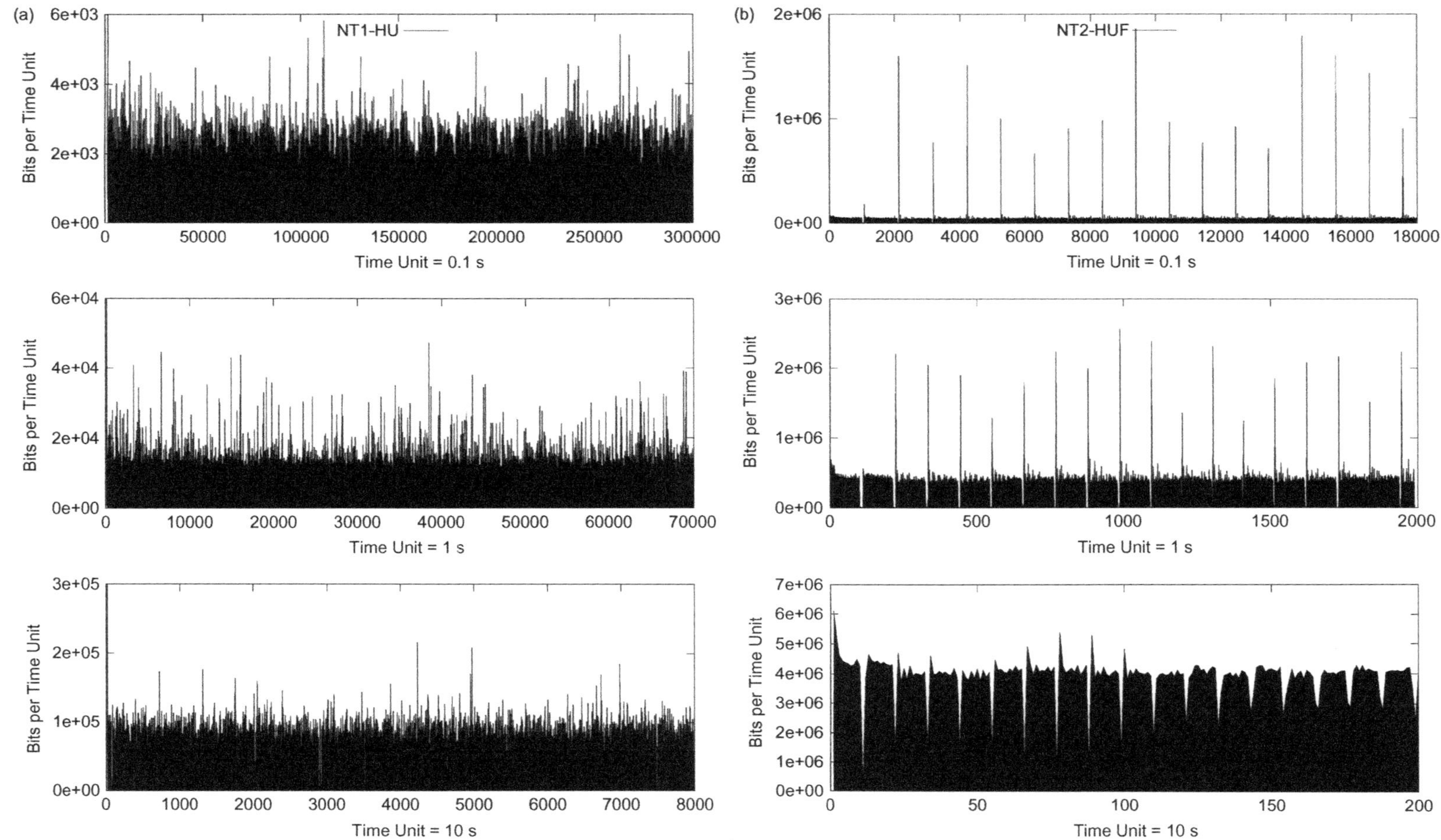

FIGURE 10.4

Bits per time unit on different time scales for a part of the datasets named NT1-HU (a) and NT2-HUF (b).

ones in Figure 10.2. The burstiness of the traffic is evidenced when the time scale is equal to a tenth of a second, and it is not lost as the time scale increases, at least up to a 10-second time unit. Although it is not depicted herein, the dataset NT1-HUS also yields a similar representation when aggregating the traffic over the same time scales. Our analysis shows that the datasets NT1-HU and NT1-HUS provide the best pictorial representations of self-similarity among all of the datasets obtained from NS3. By looking at Figure 10.4(b), it is clear that stationarity of the process is absent due to the high volume peaks, which are immediately followed by low volume peaks of traffic. The same type of representation is obtained from the remaining datasets of T2. The justification of such behavior lies with uncontrollable factors during simulation runtime, such as the way an application performs or the start of communications from one LAN to another. Generally speaking, this suggests that the artificialness associated with the simulation as well as the lack of a more unpredictable nature (from human interaction and computer-to-computer communications) may have an impact in the analysis of aggregate traffic, perhaps indicating that a simple execution of a simulated network environment may not be enough to capture the self-similarity of modeled network traffic and that some conditions should be taken into account in order to make the traffic comply with what is observed in real networks. Nevertheless, if it were not for those high and low peaks in Figure 10.4(b), the traffic actually seems to embody the visual characteristics of fGn for all the time scales considered.

Figure 10.5 now contains the outcome of the Hurst parameter estimates for all datasets concerning NS3. Although the Hurst parameter estimates for the dataset NT1-HI are within the long-range dependence boundaries, the pictorial representation of this dataset (not included in this chapter) is not compliant with self-similar network traffic. This can happen when the process under

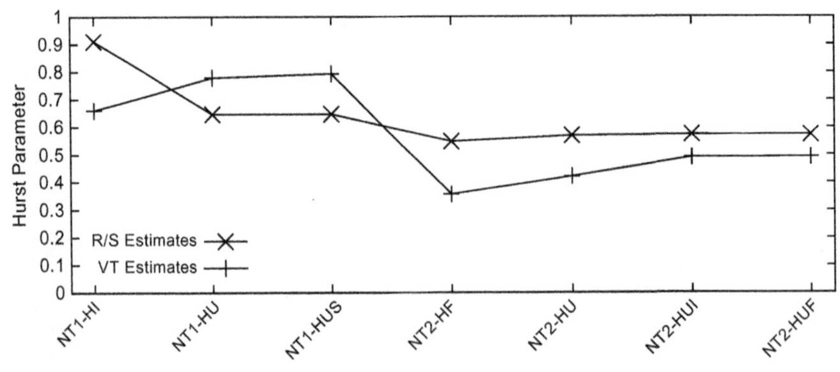

FIGURE 10.5

Hurst parameter estimates for all NS3 traces.

observation is not self-similar, and thus one needs to conduct additional analyses on additional characteristics of self-similarity, such as autocorrelation (see below). The estimates for the remaining datasets for T1 so far show that they are self-similar since their representations are valid. Regarding T2, it is observed that the VT estimates are smaller than 0.5, and that the R/S estimates are a little bigger than 0.5. Because R/S is biased, we can assume that the datasets for T2 may contain short-range dependences and take these values as indicators for the lack of self-similarity in aggregate modeled network traffic. Note that the datasets of T2 are highly condensed in terms of packet arrivals per time unit (there are several packet arrivals for a relatively small amount of simulation time). In these cases, the original process was actually built by splitting each second into 100 intervals (instead of ten intervals as in T1), each a hundredth of a second.

Some of the previous results are not very conclusive in terms of self-similarity. In this regard, Figure 10.6 depicts the empirical autocorrelation computed for all NS3 datasets along with the theoretical autocorrelation

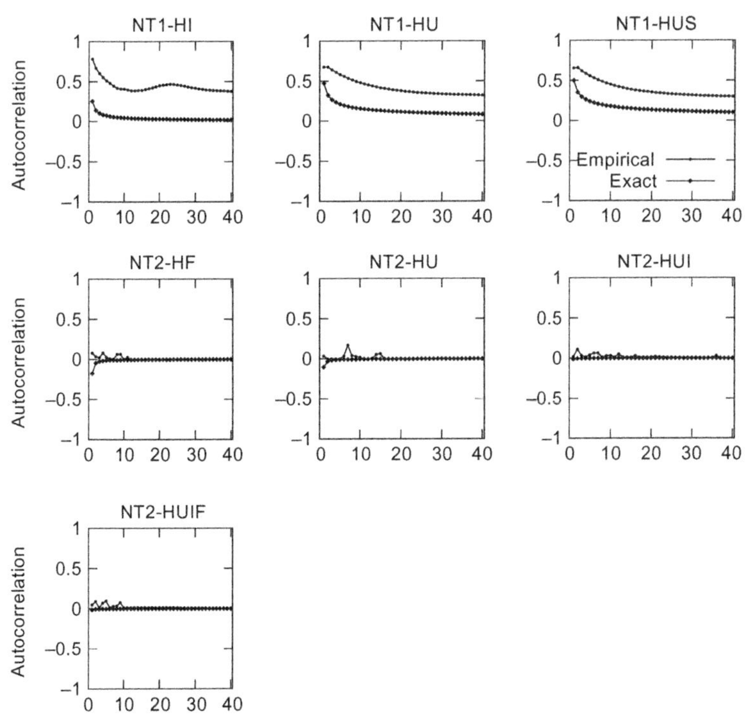

FIGURE 10.6

Empirical and exact autocorrelation for all NS3 traces. The VT method was used to provide Hurst parameter values as input to the autocorrelation function $\gamma_H(k)$.

function for an fGn, for comparison purposes, and for lags $k = 1, 2, 3, \ldots, 40$. The exact autocorrelation function of an fGn can be theoretically calculated with basis on the covariance relation between a valid Hurst parameter H and a positive lag k, such that $\gamma_H(k) = 1/2(|k-1|^{2H} - 2|k|^{2H} + |k+1|^{2H})$ [13]. The exact autocorrelation of an fGn, for all lags $k > 0$, is a hyperbolic function with positive values for $0.5 < H < 1$ and with negative values for $0 < H < 0.5$ with concavity toward the origin for both cases. Yet again, the datasets NT1-HU and NT1-HUS are the ones whose autocorrelation is most similar with the exact autocorrelation of an fGn. Except for the point where $k = 1$, their functions are hyperbolic and corroborate the results previously discussed for those datasets. The autocorrelation of the remaining datasets is not consistent with that expected for an fGn, and as such they are ruled out as self-similar. We can thus conclude that only the datasets NT1-HU and NT1-HUS from NS3 exhibited the self-similarity property imprinted into the aggregate modeled network traffic for all scales m_k considered throughout this chapter for those datasets.

4.5.3 Analysis of the network traffic of OMNeT++

The same analysis conducted for NS3 is now done for OMNeT++. With this objective, the best datasets from OMNeT++ in terms of self-similar representation are the ones named OT1-H, OT1-HF, and OT1-HU. All others showed high and low peaks of network traffic (similar to the ones in NS3), which may be due to application starting and ending execution, and made the process seem unbalanced and nonstationary. Figure 10.7 contains a pictorial view of the bits per time unit for the datasets OT1-H and OT2-HF. In Figure 10.7(a), the visual characteristics of self-similarity become clearer as the time scale increases, contrarily with what is seen on Figure 10.7(b), which shows the same periodic high and low peaks discussed in the NS3 datasets.

Figure 10.8 and Figure 10.9, respectively, depict the Hurst parameter estimates and the autocorrelation for all OMNeT++ datasets. The disparity of the Hurst parameter estimates of both estimators used suggest that the datasets are not self-similar, regardless of the estimation value, and as such they are herein included for the sake of completeness. In fact, several empirical autocorrelation functions do not agree with the underlying Hurst parameter values, and many of them are linear and not hyperbolic. Despite the apparent visual self-similarity in the datasets OT1-H, OT1-HF, and OT1-HU, the modeled network traffic is not self-similar due to the Hurst parameter estimates and their autocorrelation structures. The same reasoning is applied to the remaining datasets, thereby concluding that OMNeT++ does not perform well in terms of embedding the self-similarity property in modeled network traffic, at least for the topologies and experiments conducted in the scope of this research.

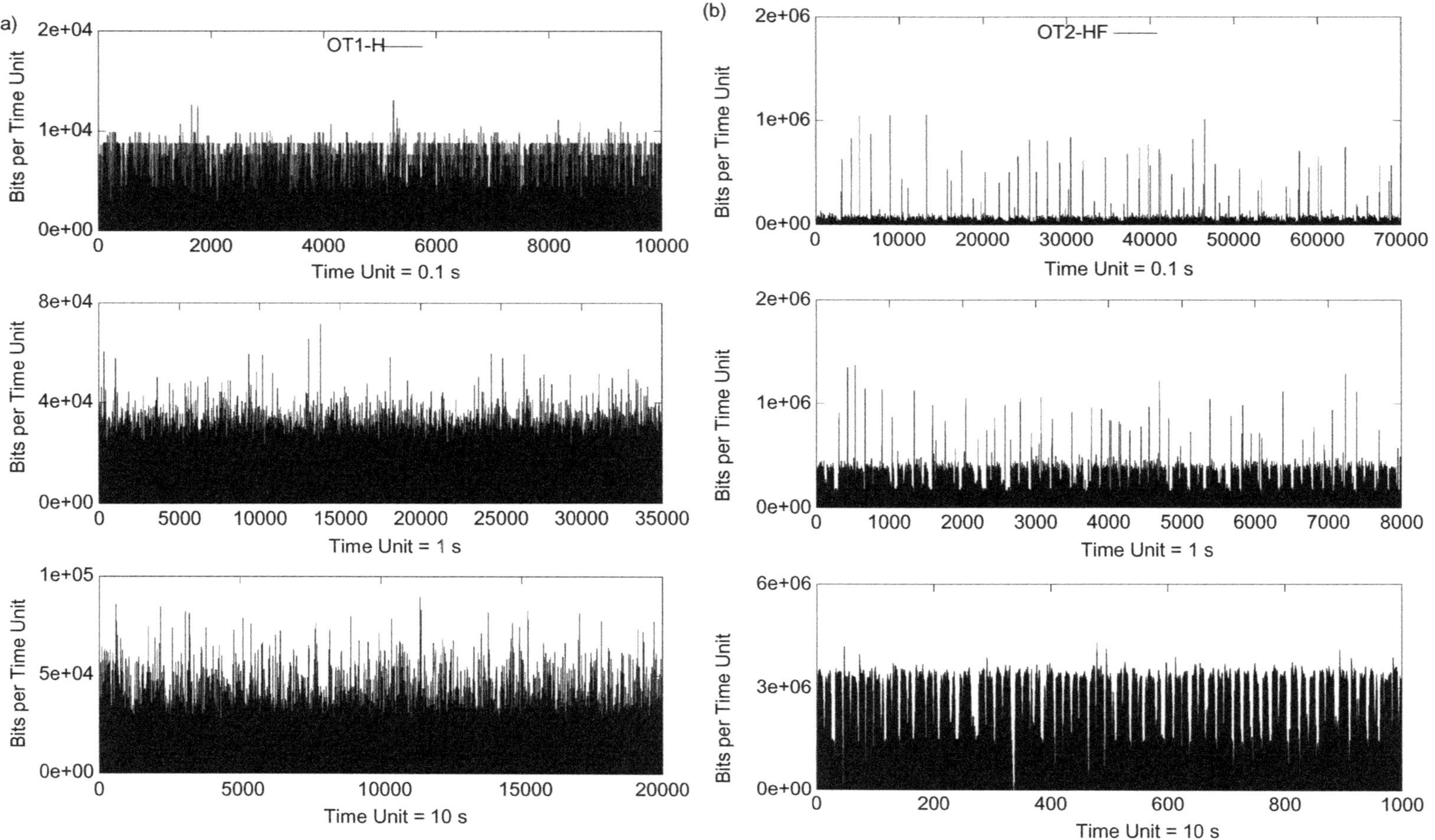

FIGURE 10.7

Bits per time unit on different time scales for a part of the datasets named OT1-H (a) and OT2-HF (b).

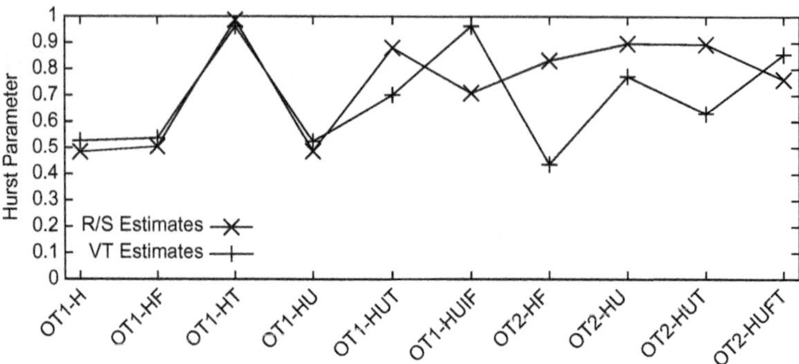

FIGURE 10.8

Hurst parameter estimates for the traces of OMNeT++.

5 CONCLUSIONS AND FUTURE WORK

The simulation of network traffic plays an essential role in research and development in the networking area by allowing abstracting of underlying network complexity. It is thus crucial to find out if the simulation primitives comply with the properties exhibited by network traffic studied under real network environments. This chapter has described the self-similar nature of network traffic and discussed its presence in simulated aggregate network traffic of various computer networking scenarios.

We have provided empirical evidence of the self-similarity property to be embedded in some NS3 modeled network traffic, but not in OMNeT++. The stationarity of the bit rate per time unit is lost because of uncontrollable events during simulation runtime, producing a direct and negative impact in the self-similarity of the traffic. It was observed that the traces varied a great deal, perhaps suggesting that the simulations are highly dependent upon the simulation engine itself, and that the behavior of the modeled applications and network nodes is directly impacted by such dependence. However, it should be emphasized that the results presented throughout this chapter should be interpreted as starting points for a more thorough study of self-similarity on simulated network traffic. The VT method, for example, only covers the variance property, and so additional estimators of the Hurst parameter that explore other properties of self-similarity should be used in the future. This work does not prove that the simulators do not generate, generally speaking, self-similar traffic. However, it shows that the property should be tested whenever it is important for research work, since it may not be naturally present. Simulating this property requires fine-tuning the simulators and aggregating more sources. Additionally, goodness of fit tests for testing Gaussianity may also be employed to assess if the distribution is Gaussian.

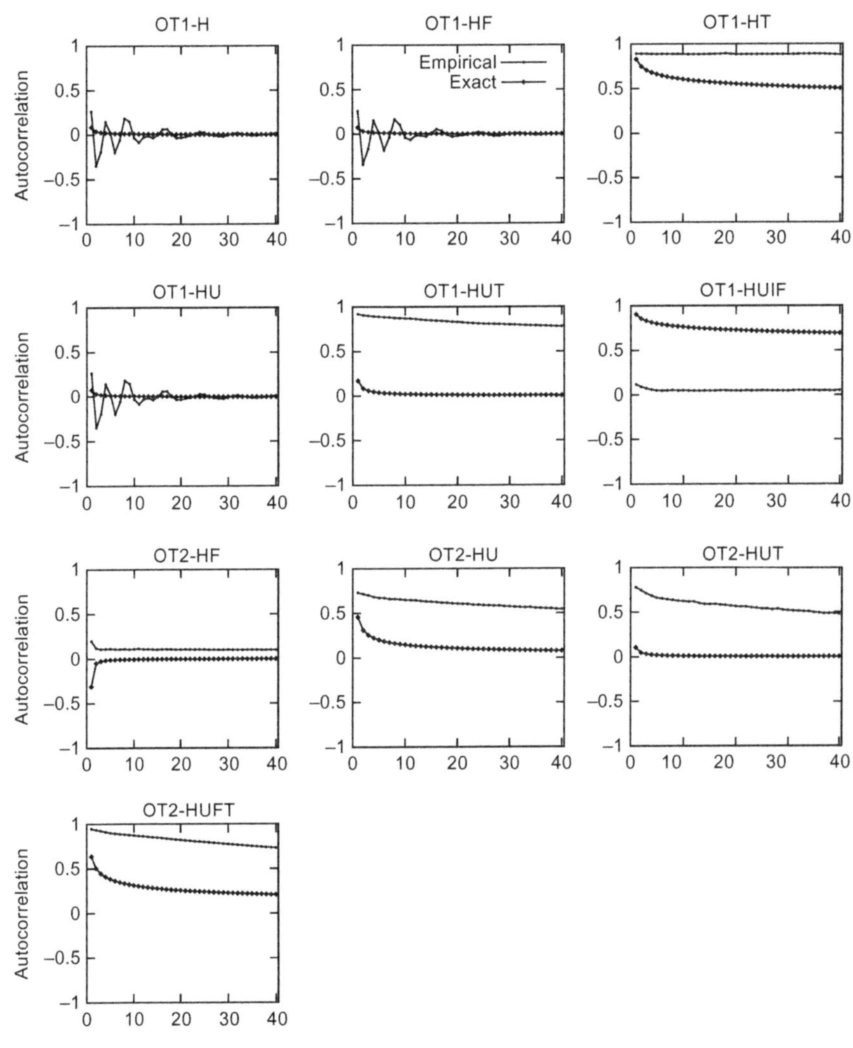

FIGURE 10.9

Empirical and exact autocorrelation for all OMNeT++ traces. The VT method was used to provide Hurst parameter values as input to the autocorrelation function $\gamma_H(k)$.

REFERENCES

[1] Gomes JVP, Inácio PRM, Lakic B, Freire MM, Da Silva HJA, Monteiro PP. Source traffic analysis. ACM Trans Multimedia Comput Commun Appl 2010;6 (3):21:1−21:23.

[2] Leland WE, Taqqu MS, Willinger W, Wilson DV. On the Self-Similar Nature of ethernet traffic (extended version). IEEE/ACM Trans Netw 1994;2(1):1−15.

[3] Hurst HE. Long-term storage capacity of reservoirs. Trans Am Soc Civ Eng 1951;116:770–99.
[4] Kenkel NC, Walker DJ. Fract Biol Sci 1996;11:77–100.
[5] Soares F, Janela F, Pereira M, Seabra J, Freire MM. Classification of breast masses on contrast-enhanced magnetic resonance images through log detrended fluctuation cumulant-based multifractal analysis. IEEE Syst J 2013;99:1–10.
[6] Leland WE, Wilson DV. High time-resolution measurement and analysis of LAN traffic: Implications for LAN interconnection, In: Proceedings of the 10th annual joint conference of the IEEE computer and communications societies (INFOCOM), vol. 3; Apr. 1991. p. 1360–6.
[7] Elbiaze H, Cherkaoui O, McGibbon B, Blais M. A Structure- preserving method of sampling self-similar traffic, in 13th IEEE International Symposium on Modeling, Analysis, and simulation of computer and telecommunication systems (MASCOTS). Atlanta, GA, USA; Sep. 2005, p. 161–8.
[8] Mandelbrot BB, Fisher AJ, Calvet LE. A multifractal model of asset returns, Cowles foundation for research in economics, Yale University, Cowles Foundation Discussion Papers 1164, Sep. 1997.
[9] Racine R. Estimating the hurst parameter, Eidgenössische Technische Hochschule Zürich, B.Sc. Thesis, 2011.
[10] Lee K, Hong S, Kim SJ, Rhee I, Chong S. SLAW: Self-similar Least-action Human Walk. IEEE/ACM Trans Netw 2012;20(2):515–29.
[11] Hosking J. Modeling persistence in hydrological time series using fractional differencing. Water Resour Res 1984;20(12):1898–908.
[12] Inácio PRM, Freire MM, Pereira M, Monteiro PP. Fast synthesis of persistent fractional brownian motion. ACM Trans Model Comput Simul 2012;22(2):11:1–11:21.
[13] Dieker T. Simulation of fractional Brownian motion, Master's thesis, University of Twente, P.O. Box 94079, 1090 GB. Amsterdam; The Netherlands; 2004.
[14] Paxson V. Fast, approximate synthesis of fractional gaussian noise for generating self-similar network traffic, SIGCOMM. Comput Commun Rev 1997;27(5):5–18.
[15] Jeong H-DJ, McNickle D, Pawlikowski K. Fast self-similar teletraffic generation based on FGN and wavelets, In: IEEE International Conference on networks (ICON). Brisbane, Queensland, Australia: IEEE Computer Society; Sep. 1999, p. 75–82.
[16] Karagiannis T, Faloutsos M. SELFIS: A tool for self-similarity and long-range dependence analysis, in 1st Workshop on fractals and Self-similarity in data mining: Issues and approaches (held in conjunction with ACM SIGKDD 2002). New York, NY, USA; ACM, Jul. 2002.
[17] Abry P, Veitch D. Wavelet analysis of long-range-dependent traffic. Inform Theory IEEE Trans 1998;44(1):2–15 on.
[18] Jones O, Shen Y. Estimating the Hurst index of a self-similar process via the crossing tree. IEEE Signal Proc Lett 2004;11(4):416–19.
[19] Mandelbrot BB, Wallis JR. Robustness of R/S in measuring noncyclic global statistical dependence, vol. 5; 1969. p. 967–88.
[20] Willinger W, Paxson V, Taqqu MS. Self-similarity and heavy tails: structural modeling of network traffic. In: Adler RJ, Feldman RE, Taqqu MS, editors. A practical guide to heavy tails. Cambridge, MA, USA: Birkhauser Boston Inc.; 1998. p. 27–53.

[21] Willinger W, Taqqu M, Sherman R, Wilson D. Self-similarity through high-variability: statistical analysis of Ethernet LAN traffic at the source level. IEEE/ACM Tran Net 1997;5(1):71–86.

[22] Cao J, Cleveland W, Lin D, Sun D. Internet traffic tends toward poisson and independent as the load increases. In: Denison D, Hansen M, Holmes C, Mallick B, Yu B, editors. Nonlinear estimation and classification, ser. lecture notes in statistics, vol. 171. New York: Springer; 2003. p. 83–109.

[23] Network Simulator, "Network Simulator Web Page," Available from: <http://www.nsnam.org/>. 2014, [accessed Apr. 2014].

[24] OMNeT++, "OMNeT++ Web Page," Available from, <http://www.omnetpp.org/>; 2013, [accessed Apr. 2014].

[25] The Internet Traffic Archive, Bellcore network traffic traces, Available from: <http://ita.ee.lbl.gov/html/contrib/BC.html>; 2008, [accessed Apr. 2014].

CHAPTER

Performances evaluation and Petri nets

11

Ousmane Diallo[1,2], Joel J.P.C. Rodrigues[1], and Mbaye Sene[3]

[1]Instituto de Telecomunicações, University of Beira Interior, Covilhã, Portugal

[2]University of Assane Seck of Ziguinchor, Ziguinchor, Senegal

[3]Université Cheikh Anta DIOP (UCAD), Dakar, Senegal

1 INTRODUCTION

Modeling and performance evaluation of discrete event systems, such as computer systems, communication network and production systems, etc. have previously and even today received much attention from different scientific communities both in academia and in modern industry [1–10]. Scientific, economic and technical progress and the increasing complexity of these systems raise more new challenges and new problems for their developers. And in addition, since imperfections in the design process can be negative factors in the development time, cost and operational efficiency in the system, it is then necessary to use effective and well-suited methods that can help the designer in the design phase, where they allow the evaluation a priori of the performance of a new uncertain system that one wants to size, as well as in the exploitation phases, where they allow, for example, the evaluation of possible changes in the system that must meet new goals.

Modeling and performance evaluation have usually been closely related research topics. This is probably due to their relationship of cause and effect. Thus, the performance evaluation of one system is most often performed a priori on a pre-established model of the system to be studied, and a good design (or not) of the underlying model affects the system analysis. The direct analysis of stochastic processes and, more generally, Markov chains were the pioneers in this field of research [5,11–17]. Thus, in 1957 the research work of Jackson proposed a new approach based on networks of queues [18–21]. Later, there was the application of operational analysis, which allowed the introduction of some simple performance criteria based solely on observations made on the system to be studied, in particular the operational formula of Little [22] that allows the definition of performance criteria such as the throughput of a system, the average number of requests in the system and the response time of the system. Many extensions of the formalism of networks of queues [19,23] have been proposed in order to study some relatively complex cases, and this research work led to powerful tools for a relatively small class of queueing systems. In 1962 Carl Adam Petri left the formalism of queueing networks and invented a new, more powerful formalism called Petri nets [24–31], in his thesis submitted to the faculty of Mathematics and Physics at the Technical University of Darmstadt, West Germany. For more history about Petri nets readers can follow [24,32]. Petri nets are graphical and mathematical tools that provide a uniform environment for modeling, formal analysis, and design of real systems with discrete events that are characterized as being concurrent, synchronous, asynchronous, distributed, parallel, nondeterministic, and stochastic, such as computer systems [33–40], communication networks [41–43], production systems, industrial automated systems [44–55], workflow management [56,57], traffic control [6,58–62], embedded systems [63–67], etc. One of the main advantages of Petri net models is that the same model can be used for the analysis of behavioral properties and performance evaluation, as well as for systemic construction of discrete-event simulators and controllers [68]. In addition, Petri nets, with their graphical notations and their simple semantics, provide a powerful communication medium between the different entities involved, e.g., researchers, practitioners, the specifications, and the customer, since complex requirement specifications can be described graphically using Petri nets. This is useful for the customer who might not well understand mathematical notations or ambiguous textual descriptions. However, the mathematical notations are also very important for researchers and practitioners, as a Petri net model can be defined by a set of linear algebraic equations, or other mathematical models that translate the behavior of the studied system. Therefore, the practitioners or researchers can perform the formal analysis of the model and verify whether the properties related to the behavior of the studied system, for instance, precedence relations among underlying events, concurrent events, appropriate synchronization, repetitive activities, and mutual exclusion of shared resources, are met. For practical applications of Petri nets, the use of computer-aided tools is necessary to help in the drawing, analysis, and/or simulation of various application systems [24]. Some tools and their application can be found in references [69–75].

The objective of this chapter is to introduce the fundamental concepts of Petri nets to the researchers and practitioners who are interested, as well as those who may become interested to work in this area of modeling and performance analysis of real systems with discrete events using Petri nets, but also to show how this tool has currently advanced in order to be convenient for modeling and analyzing the performance of complex systems involved in many and varied areas of computer science and other disciplines. Therefore, the chapter starts by defining briefly the fundamental concepts of modeling and performance evaluation of systems with their link to Petri nets, and afterwards largely describes the fundamental concepts of Petri nets, from their applications [76–79], ordinary Petri nets (graphical representation and formal definition) [26,28,80], stochastic Petri nets *(SPNs)* [81–83], generalized stochastic Petri nets *(GSPNs)* [84,85], colored Petri nets *(CPNs)* [86–89], to the stochastic well-formed Petri Nets *(SWNs)* [75,90–93], which offer high-level modeling.

The remainder of this chapter is organized as follows. Section 2 defines the fundamental concepts of modeling and performance evaluation of systems with their link to Petri nets, while Section 3 largely describes the fundamental concepts of Petri nets and their applications, as described in the previous paragraph. Finally, Section 4 concludes the chapter.

2 MODELING AND PERFORMANCE EVALUATION

2.1 MODELING

It is possible to model a system that already runs or a not yet existing system. However, the study of a real system is rarely achievable in an operational environment, an analysis consisting to represent its running in a way more or less detailed is required. Thus, to meet the criteria established by the applicant (client), a thorough and careful study is needed. After this study, the modeler must be able to know the required equipment and specifications that it requests. He must also be able to certify whether the desired results can be achieved or not. After this step, he can start the modeling process, bearing in mind the objective to achieve. The use of tools (descriptive, mathematics, etc.) is then necessary to reach this goal. The real system (or to design) is reproduced based on mathematical equations and diagrams that make abstraction of certain details. Most of the time, the model takes into account certain system parameters and abandons others. The more accurate the model is, the more its analysis is difficult and sometimes not achievable. At this point, there must be a compromise between adequacy of the model and system and an easy analysis of the model.

One can therefore say that a model is an abstraction of a real system. Depending on the degree of abstraction, a model more or less faithful is obtained. Figure 11.1 gives an illustration of the compromise between the real system and the model.

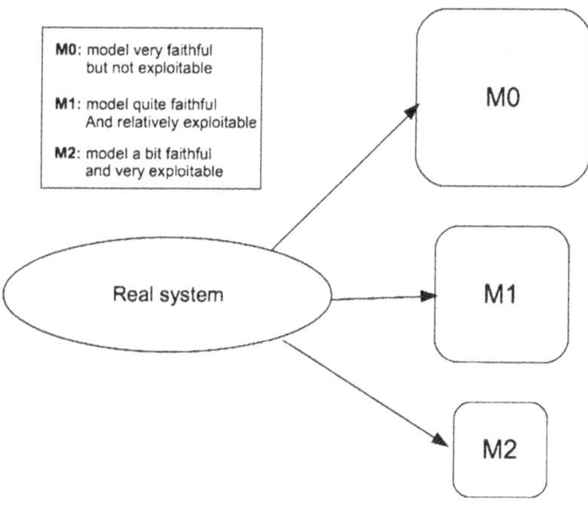

FIGURE 11.1

Model versus real system.

In the design phase, two different approaches can be adopted:

- **An empirical approach:** This approach is particularly inefficient because of the complexity of computer systems. Indeed, it is difficult to evaluate the impact of sizing the parameters (throughput of the network, server power, nature and number of transactions, etc.) on the measures identified in the system.
- **A theoretical approach:** this approach is the one used most often in modeling. In fact, the theoretical approach is a more formal method, which guarantees bounds on the performance of the system because of the difficulty of working with exact values.

2.2 PERFORMANCE EVALUATION

The performance evaluation is interested in calculating performance parameters or indices of a system. The performance parameters that one wishes to obtain are on different orders depending on the systems being studied. For example, in communication networks, an important performance parameter is the response time (routing delay) from end to end. It separates the transmission of the message and its reception by the recipient. The five performance parameters that are tremendously important to determine in many systems are: the throughput (often denoted X), the waiting time (often denoted W), the response time of a client, message or packet (often denoted R), the number of clients, messages or packets (often denoted Q) and the use rate (often denoted U). However, most of the time we are interested in average values.

2.3 WHY EVALUATE THE PERFORMANCE OF A SYSTEM?

The performance evaluation of a system is necessary when one wishes to know the performance of a system but direct measurements of the system cannot be obtained. The performance evaluation can mainly occur in two levels:

- **In Design:** in this phase the system does not yet exist; it should be then created and sized. In practice, it is necessary to perform this phase with respect to a set of specifications.
- **In exploitation:** in this phase the system does exist; however, one wishes to modify or test it outside of its normal running. It may be a goal here to design a different system that meets new objectives. This phase is generally performed to make the system more efficient.

A performance evaluation is also performed in research in order to justify interest in a new solution, such as a new architecture, algorithm, protocol, etc.

2.4 HOW TO DO A PERFORMANCE EVALUATION OF A SYSTEM?

Generally, directly measuring the performance parameters of a real system is very difficult, if not impossible. Thus, a mathematical formalism concentrating on a model, behavior and parameters that better reproduce the functioning of the system must be

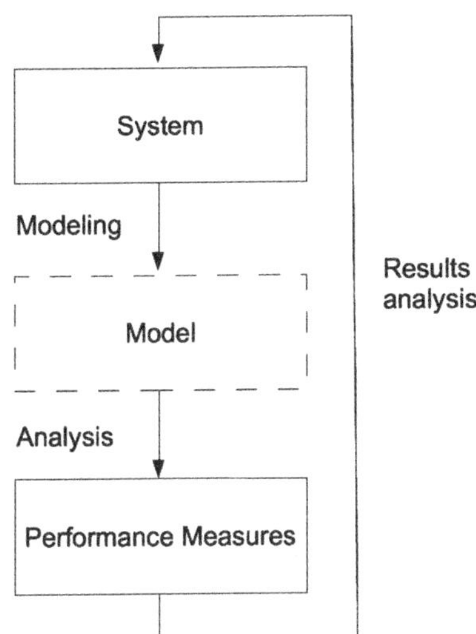

FIGURE 11.2

Illustration schema of the performance evaluation of a real system.

proposed. The model is therefore a mathematical abstraction of the real system. It can be defined by a set of equations or be described by means of formalism. Different types of formalism have been developed depending on the analysis required, such as networks of queues [18–21], stochastic automata [94–97], Petri nets [24–31], etc. Figure 11.2 gives an illustration of the performance evaluation of a real system.

The looping occurs only if the results obtained are not those that one is looking for. Note that performance analysis is done on the model; therefore, the results obtained belong to the model and not to the system.

2.5 ANALYSIS OF THE MODEL

After building the model using the required formalism, it must then be analyzed. There are two main types of analysis: qualitative analysis and quantitative analysis.

- **Qualitative analysis** consists of checking the structural and behavioral properties of the system, such as the absence of blocking (vivacity), the invariants of the system, stability, etc.
- **Quantitative analysis** is used to calculate the performance parameters of the system. It only makes sense if a qualitative analysis has already been performed. For instance, it is useless to analyze the performance of a system that is in a state of deadlock. There are two main approaches to quantitative analysis of performance: *simulation* and *analytical methods*:
    - **Simulation** consists of reproducing the evolution of the model, step by step, by studying a particular realization of the stochastic model. Various simulation languages can be used, such as the universal languages

(C++, Java, etc.), or specific languages (GPSS, Simula, sinscript), etc. The advantage of simulation is to provide a general approach to studying any model, as long as the simulation tool is adapted to the model considered. Simulation is very commonly used in industry. The big disadvantage of simulation is that it is greedy in terms of computing time and analysis of the results can be tricky.

- **Analytical methods** (stochastic automata, etc.) propose to calculate the performance indices mathematically by solving the underlying equations. Their interest lies mainly in their resolution as they are usually inexpensive in terms of computing time. Unfortunately, the class of models that can be analyzed simply and exactly is relatively limited.

In quantitative analysis, the results obtained between two executions are compared and the practitioner can modify one or more parameters depending on the targeted objectives. This phase ends when the results obtained satisfy the specifications.

2.6 ANALYSIS AND DOCUMENTATION OF THE RESULTS

The specification contains the main objectives of the model. The modeler introduces a margin of error, due to the fact that modeling is not a deterministic study.

The analysis of the results allows us to verify if the results obtained are within the range previously defined by the modeler. It is also desirable to know the parameters which have influenced the results. If the results are good, the model can be executed one last time, to be sure that the same results are obtained. If results are not those expected, testing should continue.

Saving the results of different executions along with the parameters that lead to these results is tremendously important in the validation of a model. This can help the client in the future, for example when he wants to increase the performance of the system, to know the parameters to modify, and therefore the material to change. Documentation of this process can also help in understanding the parameters that might be at the origin of an eventual loss of performance of the system during its running.

Models are made using specific formalisms. Thus, different types of formalism have been developed depending on the analysis required, such as networks of queues [18–21], stochastic automata [94–97], Petri nets [24–31], etc.

As discussed previously, this chapter is especially interested in Petri nets, which provide a uniform environment for modeling and formal analysis of real systems with discrete events.

3 PETRI NETS

As stated earlier, Petri nets [24–31] were invented in 1962 by Carl Adam Petri in his thesis submitted to the faculty of Mathematics and Physics at the Technical University of Darmstadt, West Germany. Petri nets are graphical and mathematical tools that allow a performance evaluation and analysis of real systems with discrete events that are characterized as being concurrent, synchronous, asynchronous, distributed, parallel, nondeterministic, and stochastic. Application examples with references were provided in Section 1. Initially, the Petri nets made abstraction of time. Since then, they continue to evolve.

- Time is first introduced for obtaining ***Stochastic Petri Nets (SPN)***.
- Later the immediate transitions were introduced and the ***Generalized Stochastic Petri Net (GSPN)*** arrived.
- With the introduction of colors associated with markings, the era of ***Colored Petri Nets*** and ***Stochastic Well-Formed Petri Nets (SWN)*** came into being.

3.1 OVERVIEW OF THE APPLICATIONS OF PETRI NETS

Petri nets have been used in many varieties of application areas, with various examples met in the literature, such as industrial, computer sciences, biologic, and study of organisms. Two of the most successful application areas of Petri nets have been modeling and performance evaluation [5,6,8–10,24–37,58–67,75,86–89], and communication protocols [41,43,70,93,98–102].

In industrial areas, Petri nets have been used in various sectors, such as simple production lines with buffers [103–105], automotive production systems, flexible manufacturing/industrial control systems, [45,46,49,50,52,53,55], automated assembly lines, resource-sharing systems [32,48,51,54,103,106–110], deadlock avoidance [37,111–114], discrete-event systems [115–118], multiprocessor memory systems [119–122], software development [123–131], data-flow computing systems [56,57,132–137], fault-tolerant systems [138–141], programmable logic and very-large-scale integration (VLSI) arrays [142–150], asynchronous circuits and structures [151–155], compiler and operating systems [156–158], office-information systems [159–161], formal languages [162–165], logic programs [166–169], communication networks [30,41–43,70,93,98,170–172], and so forth. For more information about the applications of Petri nets in industrial areas, readers can review [68].

Petri nets are commonly used in the computer field and are particularly suitable for modeling communication protocols [41–43,70,93,98–102,171–173]. They are used in the compilation of programs via finite automata. Moreover, with the increasing orientation of computer systems towards distributed systems, Petri nets are largely used in modeling and analysis of distributed-software systems [33–35,37,40,124–130,140,174], distributed database systems [65,66,93,173–176], and parallel and concurrent programs [33–40].

In biology, Petri nets provide a great help in modeling and analyzing biological systems, such as multiscale systems biology and the complex chemical processes that are experienced in the mechanisms of living [177–186]. Other phenomena, such as gene expression, also benefit from this modeling tool [187–192]. References [177,179,193,196] give comprehensive tutorials on Petri nets in systems biology.

Finally, in the study of organizations, the Petri nets provide a better assessment of systems. Research work in [194] proposed the use of colored Petri nets

[86–89] to model conversations between agents. The same approach is used in [195], but with the introduction of recursive colored Petri nets to facilitate the composition of conversations. In addition, with the help of timed Petri nets and stochastic Petri nets, it becomes possible to quantify, respectively, the performance of an organization and to decompose statistical studies of basic phenomena.

Generally, Petri nets are very suitable for modeling and performance evaluating many discrete event systems with organizational problems, synchronization and cooperation between processes working in parallel and sharing resources.

3.2 FORMALISM OF PETRI NETS

3.2.1 Definitions and elements of a Petri net

3.2.1.1 Petri net

A Petri net can be defined as a quadruplet:

$$R = \langle P, T, Pre, Post \rangle$$

with:

- $P = \{p_1, p_2, \ldots, p_m\}$ is a finite set of places. The places are represented by circles and model the state variables of the system (system resources);
- $T = \{t_1, t_2, \ldots, t_n\}$ is a finite set of transitions, with $P \cup T \neq \varnothing$ and $P \cap T = \varnothing$

A transition represents an event and/or an action that changes the values of the state variables of the system. Transitions are represented by a rectangle or a continuous line:

- $Pre{:}(P\ X\ T) \rightarrow N$ is the application of *precedent places*, where N is a set of nonnegative integers, and
- $Post{:}(P\ X\ T) \rightarrow N$ is the application of *following places.*

The notation below can also be used:

$$C = Post - Pre$$

with C generally known as the *incidence matrix* of the Petri net.

3.2.1.2 Marked Petri net

Each place can contain 0 or several tokens indicating the value of the variable state modeled by this place. If the place models a logical variable, then the number of tokens that can contain the place is 0 or 1. If there is one token, then the condition is true, false otherwise.

A marked Petri net is the couple:

$$R_0 = \langle R, M_0 \rangle$$

where:

- R is the Petri net and

- M_0 is the initial marking of the network *R;* it is an application

$$M_0{:}P \rightarrow N$$

$M_0\ (p)$ is the number of tokens contained in the place p.

3.2.1.3 Associated graph and matrix notations

A Petri net can be associated with a graph that possesses two types of nodes: the *places* and the *transitions*. An *arc* connects a place p to a transition t if $Pre(p,t) \neq 0$. An arc connects a transition t to a place p if $Post(p,t) \neq 0$. The nonzero values of matrix *Pre* and *Post* are associated with arcs as labels (the default value is *1*). Figure 11.3 provides an example of a graph associated with a given Petri net.

Considering Figure 11.3, the marking M_0 can be represented by a vector with dimension being the number of places. *Pre, Post* and *C* are matrices whose number of lines is equal to a number of places and the number of columns is equal to the number of transitions. Thus, in matrix notation, the Petri net of Figure 11.3 is defined as:

- $P = \{p1, p2, p3\}$,
- $T = \{t1, t2, t3, t4\}$,
- $Pre = \begin{array}{c} \begin{array}{cccc} \mathrm{t1} & \mathrm{t2} & \mathrm{t3} & \mathrm{t4} \end{array} \\ \begin{bmatrix} 0 & 1 & 0 & 0 \\ 1 & 0 & 3 & 0 \\ 0 & 0 & 0 & 1 \end{bmatrix} \end{array} \begin{array}{c} \\ p1 \\ p2 \\ p3 \end{array}$
- $Post = \begin{array}{c} \begin{array}{cccc} \mathrm{t1} & \mathrm{t2} & \mathrm{t3} & \mathrm{t4} \end{array} \\ \begin{bmatrix} 1 & 0 & 0 & 0 \\ 0 & 1 & 0 & 3 \\ 0 & 0 & 1 & 0 \end{bmatrix} \end{array} \begin{array}{c} \\ p1 \\ p2 \\ p3 \end{array}$

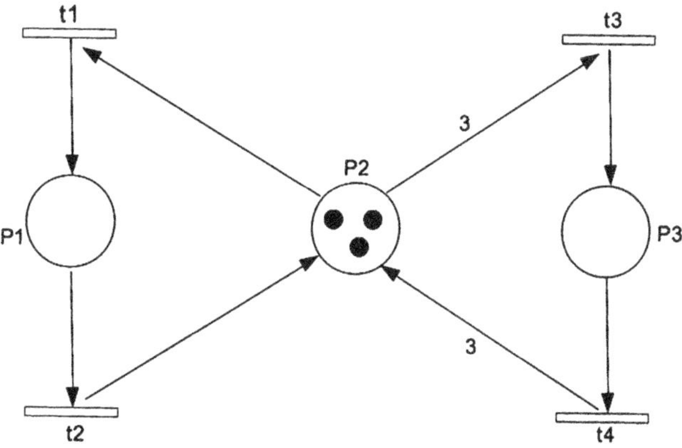

FIGURE 11.3

Example of a graph associated with a Petri net.

- $$C = \begin{array}{c} \begin{matrix} t1 & t2 & t3 & t4 \end{matrix} \\ \begin{bmatrix} 1 & -1 & 0 & 0 \\ -1 & 1 & -3 & 3 \\ 0 & 0 & 1 & -1 \end{bmatrix} \end{array} \begin{matrix} p1 \\ p2 \\ p3 \end{matrix}$$
- $$M_0 = \begin{bmatrix} 0 \\ 3 \\ 0 \end{bmatrix} \begin{matrix} p1 \\ p2 \\ p3 \end{matrix}$$

Note that, considering a place p and a transition t:

- The set of input places of t is defined as: $\bullet t = \{p \in P \backslash Pre(p,t) > 0\}$
- The set of output places of t is defined as: $t\bullet = \{p \in P \backslash Post(p,t) > 0\}$
- The set of input transitions of p is defined as: $\bullet p = \{t \in T \backslash Post(p,t) > 0\}$
- The set of output transitions of p is defined as: $p\bullet = \{t \in T \backslash Pre(p,t) > 0\}$
- *Pre(.,t)*, *Post(.,t)* and *C(.,t)* are the columns of matrices *Pre*, *Post* and *C* associated with the transition *t*.

3.2.1.4 Example of modeling with Petri nets

Consider two machines *M1* and *M2* communicating via two channels **Ch1** and **Ch2** in the following manner:

The Petri net of Figure 11.4 can model the functioning of this communication protocol.

3.2.1.5 Petri net with inhibitor arc

Petri nets have been extended with the introduction of inhibitor arcs. An inhibitor arc onnecting a place p to a transition t prevents the latter from being fired if the place p contains a number of tokens higher or equal to the value of the arc, i.e.:

$$M_0(p) \geq arc\_value$$

A Petri net with inhibitor arc is defined by:

$$R_{inh} = \langle R, \text{Inh} \rangle$$

where *Inh* is the application:

$$Inh{:}P\ X\ T \rightarrow N$$

The inhibitor arc (see Figure 11.5) is represented by a line with a small circle at the extremity.

3.2.2 Dynamic of a Petri net

The evolution of a Petri net is obtained by the firing of transitions. The conditions of enabling and firing are defined in the following section.

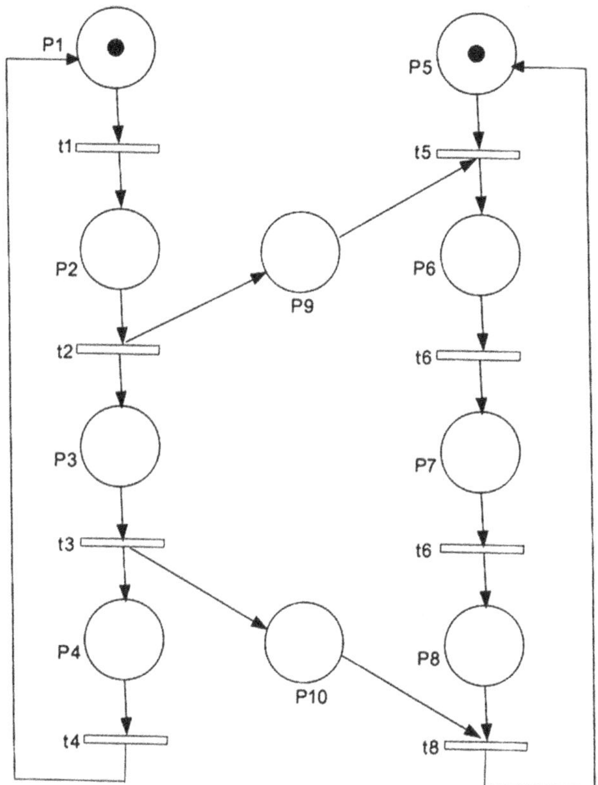

FIGURE 11.4

Petri net for modeling a communication between sender and receiver.

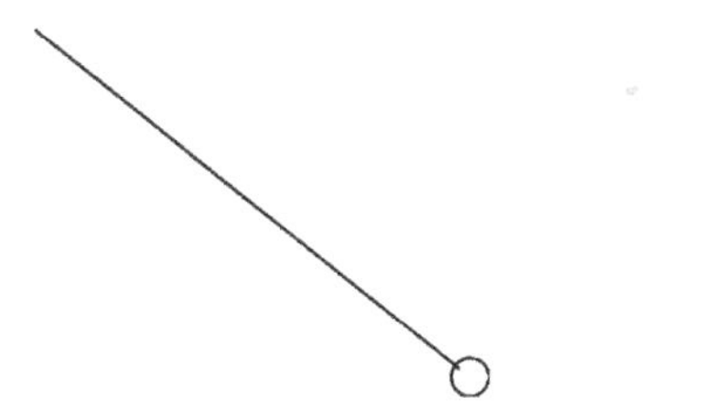

FIGURE 11.5

Inhibitor arc.

3.2.2.1 Enabling a transition

A transition t is enabled in M if and only if:

- $\forall p \in P, M(p) \geq Pre(p,t)$
- And $\forall p \in P, M(p) < Inh(p,t)$

t being enabled can be expressed by the following notations:

- $M \geq Pre(.,t)$
- $M(t>$
- $M \xrightarrow{t}$

For example, in the Petri net of Figure 11.3 and for the initial marking M_0

$$M_0 = \begin{bmatrix} 0 \\ 3 \\ 0 \end{bmatrix}$$

The transitions $t1$ and $t3$ are enabled because

$$Pre(.,\ t1) = \begin{bmatrix} 0 \\ 1 \\ 0 \end{bmatrix} \text{ and } Pre(.,\ t3) = \begin{bmatrix} 0 \\ 3 \\ 0 \end{bmatrix}$$

$$\Rightarrow M_0 > Pre(.,t1) \text{ and } M_0 = Pre(.,t3)$$

3.2.2.2 Firing of a transition

The firing of a transition t enabled in M consists of removing on each of its input places a number of tokens equal to the value of the arc connecting them and to deposit on each of its output places a number of tokens equal to the value of the arc that connects them. This firing leads to the marking M' such that:

$$\forall p \in P, M'(p) = M(p) - Pre(p,t) + Post(p,t)$$

The following notations can also be used:

$$M' = M - Pre(.,t) + Post(.,t)$$

$$M(t > M')$$

$$M \xrightarrow{t} M'$$

For example, using again the Petri net of Figure 11.3 after the firing of the transition $t1$ from the initial marking M_0, the following marking M' is obtained:

$$\begin{bmatrix} 1 \\ 2 \\ 0 \end{bmatrix} = \begin{bmatrix} 0 \\ 3 \\ 0 \end{bmatrix} - \begin{bmatrix} 0 \\ 1 \\ 0 \end{bmatrix} + \begin{bmatrix} 1 \\ 0 \\ 0 \end{bmatrix}$$

3.2.2.3 Sequence of firings

A sequence of firings σ is a succession of firings of transitions which causes the Petri net to pass from the marking M to the marking M' passing through one or more intermediate markings. At each step M_x of the sequence, an enabled transition is fired and makes the marking change from M_x to M_{x+1}.

This sequence of firings σ can be formally defined as a finite sequence of transitions $\sigma = t_1,t_2,\ldots,t_n$, $n \geq 0$ such that there are markings $M_1,\ldots, M_{n+1}$ satisfying

$$M_x(t_x > M_{x+1}) \quad \forall x = 1, 2, \ldots, n$$

Note that it is possible that the same transition can be fired many times. If all transitions of the Petri net are in the sequence, one can say that the sequence is complete.

Again using the Petri net of Figure 11.3, one has, for instance for the sequence $\sigma = t_1,t_1,t_2$ from the initial marking M_0:

$$\begin{bmatrix} 0 \\ 3 \\ 0 \end{bmatrix} \xrightarrow{t_1} \begin{bmatrix} 1 \\ 2 \\ 0 \end{bmatrix} \xrightarrow{t_1} \begin{bmatrix} 2 \\ 1 \\ 0 \end{bmatrix} \xrightarrow{t_2} \begin{bmatrix} 1 \\ 2 \\ 0 \end{bmatrix}$$

3.2.2.4 Markings graph

Considering a marked Petri net:

$$R_0 = \langle R, M_0 \rangle$$

The *reachability set*, A $(R; M_0)$, of a marked Petri net is the markings set that can be reached from the initial marking M_0 by a sequence of firings.

$$M(R; M_0) = \left\{ M_x, \exists \sigma M_0 \xrightarrow{\sigma} M_x \right\}$$

When this markings set is finite, it can be represented as a graph called a *markings graph*, GA $(R; M_0)$. In this graph, the vertices are represented by reachable markings A $(R; M_0)$ and a directed arc connects two vertices M_x and M_y, if there exists an enabled transition t such that: $M_x \xrightarrow{t} M_y$. Figure 11.6 represents the markings graph for the Petri net of Figure 11.3.

In this markings graph, the vertices are:

$$M_0 = \begin{bmatrix} 0 \\ 3 \\ 0 \end{bmatrix}, \quad M_1 = \begin{bmatrix} 1 \\ 2 \\ 0 \end{bmatrix}, \quad M_2 = \begin{bmatrix} 0 \\ 0 \\ 1 \end{bmatrix}, \quad M_3 = \begin{bmatrix} 2 \\ 1 \\ 0 \end{bmatrix}, \quad M_4 = \begin{bmatrix} 3 \\ 0 \\ 0 \end{bmatrix}$$

3.2.3 Properties of a Petri net

This section defines some properties of Petri nets.

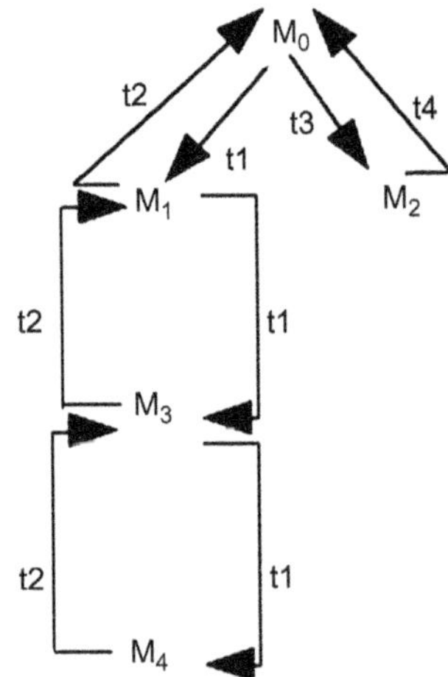

FIGURE 11.6

Illustration of a markings graph.

3.2.3.1 K-bounded place and binary

A place p of a marked Petri net R_O is *k-bounded* if and only if:

$$\forall M' \in A(R; M_0) M'(p) \leq k$$

If $k = 1$, then the place is binary (safe).

3.2.3.2 K-bounded marked Petri net and binary

A marked Petri net R_O is k-bounded if and only if all its places are k-bounded. A marked Petri net R_O is binary (safe) if and only if all its places are binary.

3.2.3.3 Almost-alive transition

A transition t of a marked Petri net R_O is almost alive if and only if there exists a sequence of firings σ such that:

$$M_0 \xrightarrow{\sigma} M' \text{ and } M' \xrightarrow{t}$$

3.2.3.4 Alive transition

A transition t of a marked Petri net R_O is alive if and only if:

$$\forall M' \in A(R; M_0) \exists \sigma M' \xrightarrow{\sigma;t}$$

R_O is alive for an initial marking M_O if and only if all its transitions are alive for this marking M_O.

3.2.3.5 Conformity

A Petri net is conformed if it is safe and alive.

3.2.3.6 Home state

A Petri net state M' is said to be a home state for an initial marking M_0 if and only if for each marking M'' in $A(R; M_0)$, M' is reachable from M'', i.e.,

$$\forall M'' \in A(R;\ M_0) \exists \sigma\ M'' \overset{\sigma;t}{\rightarrow} M'$$

3.2.3.7 Coverage

A marking M' covers a marking M'' and is noted $M' \geq M''$ if and only if:

$$\forall p \in P M'(p) \geq M''$$

3.2.3.8 Reversibility

A marked Petri net R_0 is said to be reversible for an initial marking M_0 if for each marking M' in $A(R; M_0)$, M_0 is reachable from M', i.e., if M_0 is a home state.

3.2.3.9 Invariances analysis

Consider $\overline{\sigma}$ the vector whose components $\overline{\sigma}(t)$ are the number of occurrences of transitions t in a sequence of firings σ; this vector is called a characteristic vector of σ and its dimension is equal to the number of transitions of the Petri net.

The evolutions of all the reachable markings of a Petri net can be then given by the equation below:

$$M' = M - Pre.\overline{\sigma} + Post.\overline{\sigma} \tag{11.1}$$

or

$$M' = M + C.\overline{\sigma} \text{ with } M \geq 0 \text{ and } \overline{\sigma} \geq 0 \tag{11.2}$$

This equation is called the *fundamental equation* of the Petri net and is also used to verify the marking invariance properties.

By multiplying the two members of the equation above by v^T with $v \in Z^n$, the equation below is obtained:

$$v^T.M' = v^T.M + v^T.C.\overline{\sigma} \text{ with } M \geq 0 \text{ and } \overline{\sigma} \geq 0 \tag{11.3}$$

The interesting values of the vector v are those that verify that $v^T.C = 0$; therefore the fundamental equation becomes:

$$v^T.M' = v^T.M \tag{11.4}$$

3.2.3.10 Example of analysis

Considering the Petri net of Figure 11.7:

- $M_0 = \begin{bmatrix} s \\ 0 \\ n \\ t \\ 0 \end{bmatrix}$ and $Pre(p3, t3) = Post(p3, t4) = n$

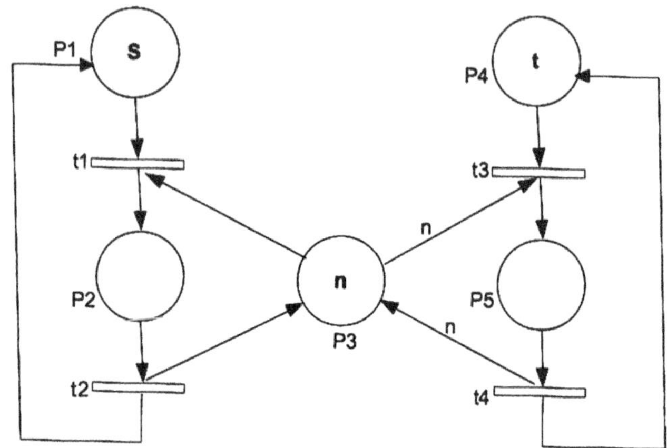

FIGURE 11.7

Analysis of a Petri net.

- $Pre = \begin{bmatrix} 1 & 0 & 0 & 0 \\ 0 & 1 & 0 & 0 \\ 1 & 0 & n & 0 \\ 0 & 0 & 1 & 0 \\ 0 & 0 & 0 & 0 \end{bmatrix}$ and $Pre = \begin{bmatrix} 0 & 1 & 0 & 0 \\ 1 & 0 & 0 & 0 \\ 0 & 1 & 0 & n \\ 0 & 0 & 0 & 1 \\ 0 & 0 & 1 & 0 \end{bmatrix}$
- $C = \begin{bmatrix} -1 & 1 & 0 & 0 \\ 1 & -1 & 0 & 0 \\ -1 & 1 & -n & n \\ 0 & 0 & -1 & 1 \\ 0 & 0 & 1 & -1 \end{bmatrix}$

In the initial marking M_0, $t1$ is enabled; its firing gives the marking M' obtained by:

$$M' = M_0 + C \begin{bmatrix} 1 \\ 0 \\ 0 \\ 0 \end{bmatrix} = \begin{bmatrix} s \\ 0 \\ n \\ t \\ 0 \end{bmatrix} + \begin{bmatrix} -1 \\ 1 \\ -1 \\ 0 \\ 0 \end{bmatrix} = \begin{bmatrix} s-1 \\ 1 \\ n-1 \\ t \\ 0 \end{bmatrix}$$

Assuming that $s \geq 3$ and $n \geq 3$ and considering the sequence $\sigma = t_1, t_1, t_1, t_2$ the characteristic vector of σ is:

$$\overline{\sigma} = \begin{bmatrix} 3 \\ 1 \\ 0 \\ 0 \end{bmatrix} \begin{bmatrix} t_1 \\ t_2 \\ t_3 \\ t_4 \end{bmatrix}$$

The marking M'' reached after the sequence σ is given by:

$$M'' = M_0 + C\begin{bmatrix}3\\1\\0\\0\end{bmatrix} = \begin{bmatrix}s\\0\\n\\t\\0\end{bmatrix} + 3\begin{bmatrix}-1\\1\\-1\\0\\0\end{bmatrix} + \begin{bmatrix}1\\-1\\1\\0\\0\end{bmatrix} = \begin{bmatrix}s-2\\2\\n-2\\t\\0\end{bmatrix}$$

3.2.4 Conflict and parallelism

3.2.4.1 Structural conflict

Two transitions *t1* and *t2* are in structural conflict if and only if they have at least one place of entry in common:

$$\exists p \; Pre(p, t1).Pre(p, t2) \neq 0$$

3.2.4.2 Effective conflict

Two transitions *t1* and *t2* are in effective conflict for a marking M if and only if they are in structural conflict and:

$$M \geq Pre(., t1)$$
$$M \geq Pre(., t2)$$

3.2.4.3 Structural parallelism

Two transitions *t1* and *t2* are structurally parallel if:

$$(Pre(., t1))^T \; X \; Pre(., t2) = 0$$

The result of this is that they have no place of entry in common.

3.2.4.4 Effective parallelism

Two transitions *t1* and *t2* are effectively parallel for a given a marking M if and only if they are structurally parallel and:

$$M \geq Pre(., t1)$$
$$M \geq Pre(., t2)$$

Considering the Petri net of Figure 11.3, the transitions *t1* and *t3* are in structural conflict because:

$$Pre(p2, t1).Pre(p2, t3) = 3$$

And *t2* and *t4* are structurally parallel because:

$$Pre(., t2) = \begin{bmatrix}1\\0\\0\end{bmatrix} \quad \text{and} \quad Pre(., t4) = \begin{bmatrix}0\\0\\1\end{bmatrix}$$

For the marking initial M_0, the transitions *t1* and *t3* are in effective conflict. Considering the marking M' below:

$$M' = \begin{bmatrix} 1 \\ 0 \\ 1 \end{bmatrix}$$

then the transitions *t1* and *t4* are effectively parallel. In fact, they can be fired independently of one another.

3.3 STOCHASTIC PETRI NETS (SPNs)

3.3.1 Introduction of time in Petri nets

In a Petri net, the time can be introduced in several levels:

- **Token**: Each token carries a time interval during which it is available to fire a given transition.
- **Place**: The tokens in a place remain there during a time interval, depending on the place, before being used.
- **Arc**: The tokens go through the arcs moving from a place to a transition (or vice versa) in a period of time corresponding to the delay of cross.
- **Transition**: The activity modeled by a transition lasts for a time corresponding to the time required for its execution. The beginning of the activity corresponds to the moment when the transition is enabled and its end corresponds to the end of the firing.

A transition has a delay of firing resulting in a random firing. At each timed transition *t*, one can associate a clock. When *t* is enabled, the clock is initialized with a value *y*; *y* is decremented with a constant speed and when its value is *0*, *t* is fired. The delay of the firing follows an exponential law:

$$F_{X_i}(t) = 1 - e^{-\lambda_i t} \tag{11.5}$$

3.3.2 Definition of a stochastic Petri net

A stochastic Petri net is defined as:

$$SPN = \{R, Inh, M_0, \lambda\}$$

where:

R = ⟨P, T, Pre, Post⟩ is a Petri net
Inh is the application: $Inh{:}PXT \rightarrow N$
M_0 is the initial marking
And $\lambda = \lambda_1, \lambda_2, \ldots, \lambda_n$ is the set of firing delays of all the transitions.

In such a Petri net, the distribution of the firing time of each transition is exponential with a rate λ *[T]*. Notice that $\lambda_k = \lambda[T]$ the firing rate of the transition T_k.

3.3.3 Performance study of a stochastic Petri net

In an SPN, if two transitions *t1* and *t2* with respective rates λ_I and λ_2 are enabled in *M*, the probability of firing *t1* before *t2* is given by:

$$P_{\lambda_1} = \frac{\lambda_1}{\lambda_1 + \lambda_2} \tag{11.6}$$

In the same way, the probability of firing *t2* before *t1* is:

$$P_{\lambda_2} = \frac{\lambda_2}{\lambda_1 + \lambda_2} \tag{11.7}$$

These formulas show that the probability of passing from the marking *M* to a marking *M′* is independent of the time spent in *M*. The time spent in *M* is:

$$P[\min(X_1, X_2) \leq x] = 1 - e^{(\lambda_1 + \lambda_2)x}$$

The quantitative analysis of an *SPN* can be performed by analyzing its associated Markov chain. Consider two reachable markings M_x and M_y such that only one transition t_k is as:

$$\mathrm{M_X} \xrightarrow{t_k} M_y$$

The firing rate of the transition from the state M_x to the state M_y is $\lambda(t_k)$ (the probability law is exponential). If two transitions t_k and t_m are allowed to pass from M_x to M_y, then the firing rate of the transition from the state M_x to the state M_y will be:

$$\lambda(t_k) + \lambda(t_m)$$

From an SPN, a Markov process can be built. The states (discrete state space and continuous time) will be the reachability markings set of the SPN, A (R; M0). From the function λ the matrix Q of transition rates are written directly. The term $q_{xy} = \lambda_k$ is the transition rate to arrive in the state M_y from the state M_x. The term q_{xx} is negative because it describes the transition rate to leave the state M_x.

The vector Π of stationary state probabilities of the associated Markov chain is obtained by solving the equations:

$$\Pi Q = 0 \text{ with } \Pi = (\Pi_1, \Pi_2, \ldots, \Pi_s) \tag{11.8}$$

$$\sum_{1}^{s} \Pi_j = 1 \tag{11.9}$$

From this moment the performances of the studied system can be calculated.

3.4 GENERALIZED STOCHASTIC PETRI NETs

3.4.1 Definition

A generalized stochastic Petri net (GSPN) is an SPN with two types of transitions: *timed transitions* and *immediate transitions* (see Figure 11.8). Timed transitions are represented by a hollow rectangle and the immediate transitions are

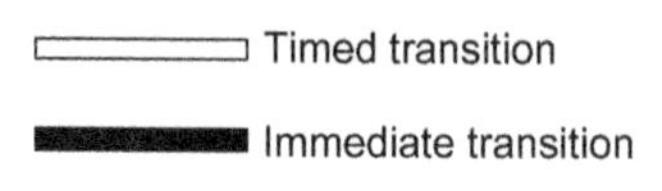

FIGURE 11.8

Timed transition and immediate.

represented by a solid rectangle. The immediate transitions are fired as soon as they are enabled (there is no waiting time), while the others are fired with a rate depending on the transition, as in SPNs.

A GSPN is characterized by $< PN, T_t, T_i, \lambda >$ where:

- $R = \langle P, T, Pre, Post, Inh, M_0 \rangle$
- $T_t \subseteq T$ is the set of timed transitions ($T_t \neq \varnothing$)
- $T_i \subseteq T$ is the set of immediate transitions ($T_i \cap T_t = \varnothing$ *and* $T = T_i \cup T_t$)
- $\lambda = (\lambda_1, \lambda_2, \ldots, \lambda_{|T|})$ with $\lambda_i \in R_+$ is the firing delay if the transition is temporized, or the probability of firing if the transition is immediate.

In the study of GSPNs, only tangible markings are considered (the reachable states set is private of evanescent markings).

The stay time of the stochastic process in the markings where at least one immediate transition is enabled (*evanescent* markings) does not follow an exponential distribution. Such markings change immediately (the process does not stop on these markings) because immediate transitions are fired with zero delay. However, the distribution of the stay time of the process in the markings where only timed transitions are enabled (*tangible* markings) is exponential. A GSPN does not directly describe a continuous time Markov chain.

3.4.2 Example of a GSPN

Figure 11.9 provides an example of a GSPN that models one system with two parallel activities.

3.5 COLORED PETRI NETS (CPNs)

3.5.1 Folding of a Petri net

One way to describe relatively complex systems with a compact Petri net is to fold a set of processes that has the same structure (or structures close to each other) in a single conservative component. The trouble is that the individuality of processes is lost and is then known in the data part. The marking of places only gives the number of processes in a given state without being able to know their identities.

The underlying idea of the colored Petri nets is precisely to perform this adjustment folding without loss of information and without losing the graphic visualization of the process structure.

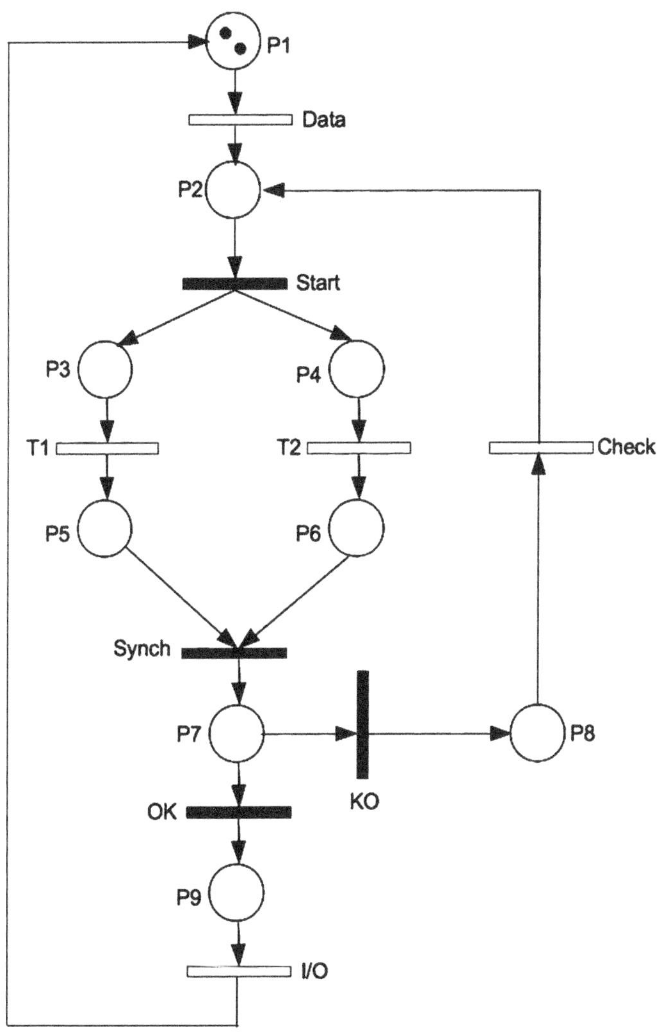

FIGURE 11.9

Example of a GSPN modeling one system with two parallel activities.

3.5.2 Association of colors to the tokens

In order to differentiate the tokens, they are associated with colors (or integers or label sets). Accordingly, each place is associated with a set of colors of tokens that can stay inside. Each transition is associated with a set of colors corresponding to the ways of firing this transition. In the simplest case, that is to say when all processes have exactly the same structure and are independent of each other,

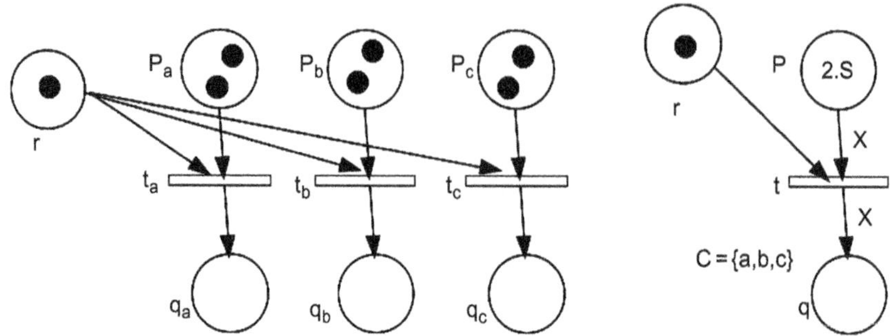

FIGURE 11.10

Example of reduction model with the colored Petri nets.

the colors of transitions are directly associated with the process and the set of colors of the places and transitions are identical. Figure 11.10 provides an example of reducing a model using the colored Petri nets.

3.5.3 Definition

A marked colored Petri net is a 6-tuple:

$$\text{CPN} = \langle \text{P, T}, C_{col}, C_{sec}, \text{W}, M_0 \rangle$$

with:

- $P = \{p_1, p_2, \ldots, p_m\}$ is a non-empty finite set of places,
- $T = \{t_1, t_2, \ldots, t_n\}$ is a non-empty finite set of transitions,
- $C_{col} = \{c_1, c_2, \ldots, c_m\}$ is a non-empty finite set of colors,
- C_{sec} is the function subset of colors that associates with each place and transition a subset of C_{col}:
  - $C_{sec}: P \cup T \rightarrow \Phi(C_{col})$ where $\Phi(C_{col})$ is the set of subsets of C_{col}
- W is the incidence function (equivalent to $C = Post - Pre$); each element $W(p,t)$ of W is itself a function:

$$W(p,t): C_{sec}(t) X\, C_{sec}(p) \rightarrow \text{N}$$

- M_0 is the initial marking. For each place and for each possible color in this place, it associates a number of tokens:

$$M_0(p): C_{sec}(p) \rightarrow \text{N}$$

Note that the definition of a *CPN* may contain the *Inh* application that allows knowing the places that are connected to transitions by an inhibitor arc. In this case, the *CPN* is defined as:

$$\text{CPN} = \langle \text{P, T}, C_{col}, C_{sec}, \text{W, Inh}, M_0 \rangle$$

3.5.4 Enabling and firing

3.5.4.1 Enabling

A transition t is enabled for the color c in the marking M and is noted $M(t(c)>$ if and only if:

$$\forall p \in P,\ M(p) \geq Pre(p,t)(c)$$

A colored Petri net can be defined with priorities. Such a Petri net is defined by a couple:

$$\langle \mathrm{CPN}, \Lambda \rangle$$

where Λ is the priority function defined from T to N. By default, it is assumed that $\Lambda(t) = 0\ \forall\ t \in T$. In a colored Petri net with priority, the validation of a transition t for a color c is given by:

- $\forall p \in P,\ M(p) \geq Pre(p,t)(c)$
- $\forall T_k \in T | \Lambda(T_k) > \Lambda(t), \forall c' \in C_{sec}(T_k), \exists\ p \in P | M(p) < Pre(p,T_k)(c')$

3.5.4.2 Firing

The firing of (t,c) leads to a new marking $M' = M(t(c)>$ with:

$$\forall p \in P, M'(p) = M(p) - Pre(p,t)(c) + Post(p,t)(c)$$

3.5.4.3 Analysis of a colored Petri net

The analysis of a colored Petri net is equivalent to analyzing the corresponding ordinary Petri net. It is necessary then to unfold the *CPN* and to apply the analysis rules of the ordinary Petri nets to an unfolding network. The disadvantage is that usually the unfolding results in a Petri net that is too large and difficult to analyze. In addition, the analytical results obtained from the unfolding network are very difficult to interpret in the original colored network. Another approach is to generalize the results of the simple Petri nets to the colored Petri nets; however, this generalization is quite heavy and difficult to achieve. Moreover, it is impossible to automate the calculation of aggregated states of the modeled system. It is therefore impossible to take advantage of symmetries of the system. So to take advantage of symmetries, it is necessary to have direct analysis tools of colored Petri nets. The problem that arises is then: "Can we define a formalism structured enough to allow defining a direct algorithm while maintaining an expressive power sufficiently high?" Thus, after extensive work on research and intermediate classes of networks (regular networks, ordered networks, networks with unary predicates/transitions), a high-level Petri net class is obtained: Well-formed Nets (*WN*).

3.6 WELL-FORMED NETS

3.6.1 Definition

A well-formed net (WN) is a CPN that respects syntactic constraints whose major objective is to develop analysis algorithms that can be applied directly to the colored Petri nets. The constraints imposed in this type of network are of three types:

1. The color areas are Cartesian products of basic areas called classes of elementary colors.
2. The color functions are built from elementary functions that are: the *identity*, the *successor* and *diffusion*.
3. The model highlights the symmetries of the system. The asymmetric behaviors are considered by partitioning the classes in static subclasses and by adding guards on the color functions and on the transitions.

3.6.2 Color areas

The color areas of places and transitions are Cartesian products of basic areas or elementary classes.

3.6.2.1 Elementary classes

An elementary class is a non-empty finite set of terminal colors. Each elementary class can be ordered and/or set. A terminal color is a color that is defined from any other color. An elementary class is generally seen as a set of elements of the same type whose behavior may be slightly different.

When you want to use the successor function, an order in the class is defined and each element of the class possesses only one successor.

The difference in behavior of elements within a same class is determined by the fact that each class can be partitioned into many static subclasses. The subclasses are pairwise disjointed. These subclasses are called static because they do not depend on markings and are fixed in the design phase of the model. Note that these subclasses can also be parameterized; in this case, the number of elements of the class is the sum of the number of elements of the subclasses that compose it.

3.6.2.2 Definition of a colors area

A colors area is a Cartesian product of finite elementary classes. Each elementary class is a finite set of elements. Notice that $Cl = \{C_1, C_2, \ldots, C_k\}$ the set of elementary classes and it requires that $\forall j \neq i \in [1 \ldots k], C_i \cap C_j = \varnothing$.

If an elementary class C_i is partitioned at its design in s_i static subclasses ($s_i > 1$), the qth subclass is noted $C_{i,q}$ and there is $C_i = \cup_q C_{i,q}$. If the subclass $C_{i,q}$ is set, it is noted $n_{i,q}$ its number of elements. In addition, $\forall r | 0 \leq r \leq s_i$ and $r \neq q, C_{i,q} \cap C_{i,r} = \varnothing$.

A colors area is usually noted by $C = C_1^{e1} X C_2^{e2}, \ldots, C_k^{ek}$ with $e1(1 \le i \le k)$ the number of times that the class C_i appears in domain C. If all e1 are zero, then the colors area is *neutral area*, noted ε. The colors area of a node *r* (*r* is either a place or a transition) is denoted by C(r).

3.6.3 Basic color functions

Arcs of a WN are labeled by color functions. As simple colored Petri nets, these functions specify, for a firing, the number and the mark color consumed or produced. In the case of WNs, these functions are built from elementary functions, which are three in number. Each of them is defined by a color area C to an elementary class C_i. These functions are:

- **Identity** (X_i^j) that allows the selection of an object among the jth instance of the class C_i

$$(\text{if } c = (c_i^j)_n^{ei(t)} \in C(t), \text{ and } C(p) = C_i \text{ is ordered, then } X_i^j(c) = c_i^j),$$

- **Diffusion** ($S_{i,q}$) used to synchronize all objects in a static subclass $D_{i,q}$ when it appears in an entering arc function of a transition, and to disseminate all objects $D_{i,q}$ when it appears in an outgoing arc function

$$(\text{if } C(p) = C_i, \text{ then, } \forall c \in C(t), S_{i,q}(c) = D_{i,q}),$$

- **Successor** ($!X_i^j$) to model the choice of the successor of a selected object from an ordered class (an immediate neighbor in a ring topology or the next message in a sequence of messages, for example).

$$(\text{if } c = (c_i^j)_n^{ei(t)} \in C(t), \text{ and } C(p) = C_i \text{ is ordered, then } !X_i^j(c) = !c_i^j).$$

3.6.4 Guards

The guards are Boolean expressions (called standard predicates) used to define the enabling of transitions: the equality between objects of the same class can be used ($X_i^j = X_i^k$) or between an object and the successor of another object ($X_i^j = !X_i^k$) in a same ordered class, membership to a static subclass $X_i^j \in D_{i,q}$ and a logical combination of these values.

3.6.5 Formalism of a WN

A WN is defined by the 7-tuple:

$$WN = \langle P, T, C_1, C, Pre, Post, Gard \rangle$$

with:

- $P = \{p_1, p_2, \ldots, p_m\}$ is a non-empty finite set of places,
- $T = \{t_1, t_2, \ldots, t_n\}$ is a non-empty finite set of transitions,
- $P \cap T = \varnothing$,

- $C_l = \{C_i | i \in \{1, 2, ..., k\}\}$ is the set of elementary classes. Each C_i is a non-empty finite set that can be partitioned into static subclasses,
- C is the colors function $P \cup T \rightarrow w$ where the subset w is a set containing a finite Cartesian product of C_l elements. An element of $C(s)$ is a tuple $\langle C_1, C_2, ..., C_k \rangle$ called the color of s. $C(s)$ is the color area of s.
- *Pre* (respectively, Post) is the before (respectively, after) incidence function which associates to each couple (P_x, T_y) of *PXT* a guarded color function from (T_y) to $Bag(C(P_x))$,
- *Gard* is a function that associates a guard to each transition. By default, *Gard* (T_x) is the constant function with a value *TRUE* $\forall T_x \in T$.

For the *successor* function to be applied to a class $C_i \in C_l$, it is necessary that this class is ordered. A *ord* function defined from C_l to True, False {True, False} is used to specify which classes are ordered (if $\text{ord}(C_i) = \text{True}$ then C_i is ordered, otherwise it is not).

It should be noted that, like any class of Petri net, the definition of a WN may involve *Inh* application defined on *PXT* that gives the places and transitions linked by an inhibitor arc. It is also possible to have a *Pri* application defined from *T* to *N* which gives priorities to transitions. In this case, the WN is defined by:

$$WN = \langle P, T, C_l, C, Pre, Post, Inh, Pri, Gard \rangle$$

By default $Pri(T_i) = 0$. Figure 11.11 provides an example of a WN.

- $P = \{P1, P2, P3\}$, $T = \{T1\}$, $C_l = \{C1\}$;
- $C1 = C_{1,1} \cup C_{1,2}$ with $C_{1,1} = \{0\}$ and $C_{1,2} = \{1...N_1\}$;
- $C(P1) = C(P2)$;
- $C(P3) = C(T1) = C1XC1$

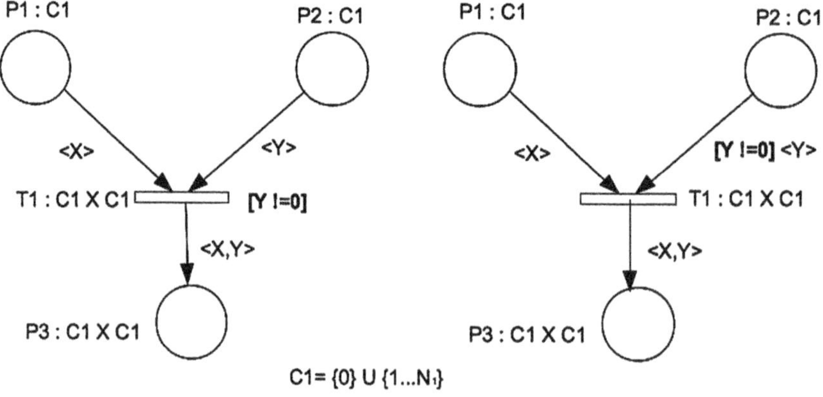

FIGURE 11.11

Example of a well-formed net.

In these WNs, the set P of places contains three elements ($P1$, $P2$, $P3$), the set T of transitions contains one element ($T1$) and the set of elementary classes C_I has a single element ($C1$). This unique class is partitioned into two static subclasses that are $C_{1,1}$ and $C_{1,2}$. The static subclass $C_{1,2}$ is set with a cardinal N_1. The class C1 is also set. The places $P1$ and $P2$ have the same area ($C1$) while the place $P3$ and the transition $T1$ have the same color area $C1XC1$. In the left WN, there is a guard on the transition $T1$, while in the right one, it is the function on the arc $P2$ to $T1$ which is guarded.

3.6.6 Enabling and firing in a WN

The firing rules are identical to those used for colored Petri nets. However on WNs, there are guards to take into account and change the rules as follows:

A transition T_i is enabled for the color $c_t \in C(T_i)$ if and only if the associated guard to T_i is true for the color c_t and T_i is normally enabled for c_t (the verified preconditions).

3.7 STOCHASTIC WELL-FORMED PETRI NETS

The need for a high-level class for direct analysis without passing through the unfolding gave birth to well-formed Petri nets. In these networks, the classes and the color areas are well structured as well as the color functions, to allow the creation of algorithms making it possible to study without using unfolding. However, these networks do not take into account the stochastic character of certain networks; this resulted in the extending of this class to the stochastic Petri nets, which led to stochastic well-formed Petri nets (SWNs).

3.7.1 Aggregation Markov

The goal of aggregation is to replace a complex system with a simpler system that keeps the same behavior as the original system had regarding certain criteria.

The aggregation is an analysis method of a Markov process, which consists of, after having grouped this process into classes, studying the new process formed from the states classes. Concerning the performance evaluation, most often the goal is to be able to calculate a synthesis of stationary measurements of the original system from stationary measurements of the reduced system.

3.7.2 Stochastic process of a well-formed Petri net

The introduction of a stochastic semantic within well-formed Petri nets has three major objectives:

1. Be consistent with the stochastic semantic of generalized stochastic Petri nets,
2. Allow the user to specify it in the well-formed Petri net,
3. Preserve the symmetry so that the symbolic reachability graph (SRG) can be used to support a quantitative evaluation.

In GSPNs, when two immediate transitions are simultaneously enabled, additional information is required to completely specify the behavior of the system. However, for SWNs, the concurrence can be between the different color instances of the same transition as well as the concurrences between transitions. So to completely specify the behavior of the system, the associated value of a transition (weight or firing delay) must be a function of the color instance of the transition and the marking. For the sake of respect for symmetry of the qualitative model, all objects of the same static subclass must be at the same level of the value associated to the transition. In the same way, the dependence of the marking will be linked to static subclasses to ensure the homogeneity of the output rate of the ordinary marking composing a symbolic marking.

3.7.3 Definition of a SWN

A stochastic well-formed Petri net is a couple S,θ such that:

- S = ⟨P, T, C_1, C, Pre, Post, Pri, Gard⟩ is a well-formed Petri net,
- θ is a function defined on T such as:

$$\theta(T_i) : \tilde{C}i(T_i)X \prod P_i \in P\ \mathrm{Bag}(C(P_i)) \rightarrow R_+$$

If $\mathrm{Pri}(T_i) > 0$, then T_i is an immediate transition and $\theta(T_i)(\tilde{C}, \tilde{M})$ represents the weight associated to the transition T_i for color c in the marking M. The probability of firing of $T_i(c)$ in M, if other immediate transitions are enabled is:

$$PB = \frac{\theta(T_i)(\tilde{C},\ \tilde{M})}{\sum_{T_{j,c'}} \theta(T_j)(\tilde{C}', \tilde{M})} \text{ with } \mathrm{Pri}(T_j) = \mathrm{Pri}(T_i) \text{ and } M(T_j(c')) > \quad (11.9)$$

If $\mathrm{Pri}(T_i) = 0$, then T_i is an exponential transition and $\theta(T_i)(\tilde{C},\ \tilde{M})$ represents the firing rate associated to a transition T_i for the color c in the marking M. This rate is a distributed random variable following the exponential law with mean $\theta(T_i)(\tilde{C},\ \tilde{M})$. Note that the firing rate (or weight) associated to a transition only depends on static subclasses of the model, and never directly on colors.

3.7.4 Example of a SWN

The network of Figure 11.12 models the operation of a multiprocessor. There are two classes of colors (C_1 and C_2), which are respectively the class of the processes and those of the processors. The class C_1 is composed of two static subclasses: $C_{1,\ 1}$ contains the slow processes and $C_{1,\ 2}$ the fast processes. Similarly C_2 is divided into two static subclasses ($C_{2,\ 1}$ and $C_{2,\ 2}$), with the cardinal 2, depending on their processing speed. The place P_1 contains the inactive processes, P_2 the processes waiting for a free processor, and P3 the processors.

After a mean time λ_1 (respectively λ_2) a slow (respectively, fast) process seeks to accede to a processor (transition T_1). When it gets it (firing of T_2), it occupies it for a duration μ_1 (respectively μ_2) before liberating it (firing of T_3). The firing of the transition T_3 depends on the type of the processor and the process.

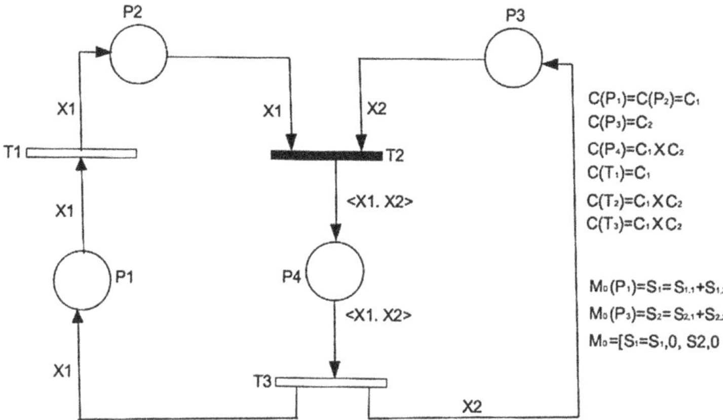

FIGURE 11.12

Modeling a multiprocessor with a SWN.

3.7.5 Analysis of a SWN

3.7.5.1 Symbolic graph

A symbolic graph of a SWN is a graph in which the states are represented as classes as well as the firings. Each node of such a graph is a class of markings represented by one of its elements. Each arc connecting two nodes is a firing class represented by an element belonging to its class.

In a symbolic graph there are marking classes (symbolic markings) and firing classes (symbolic firings). For these classes, equivalences based on the symmetries of the SWN are defined. A markings class (respectively, firings) is composed of equivalent markings (respectively, of equivalent firings).

The equivalence between the states is based on the use of symmetries preserving the structure of the model. The definition of equivalence between the states must ensure similarity of behavior of equivalent states at a firing transition.

3.7.5.1.1 Symbolic markings. Two markings of a SWN are equivalent to a close symmetry; that is to say that M is equivalent to M' if and only if there exists s such that $M' = s.M$. Therefore, the markings M' and M are equivalent (belong to the same equivalence class). The markings of the network are then grouped according to the equivalent classes. Each equivalent class corresponds to a symbolic marking of the model. Then, a symbolic marking is an ordinary markings set. In the following, the symbolic marking associated to the marking M is noted $\hat{M}$.

For the design of the symbolic graph, it is necessary to prove the compatibility of symbolic markings with the firing of transitions.

3.7.5.1.2 Symbolic firings. The contribution of well-formed networks is that they allow an easy and automatic definition of state classes. However, these state classes are meaningful only if they preserve the firings of transitions. Fortunately, the fundamental property of conservation is verified by the well-formed networks.

Proposal: The transition firing is preserved by applying a permutation on the departure markings and arrival, as well as on the color instantiating the transition.

A rule of symbolic firing is proposed. This rule acts on a representation of a symbolic marking to directly build a representation of the markings class obtained after firing. It operates in several steps:

- The first step is to divide a marking so as to isolate each of selected objects for the firing.
- Then the enabling is tested on this divided representation. If the test is positive, then the marking of dynamic subclasses is modified in accordance with incidence functions. This gives a new marking whose representation is not necessarily canonical (a canonical representation of a symbolic marking is an ordered and minimum representation).
- The last step is therefore to calculate the canonical representation of the marking obtained after firing.

3.7.5.1.3 Design of the symbolic graph. The design of the symbolic graph (see Figure 11.13) is achieved by applying the rule of symbolic firing to the initial symbolic marking, then repeatedly to the different built markings. However, as the initial marking is defined by the user, it is important to first make sure it is represented in its canonical form. If it is not, it should therefore calculate the canonical form.

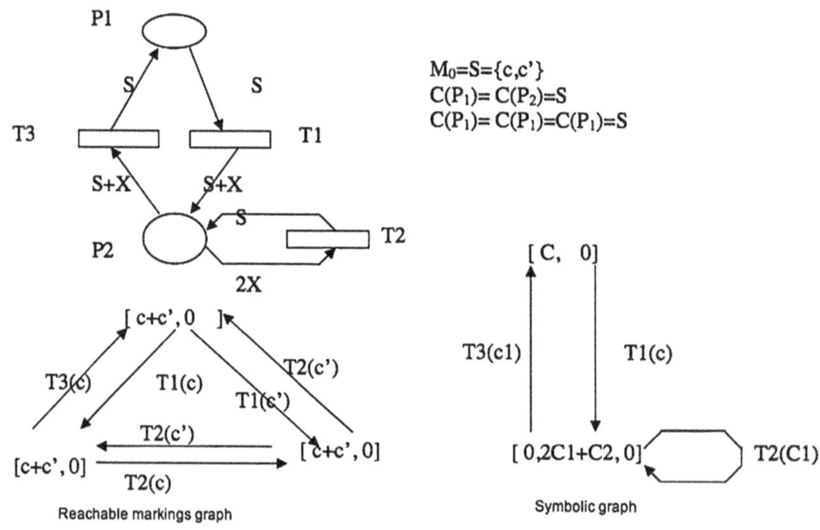

FIGURE 11.13

Example of designing of a symbolic graph.

In this example, a permutation x is applied such that $x(c') = c$ and $x(c) = c'$. Considering $M_0 = [c + c', 0]$, $M_1 = [0, 2c + c']$, and $M_2 = [0,\ c + 2c']$, one has:

$x(M_1) = x([0, 2c + c'] = [0, 2c' + c] = M_2$ in the same way $X(M_2) = M_1$, so M_1 and M_2 are equivalent and form a symbolic marking. The second symbolic marking is composed of one ordinary marking (the initial marking M_0).

By applying the permutation x to the firings, three symbolic firings, each of them formed of two ordinary firings, are obtained.

3.7.5.2 Passage from the symbolic graph to the Markov aggregation

Consider **P** the transition matrix of states of the Markov chain included in the semi-Markov process associated with the SWN.

3.7.5.2.1 Calculation of parameters of the aggregated chain. To calculate these parameters, the method of included chain is used.

Thus, this study is restricted to the calculation of coefficients $\hat{P}[\hat{M}, \hat{M}']$ of the probabilities matrix of the transition of this chain and with duration $(\hat{M})$ which is the stay time in an ordinary marking $\hat{M}$ tangible.

In the following, a symbolic firing is noted $\hat{M}(T_i(\lambda, \mu) >$ and $M(T_i(c) >$ is one of the ordinary firings corresponding to the symbolic firing.

All the ordinary firings denoted by a symbolic arc are projected on the same static subclasses; similarly, all the ordinary markings of a symbolic marking are projected to the same static partition. So the stochastic parameter of the ordinary firing $\theta(T_i)(\tilde{C}, \hat{M})$ is independent of the choice of this firing and is deducted directly from the symbolic marking and the symbolic firing. Let's note it as $\theta(T_i)\langle\lambda,\ \mu,\ \hat{M}\rangle$.

The expressions of coefficients of the matrix of the aggregated included chain and the stay time are given by these formulas:

$$\hat{P}\left[\hat{M}, \hat{M}'\right] = \frac{\sum_{\langle T_i,\lambda,\mu\rangle : \hat{M} \xrightarrow{\langle T_i,\lambda,\mu\rangle} \hat{M}'} \hat{\theta}[T_i]\langle\lambda, \mu, \hat{M}\rangle \left|\hat{M} \xrightarrow{\langle T_i,\lambda,\mu\rangle}\right|}{\sum_{\langle T_i,\lambda,\mu\rangle : \hat{M} \xrightarrow{\langle T_i,\lambda,\mu\rangle}} \hat{\theta}[T_i]\langle\lambda, \mu, \hat{M}\rangle \left|\hat{M} \xrightarrow{\langle T_i,\lambda,\mu\rangle}\right|} \tag{11.10}$$

$$\text{Duration}(\hat{M}) = \frac{1}{\sum_{\langle T_i,\lambda,\mu\rangle : \hat{M} \xrightarrow{\langle T_i,\lambda,\mu\rangle}} \hat{\theta}[T_i]\langle\lambda, \mu, \hat{M}\rangle \left|\hat{M} \xrightarrow{\langle T_i,\lambda,\mu\rangle}\right|} \tag{11.11}$$

where Equation (11.2) applies only to tangible symbolic markings and where $\left|\hat{M} \xrightarrow{\langle T_i,\lambda,\mu\rangle}\right|$ is the number of colored firings from a fixed marking $\hat{M}$ represented by the symbolic instantiation $\langle T_i,\ \lambda,\ \mu\rangle$. It is shown that:

$$\left|\hat{M} \xrightarrow{\langle T_i,\lambda,\mu\rangle}\right| = \prod_{i=1}^{h} \prod_{j=1}^{mi} \frac{\text{card}(Z_i^j)!}{(\text{card}(Z_i^j) - \mu_i^j)!} \tag{11.12}$$

with h representing the number of nonordered classes, mi representing the number of dynamic subclasses of C_i in the representation and μ_i^j the number of instantiations in Z_i^j.

Finally, the probability in equilibrium of a marking M is given by the quotient of the probability of its symbolic marking $\hat{M}$ on the cardinal of the latter. This value is then:

$$P_M = \frac{1}{|S(\hat{M})|}\left(\prod_{i=1}^{h}\prod_{j=1}^{Si}\frac{|C_{i,q}|!}{\prod_{d(Z_i^j)=q}\text{card}(Z_i^j)!}\right)\prod_{i=h+1}^{n} v(i) \tag{11.13}$$

In this formula, Si is the number of static subclasses of C_i, $v(i) = |C_i|$ if $mi > 1$ and if $Si = 1$ and $v(i) = 1$ otherwise. $S(\hat{M})$ is the set of admissible permutations of the marking $\hat{M}$, that is to say the number of permutations defined on dynamic subclasses that leaves the symbolic marking invariant.

4 CONCLUSION

The different scientific communities both in academia and in modern industry are increasingly facing the need to study the performance of a system before its implementation and even during its functioning, so that the system can be adapted to current and future users' needs. This chapter introduced the fundamental concepts of Petri nets to researchers and practitioners, but also showed how this mathematical and graphical tool has advanced in order to be convenient for modeling and analyzing the performance of real complex systems with discrete events that are characterized as being concurrent, synchronous, asynchronous, distributed, parallel, nondeterministic, and stochastic, etc.

REFERENCES

[1] Krishna CM. Performance modeling for computer architects. Wiley-IEEE Computer Society Press; September 1995. ISBN: 978-0-8186-7094-7.

[2] Kobayashi H. Modeling and analysis: an introduction to system performance evaluation methodology (The Systems programming series). Addison-Wesley; June 1978. ISBN-13: 978-0201144574.

[3] Jain R. The art of computer systems performance analysis—techniques for experimental design, measurement, simulation and modeling. John Wiley and Sons; 1991.

[4] Heidelberger P, Lavenberg SS. Computer performance evaluation methodology. IEEE Trans Comput 1984;33(12):1195–220.

[5] Labadi K. Contribution à la modélisation et à l'analyse de performances des systèmes logistiques à l'aide d'un nouveau modèle de réseaux de Petri stochastiques. PhD Thesis, UTT, November 2005.

[6] Lalouette J, Brinzei N, Malasse O, Caron R, Scherb F, Aubry J-F. Modélisation et évaluation des performances d'un système de signalisation ferroviaire intégrant BAL et ETCS par réseaux de Petri colorés. In: Sixiéme Conférence Internationale Francophone d'Automatique, CIFA 2010, Nancy: France; 2010.

[7] Calzarossa M, Ferrari D. Performance evaluation, vol. 6. North-Holland; 1986. p. 25–33.

[8] Bruno G, Biglia P. Performance evaluation and validation: tool handling in FMS using PN's. In: IEEE Int. workshop timed Petri nets, Torino, Italy; July 1–3, 1985. p. 64–71.

[9] Chen H, Amodeo L, Chu F. Modeling and performance evaluation of supply chain with batch deterministic and stochastic Petri nets. In: 13th annual European simulation symposium, simulation in industry, Marseille; October 2001, p. 415–19.

[10] Berge N, Juanolle G, Samaan M. Using stochastic timed Petri nets for modeling and analysing and industrial application based on FIP fieldbus. In: Symposium on emerging technologies and factory automation, Paris, France, ETFA 95, INRIA-IEEE; 1995.

[11] Bichteler K. Stochastic integration and stochastic differential equations, free online book. Accessed 2014. Available from: <http://www.ma.utexas.edu/users/kbi/SDE/C_1.html>.

[12] Knill O. Probability theory and stochastic processes with applications. Overseas Press; 2009.

[13] Gusak D, Kukush A, Kulik A, Mishura Y, Pilipenko A. Theory of stochastic processes. Springer; 2010. ISBN 978-0-387-87862-1.

[14] Applebaum D. Levy process stochastic calculus, 2e. Cambridge University Press; April 30, 2009.

[15] Scott M. Applied stochastic processes in science and engineering. University of Waterloo; 2013.

[16] Feller W. An introduction to probability theory and its applications. 3rd ed. John Wiley and Sons; 1968.

[17] Takas L. Processus stochastiques: problèmes et solutions. Dunod; 1964.

[18] Bolch G, Greiner S, de Meer H, Trivedi KS. Queueing networks and Markov chains: modeling and performance evaluation with computer science applications. 2nd ed. Wiley-Interscience; 2006. ISBN: 978-0-471-56525-3.

[19] Baynat B. Théorie des files d'attente: Des chaînes de Markov aux réseaux à forme produits. Edition Hermes Science; 2000.

[20] Gross D, Shortle JF, Thompson JM, Harris CM. Fundamentals of queueing theory. 4th ed. New Jersey: John Wiley and Sons; 2008.

[21] Kelly FP. Networks of queues. Adv Appl Probab 1976;8(2):416–32.

[22] Little JDC. A proof of the queueing formula $L = \lambda W$. Oper Res 1961;9:383–7.

[23] Buchholtz P. A class of hierarchical queueing networks and their analysis. Queueing Syst 1994;15:59–80.

[24] Murata T. Petri nets: properties, analysis and applications. Proc IEEE 1989;77(4): 541–80.

[25] Krings AW. Petri Nets. CS449/549 fault-tolerant systems sequence 11, 2011. Available from: <http://www2.cs.uidaho.edu/~krings/CS449/>.

[26] Vidal-Naquet G, Choquet-Geniet A. Réseaux de Petri et Systémes Paralléles. Armon Colin; 1992.

[27] Reisig W. Petri nets: an introduction. Berlin: Springer-Verlag; 1985.

[28] Choquet-Geniet A. Les Réseau de Petri, un outil de modelisation. Dunod; 2006.
[29] David R, Alla H. Du Grafcet aux réseaux de Petri. Paris: Editions Hermès; 1992.
[30] Zaitsev DA. Clans of Petri nets: verification of protocols and performance evaluation of networks. LAP LAMBERT Academic Publishing; 2013.
[31] Brams G. Réseau de Petri, Théorie et pratique. Tome 1: théorie et analyse – Tome 2: modélisation et applications. Masson; 1983.
[32] Zhou M, Wu N. System modeling and control with resource-oriented petri nets. CRC Press; December 2010. p. 6–10.
[33] Dwyer MB, Clarke LA, Nies KA. A compact Petri net representation for concurrent programs. In: Proceedings of the 17th international conference on software engineering, Seattle, Washington, USA; April 24–28, 1995. p. 147–57.
[34] Reisig W. Elements of distributed algorithms: modeling and analysis with Petri nets. Springer-Verlag; 1998.
[35] Agha GA, De Cindio F, Rozenberg G, editors. Concurrent object-oriented programming and petri nets. Berlin: Springer-Verlag; 2001, ISBN: 3-540-41942-X.
[36] Frey G, Litz L. Formal methods in PLC programming. In: Proceedings of the IEEE SMC, Nashville, TN, vol. 4; October 2000. p. 2431–36.
[37] Shatz SM, Shengru T, Murata T, Duri S. An application of Petri net reduction for Ada tasking deadlock analysis. IEEE Trans Parallel Distrib Syst Dec 1996;7(12):1307–22.
[38] Esparza J. A false history of true concurrency: from Petri to tools. In: Proc. SPIN. Ed. by Jaco van de Pol and Michael Weber, vol. 6349. LNCS. Springer; 2010, p. 180–86.
[39] Haar S, Fabre E. Diagnosis with Petri net unfoldings. In: Seatzu C, et al., editors. Control of discrete-event systems, LNCIS, 433. London: Springer-Verlag; 2013. p. 301–18.
[40] Boukala MC, Petrucci L. Towards distributed verification of Petri nets properties. Proceedings of the international workshop on verification and evaluation of computer and communication systems (VECOS'07). British Computer Society; 2007. p. 15–26.
[41] Valois F. Modélisation et Évaluation de Performances de Réseaux. Département Télécommunications Cours 4TC. <http://fvalois.insa-lyon.fr/>; [last access 2014].
[42] Song Y-Q. Evaluation de performances stochastiques des réseaux. SSR2012.
[43] Application of Petri nets to communication networks. In: Billington J, Diaz M, Rozenberg G, editors. Lecture notes in computer science, vol. 1605. Springer-Verlag; 1999.
[44] Delgadillo GM, Llano SB. Scheduling application using Petri nets: a case study: intergráficas s.a. In: Proceedings of 19th international conference on production research, Valparaiso, Chile, 2006.
[45] Sgavioli M. Modelagem de Sistemas de Manufatura Usando Redes de Petri Coloridas Fuzzy Focando a Solução de Conflitos. Msc Thesis, Universidade Federal de São Carlos; 2010.
[46] El-Tamimia AM, Abidib MH, Mianb SH, Aalamb J. Analysis of performance measures of flexible manufacturing system. J King Saud University, Eng Sci 2012;24(2): 115–29.
[47] John FR, Prasad PSS. Supply chain conflict detection with colored Petri nets. J Adv Manage Res 2012;9(2):208–16. Available from: http://dx.doi.org/10.1108/09727981211271959.
[48] Li ZW, Wu NQ, Zhou MC. Deadlock control of automated manufacturing systems based on Petri nets: A literature review. IEEE Trans Syst Man Cybern Part C Appl Rev 2012;42 (4):437–62. Available from: http://dx.doi.org/10.1109/TSMCC.2011.2160626.

[49] Uzam M, Zhou M. An iterative synthesis approach to Petri net-based deadlock prevention policy for flexible manufacturing systems. IEEE Trans Syst Man Cybern Part A Syst Humans 2007;37(3):362–71. Available from: http://dx.doi.org/10.1109/TSMCA.2007.893484.

[50] Zhou M, DiCesare F, Desrochers AA. A hybrid methodology for synthesis of Petri net models for manufacturing systems. IEEE Trans Rob Autom 1992;8(3):350–61. Available from: http://dx.doi.org/10.1109/70.143353.

[51] Jeng M, Xie X, Yu Peng M. Process nets with resources for manufacturing modeling and their analysis. IEEE Trans Rob Autom 2002;18(6):875–89. Available from: http://dx.doi.org/10.1109/TRA.2002.805655.

[52] Silva M, Teruel E. Petri nets for the design and operation of manufacturing systems. Eur J Control 1997;3(3):182–99. http://dx.doi.org.10.1016/S0947-3580(97)70077-3.

[53] Zhou M, Venkatesh K. Modeling, simulation, and control of flexible manufacturing systems: a Petri net approach. Series in intelligent control and intelligent automation, vol. 6. World Scientific Publishing Company; 1999. ISBN: 981-02-3029-X.

[54] Ramaswamy S, Valavanis KP, Barber S. Petri net extensions for the development of MIMO net models of automated manufacturing systems. J Manuf Syst 1997;16(3):175–91. http://dx.doi.org.10.1016/S0278-6125(97)88886-3.

[55] Desrochers AA, Al'Jaar RY. Applications of Petri nets in manufacturing systems: modelling, control and performance analysis. IEEE Press; 1995, ISBN: 0-87942-295-5.

[56] van der Aalst WMP. "Three good reasons for using a Petri-net-based workflow management system. In: Wakayama T, et al., editors. Information and process integration in enterprises: rethinking documents. Norwell: The Kluwer International Series in Engineering and Computer Science, Kluwer Academic Publishers; 1998. p. 161–82. [chapter 10].

[57] van der Aalst WMP. The application of Petri nets to workflow management. J Circuits Syst Comput 1998;8(1):21–66.

[58] DiCesare F, Kulp P, Gile M, List GF. The application of Petri nets to the modeling, analysis and control of intelligent urban traffic networks. In: Valette R, editor. Proceedings. of the 15th international conference application theory Petri nets. Zaragoza, Spain; June 1994. p. 2–15.

[59] Cheng Y-H, Yang L-A. A Fuzzy Petri nets approach for railway traffic control in case of abnormality: evidence from Taiwan railway system,". Expert Syst Appl 2009;36(4):8040–8. Available from: http://dx.doi.org/10.1016/j.eswa.2008.10.070.

[60] Di Febbraro A, Giglio D, Sacco N. Urban traffic control structure based on hybrid Petri nets. IEEE Trans Intell Transp Syst 2004;5(4):224–37. Available from: http://dx.doi.org/10.1109/TITS.2004.838180.

[61] List GF, Cetin M. Modeling traffic signal control using Petri nets. IEEE Trans Intell Transp Syst 2004;5(3):177–87. Available from: http://dx.doi.org/10.1109/TITS.2004.833763.

[62] Lin L, Nan T, Xiangyang M, Fubing S. Implementation of traffic lights control based on Petri nets. Proc IEEE Intell Transp Syst 2003;2:1087–90. Available from: http://dx.doi.org/10.1109/ITSC.2003.1252653.

[63] Neto PFR, Perkusich MLB, de Almeida HO, Perkusich A. A formal verification and validation approach for real-time databases. IGI Global 2009. Available from: http://dx.doi.org/10.4018/978-1-59140-851-2.ch005.

[64] Yáskara YMPF, Neto FMM, Neto PFR, Perkusich MLB, Paillard GAL, Perkusich A. QoS management for real-time DataBases in embedded systems. EATIS; 2008.

[65] Neto PFR, Perkusich MLB, Perkusich A. Real-time databases for sensor networks. ICEIS 2004;1:599–603.

[66] Neto PFR, Perkusich MLB, Perkusich A. A model in Petri nets to analyze quality of service in real-time databases. SMC; 2003. p. 300–05.

[67] Cortés LA, Eles P, Peng Z. A Petri net based model for heterogeneous embedded systems. In: Proceedings of the NORCHIP Conference, 1999. p. 248–55.

[68] Zurawski R, Zhou MC. Petri nets and industrial applications: a tutorial. IEEE Trans Ind Electron 1994;41(6):567–82.

[69] Feldbrugge F, Jensen K. Petri net tool overview 1986. In: Brauer W, Reisig W, Rozenberg G, editors. Petri nets: applications and relationships to other models of concurrency (LNCS, vol. 255). Berlin: Springer-Verlag; 1987. p. 20–61.

[70] Billington J, Wheeler G, Wilbur-Ham M. PROTEAN: a high level Petri net tool for the specification and verification of communication protocols. IEEE Trans Software Eng 1988;14(3):301–16.

[71] Holliday MA, Vernon MK. The GTPN Analyzer: numerical methods and user interface. Technical Report 639. Dept. of Computer Science, Univ. of Wisconsin–Madison; Apr. 1986.

[72] Franceschinis G, Gaeta R, Bertoncello C. WNSIM: manual. PEG, Dipart.di Informatica. Univ. di Torino (Italy); 2001.

[73] Chiola G, Franceschinis G, Gaeta R, Ribaudo M, editors. GreatSPN 1.7: graphical editor and analyzer for timed and stochastic Petri nets. Perform Eval 1995;24:47–68.

[74] Franceschinis G, Gaeta R, Bertoncello C. GreatSPN: User's Manual (version 2.0.2). PEG, Dipart. di Informatica, Univ. di Torino (Italy); 2002.

[75] Haddad S, Moreaux P, Sene M. Performance Evaluation with SWN: a technical contribution. Réseaux et Systémes Répartis—Calculateurs Parallèles 2001;13(6).

[76] He X, Murata T. High-level Petri nets – extension, analysis and applications. In: Chen WK, editor. The electrical engineering handbook. Burlington, MA: Elsevier Academic Press; 2005. p. 459–75.

[77] Kordic V. Petri net, theory and applications. I-Tech Education and Publishing, February 2008, ISBN 978-3-902613-12-7.

[78] Lectures on Petri nets II: applications, advances in Petri nets. In: Reisig W, Rozenberg G, editors. Lecture notes in computer Science, vol. 1492. Springer-Verlag; 1998. ISBN: 3-540-65307-4.

[79] Valette R. Réseaux de Petri: Théorie et Applications. Lecture. February 1999, LAAS-CNRS Toulouse.

[80] David R, Alla H. Du Grafcet aux réseaux de Petri. Hermes; 1997, 500 p.

[81] Bause F, Kritzinger PS. Stochasti Petri nets, an introduction to the theory. Wiesbaden: Verlag Vieweg; 2002.

[82] Haddad S, Moreaux P. Les réseaux de Pétri Stochastiques; December 2000.

[83] Marsan MA. Stochastic Petri nets: an elementary introduction. Advances in Petri nets 1989. Berlin: Springer; 1989. p. 1–29.

[84] Balbo G. Introduction to generalized stochastic Petri nets. Seventh international school on formal methods for the design of computer, communication and software systems: performance evaluation. May 29, 2007.

[85] Eisentraut C, Hermanns H, Katoen J-P, Zhang L. A semantics for every GSPN. In: Colom J-M, Desel J, editors. PETRI NETS 2013, LNCS 7927. Berlin Heidelberg: Springer-Verlag; 2013. p. 90–109.

[86] van der Aalst WMP, Stahl C, Westergaard M. Strategies for modeling complex processes using colored Petri nets,". In: Jensen K, et al., editors. ToPNoC VII, LNCS 7480. Berlin Heidelberg: Springer-Verlag; 2013. p. 6–55.

[87] Aly S, Mustafa K. Protocol verification and analysis using colored Petri nets. Technical Report Submitted. DePaul University; July, 2003.

[88] Jensen K. Coloured Petri nets: a high level language for system design and analysis. In: Rozenberg G, editor. Advances in Petri nets 1990, lecture notes in computer science, vol. 483. Berlin: Springer; 1990. p. 342–416.

[89] Jensen K. Coloured Petri nets. basic concepts, analysis methods and practical use. Practical use, monographs in theoretical computer science, vol. 3. Springer-Verlag; 1997. ISBN: 3-540-62867-3.

[90] Chiola G, Dutheillet C, Franceschinis G, Haddad S. Stochastic well-formed colored nets and symmetric modeling applications. IEEE Trans Comput 1993;42 (11):1343–60.

[91] Xia Y, Liu Y, Liu J, Zhu Q. Modeling and performance evaluation of BPEL processes: a stochastic-Petri-net-based approach. IEEE Trans Syst Man Cybern Part A Syst Humans 2012;42(2):503–10. Available from: http://dx.doi.org/10.1109/TSMCA.2011.2164064.

[92] Chiola G, Dutheillet C, Franceschinis G, Haddad S. On well formed colored nets and their symbolic reachability graph. In: Proceedings of the 11th international conference of application and theory of Petri net. 1990. p. 373–96.

[93] Mokdad L, Sene M, Boukerche A. Call admission control performance analysis in mobile networks using stochastic well-formed Petri Nets. IEEE Trans Parallel Distrib Syst 2011;22(8):1332–41.

[94] D'Argenio PR, Katoen J-P. A theory of stochastic systems. Part II: process algebra. Inf Comput 2005;203(1):39–74. Available from: http://dx.doi.org/10.1016/j.ic.2005.07.002.

[95] de la Higuera C, Oncina J. Learning stochastic finite automata. In: Proceedings of the seventh international colloquium on grammatical inference, LNAI, ICGI, vol. 3264. 2004. p. 175–86.

[96] Palmer N, Goldberg PW. PAC-learnability of probabilistic deterministic finite state automata in terms of variation distance. Theor Comput Sci 2007;387(1): 18–31.

[97] Verwer S, Eyraud R, de la Higuera C. PAUTOMAC: a probabilistic automata and hidden Markov models learning competition. Theoretical computer science. Springer; October 2013. Available from: http://dx.doi.org/10.1007/s10994-013-5409-9.

[98] Diaz M. Modeling and analysis of communication and cooperation protocols using Petri net based models (1976). J Comput Netw 1982;6(6):419–41. http://dx.doi.org.10.1016/0376-5075(82)90112-X.

[99] Teixeira RC, Duarte OCMB. Evaluating the impact of the communication system on distributed virtual environments. J Multimedia Tools Appl – MTA 2003;19(3): 259–78.

[100] Li D, Cui Y, Xu K, Wu J. Improvement of multicast routing protocol using petri nets. In: Rough sets, fuzzy sets, data mining, and granular computing lecture notes in computer science, vol. 3642. 2005, p. 634–43.

[101] El-Karaksy MR, Nouh AS, Al-Obaidan A. Performance analysis of timed Petri net models for communication protocols: a methodology and a package. Comput Commun 1990;13(2):73–82. http://dx.doi.org.10.1016/0140-3664(90)90174-F.

[102] Juanole G, Algayres B, Dufau J. On communication protocol modelling and design. In: Advances in Petri nets 1984, lecture notes in computer science, vol. 188; 1985, p. 267–87.

[103] Zhou M, Dicesare F. Petri net modelling of buffers in automated manufacturing systems. IEEE Trans Syst Man Cybern Part B Cybern 1996;26(1):157–64.

[104] Recaldc L, Silva M, Ezpeleta J, Teruel E. Petri nets and manufacturing systems: an examples-driven tour. In: Desel J, Reisig W, Rozenberg G, editors. Lectures on concurrency and Petri nets: advances in Petri nets, volume 3098 of lecture notes in computer science. Springer-Verlag; June 2004. p. 742–88.

[105] Praveen M, Lodaya K. Model checking counting properties of 1-safe nets with buffers in paraPSPACE, FST&TCS'09, LZI, 2009. p. 347–58.

[106] Kilincci O. A Petri net-based heuristic for simple assembly line balancing problem of type 2. Int J Adv Manuf Technol 2010;46(1):329–38. Springer.

[107] Kilincci O, Bayhan GM. A Petri net approach for simple assembly line balancing problems. Int J Adv Manuf Technol 2006;30(11):1165–73. Available from: http://dx.doi.org/10.1007/s00170-005-0154-2.

[108] Ullah H, Bohez ELJ. A Petri net model for the design and performance evaluation of a flexible assembly system. Assembly Autom 2008;28(4):325–39. Available from: http://dx.doi.org/10.1108/01445150810904486.

[109] Ullah H. Petri net versus queuing theory for evaluation of FMS. Assembly Autom 2011;31(1):29–37.

[110] Li B, Li X, Guo W, Wu S. A generalized stochastic Petri-net model for performance analysis and allocation optimization of a particular repair system. Asia-Pac J Oper Res 2013;30(01). Available from: http://dx.doi.org/10.1142/S021759591250042X.

[111] Guo JW, Li ZW. A deadlock prevention approach for a class of timed Petri nets using elementary siphons. Asian J Control 2010;12(3):347–63.

[112] Yan MM, Li ZW, Wei N, Zhao M. A deadlock prevention policy for a class of Petri nets S3PMR. J Inf Sci Eng 2009;25(1):167–83.

[113] Li ZW, Liu GY, Hanisch M-H, Zhou MC. Deadlock prevention based on structure reuse of Petri net supervisors for flexible manufacturing systems. IEEE Trans Syst Man Cybern Part A Syst Humans 2012;42(1):178–91.

[114] Uzam M, Gelen G. On a deadlock prevention policy for a class of Petri nets S3PMR. Int J Adv Manuf Technol, Springer; 2014, http://dx.doi.org.10.1007/s00170-014-5821-8.

[115] Cabasino MP. Diagnosis and identification of discrete event systems using Petri Nets. Ph.D. in Electronic and Computer Engineering, University of Cagliari; 2009.

[116] Fliss I, Tagina M. Multiple fault diagnosis of discrete event systems using Petri nets. International conference on communications, computing and control applications (CCCA). IEEE; 2011. p. 1–6, http://dx.doi.org.10.1109/CCCA.2011.6031430.

[117] Latorre JI, Jiménez E, Pérez M. Simulation-based optimization of discrete event systems with alternative structural configurations using distributed computation and

the Petri net paradigm. Simulation: Trans Soc Model Simul Int 2013; Special Issue: Advancing Simulation Theory and Practice with Distributed Computing.

[118] Latorre-Biel J-I, Jiménez-Macías E, Pérez-Parte M. Sequence of decisions on discrete event systems modeled by Petri nets with structural alternative configurations. J Comput Sci 2013. Available from: http://dx.doi.org/10.1016/j.jocs.2013.09.001.

[119] Marsan MA, Conte G, Balbo G. A class of generalized Petri nets for the performance evaluation of multiprocessor systems. ACM Trans Comput Syst 1984;2(2): 93–122.

[120] Madhukar M, Leuze M, Dowdy L. Petri net model of a dynamically partitioned multiprocessors system. In: Proceedings of the sixth international workshop on Petri nets and performance models (PNPM' 95), 1995.

[121] Laili Y, Tao F, Zhang L, Sarker BR. A study of optimal allocation of computing resources in cloud manufacturing systems. Int J Adv Manuf Technol 2012;63: 1–20.

[122] Chong Y-K, Hwang K. Performance analysis of four memory consistency models for multithreaded multiprocessors. IEEE Trans Parallel Distrib Syst 1995;6(10): 1085–99.

[123] Balsamo S, Di Marco A, Inverardi P, Simeoni M. Model-based performance prediction in software development: a survey. IEEE Trans Software Eng 2004;30: 295–310.

[124] Gold R. Petri nets in software engineering. ABWP, University of Applied Sciences; 2004.

[125] Schmietendorf A, Dimitrov E, Dumke RR. Process models for the software development and performance engineering tasks. Proceedings of the third international workshop on software and performance. Rome, Italy: ACM Press; 2002.

[126] Wirtz G. Application of Petri nets in modelling distributed software systems. In: Moldt D, editor. Workshop on modelling of objects, components, and agents. Aarhus, Denmark: 2001.

[127] Xu J, Kuusela J. Analyzing the execution architecture of mobile phone software with coloured Petri nets. Int J Softw Tools Technol Trans 1998;2(2).

[128] Saldhana J, Shatz SM. UML diagrams to object Petri net models: an approach for modeling and analysis. In: International conference on software engineering and knowledge engineering. Chicago, Illinois: 2000.

[129] Gehlot V, Way T, Beck R, DePasquale P. Model driven development of a service oriented architecture (SOA) using colored Petri nets. First workshop on quality in modeling, ACM/IEEE ninth international conference on model driven engineering languages and systems (QiM/MoDELS'06), October, 2006.

[130] Gehlot V, Pujari G. A case study in defining colored Petri nets based model driven development of enterprise service oriented architectures. In: Proceedings of the IEEE 42nd Hawaii international conference on system sciences (HICSS-42), Software Technology Track. January 2009.

[131] Aquilani F, Balsamo S, Inverardi P. Performance analysis at the software architectural design level. Perform Eval 2001;45:147–78.

[132] Bernardeschi C, De Francesco N, Vaglini G. A Petri nets semantics for data flow networks. Acta Inf, 32. 1995, Springer-Verlag.

[133] Wagner B, Dinges A, Muller P. Dataflow orchestration of image processing algorithms using high-level Petri nets. Seventh international conference on hybrid

intelligent systems. HIS 2007. IEEE; 2007. p. 344–347. Available from: http://dx.doi.org/10.1109/HIS.2007.54.

[134] Rocha J-I, Gomes L, Dias O, Petri net verification techniques on synchronous dataflow models. In: IECON 2011, 37th annual conference on IEEE Industrial Electronics Society; Nov. 2011, p. 3792–97.

[135] Rocha J-I, Gomes L, Dias O. Analysing storage resources on synchronous dataflows using Petri net verification techniques. In: IECON 2012, 38th annual conference on IEEE industrial electronics society; 2012. p. 4676–81.

[136] Rocha J-I, Dias OP, Gomes L. Exploiting dataflows and Petri nets mappings. In: 2013 11th IEEE international conference on industrial informatics (INDIN); 2013. p. 590–95.

[137] Rocha J-I, Dias OP, Gomes L. Strategies to improve synchronous dataflows analysis using mappings between Petri nets and dataflows. Technological innovation for collective awareness systems, IFIP advances in information and communication technology, vol. 423. Springer; 2014. p. 237–48.

[138] Miyagi PE, Riascos LAM. Modeling and analysis of fault-tolerant systems for machining operations based on Petri nets. Control Eng Prac, Elsevier 2006 April; 14(4):397–408. http://dx.doi.org/10.1016/j.conengprac.2005.02.002.

[139] Jian S, Shaoping W, Yaoxing S. Petri-nets based availability model of fault-tolerant server system. In: 2008 IEEE conference on robotics, automation and mechatronics, September 2008, p. 444–49.

[140] Riascos LAM, Simoes MG, Miyagi PE. Bayesian network fault diagnostic system for PEM fuel cell. J Power Sources 2007;165(1):267–78.

[141] De Cindio F, Simone C. Petri nets for modelling fault tolerant distributed systems in a modular and incremental way. In: Position paper for the fourth ACM SIGOPS European workshop on fault tolerance in distributed systems, Bologna September, 1990.

[142] Litz L, Frey G. A graduate course on logic process control based on Petri nets. In: Proceedings of the IEEE SMC'98, vol. 1. San Diego;1998. p. 274–77.

[143] Ferrarini L. An incremental approach to logic controller design with Petri nets. IEEE Trans Syst Man Cybern 1992;22:461–73.

[144] Frey G, Litz L. Correctness analysis of Petri net based logic controllers. In: Proceedings of the american control conference (ACC'2000); 2000. p. 3165–6.

[145] Frey G, Litz L. Transparency analysis of Petri net based logic controllers—a measure for software quality in automation. In: Proceedings of the American control conference (ACC'2000); 2000. p. 3182–6.

[146] Minas M, Frey G. Visual PLC-programming using signal interpreted Petri nets. In: Proceedings of the American control Conference (ACC 2002); 2002. p. 5024.

[147] Bender DF, Combemale B, Crégut X, Farines JM, Berthomieu B, Vernadat F. Ladder metamodeling and PLC program validation through time petri nets. In: Lecture notes in computer science (LNCS) vol. 5095; 2008. p. 121–36.

[148] Tsai J-I, Teng C-C. Constructing an abstract model for ladder diagram diagnosis using Petri nets. Asian J Control 2010;12:309–22. Available from: http://dx.doi.org/10.1002/asjc.187.

[149] Barghash MA, Abuzeid OM, Al-Rabadi AN, Jaradat AM. Petri nets and ladder logic for fully-automating and programmable. Am J Eng Appl Sci 2011;252–64.

[150] Gomaa MM. Petri net to ladder logic diagram converter and a batch process simulation. Asian Res Publishing Network (ARPN) J Eng Appl Sci 2011;6:67–72.

[151] Andreu D, Souquet G, Gil T. Petri net based rapid prototyping of digital complex system. In: Symposium on VLSI, IEEE computer society annual; April 2008. p. 405–10.

[152] Cortadella J, Kishinevsky M, Kondratyev A, Lavagno L, Yakovlev A. Hardware and Petri nets: application to asynchronous circuit design. Application and theory of Petri nets: 21st International Conference ICATPN 2000, Aarhus, Denmark; June 2000.

[153] Grobelna I, Grobelny M, Adamski M. Petri nets and activity diagrams in logic controller specification—transformation and verification. Mixed Design Integr Circuits Sys MIXDES 2010;607–12.

[154] Yakovlev AV, Koelmans AM, Semenov A, Kinniment DJ. Modelling, analysis and synthesis of asynchronous control circuits using Petri nets. Integration: VLSI J 1996;21(3):143–70.

[155] Koppad D, Bystrov A, Yakovlev A. Off-line testing of Asynchronous circuits. 18th International Conference on VLSI Design. Kolkata: IEEE CS Press; 2005.

[156] Baer JL, Ellis CA. Model design and evaluation of a compiler for a parallel processing environment. IEEE Trans Software Eng 1977;SE-3(6):394–405.

[157] Noe JD. A Petri net model of the CDC 6400. In: Proceedings of the ACM/ SIGOPS workshop on systems performance evaluation. p. 362–78, 1971.

[158] Valette R, Bako B. Software implementation of Petri nets and compilation of rule-based systems. Advances in Petri Nets. Springer-Verlag; 1991. p. 296–16.

[159] De Cindio F, De Michelis G, Simone C. GAMERU: a language for the analysis and design of human communication pragmatic within organizational systems. LNCS 1987;255(24):21–44.

[160] Ellis CA, Nutt GJ. Office information systems and computer science. Comput Surv 1980;12(1):27–60.

[161] Oberweis A, Sander P. Information system behavior specification by high level Petri nets. ACM Trans Inf Syst (TOIS) 1996;14(4):380–420.

[162] Darondeau P. Deriving unbounded Petri nets from formal languages. In: Sangiorgi D, de Simone R, editors. CONCUR, volume 1466 of lecture notes in computer science. Springer; 1998. p. 533–48.

[163] Lorenz R, Bergenthum R, Desel J, Mauser S. Synthesis of Petri nets from finite partial languages. In: Proceedings of the seventh international conference on application of concurrency to system design, July 10–13, 2007. p. 157–66. http://dx.doi.org.10.1109/ACSD.2007.34.

[164] van Dongen BF, Alves de Medeiros AK, Wenn L. Process mining: overview and outlook of Petri net discovery algorithms. Lecture notes in computer science. In: Jensen K, van der Aalst WMP, editors. Proceedings of the transactions on Petri nets and other models of concurrency II, vol. 5460. Berlin, Germany: Springer-Verlag; 2009.

[165] Crespi-Reghizzi S, Mandrioli D. Petri nets and szilard languages. Inf Control 1977; 33(2):177–92.

[166] Cervesato I. Petri nets and linear logic: a case study for logic programming. In: Proceedings of GULP-PRODE'95; 1995, p. 313–18.

[167] Son DT. Petri nets for modeling problem of logic programming and knowledge representation. In: Perspective for modeling Fluent Calculus; July, 2012.

[168] Behrens TM, Dix J. Model checking with logic based Petri nets. IfI Technical Report Series, Clausthal University of Technology; 2002.

[169] Darlington JL. A net based theorem prower for program verification and synthesis. Gesellschaft fur Math. und Datenverarbeitung mbH Bonn, Interner Bericht des IST 3/79 Dez; 1979.

[170] Al-Begain IAK, Kouvatsos D. Analysis of GSM/GPRS cell with multiple data service class. J Wireless Personal Comm 2002;25:41–57.

[171] Hedge N, Altman E. Capacity of multiservice WCDMA networks with variable GoS. <http://www.citeseer.ist.psu.edu/717389.html>; 2009.

[172] Mokdad L, Sene M. Performance measures of call admission control in mobile networks Using SWN. Proceedings of the first ACM international conference performance evaluation methodologies and tools (VALUETOOLS '06); 2006. p. 1–7.

[173] Sanghare OA, Sene M, Rodrigues JJPC. Distributed transactions on mobile systems: performance evaluation using SWN. In: IEEE international conference on ICC 2011. p. 1–6. http://dx.doi.org/10.1109/icc.2011.5963020.

[174] Yau SS, Caglayan MU. Distributed software system design representation using modified Petri nets. IEEE Trans Software Eng 1983;SE-9(6):733–45.

[175] Diallo O, Sene M, Sarr I. Freshness-aware metadata management: performance evaluation with SWN models". Proceedings of the 2010 IEEE/ACS international conference on computer systems and applications (AICCSA), Hammamet, Tunisia. IEEE DIGITAL LIBRARY; May 2010. p. 1–6.

[176] Haryono D, Tirtawangsa J, Erfianto B. Petri net modelling of concurrency control in distributed database system. Jurnal Sistem Komputer 2012;2(2):35–42. ISSN 2087-4685.

[177] Bertens LMF, Kleijn J, Verbeek FJ. Biomodelling and Petri nets. Eureka! Universiteit Leiden; 2012.

[178] Peleg M, Rubin D, Altman RB. Using Petri Net Tools to Study Properties and Dynamics of Biological Systems. J Am Med Inform Assoc 2005;12(2).

[179] Majumdar A. Modeling of yeast pheromone pathway using Petri nets. Msc Thesis at the faculty of the Graduate College at the University of Nebraska; December, 2012.

[180] Hamed RI, Ahson SI. Confidence value prediction of DNA sequencing with Petri net model. J King Saud University Comput Inf Sci 2011;23(2):79–89 doi:10.1016/j.jksuci.2011.05.004.

[181] Gao Q, Gilbert D, Heiner M, Liu F, Maccagnola D, Tree D. Multiscale modelling and analysis of planar cell polarity in the *Drosophila* Wing. IEEE/ACM Trans Comput Biol Bioinf 2012;99: (PrePrints):1–1.

[182] Heiner M, Gilbert D. BioModel engineering for multiscale systems biology. Prog Biophys Mol Biol 2013;111(2–3):119–28.

[183] Liu F, Heiner M. Colored Petri nets to model and simulate biological systems. In: International workshop on biological processes & Petri Nets (BioPPN). Braga, Portugal; June 21, 2010. ISBN: 978-972-8692-53-7.

[184] Ross-León R, Ramirez-Treviño A, Morales JA, Ruiz-Leon J. Control of metabolic systems modeled with timed continuous Petri nets. International workshop on biological processes & Petri nets (BioPPN). Braga, Portugal; June 21, 2010. ISBN: 978-972-8692-53-7.

[185] Machado D, Costa RS, Rocha M, Rocha I, Tidor B, Ferreira EC. Model transformation of metabolic networks using a Petri net based framework? In: International workshop on biological processes & Petri nets (BioPPN). Braga, Portugal; June 21, 2010. ISBN: 978-972-8692-53-7.

[186] Parvu O, Gilbert D, Heiner M, Liu F, Saunders N. Modelling and analysis of phase variation in bacterial colony growth. Proceedings of CMSB 2013, Vienna, Springer, LNCS, to appear; September 2013.

[187] Blätke M, Dittrich A, Rohr C, Heiner M, Schaper F, Marwan W. JAK/STAT signalling – an executable model assembled from molecule-centred modules demonstrating a module-oriented database concept for systems and synthetic biology. Mol BioSyst 2013.

[188] Blätke M, Heiner M, Marwan W. Predicting phenotype from genotype through automatically composed Petri nets. Proc. CMSB 2012, London, Springer, LNCS/LNBI 7605; 2012. p. 87–106.

[189] Gilbert D, Heiner M, Liu F, Saunders N. Colouring space – a coloured framework for spatial modelling in systems biology. In: Proc. PETRI NETS 2013, Milano, Springer, LNCS 7927; June 2013. p. 230–49.

[190] Heiner M, Gilbert D, Donaldson R. Petri nets for systems and synthetic biology. SFM 2008, Springer, LNCS 5016; 2008. p. 215–64.

[191] Heiner M, Gilbert D. How might Petri nets enhance your systems biology toolkit. In: Proceedings of the PETRI NETS 2011, Springer, LNCS 6709; 2011. p. 17–37.

[192] Marwan W, Rohr C, Heiner M. Petri nets in snoopy: a unifying framework for the graphical display, computational modelling, and simulation of bacterial regulatory networks. In: Helden JV, Toussaint A, Thieffry D, editors. Methods in molecular biology – bacterial molecular networks. Humana Press; 2012. p. 409–37.

[193] Gilbert D, Pârvu O. Petri nets for multiscale systems biology simulation and analysis. London UK: School of Information Systems, Computing and Mathematics Centre for Systems and Synthetic Biology, Brunel University; 2013.

[194] Scott Cost R, Chen Y, Finin T, Labrou Y, Peng Y. Modeling agent conversations with colored Petri nets. Third conference on autonomous agents (Agents-99), workshop on agent conversation policies, Seattle. ACM Press; May 1999.

[195] Mazouzi H, El A, Seghrouchni F, Haddad S. Open protocol design for complex interactions in multi-agent systems. Proceedings of the first international joint conference on autonomous agents and multiagent systems. ACM Press; 2002. p. 517–26.

[196] Blatke MA, Heiner M, Marwan W. Tutorial Petri nets in systems biology. Otto-von-Guericke University Magdeburg; August 2011.

CHAPTER

Towards correct and reusable Network-on-Chip architectures 12

Maryam Kamali[1], Luigia Petre[2], Kaisa Sere[2], and Masoud Daneshtalab[3]

[1]*University of Liverpool, Liverpool, UK*
[2]*Åbo Akademi University, Turku, Finland*
[3]*University of Turku, Turku, Finland*

1 INTRODUCTION

Microchips have been at the basis of our technological world for several decades now. They provide the computational core for an enormous number of embedded and cyber-physical systems that sustain the infrastructure of our society, such as in transportation with chips embedded into cars; in commerce, for instance by maintaining evidence of the retail product; in industry, with electrical grids and nuclear plants controlled by computers; in consumer electronics with DVD players, fridges, and ovens; in telecommunication with smart mobile phones, to point out only a few. Being widespread is a proof of their utility, but also signals a threat, if we cannot ensure their correct functioning.

A microchip essentially connects a computational engine to a memory system. Many dedicated technologies exist for enabling this connectivity, evolving as illustrated in Figure 12.1. Point-to-point communication provided a good performance but led to a complicated architecture. Eventually, a network was designed in between the components that need to communicate, and this proved to achieve the best communication figures. The Network-on-Chip (NoC) was thus born, where a component communicates with another one via a network of routers. The question of microchip correctness thus propagates to a question of NoC correctness. This "network" has specific features such as handling routing demands from one component to another, even though the network is "local." Another specific NoC feature is that a failed core is not replaceable. Hence, it is important to define what it means for a NoC to function correctly is extremely relevant.

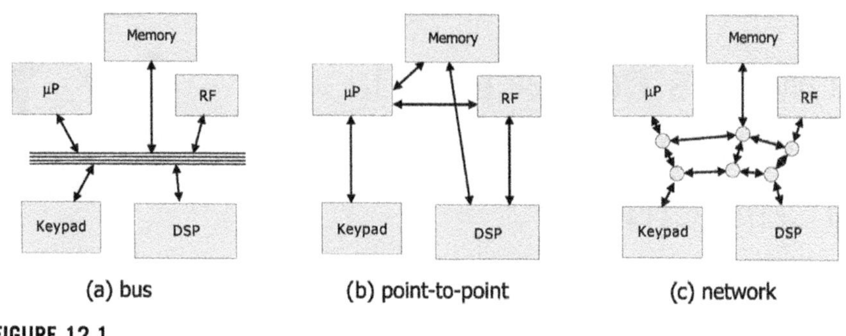

FIGURE 12.1

NoC evolution [1].

In this paper, we address the question of ensuring that a particular NoC functions correctly. Part of our goal is to propose a definition for the concept of NoC functioning. We do not claim that ours is the best or only definition for this; instead, we propose a definition we find appropriate for illustrating the main part of our goal, that of *proving* a NoC correct with respect to this definition. This means that we model a NoC mathematically and demonstrate that the model is in accordance to the definition. The advantages of proving are clear, as it offers mathematical assurance for what we are illustrating. In contrast with testing or simulating a NoC, we can be sure that in all the states and for all the scenarios, our communication definition for the NoC is respected. Proving is, however, a difficult activity, since it is typically based on an intrinsic mathematical scaffolding. Two aspects thus become instrumental for our approach, namely *reusability* and *tool support*. The former aspect refers to the possibility of reusing and extending NoC models that are already proven correct; the latter aspect refers to the possibility of modeling based on a theorem-prover tool, that offers support for editing, proof obligation generation and proof obligation discharging. For these reasons, our modeling and proving takes place with the Event-B [2] formal method and its associated Rodin platform tool [3].

Formal methods refer to the application of mathematical techniques to the design and implementation of computer hardware and software. An interesting, recent example of formal methods (application namely Separation Logic [4]) is provided by the development of the INFER tool [5] for static analysis: this ensures the safe usage of pointers, the absence of memory leaks, and the correct functioning of the dynamic heap operations. Prominent examples of applying formal methods are provided by, e.g., Intel [6,7] and IBM [8] for formally verifying hardware or systems-on-chip (SoC) [9]. By using rigorous mathematical techniques, it is possible to deliver *provably correct systems*. Formal methods are based on the capture of system requirements in a specific, precise format. Importantly, such a format can be analyzed for various properties and, if the formal method permits, also stepwise developed until an implementation is formed. By following such a formal development, we are sure that the final result correctly implements the requirements of the system.

In this paper, we propose an extensible formalization of NoC architectures, based on which we can model and analyze both functional and nonfunctional NoC properties. One of the main features of the formal method employed here (Event-B) is that system development takes place in a stepwise manner that eventually leads to a system implementation. The stepwise development is captured by the *refinement* [11,12] relation between models of the same system, so that a high-level model of a system is transformed by a sequence of correctness-preserving steps into a more detailed model that satisfies the original specification. Our models capture the general NoC architecture at a high level of abstraction. This abstraction is beneficial for constructing correct reusable models. On one hand, abstraction allows us to generalize the specification of a set of systems with common behavior. On the other hand, it allows us to prove the global properties of such general specifications.

We propose four different abstract models M_0, M_1, M_2, and M_3 for NoCs so that $M_0 \sqsubseteq M_1 \sqsubseteq M_2 \sqsubseteq M_3$, where "$\sqsubseteq$" denotes the refinement relation. Each of these abstract models can then be refined into more concrete models, for instance to describe specific NoC algorithms. When the concrete models preserve the NoC functioning definition of the abstract models, we guarantee the correctness of the concrete NoC designs. As an application of the proposed NoC architectures, we model both the *deterministic* and the *congestion-aware* XYZ routing algorithms. The deterministic XYZ routing algorithm is modeled as a refinement of the M_2 NoC architecture, while the congestion-aware XYZ algorithm refines the M_3 NoC architecture. To verify the XYZ routing algorithms, we generate the proof obligations using the Rodin platform tool and discharge them automatically or interactively.

The models include the main communication constraints that are instrumental in proving the NoC correctness. Our definition of NoC functioning at this abstract level is the property that a package injected in the network is eventually received at the destination. We also address the nonfunctional property of *performance* [13], by modeling buffer treatment in case of congestion. More precisely, at the architecture level M_0–M_3 we model that, when buffer congestion is detected, then the IP core voltage is increased, so that messages are treated faster and consequently removed faster from the buffers. In addition, the congestion-aware XYZ algorithm slightly modifies the deterministic XYZ algorithm: instead of messages being routed to their destination first on the X coordinate, then on the Y coordinate and finally on the Z coordinate, they are routed first on the coordinate where a noncongested buffer is detected.

Hence, in this paper we propose a definition for NoC functioning and we model it in several increasingly more complex NoC models. We demonstrate that the NoC models respect this definition and show the reusability of both the definition and the models. As an interesting observation, we point out some nonfunctional aspects of NoCs that fit well with our definition of correctness. These aspects are an example of the monitoring and adapting capabilities of NoCs: we demonstrate that our NoC models respect the definition even when problems (i.e., congestion in our case) occur.

The paper is organized as follows. In Section 2 we review the Network-on-Chip communication paradigm. In Section 3 we overview the Event-B formal method to the extent needed in this paper. In Section 4 we present the (abstract) NoC model we target, together with our NoC functioning definition and subsequent NoC developments. We also discuss the nonfunctional property of congestion via which we address monitoring and adaptability of NoCs in this paper. In Section 5 we describe the four increasingly more detailed models for a NoC together with the constraints for proving correctness. In Section 6 we discuss reusability and extensibility and in Section 7 we present some verification statistics. In Section 8 we review related work. In Section 9 we put forward the results of this paper and their significance. We offer some conclusions and remarks for future work in Section 10.

2 NETWORK-ON-CHIP

In this section we overview some fundamental network-on-chip (NoC) aspects.

2.1 WHAT IS A NoC?

A microprocessor essentially connects a computational engine (CPU) to a memory system. As computers evolved, both the CPU and the memory became more complex and specialized; in addition, dedicated functional units were necessary as well. The System-on-Chip [14] was thus born, consisting of a set of units named *cores* that needed to communicate with each other.

Initially, communication was handled by a bus, i.e., a long wire, globally clocked and stretched over the whole chip; only one core was able to access the bus at a time. Various problems were apparent, e.g., enabling the communication of cores working at different clock frequencies (e.g., a digital signaling core and a graphics core), latency, and complexity. It turned out that a simple bus was not very scalable, so that adding more cores on a chip made communication less effective. A first attempt to solve the problem was to replace the bus with bridged bus segments, i.e., locally clocked bundles of wires; however, this solution implied manual design and was too time-consuming and expensive. Moreover, technical issues due to several wires bundled together (such as parasitic capacitance) were also against this solution. Point-to-point links were also considered, but they resulted in a large number of connections and increased complexity. A simpler approach was then designed for the communication

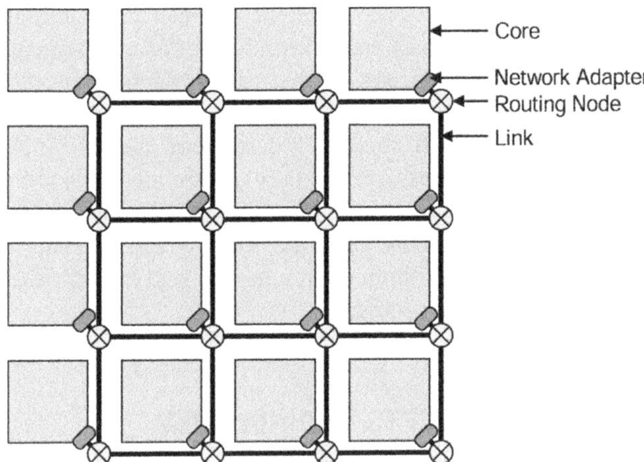

FIGURE 12.2

NoC structure [1].

paradigm on a SoC, namely a network of routers that are able to relay messages between any two cores in the SoC. This is known as a Network-on-Chip (NoC) [1] and is illustrated in Figure 12.2.

Every core has a router and a network adapter attached to it. This adapter constructs the packets from the core data to be communicated and the routers transmit the data along the network links. Typically, the routers connect the cores via copper wiring and need routing tables and algorithms much as for any usual (computer) network. Since these routers are on a chip, they have special requirements such as the need for simple logic and efficient algorithms to save the resources of the host chip. There are also circuit-switching techniques for NoCs, that ensure bandwidth efficiency and are employed in high-performance computing tasks. However, they are not as widespread as the packet-switching methods, because they are more expensive and tailor-made.

2.2 BASIC NoC CONCEPTS

NoC protocols are built as an attempt to scale down the concepts of large-scale networks and to apply them to the embedded SoC domain [14]. *Switching* in NoCs determines how data flows through the routers, defines the granularity of the transfer as well as the applied switching technique. A *phit* is the physical data unit that is transferred in one cycle on a link, while a *flit* is the logical data unit into which packets are divided, and consists of several phits. Some switching strategies are store-and-forward, where the packet is sent if the next router has buffer space for it; virtual cut through, where the first flit is sent as soon as there is space for the whole packet; wormhole, where a flit is forwarded if there is space for it in the next buffer, hence different flits can travel through different routers. The former two techniques have excessive buffer requirements, while the latter has reasonable buffer requirements, but can suffer from deadlocks, due to potential link dependencies. A *routing algorithm* is responsible for correctly and efficiently routing packets from source to destination and is typically chosen depending on the trade-offs necessary to satisfy certain metrics, such as minimizing the power consumption, increasing performance by reducing the delays and maximizing traffic utilization of the network. Routing can be static or dynamic, source or distributed routing, minimal or not, etc. and should ensure that no deadlock, livelock or starvation occurs.

2.3 EMERGING NoC CONCEPTS/TECHNOLOGIES

Although the NoC concept is quite novel as the communication paradigm of SoC, various problems attract a considerable amount of research in the area. In particular, one problem that is addressed is that of the copper cable delays on long interconnects, coupled with the resistance/capacitance increases for such cables. To overcome the problem, several other types of interconnects are studied, namely optical interconnects [15], wireless interconnects [16,17], and carbon nanotubes [18]. The research in this area is quite active and may soon lead to various innovative solutions. Another proposal for the improvement of NoC speeds consists of adopting a three-dimensional [19] topology structure for the routers instead of a two dimensional one. This would allow for massive parallelism, distributed memory architecture, easier heterogeneous layers so that one layer would consist of only one technology, and (arbitrarily) improved scalability. Not only computation but also communication can take place in parallel threads within this approach.

3 INTRODUCTION TO EVENT-B

Event-B [2] is a formal approach for the specification and development of parallel, distributed and reactive systems. Event-B comes along with the associated Rodin platform tool [3,10,20,21], which provides a platform for specifying and verifying distributed systems based on a theorem prover. Event-B allows us to formally model a system and to prove that the model fulfills certain desired properties. Models in Event-B are analyzed by proving that they simulate the desired system execution. To perform simulation and proofs, a discrete transition system formalism is used. The system simulation is represented by means of a succession of *state transitions*, called *events*. The proof is performed by demonstrating that the transitions preserve a number of desired global properties, which must be guaranteed by the states of the system components.

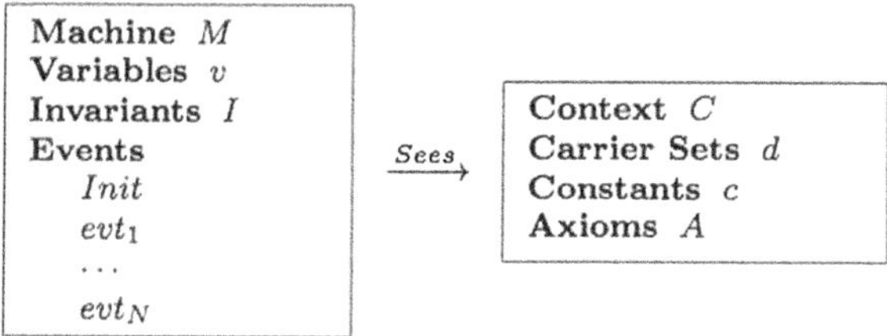

FIGURE 12.3

A machine M and a context C in Event-B.

In Event-B, a system model is organized in terms of two basic constructs: *machine* and *context*. A machine encapsulates the model state, represented as a collection of model variables, and defines operations on this state. Thus, it describes the behavior of the modeled system, also referred to as the dynamic part. A context specifies the static part of a model and isolates the parameters of a system, that hold for all instances. A context may contain user-defined carrier sets, constants and their properties, which are given as a list of model axioms. The general form of an Event-B model is illustrated in Figure 12.3. The relationship between a machine and its accompanying context is expressed by the keyword Sees, denoting a structuring technique that allows the machine access to the contents of the context.

A machine, identified by its name M, consists of state **variables**, v, defining the *state* of the model and operations called ***events***. Variables are constrained by predicates I given in the **Invariants** clause. Events specify possible state changes in a system model, i.e., define the dynamic behavior of the system. An event *evt* can be represented as follows:

$$evt \triangleq \textbf{any } vl \textbf{ where } g \textbf{ then } S \textbf{ end}$$

where vl stands for new local variables (parameters) of the event, g is the *guard*, i.e., a conjunction of predicates over the state variables v and vl, and S is the *action*, i.e., an assignment to the state variables.

The occurrence of events represents the observable behavior of the system. The guard states the necessary conditions under which the action can be executed, i.e., when the event is *enabled*. If several events are enabled at the same time, any of them can be chosen for execution nondeterministically. If some events have no variables in common and are enabled at the same time, then they can be executed in parallel since their sequential execution in any order gives the same result. For all practical purposes, this execution model is parallel and can be implemented as such when the model is refined to code. Events can be declared as *anticipated*, meaning that in the future refinements we need to set out a natural number expression called *variant* and prove that it is decreased by this event. Events can also be *convergent*, meaning that in the current machine there is a variant that decreases when this event is chosen for execution. Thus, an anticipated event is not convergent in the current machine but should become so in a future refinement of that machine.

The action of an event is a parallel composition of assignments that are in one of the forms:

$$x :=E(x,y) \quad \text{(a.1)}$$

$$x :\in Set \quad \text{(a.2)}$$

$$x :| P(x,y,x') \quad \text{(a.3)}$$

where $x,y \subseteq v$, $E(x,y)$ is an expression on variables x, y, *Set* is a set of values and $P(x,y,x')$ is a before-after predicate relating initial values of x, y (before the action) to some final value x' (afterwards). Assignment a.1 is a deterministic assignment whereas a.2 and a.3 are nondeterministic. In assignment a.2, x is assigned an element of *Set* and in assignment a.3, x is assigned an after-state x' satisfying P.

The semantics of a whole Event-B model is formulated as a number of *proof obligations*, expressed in the form of logical sequents. Below we describe only the most important proof obligations that should be verified (proved) for the initial and refined models. The full list of proof obligations can be found in [22].

Invariant property: A primary element of an Event-B model is the invariant. It models properties that should hold in every reachable state of the model. To prove an invariant property for an Event-B model, invariant establishment and preservation rules should be proved. Invariant establishment states that any possible state after initialization characterized by the so called before-after predicate $BA_{Init}(d,c,v')$ must satisfy the invariant $I(d,c,v')$, also assuming the axioms. The proof obligation rule is as follows:

$$A(d,c), BA_{Init}(d,c,v') \vdash I(d,c,v') \quad \text{(INIT)}$$

where A stands for the conjunction of the model axioms, I is the conjunction of the model invariants, d stands for the model sets, c are the model constants, and v, v' are the variable values before and after event execution.

Invariant preservation states that the property is maintained whenever variables change their values. Invariant preservation shows that for every event of the model, evt_i, the invariants still hold in any possible state after event execution.

$$A(d,c), I(d,c,v), g_i(d,c,v), BA_i(d,c,v,v') \vdash I(d,c,v') \quad \text{(INV)}$$

where g_i stands for the event guard.

Feasibility property: Every Event-B model should satisfy the event feasibility. Feasibility states that whenever an event of the Event-B model, evt_i, is enabled, the action of the event is always feasible, i.e., there exists some reachable after-state:

$$A(d,c),\ I(d,c,v),\ g_i(d,c,v) \vdash \exists v' \cdot BA_i(d,c,v,v') \quad \text{(FIS)}$$

The proof obligation rules provide us with a foundation for establishing correctness of Event-B specifications. In particular, to verify correctness of a specification, we need to prove that its initialization and all events preserve the given invariant.

Event-B employs the concept of refinement, developing a system model through a number of correctness preserving steps. The refinement concept provides for a top-down approach for constructing systems following rules for gradually introducing details to an initial abstract specification. While capturing more detailed requirements, each refinement step typically introduces new events and variables into an abstract specification. These new events correspond to stuttering steps that are not visible in the abstract specification. We call such model refinement *superposition refinement* [23,24]. Moreover, Event-B formal development supports *algorithmic refinement* [25], allowing us to refine an event of an abstract machine with several corresponding events in a refined machine. This will model different branches of execution that can, for instance, take place in parallel and thus can improve the algorithmic efficiency. Out of these refinement approaches, in this paper we employ the superposition refinement.

When presenting events in a refined model, we often use the shorthand notation "*refined_event* **extends** *abstract_event*." The meaning of this notation is that the refined event is created from the abstract one by simply adding new guards and/or new actions. Only the added elements are shown in the extended event, while the old guards and actions are implicitly present.

To verify the correctness of a refinement step, we need to prove a number of proof obligations for a refined model. For brevity, here we show only a few essential ones.

Let us first introduce a shorthand $H(d,c,v,w)$ to stand for the hypotheses $A(d,c), I(d,c,v), I'(d,c,v,w)$, where I, I' are respectively the abstract and refined invariants, and v, w are respectively the abstract and concrete variables. Then the feasibility refinement property for an event evt_i of a refined model can be presented as follows:

$$H(d,c,v,w), g_i'(d,c,w) \vdash \exists w' \,.\, BA_i'(d,c,w,w') \quad \text{(REF\_FIS)}$$

where g_i' is the refined guard and BA_i' is a before-after predicate of the refined event.

The event guards in a refined model can only be strengthened in a refinement step:

$$H(d,c,v,w), g_i'(d,c,w) \vdash g_i(d,c,v) \quad \text{(REF\_GRD)}$$

where g_i, g_i' are respectively the abstract and concrete guards of the event evt_i.

Finally, the *simulation* proof obligation requires us to show that the "execution" of a refined event is not contradictory to its abstract version:

$$H(d,c,v,w), g_i'(d,c,w), BA_i'(d,c,w,w') \vdash \exists v' \,.\, BA_i(d,c,v,v') \wedge I'(d,c,v',w') \quad \text{(REF\_SIM)}$$

where BA_i, BA_i' are respectively the abstract and concrete before-after predicates of the same event evt_i.

The Event-B refinement process allows us to gradually introduce implementation details, while preserving functional correctness during stepwise model transformation. The model verification effort and, in particular, automatic generation and proving of the required proof obligations, are significantly facilitated by the provided tool support-the Rodin platform tool [3,10,20,21].

Let us note here the quintessential feature of Event-B modeling with the associated Rodin platform. Modeling in Event-B is semantically justified by proof obligations. Every update of a model generates a new set of proof obligations in the background. It is this interplay between modeling and proving that sets Event-B apart from other formalisms. Without proving the required obligations, we cannot be sure of the correctness of a model. The proving effort thus encourages the developer to structure formal model development in such a way that manageable proof obligations are generated at each step. This leads to very abstract initial models upon which we can gradually introduce into a system model various facets of the system. Such a development method fits well when we have to describe complex algorithms.

4 NoC MODELS AND PROPERTIES

Our object of study is described by the NoC model of computation.

4.1 THE NoC STRUCTURE

Our most abstract model for the NoC model of computation, M_0, contains:

- A set of nodes, referred to as routers
- A binary relation on the set of nodes, denoting the neighbor structure
- A set of data, modeling the possible contents of messages
- A set of identifiers pointing to triples that model messages, i.e., each triple contains a data content, a source node, and a destination node.

The next model, M_1, refines the general network structure in M_0 to a three-dimensional NoC. This is achieved by associating a 3D-mesh structure to the set of nodes and their neighboring relation, as well as associating channels to the new 3D neighbor relation and fourteen ports to each node. The following model, M_2, adds buffers to each port in M_2. In M_3 we model congestion and voltage levels.

Models M_0–M_2 together with their refinement relation and derived algorithms for a deterministic XYZ routing and a congestion-aware XYZ routing are illustrated abstractly in Figure 12.4. In the following section, this figure is more detailed.

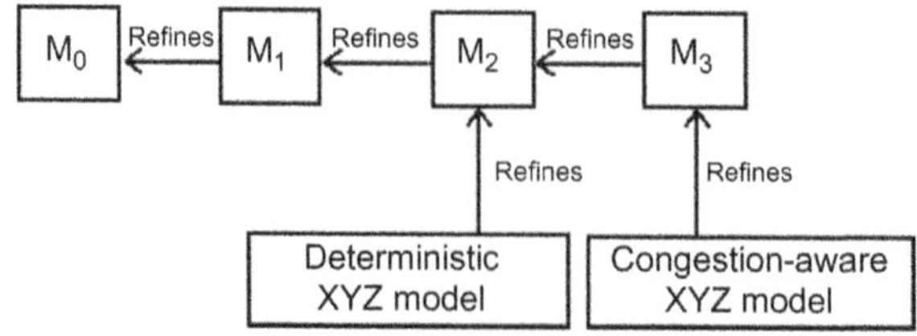

FIGURE 12.4

The abstract structure of NoC modeling.

4.2 THE NoC COMMUNICATION MODEL

For transmitting messages in M_0, we essentially copy them from node to node until the destination is reached. In M_1, a message is stored in a port of the node and when we model transmission, essentially we copy the message in a corresponding channel. In M_2, the messages are stored in the buffers of ports.

4.3 THE NoC FUNCTIONING DEFINITION

For our NoC models, we use the following definition for modeling successful NoC functioning: the messages injected in the network will eventually reach their destinations. A *sent_pool* set denotes the list of messages injected into the network. The *sent_pool* set is updated whenever a new message is injected into the network. A *moving_pool* set denotes the current position of travelling messages. All the messages injected into the network are added into the *moving_pool* and whenever a message is routed from a node to another one, the current position of that message is updated in *moving_pool*. A *received_pool* set denotes the list of messages received from the network by destination nodes. Whenever a message is received at its destination, it will be added to *received_pool* and removed from *moving_pool*. This behavior of the message pools is illustrated in Figure 12.5, where a message is denoted as a triple (data, src, des), with data content "data," source node "src" and destination node "des."

For demonstrating the correctness of our NoC model, we need to show that *sent_pool* eventually becomes equal to *received_pool*, at which point *moving_pool* is empty. We model this by first defining the invariant $sent\_pool = received\_pool \Leftrightarrow \text{moving\_}pool = \varnothing$. The events that manipulate the contents of these pools are then specified as either anticipated or convergent, meaning that they will only execute a finite number of times. This means that, at some point, these events are not enabled anymore. As the guards of these events rely on $\text{moving}_{pool} \neq \varnothing$, the fact that they are not enabled anymore implies that $\text{moving\_}pool = \varnothing$. As ensured by the invariant, at this point *sent_pool* is equal to *received_pool*, hence, every message injected in the network has reached its destination.

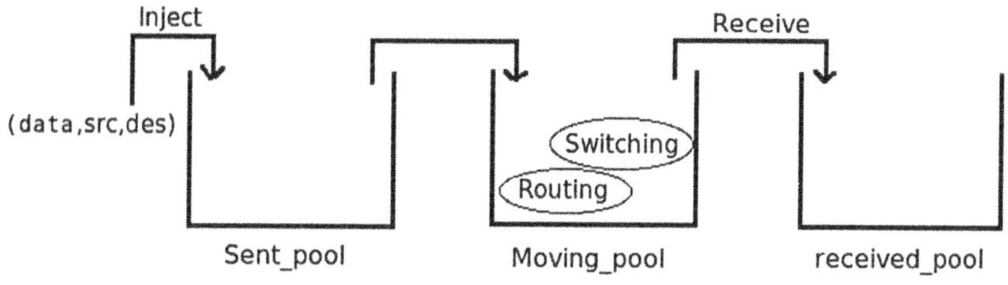

FIGURE 12.5

Message pools.

The switching event is specified throughout M_0–M_3 models as anticipated and then it becomes convergent in the two XYZ models. Due to the nondeterministic choice involved in the switching event, we cannot attach, at the M_0–M_3 level of abstraction, a similar variant as in the XYZ models. This is because we cannot ensure that the switching event decreases the distance between the message position and its destination, due to the lack of fairness in our modeling. We did not model fairness because it is a rather concrete feature of an algorithm, but obviously, this needs to be considered when implementations are developed.

4.4 CONGESTION

The NoC models described here, together with the associated NoC functioning definition, supply the fundamental infrastructure of specifying a correct 3D NoC with respect to the functional properties. In M_3 we aim to also model a nonfunctional property, i.e., we add a property used for *optimizing* the 3D NoC architecture. The purpose of this model is to demonstrate that our correct-by-construction specification of (3D) NoCs can be employed for modeling both functional and nonfunctional properties.

One of the main concerns in design optimization is power consumption. One optimization technique for NoC power consumption consists of scaling the speeds of the communication links via a corresponding *voltage* level. We employ this technique here, by modeling different voltage rates in every node; in certain conditions, the current voltage rate of a node can be changed. For instance, if the load in a node is high, the node automatically adjusts its voltage level to *high*, so that the node performs faster. Here we model the node load by the *congestion level* of that node. When the local congestion level is high in a node, the node may not be able to transmit packets successfully, thus eventually leading to a high queue build-up, a high channel loading, or a high packet drop rate. Possible reasons for congestion level increases in nodes can be the increase of the incoming traffic rate, the low speed of transformation, improperly managed buffers, etc. Therefore, by adapting the voltage level in each node, according to the local congestion level, we can avoid the large packet delay and the dropping of packets due to queue overflows; overall, we avoid throughput degradation.

One approach to measuring the local congestion level is by using a buffer utilization metric. Here, we employ this approach.

5 FOUR ABSTRACT MODELS FOR THE 3D NoC: M_0, M_1, M_2, M_3

In this section we formally develop four high-level models M_0, M_1, M_2, and M_3 for the 3D NoC. Our models are at different and increasing levels of detail so that each model is a refinement of the previous one: $M_0 \sqsubseteq M_1 \sqsubseteq M_2 \sqsubseteq M_3$. In the initial model, we specify a network of nodes and define the communication property of this network based on a specific data structure called *pool*, as suggested by Abrial [22]. In the second model, we add new data and events to model the 3D mesh-based NoC architecture; besides, we specify the channels between nodes. In the third model, we model buffers for nodes and refine the communication model. In the fourth model we add the specification of a nonfunctional property of NoC by modeling voltage adaptation according to the congestion level in NoC. The structure and steps of the modeling are shown in Figure 12.6.

By starting from an initial model that is rather abstract, i.e., without detailing the communication topology, we obtain a rather general starting point that can later be refined to various topologies. Moreover, adding channels and ports only in the second model leads to a clean modeling of the basic communication mechanism (via routing and switching) in the initial model; the required detail (of channels and ports) are not needed for understanding the communication

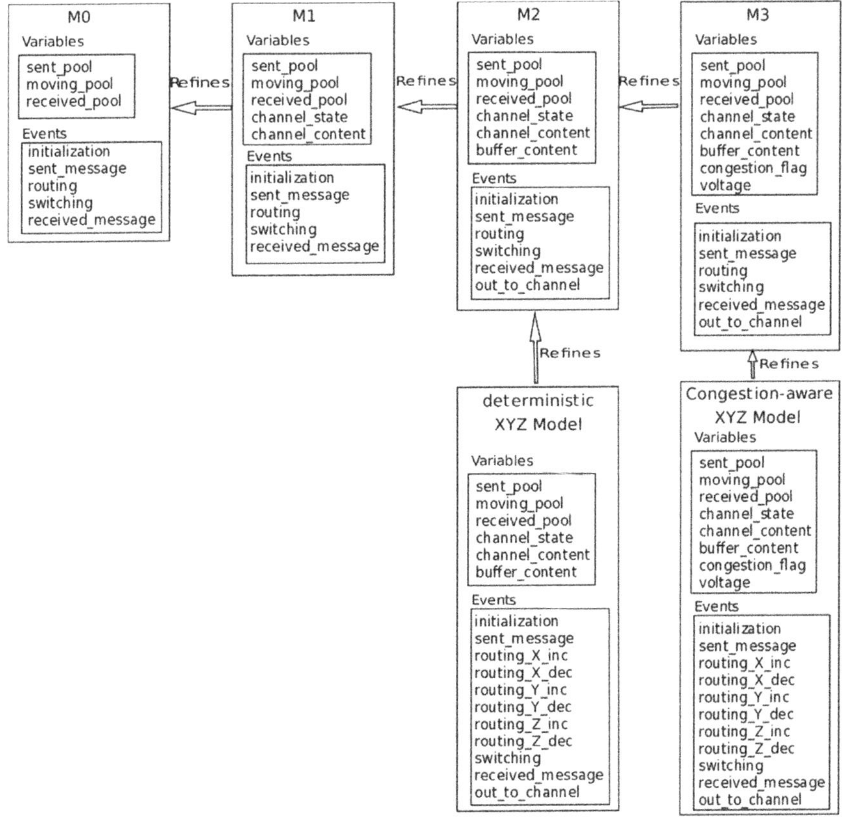

FIGURE 12.6

The structure of NoC modeling.

mechanism. Adding buffers in the third model illustrates an extra level of detail. Networks where the nodes have no buffers for communication will, therefore, employ the second model as their abstraction and not the third. By adding buffers to the model, we can then capture buffer congestion and based on that, the adaptation of voltage. Hence, by modeling the primitive functional architecture of NoC in the first three abstract models, the non-functional properties can also be specified by refining the most concrete one, i.e., the third model.

5.1 MODEL M_0

The first model M_0 that we construct is rather abstract: we do not consider numerous parts of the network such as channels or buffers; they will be introduced in subsequent refinements. This will thus allow us to reason about the system very abstractly [22]. The model is formed of the static part and the dynamic part, as follows.

5.1.1 The static part

The static part of our model is described in Figure 12.7 and contains the sets *MESSAGES*, *ROUTER*, *DATA* and the constants *data*, *des*, *src* and *Neigh*. The message identifiers are modeled by the non-empty and finite *MESSAGES* set (@**axm1**). We use the following modeling idea for messages. A message id in the *MESSAGES* set relates to a triple (*data, src, des*) where *data* is an element of the *DATA* set (@**axm4**), *src* models the source node where a message is injected (@**axm5**), and *des* models the destination node where a message should be received (@**axm6**). A message should not be destined to its source node (@**axm7**). The set of network nodes and data are modeled by the sets *ROUTER* (finite and non-empty) (@**axm2**) and *DATA* (finite and non-empty) (@**axm3**), respectively. The relation Neigh (non-empty, symmetric, and non-reflexive) (@**axm9**) models the neighbor structure i.e., which node can communicate with which node (@**axm8**).

SETS *MESSAGES ROUTER DATA*
CONSTANTS *des src data Neigh*
AXIOMS
@axm1 $MESSAGES \neq \varnothing \land finite(MESSAGES)$
@axm2 $ROUTER \neq \varnothing \land finite(ROUTER)$
@axm3 $DATA \neq \varnothing \land finite(DATA)$
@axm4 $data \in MESSAGES \rightarrow DATA$
@axm5 $src \in MESSAGES \rightarrow ROUTER$
@axm6 $des \in MESSAGES \rightarrow ROUTER$
@axm7 $\forall m, sp, dp \cdot m \in MESSAGES \land sp \in ROUTER \land dp \in ROUTER$
$\land m \mapsto sp \in src \land m \mapsto dp \in des \Rightarrow sp \neq dp$
@axm8 $Neigh \in ROUTER \leftrightarrow ROUTER$
@axm9 $Neigh \neq \varnothing \land Neigh = Neigh^{-1} \land dom(Neigh) \lhd id \cap Neigh = \varnothing$

FIGURE 12.7

M_0: Static part.

To define structure types such as records in Event-B, we use functions to represent attributes. Therefore, our modeling idea translates to the functions *data*, *src* and *des* with ranges *DATA*, *ROUTER*, and *ROUTER*, respectively.

5.1.2 The dynamic part

For expressing NoC communication, we define two message subsets and one partial message-to-node map as machine variables: $sent_{pool} \subseteq MESSAGES$, $received\_pool \subseteq MESSAGES$ and $moving\_pool \in sent\_pool \rightarrow ROUTER$. We have illustrated these in Figure 12.5. To model the communication and the message pool functions, we define three events as explained below. The $sent_{message}$ event described in Figure 12.8 handles the injection of a new message into the network. Whenever a message is injected into the network, both *sent_pool* as well as *moving_pool* are updated.

A message in *moving_pool* should be routed toward its destination. This is composed of two actions: one for deciding which node would be the next one (routing) and the other for transferring the message to that node (switching). These two actions are available for all the nodes, including the source, the destination as well as all the intermediate nodes and are modeled respectively by the *routing* and *switching* events shown in Figure 12.9. In this abstract model we do not have any routing decisions; hence, the *routing* event is modeled by *skip* (@**act1**). The *switching* event in the model only transfers a message from the current node to one of its neighbors, provided as an argument, and updates the *moving_pool* by changing the current position of a message (@**act1**). To avoid cycling, we do not allow a message to return to its source (@**grd4**). The reason for not considering a specific routing algorithm is that it makes our initial model more general and reusable for a wide variety of routing algorithms implementations. The *switching* event has the status *anticipated.*

The *received_message* event shown in Figure 12.10 adds a message received at its destination to *received_pool* (@**act2**) and removes the message from *moving_pool* expressed by using the domain subtraction operator in the @**act1**. This event is convergent: if new messages are not injected to the network for a certain time, all the messages will be received at their destinations. This is proved based on the ($sent\_pool \setminus received\_pool$) variant denoting the difference between the sets *sent_pool* and *received_pool.*

```
event sent_message
Any
   current_msg
Where
   @grd1 current_msg ∈ MESSAGES
   @grd2 current_msg ∉ sent_pool
Then
   @act1 sent_pool := sent_pool ∪ {current_msg}
   @act2 moving_pool := moving_pool ∪ {current_msg ↦ src(current_msg)}
End
```

FIGURE 12.8

M_0: sent_message event.

```
event routing
Then
   @act1 skip
End

event switching
status anticipated
Any
   current_msg   new_position
Where
   @grd1 current_msg ∈ dom(moving_pool)
   @grd2 des(current_msg) ≠ moving_pool(current_msg)
   @grd3 new_position ↦ moving_pool(current_msg) ∈ Neigh
   @grd4 new_position ≠ src(current_msg)
Then
   @act1 moving_pool(current_msg):=new_position
End
```

FIGURE 12.9

M_0: switching and routing event.

```
event   received_message
status convergent
Any
   current_msg
Where
   @grd1 current_msg ∈ dom(moving_pool)
   @grd2 des(current_msg) = moving_pool(current_msg)
Then
   @act1 moving_pool := {current_msg} ⩤ moving_pool
   @act2 received_pool := received_pool ∪ {current_msg}
End
```

FIGURE 12.10

M_0: received_message event.

In order to prove the model correctness, we need to prove that the *sent_pool* subset eventually becomes equal with the *received_pool* subset (**@inv1-2**, **@inv4-6**) and the *moving_pool* subset is empty when all the messages are received at their destinations (**@inv3**). These properties are formulated in Figure 12.11 as invariants.

M_0 is a general specification of a general network and will be refined to model 3D NoC communication designs in the following. Moreover, the model provides the necessary properties that should be preserved by refinement. These properties, which guarantee the overall NoC correctness, are defined as the list of invariants.

5.2 MODEL M_1

Transferring a message from a node to its neighbor in model M_0 is achieved simply by copying the message from a node to another. In this section we refine the

INVARIANTS

@inv1 $dom(moving\_pool) \subseteq sent\_pool$
@inv2 $received\_pool \cap dom(moving\_pool) = \varnothing$
@inv3 $sent\_pool = received\_pool \Leftrightarrow moving\_pool = \varnothing$
@inv4 $\forall msg \cdot msg \notin sent\_pool \Rightarrow msg \notin received\_pool$
@inv5 $sent\_pool \setminus dom(moving\_pool) = received\_pool$
@inv6 $sent\_pool \setminus received\_pool = dom(moving\_pool)$

FIGURE 12.11

M_0: Invariants (pool modeling).

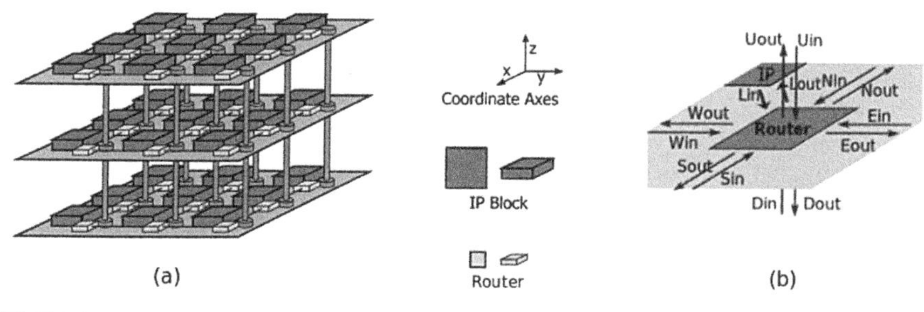

FIGURE 12.12

(a) 3D mesh-based NoC architecture; (b) router channels.

initial model M_0 to also specify channels specific to the 3D NoC. To specify channels, we need 3D NoC architecture, such as mesh-based [19], or tree-based [26]. We consider here NoC with 3D mesh topologies. The 3D mesh-based NoC (Figure 12.12(a)) consists of $N = m^*n^*k$ nodes; each node has an associated integer *coordinate triple*, (x, y, z), so that $0 < x \leq m, 0 < y \leq n, 0 < z \leq k$.

Our 3D NoC architecture employs seven-port routers: one port to the (**L**ocal) IP block, one port to above (**U**p) and below (**D**own) routers, and one in each cardinal direction (**N**orth, **S**outh, **E**ast and **W**est), as shown in Figure 12.12(b). Each of these ports has two channels, *in* and *out*, resulting in a total of 14 channels.

5.2.1 The static part

We extend the static part of the initial model M_0 in three ways: we map routers to coordinate triples, we add new properties for the *Neigh* relation based on the coordinate triples, and we model ports and channels for the 3D NoC. In order to map routers to the coordinate triples, we define four constants: *coordX*, *coordY*, *coordZ* and *mk_position* as shown in Figure 12.13. The *coordX*, *coordY* and *coordZ* constants represent coordinate triples *(x,y,z)* (**@axm12-14**) and the *mk_position* constant is a map associating each router to a position in space given by the coordinates (**@axm11**). We define **@axm15** to establish the one-to-one relation

SETS *CHANNEL PORTS*
CONSTANTS *coordX coordY coordZ mk_position crossbarX crossbarY crossbarZ mk_channel*
AXIOMS
@axm10 $crossbarX \in \mathbb{N}_1 \land crossbarY \in \mathbb{N}_1 \land crossbarZ \in \mathbb{N}_1$
@axm11 $mk\_position \in (1\,..\,crossbarX) \times (1\,..\,crossbarY) \times (1\,..\,crossbarZ) \rightarrowtail\!\!\!\!\twoheadrightarrow ROUTER$
@axm12 $coordX \in ROUTER \twoheadrightarrow (1\,..\,crossbarX)$
@axm13 $coordY \in ROUTER \twoheadrightarrow (1\,..\,crossbarY)$
@axm14 $coordZ \in ROUTER \twoheadrightarrow (1\,..\,crossbarZ)$
@axm15 $\forall xx, yy, zz \cdot xx \in 1\,..\,crossbarX \land yy \in 1\,..\,crossbarY \land zz \in 1\,..\,crossbarZ$
$\Rightarrow coordX(mk\_position(xx \mapsto yy \mapsto zz)) = xx$
$\land coordY(mk\_position(xx \mapsto yy \mapsto zz)) = yy$
$\land coordZ(mk\_position(xx \mapsto yy \mapsto zz)) = zz$
@axm16 $\forall pos1, pos2 \cdot pos1 \in ROUTER \land pos2 \in ROUTER \land pos1 \neq pos2$
$\Rightarrow coordX(pos1) \neq coordX(pos2) \lor coordY(pos1) \neq coordY(pos2)$
$\lor coordZ(pos1) \neq coordZ(pos2)$
@axm17 $\forall r1, r2 \cdot r1 \mapsto r2 \in Neigh \Leftrightarrow abs(coordX(r1) - coordX(r2)) + abs(coordY(r1)$
$- coordY(r2)) + abs(coordZ(r1) - coordZ(r2)) = 1$

FIGURE 12.13

M_1: Static part (1).

AXIOMS
@axm18 $def\_channel \in (ROUTER \times PORTS) \rightarrow (ROUTER \times PORTS)$
@axm19 $partition(PORTS, \{Ein\}, \{Eout\}, \{Win\}, \{Wout\}, \{Nin\}, \{Nout\}, \{Sin\}$
$, \{Sout\}, \{Uin\}, \{Uout\}, \{Din\}, \{Dout\}, \{Lin\}, \{Lout\})$
@axm20 $mk\_channel \in def\_channel \rightarrowtail\!\!\!\!\twoheadrightarrow CHANNELS$

FIGURE 12.14

M_1: Static part (2).

between coordinate triples and routers and **@axm16** to guarantee that each coordinate triple is assigned to only one router. The *crossbarX, crossbarY* and *crossbarZ* constants (**@axm10**) model the number of nodes in X, Y and Z coordinates in the network, respectively.

Two nodes with coordinates (x_i, y_i, z_i) and (x_j, y_j, z_j) are connected by a communication channel if and only if $|x_i - x_j| + |y_i - y_j| + |z_i - z_j| = 1$. To model this neighbor structure, the *Neigh* relation in the initial model M_0 is restricted in this model by adding **@axm17** in Figure 12.13.

We define the *CHANNEL* set to model the communication channels between routers and we define the *PORTS* set to define the input and output ports of nodes (**@axm19**) in the static part of the second model. To show how two neighbors are connected to each other through channels, we define the *def_channel* and *mk_channel* relations with the help of **@axm18** and **@axm20**, as shown in Figure 12.14. The *def_channel* relation models the relation of a port of a node to the corresponding port of its neighbor and the *mk_channel* relation maps the port relations to channels.

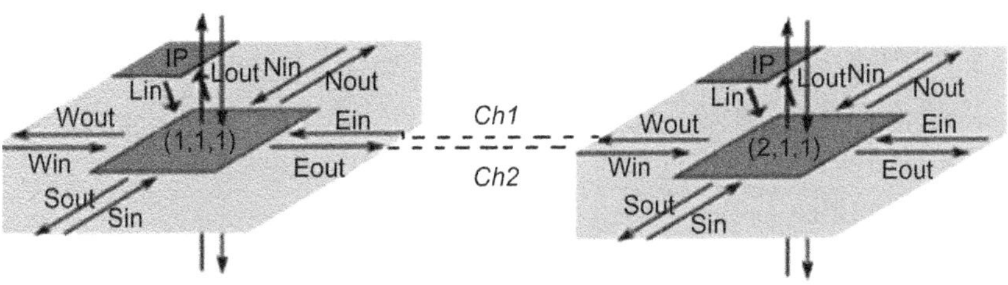

FIGURE 12.15

Channels in 3D mesh-Based NoCs.

AXIOMS

@**axm21** $\forall n, m, i, j \cdot (n \mapsto i) \mapsto (m \mapsto j) \in def\_channel \wedge i = Wout \wedge j = Ein$
$\Leftrightarrow coordX(n) - coordX(m) = 1 \wedge coordY(n) = coordY(m) \wedge coordZ(n) = coordZ(m)$

@**axm22** $\forall n, m, i, j \cdot (n \mapsto i) \mapsto (m \mapsto j) \in def\_channel \wedge i = Eout \wedge j = Win$
$\Leftrightarrow coordX(n) - coordX(m) = -1 \wedge coordY(n) = coordY(m) \wedge coordZ(n) = coordZ(m)$

FIGURE 12.16

M_1: Static part (3).

East and west ports of neighbor nodes with different X coordinate are related to each other through a channel. For instance, as shown in Figure 12.15, *Ein* and *Eout* ports of node *(1,1,1)* are connected to *Wout* and *Win* ports of node (2,1,1) through a channel $((1,1,1) \mapsto Eout) \mapsto ((2,1,1) \mapsto Win)$ and $((2,1,1) \mapsto Wout) \mapsto (1,1,1) \mapsto Ein)$ relations in *def_channel.* This connection of the ports of the neighboring nodes on the X coordinate is modeled by the **@axm21** and **@axm22** shown in Figure 12.16. The port relation between neighbors on other coordinates is defined by similar axioms, not shown here for simplicity.

5.2.2 The dynamic part

In the static part of the model M_1, we define the 3D mesh NoC architecture with the triple coordinate of nodes and their channels. In the dynamic part of the model M_1, we refine the dynamic part of the model M_0 to specify the transferring of data through the communication channels, so that the overall NoC correctness holds.

The communication channels between routers are asynchronous, transferring data upon request. Each channel propagates data as well as control values. In our case, a control value models the fact that a channel is occupied by a message. When a message is injected to a channel, the control value of that channel is set to *busy* and when the message is received at the other side of channel, the control value of that channel is set to *free*.

In order to model the transferring of messages through the communication channels, the variables *channel_state* (**@inv9**) and *channel_content* (**@inv7**)

VARIABLES *channel_content* *channel_state*
INVARIANTS
@inv7 $channel\_content \in CHANNELS \twoheadrightarrow\!\!\!\!\rightarrowtail MESSAGES$
@inv8 $dom(channel\_content) = channel\_state^{-1}[\{busy\}]$
@inv9 $channel\_state \in CHANNELS \rightarrow state$
@inv10 $dom(channel\_state) = ran(mk\_channel)$
@inv11 $ran(channel\_content) \subseteq dom(moving\_pool)$
@inv12 $\forall msg \cdot msg \in dom(moving\_pool) \land des(msg) = moving\_pool(msg) \Rightarrow msg \notin ran(channel\_content)$

FIGURE 12.17

M_1: Invariants (gluing and channels).

event *out_to_channel*
Any
new_position *current_msg* *out_p* *in_p*
Where
@grd1 $current\_msg \in dom(moving\_pool) \land new\_position \in ROUTER$
@grd2 $moving\_pool(current\_msg) \mapsto new\_position \in Neigh$
@grd3 $out\_p \in \{Nout, Sout, Wout, Eout, Uout, Dout\} \land in\_p \in \{Nin, Sin, Win, Ein, Uin, Din\}$
@grd4 $(moving\_pool(current\_msg) \mapsto out\_p) \mapsto (new\_position \mapsto in\_p) \in dom(mk\_channel)$
@grd5 $channel\_state(mk\_channel((moving\_pool(current\_msg) \mapsto out\_p) \mapsto (new\_position \mapsto in\_p))) = free$
@grd6 $moving\_pool(current\_msg) \neq des(current\_msg) \land current\_msg \notin ran(channel\_content)$
Then
@act1 $channel\_state(mk\_channel((moving\_pool(current\_msg) \mapsto out\_p) \mapsto (new\_position \mapsto in\_p))) := busy$
@act2 $channel\_content(mk\_channel((moving\_pool(current\_msg) \mapsto out\_p) \mapsto (new\_position \mapsto in\_p))) := current\_msg$
End

FIGURE 12.18

M_1: out_to_channel event.

are defined in M_1 to represent the control and the data value on each channel. Each channel can have the *busy* or *free* state. When the channel receives data, its state switches from *free* to *busy* and the message is added to the *channel_content*. When the channel transfers data to the end, *channel_state* changes to *free* and the channel is released by removing the message from *channel_content*. The invariants of M_1 model that, when a channel is released, then its content is empty and can thus receive the next message; when a channel is busy, the message is in the channel. We illustrate these invariants (**@inv8**, **@inv10–12**) in Figure 12.17. In order to maintain the consistency between model M_1 and M_0, we define the gluing invariant **@inv11**, expressing that the content of channels are a subsets of traveling messages. It guarantees that adding channels to the concrete model preserve the properties of the abstract model.

The *switching* event is now refined to transfer a message to the next router through channels. In order to model this, we add a new event *out_to_channel as* shown in Figure 12.18 to model pushing a message in the channel (**@act1-2**). This event is enabled when there is a message for transferring in a node

```
event  switching  extends   switching
status  anticipated
Any
   p1  p2
Where
   @grd5 p1 ∈ {Nout, Sout, Wout, Eout, Uout, Dout}
   @grd6 p2 ∈ {Nin, Sin, Win, Ein, Uin, Din}
   @grd7 (moving_pool(current_msg) ↦ p1) ↦ (new_position ↦ p2) ∈ dom(mk_channel)
   @grd8 channel_state(mk_channel((moving_pool(current_msg)↦p1)↦
            (new_position ↦ p2))) = busy
   @grd9 current_msg =
      channel_content(mk_channel((moving_pool(current_msg)↦p1)↦(new_position↦p2)))
Then
   @act2 channel_state(mk_channel((moving_pool(current_msg) ↦ p1) ↦
            (new_position ↦ p2))) := free
   @act3 channel_content := channel_content ⩥ {current_msg}
End
```

FIGURE 12.19

M_1: switching event.

(**@grd1**) and the node is not the destination of the message (**@grd6**). Another condition in the event that should be hold is that the channel between the node and the next node (**@grd2-4**) is free (**@grd5**).

In addition, we refine the *switching* event as shown in Figure 12.19 to model releasing the channel by receiving the message at the end of the channel (**@act2-3**). This event is enabled when a message is in the channel (**@grd8-9**). To check a channel between two neighbors (**@grd7**), we define two parameters *p1* and *p2*, which are output and input channels, respectively (**@grd5-6**).

5.3 MODEL M_2

In this model, we define buffers for the ports of the nodes and refine M_1 to model the communication in 3D NoC by considering these buffers.

5.3.1 The static part

The context of the third model contains a single constant *buffer_size*, which is a strict natural number denoting the maximum number of messages allowed in a buffer.

5.3.2 The dynamic part

Each node has 14 buffers, each assigned to node ports; those assigned to output ports are called *output buffers* and those assigned to input ports are called *input buffers*. When there is a message in an output buffer of a node, the node can transfer it to the channel provided that the channel is free. If in the other side of the channel the input buffer has an empty place, the message is transferred to the input buffer of the next node and the channel is released; otherwise, the channel will be busy until an empty place appears in the input buffer. To model the buffer

VARIABLES *buffer_content*
INVARIANTS
  @inv13 $buffer\_content \in MESSAGES \rightarrowtail\!\!\!\!\!+ (ROUTER \times PORTS)$
  @inv14 $dom(buffer\_content) \cup ran(channel\_content) = dom(moving\_pool)$
  @inv15 $dom(buffer\_content) \cap ran(channel\_content) = \varnothing$
  @inv16 $\forall b \cdot b \in ran(buffer\_content) \Rightarrow card(buffer\_content \rhd \{b\}) \in 1\,..\,buffer\_size$

FIGURE 12.20

M$_2$: Invariants (gluing and buffer).

event *switching* **extends** *switching*
status *anticipated*
Where
  @grd10 $card(buffer\_content \rhd \{new\_position \mapsto p2\}) < buffer\_size$
Then
  @act4 $buffer\_content := buffer\_content \cup \{current\_msg \mapsto (new\_position \mapsto p2)\}$
End

event *out_to_channel* **extends** *out_to_channel*
Where
  @grd6 $card(buffer\_content \rhd \{new\_position \mapsto in\_p\}) > 0$
Then
  @act3 $buffer\_content := buffer\_content \setminus \{current\_msg \mapsto (moving\_pool(current\_msg) \mapsto out\_p)\}$
End

FIGURE 12.21

M$_2$: switching and out_to_channel events.

structure in M_2, we add the new variable *buffer_content* (@**inv13**) that models the current content of all buffers. Indeed, adding and removing messages in/from buffers is modeled by the *buffer_content* variable.

In order to guarantee the correctness of the buffer modeling, we need the invariants shown in Figure 12.20. They model that the content of a buffer never becomes more than its size (@**inv16**). In addition, while a message is in the *moving_pool*, i.e., it has not reached to its destination; it must be either in a channel or in a buffer (@**inv14-15**). In fact, @**inv14-15** are gluing invariants that establish connection between the model M_2 and M_1.

The *switching* event, as shown in Figure 12.21, is refined to be enabled when there is an input buffer with at least one empty place at the end of the channel (@**grd10**). Then, besides releasing the channel, the message in the channel is transferred to the input buffer (@**act4**). The event, as shown in Figure 12.21, is refined to be enabled when there is a message in an output buffer (@**grd6**) meaning that the message is removed from buffer (@**act3**).

The *sent_message* and *received_message* events are refined so that they update the *buffer_content* variable as shown in Figure 12.22. In the *sent_message* event when a new message (@**grd3**) is injected to the network, it should be added to the local input buffer of the original node (@**act3**). In the *received_message* event

event *sent_message* **extends** *sent_message*
Where
  @grd3 $current\_msg \notin dom(buffer\_content)$
Then
  @act3 $buffer\_content := buffer\_content \cup \{current\_msg \mapsto (src(current\_msg) \mapsto Lin)\}$
End

event *received_message* **extends** *received_message*
Where
  @grd3 $current\_msg \mapsto (des(current\_msg) \mapsto Lout) \in buffer\_content$
Then
  @act3 $buffer\_content := buffer\_content \setminus \{current\_msg \mapsto (des(current\_msg) \mapsto Lout)\}$
End

FIGURE 12.22

M_2: send_message and received_message events.

event *routing* **refines** *routing*
Any
  msg router in_p out_p
Where
  @grd1 $in\_p \in \{Win, Ein, Sin, Nin, Uin, Din, Lin\}$
  @grd2 $out\_p \in \{Wout, Eout, Sout, Nout, Uout, Dout, Lout\}$
  @grd3 $(in\_p = Win \land out\_p \neq Wout) \lor (in\_p = Ein \land out\_p \neq Eout)$
  $\lor(in\_p = Sin \land out\_p \neq Sout) \lor (in\_p = Nin \land out\_p \neq Nout)$
  $\lor(in\_p = Uin \land out\_p \neq Uout) \lor (in\_p = Din \land out\_p \neq Dout)$
  $\lor(in\_p = Lin \land out\_p \neq Lout)$
  @grd4 $msg \mapsto (router \mapsto in\_p) \in buffer\_content$
  @grd5 $card(buffer\_content \rhd \{router \mapsto out\_p\}) < buffer\_size$
Then
  @act1 $buffer\_content(msg) := router \mapsto out\_p$
End

FIGURE 12.23

M_2: routing event.

when a message is received to its destination (@**grd3**), it should be removed from the local output buffer of the destination node (@**act3**).

At this level of abstraction, we refine the *routing* event (modeled as *skip* in the previous models) as shown in Figure 12.23. A routing algorithm decides on choosing an output channel (@**grd2**) for a message in an input channel (@**grd1**). We are at an abstract level here; hence we do not consider any specific routing algorithm but model the routing decision nondeterministically. That is, when there is a message in an input buffer of a node (@**grd4**), it can be routed to any output buffer of the node except the output buffer in the same direction with the input buffer (@**grd3**) e.g., a message in the northern input buffer cannot be routed to the northern output buffer. We also check that there is enough space in the chosen buffer (@**grd5**). We have this constraint to prevent a cycling problem in the communication that would lead to deadlock in the interconnection network. If the condition in the *routing* event holds, one of the output buffers of the node, modeled by *out_p* can nondeterministically be chosen (@**act1**).

5.4 MODEL M_3

We now continue our modeling with one more step for capturing the congestion level of a node and modeling the voltage levels of the nodes. The later will then be adjusted based on the values of the former.

5.4.1 The static part

To model different levels of voltage, we define the set *VOLTAGE* = *{low, high}* (**@axm17**) as shown in Figure 12.24. The exact values of the voltage level do not affect the specification and verification in this step hence; we abstractly define only these two levels. These levels can be refined to specific voltage levels in the next refinements when a particular voltage table is assigned to the system.

To recognize congestion in buffers, we first need to define a threshold used to reason about the state of buffers; we define this level in the static part as the constant *threshold_buffer*. If the total number of used cells in any buffer is greater than the *threshold_buffer*, it means the congestion level in the buffer is high. We restrict in our modeling the value of *threshold_buffer*: it should be less than or equal to *buffer_size* and greater than or equal to half of the *buffer_size* (**@axm18**). The exact value of threshold is a parameter of modeling.

5.4.2 The dynamic part

We define two new variables in this refinement step. The *congestion_flag* variable is a function denoting the congestion level of every buffer of a node (**@inv17**). The *congestion_flag* for every buffer of a node should be set if the total number of used cells in the buffer is greater than the (**@inv18**); otherwise the *congestion_flag* of the buffer should be reset (**@inv19**). The *voltage* variable denotes the voltage level of every node (**@inv20**). If more than half of the node buffers have passed the threshold, then the voltage level of the node should be set to *high* (**@inv21**); otherwise the voltage level should be set to *low* (**@inv22**). This is modeled via the invariants to preserve throughout the model. In other words, by proving this invariant we guarantee that the system model is correct and corresponds to the requirements of the system design. The definition of the variables and the invariants related to voltage modeling are shown in Figure 12.25.

SETS $VOLTAGE$
CONSTANTS $low \quad high \quad threshold\_buffer$
AXIOMS
  @axm17 $partition(VOLTAGE, \{low\}, \{high\})$
  @axm18 $threshold\_buffer \in [buffer\_size \div 2] \ .. \ buffer\_size$

FIGURE 12.24

M_3: Static part.

VARIABLES
  $congestion\_flag$ $voltage$
INVARIANTS
  @inv17 $congestion\_flag \in (ROUTER \times PORTS) \to 0\,..\,1$
  @inv18 $\forall r, b \cdot r \mapsto b \in ran(buffer\_content) \wedge$
    $card(buffer\_content \rhd \{r \mapsto b\}) \geq threshold\_buffer \Rightarrow congestion\_flag(r \mapsto b) = 1$
  @inv19 $\forall r, b \cdot r \mapsto b \in ran(buffer\_content) \wedge$
    $card(buffer\_content \rhd (\{r \mapsto b\})) < threshold\_buffer \Rightarrow congestion\_flag(r \mapsto b) = 0$
  @inv20 $voltage \in ROUTER \to VOLTAGE$
  @inv21 $\forall r \cdot card((\{r\} \times PORTS) \lhd congestion\_flag \rhd \{1\}) > (card(PORTS)/2)$
  $\Leftrightarrow voltage(r) = high$
  @inv22 $\forall r \cdot r \in ROUTER \wedge$
   $card((\{r\} \times PORTS) \lhd congestion\_flag \rhd \{1\}) \leq (card(PORTS)/2) \Leftrightarrow voltage(r) = low$

FIGURE 12.25

M_3: Invariants (congestion and voltage).

event $out\_to\_channel$ **extends** $out\_to\_channel$
Any
  CF_out_p vol
Where
  @grd7 $card(buffer\_content \rhd \{moving\_pool(current\_msg) \mapsto out\_p\}) - 1 \geq$
    $threshold\_buffer \Rightarrow CF\_out\_p = 1$
  @grd8 $card(buffer\_content \rhd \{moving\_pool(current\_msg) \mapsto out\_p\}) - 1 <$
    $threshold\_buffer \Rightarrow CF\_out\_p = 0$
  @grd9 $card((\{moving\_pool(current\_msg)\} \times (PORTS \setminus \{out\_p\})) \lhd congestion\_flag \rhd \{1\})$
    $+CF\_out\_p < (card(PORTS)/2) \Rightarrow vol = low$
  @grd10 $card((\{moving\_pool(current\_msg)\} \times (PORTS \setminus \{out\_p\})) \lhd congestion\_flag \rhd \{1\})$
    $+CF\_out\_p \geq (card(PORTS)/2) \Rightarrow vol = high$
Then
  @act4 $congestion\_flag(moving\_pool(current\_msg) \mapsto out\_p) := CF\_out\_p$
  @act5 $voltage(moving\_pool(current\_msg)) := vol$
End

FIGURE 12.26

M_3: out_to_channel event.

Initially, the congestion flags of all the buffers are set to *0*, because the buffers are empty; accordingly, the voltage of all the nodes is set to *low*. Now, in each event of the model, we update the congestion flag of involved buffers by comparing the number of occupied cells in the buffer with *threshold_buffer*. Moreover, we update the voltage level of involved nodes by calculating the congestion level in nodes.

In order to show these updates in events, we show the refinement of the *out_to_channel* event in Figure 12.26. We add two local parameters: *CF_out_p* and *vol* to set the value of congestion flag of *out_p* buffer and the voltage level of the corresponding node. The guards **@grd7** and **@grd8** express comparison of the occupied cells of the buffer after removing a node from it with *threshold_buffer* and updates the value of *CF_out_p* that replaces in *congestion_flag* (**@act4**). If the occupied cells of the buffer have passed the *threshold_buffer* value, the congestion flag of the buffer is set, i.e., the *CF_out_p* is set to *1*. If the occupied cells of the buffer have not passed the *threshold_buffer* value, the congestion flag of the buffer is reset, i.e., the *CF_out_p* is set to *0*. According the congestion flags of a node, the voltage level of the node (**@grd9-10**) is adjusted (**@act5**).

6 REUSABILITY AND EXTENSIBILITY OF OUR MODELS

To demonstrate the reusability feature of our models, in this section we extend both M_2 and M_3 described in Section 5, in order to specify two concrete routing algorithms. The NoC communication definition is respected by both algorithms, as they correctly refine (reuse and extend) the previously proven models M_2 and M_3.

Both these algorithms provide an optimal routing and therefore in these models the previously anticipated event becomes convergent. This is because we describe as variant the distance from the current position of the switched message to its destination that decreases in each switching step.

6.1 THE DETERMINISTIC XYZ ROUTING

In this section, we formally model a dimension-order routing (DOR) algorithm, which is a deterministic routing scheme widely used for NoC [27]. To make the best use of the regularity of the topology, the dimension-order routing transfers packets along minimal paths in the traversing of the low dimension first until no further move is needed in this dimension. Then, they go along the next dimension, and so forth, until they reach their destination. For example, the dimension-order routing in the 3D NoC called the XYZ *routing algorithm* uses Z dimension channels after using Y and X dimension channels. Packets travel along the X dimension, then along the Y dimension and finally along the Z dimension. Thus, if $current_{node} = (c_x, c_y, c_z)$ is a node containing a message addressed to the node $destination = (d_x, d_y, d_z)$, then the XYZ routing function $R_{xyz}(,)$ is defined as follows:

$$R_{xyz}((c_x, c_y, c_z), (d_x, d_y, d_z)) = \begin{cases} (c_{x-1}, c_y, c_z) & \text{iff} \quad cx > d_x \\ (c_{x+1}, c_y, c_z) & \text{iff} \quad cx < d_x \\ (c_x, c_{y-1}, c_z) & \text{iff} \quad c_x = d_x \wedge c_y > d_y \\ (c_x, c_{y+1}, c_z) & \text{iff} \quad c_x = d_x \wedge c_y < d_y \\ (c_x, c_y, c_{z-1}) & \text{iff} \quad c_x = d_x \wedge c_y = d_y \wedge c_z > d_z \\ (c_x, c_y, c_{z+1}) & \text{iff} \quad c_x = d_x \wedge c_y > d_y \wedge c_z > d_z \end{cases}$$

We model the XYZ routing algorithm based on the model M_2; for this, we have to refine the routing event which is nondeterministically defined in M_2. As shown in the above formula, a message can be transferred to six different directions based on its current position and its destination. Therefore, we refine the *routing* event in the previous model to six *routing* events so that their guards are based on the routing formula. As an example of the *routing* event, we show in Figure 12.27 the situation where c_x is greater than d_x (**@grd5**): the message is routed to the $x - 1$ coordinate, i.e., to the west ($out\_p = Wout$). All the correctness properties defined for the abstract models are proved. Hence, the XYZ routing algorithm guarantees the overall communication correctness.

```
event  routing_X_dec extends  routing
Where
  @grd5 coordX(router) > coordX(des(msg)) ∧ out_p = Wout
Then
  @act1 buffer_content(msg) := router ↦ out_p
End
```

FIGURE 12.27

The XYZ Model: routing event ($c_x > d_x$).

6.2 THE CONGESTION-AWARE XYZ ROUTING

We now apply the detection of congestion in a buffer in a different manner than in M_3: we model an adaptive routing path selection that balances traffic load distribution and alleviates congestion caused by heavy NoC traffic. We achieve this by specifying an adaptive XYZ routing algorithm, which switches between deterministic routing at low congestion rates and dynamic routing when the network congestion increases. In this congestion-aware routing algorithm, routers make decisions locally by monitoring the congestion status in each direction to the neighboring routers. Moreover, the deadlock-free feature of XYZ routing is still incorporated by limiting a packet to traverse the network only by following one of the shortest paths between the source and destination.

We model the following algorithm. The destination of a message is compared with the current router. If the destination is equal with the current router, the message is routed to the local output buffer. Otherwise, if the destination of the message has the same x and y (or x and z, or y and z respectively) coordinate address as the current router, the message should be routed to the neighboring router on the z-axis (or y-axis, or x-axis, respectively) towards the destination. Thus, the message in an input buffer of the current router is routed to z (or y, or x, respectively) output buffers of the current router. If the destination of the message has the same x (or y, or z, respectively) coordinate address as the current router, the message should be routed to the neighboring router with the smallest congestion level on the y or z axes, (or with the smallest congestion level on the x or z axes, or with the smallest congestion level on the x or y axes, respectively) towards the destination. This is modeled by routing the message toward the y or z (or x or z, or x or y, respectively) output buffers of the current node. Else, the congestion level of the output buffer of current router towards the destination is checked and the message is sent to the output with the smallest congestion level.

We show in Figure 12.28 the situation when the west output buffer is chosen for routing the message in a router. Instrumental in modeling is the constant *candidate_output*, defined in the static part of the model. This constant function takes as arguments the address of a router and the destination address of a message and returns the possible directions that the message could be routed on, when at that router. The *routing_X_dec* event is enabled in three different situations modeled

```
event routing_X_dec extends routing
Where
  @grd5 coordX(router) > coordX(des(msg))
  @grd6 out_p = Wout
  @grd7 (congestion_flag(router ↦ Wout) = 0)∨
    (∀p·p ∈ candidate_output(router ↦ des(msg))⇒
    card(buffer_content ▷ {router ↦ Wout}) ≤ card(buffer_content ▷ {router ↦ p}))
    ∨(coordY(router) = coordY(des(msg)) ∧ coordZ(router) = coordZ(des(msg)))
Then
  @act1 buffer_content(msg) := router ↦ out_p
End
```

FIGURE 12.28

The congestion-aware XYZ model:routing event ($c_x > d_x$).

by the **@grd7** of the event. The guard consists of the disjunction of the following three different situations. The first disjunction models the situation with no congestion in the west output buffer and c_x greater than d_x. The second disjunction models the situation when there is congestion in the west output buffer, but still the congestion level in this output buffer is smaller than the congestion levels of other possible outputs, returned by *candidate_output*. The third disjunction models the situation when the y and z coordinates of the destination are the same as those of the current node; hence, the message can only be routed on the coordinate despite the congestion level in the west output buffer.

7 VERIFICATION OF MODELS

In order to prove the correctness of our modeling, we have to discharge two types of proof obligations, as detailed in Section 3. First, we need to discharge properties related to the semantics of each model, such as (FIS), (INV), (INIT), etc. Second, we need to discharge properties related to the refinement relation between various models, such as (REF_FIS), (REF_GRD), (REF_SIM), etc. Some of these properties are automatically discharged by the supporting Rodin platform tool, while others need to be discharged by supplying the tool with various hypotheses in order to carry the proving out. The latter category of proofs is referred to as *interactively discharged.*

The proof statistics for our models are shown in Table 12.1. These figures express the number of proof obligations generated by the Rodin platform tool as well as the number of obligations automatically discharged by the platform and those interactively proved. We note that these proof statistics concern the finished models discussed in the paper. However, the process of trying to discharge proof obligations on earlier versions of the model led to various developments that in some cases meant the proofs were afterwards completed automatically. This is one of the biggest advantages of the Event-B development with the Rodin platform: the integrated approach for modeling and proving in an interleaving

Table 12.1 Proof Statistics

Model	Number of Proof Obligations	Automatically Discharged	Interactively Discharged
Context	21	6(28%)	15(72%)
The M_0 Model	38	34(89%)	4(11%)
The M_1 Model	33	11(33%)	22(67%)
The M_2 Model	33	7(21%)	26(79%)
The M_3 Model	80	62(%)	18(%)
Deterministic XYZ Model	13	13(100%)	0(0%)
Congestion-aware XYZ Model	13	10(%)	3(%)
Total	231	143	88

manner. Upon conceiving a model, we are immediately shown which proofs are not obvious to discharge, thus prompting one to reconsider the models.

All the interactive proofs we were left to discharge are in fact concerned with several proving technicalities specific to the current version of the Rodin platform tool platform. A high number of interactive proofs were due to reasoning about set comprehension and unions, not currently supported automatically in Rodin. In addition, the interactive proving often involves manually suggesting values to discharging various properties containing logical disjunctions or existential quantifiers. Extra proving was due to the fact that we cannot currently create proof scripts and reuse them whenever needed in Rodin. Thus, in some cases we had to manually repeat very similar or almost identical proofs.

8 RELATED WORK

Three-dimensional Network-on-Chip (3D NoC) architectures [19] provide more reliable interconnections than two-dimensional NoC due to the increased number of links between components. Due to their promise of parallelism and efficiency, 3D NoC have a critical role in leading towards reliable computing platforms. However, the majority of their evaluation approaches are simulation-based tools, such as XMulator [28], Noxim [29], etc. Simulation-based approaches are usually applied in the late stages of design and are limited, e.g., by the length of time that a system is simulated. This means that exhaustive checking of all the system states is impossible in practice for complex 3D NoC and thus, simulation is not suitable for verifying the correctness of a NoC design. In this paper we handle a proving approach in order to address the verification problem for a NoC design.

The high complexity of NoC architectures is a trade-off to be handled in exchange for them being able to perform faster and with an increased parallelism level for both communication and computation. Formal techniques are one increasingly important candidate in addressing the complexity problems, essentially via abstraction. By applying formal techniques to designing NoC, errors can be detected and corrected during the design and the resulting models are proven to be correct with respect to the specification. We have applied the formal technique of refinement in modeling several other architectures based on communication networks [30], wireless sensor networks [31], Network-on-Chip [32,33], video decoding on multiple cores [34], peer-to-peer networks [35], etc. In all of these correct-by-construction developments, a system is initially modeled abstractly in order to reason about correctness properties and gradually more details of the system are added to the model in such a way that a more concrete model preserves the correctness properties of the system.

Our approach to exploring NoC properties is based on a theorem prover. A formal model and development flow for GALS systems is proposed via Petri nets in [36]. An alternative formal approach used for exploring NoC properties is that of model checking. In model checking techniques, a specific design for a finite state space system is specified and verified. In [37], the authors use the SMV model checker to verify design invariants of a SoC bus protocol. In [38,39], model-checking tools are applied to asynchronous hardware designs. In contrast to model checkers, theorem provers are not restricted to a specific architecture and they can be used for modeling and verifying infinite state space. Theorem provers have been used for developing general models for specifying and verifying specific designs. In [40], the authors propose a general formal model of two-dimensional NoC based on the ACL2 theorem prover, in order to verify a specific design. A general formal model of asynchrony in the Boyer-Moore logic is presented in [41]. All these approaches do not consider nonfunctional properties.

Much of the research concerning the 3D NoC design is concentrated on various bottom-up approaches, such as the study of routing algorithms [42] or the design of dedicated 3D NoC architectures [43] where parameters such as hop count or power consumption are improved. Here we are concerned with a reverse, top-down approach where we start from simple models and add complexity later. There are already research results regarding the detection of faults as well as debugging in the early stages of NoC design. A generic model for specifying and verifying NoC systems is presented in Borrione et al. [40], where the formal verification is addressed with the ACL2 theorem prover, a mechanized proof tool. This tool produces a set of proof obligations that should be discharged for particular NoC instances, but does not handle refinement and thus cannot propose a reliable and reusable approach. Another formal approach to the development of the NoC systems employing the B-action systems formalism has been described in [44], where the focus is on the formal specification of communication routers. A framework for modeling 2D-NoC systems by composing more advanced routing components out of simpler ones is proposed there.

One proposal for congestion-awareness in NoC is overviewed in [45], where a deflective routing algorithm is employed. In order to avoid congestion in hot spots (such as in the center of a two-dimensional NoC mesh), the study of all the FIFO buffers in the network mesh is described. Our proposal offers a distributed approach, where individual nodes detect and attempt to resolve congestion. In [46], the authors discuss an optimization of the best-effort communication services in NoC and propose an algorithm to solve the optimization problem. This algorithm can then be implemented by a centralized controller, again a different approach with respect to our proposal. In [47], the authors consider fairness in NoC routing and propose a control scheme based on a mathematical model and simulation. As we base our modeling on Event-B and its nondeterministic execution model, we do not attempt to address fairness in this paper.

Some other interesting approaches to address the modeling and verification of NoC properties are provided in [48,49]. In [48], the authors propose an extension of a functional formalization of on-chip communications [50], referred to as GeNoC, for modeling various features of Nostrum, a two-dimensional NoC architecture. They prove various communication properties in this framework. In [49], runtime verification is proposed for ensuring the functional correctness of NoC communications. The idea is to have a separate, lightweight checker that informs destination nodes of incoming packets ahead of time. If these packets do not arrive, an error is signaled and the packets are safely delivered via the checker network. Although centralized, this approach addresses the self-treatment of errors and thus provides an interesting alternative to modeling and verifying NoC communications.

9 RESULTS

The communication model proposed by the NoC architecture is at the basis of our technological world; hence understanding it well is essential. One question we ask in this paper is how to determine if a NoC model functions correctly. For this, we have proposed a NoC functioning definition and then proved that several increasingly more detailed NoC models respect this property. We have handled proving via a mathematical based formalism and its associated methodology. Due to the required expertise for this as well as the advent of manycore architectures making the NoC models ever more complex, we can extremely benefit from a reusable approach to model NoC correctness. This is a problematic research gap in the literature, namely correctness and extensibility of hardware platform verification.

Our approach in this paper addresses this gap by proposing several NoC models for on-chip communication, at different levels of abstraction. Our main design tools are abstraction and refinement, as they are suitable in managing complexity. We also validate our models by applying them to derive algorithms.

In this chapter we have proposed an approach to be seen as a proof-of-concept for demonstrating reusable NoC correctness. Obviously, more complex models than those that can be illustrated in a paper are needed in the real world. However, we expect that tool support such as the Rodin platform tool will be a considerable aid in applying our approach to construct complex models as well as to construct NoC models that verify different communication properties. Refinement patterns can also be defined in order to further increase the reusability of such an approach, an example of which having been proposed, for instance in [31] to develop wireless sensor-actor networks.

10 CONCLUSIONS

In this paper, we have proposed a correct-by-construction formalization of NoC architectures that we employ in modeling and proving various NoC properties. The reusability of the NoC models (expressed using a special data structure called *pool* [22]) is guaranteed. Moreover, we have modeled two aspects of congestion-awareness. Upon detecting buffer congestion, a node self-updates its voltage level, thus performing operations faster and consequently freeing the buffers faster. We note that, when congestion disappears, the node self-updates the voltage as well, switching to lower voltage and thus performing in a more energy-efficient manner. Moreover, we have shown how to adapt a routing algorithm so that it attempts to route messages on noncongested routes. All these properties are ensured via invariants. In order for the invariant to be satisfied by a model, a number of proof obligations need to be discharged. Moreover, in order for the models to respect the refinement relation $\sqsubseteq$, i.e., to develop each other in a provably correct manner, some other proof obligations need to be generated. As we have employed the Rodin platform tool to specify our 3D NoC modeling, many of these proof obligations have been automatically discharged, while for the rest it was possible to discharge them interactively. We note an interesting property of our proposed NoC functioning definition that essentially reduces to the fact that all the messages will eventually reach their destinations. This is a typical liveness property that we model here as an invariant, also based on variant expressions ensuring that our models will eventually terminate. The liveness property can also be verified via a model checker, for instance Pro-B [51], which is associated with the Rodin platform tool.

The NoC communication can be either unicast or multicast [52]. In the unicast communication a message is sent from a source node to a single destination node, while in the multicast communication a message is sent from a source node to an arbitrary set of destination nodes. We have considered here sending a message from a source to a single destination, hence modeled unicast communication. We have also extended our modeling to specify multicast communication [32].

All the refinements of our proposed models are superposition-based: we add data and events or strengthen the guards of and add actions to existing events so that the original behavior is not overtaken or influenced by the new behavior. An interesting property of our multicast modeling [32] is that the concrete algorithm employs an algorithmic refinement instead of a superposition one like in this paper. This is because we can have several messages that could be routed in parallel to several destinations using different events via several channels. Our XYZ routing in this paper employs already several events for the routing instead of the abstract routing event of the model M_2, but only one of them is enabled at all moments.

By strengthening the invariants we can verify more diverse properties of the 3D NoC designs, for instance we could prove deadlock-freedom for routing algorithms-currently, one of the most challenging properties for the 3D NoC. For this, we envision an extension of the abstract 3D NoC models with an extra channel dependency graph to reason about deadlock-freedom. We also plan to investigate the optimality of our proposed modeling, for instance via simulation, and compare it to other approaches.

REFERENCES

[1] Bjerregaard T, Mahadevan S. A survey of research and practices of network-on-chip. ACM Computing Survey; 2006.

[2] Abrial J-R. Modeling in Event-B: system and software engineering. New York: Cambridge University Press; 2010.

[3] Rodin Tool Platform. Event-B.org. tool platform. Available from: http://www.event-b.org/platform.html>; 2014.

[4] Ishtiaq SS, O'Hearn PW. Bi as an assertion language for mutable data structures. POPL 2001;14–26.

[5] Calcagno C, Distefano D. Infer: an automatic program verifier for memory safety of c programs. In: Bobaru MG, Havelund K, Holzmann GJ, Joshi R, editors. NASA Formal Methods – Third International Symposium, NFM 2011, Pasadena, CA; April 18–20, 2011. Proceedings; 2011. p. 459–65.

[6] Harrison J. Formal verification at intel. Logic in computer science, symposium on; 0:45; 2003.

[7] Kaivola R, Ghughal R, Narasimhan N, Telfer A, Whittemore J, Pandav S, et al. Replacing testing with formal verification in intel coretm i7 processor execution engine validation. In: Proceedings of the 21st international conference on computer aided verification, CAV '09, Berlin, Heidelberg, Springer-Verlag; 2009. p. 414–29.

[8] Liao WS, Hsiung PA. Creating a formal verification platform for ibm coreconnect-based soc. In: Proceedings of the 1st international workshop on automasted technology for verificatin and analysis (ATVA2003); 2003. p. 7–18.

[9] Le Guernic P, Gupta R, Skuhla SK. Formal methods and models for system design: a system level perspective (The Kluwer international series in video computing). 1st ed. Springer; 2004.

[10] Abrial J-R. A system development process with Event-B and the Rodin platform. Formal Methods and Software Engineering 2007;1–3.

[11] Abrial J-R, Cansell D, Méry D. Refinement and reachability in Event-B. ZB'05. Springer; 2005. p. 222–41.

[12] Abrial J-R, Hallerstede S. Refinement, decomposition, and instantiation of discrete models: application to event-B. Fundam Inf 2007;77:1–28.

[13] Shin D. Power-aware communication optimization for networks-on-chips with voltage scalable links. In: Proceedings CODES + ISSS'04; 2004. p. 170–75.
[14] Pasricha S, Dutt N. On-Chip communication architectures system on chip interconnect. Morgan Kaufmann; 2008.
[15] Shacham AA. Photonic networks-on-chip for future generations of chip multiprocessors. IEEE Trans Comput 2008;1246–60.
[16] Wang C, Hu W-H, Bagherzadeh N. A wireless network-on-chip design for multicore platforms. PDP'11. IEEE; 2011. p. 409–16.
[17] Chang MF, Cong J, Kaplan A, Naik M, Reinman G, Socher E&-W. CMP network-on-chip overlaid with multi-band RF-interconnect. HPCA. IEEE Computer Society; 2008. p. 191–202.
[18] Kempa KA. Carbon nanotubes as optical antennae. Adv Mater 2007;421–6.
[19] Feero BS, Pande PP. Networks-on-chip in a three-dimensional environment: a performance evaluation. IEEE Trans Comput 2009;58:32–45.
[20] Rodin. Rigorous open development environment for complex systems. IST FP& STREP project. Documentation at: http://rodin.cs.ncl.ac.uk/; 2011. (accessed 26 october 2011).
[21] Abrial J-R, Butler M, Hallerstede S, Hoang TS, Mehta F, Voisin L. Rodin: an open toolset for modelling and reasoning in Event-B. Int J Softw Tools Technol Trans 2010;447–66.
[22] Abrial J-R. The B-book: assigning programs to meanings. New York: Cambridge University Press; 1996.
[23] Back RJR, Sere K. Superposition refinement of reactive systems. Formal Aspects Comput 1993;8(3):324–46.
[24] Katz S. A superimposition control construct for distributed systems. ACM Trans Program Lang Syst 1993;15:337–56.
[25] Back R-J, Sere K. Stepwise refinement of action systems. In: van de Snepscheut JLA, editor. Mathematics of program construction, 375th Anniversary of the Groningen University, International conference, Groningen, The Netherlands, June 26–30, 1989. Proceedings, volume 375 of lecture notes in computer science. Springer; 1989. p. 115–38.
[26] Grecu C, Pande PP, Ivanov A, Saleh R. A scalable communication-centric soc interconnect architecture. In: Proceedings of the 5th international symposium on quality electronic design, ISQED'04, Washington, DC, IEEE Computer Society; 2004. p. 343–48.
[27] Nayebi A, Meraji S, Shamaei A, and Sarbazi-Azad H. Xmulator: a listener-based integrated simulation platform for interconnection networks. In: Asia international conference on modelling and simulation; 2007. p. 128–32.
[28] Montanana JM, Koibuchi M, Matsutani H, Amano H. Balanced dimension-order routing for k-ary n-cubes. In: Proceedings of the 2009 international conference on parallel processing workshops, ICPPW'09, Washington, DC, IEEE Computer Society; 2009. p. 499–06.
[29] Palesi M, Holsmark R, Kumar S, Catania V. Application specific routing algorithms for networks on chip. IEEE Trans Parallel Distributed Syst 2009;20:316–30.
[30] Kamali M, Petre L, Sere K, Daneshtalab M. Correcomm: a formal hierarchical framework for communication designs. In: 2nd IEEE International conference on embedded systems for enterprise applications. IEEE; 2011.
[31] Kamali M, Laibinis L, Petre L, Sere K. Formal development of wireless sensor–actor networks. Sci Comput Program 2014;25–49.
[32] Kamali M, Petre L, Sere K, Daneshtalab M. Formal modeling of multicast communication in 3d nocs. In: 14th Euromicro conference on digital system design. IEEE; 2011.

[33] Kamali M, Petre L, Sere K, Daneshtalab M. Refinement-based modeling of 3d nocs. In: 4th IPM International conference on fundamentals of software engineering. LNCS; 2011.

[34] Sandvik P, Sere K. Formal analysis and verification of peer-to-peer node behaviour. In: AP2PS 2011, Proceedings of the third international conference on advances in P2P systems; 2011. p. 47–52.

[35] Lumme K, Petre L, Sandvik P, Sere K. A formal approach to h.264 video decoding on multicore systems. In: International journal of critical computer-based systems (IJCCBS). Submitted; 2011.

[36] Moutinho F, Gomes L, Barbosa P, Barros J, Ramalho F, Figueiredo J, et al. Petri net based specification and verification of globally-asynchronous-locally-synchronous system. In: Camarinha-Matos L, editor. Technological Innovation for Sustainability, volume 349 of IFIP Advances in Information and Communication Technology. Boston: Springer; 2011. p. 237–45.

[37] Roychoudhury A, Mitra T, Karri SR. Using formal techniques to debug the AMBA system-on-chip bus protocol. Design, automation and test in europe conference and exhibition; 2003. p. 828–33.

[38] Wang X, Kwiatkowska M. On process-algebraic verification of asynchronous circuits. Fundam Informaticae 2007;80(1–3):283–310.

[39] Kapoor HK, Josephs MB. Modelling and verification of delay-insensitive circuits using CCS and the concurrency workbench. Inform Process Lett 2004;293–6.

[40] Borrione D, Helmy A, Pierre L, Schmaltz J. A formal approach to the verification of networks on chip. EURASIP J Embedded Syst 2009;2(1–2):14.

[41] Moore JS. A formal model of asynchronous communication and its use in mechanically verifying a biphase mark protocol. Formal Aspects Comput 1993;6:60–91.

[42] Andreasson D, Kumar S. Slack-time aware routing in NoC systems. In: IEEE international symposium on circuits and systems; 23–26 May, 2005. p. 2353–6.

[43] Yan S, Lin B. Design of application-specific 3D networks-on-chip architectures. IEEE; 2008 (0407):142–49.

[44] Tsiopoulos L, Waldén M. Formal development of NoC systems in B. Nord J Comput 2006;13:127–45.

[45] Nilsson E, Millberg M, Oberg J, Jantsch A. Load distribution with the proximity congestion awareness in a network on chip. In: Proceedings of the conference on design, automation and test in europe – vol. 1, DATE'03. IEEE Computer Society; 2003.

[46] Talebi MS, Jafari F, Khonsari A. A novel flow control scheme for best effort traffic in NoC based on source rate utility maximization. IEEE 15th international symposium on modeling, analysis, and simulation of computer and telecommunication systems. MASCOTS '07. IEEE Computer Society; 2007. p. 381–86.

[47] Jafari F, Yaghmaee MH. A novel flow control scheme for best effort traffics in networkon-chip based on weighted max-min-fairness. International symposium on telecommunications. IST 2008. IEEE Computer Society; 2008. p. 458–63.

[48] Helmy A, Pierre L, Jantsch A. Theorem proving techniques for the formal verification of NoC communications with non-minimal adaptive routing. IEEE 13th international symposium on design and diagnostics of electronic circuits and systems (DDECS), 2010. IEEE Computer Society; 2010. p. 221–24.

[49] Parikh R, Bertacco V. Formally enhanced runtime verification to ensure NoC functional correctness. In: 44th Annual IEEE/ACM international symposium on microarchitecture. ACM; 2011.

[50] Schmaltz J, Borrione D. A functional formalization of on chip communications. Formal Aspects Comput 2008;20(3):241–58.

[51] ProB Model Checker [online]. Available at: <http://www.stups.uniduesseldorf.de/ProB/overview.php> [accessed 26.10.11.].

[52] Lu Z, Yin B, Jantsch A. Connection-oriented multicasting in wormhole-switched networks on chip. In: Proceedings of the IEEE computer society annual symposium on emerging VLSI technologies and architectures, Washington, DC, IEEE Computer Society; 2006. p. 205.

FURTHER READING

Ebrahimi M, Daneshtalab M, Liljeberg P, Tenhunen H. Hamum - a novel routing protocol for unicast and multicast traffic in mpsocs. In: Danelutto M, Bourgeois J, Gross T, editors. PDP. IEEE Computer Society; 2010. p. 525–32.

Gerhart S, Craigen D, Ralston T. Case study: Paris Metro signaling system. January 1994;11(1):32–35.

Kim YB, Kim Y-B. Fault tolerant source routing for network-on-chip. In: Proceedings of the 22nd IEEE international symposium on defect and fault-tolerance in VLSI systems, Washington, DC, IEEE Computer Society; 2007. p. 12–20.

Loi I., Benini L. An efficient distributed memory interface for many-core platform with 3d stacked dram. In: Proceedings of the conference on design, automation and test in europe, DATE'10, 3001 Leuven, Belgium. European Design and Automation Association; 2010. p. 99–104.

Park D, Eachempati S, Das R, Mishra AK, Xie Y, Vijaykrishnan N, et al. Mira: a multi-layered on-chip interconnect router architecture. SIGARCH Comput Archit News 2008;36:251–61.

CHAPTER

Markov chain models and applications

13

Kishor S. Trivedi[1], **Kalyanaraman Vaidyanathan**[2], **and Dharmaraja Selvamuthu**[3]

[1]*Duke University, Durham, NC, USA*

[2]*Oracle Corporation, San Diego, CA, USA*

[3]*Indian Institute of Technology Delhi, New Delhi, India*

1 INTRODUCTION

Modeling is a fundamental aspect in the design process of a complex system, as it allows the designer to compare different architectural choices as well as to predict the behavior of the system under varying input traffic, service, fault and prevention parameters. Depending on the final target of the analysis and on the complexity of the system, different modeling techniques can be adopted. When a more detailed description of the system is required, Markov models can be adopted as a very powerful modeling framework. The analytical tractability of Markov models is based on the exponential assumption of the distribution of the holding time in a given state. This implies that the future evolution of the system depends only on the current state and, based on this assumption, simple and tractable equations can be derived for both transient and steady-state analysis. The hypothesis of exponential distribution allows the definition of models which can give a more qualitative rather than quantitative analysis of real systems [1].

The main purpose of this chapter is to review the existing importance measures and their interrelationships, and to introduce novel techniques on how to obtain these measures using reward rate functions in a Markov reward model. In this chapter, we discuss the various analytical modeling paradigms for quantitative analysis of various preventive maintenance techniques. Further, the features of the Markov regenerative process (MRGP) for preventive maintenance analysis of systems are also explained. The theoretical framework is described, along with an in-depth analysis of solution techniques. Computational methods for systems with deterministically and exponentially distributed event times are discussed in particular and some interesting applications are presented.

2 STRENGTHS OF MARKOV MODELS

The essential necessity for a stochastic process to be a homogeneous continuous time Markov chain (CTMC) is that the sojourn time in each state must be exponentially distributed. However, the sojourn times in each state may not follow exponential distribution while modeling practical or real-time situations. The existence of the non-exponentially distributed event time gives rise to non-Markovian models. A non-Markovian model can be modeled using phase-type approximation. However, phase-type expansion increases the already large state-space of a real system model. The problem becomes really severe when mixing deterministic times with exponential ones. The strict Markovian constraints are relaxed by using Markov regenerative processes (MRGP). A generalization of CTMC where the time spent by the process in a given state is allowed to follow non-exponential (general) distribution is a semi-Markov process (SMP). Further generalization is provided by MRGP. This concept is used in extending the CTMC model by allowing general distribution for all the event times other than failure times in the given examples. As a result the stochastic process under consideration becomes MRGP.

A Markov renewal process becomes a Markov process when the transition times are independent exponential and are independent of the next state visited. It becomes a Markov chain when the transition times are all identically equal to 1. It reduces to a renewal process if there is only one state and then only transition time becomes relevant. Renewal theory is used to analyze stochastic processes that regenerate themselves from time-to-time. The long-run behavior of a regenerative stochastic process can be studied in terms of its behavior during a single regeneration cycle. Semi-Markov processes are used in the study of certain queuing systems.

Numerous studies have described and reported the occurrence of "software aging" [2–4] in which the state of software degrades with time. This degradation is caused primarily by the exhaustion of operating system resources, data corruption and numerical error accumulation. If untreated, this may lead to performance degradation of the software or crash/hang failure, or both in the long run. Examples of software aging are memory bloating and leaking, unreleased file-locks, data corruption, storage space fragmentation and accumulation of round-off errors [3,4]. Aging has not only been observed in software used on a mass scale but also in specialized software used in high-availability and safety-critical applications [2,5]. To counteract software aging, a preventive maintenance technique called "software rejuvenation" has been proposed [2,6,7], which involves periodically stopping the system, cleaning up, and restarting it from a clean internal state. This "renewal" of software prevents (or at least postpones) a crash failure. The internal state of the software can be cleaned by techniques like garbage collection, flushing operating system kernel tables and reinitializing internal data structures.

Rejuvenation has been implemented in various types of systems, from telecommunication systems [2,8,9], operating systems [10], transaction processing systems [11], web servers [12–14], cluster servers [15–17], cable modem systems [18], spacecraft systems [19], safety-critical systems [5,20], to biomedical applications [21]. Preventive maintenance, however, incurs an overhead (lost transactions, downtime, additional resources, etc.) which should be balanced against the cost incurred due to unexpected outage caused by failure. This in turn demands a quantitative analysis, which in the context of software systems has only recently started to receive attention.

3 ANALYTICAL MODELING TECHNIQUES

Performance and availability attributes of a system can be predicted in several ways: (1) building prototypes and taking measurements, (2) using discrete event simulation to model the system, or (3) constructing an analytical model for evaluating the attributes of the system. Measurement is the direct method for assessing an existing system, but it is not a feasible option during system design and implementation phases. Discrete event simulation techniques are a commonly used modeling approach. Simulation can capture system characteristics to a high degree of fidelity. Many software packages are available to facilitate the construction and solution of discrete event simulation models. However, this method tends to be expensive since it takes a longer time to execute the models, particularly when highly accurate results are required. Analytical modeling, on the other hand, has been proven to be a cost-effective and quite accurate approach for system analysis. A model is an abstraction of a system that includes sufficient details to facilitate the understanding of system behavior. Due to recent developments in model generation and solution techniques, and due to the availability of software packages, large and realistic models can be developed and studied. An analyst can choose from different types of analytical models based on the accessibility, construction, efficiency, and accuracy of solution algorithms, and the availability of suitable software packages.

Analytical models can be broadly classified into non-state space and state space models. Non-state space models, also known as combinatorial models, do not enumerate all possible system states to solve for the required attributes. Some of the commonly used combinatorial models are reliability block diagrams, reliability graphs, fault trees, and attack trees [22]. However, system features, such as dependent behavior, imperfect coverage, and nonzero reconfiguration delays, are not easily captured by these models. These limitations can be overcome by state space models, since they enable us to model the complicated interactions between the components and the trade-offs between different measures of interest. State space models are much more comprehensive. They allow explicit modeling of complex relationships and their transition structure can encode important sequencing information. The most commonly used state space models are Markov chains. They provide great flexibility for modeling performance, reliability, availability, survivability, and security. But the size of the state space of a Markov chain grows much faster than the number of system components, thereby making the model specification difficult and an error-prone process. Some of the techniques that have been used extensively to reduce the size of the models include state truncation methods, fixed-point iteration, and hierarchical models. Another state space modeling formalism is stochastic Petri nets and their extensions that allow for the concise specification and automated generation of the underlying Markov chain [23]. In this research work, models are analyzed by a software package like SHARPE [24,25], which was developed by the researchers at Duke University.

In this section, we present Markov chains and Markov reward models. Later, we see how strict Markovian constraints are relaxed by using Markov regenerative processes. The notations are as follows:

$P_i(t)$	probability of being in the state i at time t
r_i	reward rate assigned to the state i
$V(t)$	transition probability matrix
$K(t)$	global kernel matrix
$E(t)$	local kernel matrix
k	total number of deterioration stages
i	deterioration stage
g	minimal preventive maintenance threshold
b	major preventive maintenance threshold
λ_i	transition rate at stage i
$1/\lambda_{in}$	mean time between inspections (MTBI)
$1/\mu_{in}$	mean time to carry out maintenance inspection of the device
$1/\mu_m$	mean duration of preventive minimal maintenance
$1/\mu_M$	mean duration of preventive major maintenance
$1/\mu_R$	mean time to repair after a deterioration failure
F	deterioration failure state.

3.1 MARKOV MODELS

Consider a stochastic process $\{X\ (t),\ t \geq 0\}$ with state space $\Omega = \{1, 2, \ldots, N\}$. Here, the random variable $X(t)$ represents a state of the system at time t. The stochastic process $\{X\ (t),\ t \geq 0\}$ turns into a homogeneous continuous-time Markov chain (CTMC) with the assumption that the sojourn time in each state is exponentially distributed. In a CTMC, the Markov property is satisfied, i.e., the past history of the process is completely summarized by the current state of the process. Hence, for any $t_0 < t_1 < t_2 < \ldots < t_n < t$,

$$P[X(t) \leq x | X(t_n) = x_n, \ldots, X(t_0) = x_0] = P[X(t) \leq x | X(t_n) = x_n].$$

A CTMC is said to be irreducible if every state is reachable from every other state. Define $Q = [q_{ij}]$ as the infinitesimal generator matrix of the CTMC $\{X(t), t \geq 0\}$, where $q_{ii} = -\sum_{j \neq i} q_{ij}$. Also, define $P_i(t) = P\{X(t) = i\}$ as the unconditional probability of the CTMC being in the state i at time t. Then $P(t) = [P_1(t), P_2(t), \ldots, P_N(t)]$ is the row vector of the time-dependent state probabilities, which is computed by solving the Kolmogorov forward equation [26]:

$$\frac{dP(t)}{dt} = P(t)Q \tag{13.1}$$

given $P(0)$, where $P(0)$ is the initial state probability vector (at time $t = 0$). If the CTMC is irreducible with all states recurrent non-null, then the limit $\pi_j = \lim_{t\to\infty} P_j(t), j \in \Omega$ always exists and the unique steady-state probability vector π, defined as $\pi = [\pi_1, \pi_2, \ldots, \pi_N]$, satisfies Equation 13.2 [26]:

$$\pi Q = 0, \sum_{i\in\Omega} \pi_i = 1. \tag{13.2}$$

Once the state probabilities $P_i(t)$ and π_i have been computed, the measures of interest as the weighted averages of these quantities can be obtained easily. On assigning reward rates to the states of the CTMC, the CTMC turns into a Markov reward model (MRM). Assume that a weight or reward rate r_i is assigned to the state i. The reward rate is assigned on the basis of desired measures. Define $Z(t) = r_{X(t)}$ as the instantaneous reward rate of the MRM at time t. Then the expected instantaneous reward rate at time t is given by [26]:

$$E[Z(t)] = \sum_{i\in\Omega} r_i P_i(t).$$

The expected reward rate in steady-state is given by:

$$E[Z] = \sum_{i\in\Omega} r_i \pi_i.$$

The essential necessity for a stochastic process to be a homogeneous CTMC is that the sojourn time in each state must be exponentially distributed. However, the sojourn times in each state may not follow exponential distribution while modeling the practical situations. A generalization of CTMC where the time spent by the process in a given state is allowed to follow nonexponential (general) distribution is SMP. Further, generalization is provided by MRGP. We have discussed MRGP in detail in this section.

3.2 MARKOV REGENERATIVE PROCESS MODELS

In this section, we provide a brief overview of MRGP theory and the solution techniques. For a more detailed description, the reader is referred to [27].

Consider a stochastic process such that there exist few time points where the process satisfies Markov property. These time points are called regeneration points. In an MRGP, the stochastic evolution between two successive regeneration points depends only on the state at regeneration, and not on the evolution before regeneration. Furthermore, the evolution of the MRGP becomes a probabilistic replica after each regeneration because of the time homogeneity of the embedded Markov renewal process. As a consequence, all memory other than the state is reset at a regeneration point [27]. The concepts of MRGP are explained in the next two definitions.

3.2.1 Definition 1

Markov Renewal Sequence: A sequence of bivariate random variables $\{(Y_n, S_n), n \geq 0\}$ is called a *Markov renewal sequence* if:

(i) $S_0 = 0$, $S_{n+1} \geq S_n$; $Y_n \in \Omega' = \{0, 1, 2, \ldots\}$, where $\Omega' \subseteq \Omega$, and
(ii) for all $n \geq 0$,

$$\begin{aligned} &P\{Y_{n+1} = j, S_{n+1} - S_n \leq t | Y_n = i, S_n, Y_{n-1}, S_{n-1}, \ldots, Y_0, S_0\} \\ &\quad = P\{Y_{n+1} = j, S_{n+1} - S_n \leq t | Y_n = i\} \quad (\textit{Markov Property}) \\ &\quad = P\{Y_1 = j, S_1 \leq t | Y_0 = i\}. \quad (\text{Time Homogeneity}) \end{aligned}$$

3.2.2 Definition 2

Markov Regenerative Process: A stochastic process $\{Z(t), t \geq 0\}$ on Ω is called an MRGP if there exists a Markov renewal sequence $\{(Y_n, S_n), n \geq 0\}$ of random variables such that all conditional finite dimensional distributions of $\{Z(S_n + t), t \geq 0\}$ given $\{Z(u), 0 \leq u \leq S_n, Y_n = i\}$ are the same as those of $\{Z(t), t \geq 0\}$ given $Y_0 = i$, $i \in \Omega' \subseteq \Omega$. Note that the above definition implies that in this case $\{Z(S_n^+), n \geq 0\}$ or $\{Z(S_n^-), n \geq 0\}$ is an *embedded DTMC* or just the *embedded Markov chain* (EMC) in $\{Z(t), t \geq 0\}$, and also that S_n are regeneration points of $\{Z(t), t \geq 0\}$. From this definition, it is obvious that every SMP is an MRGP. In an SMP, every state transition time instant is a regeneration point; however, this is not necessarily true for an MRGP.

To provide an analytical formulation of MRGP, we define $V(t) = [V_{i,j}(t)]$, $K(t) = [K_{i,j}(t)]$ and $E(t) = [E_{i,j}(t)]$ as the following matrix valued functions.

$$\begin{aligned} V_{i,j}(t) &= P\{Z(t) = j | Y_0 = i\} \\ K_{i,j}(t) &= P\{Y_1 = j, S_1 \leq t | Y_0 = i\} \\ E_{i,j}(t) &= P\{Z(t) = j, S_1 > t | Y_0 = i\}. \end{aligned}$$

$V(t)$ is the transition probability matrix and the entry $V_{i,j}(t)$ provides the probability that the stochastic process $Z(t)$ is in state j at time t given that it was in state i at $t = 0$. Thus $V(t)$ captures the transient behavior of the process. The matrix $K(t)$ is called the global kernel matrix and the entry $K_{i,j}(t)$ provides the probability that the system will be in state j at the time of the next regeneration instant which occurs on or before time t given that the system was in state i just after the previous regeneration instant. Finally, the matrix $E(t)$ is called the local kernel since it describes the behavior of the MRGP between two consecutive regeneration time points. The element $E_{i,j}(t)$ of the matrix $E(t)$ is the probability that the system will be in state $j \in \Omega$ at time t and the next regeneration instant occurs after t given that the system was in state $i \in \Omega'$ just after the previous regeneration instant. From the above definitions:

$$\sum_{j \in \Omega'} K_{i,j}(t) + \sum_{j \in \Omega} E_{i,j}(t) = 1, \forall i \in \Omega'.$$

The transient behavior of the MRGP can be studied by solving the following generalized Markov renewal equation (in matrix form) [27]:

$$V(t) = E(t) + K^{*}V(t) \tag{13.3}$$

where $K^{*}V(t)$ is a convolution matrix.

Steady-state Solution: If the DTMC defined at regeneration points $\{Z(S_n), n \geq 0\}$ is finite, aperiodic and irreducible, then its steady-state probability vector ν given by the solution of the linear system $\nu = \nu K(\infty)$, under the condition $\sum_{k \in \Omega'} \nu_k = 1$ can be obtained. Let $\alpha_{i,j}$ be the mean time that the MRGP spends in state j between two successive regeneration instants, given that it started in state i just after the last regeneration.

$$\alpha_{i,j} = E[\text{time in } j \text{ during } (0, S_1) | Y_0 = i] = \int_0^{\infty} E_{i,j}(t)\mathrm{d}t. \tag{13.4}$$

Then the steady-state probabilities of the MRGP are given by [27]

$$\pi_j = \frac{\sum_{k \in \Omega'} \nu_k \alpha_{k,j}}{\sum_{k \in \Omega'} \nu_k \beta_k}, \text{ where } \beta_k = \sum_{l \in \Omega'} \alpha_{k,l}. \tag{13.5}$$

Transient Solution: The coupled integral equation (Equation (13.3)) can be numerically solved in the transform domain. The transient probabilities in Laplace-Stieltjes Transform (LST) domain are obtained as

$$V^{*}(s) = [I - sK^{*}(s)]^{-1}E^{*}(s). \tag{13.6}$$

4 MARKOV MODELING

We discuss two schemes of preventive maintenance: (a) condition-based and (b) time-based [28]. In the former the action taken after each inspection is dependent on the state of the system: either (1) no action, or (2) a minimal maintenance (preventive maintenance with minimal effort and effect) to recover the system to the last good state (i.e., the state in which it was just before failure), or (3) a major maintenance to bring the system to as good as the initial state. In the time-based approach, maintenance is carried out at predetermined time intervals and the system is restored to its initial state [29]. In this section, we present a closed form solution of the Markov chain for condition-based approach including minimal and major maintenance. This is helpful in obtaining an optimal inspection interval to maximize the availability or minimize the operation cost [30].

4.1 SYSTEM WITH THRESHOLD *B* AND *G*, UNDER DEVICE DETERIORATION FAILURE

The CTMC model for the major and minimal maintenance problem is shown in Figure 13.1 [9]. In this figure, state $(i, 0)$ represents the state in which the device is operational but at the ith deterioration stage, $(i, 1)$ represents the state in which

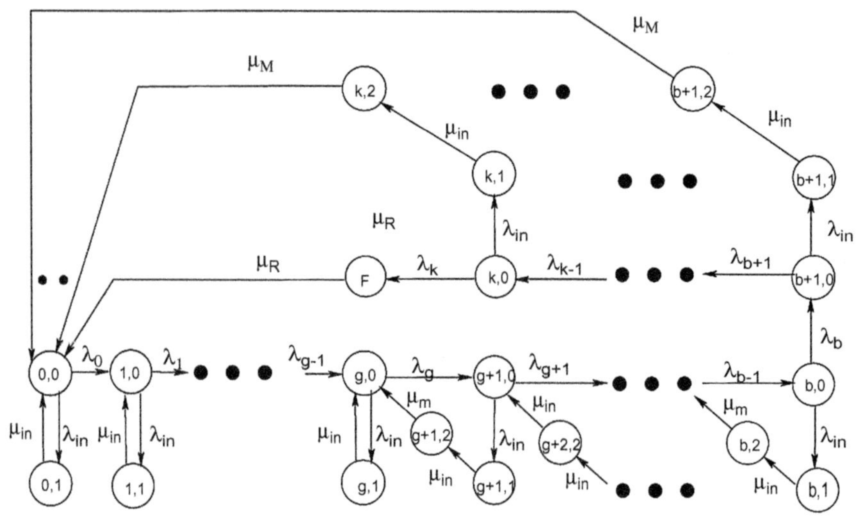

FIGURE 13.1

CTMC model for preventive major and minimal maintenance problem.

the device is in the ith deterioration stage and under inspection, $(i, 2)$ is the state where the device is in the ith deterioration stage and is under maintenance, and state F is the deterioration failure state. The device is inspected after a random period that is exponentially distributed with mean $1/\lambda_{in}$.

By this model, the device experiences no maintenance when the deterioration stage is determined by inspection to be $i \le g$; experiences minimal maintenance if the device is found to be in one of its deterioration stages $(i, 0)$ with $g < i \le b$ by which the device is restored to $(i-1, 0)$ with a mean duration of $1/\mu_m$, and undergoes major maintenance when the deterioration stage is found to be in $(i, 0)$ with $b < i \le k$ by which the device is brought to the state that is as good as new (state $(0, 0)$). When the device is in deterioration failure state F, a major repair is carried out to bring the device to state $(0, 0)$, with a mean duration $1/\mu_R$. Thus, the minimal maintenance threshold is g, and the major maintenance threshold is b.

4.2 STEADY-STATE SOLUTION

By solving the CTMC model in Figure 13.1, the steady-state probability of the system being in state (0, 0) has the following closed form expression [30]:

$$\pi_{0,0} = \left[\begin{array}{l} \sum_{i=0}^{g-1} \left(1+\frac{\lambda_{in}}{\mu_{in}}\right)\frac{\lambda_0}{\lambda_i} + \left(1+\frac{\lambda_{in}}{\mu_{in}}\right) E_g + \sum_{i=g+1}^{b} \left(1+\frac{\lambda_{in}}{\mu_{in}}+\frac{\lambda_{in}}{\mu_m}\right) E_i \\ + \sum_{i=b+1}^{k} \left(1+\frac{\lambda_{in}}{\mu_{in}}+\frac{\lambda_{in}}{\mu_M}\right) \frac{\lambda_0}{\lambda_i+\lambda_{in}} + \frac{\lambda_0 \lambda_k}{\mu_R(\lambda_i+\lambda_{in})} \end{array} \right]^{-1}.$$

where

$$E_i = \frac{\lambda_0 \sum_{j=0}^{b-i} \lambda_{in}^j \prod_{m=0}^{b-i-j-1} \lambda_{b-m}}{\prod_{j=0}^{b-i} \lambda_{b-j}}.$$

Note that when the transition rates at each deterioration stage are the same, say λ, then $E_i = \sum_{j=0}^{b-i} \left(\frac{\lambda_{in}}{\lambda}\right)^j$. Further the system is available when it is in any of the states $(i,0)$. Hence the steady-state availability A_s of the system is given by:

$$A_s = \sum_{i=0}^{k} \pi_{i,0} = \sum_{i=0}^{g-1} \frac{\lambda_0}{\lambda_i} \pi_{0,0} + \sum_{i=g}^{b} E_i \pi_{0,0} + \sum_{i=b+1}^{k} \frac{\lambda_0}{\lambda_i + \lambda_{in}} \pi_{0,0}.$$

The system Mean Time to Failure (MTTF) is obtained by making the system failure state F as absorbing State and is given by [30]:

$$\text{MTTF} = \frac{1}{\lambda_k} \left[\begin{array}{l} \sum_{i=0}^{g-1} \left(1 + \frac{\lambda_{in}}{\mu_{in}}\right) \frac{\lambda_k + \lambda_{in}}{\lambda_i} + \left(1 + \frac{\lambda_{in}}{\mu_{in}}\right) \frac{\lambda_k + \lambda_{in}}{\lambda_0} E_g + \sum_{i=g+1}^{b} \left(1 + \frac{\lambda_{in}}{\mu_{in}} + \frac{\lambda_{in}}{\mu_m}\right) \frac{\lambda_k + \lambda_{in}}{\lambda_0} E_i \\ + \sum_{i=b+1}^{k} \left(1 + \frac{\lambda_{in}}{\mu_{in}} + \frac{\lambda_{in}}{\mu_M}\right) \frac{\lambda_k + \lambda_{in}}{\lambda_i + \lambda_{in}} \end{array} \right]$$

The results can be verified by comparing the closed form solution and the results obtained numerically by SHARPE [23]. The following parameters are chosen for the purpose:

$$k = 4, g = 1, b = 3, \lambda_0 = 0.03, \lambda_1 = 0.03, \lambda_2 = 0.03, = 0.03, \lambda_4 = 0.03,$$
$$\mu_R = 0.025, \mu_{in} = 2, \mu_M = 0.2, \mu_m = 0.5.$$

The variation of steady-state availability with various Mean Time between Inspection (MTBI) is shown in the Figure 13.2(a) and that of system MTTF in Figure 13.2(b). It can be observed from Figure 13.2(b) that the system MTTF is a monotonically decreasing function of MTBI, which indicates that more frequent inspection will result in longer system MTTF. But a too frequent inspection will decrease the system availability as shown in Figure 13.2(a). Therefore a balance between system MTTF and availability should be pursued.

4.3 SYSTEM WITH THRESHOLD *B* AND *G* UNDER BOTH DEVICE RESTORATION FAILURE AND POISSON FAILURE

Now, to the CTMC model in Figure 13.1, one Poisson failure state $(i, 3)$ is added corresponding to each state $(i, 0)$ with failure rate λ_{pf} and repair rate μ_r. Then we have the added balance equations in the steady state for $0 \leq i \leq k$:

$$\mu_r \pi_{i,3} = \lambda_{p,f} \pi_{i,0} \Rightarrow \pi_{i,3} = \frac{\lambda_{p,f}}{\mu_r} \pi_{i,0}.$$

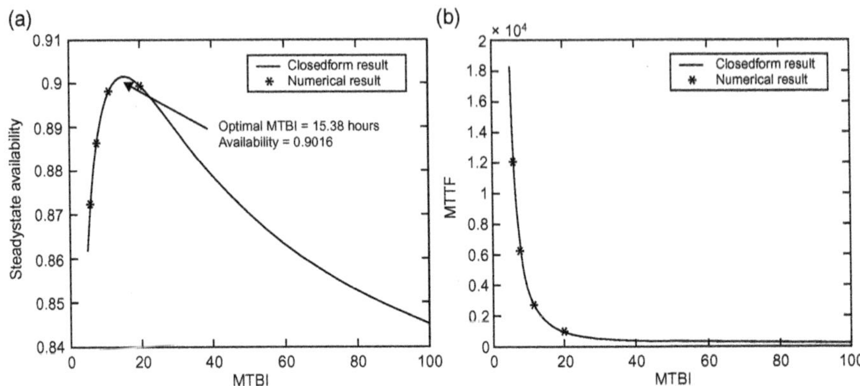

FIGURE 13.2

Steady-state availability MTTF under device restoration failure.

The closed form solution of $\pi_{0,0}$ is then [30]:

$$\pi_{0,0}=\left[\begin{array}{l}\sum_{i=0}^{g-1}\left(1+\frac{\lambda_{pf}}{\mu_r}+\frac{\lambda_{in}}{\mu_{in}}\right)\frac{\lambda_0}{\lambda_i}+\left(1+\frac{\lambda_{pf}}{\mu_r}+\frac{\lambda_{in}}{\mu_{in}}\right)E_g+\sum_{i=g+1}^{b}\left(1+\frac{\lambda_{in}}{\mu_{in}}+\frac{\lambda_{in}}{\mu_m}+\frac{\lambda_{pf}}{\mu_r}\right)E_i\\+\sum_{i=b+1}^{k}\left(1+\frac{\lambda_{in}}{\mu_{in}}+\frac{\lambda_{in}}{\mu_M}+\frac{\lambda_{pf}}{\mu_r}\right)\frac{\lambda_0}{\lambda_i+\lambda_{in}}+\frac{\lambda_0\lambda_k}{\mu_R(\lambda_i+\lambda_{in})}\end{array}\right]^{-1}$$

and the steady-state availability A_s is given by:

$$A_s=\sum_{i=0}^{k}\pi_{i,0}=\sum_{i=0}^{g-1}\frac{\lambda_0}{\lambda_i}\pi_{0,0}+\sum_{i=g}^{b}E_i\pi_{0,0}+\sum_{i=b+1}^{k}\frac{\lambda_0}{\lambda_i+\lambda_{in}}\pi_{0,0}. \tag{13.7}$$

The system MTTF is given by [30]:

$$\text{MTTF}=\frac{1}{\lambda_k}\left[\begin{array}{l}\sum_{i=0}^{g-1}\left(1+\frac{\lambda_{in}}{\mu_{in}}+\frac{\lambda_{pf}}{\mu_r}\right)\frac{\lambda_k+\lambda_{in}}{\lambda_i}+\left(1+\frac{\lambda_{in}}{\mu_{in}}+\frac{\lambda_{pf}}{\mu_r}\right)\frac{\lambda_k+\lambda_{in}}{\lambda_0}E_g\\+\sum_{i=g+1}^{b}\left(1+\frac{\lambda_{in}}{\mu_{in}}+\frac{\lambda_{in}}{\mu_m}+\frac{\lambda_{pf}}{\mu_r}\right)\frac{\lambda_k+\lambda_{in}}{\lambda_0}E_i\\+\sum_{i=b+1}^{k}\left(1+\frac{\lambda_{in}}{\mu_{in}}+\frac{\lambda_{in}}{\mu_M}+\frac{\lambda_{pf}}{\mu_r}\right)\frac{\lambda_k+\lambda_{in}}{\lambda_i+\lambda_{in}}\end{array}\right].$$

The results can be verified numerically, by choosing the parameters as chosen above and in addition we choose: $\lambda_{pf}=0.03$, $\mu_r=1$. The steady-state availability is shown in Figure 13.3(a) and system MTTF in Figure 13.3(b).

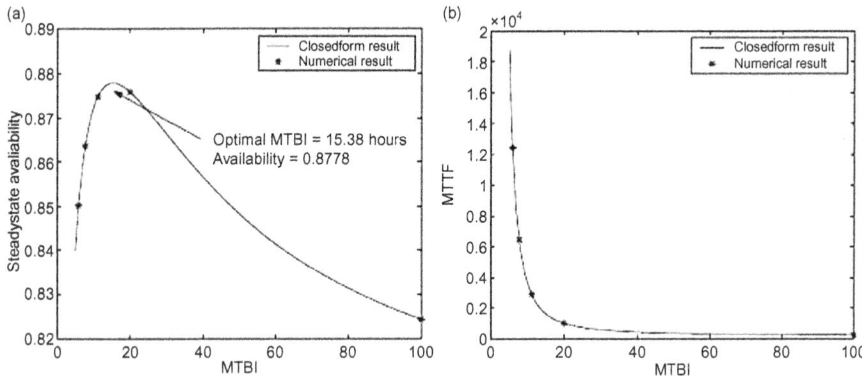

FIGURE 13.3

Steady-state availability and MTTF under device restoration failure and Poisson failure.

4.4 OPTIMIZATION

With the objective to maximize the availability A_s, the optimal inspection interval can be found by solving the polynomial equation:

$$\frac{dA_s}{d\lambda_{in}} = \sum_{i=0}^{g-1} \frac{\lambda_0}{\lambda_i}\frac{d\pi_{0,0}}{d\lambda_{in}} + \sum_{i=g}^{b}\left(\frac{dE_i}{d\lambda_{in}}\pi_{0,0} + E_i\frac{d\pi_{0,0}}{d\lambda_{in}}\right) + \sum_{i=b+1}^{k}\left(\frac{\lambda_0}{\lambda_i+\lambda_{in}}\frac{d\pi_{0,0}}{d\lambda_{in}} - \frac{\lambda_0}{(\lambda_i+\lambda_{in})^2}\pi_{0,0}\right) = 0.$$

Assuming the failure rate at each deterioration stage is the same, i.e. $\lambda_i = \lambda$, $\forall$ $i = 0, 1, \ldots, k$, we obtain the optimal time between inspections corresponding to different λ, by numerically solving the above equation.

The optimal time between inspection corresponding to different λ is presented in Figure 13.4(a) and Figure 13.4(b) shows the maximal availability for the optimal inspection time interval, corresponding to different λ. We observe from Figure 13.4(a) that the optimal inspection interval decreases exponentially with the increase in the failure rate λ, at each stage, which may be clearly understood from the intuition that the faster a failure occurs, the more frequently we need to conduct inspection and maintenance. By comparing the maximal availability obtained by preventive maintenance and the availability obtained without any maintenance, we observe that the improvement in availability by preventive maintenance is more significant when the failure rate at each deterioration stage is larger. Further we calculate the expected cost. The system operational cost is given by:

$$C_s = c_m\left(\sum_{i=0}^{g-1}\pi_{i,1} + \sum_{i=g}^{k}(\pi_{i,1}+\pi_{i,2})\right) + c_d\left(\sum_{i=0}^{k}\pi_{i,3} + \pi_F\right) + c'_m\sum_{i=g}^{b}(\mu_m\pi_{i,2}) + c'_M\sum_{i=b+1}^{k}(\mu_M\pi_{i,2}) + c'_r\sum_{i=0}^{k}(\mu_r\pi_{i,3}) + c'_R\mu_R\pi_F$$

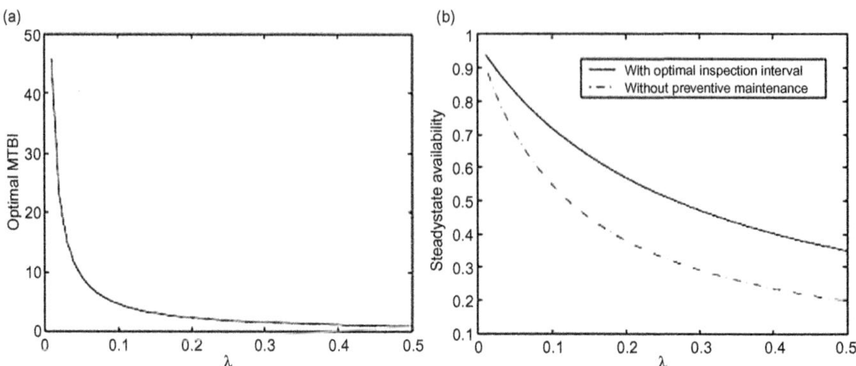

FIGURE 13.4

Optimal MTBI and steady-state availability under preventive maintenance for different λ.

Table 13.1 System Operational Cost Assignment

Cost Meaning		Value
c_m	Cost per unit time in down states due to maintenance	10
c_d	Cost per unit time in down states due to repair	50
c'_m	Cost for each minimal maintenance	10
c'_M	Cost for each major maintenance	20
c'_r	Cost for each minimal repair	10
c'_R	Cost for each major repair	50

where the costs assigned are mentioned in Table 13.1, and the optimal inspection interval to minimize the operational cost can be found by numerically solving the equation:

$$\frac{dc_s}{d\lambda_{in}} = 0.$$

Figure 13.5(a) and Figure 13.5(b) show the optimal inspection interval and the minimal operational cost under varying rate λ. It can be observed from these figures that significant reduction in operational cost can be achieved by the choice of optimal inspection intervals. As shown in Figures 13.3(a) and 13.3(b), the system can be made more reliable by decreasing the inspection interval, which will result in an increase in the MTTF, at the cost of lower availability. When the objective is to maximize the MTTF while still meeting the target availability, the optimization problem could be written as:

$$max\ MTTF\ such\ that\ A_s > A_{th}$$

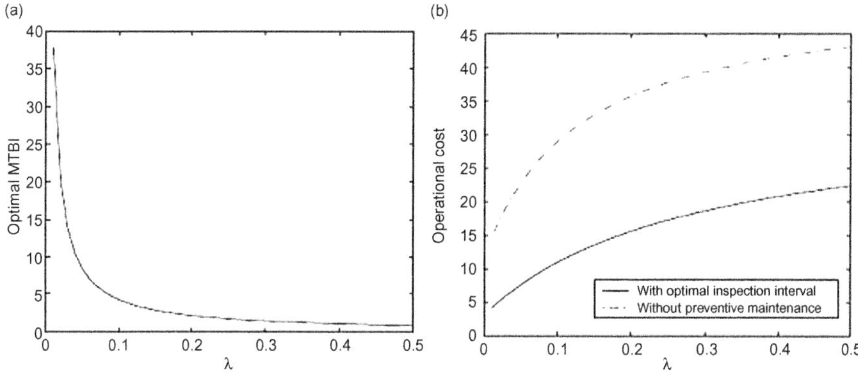

FIGURE 13.5

Optimal MTBI to minimize operational cost and minimal cost achieved by preventive maintenance for different λ.

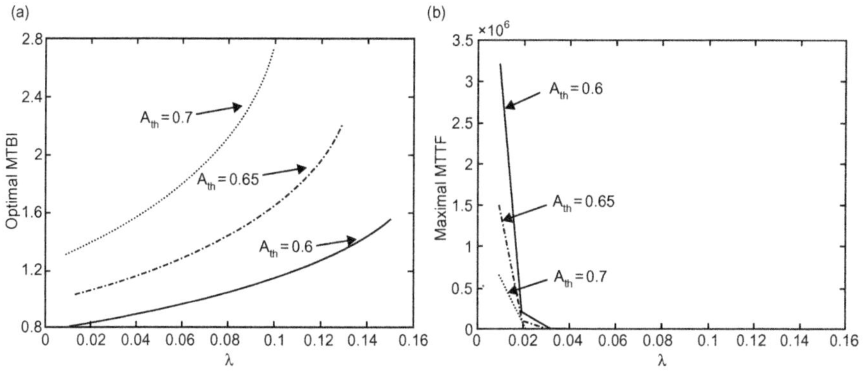

FIGURE 13.6

Optimal MTBI to maximize MTTF under steady-state availability constraint and maximal MTTF achieved by preventive maintenance.

where A_{th} is the target availability. It can be observed that the MTTF monotonically decreases with increasing MTBI, as in Figure 13.3(b), and between the steady-state availability and the MTBI follows a bell-shape curve, as shown in Figure 13.3(a). For this reason, the optimal inspection rate, λ_{in}^{*} that satisfies the above equation, is thus the smallest positive real root of the equation $A_s(\lambda_{in}) = A_{th}$, where A_s is as in Equation (13.7). Then, the optimal MTBI is $1/\lambda_{in}^{*}$. As a numerical example, Figure 13.6(a) shows the optimal MTBI under varying failure rate λ, while Figure 13.6(b) shows the maximal MTTF achievable by selecting optimal MTBI.

5 MARKOV REGENERATIVE PROCESS MODELING

In this section, we extend the CTMC model of Section 3 by allowing general distributions for all event times other than failure times. As a result, the stochastic process under consideration becomes a MRGP [31].

5.1 MODEL DESCRIPTION

Consider that the software system starts in a "robust" (or new) working state, D_0 (Figure 13.7). As time progresses, it transits through several deterioration failure stages (D_1 through D_k) and ultimately suffers a major failure (state F). Each deterioration stage, D_i, is an exponential stage and the deterioration rate in state i is λ_i. A series of deterioration failures leads to a major failure. Hence the time to failure for the software system starting from the initial state (ignoring Poisson failures and preventive maintenance) is hypo-exponentially distributed [26]. This facilitates the modeling of the phenomenon of software aging as it follows an increasing failure rate distribution. From the failure state F, a full restart (hardware reboot) is required to bring the system back to the "robust" state, D_0. Consider that the time taken for the full restart has the distribution $F_R(t)$. The system can also experience Poisson failures (constant failure rate, λ_p) at any stage (states D_0 through D_k). These failures occur abruptly unlike the gradually worsening deterioration failures and lead the system to the corresponding states, P_0 through P_k. This could correspond to sudden failures from causes other than software aging in a real software system. A deterioration failure is a "soft" failure in which the system can still provide service (possibly at a degraded level), whereas a Poisson failure is a "hard" failure in which the system cannot perform any given work. A simple retry [32] can bring back the system from the Poisson failure to the deterioration stage from which it failed. Denote the distribution of the time to perform a simple retry by $F_r(t)$.

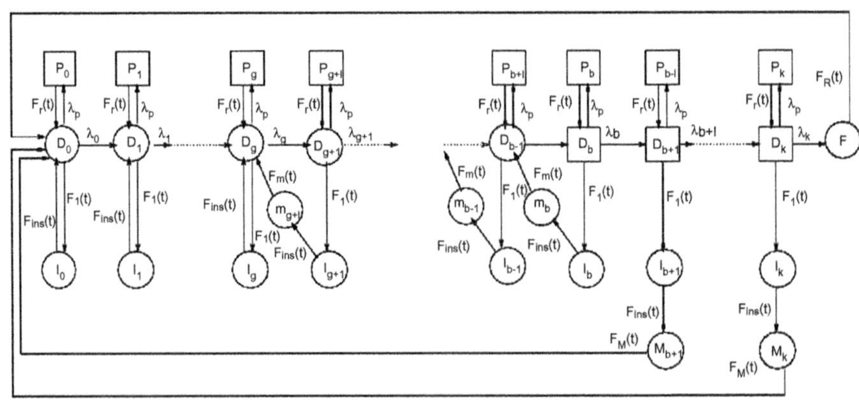

FIGURE 13.7

State transition diagram for inspection-based preventive maintenance.

Two kinds of preventive maintenance are performed on this system. A minimal maintenance (corresponding to a partial system cleanup) [33] performed at a deterioration stage restores the system to the previous deterioration stage and a major maintenance (full cleanup/reboot) performed at any of the deterioration stages restores the system to the "as good as new" state. The system is inspected so that the time period between two inspections is generally distributed with CDF $F_I(t)$. States I_0 through I_k denote the states where the actual inspection takes place and the time to perform the inspection is generally distributed ($F_{ins}(t)$). At the end of inspection, one of two kinds of preventive maintenance of system is carried out based on the following inspection-based policy. There are k deterioration stages in the system and that the current deterioration stage, i, is observable through some system parameter(s) [3,16,34]. Thresholds g and b are set up so that (i) no maintenance is done if the inspection finds the system in state D_i, $i \leq g$; (ii) a minimal maintenance (CDF $F_m(t)$) is performed when $g < i \leq b$, where the system is restored to degradation stage $i-1$; (iii) a major maintenance (CDF $F_M(t)$) is performed when $b < i \leq k$, whereupon the system is restored to the initial state D_0. States m_n correspond to states where a minimal maintenance is being performed and states M_n correspond to states where a major maintenance is being performed. The system is operational only in states $\{D_n,\ n = 0,\ 1,\ \ldots,\ k\}$. Next we describe the methodology for obtaining the steady-state availability. The optimal value of the inspection interval which minimizes expected downtime or the expected cost for a given set of parameter values can also be then obtained.

5.2 KERNEL MATRICES

Let the underlying stochastic process of Figure 13.7 be $\{Z(t),\ t \geq 0\}$, where $Z(t)$ is the state of the system at time t. Noticeably, this is a continuous-time discrete state process. It should be noted that since sojourn times in all the states are not exponentially distributed, the process $\{Z(t),\ t \geq 0\}$ is not a homogeneous CTMC. Moreover, $Z(t)$ is not even a semi-Markov process since whenever the system is in a deterioration stage and inspection has not yet triggered, there is a need to keep track of the remaining time in the inspection clock in order to predict the future behavior of the system. The underlying process satisfies the Markov property only at the following instants (regeneration time points)—an inspection trigger occurs, an inspection ends, a maintenance (major or minimal) action is completed, the system enters the failure state F, or full reboot is completed and the system enters the state D_0.

Observe that the regeneration instances exactly correspond to time points of entering the states in $\Omega' = \{D_0,\ \ldots,\ D_{b-1},\ I_0,\ \ldots,\ I_k,\ m_{g+1},\ \ldots,\ m_b,\ M_{b+1},\ \ldots,\ M_k,\ F\}$ (circles in Figure 13.7). Note that entering the states $D_0,\ \ldots,\ D_{b-1}$ from the Poisson failure states $P_0,\ \ldots,\ P_{b-1}$ and from the previous deterioration stages (squares in Figure 13.7) are not regeneration time points since the inspection clock has not been reset. The inspection clock is also not reset when the system enters states $P_0, \ldots, P_k$ but the clock is stopped till it returns to one of the deterioration states $D_0, \ldots, D_k$. Consider the sequence of regeneration epochs $\{S_n, n \geq 0\}$ at which the process $\{Z(t), t \geq 0\}$ is observed. The state of the process $Z(t)$ can change between S_n and S_{n+1}, because of Poisson failure and deterioration failure at each stage. We observe that the sequence $(Z(S_n^+), S_n)$ is a Markov renewal sequence and $\{Z(t), t \geq 0\}$ is a Markov regenerative process. We now proceed to determine the local kernel $E(t)$ and the global kernel $K(t)$ matrices in the LST domain. The global kernel, $K(t)$, forms a square matrix of order $(2k+b-g+2)$ and its LST, $\tilde{K}(s)$, can be divided into submatrices of the form:

$$\tilde{K}(s) = \begin{pmatrix} 0 & \tilde{K}_1(s) & \tilde{K}_2(s) \\ \tilde{K}_3(s) & 0 & 0 \\ 0 & 0 & \tilde{K}_4(s) \\ \tilde{K}_5(s) & 0 & 0 \end{pmatrix}.$$

Submatrices $\tilde{K}_1(s)$ and $\tilde{K}_2(s)$ are shown in Table 13.2 and $\tilde{K}_3(s)$, $\tilde{K}_4(s)$ and $\tilde{K}_5(s)$ are shown in Table 13.3. The determination of the elements of $\tilde{K}_1(s)$ and $\tilde{K}_2(s)$ is described in Section 5.3.

The local kernel, $E(t)$ forms a $(2k+b-g+2) \times (4k-g+4)$ matrix and its LST, $\tilde{E}(s)$ can be divided into submatrices of the form:

$$\tilde{E}(s) = \begin{pmatrix} \tilde{E}_1(s) & \tilde{E}_2(s) & 0 \\ 0 & 0 & \tilde{E}_3(s) \end{pmatrix}.$$

Submatrices $\tilde{E}_1(s)$, $\tilde{E}_2(s)$ and $\tilde{E}_3(s)$ are shown in Table 13.4. With some minor computations, it can be shown that

$$J_i(s) = \frac{1}{s}\lambda_p\left[1 - \tilde{F}_r(s)\right]H_i(s), \quad i = 0, \ldots, k.$$

The determination of the elements of the $\tilde{E}_1(s)$, $H_0(s), \ldots, H_k(s)$, is described in Section 5.3. The elements of $\tilde{E}_2(s)$ can be computed from $\tilde{E}_1(s)$, where, $\overline{F}(s) = 1 - F(s)$.

5.3 SUBORDINATED SEMI-MARKOV REWARD PROCESS

Next we describe the approach to determine the elements of submatrices $\tilde{K}_1(s)$, $\tilde{K}_2(s)$, $\tilde{E}_1(s)$ and $\tilde{E}_2(s)$. Let Ω (D_i) be the set of all states reachable from state D_i $(i = 0, \ldots, b-1)$ in which the subordinated semi-Markov process (since sojourn time in state P_j, $j = 0, 1, \ldots, k$ are generally distributed) can spend a non-zero time before the next regeneration instant occurs. Therefore, $\Omega(D_i) = \{D_i, \ldots, D_k, P_i, \ldots, P_k\}$.

Let $\{Z^{Di}(t), t \geq 0\}$ be the subordinated semi-Markov process defined over Ω (D_i) (Figure 13.8). The inspection clock can only start when system is in states from D_i through D_{b-1} and end only by transitions from states D_0 through D_k to states I_0 through I_k, respectively. The clock is disabled in the Poisson failure states P_0 through P_k and resumes in states D_0 through D_k, from where it

Table 13.2 Submatrices $\tilde{K}_1(s)$ and $\tilde{K}_2(s)$

	I_0	I_1	...	I_{b-1}	...	I_k
D_0	$G_0(s)$	$G_1(s)$	...	$G_{b-1}(s)$	...	$G_k(s)$
D_1		$G_0(s)$	...	$G_{b-2}(s)$	...	$G_{k-1}(s)$
$\vdots$			$\ddots$	$\vdots$	$\vdots$	$\vdots$
D_{b-1}				$G_0(s)$	...	$G_{k-b+1}(s)$

	m_{g+1}	...	m_b	M_{b+1}	...	M_k	F
D_0							$1-\sum_0^k[G_i(s)+H_i(s)+J_i(s)]$ 0
D_1							$1-\sum_0^{k-1}[G_i(s)+H_i(s)+J_i(s)]$ 0
$\vdots$							$\vdots$
D_{g-1}							$1-\sum_0^{k-g+1}[G_i(s)+H_i(s)+J_i(s)]$ $0 \quad [G_i(s)+H_i(s)+J_i(s])$
D_g							$1-\sum_0^{k-g}[G_i(s)+H_i(s)+J_i(s)]$ $1-\sum_0 \quad [G_i(s)+H_i(s)+J_i(s)]$
D_{g+1}							$1-\sum_0^{k-g-1}[G_i(s)+H_i(s)+J_i(s)]$ $1-\sum_0 \quad [G_i(s)+H_i(s)+J_i(s)]$
$\vdots$							$\vdots$
D_{b-1}							$1-\sum_0^{k-b+1}[G_i(s)+H_i(s)+J_i(s)]$

left off before the system entered the Poisson failure states. The clock also continues as the system moves through the deterioration stages, D_0 through D_k. States D_0, D_1, ..., D_k and $P_0, P_1, \ldots, P_k$ are therefore *prs* (preemptive resume) states [35].

Assume that the generally distributed inspection clock time, with CDF $F_I(t)$ begins in state D_i, $i=0, \ldots, b-1$ at $t=\tau_0^*=0$. Let x denote a sample of the distribution $F_I(t)$. The next regeneration time point, τ_1^*, will be either the epoch of actual inspection starting when the system enters one of the states I_0 through I_k (this event occurs when the cumulative amount of time spent in the deterioration states reaches x), or the epoch at which the system enters the state F. The problem of determining the elements of submatrices $\tilde{K}_1(s)$, $\tilde{K}_2(s)$, $\tilde{E}_1(s)$ and $\tilde{E}_2(s)$ is transformed into the completion time (task completion) problem [35,36]. We need to determine the distribution of "task completion" time given the CDF of the "work" requirement X , which in this case is $F_I(t)$. Define reward rates r_n, as equal to 1 when $n=D_0$, $D_1, \ldots, D_k$ and 0 when $n=P_0$, P_1, ..., P_k. Let $Q^{D_i}(t)$ be the kernel of the semi-Markov process $Z^{D_i}(t)$ and let R be the diagonal matrix of the corresponding reward rates (r_n). Let $C(x)$ denote the task completion time for a fixed

Table 13.3 Submatrices, $\tilde{K}_3(s)$, $\tilde{K}_4(s)$ and $\tilde{K}_5(s)$

	D_0	D_1	...	D_{g-1}	D_g	D_{g+1}	...	D_{b-1}
I_0	$\tilde{F}(s)$ *Ins*							
I_1		$\tilde{F}(s)$ *Ins*						
⋮			⋱					
I_{g-1}				$\tilde{F}(s)$ *Ins*				
I_g					$\tilde{F}(s)$ *ins*			

	m_{g+1}	...	m_b	M_{b+1}	...	M_k	F
I_{g+1}	$\tilde{F}_{ins}(s)$ $_{ins}(s)$						
⋮		⋱					
I_b			$\tilde{F}_{ins}(s)$ $_{ins}$				
I_{b+1}				$\tilde{F}_{ins}(s)$ $_{ins}(s)$			
⋮					⋱		
I_k						$\tilde{F}_{ins}(s)$ $_{ins}$	

	D_0	D_1	...	D_{g-1}	D_g	D_{g+1}	...	D_{b-1}
m_{g+1}					$\tilde{F}_m(s)$ $_m(s)$			
m_{g+2}						$\tilde{F}_m(s)$ $_m(s)$		
⋮							⋱	
m_b								$\tilde{F}_m(s)$ $_m$
M_{b+1}	$\tilde{F}_M(s)$ $_M$ (s)							
⋮	⋮							
M_k	$\tilde{F}_M(s)$ $_M$							
F	$\tilde{F}_R(s)$							

work requirement, x. We now define two matrices $M(t)$ and $N(t)$ whose entries are defined by:

$$M_{p,q}(t, X=x) = P\{Z^{D_i}(\tau_1^*) = q, C(x) \le t | Z^{D_i}(0) = p\}$$
$$N_{p,q}(t,\ X=x) = P\{Z^{D_i}(t) = q, C(x) > t | Z^{D_i}(0) = p\}.$$

where $p,\ q \in \Omega\ (D_i)$. $M_{p,q}(t,\ x)$ gives the probability of x amount of inspection clock time (which is different from absolute time since the clock is disabled in the P_j states) elapsing until state q is reached before time t, starting at state p. $N_{p,q}(t,\ x)$ gives the probability of being in state q at time t before x amount of inspection clock time has elapsed, starting in state p. From the above definitions it is clear that

$$\sum_{\Omega(D_i)} M_{p,q}(t,x) + \sum_{\Omega(D_i)} N_{p,q}(t,x) = 1.$$

Table 13.4 Submatrices $\tilde{M}^{*}(s,w) = [sI + wR - Q^{D_i}]^{-1}R$, $\tilde{E}_2(s)$ and $\tilde{E}_3(s)$

	D_0	D_1	...	D_{b-1}	...	D_k
D_0	$H_0(s)$	$H_1(s)$	...	$H_{b-1}(s)$	...	$H_k(s)$
D_1		$H_0(s)$	...	$H_{b-2}(s)$	...	$H_{k-1}(s)$
⋮			⋱	⋮	⋮	⋮
D_{b-1}				$H_0(s)$	...	$H_{k-b+1}(s)$

	P_0	P_1	...	P_{b-1}	...	P_k
D_0	$J_0(s)$	$J_1(s)$	...	$J_{b-1}(s)$	...	$J_k(s)$
D_1		$J_0(s)$	...	$J_{b-2}(s)$	...	$J_{k-1}(s)$
⋮			⋱	⋮	⋮	⋮
D_{b-1}				$J_0(s)$	...	$J_{k-b+1}(s)$

	I_0	...	I_k	m_{g+1}	...	m_b	M_{b+1}	...	M_k	F
I_0	$\tilde{\tilde{F}}_{ins}(s)$									
⋮		⋱								
I_k			$\tilde{\tilde{F}}_{ins}(s)$							
m_{g+1}				$\tilde{\tilde{F}}_{m}(s)$						
⋮					⋱					
m_b						$\tilde{\tilde{F}}_{m}(s)$				
M_{b+1}							$\tilde{\tilde{F}}_{M}(s)$			
⋮								⋱		
M_k									$\tilde{\tilde{F}}_{M}(s)$	
F										$\tilde{\tilde{F}}_{R}(s)$

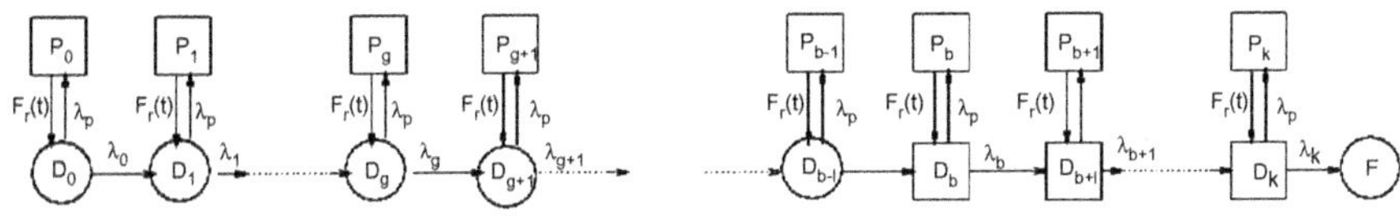

FIGURE 13.8

Subordinate process starting in states D_0 through D_{b-1}.

The elements of these matrices in the LST-LT domain is given by (refer to [35,36] for derivation)

$$\tilde{M}^*_{p,q}(s,w) = \delta_{p,q}\frac{r_p}{s+wr_p}\left[1 - Q_p^{\sim D_i}(s+wr_p)\right] + \sum_{u\in\Omega(D_i)} Q_{p,u}^{\sim D_i}(s+wr_p)\tilde{M}^*_{u,q}(s,w)$$

$$\tilde{N}^*_{p,q}(s,w) = \delta_{p,q}\frac{r_p}{w(s+wr_p)}\left[1 - Q_p^{\sim D_i}(s+wr_p)\right] + \sum_{u\in\Omega(D_i)} Q_{p,u}^{\sim D_i}(s+wr_p)\tilde{N}^*_{u,q}(s,w)$$

where $\delta_{p,q}$ is the Kronecker delta, and

$$Q_p^{\sim D_i}(s) = \sum_q Q_{p,q}^{\sim D_i}(s), p, q \in \Omega(D_i).$$

If the subordinated process, $\{Z^{D_i}(t), t \geq 0\}$, is a CTMC, then the above equations can be written in the matrix form as

$$\tilde{M}^*(s,w) = [sI + wR - Q^{D_i}]^{-1}R$$

and

$$\tilde{N}^*(s,w) = (s/w)[sI + wR - Q^{D_i}]^{-1}$$

where Q^{D_i} is the generator matrix of the CTMC.

Inverting the Laplace transform with respect to w and then unconditioning, we get

$$\tilde{M}_{p,q}(s) = \int_0^\infty \tilde{M}_{p,q}(s,x)dF_I(x)$$

and

$$\tilde{N}_{p,q}(s) = \int_0^\infty \tilde{N}_{p,q}(s,x)dF_I(x).$$

By the time inspection clock expires, if the subordinated SMP is in state D_j, then the corresponding state of the MRGP after the corresponding regeneration instant will be I_j. Hence, the elements of $\tilde{K}_1(s)$ and $\tilde{E}_1(s)$ are given by $G_j(s) = \tilde{M}_{D_0,D_j}(s)$ and $H_j(s) = \tilde{N}_{D_0,D_j}(s)$. Submatrices $\tilde{K}_2(s)$ and $\tilde{E}_2(s)$ can be determined once $\tilde{K}_1(s)$ and $\tilde{E}_1(s)$ are known.

5.4 STEADY-STATE SOLUTION

To obtain the MRGP steady-state probabilities, we first need to solve the embedded DTMC given by

$$v = vK(\infty) = v\lim_{s\to 0} s\tilde{K}(s) \text{ and } \sum_i v_i = 1$$

where

$$v = [v_{D_0}, \ldots, v_{D_{b-1}}, v_{I_0}, \ldots, v_{I_k}, v_{m_{g+1}}, \ldots, v_{m_b}, v_{M_{b+1}}, \ldots, v_{M_k}, v_F].$$

The above can be rewritten as

$$v_{I_n} = \sum_{r=0}^{\min(n,b-1)} v_{D_r} K_{D_r,I_n}, n = 0, 1, \ldots, k \text{ and } \sum_i v_i = 1$$

where $K_{i,\, j}$ denotes the entry of $K_{i,\, j}\,(\infty)$ and

$$\begin{aligned}
v_{D_0} &= v_F + v_{I_0} + v_{I_{b+1}} + \cdots + v_{I_k} \\
v_{D_r} &= v_{I_r}, r = 1, 2, \ldots, g-1 \\
v_{D_g} &= v_{I_g} + v_{I_{g+1}} \\
v_{D_n} &= v_{I_{n+1}}, n = g+1, g+2, \ldots, b-1 \\
v_{m_n} &= v_{I_n}, n = g+1, g+2, \ldots, b \\
v_{M_n} &= v_{I_n}, n = b+1, b+2, \ldots, k.
\end{aligned} \tag{13.8}$$

The above equations can by written in matrix form as

$$[v_{I_0}, \ldots, v_{I_k}, v_F]A = B$$

where A is a square matrix of order $(k+2)$ and B is a vector of order $(k+2)$. The computation reduces the system of $(2k+b-g+2)$ equations to $(k+2)$ equations. Once $v_{I_0}, \ldots, v_{I_k}, v_F$ are obtained by solving the matrix equation, the other probabilities can be determined from Equation (13.8).

From Equation (13.4), we find $\alpha_{i,j}$ as

$$\alpha_{i,j} = \begin{cases}
\frac{1}{s}\tilde{N}_{i,j}(s)|_{s=0}, i = D_r, r = 0, \ldots, b-1, j = D_r, \ldots, D_k, P_r, \ldots, P_k \\
\frac{1}{s}(1 - \tilde{F}_{ins}(s))|_{s=0}, i = j = I_0, \ldots, I_k \\
\frac{1}{s}(1 - \tilde{F}_m(s))|_{s=0}, i = j = m_{g+1}, \ldots, m_b \\
\frac{1}{s}(1 - \tilde{F}_M(s))|_{s=0}, i = j = M_{b+1}, \ldots, M_b \\
\frac{1}{s}(1 - \tilde{F}R_R(s))|_{s=0}, i = j = F \\
0, \textit{otherwise}.
\end{cases}$$

From the above equations, we find

$$\begin{aligned}
\beta_{D_n} &= \textstyle\sum_j \alpha_{D_n,j}, n = 0, 1, \ldots, b-1, j = D_n, \ldots, D_k, P_n, \ldots, P_k \\
\beta_{I_n} &= \alpha_{I_n,I_n}, n = 0, 1, \ldots, k \\
\beta_{m_n} &= \alpha_{m_n,m_n}, n = g+1, g+2, \ldots, b \\
\beta_{M_n} &= \alpha_{M_n,M_n}, n = b+1, b+2, \ldots, k \\
\beta_F &= \alpha_{F,F}.
\end{aligned}$$

Now, the MRGP steady-state probabilities are computed as [27]

$$\pi_j = \frac{\sum_{k \in \Omega'} \nu_k \alpha_{k,j}}{\sum_{k \in \Omega'} \nu_k \beta_k}, j \in \Omega$$

where v_k's, $\alpha_{k,j}$'s and β_k's are obtained from the previous computation.

5.4.1 Expected downtime

The steady-state availability of the system is given by

$$A = \sum_{j=0}^{k} \pi_{D_j} = \frac{\sum_{i=0}^{b-1} \nu_{D_i} \sum_{j=i}^{k} \alpha_{D_i,D_j}}{\sum_{k \in \Omega'} \nu_k \beta_k}.$$

The expected downtime over a time interval [0, T], is then given by, $D = T^*(1 - A)$.

5.4.2 Expected cost

The expected cost incurred per unit time is given by

$$c = c_r \sum \pi_{P_j} + c_I \sum \pi_{I_j} + c_m \sum \pi_{m_j} + c_M \sum \pi_{M_j} + c_F \pi_F$$

which is equal to

$$c = \frac{1}{\sum_{k \in \Omega'} \nu_k \beta_k} \left[c_r \sum_{0}^{b-1} \nu_{D_i} \sum_{i}^{k} \alpha_{D_i,P_j} + c_I \sum_{0}^{k} \nu_{I_j} \alpha_{I_j,I_j} + \right.$$

$$\left. c_m \sum_{g+1}^{b} \nu_{m_j} \alpha_{m_j,m_j} + c_M \sum_{b+1}^{k} \nu_{M_j} \alpha_{M_j,M_j} + c_F \nu_F \alpha_{F,F} \right]$$

where c_r, c_I, c_m, c_M and c_F are the costs/unit time of Poisson failure, inspection, minimal maintenance, major maintenance and major failure, respectively. (It is also possible to include one-time costs for each of these events through an impulse reward function.) The expected cost over a time interval [0, T], is given by:

$$C = T^* c.$$

5.5 TRANSIENT SOLUTION

The time-dependent probabilities are obtained by numerically inverting Equation (13.6) by Jagerman's method [37]. The transient availability is computed as

$$A(t) = \sum_{0}^{k} \pi_{D_j}(t).$$

5.6 NUMERICAL RESULTS

We present numerical examples in this section to illustrate the applicability of the model and solution approach. Consider an inspection-based maintenance model with four deterioration stages ($i = 0, \ldots, 3$). No maintenance is done in stages 0 and 1. A minimal maintenance is done when the system is in deterioration stage 2

Table 13.5 Parameter Values

Parameter	Description	Value
λ_i	Deterioration failure rate	0.1/hr
λ_p	Poisson failure rate	0.05/hr 2 min
$1/\mu_r$	Mean time for retry	1 hr
$1/\mu_R$	Mean time for full reboot	5 min
$1/\mu_m$	Mean time for minimal maint. Mean time for major maint.	15 min
$1/\mu_M$ $1/\mu_{ins}$ C_r	Mean time for actual inspection	40 sec
C_R C_m C_M C_{ins}	Cost of retry	$500/hr
	Cost of full reboot	$5000/hr
	Cost of minimal maint.	$1500/hr
	Cost of major maint.	$3000/hr
	Cost of actual inspection	$1000/hr

and a major maintenance is done in stage 3. Therefore, the parameters g, b and k are 1, 2 and 3, respectively (Section 5.1). The time taken for failure, reboot, retry, maintenance and actual inspection are considered to be exponentially distributed with rate parameters given in Table 13.5. The inspection interval δ is considered deterministic.

5.6.1 Steady-state results

The following steady-state measures are computed: expected downtime, D, and the expected cost, C, over an interval $[0, T]$. The value of T is fixed to be 1000 hours. The costs incurred per unit time due to maintenance, reboot, retry and actual inspection are assumed to be fixed and are given in Table 13.5. It is assumed that the cost incurred due to maintenance is less than that incurred due to system failure (full reboot), since maintenance can be done gracefully at predetermined or scheduled times. Minimal maintenance costs less than major maintenance. Based on the measures D and C, optimal inspection intervals can be obtained for a given set of parameters. We may not obtain the same inspection interval which minimizes both expected downtime and expected cost. Therefore, there is a trade-off involved based on what the user/operator considers important.

Figure 13.9 shows the plot for expected downtime (in hours) versus inspection interval (in hours). The solid line corresponds to the plot when the inspection interval is deterministic and the dashed line corresponds to the plot when the deterministic interval is approximated using an exponential distribution with mean equal to the deterministic interval. If the inspection interval is close to zero, the system is performing preventive maintenance very often and thus incurs high downtime. As the inspection interval increases, the expected downtime decreases and reaches an optimum value. If the inspection interval goes beyond the optimal

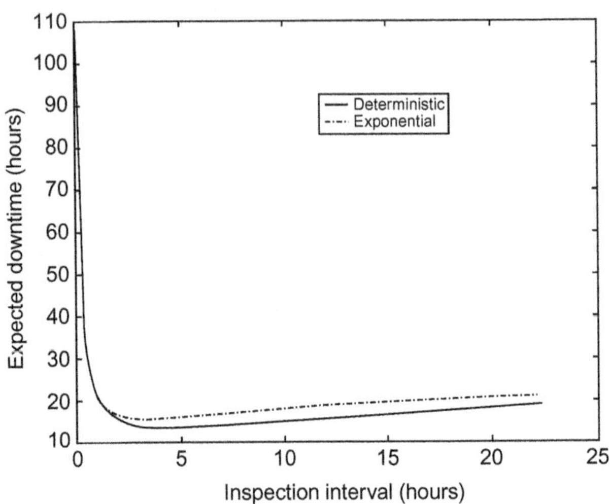

FIGURE 13.9

Expected downtime versus δ.

value (i.e., we perform preventive maintenance less and less frequently), the system failure has more influence on the expected downtime than maintenance. Hence the expected downtime begins to increase. In the deterministic inspection interval case, the minimum for the expected downtime (13.6119 hours) occurs at 4.85 hours. In the case of the exponential approximation, the minimum value of downtime is 15.7475 hours which occurs when the inspection interval is 3.35 hours. Hence if we approximate the deterministic interval using an exponential distribution, a higher downtime would be incurred (which could be avoided) by inspecting the system every 3.35 hours.

Similarly, Figure 13.10 shows the plot of expected cost incurred versus the inspection interval (in hours), both for the deterministic case as well as the exponential approximation. For the deterministic inspection interval case, the minimum expected cost is $23,669, which occurs at 1.85 hours. For the exponential approximation, the minimum cost is $29,232, which occurs when the inspection interval is 1.35 hours. We find that the optimal inspection interval is not the same for the expected downtime and the expected cost. Therefore a compromise has to be made regarding the inspection interval based on whether the expected downtime or the expected cost incurred is more important.

We now compare the previously described inspection-based system with a non- inspection system. The basic system is the same as that described earlier. Instead of inspecting the system after every δ hours and then deciding on the type of preventive maintenance (major/minor), here, no inspection is done and a major maintenance (full rejuvenation) is performed after every δ hours. Hence, for the non-inspection system, we will call δ the rejuvenation trigger interval.

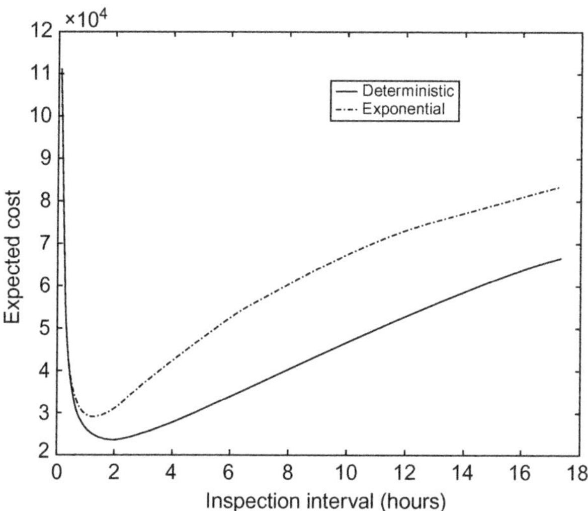

FIGURE 13.10

Expected cost versus δ.

Figure 13.11 shows the plot of expected downtime versus the rejuvenation trigger interval for the non-inspection system (solid line). The minimum expected downtime of 19.69 hours occurs when the rejuvenation trigger interval is 22.35 hours. For the inspection-based system, we varied the mean time for actual inspection (mean time to carry out inspection) from 10 minutes to 28 seconds. The plots are also shown in Figure 13.11. When the mean time for actual inspection is 10 minutes, the minimum expected downtime is more than that for the non-inspection system. For other cases, the minimum expected downtime is less than that for the non-inspection system. As expected, the downtime decreases as the mean time to actual inspection decreases, but beyond a certain limit, the increase in gain becomes progressively smaller. Hence we can conclude that generally, an inspection-based system performs better than a non-inspection system and that the lesser the time taken for actual inspection, the greater is the benefit.

5.6.2 Transient results

For the transient analysis, we assume the value of the inspection trigger interval, δ to be 5 hours. The transient availability, $A(t)$, is plotted in Figure 13.12 for both the deterministic inspection trigger interval as well as for exponentially distributed inspection trigger interval. We can see from the plot that for the deterministic interval, transient availability ripples for some time before it settles down to a constant value. Each ripple has a local minimum which occurs every 5 hours (the inspection trigger interval). The steady-state availability in the exponential case is lesser than the steady-state availability in the deterministic case.

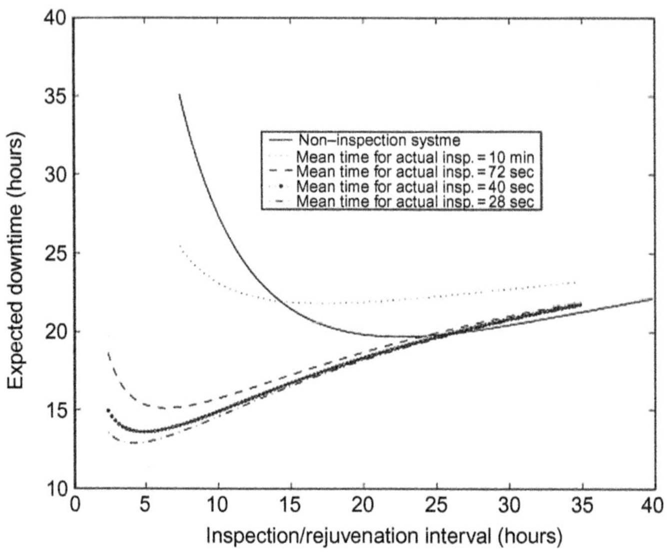

FIGURE 13.11

Comparison of inspection-based systems with non-inspection system.

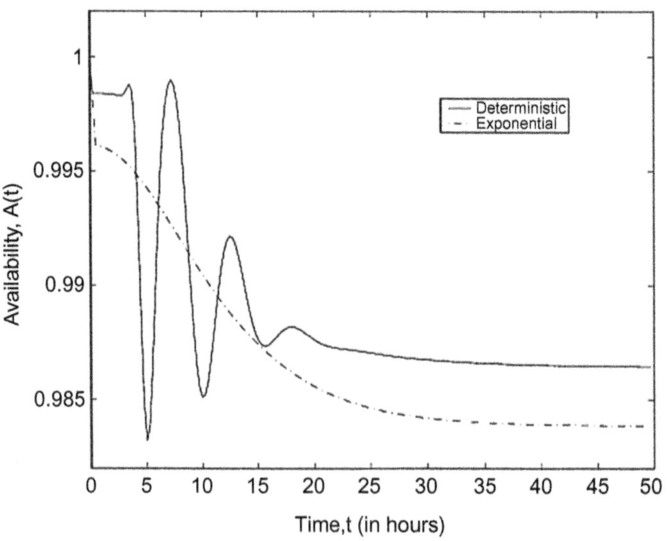

FIGURE 13.12

Transient availability.

6 CONCLUSIONS

In computer systems, the existence of deterministic and exponentially distributed parameters gives rise to stochastic models that are non-Markovian in nature. A non-Markovian model can be studied, analyzing its underlying process.

In many cases this process can be shown to be a MRGP and therefore Markov renewal theory can be applied for its long-run as well as time-dependent behavior.

In this chapter, we analyzed preventive maintenance in software systems using analytical modeling paradigms using Markov and Markov regenerative process models. Using Markov models, we presented closed-form solutions for preventive major and minimal maintenance problems under both device deterioration failure and Poisson failures. We then described the method to determine the optimal inspection interval. And finally, using MRGP modeling, we introduced and analyzed inspection-based preventive maintenance in operational software systems, in which two levels of maintenance are performed. We then numerically obtained the optimal value of the inspection interval that minimizes expected downtime and expected cost for an assumed set of parameter values. We show that inspection-based maintenance is advantageous in many cases over non-inspection based maintenance. The application of two levels of preventive maintenance in software systems, based on the degradation stage, is a novel approach. Our model is very general and captures general distribution for the inspection clock, time to actual inspection, time for maintenance and restart.

ACKNOWLEDGEMENT

The research work of D. Selvamuthu is supported by grant number MHRD 3596, and the research work of K. Trivedi was supported by a US NSF grant NSF-CNS-08- 31325.

REFERENCES

[1] Logothetis D, Trivedi K, Puliafito A. Markov regenerative models. In: Proceedings of the international computer performance and dependability symposium, Erlangen, Germany; 1995. p. 134–43.

[2] Huang Y, Kintala C, Kolettis N, Fulton ND. Software rejuvenation: analysis, module and applications. In: Proceedings of FTCS-25, Pasadena, CA; June 1995. p. 381–90.

[3] Trivedi KS, Vaidyanathan K, Goševa-Popstojanova K. Modeling and analysis of software aging and rejuvenation. Proceedings of 33rd annual simulation Symposium, Greece; 2000. p. 270–79.

[4] Machida F, et al. Software life-extension: a new countermeasure to software aging. Software Reliability Engineering (ISSRE), IEEE 23rd international symposium on; 2012. p. 131–40.

[5] Sabino MEG, Merabti M, Llewellyn-Jones D, Bouhafs F. Detecting software aging in safety-critical infrastructures p. 78–85 Science and information conference (SAI), 2013. IEEE; 2013.
[6] Duke University. Software Rejuvenation website. Available from: <http://srejuv.ee.duke.edu/>; 2015.
[7] Bao Y, Sun X, Trivedi KS. A workload-based analysis of software aging, and rejuvenation. IEEE Trans Reliab 2005;54(3):541–8.
[8] Avritzer A, Weyuker EJ. Monitoring smoothly degrading systems for increased dependability. Empir Softw Eng J 1997;2(1):59–77.
[9] Hanmer RS, Mendiratta VB. Rejuvenation with workload migration. In: DSN workshops. 2010. p. 80–5.
[10] Cotronco D, Natella R, Pietrantuono R, Russo S. Software aging analysis of the Linux operating system. In: Software reliability engineering (ISSRE), IEEE 21st international symposium on; 2010. p. 71–80.
[11] Cassidy K, Gross K, Malekpour A. Advanced pattern recognition for detection of complex software aging in online transaction processing Servers. In: Proceedings of DSN 2002, Washington D.C; 2002. p. 478–82.
[12] Grottke M, Li L, Vaidyanathan K, Trivedi KS. Analysis of software aging in a web server. IEEE Trans Reliab 2006;55(3):411–20.
[13] Microsoft. Enterprise website. 2014. Available from: <http://www.microsoft.com/serviceproviders/deployment/iishostingP74416.asp>.
[14] Zhao J, Trivedi KS. Performance modeling of apache web server affected by aging. In: 2011 IEEE third international workshop on software aging and rejuvenation (WoSAR), 2011. p. 56–61.
[15] Castelli V, Harper RE, Heidelberger P, Hunter SW, Trivedi KS, Vaidyanathan K, et al. Proactive management of software aging. IBM JRD 2001;45(2):311–32.
[16] Vaidyanathan K, Harper RE, Hunter SW, Trivedi KS. Analysis and implementation of software rejuvenation in cluster systems. In: Proceedings of the ACM SIGMETRICS 2001/Performance 2001, Cambridge, MA; June 2001. p. 62–71.
[17] Yang M, Li Z, Yang W, Li T. Analysis of software rejuvenation in clustered computing system with dependency relation between nodes. In: 2010 IEEE tenth international conference on computer and information technology (CIT); 2010. p. 46–53.
[18] Liu Y, Ma Y, Han J, Lavender H, Thrived KS. A proactive approach towards always-on availability in broadband cable networks. Comput Commun 2005;28(1):51–64.
[19] Tai AT, Alkalaj L, Chau SN. On-board preventive maintenance: a design-oriented analytic study for long-life applications. Perform Eval 1999;35(3–4):215–32.
[20] Marshall E. Fatal error: how patriot overlooked a scud. Science 1992;Mar(13):1347.
[21] Mansoor A, Patsekin V, Scherl D, Robinson J, Rajwa B. A statistical modeling approach to computer-aided quantification of dental biofilm. IEEE J Biomed Health Informatics 2014;99(1):1.
[22] Nicol D, Sanders WH, Trivedi KS. Model-based evaluation: from dependability to security. IEEE Trans Dependable Secure Comput 2004;1(1):48–65.
[23] Marsan MA, Balbo G, Conte S, Donatelli G, Franceschinis M. Modeling with generalized stochastic Petri nets. West Sussex, England: John Wiley and Sons; 1995.
[24] Trivedi KS, Sahner RA. SHARPE at the age of twenty two. ACM SIGMETRICS Perform Eval Rev 2009;36(4):52–7.

[25] Sahner RA, Trivedi KS, Puliafito A. Performance and reliability analysis of computer systems. Dordrecht/Boston: Kluwer Academic Publishers; 1996.
[26] Trivedi KS. Probability and statistics, with reliability, queuing and computer science applications. 2nd ed. New York: John Wiley and Sons; 2001.
[27] Kulkarni VG. Modeling and analysis of stochastic systems. London: Chapman & Hall; 1995.
[28] Legat V, Zaludova AH, Cervenka V, Jurca V. Contribution to optimization of preventive maintenance. Reliab Eng Syst Saf 1996;51:259–66.
[29] Auriol JK. On time-dependent availability and maintenance optimization of standby units under various maintenance policies. Reliab Eng Syst Saf 1997;56:79–89.
[30] Chen D, Trivedi KS. Closed-form analytical results for condition-based maintenance. Reliab Eng Syst Saf 2002;76:43–51.
[31] Vaidyanathan K, Dharmaraja S, Trivedi KS. Analysis of inspection-based preventive maintenance in operational software systems. In: Proceedings of international symposium on reliable distributed systems (SRDS 2002), October, Japan; 2002. p. 286–95.
[32] Wang Y, Huang Y, Fuchs WK, Kantilla C, Sure G. Progressive retry for software failure recovery in message passing applications. IEEE Trans Comput 1997;46(10):1137–41.
[33] Candea G, Fox A. Recursive restartability: turning the reboot sledgehammer into a scalpel. Proceedings of eighth HOT-OS. Germany: Schloss Elmau; May 2001. p. 110–15.
[34] Bobbio A, Sereno M, Anglano C. Fine grained software degradation models for optimal rejuvenation policies. Perform Eval 2001;46:45–62.
[35] Kulkarni VG, Nicola VF, Trivedi KS. On modeling the performance and reliability of multi-mode computer systems. J Syst Softw May 1986;175–82 North-Holland.
[36] Bobbio A, Telek M. The task completion time in degradable systems. In: Haverkort B, Marie R, Rubino G, Trivedi KS, editors. Performability modeling tools and techniques. New York: John Wiley & Sons; 2001.
[37] Jagerman DL. An inversion technique for the Laplace transform. Bell Syst Tech J 1982;61:1995–2002.

PART

4

Simulation methodologies in computer networks and systems

CHAPTER

14 A model-driven method for the design-time performance analysis of service-oriented software systems

Paolo Bocciarelli and Andrea D'Ambrogio

University of Roma "Tor Vergata", Roma, Italy

1 INTRODUCTION

The development of modern software systems usually targets the adoption of distributed architectures, due to both the intrinsic openness of current systems and the availability of powerful middleware and integration technologies. In this respect, Service-oriented Architecture (SOA) is a widely used software building paradigm, according to which an application results from the composition of a set of services in execution onto networked server hosts [1]. A service can be seen as an open and self-describing component providing a well-defined set of operations. *Web Services* is the most used technology for implementing the service concept [2].

In an SOA perspective, a *service-oriented software system* can be seen as distributed software implemented as an *orchestration* of Web services that cooperate to execute the process defining the collaboration workflow.

Service-oriented software systems are often at the core of mission- or business-critical systems, and thus advanced quantitative analysis techniques are needed to assess, from the early development stages, whether or not the system accomplishes the stakeholder requirements and constraints [3]. In this respect, simulation is an effective and widely adopted approach for carrying out a quantitative early evaluation of the system behavior. Simulation-based techniques provide a valuable strategy both to cut the cost of developing experimental prototypes and to mitigate the risk of time/cost overrun due to re-design and re-engineering of a system that does not provide the required quality.

Unfortunately, the use of simulation-based analysis approaches is still limited in practice due to [4]:

- inadequate technical skills of software analysts and designers in charge of conducting the simulation-based analysis;
- costs and difficulties to retrieve and analyze the data required for the simulation model parameterization;
- use of models that may be (partially) incorrect or may not include an adequate level of detail.

Such obstacles are even exacerbated when it comes to the use of distributed simulation (DS), which naturally addresses the inherent complexity and the distributed nature of service-oriented software systems. A DS is built by integrating into a so-called *federation* the several simulation components (i.e., the *federates*) that are executed onto a set of nodes interconnected by a network infrastructure of LAN or WAN type. DS has been traditionally used as a means to achieve scalability, reusability and parallelism. Nevertheless, the adoption of DS techniques requires a significant expertise and a considerable effort, due to the complexity of currently available DS standard and technologies.

To reduce the effort and make DS easier to use, this chapter introduces a model-driven automated method for enacting at design-time the DS-based performance analysis of service-oriented software systems.

The proposed method consists of two model transformations that take as initial input the UML model of the system under study, generate the corresponding performance model, and yield as final output the DS code ready to be executed.

Specifically, this work adopts the EQN (Extended Queueing Network) formalism to define the performance model and makes use of the jEQN language [5] to implement and execute the performance model. jEQN is part of SimArch, a layered architecture that eases the use of DS techniques by hiding all the implementation details of the distributed execution platform [6].

The several model transformations at the core of this chapter contribution have been implemented according to principles and standards introduced in the model-driven engineering field, specifically by use of MDA (Model Driven Architecture), the Object Management Group's incarnation of model-driven engineering principles [7]. The proposed approach makes use of *model-to-model* and *model-to-text* transformations that have been implemented in QVT (Query/View/Transformation) and MOFM2T (MOF Model to Text) transformation language, respectively [8].

The availability of such automated model transformations allows software engineers to predict system behavior with no extra effort and without being required to own specific skills of DS or performance theory.

In order to properly describe the service-oriented system, the input UML model is annotated with stereotypes provided by the UML profile for the Service-oriented architecture Modeling Language (SoaML) [9]. Moreover, the UML

Profile for Modeling and Analysis of Real-Time Embedded systems (MARTE) [10] is also used to characterize the input UML model in terms of performance-related properties and requirements.

In an SOA, the service provider creates a WSDL (Web Service Description Language) document and publishes it to one or more discovery registries (such as UDDI, Universal Description Discovery and Integration), so that service consumers can find the service using a wide variety of search criteria and then use the WSDL description to develop or configure a client that will interact with the service.

Unfortunately, a WSDL document only addresses the functional aspects of a Web service without containing any useful description of nonfunctional (e.g., performance-related) characteristics. Different Web services may provide similar functionality, but with distinct performance properties.

To this purpose, the model-driven method discussed in this chapter makes use of Q-WSDL (Quality-enabled WSDL), a lightweight WSDL extension for the description of the performance properties of a Web service [11].

The rest of this work is organized as follows: the related work section reviews various contributions dealing with the performance analysis of service-oriented software systems, while the background section summarizes the main concepts at the basis of this chapter contribution. Then, the implementation of the proposed model-driven method is illustrated. Finally, an example application of the method is described and the concluding remarks are drawn.

2 RELATED WORK

This section reviews the several contributions dealing with the performance analysis of SOA-based software systems implemented as orchestration of Web services.

Concerning the performance analysis of service-oriented systems based on the use of simulation-based techniques, a relevant contribution has been proposed in [12], which obtains the service time and communication latency for each Web service in the scenario by costly and time-consuming load testing rather than by use of predictive models.

The queueing network formalism is widely used in the performance analysis domain as it is supported by several tools and framework, such as JMT [13] or OMNet++ [14].

Specifically, the use of queueing-based models to analyze performance-related issues in the SOA domain has been widely investigated in [15–18].

In [15] parallel and concurrent structures adopted in open and closed queuing networks are analyzed to provide a clear understanding of the impact of parallelism on system performance. Specifically, response time approximations are proposed for parallel constructs modeled as fork and join queues, which allow one to effectively compare the several configurations from a performance-related perspective.

In [16] an approximate two-level single-class queueing network model is proposed to predict the execution time of applications executed on multicore systems. The model captures the memory contention due to multiple cores access and incorporates it into an application-level model.

In [17] a method is introduced to estimate the service response time of composite Web services. The proposed method takes as input the model of the system and derives a family of bounding models for the composite Web service response time. A valuable strength of such an approach is that the use of bounding models allows one to obtain an appropriate trade-off between accuracy and computational complexity.

In [18] the response time of composite Web services is addressed. The proposed approach, which aims at overcoming the state explosion problem, consists of the definition of bounding models providing upper and lower bounds on response time. Both single and multiple composite Web service execution instances are studied. Such an approach, similarly to the one proposed in [16], allows one to find the right trade-off between accuracy and computational complexity.

The above-mentioned contributions propose approaches to predict performance-related characteristics (e.g., the service time) of composite Web services by use of analytical-based techniques, providing valuable solutions for addressing relevant and challenging issues (e.g., the impact of using parallel/concurrent structures, the evaluation of resource contention, the attention to the computational complexity). Beside their relevance and effectiveness, such approaches show some limitations: the definition of the required analytical performance models needs a nonnegligible effort and the analysis of such models exhibits a high computational cost due to the state explosion problem. In this respect, most of the above-mentioned contributions adopt bounding models to manage the computational complexity at the expense of the prediction accuracy.

Alternatively, the approach proposed in this chapter addresses the simulation-based performance prediction of SOA-based software by introducing an automated method that does not require specific skills of performance modeling theory. The proposed method has been validated in order to ensure the correctness and the accuracy of the provided outcomes.

Approaches that face the performance of service-oriented systems from a wider perspective (i.e., not limited to queueing-based modeling) can be found in [19,20].

In [19] an approach that discusses a QoS prediction algorithm consisting of a set of reduction rules applied to Web service processes is proposed. The main limitation of such an analytical approach is that the use of reduction rules for the prediction of performance-related QoS attributes (e.g., the overall response time) does not take into consideration the contention experienced by different service consumers for accessing shared services and can thus be considered as a best-case (or stand-alone) estimation.

In [20] a UML-based approach for integrated security and performance analysis of service-based systems is proposed. The performance analysis is carried out by generating and solving a PEPA (Performance Evaluation Process Algebra) model. The approach is applied to the development of Web service-based portals and the authors claim it can be used to assess service discovery protocols as well, without mentioning composite Web service scenarios.

Differently from the above illustrated contributions, this work makes use of the EQN notation for specifying the performance model and adopts the jEQN language for implementing and executing the EQN model. The use of a model-driven approach combined with jEQN brings several benefits, which contribute to overcome most limitations of existing works:

- the proposed method automatically builds the performance model by use of model transformations and thus does not require any specific expertise in the performance theory field;
- the proposed method is open and fully customizable, meaning that the proposed performance model can be easily improved or extended in order to better model distinctive characteristics of specific domains;
- jEQN supports the representation of Allocate-Release or Fork-Join primitives, in order to effectively model access control and concurrency, respectively;
- jEQN supports the transparent execution of simulation models into local or distributed simulation environments.

3 BACKGROUND

3.1 MODEL-DRIVEN ARCHITECTURE

Model-driven engineering (MDE) is an approach to software design and implementation that addresses the raising complexity of execution platforms by focusing on the use of formal models [21,22]. According to this paradigm, a software system is initially specified by use of high-level models. Such models are then used to generate other models at a lower level of abstraction, which are used in turn to generate other models, until stepwise refined models can be made executable.

One of the most important initiatives driven by MDE principles is MDA (Model-driven Architecture), the OMG's incarnation of MDE [7]. MDA-based software development is founded on the principle that a system can be built by specifying a set of model transformations, which allow one to obtain models at lower abstraction levels from models at higher abstraction levels.

To achieve such an objective, MDA provides the following standards:

- **Meta Object Facility (MOF):** for specifying technology neutral metamodels (i.e., models used to describe other models) [23];
- **XML Metadata Interchange (XMI):** for serializing MOF metamodels/models into XML-based schemas/documents [24];
- **Query/View/Transformation (QVT):** for specifying model transformations [25].

The relationship among MDA standards is summarized by the scenario depicted in Figure 14.1. Model M_A and model M_B are instances of metamodel

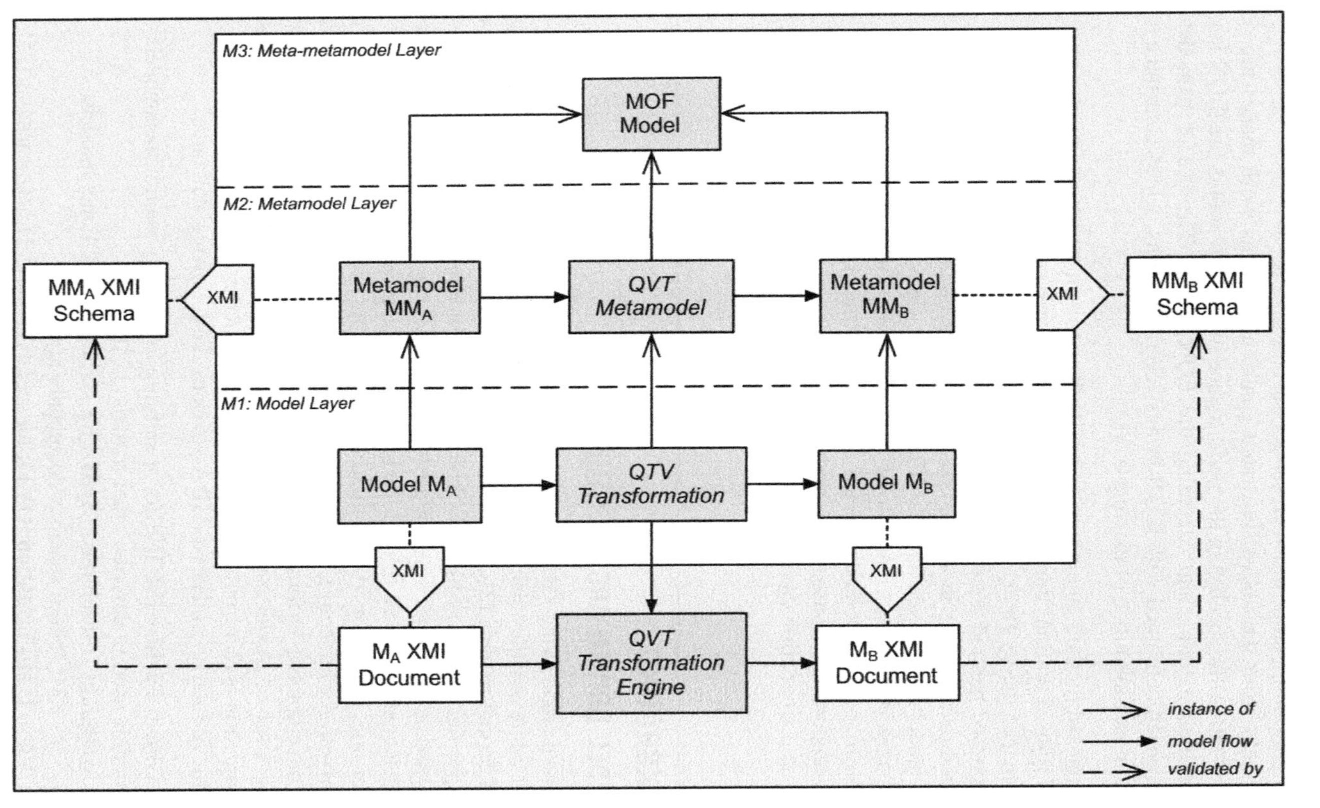

FIGURE 14.1

Relationship among MDA standards.

MM_A and metamodel MM_B, respectively, which in turn are defined in terms of MOF constructs. A model transformation, i.e., an instance of the QVT metamodel, gives the rules to map model M_A into model M_B. The so-specified transformation is then executed by a QVT transformation engine.

Both model M_A and model M_B can be serialized by use of XMI rules, to obtain the corresponding XMI-based documents. XMI rules can also be used at the metamodel layer to serialize metamodels and obtain XMI schemas, which are used to validate XMI documents.

In this chapter MDA has been used to design and develop the method to enact the design-time performance analysis of service-oriented software systems.

3.2 SIMULATION EXECUTION PARADIGMS

A given simulation model is implemented into a simulation program that can be executed according to three different paradigms: local, parallel and distributed.

In a local simulation the simulation program is deployed onto a single processor platform that completes its execution. According to a parallel/distributed approach, the simulation program may be executed over multiple processors. In this chapter the following taxonomy is considered [26]:

- **Parallel Simulation (PS)**, which is concerned with the execution on multiprocessor computing platforms containing multiple CPUs that interact frequently, e.g., thousands of times per second.
- **Distributed Simulation (DS)**, which is concerned with the execution on loosely coupled computing platforms, in which interactions take much more time, e.g., milliseconds or more, and occur less often. Its execution relies on a distributed system consisting of a set of hosts interconnected through a LAN or a WAN (e.g., the Internet) network infrastructure.

This chapter specifically addresses the DS case. Simulation of modern and complex service-oriented software systems requires computational resources that might not be available on a single host, and thus DS is often used to enact a scalable way to simulate a complex system by partitioning the simulation model into submodels, each simulated on an independent host [27].

More specifically, the following benefits can be obtained by the adoption of DS techniques, as underlined in [27, 6]

- reduced execution time: DS allows one to split a large simulation computation into a set of subcomputations, each executed concurrently on a different host;
- geographical distribution: DS involves distributed hosts enabling the creation of virtual worlds with multiple participants that are physically located at different sites;
- interoperability and reusability: a complex system may involve several subsystems, developed by different manufacturers. Rather than porting a simulation program for each submodel to a single host, it may be more cost

effective to integrate the existing simulators, each responsible to simulate a specific submodel and each executing on a different host;

- fault tolerance: DS increases tolerance to failures. If one host goes down, it may be possible for other hosts to pick up the work of the failed machine, allowing the simulation computation to proceed despite the failure.

3.3 SIMARCH AND JEQN

SimArch is a layered architecture that eases the development of local and distributed simulation systems by removing the developers from all the details concerning the execution environment, which can be either a conventional local execution platform or a distributed execution platform, e.g., one based on the HLA (High Level Architecture) standard [28].

The simulation model is specified in terms of the adopted domain-specific language (DSL), defined at the upper layer of SimArch. As aforementioned, this work exploits the jEQN language, a DSL for the specification of the EQN models [5]. The language provides several simulation components (i.e., *jEQN components*) whose implementation exploits services provided by the underlying SimArch layers. Such jEQN components are classified into *simulation entity components* and *support components*. The simulation entity components identify the simulation logic and are named using the EQN standard taxonomy (e.g., user sources, waiting systems, service centers, routers and special nodes). The support components identify all the objects that do not affect the simulation logic. These components give the structures for the entity components parameterization, e.g., policy frameworks, and for the data definition, e.g., users and queues. For a detailed description of SimArch and jEQN the reader is sent to [6] and to the SimArch project web site [29].

4 MODEL-DRIVEN METHOD

This section illustrates the model-driven method for the performance-based analysis of service-oriented software systems. In order to benefit from the distributed nature of service-oriented systems, the method exploits the SimArch architecture and the jEQN language to conduct a DS-based analysis of the system under study. The method effectively supports the effortless development of a distributed simulation, starting from an UML model of the system. The rationale of the method is outlined in Figure 14.2.

At the initial step of the method, the functional requirements are used to specify a design model of the service-oriented system by use of the UML notation. Such a model, which is annotated with stereotypes provided by the

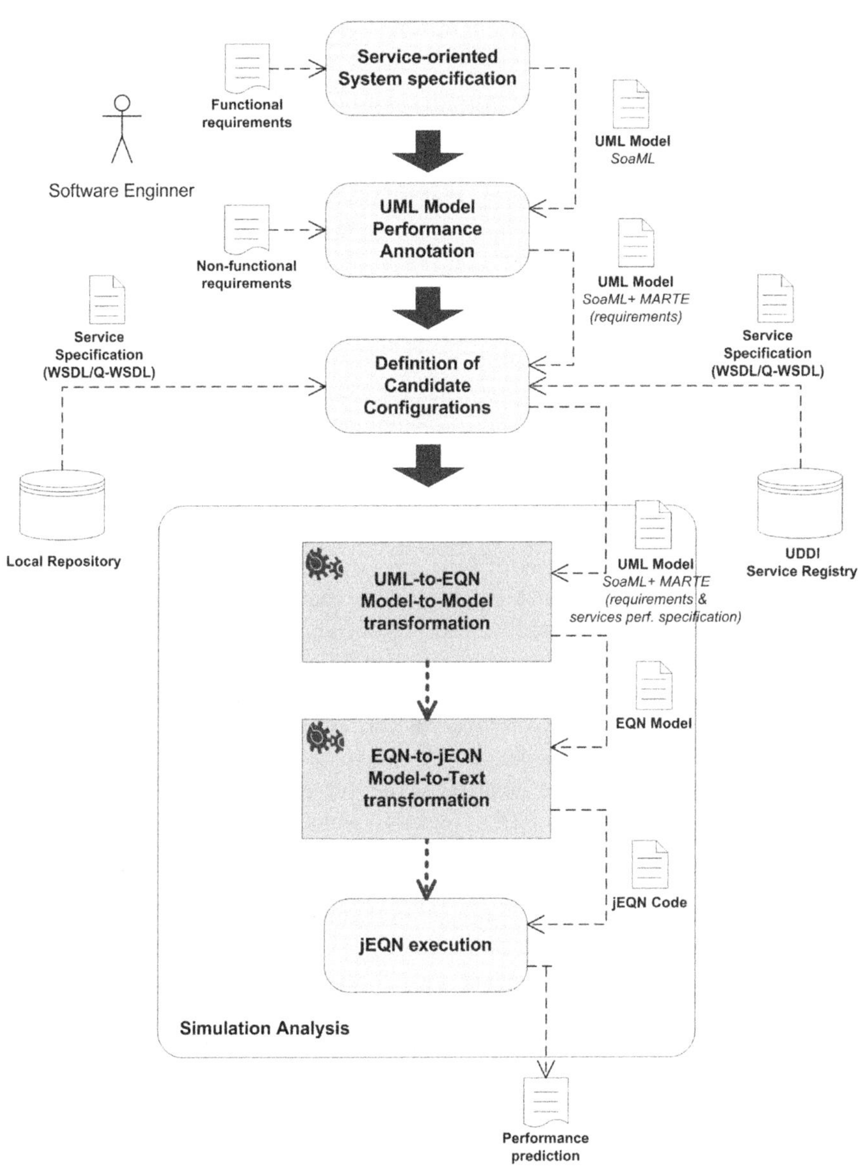

FIGURE 14.2

Model-driven method for the performance prediction of service-oriented software systems.

SoaML profile [9], consists of the following set of packages, according to the modeling conventions for the application of the SoaML profile [30]:

- *Service Interface Package*: specifies the capabilities and the related interfaces that the software services must implement in order to be integrated in the service-oriented system;
- *Service Contract Package*: defines the Service Contracts (i.e., the specification of agreements between interacting parties, in SoaML terms) that specify how the interfaces are used;
- *Participant Package*: contains the Participants (i.e., a person, a system or an organization that provides or consumes a service, in SoaML terms) that implement the abstract interfaces;
- *Service Architecture Package*: specifies the Service Architecture (i.e., the network of interacting participants providing and consuming services, in SoaML terms), by means of a UML collaboration enriched with a nested UML activity diagram, which specifies the system behavior.

Beside such a functional specification, software engineers may want to describe the system from a nonfunctional point of view, e.g., to represent the performance-related requirements that the system must satisfy when executed in its operational environment. In this respect, the proposed method makes use of the MARTE profile, whose stereotypes are used to further annotate the initial UML model [10].

As the service-oriented system is implemented by an orchestration of Web services, at the next step a service discovery is carried out to bind each service interface included in the UML model to one of the several concrete Web services matching the relevant service interface. The set of concrete services can be retrieved both from a public UDDI registry, which stores the specification of third-party services, and from a local repository, where organizations store the description of ad-hoc services specifically provided by their known suppliers. The method has been designed to be compliant to both standard WSDL and Q-WSDL. In the former case, as WSDL does not include any nonfunctional description of Web services, the software designer must provide the required performance-related information. In the latter case, the performance characterization of each Web service is included in the Q-WSDL document. Such performance-related information is then included in the UML model by use of the MARTE profile. Indeed, at this step the UML model contains both the performance requirements and the performance-related characterization of the orchestrated Web services.

At the next step, the *model-to-model* `UML-to-EQN` transformation is executed to derive the EQN performance model of the system under study. This EQN model is based on an enhanced EQN metamodel, which allows one to represent the partitioning of an EQN model into a set of submodels (or subnetworks). The UML-to-EQN model transformation allows to specify how EQN subnetworks must be run onto the independent hosts of the underlying distributed simulation infrastructure.

Finally, the *model-to-text* `UML-to-jEQN` transformation is executed to generate the jEQN code that implements the DS simulation.

The so-obtained jEQN code is deployed on a SimArch-based distributed platform and is executed for ultimately yielding the performance predictions.

The next sections describe the EQN metamodel and the method implementation, respectively.

4.1 EQN METAMODEL

Figure 14.3 illustrates the EQN metamodel used to build EQN-based performance models. The metamodel specifies the relationships among EQN entities, such as *sources*, *service centers*, *waiting systems*, etc. Specifically, the abstract metaclass `BaseElement` defines the basic element of the EQN metamodel. This metaclass is characterized by `name` and `id` attributes, which specify the element name and the unique element identifier, respectively. The `EQNModel` metaclass represents the root element of the EQN model. As an EQN model is specified in terms of centers and links that constitute the queueing network, the metamodel also includes the metaclasses `Center` and `Link`, that are connected to the `EQNModel` metaclass by aggregation associations.

The EQN model constitutes the input of the model-to-text transformation that generates the jEQN code and thus it must include information needed to drive the generation of the several components of the distributed simulation. To this purpose, in order to model the partitioning of an EQN model into several subnetworks, the `Subnetwork` metaclass has been added to the metamodel. Moreover, as a subnetwork consists of a collection of *centers* and *links*, the metamodel also includes the aggregation relationships between the `Subnetwork` metaclass and `Center` and `Link` metaclasses, respectively, as illustrated in the portion of Figure 14.3 bounded by a dotted line.

For a complete description of the EQN metamodel interested readers may refer to [31].

4.2 METHOD IMPLEMENTATION

The core of the proposed model-driven method consists of two model transformations. The first, namely `UML-to-EQN`, is the *model-to-model* transformation that generates the EQN-based performance model. Specifically, it maps an input UML Activity Diagram (AD) to an output EQN model. The second, namely `EQN-to-jEQN`, is the *model-to-text* transformation that takes as input the EQN model and yields as output the jEQN code implementing the distributed simulation. The next two sections give a detailed view of such model transformations.

4.2.1 Generation of the EQN performance model

The `UML-to-EQN` model transformation has been specified by use of QVT [24] and has been implemented and executed by use of the Eclipse MMT plugin [32].

QVT prescribes that both the source and the target models used in a transformation must be instances of a MOF compliant metamodel [23]. To this respect, the target EQN model is an instance of the EQN metamodel in Figure 14.3.

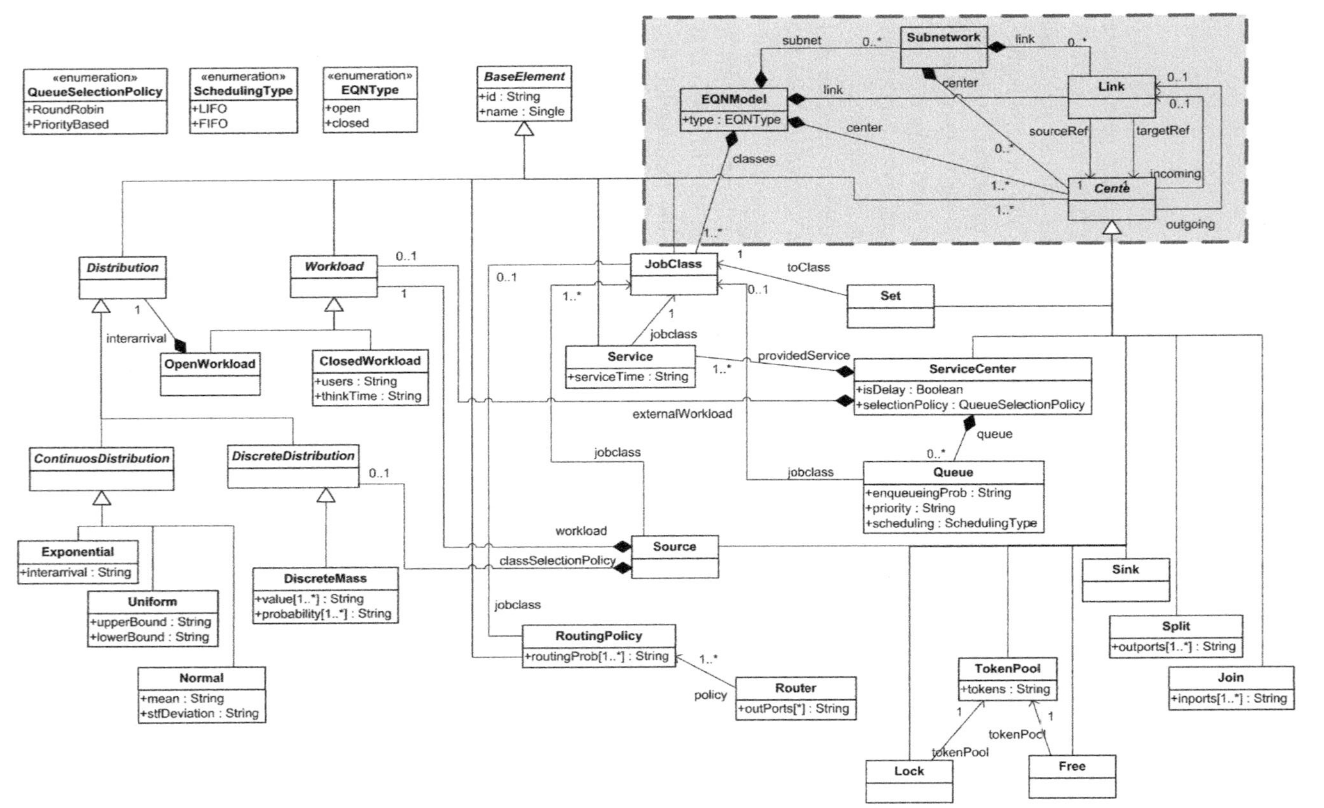

FIGURE 14.3

EQN Metamodel.

The service-oriented system has been designed as an orchestration of services implemented by use of the *Web Services* technology. It is assumed that the related AD includes a detailed view of the party that acts as the coordinator of the service composition (the *orchestrator*, in SOA terminology), while other parties, which act as *black-box* entities providing a single service, are represented by *swimlanes* including an *activity* for each provided operation.

In this perspective, a service invocation is represented in the UML model by an *activity edge* that connects a pair of *activity nodes*, the first one belonging to the *swimlane* associated to the *orchestrator*, and the second one belonging to the *swimlane* associated to the service provider.

In order to give the rationale of the model transformation process, Figure 14.4 outlines the mapping of the UML fragment representing a service invocation into the corresponding portion of the EQN model.

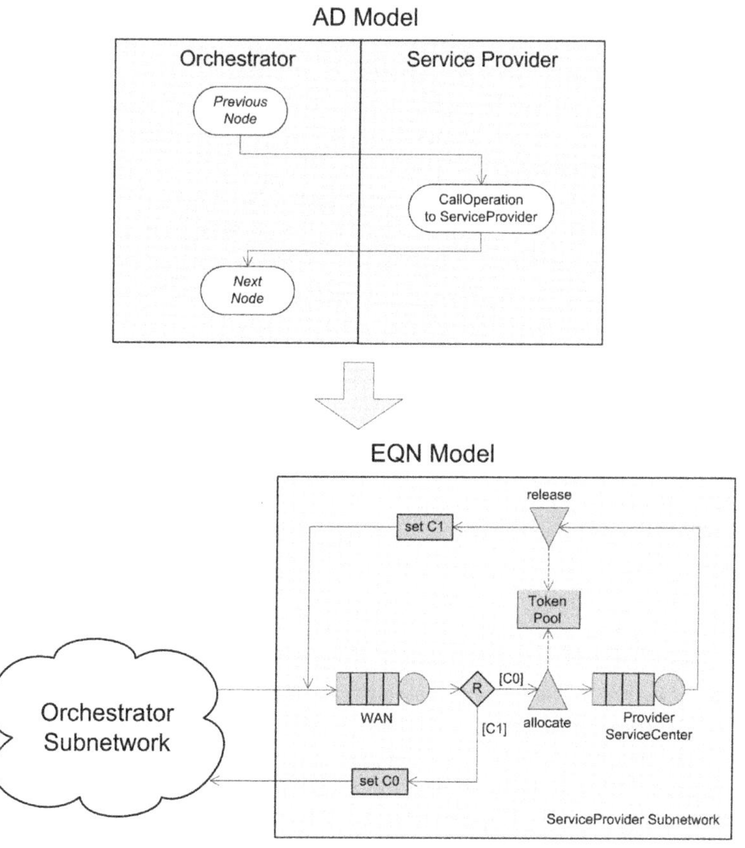

FIGURE 14.4

EQN mapping of the UML-based service invocation pattern.

The EQN model includes a subnetwork associated with each participant (e.g., the orchestrator and the set of service providers). Each subnetwork can be simulated by different simulation components of the distributed simulation infrastructure, as detailed later on. Two classes of jobs are introduced: the first one, named "`toServe`", is used to represent jobs that have to be served by a participant; the other one, named "`Served`", is used to model a job just served by a participant. The pattern shown in the figure is related to jobs processed by the orchestrator service center that have to be served by the next service center, according to the orchestration paradigm. The job class is initially set to "`toServe`", briefly denoted as "`C0`". A request to the next service center is structured as follows:

- the job passes through the *WAN Service Center*, to model the request message that the orchestrator sends to the service provider;
- the job passes through the *Provider Service Center*, to model the service execution performed by the participant. It should be noted that the router `R` forwards jobs of class "`C0`" to the *Provider Service Center*. Moreover, the access to the *Provider Service Center* is controlled by `Allocate/Release` nodes, in order to model the capacity of the server that provides the requested service. Finally, as the job leaves the `Release` node, its class is changed to "`Served`", briefly denoted as "`C1`";
- the job passes through the *WAN Service Center*, to model the response message that the service provider sends to the orchestrator;
- the job returns to the *Orchestrator Service Center*. It should be noted that the router `R` forwards all jobs belonging to class "`C1`" to the "`Set C0`" node and then to the *Orchestrator Service Center*.

The parameterization of the EQN model has been carried out by implementing an algorithm based on the one specified in [33], which takes into account the MARTE annotations included in the input UML model.

A complete outline of the mapping rules between UML elements and EQN elements is provided by Table 14.1.

The next section describes some implementation issues regarding the `EQN-to-jEQN` *model- to-text* transformation.

4.2.2 Generation of the jEQN-based DS implementation

Once the EQN model has been obtained from the `UML-to-EQN` transformation, the `EQN-to-jEQN` transformation is executed to derive the jEQN code. The `EQN-to-jEQN` transformation, which has been specified as a MOFM2T model-to-text transformation [8], has been implemented and executed by use of the Acceleo transformation language [34], which is itself provided as an Eclipse plugin. Such a transformation also provides the required set of property files for configuring the simulation environment.

Table 14.1 Mapping of UML Elements to EQN Elements

UML Element	EQN Element
Swimlane	Subnetwork
MARTE annotation *(associated to swimlane)*	Users and think time parameters *(for closed EQN)*
MARTE annotation *(associated to swimlane)*	Distribution of interarrival time *(for open EQN)*
Start/Final Node	Terminal node *(for closed EQN)* Source/Sink node *(for open EQN)*
Opaque Action Node	request to Orchestrator Service Center
Call Operation Node	see Figure 14.4
Fork Node	Fork Node
Join Node	Join Node
Decision Node Decision Node	Router Node
Control Flow	used for defining the routing within the EQN

The transformation is structured in the following steps:

Step 1. generation of both the software components managing the execution of the distributed simulation systems and the data structures exchanged by simulation components;
Step 2. generation of simulation scenario settings;
Step 3. generation of jEQN simulation components, according to the EQN partitioning;
Step 4. generation of the batch file to start the Java programs containing the jEQN components.

As the currently adopted jEQN implementation makes use of the HLA standard [27], the software components created at Step 1 refer to the HLA *Federation Manager* federate, whose role is the coordination of the simulation lifecycle [35]. This steps also produces the data definition files for data exchange among federates.

Step 2 produces the simulation environment configuration files, which are obtained by tailoring the templates already defined in SimArch. In particular, these files define the configuration of both the HLA environment (i.e., hostname and port number of the HLA server) and the simulation scenario (i.e., number of jEQN components, simulation length, etc.).

Step 3 derives the code of the required jEQN components. This step is further specified as follows:

3.1. Generation of import statements and Class skeleton

3.2. Generation of `main` method
  3.2.1. Generation of statements for SimArch and jEQN initialization
  3.2.2. Generation of statements for the subnetwork's EQN (local) entities
  3.2.3. Generation of statements for the remote stubs for adjancent subnetwork's EQN entities
  3.2.4. Generation of statements for the connections of the declared entities (both local and remote stubs)
  3.2.5. Generation of the statements to activate the simulation execution

The mapping makes a distinction among *local* and *remote* EQN elements. Specifically, for each EQN subnetwork S:

- all the EQN elements e_i that belong to S are *local* elements;
- all the EQN elements e_j that belong to a subnetwork $S' \neq S$, and that are directly connected to EQN elements e_i in S, are *remote* elements.

For each subnetwork, the mapping of local EQN elements to jEQN classes is straightforward and is outlined in Table 14.2.

Alternatively, for managing the interaction among elements that belong to different subnetworks (e.g., to different jEQN remote components) the mapping algorithm generates local stubs for the remote referenced entities.

Once all the stubs for the remote entities are defined, the `EQN-to-jEQN` transformation generates the jEQN code that implements the connections among jEQN elements, according to the EQN topology.

Finally, Step 4 generates the batch files for the execution on the SimArch environment, including the HLA server, the Federation Manager and the individual jEQN simulators. These batch files also refer to the above generated configuration files for the considered analysis scenario.

Table 14.2 Mapping of EQN Elements to jEQN Classes

EQN UML Class	jEQN Class
Terminal	Source Class InfiniteServer Class
Source	Source Class
Sink	Sink Class
Queue	WaitingSystem Class
Passive Queue	PassiveQueue Class
Split Node	SplitNode Class
Join Node	JoinNode Class
Router	Router Class

5 EXAMPLE APPLICATION

This section presents an example method application to a service-oriented application for reserving cars provided by a *car sharing* company. Let us suppose that the reservation process includes the following main steps:

1. the user provides the authentication credentials;
2. according to the existing user profile, the system retrieves and shows to the user a set of tailored commercial offers. The system also shows on a map the location of available cars ready for renting;
3. the user selects the car to be rented;
4. the system retrieves and shows the list of fuel stations and parking areas;
5. the user confirms the reservation.

As regards step 4, it is assumed that with a 70% probability the user prefers a public parking area. Depending on the user's choice the system retrieves and shows the desired type of parking.

According to the method shown in Figure 14.2, the first step deals with the specification of functional and nonfunctional requirements of the service-oriented software system, by use of a UML description annotated with stereotypes provided by SoaML and MARTE profiles.

The system is designed as an orchestration of the following services:

- a service providing commercial services, denoted as *AdvertisingManager* service;
- a service managing the car fleet, denoted as *CarFleetManager* service;
- a service managing points of interests (POI), denoted as *PointsOfInterestsManager* service.

As discussed above, the UML model consists of four diagrams that provide the following views:

- Service interfaces and capabilities, which specify the capabilities and the related interfaces that orchestrated services must implement, as shown in Figure 14.5;
- Participants, which describe the parties involved in the implementation of the service-oriented systems, as shown in Figure 14.6;
- Service contracts, which specify the agreements between interacting parties, as shown in Figure 14.7;
- Service architecture, which gives both an architectural view, by specifying how participants provide and consume services, and a dynamic understanding of the system behavior. The UML Composite Structure diagram, which specifies the service architecture, is shown in Figure 14.8, while the nested UML AD is depicted in Figure 14.9.

At the next step, a service discovery is carried out to find a set of concrete services that match the abstract service interfaces specified in the UML model. For

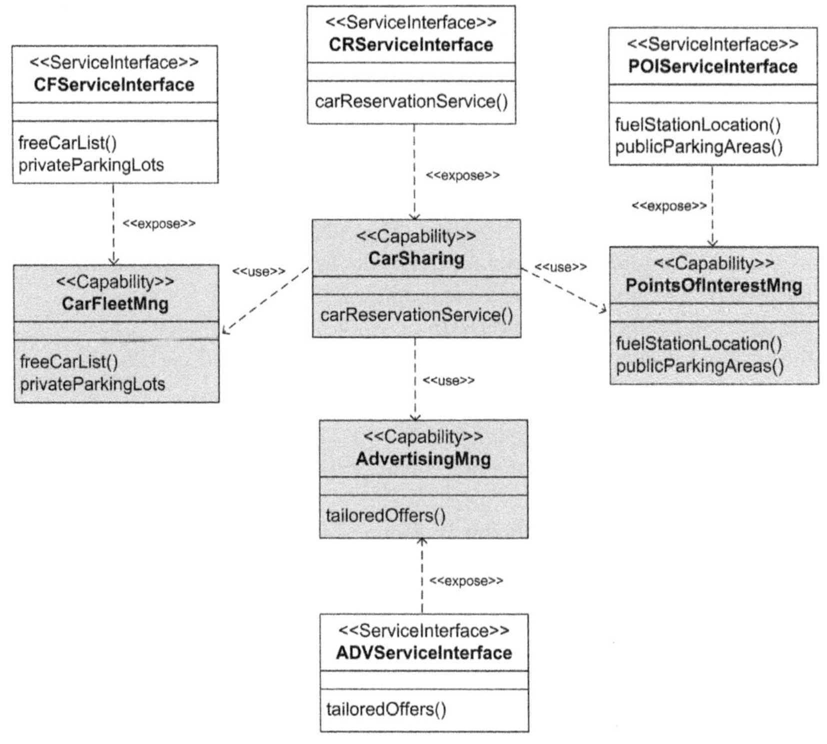

FIGURE 14.5

Service interfaces and capabilities.

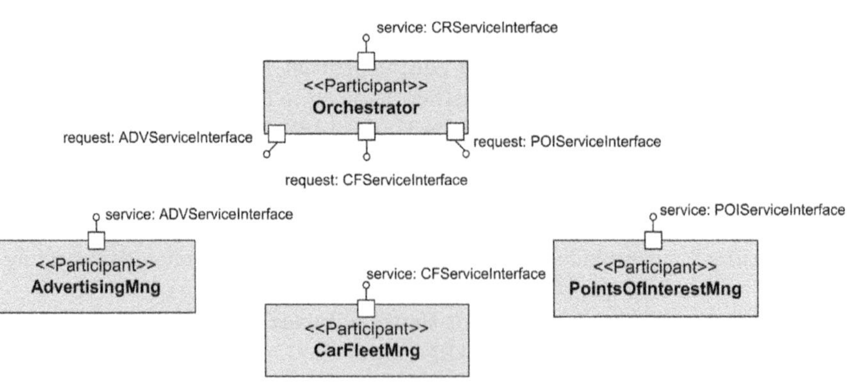

FIGURE 14.6

Participants.

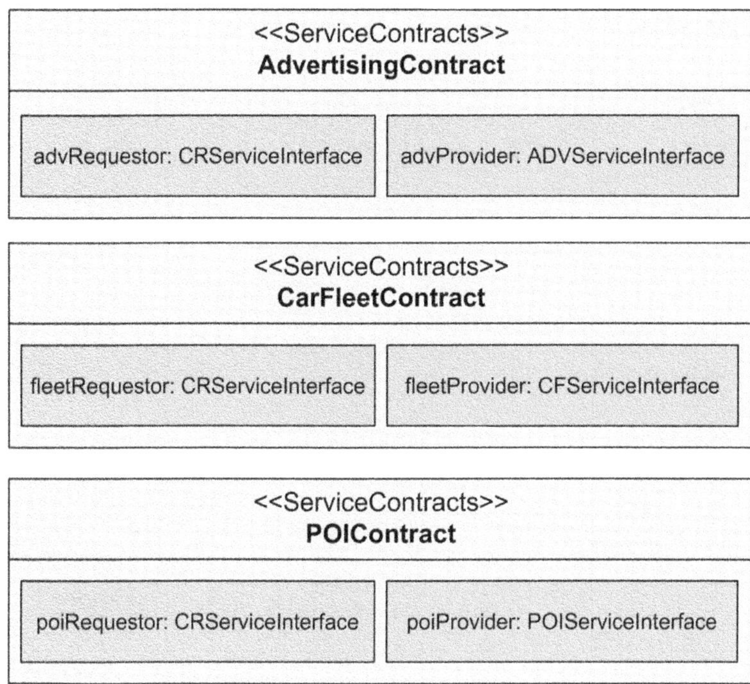

FIGURE 14.7

Service contracts.

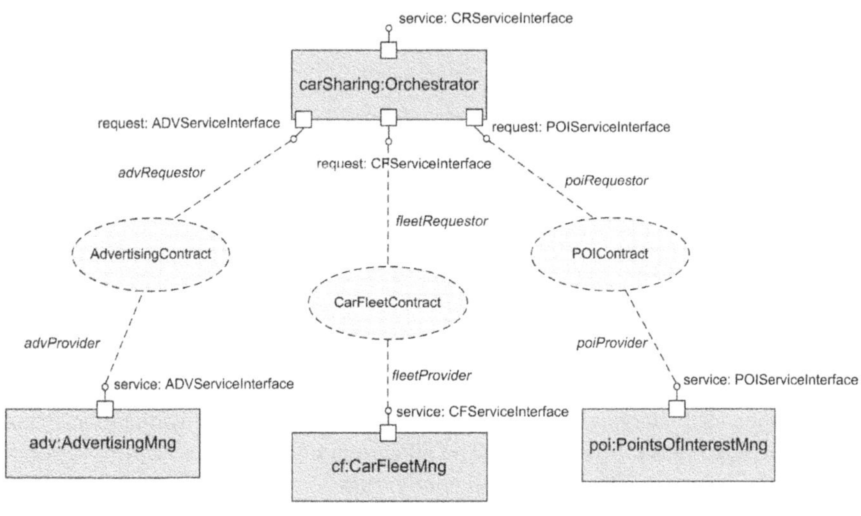

FIGURE 14.8

Composite structure diagram.

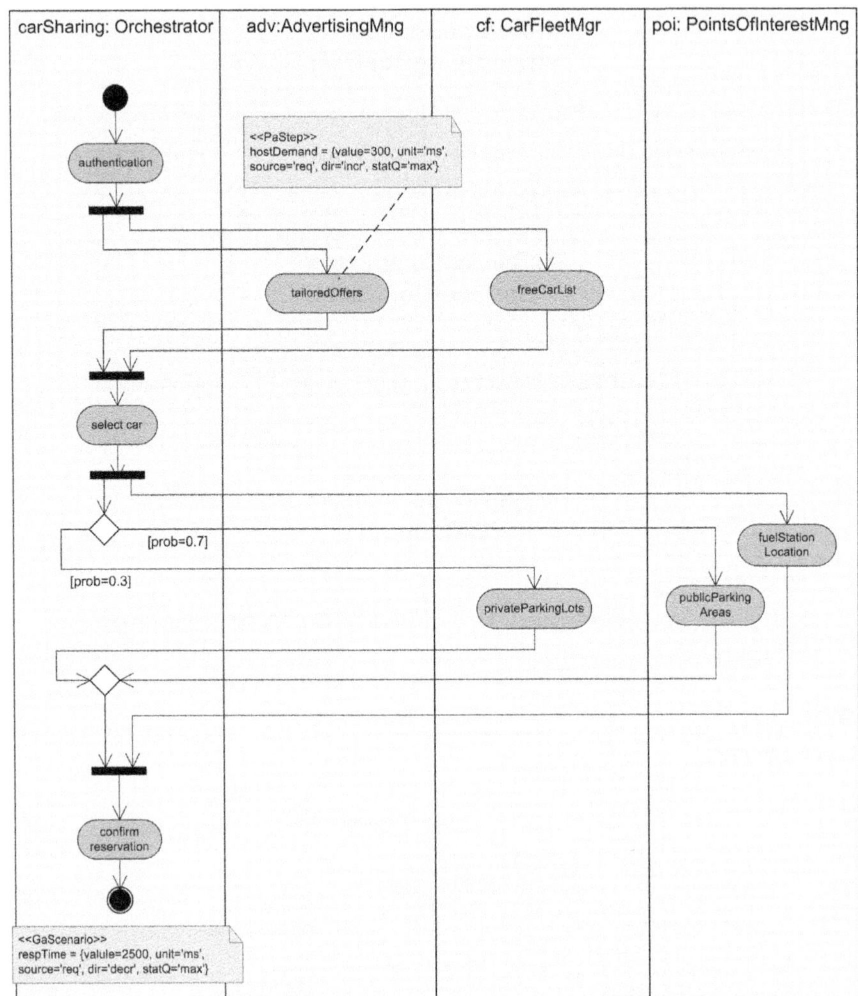

FIGURE 14.9

Activity diagram with MARTE annotations (requirements).

the sake of simplicity, it is supposed that a single concrete service is found for each required service interface. After the execution of a service discovery, the performance characteristics of the candidate service are finally available and thus can be included in the UML model. The allocation of abstract services to concrete services defines a so-called *candidate configuration*, which specifies both the performance requirements of the service-oriented system and the performance characteristics of the selected concrete services. The resulting extended UML AD is shown in Figure 14.10.

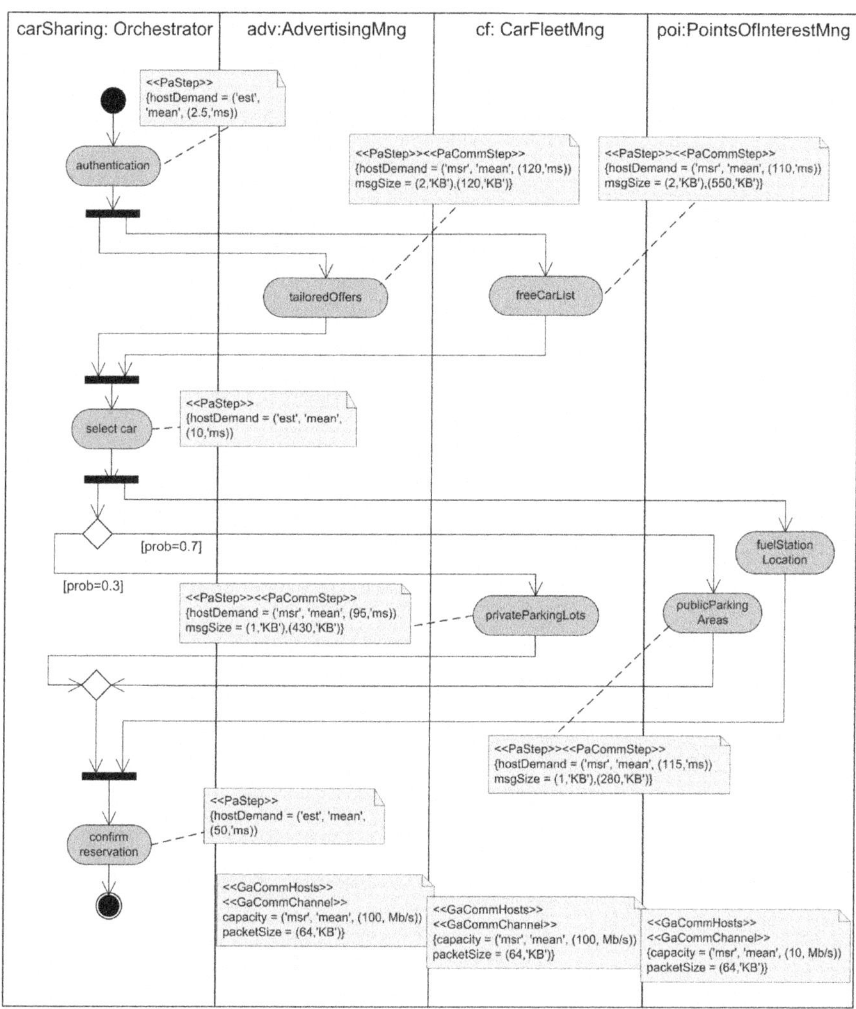

FIGURE 14.10

Activity diagram with MARTE annotations (service characteristics).

The performance prediction is then carried out by first generating the EQN model and then analyzing it by deriving and executing the corresponding jEQN code. The resulting EQN model is shown in Figure 14.11, while Listing 14.1 illustrates a fragment of the jEQN code implementing Subnet 1.

Even though this chapter is focused on an automated model building method, without taking into explicit consideration the model evaluation activity, for the

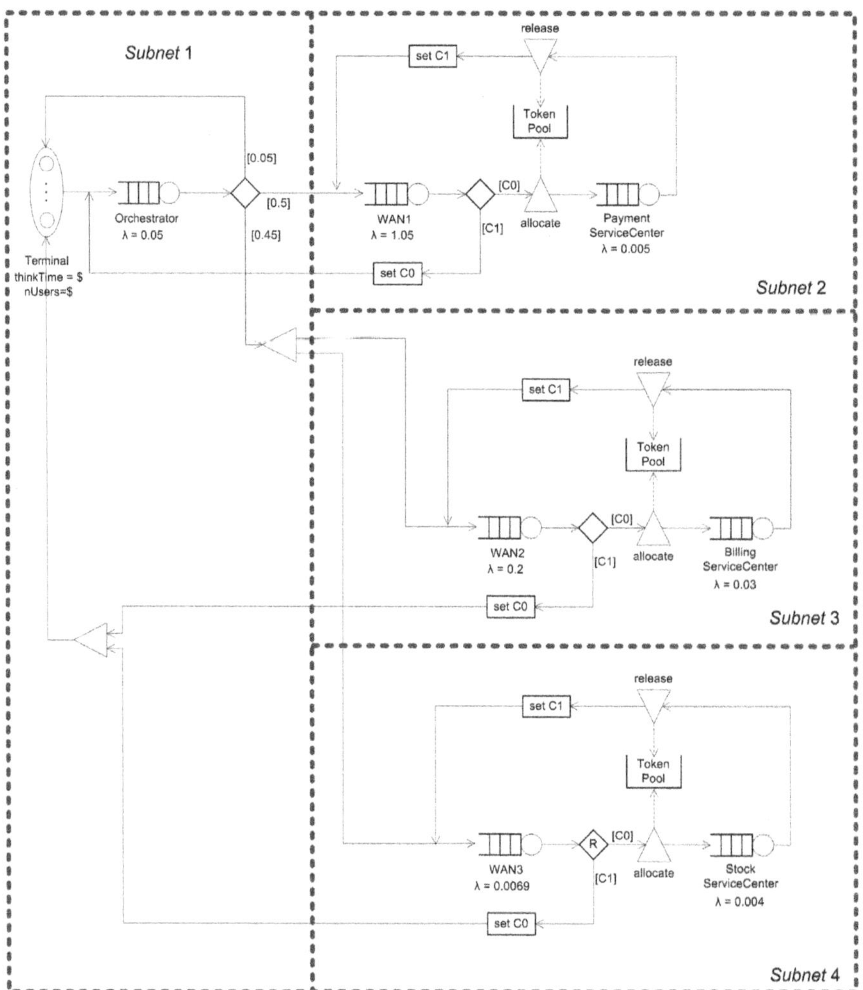

FIGURE 14.11

EQN model.

sake of completeness Figure 14.12 summarizes the results of an example performance prediction, by plotting the session mean response time with respect to the number of users, with a confidence interval of 95%.

The model evaluation activity can be used to check whether the prediction satisfies the given performance requirements and, in the negative case, take corrective actions (e.g., predicting the performance of additional candidate configurations) before system implementation.

```
...
#Initialization statement
NumericStream ns = new ExponentialStream(new JavaSimGenerators(), 0.08);
MaskBasePolicy<List<User>,User,?,Integer> insertionPolicy = new
      MaskImplicitButNotExplicitInputDependentPolicy<List<User>, User,
      Integer>(new FIFOEnqueuingPolicy(new ArrayList<User>()));
UserQueue userQueue = new InfiniteUserQueue(insertionPolicy,
userListImplementation);
...
#Creation of local components
NonPreemptiveWaitingSystem ws = new NonPreemptiveWaitingSystem(new
      JEQNName("Orchestrator"), timeFactory, layer2Factory,new InfiniteUser
      Queue( new UserDecisionDataFactory(),insertionPolicy,
      userListImplementation),new SingleCatServiceRequestGenerator(ns));
...
#Creation of stubs to remote components
RemoteEntity WAN1 = new BasicRemoteEntity(new JEQNName("Subnet2"), new
      JEQNName("WAN1"));
InPort wan1InPort = new InPort(new JEQNName(WaitingSystem.IN_PORT_NAME),
      WAN1);
...
#Creation of links
BasicLink l = new PointToPointLink(<senderEntityName>.getOutPort(),
<recipientEntityName>.getInPort());
...
```

LISTING 14.1

Fragment of jEQN code (Orchestrator subnetwork).

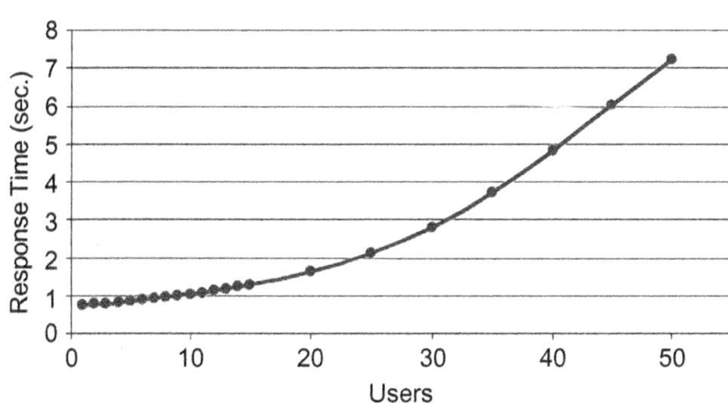

FIGURE 14.12

Results of the performance prediction.

6 CONCLUSIONS

The ability to include accurate and timely predictive performance analysis techniques in the development process of service-oriented software systems is essential to achieve the required quality and to ensure compliance with budget and time constraints.

In this respect, this chapter has introduced a model-driven method to automate the design time performance analysis of service-oriented systems. The method makes use of a distributed simulation approach that, differently from conventional (i.e., local) simulation approaches, replicates the distributed architecture of the addressed systems and provides several advantages, the most relevant ones being the increased computational capability and the ability to reuse and integrate already existing simulation components.

The proposed method consists of two model transformations that take as initial input the UML model of the system under study and yield as final output the Java code of the distributed simulation based on the jEQN domain specific language and the SimArch architecture. The execution of the so-obtained simulation code allows one to predict of system behavior and obtain the required performance-related indices.

An example application has been provided, to show how a jEQN-based simulation system can be easily derived from a UML model of a scenario dealing with a car-sharing facility. The example, which has been kept simple enough to illustrate the proposed approach without dealing with unnecessary and lengthy details, has revealed how the proposed method can be easily used even by software analysts who are not familiar with performance theory and/or distributed simulation standards and technologies.

REFERENCES

[1] Papazoglou MP, Georgakopoulos D. Service-oriented computing. Commun ACM 2003;46(10):25–8.

[2] Alonso G, Casati F, Kuno H, Machiraju V. Web services. Springer-Verlag; 2004.

[3] Menascé DA, Almeida VAF. Capacity planning for web services: metrics. Models and methods. Prentice Hall; 2001.

[4] Ryan J, Heavey C. Process modeling for simulation. Comput Ind 2006;57:437.

[5] D'Ambrogio A, Gianni D, Iazeolla G. (2006) jEQN: a Java-based language for the distributed simulation of queueing networks. Proceedings of the 21th international symposium on computer and information sciences. volume 4263 of LCNS, p. 854–65.

[6] D'Ambrogio A, Iazeolla G, Gianni D. A software architecture to ease the development of distributed simulation systems. Simulation 2011;87(9):813.

[7] Object Management Group (OMG). MDA Guide, v. 1.0.1. Miller J, Mukerji J, editors. Available from: <http://staffwww.dcs.shef.ac.uk/people/A.Simons/remodel/papers/MDAGuide101Jun03.pdf>; 2003.

[8] Object Management Group (OMG). MOF Model to text transformation language (MOFM2T), 1.0. Available from: <http://www.omg.org/spec/>; 2008b.

[9] Object Management Group (OMG). Service oriented architecture modeling language (SoaML), v. 1.0.1 2012. Available from: <http://www.omg.org/spec/SoaML/>; 2012.

[10] Object Management Group (OMG). A UML Profile for modeling and analysis of real-time embedded systems (MARTE), v. 1.1. 2011. Available from: <http://www.omg.org/spec/MARTE/>; 2009.

[11] D'Ambrogio A. A model-driven WSDL extension for describing the QoS of Web services. Proceedings of the IEEE international conference on Web Services (ICWS'06). IEEE Computer Society. Washington, DC, USA; 2006. p. 789–96.

[12] Song H, Ryu Y, Chung T, Jou W, Lee K. Metrics, methodology, and tool for performance-considered web service composition. In: Yolum P, Gungor T, Gurgen F, Ozturan C, editors. Computer and information sciences–ISCIS 2005. Berlin/Heidelberg: Springer; 2005. p. 392–401. volume 3733 of LNCS.

[13] Bertoli M, Casale G, Serazzi G. JMT: performance engineering tools for system modeling. SIGMETRICS Perform Eval Rev 2009;36(4):10–15.

[14] Varga A. The OMNeT++ discrete event simulation system. Proceedings of the european simulation multiconference (ESM'2001), vol. 9; 2001. p. 185.

[15] Alomari F, Menascé DA. Efficient response time approximations for multiclass fork and join queues in open and closed queuing networks. IEEE Trans Parallel Distrib Syst 2014;25(6):1437–46.

[16] Bardhan S, Menascé DA. Analytic performance models of applications in multi-core computers. Proceedings of the 21st international symposium on modeling, Analysis and simulation of computer systems and telecommunication systems (MASCOTS 2013). San Francisco; August 14–16; 2013, p. 318–22.

[17] Haddad S, Mokdad L, Youcef S. Bounding models families for performance evaluation in composite web services. Comput Sci J 2013;4(4):232–41. Elsevier.

[18] Mokdad L, Youcef S. Stochastic bounds for composite Web services response times. J Cluster Comput 2012;15(4):363–71. Springer.

[19] Cardoso J, Sheth AP, Miller JA, Arnold J, Kochut K. Quality of service for workflows and web service processes. J Web Semant. 2004;1(3):281–308.

[20] Gilmore S, Haenel V, Kloul L, Maidl M. Choreographing security and performance analysis for web services. In: Bravetti M, Kloul L, Zavattaro G, editors. Formal techniques for computer systems and business processes. Berlin/Heidelberg: Springer; 2005. p. 200–14. Volume 3670 of LNCS.

[21] Atkinson C, Kuhne T. Model-driven development: a metamodeling foundation. IEEE Software 2003;20(5):36–41.

[22] Schmidt CD. Model-driven engineering. IEEE Comput 2006;39(2):25–31.

[23] Object Management Group (OMG). Meta object facility (MOF) Specification, version 2.0. Available from: <http://www.omg.org/spec/>; 2004.

[24] Object Management Group (OMG). XML metadata interchange (XMI) Specification, version 2.1.1. Available from: <http://www.omg.org/spec/>; 2007.

[25] Object Management Group (OMG). Meta object facility (MOF) 2.0 Query/View/Transformation, version 1.0. 2008. Available from: <http://www.omg.org/spec/>; 2008a.

[26] Fujimoto RM. Parallel simulation: parallel and distributed simulation systems. Proceedings of the 33nd conference on winter simulation (WSC'01). Washington, DC, USA: IEEE Computer Society. p. 147 –57; 2001.

[27] Fujimoto RM. Parallel and distribution simulation systems. New York, NY, USA: John Wiley & Sons, Inc.; 1999.

[28] IEEE Standard for modeling and simulation (M&S) High level architecture (HLA)–frameworks and rules. Technical Report 1516; 2000.

[29] SimArch and jEQN Projects. Official website hosted by google code. Available at: <https://sites.google.com/site/simulationarchitecture/>; 2015.

[30] Amsden J. Modeling with SoaML, The service-oriented architecture modeling language. Technical article, IBM. Available at: <http://www.ibm.com/developerworks/rational/library/09/modelingwithsoaml-1/index.html>; 2010.

[31] Bocciarelli P, Pieroni A, Gianni D, D'Ambrogio A. A model-driven method for building distributed simulation systems from business process models. In: Rose O, Uhrmacher AM, editors. Proceedings of the winter simulation conference. Berlin, Germany; 2012. p. 227.

[32] Eclipse Foundation. Eclipse QVT Model-to-model transformation (MMT). Available at: <http://projects.eclipse.org/projects/modeling.mmt>; 2014a.

[33] D'Ambrogio A, Iazeolla G. Design of XMI-based Tools to build EQN Models of Software Systems. Proceedings of the IASTED international conference on software engineering. Innsbruck, Austria; 2005.

[34] Eclipse Foundation. Acceleo transformation language. Available at: <http://www.acceleo.org/pages/home/en>; 2014b.

[35] Khul F, Weatherly R, Dahmann J. Creating computer simulation systems: an introduction to high level architecture. Prentice Hall; 1999.

CHAPTER

Current and future trends in open source network simulators for wireless systems

15

Lorenzo Carlà, Francesco Chiti, Romano Fantacci and Tommaso Pecorella
University of Florence, Florence, Italy

1 NETWORK SIMULATION ISSUES

Over the last few years, there has been an increased interest in the design of *smart environments*, often conceived of as *pervasive ecosystems*, whose enabling technology is represented by the so-called Internet of Things. Most of the foreseeable scenarios (e.g., Smart Cities, Smart Grids, Building Automation) rely on the adoption of wireless sensor networks (WSNs). Ideally, those devices should be small, battery-operated, low cost and necessarily connected to the Internet.

The need for interoperability is leading to open standards, specifically IETF and IEEE ones, i.e., Internet and IEEE 802.15.4. The benefits of this are quite obvious. However, there are some aspects that need to be carefully evaluated: the system is more complex, the optimal system's parameters have to be tailored to the specific scenario and the protocol interactions are larger. For these reasons, the protocol suite optimization is still an open issue.

A weak design can easily hinder WSN operations, leading to early battery discharge, connection problems, low overall system reliability or even complete instability. These issues are of particular relevance in WSNs as they are inherently resource-constrained systems. This often implies protocols being hardcoded directly into the device's firmware. Therefore, it is essential to be able to investigate the overall system's behavior before industrial development and practical deployment. This represents a critical point, as WSN operating systems' (OS) (for example, TinyOS, Contiki) main design goals are about memory footprint and energy efficiency rather than code clarity, expandability and maintainability. Moreover, a real testbed is often extremely costly. As a consequence, Open Source network simulators can represent a valid alternative to real device development and testbed deployment for academic and industrial research goals.

A simulation framework is normally meant to be an effective support in the protocol/algorithm design, evaluation and optimization phases, whereas a real testbed may be either impossible, or too costly, or unable to provide experiment reproducibility if that is a key factor.

Any simulation tool should be based on a real-world scenario model, which should be either analytical or functional. The analytical approach is usually necessary to build an approximate model of the aspects that cannot be functionally simulated, e.g., the channel error patterns. On the other hand, the functional model is useful to characterize the protocols and their interactions, e.g., TCP, IP, routing protocols, etc. Even though it is possible to functionally describe all the network components, this is often unnecessary and detrimental, as it could lead to an extremely long simulation time or even hinder the simulator's capabilities. As a matter of fact, an over-detailed model will produce detailed data for very specific scenarios that are hard to generalize. A simulator based on analytical assumptions might lack some details, but it is more useful for research purposes.

The network simulator framework should allow the implementation of an abstract model, possibly alleviating the problems arising in a real implementation (e.g., device memory constraints, etc.). However, before being adopted in a simulation campaign, any model should be validated. To this purpose, there are three possible approaches:

- Qualitative: tests assessing whether the system behaves as expected;
- Quantitative: tests evaluating the performance at component or system level through a set of suitable metrics, possibly introducing several reference scenarios; and
- Cross-validation: tests evaluating the model behavior against a reference implementation.

The accuracy degree of validation is strictly connected with granularity and flexibility of the adopted simulation tool, along with the capability of data gathering and processing. In our opinion, WSNs represent an appropriate case study since their hardware has extremely limited resources, their operating systems are very small and optimized, collecting data from the implementation might be very difficult, and deploying large-scale testbeds could be quite a challenge. Summarizing, a WSN simulation framework should have the following capabilities:

- To implement standard protocols (such as IETF and IEEE ones) in order to optimize the WSN for a specific scenario;
- To allow a system's parameters modification, particularly the ones suggested by the standards;
- To be capable of going beyond the standards, in order to propose amendments or new standards;
- To test and evaluate application-level protocols (e.g., CoRE, CoAP) or, more generally, applications.

This chapter is organized as follows. First, we describe the open issues concerning wireless system simulation and we present a brief overview and comparison of the main available open source simulation frameworks. Afterwards, we focus on ns-3, describing its actual simulation models and pointing out the most relevant features of the framework.

2 SIMULATION FRAMEWORKS OVERVIEW

In simulating (or emulating) a wireless network, several alternatives need to be considered:

1. A generic network simulation framework can be extended with necessary modules; or
2. A specific simulator can be used;
3. A device simulator might be included in most operating systems of the device under investigation.

OS-based simulators reproduce almost exactly the behavior of a real node. On one hand, this might be an advantage. On the other, they usually lack detailed and flexible channel models and they force the user to cope with OS-specific constraints. Moreover, the user must adapt to the programming language and model used in the OS, often not a standard one.

Even though generic simulation frameworks have the advantages of being extremely flexible, well documented and allow the user to leverage a number of already existing models to enhance their simulations, they often require particular attention to develop and validate the WSN models.

As for specialized simulation tools, they may seem a good alternative. However, they are rarely used and lack user-based verification. Additionally, the learning curve and the simulator specific focus make it hard to reproduce the results.

Further on, the most relevant alternatives are presented and discussed. For a survey of existing simulators, see [1,2].

3 OPEN SOURCE NETWORK SIMULATORS

3.1 OMNeT++

OMNeT++ is a C++ based, open-source, component-based, modular, open architecture and discrete-event network simulator with strong GUI support and an embeddable simulation kernel [3]. An OMNeT++ model is composed by hierarchically nested modules communicating through message passing. Modules can send messages directly to their destination or along a predefined path using gates and connections. OMNeT++ has a Mobility Framework (MF) supporting node mobility, dynamic connection management and wireless channel models.

Several models can be used for WSNs simulations, such as:

- INET: this framework can be considered the standard protocol model library of OMNeT++. INET contains models for the Internet stack (TCP, UDP, IPv4, IPv6, OSPF, BGP, etc.), wired and wireless link layer protocols (Ethernet, PPP, IEEE 802.11, etc.), support for mobility, MANET protocols, DiffServ, MPLS with LDP and RSVP-TE signaling, several application models, and many other protocols and components. The INET Framework is maintained by the OMNeT++ team for the community, utilizing patches and new models contributed by members of the community.
- MiXiM: conceived for mobile and fixed wireless networks (wireless sensor networks, body area networks, ad-hoc networks, vehicular networks, etc.). MiXiM focus is on the lower layers of the protocol stack and it offers detailed models of radio wave propagation, interference estimation, radio transceiver power consumption and wireless MAC protocols.
- Castalia: a WSNs/BANs simulator that can be used also for generic networks made of low-power embedded devices. Castalia's salient features include: model for temporal variation of path loss, fine-grain interference and RSSI calculation, physical process modeling, node clock drift and several popular MAC protocols. Additionally, Castalia is highly parametric. It provides tools to help run large parametric simulation studies and to process and visualize the results.
- Veins: it is an open-source framework for running vehicular network simulations. It is based on two well-established simulators: OMNeT++, an event-based network simulator, and SUMO, a road traffic simulator. It extends these to offer a comprehensive suite of models for Infrastructure-to-Vehicles Communications (IVC) simulation.

3.2 Ns-3

Ns-3 is an open-source, discrete-event network simulator with excellent features for network research and educational purposes [4]. The ns-3 project was started around 2004–05 in order to overcome the limitations of the previous ns-2 simulator. At the time, the developers decided to rewrite, almost from scratch, a novel tool based on solid foundation, with the aim of maintainability, expandability and execution speed. The simulator is completely in C++ with optional Python bindings, allowing the writing of a simulation in Python as well (modules must be in C++).

Over the years, ns-3 has been extended with a number of features, such as Wi-Fi, LTE, parallel execution, real code integration (Direct Code Execution (DCE)) and real network integration (real-time support and real network integration), just to name a few. All the code is verified using tests and any bug or new feature is handled by code maintainers. Bugs and new features are tracked using online tools and discussed using code review tools. Moreover, a continuous integration system provides reports about code building capabilities and memory leaks, which are handled as priority bugs. The simulator can run on Linux, OS X and Windows, with different features according to the platform (e.g., OS X does not support real-time execution, etc.).

A simulation is typically made by nodes connected through one or more channels. A channel is only responsible for packet delivery, delays (including jitter) and, in the case of wireless links, for signal attenuation. A node represents a physical device, and an aggregation system allows installation into the node of the required models, e.g., TCP, UDP, IPv4/v6, routing protocols, applications and NetDevices. NetDevices are the abstract mechanism to define a network device model (e.g., Wi-Fi, LTE, etc.). Models are usually built with particular attention to the real system's behavior and APIs. As a consequence, using the models is not much different than using a real system. This feature is quite important as it encourages users to adopt good implementation practices.

Designing and implementing a new ns-3 model is quite straightforward. The framework does provide memory management, sophisticated parameters definition (able to dynamically change a parameter at run-time), extensive debugging and tracing capabilities. Additionally, ns-3 provides a strict correspondence with real packet structures allowing easy definition of headers, footers and their processing in a node. These features also enable one to trace a network using standard Pcap files and to visualize them using Wireshark.

4 OS-ORIENTED TOOLS

In this section we compare several open source operating systems with a special focus on wireless sensor networks, as a special and remarkable case in which they are adopted.

4.1 TinyOS

TinyOS [5] is an open source operating system (OS) with a large developer community. It is one of the most popular operating systems for WSNs and runs on various hardware platforms such as Mica2, Micaz, and TelosB. It is inherently an event-driven OS.

WSN nodes are normally in low-power mode and, whenever an event occurs, TinyOS wakes up the node (normal power mode) and calls the event handler. The handler may schedule tasks that are executed afterwards by the TinyOS kernel, putting the node in a low-power mode once the tasks are completed. Both TinyOS and applications are written in nesC, an extension of the C programming language. The programs are built using statically linked components, resulting in a small memory footprint. Components may be nested in order to make other components.

TinyOS is endowed with a WSN simulator, namely TOSSIM; it has the advantage that the code used for simulation also runs on the real node, reducing the effort to rewrite the code for the sensor node and allowing better comparison between experimental and simulation results.

Extensions to TOSSIM have been proposed to include other features like the evaluation of power consumption and realistic radio models.

4.2 CONTIKI

Contiki OS is built around an event-driven kernel [6]. An event scheduler that dispatches events to the running processes is included in the Contiki kernel. Contiki maintains a queue of pending events and the events are dispatched to target processes according to a First In First Out (FIFO) policy. In addition, a preemption mechanism allows managing and prioritizing events. Concurrent events can be managed by resorting to Protothreads. Interrupts can preempt an event handler, but to avoid synchronization issues, interrupts cannot post an event. Contiki and its applications are written in Embedded C (a special C for embedded systems).

Contiki OS has its own simulator (Cooja), which allows the simulation of Contiki motes. Motes can be emulated at hardware level (slower but allowing precise inspection of the system behavior) or at a less detailed level (faster and allowing simulation of larger networks).

4.3 RIOT

RiOT OS aims at bridging the gap between OSes for WSNs and "normal" OSes, currently running on Internet hosts [7]. Among the goals of RiOT are included energy-efficiency, small memory footprint, modularity and uniform, hardware-independent, API access. RiOT implements a microkernel architecture inherited from FireKernel, thus supporting multi-threading with standard API. In addition to the original FireKernel features, RiOT adds support to C++ and provides a TCP/IP network stack.

The advantages of the RiOT architecture include: (i) high reliability and (ii) a developer-friendly API. The RiOT modular microkernel structure makes it robust against bugs in single components: failures in the device driver or the file system do not harm the whole system. RiOT allows developers to create as many threads as necessary and distributed systems can be easily implemented by using the kernel message API. The number of threads is only limited by the available memory and the stack size for each thread, while the computational and memory overhead is minimal.

RiOT does not have a specific node simulator. However, it can be run on a normal Virtual Machine, so as to emulate complex networks.

5 NS-3 FRAMEWORKS

In this section we will introduce the main wireless network ns-3 models, their current development status and capabilities.

As outlined in the previous section, ns-3 is based on modules, where each module provides the functionalities needed to simulate a particular network part. In order to understand the software capabilities, it is worthwhile to explain exactly what an ns-3 module is, and the software organization. The ns-3 source directory is divided into multiple folders, each of them representing a *module*. The module's organization reflects multiple (sometimes conflicting) goals:

1. Modules should be independent from each other (as much as possible);
2. Each module should have a designated maintainer;
3. Each module should have a limited scope, i.e., provide a single, well-defined, functionality.

The modules' independency allows elimination of all the unnecessary modules and reduction of the compilation time. Moreover, modules with particular requirements (e.g., modules requiring unavailable external libraries) can be excluded from the compilation at configuration time.

The basic simulation components are grouped in the *core* and *network* modules. The first implements the very basic building blocks of the framework, like the event schedulers, smart pointer, attribute system, etc.

The second module groups the classes defining generic network concepts, like a node, headers and trailers, packets, etc. Most of those classes are meant to be further specialized, e.g., an Address into a Mac48Address.

The *internet* module contains all the Internet protocols, e.g., IPv4, IPv6, TCP (Vegas, Reno, NewReno, Westwood, etc.), UDP, etc.

It is worth considering that the basic routing protocols are in the internet module. However, more specialized routing protocol (e.g., AODV, DSR, DSDV, etc.) are provided in separate modules. This difference allows a better maintainability of each module, and keeps the internet module as small as possible.

In a wireless simulation, several blocks are needed. As shown in Figure 15.1, the main simulation components are:

1. The channel model, which can be further split into
   a. The propagation model
   b. The spectrum interference model
2. The mobility model
3. The wireless network device model
4. Any additional model above layer 2, including
   a. TCP/IP
   b. Routing
   c. Traffic generators
   d. Etc.

In order to fully understand the various modules and their relationship it is worth explaining some basic concepts and objects used in ns-3.

In a typical ns-3 simulation there are some *nodes*, exchanging packets through channels. A node is a particular object acting as a container for other objects, in particular *Applications, Protocols*, and *NetDevices*:

- An *Application* generates and receives traffic, typically using a socket (ns-3 sockets are very similar to Linux sockets).

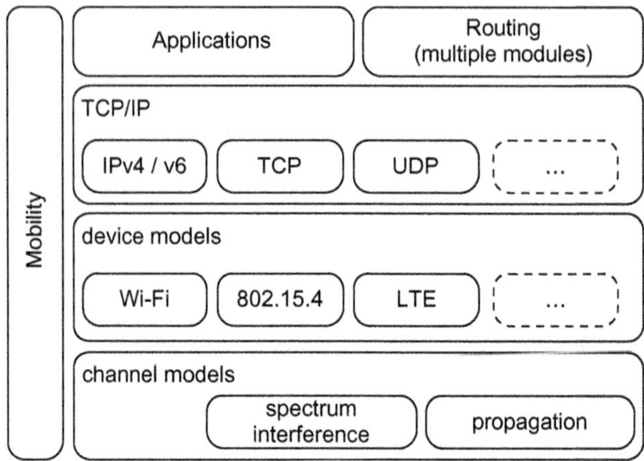

FIGURE 15.1

Ns-3 simplified module layout.

- *Protocols* are responsible for any packet processing, e.g., TCP, UDP, IP, ICMP, etc. Each protocol *registers* its protocol number in the node, allowing to dispatch incoming packets to the proper protocol.
- *NetDevices* are the ns-3 representation of a network device. A NetDevice must simulate/emulate both L2 and L1, with the appropriate simplifications to allow a satisfactory trade-off between simulation precision and execution speed. A NetDevice must be *aggregated* to a node.

Summarizing, ns-3 nodes are not just object containers. They are the logical representation of the relationship between the Applications, Protocols and Network Devices (NetDevices) of a real network node.

In the following subsections we will outline the most relevant modules for wireless networks simulations.

5.1 CHANNEL MODELS

Channel models represent one of the fundamental building blocks for wireless simulations. A network simulator should be able to take into account the main channel impairment effects over the transmitted signal. It is obvious that the channel model precision can severely affect the simulation performances. As a consequence, ns-3 uses a packet-level channel model, even though more precise (chunk-level) models are possible.

In ns-3, the channel effects are typically evaluated at packet reception. Once a packet is received, the sender and receiver position are evaluated, and a specific channel model is used to derive the received signal power, along with any possible interfering signals. The received power is used to calculate the bit and packet error probabilities, and in turn to accept or reject the packet.

The *propagation* and *spectrum* models can help in this task, allowing the simulation of various channel models (e.g., Friis, Jakes, Nakagami, Okumura Hata, etc.) and the composition of signals from various sources.

It should be noticed that the use of such models is not mandatory: the user can define their own or use a completely different channel model. As an example the Wi-Fi channel model uses the models in the *propagation* module, but it doesn't use the *spectrum* module.

5.2 MOBILITY MODELS

The node's mutual position is important in wireless simulations. The ns-3 *mobility* model is used to keep track of a node's position over time, even if the node is stationary (the ConstantPosition model is a special case of mobility). As a consequence, any wireless node will have to have a mobility model.

The current ns-3 mobility module already provides some mobility models (e.g., random waypoint, Gauss-Markov, random walk, etc.), and the user can define new models as well. However, one interesting characteristic is the ability to draw the node's mobility from a user-supplied mobility trace. This trace can be generated by popular software (SUMO[1], BonnMotion[2], TraNS[3]) and can be used to precisely mimic specific scenarios, e.g., vehicular mobility in a city.

It is worth pointing out that the ns-3 mobility models are 3D. However, at the moment of writing the position is associated with the node, and not with the objects aggregated to the node. This limitation may be removed in the future, as it may be relevant for specific scenarios (e.g., two radio devices in the head and in the tail of a train).

5.3 IEEE 802.11 MODELS

The 802.11 model is, most probably, the oldest and most used ns-3 model. The model is very complex, as it grew over the years to include the different extensions to the standard, up to 802.11n. It must be observed that not every part of the standard is fully modeled, and the standard features are not implemented at all. This does not undermine the model's usefulness, and it fully reflects the scientific interest (or lack of) for some specific features. As an example, the node's association phase in Wi-Fi infrastructure mode is not available. The missing functionalities may, of course, be developed if the scientific community shows interest in them.

---

[1]http://sourceforge.net/apps/mediawiki/sumo/index.php?title = Main_Page
[2]http://net.cs.uni-bonn.de/wg/cs/applications/bonnmotion/
[3]http://trans.epfl.ch/

Currently, the *wifi* model supports 802.11a, b, g, n, p and s. The models can be used for both infrastructure and ad-hoc modes, with and without QoS and frame aggregation support. The model has been used for a number of publications, and it is widely regarded as a very precise Wi-Fi simulation tool.

At the time of writing, the development is focused on two main areas:

1. Fully support 802.11n multiple antennas and MIMO, and
2. Extend the vehicular and ad-hoc models.

A possible future development could be the channel model harmonization between the *wifi* model and the other wireless models. As a matter of fact, the *wifi* model is older than the spectrum model, and it does not use it. The drawback is that it is not possible to use the spectrum model to simulate the cross-interference between Wi-Fi and other models, e.g., 802.15.4. This limitation may be removed in the future, but it will require a particular care to maintain the performance and precision of the current model.

5.4 IEEE 802.15.4 MODELS

The IEEE 802.15.4 standard is foreseen to become widespread, thanks to its low energy consumption and simplicity, making it ideal for wireless sensor networks [8–10]. The standard is much simpler than IEEE 802.11, but there are various elements which are implementation-dependent. As an example, the 802.15.4 MAC is not fully standardized, and a number of 802.15.4 compliant MACs have been proposed.

The *lr-wpan* ns-3 module has been recently added to ns-3. The module is based on IEEE 802.15.4-2006 standard, and focuses on ad-hoc network types.

Thanks to a rigorous design, the module tightly follows the standard defined primitives, allowing it to precisely mimic a real system. The main limitations, at the time of writing, are represented by the lack of multiple MAC protocols. However, according to feedback received, the model will be expanded with other MAC models and realistic energy consumption models soon.

For what concerns the channel models used by the module, they are based on the data provided by the standard.

The model can be used to carry IP and non-IP traffic, thanks to the *sixlowpan* module and a protocol dispatcher similar to the one used by ZigBee. As a consequence, the module can be used for IoT simulations. The routing models for 802.15.4 are being developed, and they will be publicly available as soon as they are fully validated [11].

5.5 LTE/LTE-A MODELS

The LTE/LTE-A model, known also as LENA (LTE/EPC Network simulator), was first developed by the Centre Tecnologic de Telecomunicacions de Catalunya (CTTC) [12]. The framework has been designed in order to provide a tool for the test and the evaluation of several aspects of the LTE network, such as Radio

Resource Management (RRM), MAC scheduler, Inter-cell Interference Coordination and mobility management. In order to fulfill this purpose, multiple issues were considered throughout the development of the framework, though several points are open to future development, which will be analyzed in the rest of the section.

The simulator allows the deployment of network scenarios composed by multiple eNodeB (eNB) and User Equipment (UE), where each device can be configured with different simulation parameters. In particular, thanks to the *Buildings module*, the LTE module allows the simulation of real network scenarios characterized by both outdoor and indoor propagation conditions, fading, antenna model, and, in general, the propagation model typical of scenarios that involve macro eNB, pico and femtocells.

5.5.1 LTE/LTE-A model architecture

For what concerns the PHY layer, the minimum level of granularity is represented by the Resource Block (RB). Hence, the simulator does not allow an evaluation of the radio interference with an accuracy up to the symbol level. However, since all the operation-concerning packet scheduling takes place on a per-RB basis, this choice allows a correct modelling of the packet scheduling algorithms and inter-cell interference. In addition, the complexity of the simulation is significantly reduced.

In addition, in order to support the different frequency and bandwidth configurations, it was necessary to implement a new channel interface, namely *MultiModelSpectrumChannel*, within the spectrum module.

The MAC scheduler of the LTE/LTE-A model follows the guidelines proposed by the SmallCell Forum [13], providing an interface that can be used by researchers and manufacturers for the implementation of scheduling and Radio Resource Management algorithms. In addition, the simulator comes with a good number of schedulers, such as Round Robin, Maximum Throughput, Blind Average Throughput, Token Bank Fair Queue, Priority Set and Channel and QoS Aware.

The LENA simulator can be divided into two main components, namely the EPC model and the LTE model.

The objective of the EPC model is to provide the end-to-end IP connectivity over the LTE model, allowing the simulation and performance evaluation of realistic applications. The simulator is compatible with all the TCP and UDP applications within ns-3. It is worth to note that the EPC currently supports only IPv4 connectivity.

The EPC model is composed of the SGW, PGW, MME and, partially, the eNB and it is mainly focused on the data plane aspects. For this reason, all the functionality concerning the SGW and PGW are gathered in one single node. Moreover, the protocols that compose the control plane are not fully implemented and, in many cases, the signaling between the nodes is modelled with direct function call.

To this end, several points are left to future contributions. First, future release of LENA will support IPv6 connectivity, an accurate modelling of the control plane signaling and the introduction of the HSS and PCRF. In particular, a real implementation of the interfaces that compose the EPC provides the tools for the research and development of algorithms that spans several areas of interests such as UE attachment, set-up of bearers, mobility management, measurement configuration and reporting, security, paging and cell selection. In addition, the implementation of logical modules like HSS is expected.

On the other hand, the LTE model is composed of the UEs and eNBs. In particular, within these entities each level of the protocol stack (i.e., RRC, PDCP, RLC, MAC and PHY) is modelled. Starting from the RRC module, it currently provides part of the System Information message, the initial cell selection, a procedure for the establishment and reconfiguration of the RRC connection and an algorithm in case of handover. The messaging of the RRC module is modelled following [14].

Since the LTE scheduling and Radio Resource Management do not work with IP packets, a great effort has been made in the modelling of the RLC and PDCP entities. As a matter of fact, the LTE standard defines that the RLC module segments or concatenates IP packets before the transmission according to the size of the packet and the modulation scheme adopted. To this end, the RLC entity is modelled following [15]. In particular, the simulator provides the three different types of RLC included in the standard, namely Transparent Mode (TM), Unacknowledge Mode (UM) and Acknowledge Mode (AM).

As for the PDCP entity, the reference document is represented by [16]. Currently, the main feature of the PDCP entity is represented by the maintenance of the PDCP Sequence Number (SN). A possible future development could be represented by the implementation of other functionality of the PDCP entity such as header compression, cyphering, integrity protection and duplicate discarding.

Finally, it is worth noting that the interfaces between the RLC and PDCP are realized by means of the use of direct function call.

From the point of view of the PHY layer, though the RB represents the fundamental unit for resource allocation, the simulator is able to simulate the different behavior of the control channel and data channel within the same RB. In addition, in order to reduce the computational complexity, the simulator adopts the *link-to-system mapping* (LSM) technique, which is based on the mapping of the PHY layer link performance to a system simulator. The link layer performance is obtained by means of the use of the Vienna LTE Simulator [17]. Future enhancement at the PHY layer could be represented by algorithms that allow Carrier Aggregation (CA).

At the time of writing, the main limitation is represented by the lack of the Multimedia Broadcast/Multicast Service (MBMS) framework. The implementation of the MBMS feature involves changes among all the LTE protocol stack. In fact, several modules, such as MBMS-GW and Multi-cell/multicast coordination Entity (MCE), shall be introduced. In addition, at the PHY layer, the simulator should be able to handle the multiple versions of the signal coming from different cells. In fact, one of the main features of the MBMS framework is the ability to transmit a multicast (or broadcast) service by means of synchronized eNBs, which form a *single frequency network* (SFN). In particular, the transmission is called *Multimedia Broadcast Single Frequency Network* (MBSFN). Hence, in a MBMS transmission, the signals from different cells, belonging to the SFN, contribute to increase the signal power perceived by the user. Finally, the PHY layer of the simulator should be able to discern the signals that contribute to inter-cell interference (i.e., unicast transmission) from the signals belonging to the same multicast transmission.

5.6 TCP/IP, ROUTING AND TRANSPORT PROTOCOLS (L3 AND L4 MODELS)

The ns-3 *internet* module provides support for TCP/IP standard protocols, i.e., IPv4, IPv6, TCP, UDP, along with most of the accessory protocols, e.g., ARP, NDP, etc.

The routing protocols are, usually, particularly important for wireless networks. The ns-3 framework allows use of both proactive and reactive routing. At the moment of writing, the following wireless mesh routing protocols are supported: AODV, DSR, DSDV, and OLSR. Additional protocols are being developed or are in review phase, e.g., RPL, CTP, CLWPR, etc.

Moreover, it is possible to use quagga[4] and click[5] to leverage already existing and widely deployed routing demon implementations.

Current and future developments with respect to the L3 and L4 models are foreseen to be focused on extending routing protocols (e.g., IPv6 support for OSPF and AODV, new routing protocols, etc.) and TCP extensions (e.g., L4 Path MTU discovery, extra TCP flavors support, etc.). Moreover, some efforts are directed toward support for satellite-specific protocols, like LTP and Bundle Protocols.

5.7 DIRECT CODE EXECUTION

Direct Code Execution (DCE) is an interesting and powerful feature of ns-3. It allows the experimenter to use a slightly modified Linux system while maintaining the ns-3 framework as a communication system. As a consequence, the experimenter can use a real Linux stack (down to the IP level), allowing a greater precision in the simulations, and also can use native programs, e.g., video and audio codecs.

The DCE environment is not limited to wireless network simulations, as it can be used in any simulated scenario. In wireless systems it can be especially useful to simulate protocols not yet available as native ns-3 implementations or to perform cross-implementation validations.

---

[4]http://www.nongnu.org/quagga

[5]http://read.cs.ucla.edu/click/click

REFERENCES

[1] Singh C, Vyas O, Tiwari M. A survey of simulation in sensor networks, In: International conference on computational intelligence for modelling control automation; 2008, p. 867–72.

[2] Korkalainen M, Sallinen M, Karkkainen N, Tukeva P. Survey of wireless sensor networks simulation tools for demanding applications, In: Fifth international conference on networking and services. ICNS '09; 2009, p. 102–6.

[3] OMNeT++. [Online]. Available from: <http://www.omnetpp.org>; 2015.

[4] ns-3 Consortium. ns-3. [Online]. Available from: <http://www.nsnam.org>; 2014.

[5] TinyOS. [Online]. Available from: <http://www.tinyos.net>; 2015.

[6] Contiki–The open source OS for the internet of things. [Online]. Available from: <http://www.contiki-os.org>; 2015.

[7] RiOT–The friendly operating system for the internet of things. [Online]. Available from: <http://www.riot-os.org>; 2015.

[8] Institute of Electrical and Electronics Engineers (IEEE). IEEE Standard for Information technology – Local and metropolitan area networks – Specific requirements – Part 15.4: Wireless medium access control (MAC) and Physical layer (PHY) Specifications for low rate wireless personal area networks (WPANs). IEEE Std 802.15.4-2006; 2006. p. 1–320.

[9] IPv6 over low power WPAN (Active IETF WG). [Online]. Available: <http://tools.ietf.org/wg/6lowpan/>; 2014.

[10] Routing over low power and lossy networks (Active IETF WG). [Online]. Available: <http://tools.ietf.org/wg/roll/>; 2014.

[11] Bartolozzi L, Pecorella T, Fantacci R. ns-3 rpl module: Ipv6 routing protocol for low power and lossy networks. In: Proceedings of the 5th international ICST conference on simulation tools and techniques, ser. SIMUTOOLS '12; 2012, p. 359–66.

[12] LENA–LTE-EPC Network SimulAtor. [Online]. Available from: <http://networks.-cttc.es/mobile-networks/software-tools/lena/>; 2014.

[13] Small Cell Forum. [Online]. Available from: <http://www.smallcellforum.org/>.

[14] 3GPP: The Mobile Broadband Standard. "3GPP, TS 36.301". Evolved universal terrestrial radio access (E-UTRA); User Equipment (UE) radio transmission and reception. Available from: <http://www.3gpp.org/DynaReport/36101.htm>; 2015.

[15] 3GPP: The mobile broadband standard. "3GPP, TS 36.322". Evolved universal terrestrial radio access (E-UTRA); Radio link control (RLC) protocol specification; 2015.

[16] 3GPP: The Mobile Broadband Standard. "3GPP, TS 36.323". Evolved Universal Terrestrial Radio Access (E-UTRA); Packet data convergence protocol (PDCP) specification. Available from: <http://www.3gpp.org/dynareport/36323.htm>.

[17] "Vienna LTE Simulator." [Online]. Available from: <http://www.nt.tuwien.ac.at/research/mobile-communications/lte-simulators/>; 2014.

CHAPTER

16

Simulating wireless and mobile systems

The Integration of DEUS and Ns-3

M. Amoretti, M. Picone, F. Zanichelli, and G. Ferrari

Università degli Studi di Parma, Parma, Italy

1 INTRODUCTION

Mobile and distributed systems are the result of the interconnection of several nodes, characterized by decentralized goals and control, that as a whole exhibit one or more properties (i.e., behavior) which are not easily inferred from the properties of the individual parts. Such systems are complex, because the interactions of the nodes determine their future individual states and that of the system [1]. Moreover, they usually exhibit high levels of concurrency and asynchrony and their performance may be highly influenced by the changing environmental conditions of the environment (e.g., if they move).

For the qualitative and quantitative analysis of such systems, discrete event modeling and simulation (in which time jumps from event to event) are usually adopted [2]. In order to choose the proper simulation environment, the following criteria are taken into account: simulation architecture (the operation and the design of the simulator), usability (how easy the simulator is to learn and use), extensibility (the possibility to modify the standard behavior of the simulator in order to support specific protocols), configurability (how easily the simulator can be configured and with what level of detail), scalability (the ability to simulate how a decentralized protocol scales with thousands, or more, nodes), statistics (how meaningful and easy to manipulate the results are), reusability (the possibility to use the simulation code to write the real application). Moreover, the design of mobile ubiquitous applications can be achieved efficiently only by taking into account multiple aspects: networking, user behavior, environment dynamics. Depending on the problem to be studied, omitting some of these points of view may lead to less-than-useful simulation results.

By looking at the state of the art, it is evident that almost every simulation tool targets a specific problem class. Only a few of them are truly general-purpose. Among these, in our opinion, the most advanced is CD++ [3], a modeling environment that enables the definition and execution of Discrete

Event System Specification (DEVS) models [2]. OMNeT++ is another well-known general-purpose discrete event simulation tool, which has been publicly available since 1997 [4]. Like CD++, OMNeT++ is based on the concept of simple and compound modules. The user defines the structure of the model (the modules and their interconnection) using a topology description language called NED. OMNeT++ has been used in numerous domains from queueing network simulations to wireless and ad-hoc network simulations, from business process simulation to peer-to-peer network, optical switch and storage area network simulations.

Unfortunately, all of these simulation tools are not particularly suitable for the analysis of distributed systems with thousands of nodes, characterized by a high level of churn (node joins and departures) and reconfiguration of connections among nodes. Trying to fill this gap, in 2009 we started a project for the development of an open source, Java-based, general-purpose discrete event simulation tool, called DEUS [5]. To simulate a distributed system at the application level, DEUS is particularly convenient, because of its extreme ease of use and flexibility. However, it does not provide packages for simulating networking layers, and we do not foresee implementing them. For this reason, until this point the scheduling of application-level events to simulate the exchange of messages among nodes has been necessarily configured by the user, using reasonable values—which could be considered a naive approach.

In this chapter, we present a general co-simulation methodology to obtain realistic DEUS-based simulations of mobile and distributed systems, leveraging on a highly reliable and complete open source tool for the discrete event simulation of Internet systems, namely ns-3 [6]. Such a tool relies on high-quality contributions of the community to develop new models, debug or maintain existing ones, and share results. As a proof of concept, we describe our positive experience in integrating ns-3's LENA LTE-EPC package (see Section 4) to support the network-aware simulation of a peer-to-peer overlay scheme called Distributed Geographic Table (DGT), which allows mobile nodes to efficiently share geo-referenced information without centralized control. To the best of our knowledge, OVNIS [7] is the only other tool which integrates ns-3 with a higher level discrete event platform, namely the SUMO road traffic simulator [8]. However, the only available release of OVNIS is the initial one, which includes an outdated version of ns-3.

The chapter is organized as follows. Section 2 analyzes related work on wireless and mobile systems co-simulation. Section 3 recalls the main features of DEUS. Section 4 is devoted to ns-3. Section 5 illustrates the methodology we propose to use ns-3 to improve the realism of DEUS-based simulations. Section 6 describes a challenging case study (regarding mobile nodes that form a peer-to-peer overlay network operating on top of LTE), and compares the results obtained with the proposed methodology with those obtained with a naive approach that models only the application layer. Finally, Section 7 concludes the chapter with a discussion of open problems and future work.

2 CO-SIMULATION OF WIRELESS AND MOBILE SYSTEMS

Although an intuitive way to simulate complex systems is to engineer a new tool from scratch that would contain building modules for communication components and others for physical components (such as mobile devices and vehicles), a good practice and a basic principle in engineering is to avoid reinventing the wheel and to rely on well-developed ideas as much as possible. Thus, adapting and integrating existing simulation tools provides a practical and convenient approach. In particular, co-simulation (cooperative simulation) is a methodology that allows individual components to be simulated by different simulation tools running sequentially or simultaneously, and exchanging information in a collaborative manner. In general, the type of information exchanged during co-simulation may be boundary conditions such as pressure, flow rate and temperature, or simulation parameters such as time steps or control signals. In the context of wireless and mobile systems, co-simulation is implemented by integrating a network simulator, producing information like accurate packet delays and transmission ranges, with other simulation tools, either specialized or general-purpose.

A brief description of co-simulation tools for network control systems (NCSs)—e.g., MATLAB®, Jitterbug, TrueTime—has been proposed by Årzén and Cervin [9]. For wireless network control systems (WNCS) simulations, a network simulator has been implemented as C MEX S-functions, to execute simultaneously with the SIMULINK control system [10]. Co-simulation of control and network based on MATLAB/SIMULINK has been proposed in several research works [11–14] that investigated NCS performance for various data rates, traffic, loads, network delays, networked predictive control, and compensation of transmission delay. However, the MATLAB/SIMULINK environment does not provide sufficient support for simulation of real-time implementation issues. MATLAB is also limited in simulating important aspects of wireless networks, such as node movement models and wireless signal propagation models. Jitterbug and TrueTime have been used to investigate the effects on system performance of the sampling period, communication delay, jitter, control-task scheduling, and blocking of real time tasks [15,16]. However, TrueTime does not support wireless networks and uses simplified network models. Moreover, it is not possible to use Jitterbug to evaluate the performance of a feedback scheduling system where the CPU loads change, and where the sampling periods of the controllers are changing over time. Another limitation of Jitterbug is that only linear systems can be analyzed [15].

Other research works [17–20] combined two simulation packages to achieve a more efficient co-simulation approach. A co-simulation platform that combines the ns-2 network simulator with the Modelica framework has been presented by Al-Hammouri et al. [17], where ns-2 models the communication network and Modelica simulates sensors and actuators. SIMULINK and OPNET co-simulation for WNCS over MANET has been considered by Hasan et al. [18], to investigate the situation where the controller communicates with the simulated stationary

MANET and plant nodes, over a real wireless link. In a more recent work [19], Hasan et al. have presented a SIMULINK-OPNET co-simulation methodology, with comprehensive simulation results, also considering the impact of different network sizes with stationary and mobile nodes. Leclerc et al. have developed a multi-modeling platform called AA4MM (Agent and Artefact for Multiple Models) [20]. Its main goal is reusability and interoperability of different simulators with a software architecture that is completely decentralized and based upon the multi-agent paradigm. Each simulator is controlled by a simulator manager (formally an agent) which is an autonomous entity. All of these manager/agents cooperate in order to run the whole simulation and to take care of the interaction problematics. Such an approach has been validated by coupling a user behavior simulator (MASDYNE) with a MANET simulator (JANE [21]). The most interesting result of such an approach is the ability to take mutual influences of user behaviors and network performances into account. However, this is not always possible. For example, ns-3 is not designed for being used as an on-demand provider of data items (e.g., the delay of a specific packet transmitted wirelessly in a complex environment); instead, it is particularly suitable to collect general network statistics. Fortunately, co-simulation can be also implemented by means of sequential integration of powerful, independent tools. In Section 5 we present our integration of DEUS and ns-3.

3 DEUS

DEUS is a multi-platform tool, developed in Java language (the code can be downloaded from the official site [22]). Its API, by subclassing, enables the implementation of (i) *nodes*, i.e. the entities which interact in a complex system, leading to emergent behaviors such as humans, pets, cells, robots or intelligent agents; (ii) *events*, e.g., node births and deaths, interactions among nodes, interactions with the environment, logs and so on; and (iii) *processes*, either stochastic or deterministic ones, constraining the timeliness of events.

Once specific Java classes have been implemented, it is possible to configure a simulation by means of the DEUS graphical user interface, which includes:

- the Visual Editor, for the generation of XML documents describing specific simulations;
- the Automator, for the execution of parametric simulations and the automatic generation of statistics in a Gnuplot-compliant format.

Figure 16.1 illustrates how DEUS simulation models are created (using also a Visual Editor), and then executed by the Engine, which is the core of DEUS, managing the event queue and the simulation loop. The Automator allows sensitivity analysis to be performed, by setting ranges for node and process parameters.

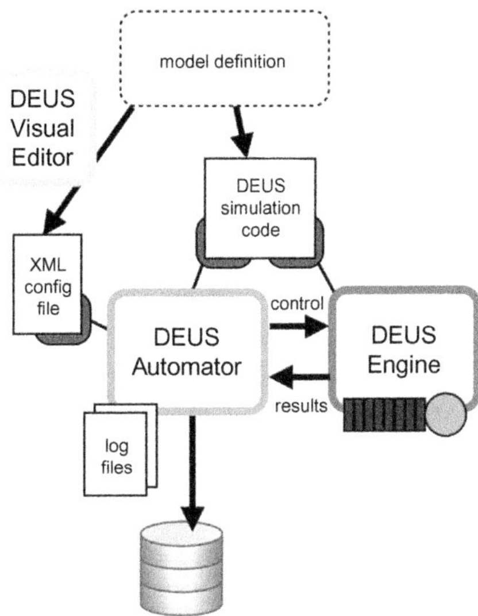

FIGURE 16.1

Discrete event simulation with DEUS.

A node may represent a dynamic system characterized by a set of possible states, whose transition functions may be implemented either in the source code of the events associated to the node, or in the source code of the node itself. Multi-scale modeling of complex systems can be achieved by defining nodes of different complexity, and connecting them. DEUS comes with a library of predefined, common processes, and many others can be implemented by the user.

3.1 DEUS API STRUCTURE AND FEATURES

Since DEUS is a general-purpose simulator, basic interfaces and classes are kept separated from more specific ones. By means of subclassing, it is possible to create specific modules for the simulation of any type of complex system. An extension package related to peer-to-peer resource sharing networks is provided by default.

The experience we acquired during the development of other simulation code (mainly using ns-2 and PeerSim [23]) showed us how difficult it is to manage memory when it comes to the simulation of systems with a large number of interacting parts (nodes, if systems are described as graphs). Java is an extremely powerful language and the flexibility of its object orientation, plus the reflection mechanism, make it highly suitable to build such a type of project. However, the difficulties in managing the garbage collection mechanism require a good design of the memory management. For these reasons, as we describe in more detail later on, DEUS relies on an efficient cloning mechanism: the initial process loads configuration objects into memories and new instances of those objects are obtained through deep cloning.

3.2 SIMULATION OBJECTS AND BEHAVIOR

The development of DEUS started from the definition of the basic simulation objects and the design of the configuration procedure, having in mind all the dynamics of complex systems that one may need to simulate. The goal was to achieve high flexibility and usability, allowing developers to specify a section with simulation objects and another one with simulation behavior, maximizing the possibility to reuse components and providing self-validation constraints so that the engine could process the configuration file through reflection and without any further validation. In particular, we have recently demonstrated that DEUS allows testing of deployment software on simulated devices and environments [24]. Simulation objects are events, nodes, and resources, while simulation behavior is managed through processes and engine objects.

An *event* represents the base simulation unit: i.e., the piece of code that is going to be scheduled by the system. Moreover, as complex systems are made by interacting components, we introduced the concept of *node*, which also corresponds to a data structure collector the event could rely on. Each node can have a set of *resources*, a structured way to represent objects the node can share or use through the event code. The association between events and nodes is given by *process* objects, which are responsible for event schedule timing calculation. The *engine* object puts everything together by linking events that are scheduled at the beginning of the simulation.

The simulation behavior follows the standard model of discrete event simulations: initialization of system state variables and clock, scheduling of initial events and, until the ending condition is true, calculation of next clock time and processing of the next event in the scheduling queue. However, a few additions have been made to make the model more flexible. For each event, it is possible to specify whether its execution is *one-shot*, so that the event will be removed from the schedule after its completion, or not, so that the event will be rescheduled according to the timing given by its associated process. Moreover, each event is provided with a listening mechanism over the scheduling process so that the latter will be able to schedule other events, namely *referenced events*, right after the event's execution. The ending condition of the simulator happens once the maximum simulation time has been reached or the scheduling queue is empty.

3.3 DEUS CORE

DEUS has been divided into packages, each one addressing a specific aspect of the simulation. The root package is `it.unipr.ce.dsg.deus`, containing the following subpackages:

- `core` – base system components including simulation object interfaces, configuration parser and engine;
- `schema` – object model representing the configuration file;

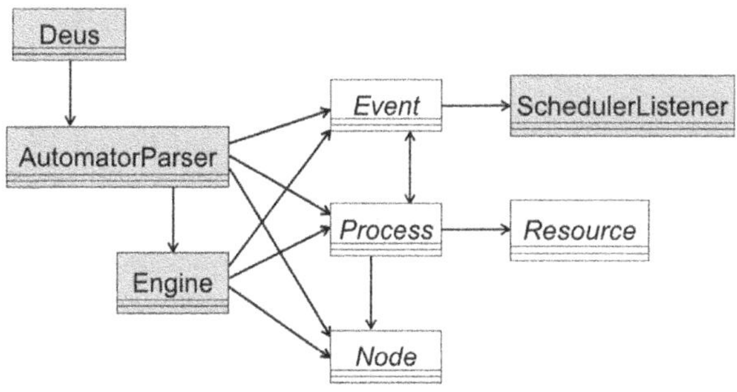

FIGURE 16.2

Class diagram of DEUS `core` package.

- `util` – support classes for the simulation engine;
- `impl.event` – reference implementations of the event object;
- `impl.node` – reference implementations of the node object;
- `impl.resource` – reference implementations of the resource object;
- `impl.process` – reference implementations of the process object.

In the following we provide a detailed description of the main classes contained in each package. A class diagram of the core package is shown in Figure 16.2.

3.3.1 `core` *package*

The `Event` class represents the simulation object being scheduled by the Engine. Each event is identified by a configuration id, a set of properties, a flag indicating if the event should be executed only once, a set of referenced events, a parent process, the triggering time and a listener to handle the execution of referenced events. In order to keep the simulation memory area as small as possible, each event is created by cloning the original event obtained from the simulation configuration parser; therefore, each implementing class should provide the code for cloning the event ensuring that its internal state is consistent, by reinitializing the event members that do not have to be cloned.

The `Node` class represents a generic data structure collector inside the simulation, so the main use is to store, read and delete information useful to characterize the simulation state. Each node is identified by a configuration id, a set of properties and a set of resources. Similarly to the `Event` class, there is the same cloning mechanism to keep the memory requirements small for the simulation execution.

The `Resource` class represents a generic resource associated to a node, with getter and setter methods.

The `Process` class represents the simulation object responsible to determine the timestamps of the events to be scheduled. Each process is identified by a configuration id, a set of properties, a set of referenced nodes and a set of referenced events.

The `Engine` class represents the simulation engine of DEUS. After the configuration file has been parsed, the obtained configured simulation objects (nodes, events and processes) are passed to the Engine, to let it properly initialize the queue of events to be run. The simulation is a standard discrete event simulation where each event has an associated triggering time, used as a sorting criteria. The events inserted into the simulation queue are processed individually one after the other, each time updating the current simulation virtual time. The run method of the engine will process each event in the event queue until a maximum virtual time is reached or the queue is empty. In each cycle the first event of the queue is removed (the one with the lowest triggering time), the virtual time of the simulation is updated and the event is executed. If the event has some referenced events, those will be scheduled right after the event execution. If the event is not one-shot and has a parent process, then it will be scheduled for execution with a triggering time calculated according to the parent process strategy.

The `AutomatorParser` class is responsible for handling the simulation configuration file, according to the *DEUS XML schema*. The configuration can be seen as a set of nodes, resources, events, processes and engine parameters. The `AutomatorParser` class handles the configuration of each simulation object and stores them in a set of array data structures. Each simulation object has a set of base features, plus references to other simulation objects: nodes can have a set of resources, events can have a set of referenced events, and processes can have references to both nodes and events. At the end of the configuration file parsing process, the `AutomatorParser` initializes the Engine object enabling the simulation execution.

3.3.2 `impl.event` package

The `BirthEvent` class represents the birth of a simulated node. During its execution, an instance of the node associated to the event is created.

The `DeathEvent` class represents the death of a simulated node. During the execution of the event the associated node is killed or, if nothing is specified, a random node is chosen instead.

The `LogPopulationSizeEvent` class is used to simulate a logging event that stores the number of nodes in the simulation, each time it is scheduled. It demonstrates that an event can really be anything, in the context of the complex system to be simulated.

3.3.3 `impl.node` package

The `BasicNode` class is the default implementation of the node abstract class, without any specific properties. A specific implementation is provided in the p2p package, which is described later in the chapter.

3.3.4 `impl.resource` package

The `AllocableResource` class represents a generic allocable resource, having a type/amount pair parameter which must be specified through the configuration file.

The ResourceAdv class represents a resource advertisement, i.e., a document that describes a ConsumableResource (with a name and an amount), and the interested node. Once the resource described by a ResourceAdv has been discovered, the owner of the resource should be registered into the ResourceAdv, and the found flag set to true.

3.3.5 impl.process *package*

The PeriodicProcess represents a generic periodic process. It has a parameter called *period*, which is used to generate the triggering time. Each time the process receives a request to generate a new triggering time, it computes it by adding the period value to the current simulation virtual time. An extension of this class is provided through the TwoSpeedsPeriodicProcess class that allows the specification of two different periods; the switch between first period and second period is made using a virtual time threshold.

The PoissonProcess represents a generic Poisson process. It has one parameter called *meanArrival*, which is used to generate the triggering time. Each time the process receives a request to generate a new triggering time, it computes it by adding the current simulation virtual time to the value of a homogeneous Poisson process with the rate parameter calculated as 1/*meanArrival* time.

Similarly to the TwoSpeedPeriodicProcess, there is the TwoSpeedPoissonProcess class to provide a Poisson Process that changes its speed after a virtual time threshold has been reached. Other classes allow for limiting the event scheduling to a time period, by specifying a starting time and an ending time.

3.4 EXTENSION PACKAGE FOR THE SIMULATION OF PEER-TO-PEER SYSTEMS

To simulate a particular class of complex systems, namely peer-to-peer resource sharing networks, we implemented the it.unipr.ce.dsg.deus.p2p package, which contains the following subpackages:

- node – the model of peer;
- event – the events characterizing a P2P network.

In the following we provide a detailed description of the main classes contained in each package. The related class diagram is illustrated in Figure 16.3.

3.4.1 node *package*

The Peer class is an extension of the Node class that represents the concept of peer in a network. Each peer is identified by a unique key generated by the engine (in the given key space) and is characterized by a list of neighbors, i.e., peers with whom it has an active link connection, and a status regarding peer connection to the network (whether is connected or not). Some methods have been implemented to manage neighborhood and notification messages.

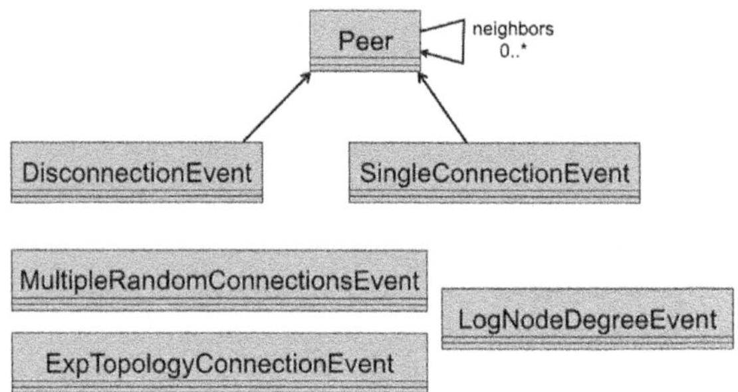

FIGURE 16.3

Class diagram of the `p2p` package.

3.4.2 `event` *package*

The `SingleConnectionEvent` class simulates the connection event of a peer in the network. The peer can connect to a randomly chosen node, or to a specific one.

An extension of this class is provided through the class called `MultipleRandomConnectionsEvent`, which enables a connection to more than one node, randomly chosen in the network.

The `DisconnectionEvent` class is used to disconnect a specific node from the network. Alternatively, it can be used to disconnect a random node from the network.

The `LogNodeDegreeEvent` class provides a logger that computes the node degree distribution for each peer of the network. The result is an array, whose index represents the node degree, and each value is the number of nodes that have the node degree corresponding to the considered index.

4 NS-3

Ns-3 is a discrete-event network simulator for Internet systems. It is a free, open source software project (with GPLv2 licensing) organized around research community development and maintenance. Like its predecessor ns-2, ns-3 relies on C++ for the implementation of the simulation models. However, ns-3 no longer uses oTcl scripts to control the simulation, thus overcoming the problems which were introduced by the combination of C++ and oTcl in ns-2. Instead, network simulations in ns-3 can be implemented in pure C++, while parts of the simulation optionally can be realized using Python as well.

Moreover, ns-3 integrates architectural concepts and code from GTNetS [25], a simulator with good scalability characteristics. Such design decisions were made at the expense of compatibility—porting ns-2 models to ns-3 must be done in a manual way. Besides performance improvements, the simulator has an extended feature set. For example, ns-3 supports the integration of real implementations code by providing standard APIs, such as Berkeley sockets or POSIX threads, which are transparently mapped to the simulation.

Among the packages being developed for ns-3, the LENA LTE-EPC is particularly rich and efficient [26]. In the LTE-EPC simulation model, there are two main components. First, the LTE Model, which includes the LTE Radio Protocol stack (RRC, PDCP, RLC, MAC, PHY). Such entities reside entirely within the User Equipment (UE) and the E-UTRAN Node B (eNB) nodes. Second, the EPC Model, including core network interfaces, protocols and entities, which reside within the SGW, PGW and MME nodes, and partially within the eNB nodes.

5 INTEGRATION OF DEUS AND NS-3

As illustrated in Section 3, to simulate a distributed system with DEUS, it is necessary to write the classes that represent nodes, events and processes. Node may represent devices, servers, virtual machines, applications, etc. Events may be associated to specific nodes (e.g., start, connection, disconnection, internally/externally triggered state change, stop, etc.), or involving several nodes (it is the case of logging events). To simulate a message delivery from one node to another, it is necessary to define the sender, the destination and to schedule a "delivered message" event in the future (in terms of virtual time of the simulation). The scheduling time of such an event must be set using a suitable process, selected among those that are provided by the DEUS API, or defined by the user, possibly.

For example, if the purpose of the simulation is to measure the average delay of propagating multi-hop messages within a network of nodes (e.g., a peer-to-peer network), the value of each link's delay must be realistic, taking into account the underlying networking infrastructure. In particular, if the communication is wireless, estimating the delay of point-to-point communication is a challenging task.

The direct integration of DEUS with ns-3, with the former that "calls" the latter to compute a delay value every time a node must send a message to another node, taking into account current surrounding conditions, is unpractical and would highly increase the duration of the simulation. Instead, a more effective and efficient solution (illustrated in Figure 16.4) includes the following steps:

1. given a complex system to be simulated, identify the main subsystem types, each one being characterized by specific networking parameters;
2. with ns-3: create detailed simulation models of the subsystems (i.e., submodels) and measure their characteristic transmission delays, taking into account both message payloads and proper headers;
3. with DEUS: simulate the whole distributed system, with refined scheduling of communication events, taking into account the transmission delays computed at step 2.

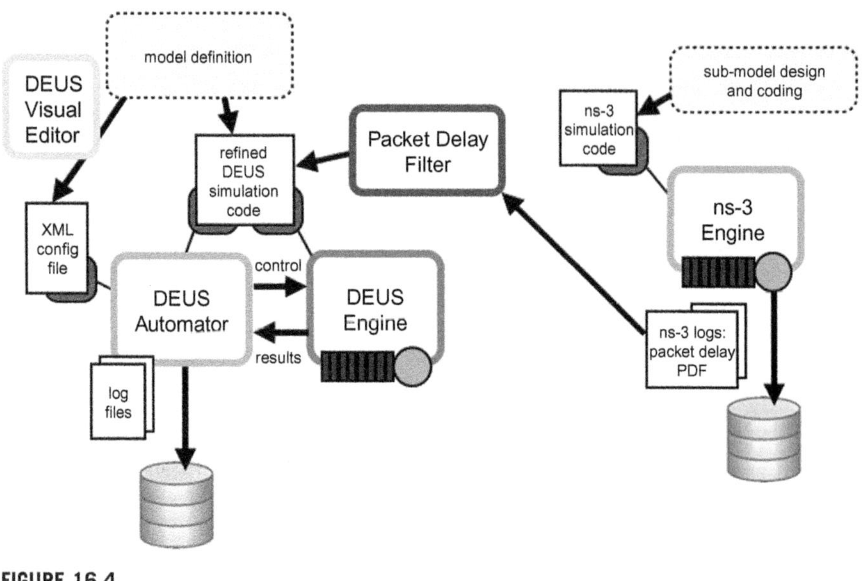

FIGURE 16.4

Discrete event simulation with DEUS and ns-3.

For example, if the overlay network relies on a cellular network, the submodel to be characterized with ns-3 could be a set of cells. Multicell communication may be very fast, if base stations are connected by optical fibers [27]. However, intercell interference and horizontal handover could be taken into account, when simulating mobile nodes. Moreover, the simulation of each cell should take into account the presence of other mobile nodes, which are not directly involved in the distributed application of interest, but consume significant resources. Finally, the same subsystem could be simulated with different geographic conditions, e.g., in a city (with small cells, buildings, and noisy channel), or in a rural area (with larger cells and a less disturbed channel).

Regarding step 2, with reference to the LTE package, it is necessary to modify the C++ class that logs the uplink and downlink delays. The modified class must log a discretized probability density function (PDF) of the RLC packet delay. Such a discretized PDF is then used to generate realistic packet delays in the DEUS-based simulations, using the well-known *inversion method* [28], which is based on the inverse probability theorem:

- choose the cumulative distribution function $F(x)$ of the random variable to be sampled;
- generate a set of uniform random numbers such that $R \sim U(0,1)$;
- compute the random variate $X_i = F^{-1}(R_i)$.

The Packet Delay Filter, illustrated in Figure 16.4, is a Java module that approximates the discretized PDF by a piecewise constant function, whose numerical inversion is straightforward and computationally inexpensive. The Packet Delay Filter implements the following algorithm:

1. put the points of the discretized PDF in a list L
2. divide L into n sub-lists
3. for each sub-list, compute the mean value of the points
4. merge two neighbor sub-lists, if the difference of their mean values is below t
5. repeat from step 2 until the set of sub-lists converges

The algorithm is repeated several times, for different values of t, in order to find the best set of sub-lists—i.e., the one that corresponds to a piecewise constant function whose integral is closer to 1. An example PDF approximated with the aforementioned algorithm is shown in Figure 16.7(b).

6 EVALUATION

We have applied the proposed methodology to model and simulate the Distributed Geographic Table (DGT), which is a peer-to-peer overlay scheme with the main objective to provide support for mobile node localization. Compared to centralized localization approaches, the DGT is more scalable, since its performance (in terms of responsiveness, completeness and robustness) remains valuable also for a large number of nodes, and when the nodes' dynamics are very high [29]. In a DGT-based system, the responsibility for maintaining information about the position of active peers is distributed among nodes, for which a change in the set of participants causes a minimal amount of disruption.

Every peer maintains a set of geo-buckets (GB), each one being a regularly updated list of known peers, sorted by their distance from the Global Position of the peer itself. GBs can be represented as concentric circles, each one having a different (application-specific) radius and thickness. The distance between two DGT peers is defined as the actual geographic distance between their locations in the world. The neighborhood of a geographic location is the group of nodes located inside a given region surrounding that location.

The main service provided by the DGT overlay is to route requests to find available peers in a specific area, i.e., to determine the neighborhood of a generic global position (Figure 16.5). The routing process is based on the evaluation of the region of interest centered in the target position. The idea is that each peer involved in the routing process selects, among its known neighbors, those that presumably know a large number of peers located inside or close to the chosen area centered in the target point. If a contacted node cannot find a match for the request, it does return a list of closest nodes, taken from its routing table. This procedure can be used both to maintain the peer's local neighborhood and to find available nodes close to a generic target.

Further details about the DGT can be found, for example, in recent articles by Picone et al. [29,30]. Simulation results presented there were obtained by means

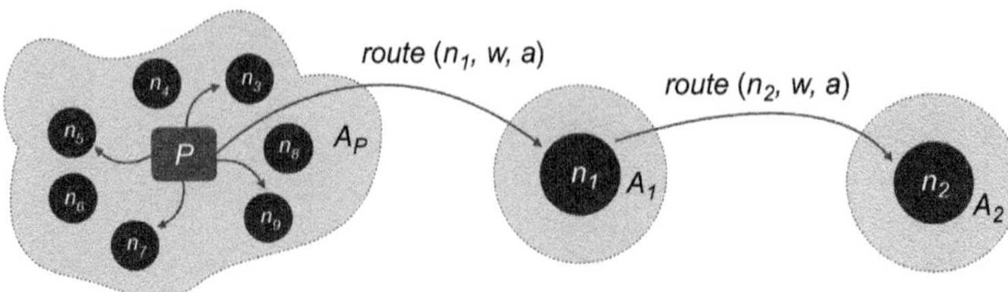

FIGURE 16.5

Propagation of a DGT query between nodes to retrieve the neighborhood of a local or remote regiòn of interest.

of a DEUS simulation model, integrated with Google Maps to have a realistic characterization of the urban environment (the city of Parma). However, simplistic assumptions on the packet transmission delay were made. In the following we illustrate how the methodology illustrated in this chapter has been used to simulate the DGT with more realistic packet transmission delays.

The simulation considers a number of vehicles that move over 100 km of realistic paths, generated using the Google Maps API. Each simulated vehicle selects a different path and starts moving over it. Using the features provided by the Google Maps API, we created a simple HTML and Javascript control page, which allows the monitoring of the time progression of the simulated system, where any node can be selected to view its neighborhood (demo videos are available online [31]).

The simulation covers 10 hours of DGT system life, with 500 to 2000 mobile nodes, 5 virtual paths with bad road surface (due to either ice, water, snow, or pothole), accident events scheduled during the simulation according to a Poisson stochastic process and with different message types to disseminate information about sensed data and traffic situation. Simulations with DEUS have been repeated with 20 different seeds for the random number generator, which are sufficient to obtain a narrow I95 confidence interval (5% of the steady state value, in the worst case). Obtained graphs consider means and standard deviations obtained by averaging over the whole set of simulated nodes, and over the 20 different simulation runs.

The considered DGT configuration is the one that gives the best performance in urban scenarios [29]. Each node has 4 GBs with a 0.5 km thickness and a peer discovery limiting number equal to 10 nodes, covering a region of interest of 12.5 km^2 and an adaptive discovery period ranging from 1.5 min to 6 min, depending on the number of new discovered nodes during each lookup process. The period increases with the knowledge degree of the node neighborhood, corresponding to the decrement of the number of new discovered peers in the same area of interest.

The transmission delay of a DGT packet has been computed by simulating with ns-3 the subsystem illustrated in Figure 16.6 (by averaging over 20 simulation runs), using the LTE package illustrated in Section 4 [32]. To match the previously described DGT configuration (i.e., DGT peers having GBs with radius of 2 km), we defined a square area having side length $l = 2$ km, with a grid of $r = 10$

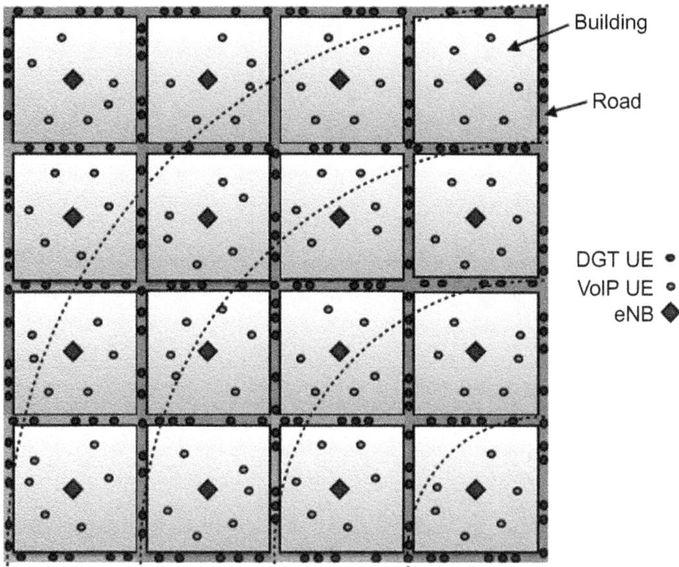

FIGURE 16.6

Bird's-eye view of the simulated scenario, with $n = 200$ DGT nodes and $n_v = 96$ other UEs randomly placed within the buildings. The geo-buckets of the DGT node in the bottom right corner of the map are also drawn, to show that the side length of the considered area equals the GB radius.

roads (5 in the N-S direction, and 5 in the W-E direction) and vehicles running over them (with linear density δ). The total amount of DGT User Equipments (UEs) is $n = r\ \delta\ l$. Parallel roads are spaced by $l/4 = 0.5$ km. In the map, there are 16 large buildings with square footprint, each one having seven floors. Randomly located within each building, there are $n_v/16$ other UEs, where n_v is their total amount. The path loss model is `ns3::BuildingsPropagationLossModel`. On top of each building, exactly in the middle, there is an eNB, i.e., a base station that serves a subset of the $n + n_v$ UEs. Such a dense deployment of eNBs may appear to be quite optimistic. We plan to test other models with 500–1000 m radius cells, and 200 active users each, which should be the best estimates for near-term LTE deployment.

The configuration of the eNBs includes FDD paired spectrum, with 50 Resource Blocks (RBs) for the uplink (which corresponds to a nominal transmission rate of 50 Mbps) and the same for the downlink—this is coherent with currently deployed LTE systems. DGT UEs use UDP to send four types of DGT packets to each other. The first type, called `Descriptor` (24 bytes), is for neighborhood consistency maintenance purposes. The second type of packet is the `Lookup Request` (20 bytes), which is used to search for remote nodes placed around a specified location. The third packet type is the `Lookup Response`

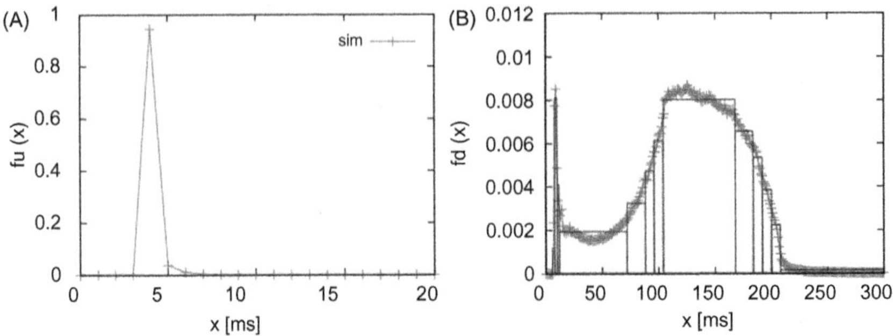

FIGURE 16.7

PDFs of the uplink (left) and downlink (right) delays for DGT packets (for the case with $\delta = 10$ vehicles/km), obtained with ns-3. The downlink PDF produced by means of ns-3 has 300 points. Its approximation obtained from the Packet Delay Filter has 13 levels > 0.

(500 bytes), which is sent by a DGT node as a reply to a lookup request, if the node owns the searched resource or information. Finally, there are traffic information packets (66 bytes). All packet types have also a 12 byte header. We set an interpacket interval of 50 ms for all types of DGT messages. Thus, the maximum rate is about 10 kB/s, while the minimum is $32 \times 20 = 0.64$ kB/s.

In a dynamic DGT scenario (the one simulated with DEUS), packets are not sent periodically—descriptors are sent only every ε meters; lookup requests are sent only when necessary, as well as lookup responses; traffic information messages are sent only when something interesting can be communicated to the other nodes (for example, a traffic jam or an incident). To simulate the presence of non-DGT traffic over LTE networks, we also included $n_v = 96$ other UEs, transmitting and receiving VoIP packets (using UDP) with a remote host located in the Internet. Such packets have a 12-byte header and a 13-byte payload, with interpacket interval set to 20 ms (we considered the AMR 4.75 kbps codec). The PDF of the resulting uplink delay is basically a Dirac delta function, shown in Figure 16.7a. Instead, the PDF of the downlink delay can be approximated with a corresponding piecewise constant function, with 13 levels, shown in Figure 16.7b.

Such delay profiles scale from small scenarios to larger ones, as they refer to intra-GB communications only. A DGT message could be propagated across the whole city, from one peer to another, relayed by intermediate peers. Each message propagation would be affected only by the data traffic within the GB of the forwarding peer, where the obtained delay profiles apply.

Such a packet delay model is a considerable improvement with respect to the one we used in our previous DEUS-based DGT simulations, which used, for every transmission, an exponential delay with mean value obtained by considering the nominal uplink and downlink.

Then, while simulating the whole overlay network with DEUS, we logged the average packet delay and amount of sent data per node. Figures 16.8 to 16.10

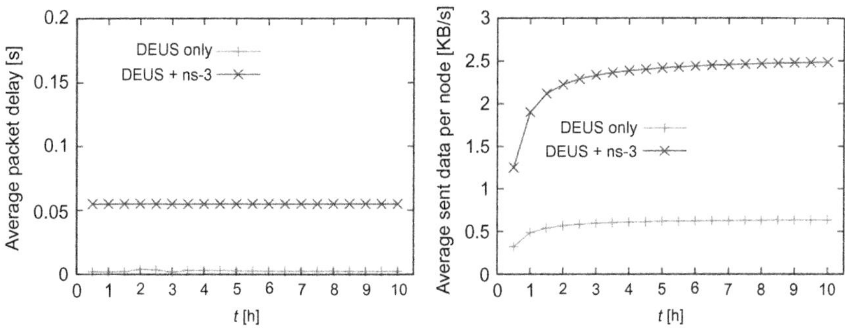

FIGURE 16.8

Average packet delay (left) and amount of sent data per node (right), measured with DEUS, for the simulated DGT overlay network with 500 vehicles.

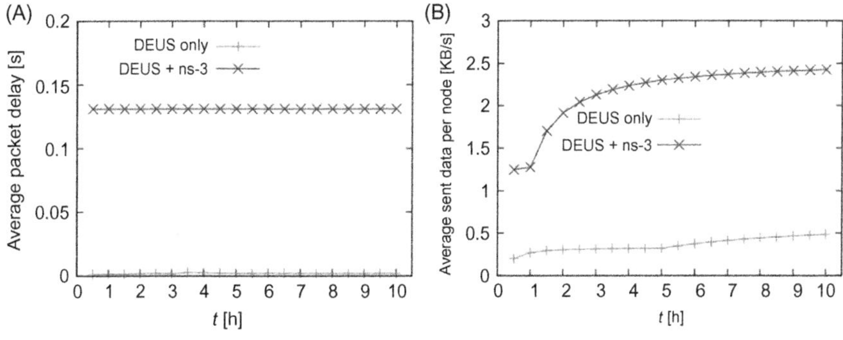

FIGURE 16.9

Average packet delay (left) and amount of sent data per node (right), measured with DEUS, for the simulated DGT overlay network with 1000 vehicles.

compare the results obtained with the old simulation model, and those obtained with the refined one, for different network sizes.

As we expected, in the refined model the average delay is higher than the one obtained with the naive model, which is based on nominal uplink and downlink values. Also the average amount of sent data is higher, as the refined model takes into account also packet headers.

7 CONCLUSION

In this chapter we have presented an effective and efficient co-simulation solution for wireless and mobile systems, based on the general-purpose simulation tool called DEUS, and the network-specific simulation tool called ns-3. With respect

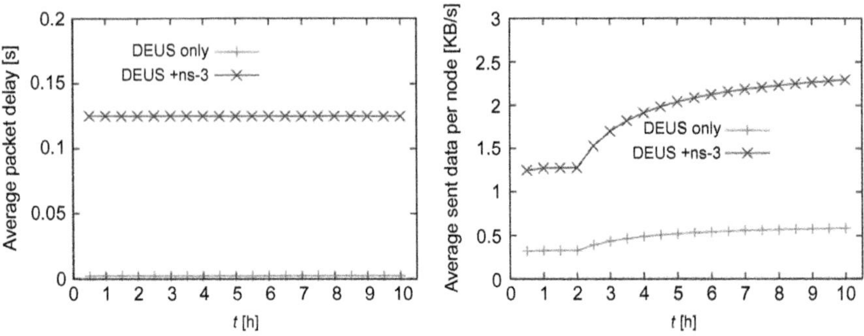

FIGURE 16.10

Average packet delay (left) and amount of sent data per node (right), for the simulated DGT overlay network with 2000 vehicles.

to the state of the art, such a solution has two main advantages. First, ns-3 allows us to obtain highly detailed statistics, encompassing all network layers. Second, DEUS is highly flexible, allowing us to simulate any type of mobility model, and to use deployment software on simulated devices and environments. The proposed approach has been successfully applied to the simulation of a peer-to-peer overlay with mobile nodes, associated to vehicles in a urban scenario.

REFERENCES

[1] Gershenson C, Heylighen F. How can we think the complex? In: Richardson K, editor. Managing organizational complexity: philosophy, theory and application. Information Age Publishing; 2005, Chapter 3.

[2] Zeigler BP, Praehofer H, Kim TG. Theory of modeling and simulation. 2nd ed. Academic Press; 2000.

[3] Wainer G. CD++: a toolkit to develop devs models. Software – Pract Exp 2002;32 (13):1–46.

[4] Varga A, Hornig R. An overview of the OMNeT++ simulation environment. In: First international conference on simulation tools and techniques for communications, networks and systems (SIMUTools 2008), Marseille, France, Mar. 2008.

[5] Amoretti M, Agosti M, Zanichelli F. DEUS: a Discrete Event Universal Simulator, 2nd ICST/ACM International conference on simulation tools and techniques (SIMUTools 2009), Roma, Italy; March 2009. ISBN 978-963-9799-45-5.

[6] NS-3 Consortium. ns-3. Official website. Available from: <http://www.nsnam.org>; 2014.

[7] University of Luxembourg. The OVNIS platform. Website. Available from: <http://ovnis.gforge.uni.lu>; 2014.

[8] Behrisch M, Bieker L, Erdmann J, Krajzewicz D. SUMO – simulation of urban mobility: an overview, in SIMUL 2011. In: Third international conference on advances in system simulation, Barcelona, Spain; October 2011. p. 63–8.

[9] Årzén KE, Cervin A. Control and embedded computing: survey of research directions, presented at the 16th IFAC World Congress, Prague, Czech Republic; 2005.
[10] Colandairaj J, Irwin GW, Scanlon WG. An integrated approach to wireless feedback control, presented at the UKACC international control conference, Glasgow, UK; 2006.
[11] Colandairaj J, Irwin GW, Scanlon WG. Analysis and co-simulation of an IEEE 802.11B wireless networked control system, in 16th IFAC world congress, Prague, Czech Republic; 2005.
[12] Chen Z, Liu L, Zhang J. Observer based networked control systems with network-induced time delay. In: IEEE International conference on systems, man and cybernetics, Hague, The Netherlands; 2004. p. 3333–7.
[13] Liu GP, Rees D, Chai SC. Design and practical implementation of networked predictive control systems. In: International conference on networking, sensing and control, Arizona, USA; 2005. p. 336–41.
[14] Yang Y, Wang Y, Yang SH. A networked control system with stochastically varying transmission delay and uncertain process parameters. In: 16th IFAC world congress, Prague, Czech Republic; 2005.
[15] Cervin A, Hanriksson D, Lincoln B, Eker J, Arzen KE. How does control timing affect performance? Analysis and simulation of timing using Jitterbug and TrueTime. IEEE Control Syst Mag 2003;23(3):16–30.
[16] Andersson M, Henriksson D, Cervin A, Årzén KE. Simulation of wireless networked control systems, presented at the 44th IEEE conference on decision and control and European control conference (ECC); 2005. p. 476–81.
[17] Al-Hammouri A, Liberatore V, Al-Omari H, Al-Qudah Z, Branicky MS, Agrawal D. A co-simulation platform for actuator networks. In: Fifth international conference on embedded networked sensor systems, Sydney, Australia; 2007. p. 383–4.
[18] Hasan MS, Yu H, Griffiths A, Yang TC, Interactive co-simulation of MATLAB and OPNET for networked control systems. In: 13th international conference on automation and computing, Stafford, UK; 2007. p. 237–42.
[19] Hasan MS, Yu H, Carrington A, Yang TC. Co-simulation of wireless networked control systems over mobile ad hoc network using SIMULINK and OPNET. IET Commun 2009;3(8):1297–310.
[20] Leclerc T, Siebert J, Chevrier V, Ciarletta L, Festor O. Multi-modeling and co-simulation-based mobile ubiquitous protocols and services development and assessment. In: Seventh international ICST conference on mobile and ubiquitous systems – mobiquitous 2010. Sydney, Australia; 2010.
[21] Gorgen D, Frey H, Hiedels C. Jane - the Java Ad hoc network development environment, In: ANSS '07; 2007. p. 163–76, USA.
[22] Deus: a simple tool for complex simulations. Official website. Accessed at: <https://code.google.com/p/deus/>; 2015.
[23] Montresor A, Jelasity M. PeerSim: a scalable P2P simulator. In: Ninth IEEE international conference on Peer-to-Peer (P2P'09), Seattle, WA, USA; September 2009.
[24] Brambilla G, Grazioli A, Picone M, Zanichelli F, Amoretti M. A cost-effective approach to software-in-the-loop simulation of pervasive systems and applications, in PerCom 2014, WiP Session, Budapest, Hungary, March 2014.
[25] Riley G. Large scale network simulations with GTNetS. In: Winter simulation conference, New Orleans, Louisiana, USA; Dec. 2003.

[26] Baldo N, Requena-Esteso M, Nin-Guerreo J, Miozzo M. A new model for the simulation of the LTE-EPC data plane, in fifth ICST/ACM international conference on simulation tools and techniques (SIMUTools 2009), Sirmione, Italy; Mar. 2012.

[27] Nagate A, Hoshino K, Mikami M, Fujii T. A field trial of multicell cooperative transmission over LTE system. In: IEEE international conference on communications (ICC 2011), Kyoto, Japan; Mar. 2011.

[28] Papoulis A. Probability, random variables, and stochastic processes. 3rd ed. McGraw Hill; 1991.

[29] Picone M, Amoretti M, Zanichelli F. Proactive neighbor localization based on distributed geographic table. Int J Pervasive Comput Commun 2011;7(3):240–63.

[30] Picone M, Amoretti M, Zanichelli F. Evaluating the robustness of the DGT approach for smartphone-based vehicular networks. In: Fifth IEEE workshop on user mobility and vehicular networks, Bonn, Germany; Oct. 2011.

[31] Distributed Systems Group (DSG). DGT – Distributed Geographic Table. Demo videos. Available from: <http://dsg.ce.unipr.it/?q=node/38#media>; 2015.

[32] Amoretti M, Picone M, Zanichelli F, Ferrari G. Simulating mobile and distributed systems with DEUS and Ns-3, international conference on high performance computing and simulation 2013, Helsinki, Finland; July 2013. Proceedings published by IEEE.

CHAPTER

17 Simulation methods, techniques and tools of computer systems and networks

Eugene David Ngangue Ndih and Soumaya Cherkaoui

Université de Sherbrooke, Sherbrooke, QC, Canada

1 INTRODUCTION

Simulations represent the evolution of a physical system over time. The main purpose of simulations is to mimic real systems or systems in the design stage to gain a deep insight into their behavior. In order to study a system, one might use the real system itself to perform various experimental test scenarios. However, in many cases, reproducing test scenarios on the physical system might be highly costly, or dangerous. In other cases, either testing the physical system is not possible because the latter is still under design, or else reproducing some of the conditions of the tests in the real world is not possible. The study of a scaled version of the system can certainly give an idea of the behavior of its full-scale counterpart. However, one must make sure that all details of the real physical system are taken into account when building its scaled version. This can consume lots of time and effort, depending on the complexity of the original system. When testing with the real system or a scaled version of it is not possible, the use of a model of the system is more practical. Usually, a system model is composed of a number of mathematical relationships binding quantitatively and logically parts of the system to each other under some assumptions.

For example, we can cite the Markovian models for the Aloha Slotted network and their extension for the case of 802.11 DCF [1].

Working with a model of a system has the advantage of allowing an evaluation of the system without the cost of physical experimentation. It also allows redesigning parts of the system prior to its deployment. Moreover, it is easier to reproduce the same test conditions many times with a model than with the system itself.

If the model of the system is simple enough, the behavior of the system can be deduced through an analytical study of the mathematical equations governing its behavior and interactions. However, often, when the system is too complex, its

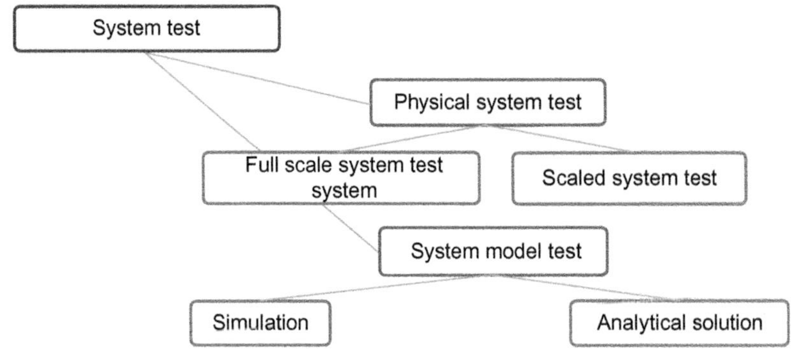

FIGURE 17.1

Options for testing a system.

model itself is also complex, making it very difficult, if not impossible, to find analytical solutions for it without using strong assumptions for simplification or constraining solutions to specific case studies. For example, in modeling the 802.11 performance, one may consider the hidden node problem which is difficult to model and which could lead to inaccurate results when modeled inappropriately.

An alternative way to take advantage of a system model is to perform simulations. Simulations can be used to validate an analytical model. However, in contrast to solving the analytical model, which usually requires simplifying assumptions, the description of the system using a simulation language could be very detailed and accurate with regard to the real system. Simulations make it possible to test scenarios that are too complex or too expensive to test on the real system. For instance, simulations can be used to simulate complex economic systems. Simulations also have the advantage of allowing reproducible test situations with a better control on the test environment than would otherwise be possible on the real system. Figure 17.1 summarizes the different options for testing a system.

Nevertheless, running simulations does not come without disadvantages and challenges. First, in order for simulations to be meaningful, one must reproduce the system model as closely as possible in the simulation environment. This can be costly in terms of time and effort. Second, one also needs to take particular care in reproducing the working environment of the real-life system. This usually translates into a number of inputs and physical constraints under which the real system will be evolving. Reproducing the working environment closely may thus require considerable effort too. This is particularly true when realistic input data cannot be simulated and experimentation is needed to obtain sample data to be fed to simulations as input. Third, as the sophistication of systems increases, the number of components that compose a system and the number of interactions among these components and with their working environment increases exponentially. Therefore, studying the exact behavior of such systems with simulations becomes a complex task requiring considerable computing memory and processing time. Finally, most of the time one must work with an existing simulation environment which might present limited capabilities.

In the following we will first introduce some of the best-known simulation methods, techniques and tools, with a special emphasis on simulation of discrete systems and computer networks.

2 SIMULATION TECHNIQUES, MODELS AND TOOLS

We distinguish two distinct categories of systems usually simulated: continuous systems whose state varies continuously in time, and discrete systems whose state varies only at discrete times. In addition, a system model can either be deterministic or stochastic, static or dynamic. In the deterministic case, a given set of input parameters produces a unique set of outputs, while in the stochastic case, the output obtained is an estimate of the real output. Most of the time, stochastic systems are systems whose model contains some probabilistic elements. For stochastic systems, randomness of some of the parameters is an important aspect which has to be implemented within simulations. Finally, a simulated model is said to be dynamic when the evolution of the system depends on time, while it is static when the system representation does not depend on time. Computer networks, wired and wireless, constitute an important family of *dynamic stochastic discrete* event systems as the states change (randomly in a huge number of applications) only at discrete instants of time, upon the occurrence of an event such as the reception or the transmission of a message. Their behavior can be described by discrete states.

For simulation purposes, a system can be modeled by a logical process (LP) consisting of virtual entities communicating with each other and executing some tasks or events. The evolution of the system is caused by the virtual entity state changes, or event execution, over time. The execution of an event might result in the creation of new events until the simulation end where no furthers events are created. For example, the reception of a frame in a wireless network leads to the generation of a reply (ACK) frame that the receiver schedules for transmission after a fixed delay depending on the (simulated) MAC algorithm. Each event has an associated time-stamp that indicates the time at which the event occurs (i.e., has to be executed). Basically, an event in simulations represents an indication of an update to the simulation system state at a specific time, referred to as time-stamp. In general, we distinguish three notions of time in simulations: the physical time, which is the time of the modeled physical system; the simulation time, representing the physical time in the simulation; and the wall-clock time, which is the real time elapsed during the execution of the simulation. It is possible to switch from one time to another depending on the mode of execution used in the simulation (real-time simulation or as fast as possible simulation); that is, it is possible, for example, to know the simulation time given the wall-clock time.

To have a correct simulation, synchronization algorithms might be required in general in order to avoid undesirable behaviors such as deadlocks or live-locks, and to satisfy the local causality constraint [2]—that is, the execution by LPs of the events in a non-decreasing time-stamp order.

Because of the proliferation of discrete event systems and their representation of many systems in real-life, especially computer networks, we put a special emphasis on these systems in this chapter. Several techniques of discrete event simulation can be found in the literature. We highlight some of them. Specifically, we describe sequential simulation techniques and parallel simulation techniques of discrete event systems when central processing units are the sole involved computing entities considered. In the next section, we present some simulation techniques that introduce the use of other computing entities such as graphic processing units together with central processing units in order to reduce the cost of computing operations. Finally, we mention the case of using multi-agent–based simulations for wireless ad hoc networks.

3 DISCRETE EVENT SIMULATION

Discrete-event simulation (DES) is an extensive field of simulation which has been widely used in the past [2,3]. DES allows representing and studying the behavior of physical systems, i.e., collections of entities interacting over time, whose states evolve in time as is the case in computer networks. DES is built upon two fundamental building blocks: the *simulation objects*, and the *events*. The simulation objects map the real physical objects (entities), while the events have potentially two functions: modify the state of a simulation object, or schedule future events.

In DES, the events to be processed are maintained in a *list* or *event calendar*. We distinguish *sequential* and *parallel* DES.

3.1 SEQUENTIAL DISCRETE SIMULATION

Sequential simulations typically use three data structures: *state variables*, a single *calendar*, and a single *global clock*. They require three basic operations: *removal* of events from the calendar, the *execution* of events removed, and the *insertion* of new events resulting from the execution of an event, in the calendar.

Basically, the events are executed in a non-decreasing time-stamp order in a loop by removing the smallest time-stamped event from the event calendar, and when some events have the same time-stamp, they are executed concurrently since they might share resources. More specifically, we distinguish *time-driven* and *event-driven* sequential simulations.

In a time-*driven sequential simulation*, the events are grouped into lists of activities corresponding to timing cycles, and the time is updated at the end of each cycle; that is, the starting time of the next timing cycle reflects the ending time of the current timing cycle, whereas in the *event-driven sequential simulation*, time is dynamically updated to reflect the time of the next scheduled event in the calendar. In other words, in event-driven sequential simulation, the time is updated for each event.

In general, sequential simulations are deterministic; they easily obey the local causality constraint, and are simple to implement. They have the disadvantage, however, that large discrete event simulations consume a considerable amount of processing time.

3.2 PARALLEL DISCRETE EVENT SIMULATION

Despite the increase in computation capabilities of computers used for sequential simulations, the use of multiple processing nodes is an increasingly appealing approach to speed up simulation of discrete event systems. With multiprocessor computers, a sequential simulation execution can be distributed or run over multiple processors to take advantage of parallel processing capabilities, thus gaining time and resources. Techniques of parallel discrete event simulation (PDES) substantially decrease run times of sequential simulations and improve both the performance and the scalability of simulations in general, and computer network simulations in particular, and the results achieved must match those obtained by sequential simulation.

Basically, in PDES, a simulation is divided across the number of available LPs. An LP has a single or multiple queues to store events to be executed or messages to be processed. It also contains several objects such as its *Local Virtual Time*, its *Future Event List* containing internal events posted within the LP itself, *events* and *messages*. Messages are events received from other LPs or to be sent to other LPs. They are stored in the input/output queues according to first-in/first-out (FIFO) protocol.

With this distribution of events throughout the set of LPs, the choice of the next event is no longer trivial as was the case in sequential simulations. Hence, for parallel simulation, LPs need to communicate with each other using exchange of time-stamped messages (message passing) delivered to other LPs at the right order in order to preserve the causality constraint. In parallel simulation, the main challenge is to guarantee causality dependency among events at different LPs while reducing the overhead introduced for synchronization between the LPs.

For this aim, we distinguish two main approaches for parallel simulation: parallel simulation based on *replication*, and parallel simulation based on *model decomposition*.

3.2.1 Parallel simulation based on replication

Simulations of discrete event systems are time consuming and may take, for example, several days to obtain accurate estimates in communication network simulations. In some applications, the analysis under certain input parameters requires a given confidence interval, and in these cases, performing multiple replications of the same simulation is necessary [5]. Also known as Multiple Replications in Parallel (MRIP) [6] or Parallel Independent Replicated Simulation

(PIRS) [7], this simulation technique basically consists of running the same sequential simulation from the beginning to the end in several LPs in parallel with different input parameters which send intermediate outputs to a central master process that merges and analyzes the different results.

PIRS is technically simple and easy to implement, and the efficiency achieved is good since message passing and synchronization are not required among the involved LPs. This technique has been used, for example, to simulate non-Markovian stochastic Petri networks in [8]. An important point with PIRS is to schedule the LPs for parallel replication executions. Several policies can be found in the literature (see [7] and the references therein). A simple scheduling policy is the *Fixed number of Replications per Processor* (FRP) that consists of assigning a fixed number of replications to each processor involved in the simulation. While FRP is simple to implement, processors that have finished the execution of the replications assigned to them before the other processors will remain idle until the last processor finishes executing its assigned replications. Better scheduling policies consist of dynamically assigning replications to the involved LPs. For example, at the end of its replications execution, a processor returns the results to the central master process (scheduler) that checks the results, and then either initiates for execution new replications to the corresponding processor if additional results are required, or aborts the execution of all LPs and terminates the simulation if the results satisfy the termination condition.

The applicability of the replication-based simulation is restricted, and depends on the set of input parameters of the simulation and on the initial transient strength [9,10], and the scalability is bounded by the centralized architecture. Another disadvantage of PIRS is that each processor must have enough memory to run the entire sequential simulation. Replication is very useful when the simulation is largely stochastic, and may be useless when the results of one sequential simulation depend on some results of another sequential simulation.

3.2.2 Parallel simulation based on model decomposition

Despite the wide use of PIRS techniques, the most commonly used approach for performing parallel simulation is model decomposition. For example, the simulation of wireless ad hoc networks could be computationally expensive, and difficult to parallelize. However, one can exploit, for instance, the relative dependency among the wireless transmissions in the time domain to decompose the simulation model for fast execution.

Basically, model decomposition-based parallel simulation consists of the decomposition of the set of state variables into several subsets, each assigned to an LP which executes it concurrently. In other words, model parallelization decomposes the main task of a simulation into multiple subtasks, executed each in parallel by an LP.

Considering the representation of the simulation computation as a space-time model [11,12], PDES techniques are broadly categorized into *spatial parallel*, *time parallel* and *space-time parallel*. Next, we present spatial parallel and time parallel simulation. Readers interested in space-time parallel can refer to [12].

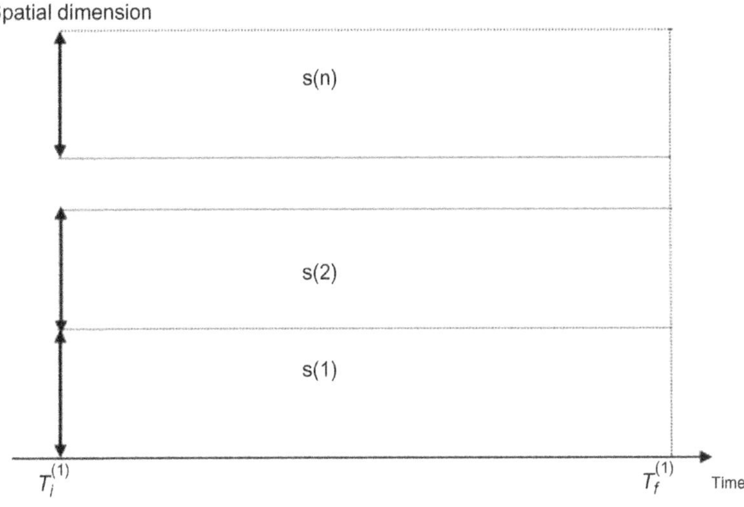

FIGURE 17.2

All the subsets $S(j)$ of state variables have the same initial state $T_i^{(j)}$ and the same ending state $T_f^{(j)}$

3.3 SPATIAL PARALLEL SIMULATION

In spatial parallel simulation, the simulation is divided along the spatial dimension underlying the model [13] into subsets of state variables. Figure 17.2 naively shows how the decomposition is performed along the spatial dimension. Each subset $S(j)$ is assigned to the LP j for concurrent execution with respect to the temporal order. The temporal order is either maintained at every simulation time or is achieved asymptotically [13]. Drawbacks of this technique are either the limited amount of parallelism achievable due to the available decomposability of the model [10,14], or the difficulty of synchronizing the events in the concurrent LPs. There are different techniques that ensure local causality constraint and coherence between simultaneous executions of the events with different time-stamps. In this chapter, we briefly present *synchronous* simulation and *asynchronous* simulation. Further readings can be found in [5].

3.3.1 Synchronous PDES

In synchronous PDES, all the LPs involved in the parallel simulation share a common view of the clock in order to synchronize their actions. In these simulations, most events occur during relatively short intervals containing the clock edges, such that there are only a few events to simulate for the rest of the time. Two different approaches can therefore be used to address the key issue of synchronous simulation—that is, exploit as much as possible the available parallelism with the least possible overhead. These approaches are the *time-driven* synchronous simulation, and the *event-driven* synchronous simulation.

3.3.1.1 Time-driven synchronous PDES

In time-driven synchronous simulation, the different LPs execute only the events whose time-stamps correspond to the value of the clock [5]. Then, the time is incremented in one time unit, and using a barrier operation, the synchronization is performed among the LPs to prepare the next iteration.

3.3.1.2 Event-driven synchronous PDES

When event-driven synchronous simulation is used, instead of incrementing the clock in one unit time at the end of each iteration as is the case in time-driven simulation, the clock of all the LPs is advanced to the minimum time-stamp of the pending events among all the LPs at each iteration termination [5]. In general, in event-driven simulations, the simulation system is organized as a series of active components connected by nets, and whose values are changed by the execution of an event. Further readings and examples of event-driven simulation can be found in [5,15–17].

3.3.2 Asynchronous PDES

In asynchronous PDES, the different LPs involved in the simulation do not necessary share a common clock. They operate asynchronously with different progress rates and potentially different clocks and use message passing as means to communicate with other LPs for synchronization purposes. Hence, at the cost of additional complexity in the synchronization mechanisms between the LPs, asynchronous PDES lead to fast simulations.

Asynchronous PDES have been widely studied (see [18,19] and the references therein), and different categories are proposed in the literature. We present in this chapter the main ones used: *conservative* and *optimistic*.

3.3.2.1 Conservative PDES

In conservative PDES, the events at each LP are executed in time-stamp order, and the causality constraint is always respected. Basically, before executing an event with a given time-stamp, each LP makes sure that there is no incoming message events from other LPs with smaller time-stamps pending [20] in its event queue. In other words, in conservative PDES, causality errors are not possible, and the events are processed safely—that is, if an LP does not have enough information in order to predict future interactions with other LPs, it suspends the execution of its events.

For this end, the conservative PDES method makes use of the look-ahead [21] model property (which is an a priori knowledge of some properties of the model) to ensure safety predictions.

Basically, in the look-ahead-based conservative synchronization PDES, also known as Chandy-Misra-Bryant (CMB) [22] simulation techniques, all the LPs involved in the parallel simulation are interconnected from the beginning to the end of the simulation, so that each LP is aware of the events executed/sent in non-decreasing time-stamp order by the other LPs. Therefore, in addition to its own queue, an LP has multiple other FIFO-based (First-In/First-Out) input queues, each associated to the incoming messages for each specific LP involved in the global simulation.

To execute an event from one of its queues (its own queue, plus all the input queues associated each to the incoming messages of a specific LP), an LP selects the event with the minimum time-stamp, and compares it to a threshold known as the acceptance horizon or the lower bound time stamp [23]; below the acceptance horizon, the event is executed, and above, the event is blocked until the LP receives additional information (events) from the other LPs. The acceptance horizon is not constant, and has to be updated as long as the simulation progresses. Note that in order to preserve causality, events extraction is performed only if all the queues present a message; otherwise extraction is delayed.

To avoid long blocking periods (deadlocks) due to the presence of empty input queues (obsolete acceptance horizon), or system memory overflow, one can use the standard null-message-based protocol, which consists of an exchange of null-messages (message without event but with time-stamp) between LPs in order to update the acceptance horizon without creating additional events. Further details can be found in [5,23].

Due to the considerable amount of null messages sent to avoid deadlock in the standard approach, especially in distributed systems, several alternative conservative techniques which implement null-message avoidance/reduction mechanisms have been proposed in the literature.

For example, in the deadlock avoidance protocol [24], an LP sends null-messages only if it is to suspend the execution of an event.

In the timeout-based protocol [22], the null-messages are sent only after the expiration of a specific timer. The reduction of null-messages is effective since the null-messages are sent only when the normal update of the acceptance horizon fails.

In some scenarios, it is useful to combine both deadlock avoidance and timeout-based protocols [23] to further reduce the amount of null-messages exchanged.

3.3.2.2 Optimistic PDES

The first optimistic parallel simulation is known as *Time Warp* (TW) and can be found in [25]. In this technique, the time used to model the time in the simulation is a *virtual time*. Each LP has its own Local Virtual Time (LVT) and executes its events independently of the other LPs, that is, without synchronization. However, contrary to conservative simulations where the LPs only need to keep a record of its current state, in TW, each LP keeps a record of its past input and output events in order to manage causality errors, until it is sure that no event with smaller time-stamp can be received elsewhere in the simulation.

Indeed, with optimistic PDES, the local causality constraint can be violated. But when this occurs—that is, if some events, known as stragglers, are later detected to have been executed in an incorrect order—the system restoration or a reverse computation is invoked, and the affected LPs undo the incorrect computations, and roll back to the last point in the past where the causality constraint was respected and free of error.

In order to cancel all the messages sent to other LPs during the erroneous computation phase, affected LPs send to other LPs an *anti-message* for each erroneous message previously sent. Note that the purpose of an anti-message is to annihilate its corresponding message. This procedure is invoked at each affected LP. All the erroneous messages pending in the input queues are deleted once detected. While this technique avoids look-ahead delays of the conservative simulation, it has to enable techniques for ease *rollback*—that is, the recovery of the previous error-free states of the simulation—and to reduce the amount of memory required due to the large number of events recorded [26,27].

Rolling back is computational expensive and can have a considerable effect on the performance of a TW simulation. In a TW simulation, there exist two types of rollback: the *primary rollbacks*, and the *secondary rollbacks*. The primary rollbacks are caused by stragglers, and the secondary rollbacks are caused by the anti-messages. Thus, the main purpose of the rollback-handling techniques in TW simulations is to control as much as possible the rollback mechanism to avoid the *domino effect* as a consequence of uncontrolled rollback behavior. To ease or reduce the rollback procedure in a TW simulation, different schemes have been developed and proposed in the literature. In the following, we briefly mentioned some of them, but more examples can be found in [28]. According to [28], these techniques can be grouped as *optimism control* techniques, *event cancellation* techniques, and *LP aggregation* techniques.

The purpose of optimism control techniques, such as the Breathing Time Warp [29] or the Switch Time Warp [30] algorithms as examples, is to reduce the rollback by improving in a TW simulation the *temporal locality* of the event, thus reducing as much as possible the relative difference between the time-stamps of the concurrent processed events. However, these techniques require *a priori* a global knowledge of some aspects of the simulation. With event cancellation techniques, such as the Throttled Lazy Cancellation [31] or the Early Cancellation [32], the communication overhead due to the exchange of anti-messages is reduced by either controlling the broadcast of the anti-messages, or by cancelling wrong messages with an appropriate mechanism. To reduce the anti-messages, event cancellation techniques in general introduce extra overhead, which is further reduced using LP aggregation techniques (see [28] for further details).

To reduce the amount of memory used in the TW simulation, several techniques have been proposed in the literature and are classified, according to [28], as *fossil collection* techniques [33,34], *memory stall recovery* techniques [27], or *checkpointing* techniques [35].

Basically, fossil collection techniques consist of reclaiming memory from the state and event histories in order to reduce memory stalls in TW simulations. However, fossil collection techniques suffer from generating extra significant overhead, and from not being able to guarantee absence of memory stall in certain cases.

In memory stall recovery, memory recovery can be performed either upon shared-memory LP's requests to a global pool or by releasing past state memories.

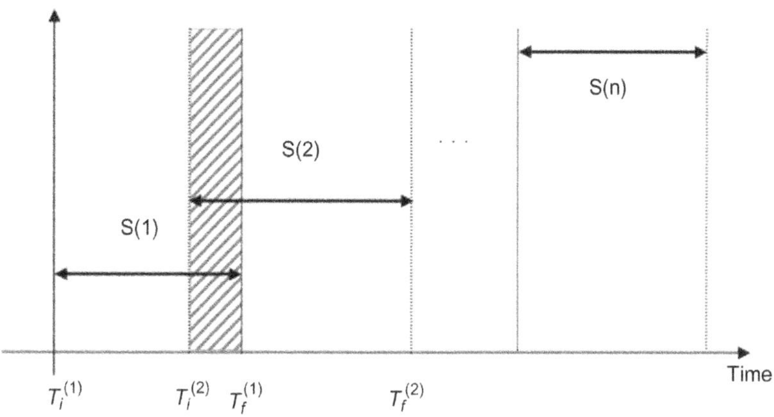

FIGURE 17.3

State matching problem in time-parallel simulation.

At the cost of increasing the computational overhead, checkpointing techniques allow rolling back by reducing state-saving overhead either by reducing the number of states saved in the simulation, or by reducing the amount of data saved in each state.

3.4 TIME-PARALLEL SIMULATION

In time-parallel simulation, it is possible to achieve a huge parallelism by decomposing the space-time diagram along the time dimension, such that the resulting subsets of state variables are executed in parallel, each by an LP. There are two main advantages of the time-parallel simulation: the decomposition along the time dimension is straightforward and can be applied even in the case where a reasonable partitioning of the state variables is not possible, and the simulations of LPs are (closely) independent, that is, there is no need a priori for communication between LPs during the simulation.

However, in time-parallel simulation techniques, it is required to address the *state matching decomposition* issue—that is, to ensure matching at boundaries of the initial state and the ending state of two adjacent subsets of state variables [12]. Figure 17.2 illustrates the state matching decomposition problem that occurs in time-parallel simulations.

In Figure 17.3 the ending state $T_f^{(1)}$ of the state variable subset S (1) does not coincide with the initial state $T_i^{(2)}$ of the state variable subset S (2).

To solve the state matching problem, we distinguish two main approaches: *regeneration point* and *fix-up computation.*

3.4.1 Regeneration point-based temporal decomposition

Also known as *precomputation* of state at specific time division points for time-parallel simulation [36], the regeneration point-based temporal decomposition consists of reoccurring states (regeneration points) throughout the simulation. Then, given that the regeneration points are identified a priori, the decomposition along the time axis of the global set of state variables is such that starting and ending states of each subset are regeneration points. Afterwards, all the parallel simulation traces are threaded to obtain a correct trace of the entire simulation. The performance of the regeneration point-based approach depends on the occurrence frequency of the regeneration points in the decomposition of the set of state variables.

Several examples of regeneration point-based parallel simulations can be found in the literature. For instance, the authors in [37] propose a regeneration points-based scheme for parallel simulation of multiple access protocols for medium access.

In general, the disadvantage of the regeneration point-based parallel simulation is the difficulty in identifying the regeneration points. For instance, for nonregenerative systems such as wireless networks in which the regeneration point would correspond to a state in which all the state variables and the routing tables are flushed and the nodes are off [38], an approximate solution [39] can be used instead.

3.4.2 Fix-up computations-based temporal decomposition

With fix-up computation-based simulation, each LP first guesses the initial state of its corresponding subset of state variables, and then executes the simulation accordingly [40]. After message passing, if the final state executed in the adjacent LP does not match its guessed initial state, it performs a fix-up computation by replacing its initial state by the final state of the adjacent subset. This procedure is performed repeatedly at each LP, and the simulation is completed once all the initial states/final states of adjacent subsets of state variables matches.

3.5 OTHER TIME-PARALLEL SIMULATIONS

Other time-parallel-based simulations have been proposed to simulate specific networks. For example, [41] proposed a combination of a priori initial state approximation and a simulation warm-up interval.

In the warm-up interval technique, two intervals (the warm-up interval and the measurement interval) are considered for each simulation segment. In this technique, no measurement is performed during the warm-up phase (first phase); the measurements are performed during the second phase. During this second phase, the initial state of the simulation is "approximated" by the final state of the warm-up interval [38,41] which takes into account perturbing events that take place during the warm-up phase. The accuracy of the simulation depends on the duration of the warm-up phase; the longer the warm-up, the better the simulation accuracy. The simulation stops when the desired accuracy is obtained.

To speed up the simulation, the authors replace completely/partially the warm-up phase with a *compressed history* [38], which represents a simplified scenario that will leave the network in the same final state even if the simulation traces are different. This simplified scenario is executed during the first phase before the measurement phase.

4 GPU-BASED SIMULATIONS

As seen in the previous sections, various approaches proposed to speed up simulation of discrete event systems synchronize the asynchronous LP in order to respect the local causality constraint. In these techniques, the simulations are executed on central processing units (CPU), which become costly and thus inappropriate for scenarios requiring a large number of parallel processing units.

Recently, graphics processing units (GPU) have gained more interest as an attractive alternative to replace CPUs due to their ubiquitousness (low cost) and their considerable computing power. However, despite the highly parallel architecture of a GPU, executing GPU-based discrete event parallel simulations still remains a big challenge. This is due to the fact that, on the one hand, GPU hardware is Single Instruction, Multiple Data (SIMD)-based hardware, and is more adapted for time-synchronous simulations [42]. On the other hand, discrete event systems behave generally asynchronously, thus the times of the events have to be modified in order to be synchronized [43].

Various approaches have been proposed in the literature to improve SIMD-based simulation of discrete event systems. According to [42], in order to execute discrete event simulations on SIMD-based platforms, traditional events and LPs have to be cast into a stream processing paradigm which is supported by the GPU-based hardware. Hence, Perumalla in [42] proposed a hybrid algorithm for GPU-based discrete event simulation, which is a combination of the time-stepped simulation and the discrete event simulation. In this technique, all the elements of the grid of events are updated synchronously, and this updating is achieved using a time-step corresponding to the minimum event time chosen from the list of update times. He showed that the performance of GPU-based simulation becomes more effective compared to CPU-based simulations for problems of larger size exceeding the L2 cache size of the CPU.

In [43], Park and Fishwick proposed also a hybrid time-synchronous/event approach to parallel discrete event simulation of queuing networks. Their technique is SIMD-based and combines CPU and GPU into a master-slave paradigm, where the CPU acts as the master and the GPUs act as the slaves. At the beginning of the simulation, the CPU extracts from the Future Event List (FEL) the minimum time-stamp event as long as it encounters events with the same time-stamp. Events with the same time-stamp constitute a Current Event List (CEL) and are sent each for execution to a GPU. At the end of the execution, each GPU sends the results and the time-stamp increments for the next execution to the CPU which schedules the next corresponding vent to the FEL with the time-stamp increment. For synchronization purpose, the CPU waits for the responses of each GPU before continuing the extraction of the FEL events.

5 MULTI-AGENT–BASED SIMULATION

In multi-agent–based simulation (MABS) [44–47], the simulated entities are modeled as *agent.* Each agent knows its capabilities and is capable to function on its own. Hence, with little information about the other agents, device agents cooperate and good decisions regarding the system can be made in reasonable time. Thus, MABS techniques are preferred to simulate complex systems, and can be used together with other object-oriented simulation techniques.

Using or designing a specific MABS technique requires appropriate tools such as the communication standard between the agents, the simulation platform, the simulation language, etc. There are different methodologies proposed in the literature (for example, Gaia [48] and Prometheus [49]) to develop multi-agent systems (MAS)). However, each methodology is in general appropriate for specific problems, and can be used only in certain platforms with specific languages.

As already mentioned, the simulation platform and the languages for MAS are other important aspects to be considered in MABS techniques. There are several platforms proposed in the literature. For instance, the open-source platform JADE (Java Agent Development) [45] allows the implementation of MAS for peer-to-peer agent-based applications. JADE acts in accordance with the specifications of the Foundation for Intelligent Physical Agents (FIPA) [50], an IEEE organization that actively encourages agent-based technology, standards and interoperability with other technologies. Multi-Agent Simulator Of Networks (MASON) [46,51] is another well-known open-source platform used for modeling and simulating MAS. With MASON, the system is written in Java.

Traditionally, MABS has been used mostly in purely social contexts; it supports a wide range of scenarios including proactive behavior and dynamic systems, in comparison to discrete event simulations widely used in computer network simulations [44]. Nowadays, due to the increasing complexity of nodes in computer-based networks, especially in wireless ad hoc networks which are created on-the-fly without any central coordinator, the presence of agents in such systems has become very popular and MABS might provide adequate simulation environments.

For example, in vehicular networks, intelligent vehicles communicate intensively with each other under the Dedicated Short Range Communications (DSRC) [52] standard for safe and efficient traffic. To be able to capture the behavior of both the system and each entity of the system, and due to various real-time constraints in wireless communication, network dynamics, etc., Fernandes and Nunes [47] proposed an architecture using MAS to simulate vehicular traffic with communications. Their architecture includes multi-agents: the network simulator (NS) is in charge of simulating the network transmission between Communication Manager agents; the Communication Manager (CM) agent manages the communications between the vehicle and the other vehicles or infrastructures; the Intersection Traffic Rules Arbiter (ITRA) agent is in charge of recording traffic events, and resolves conflicts between User Managers; the User Manager (UM)

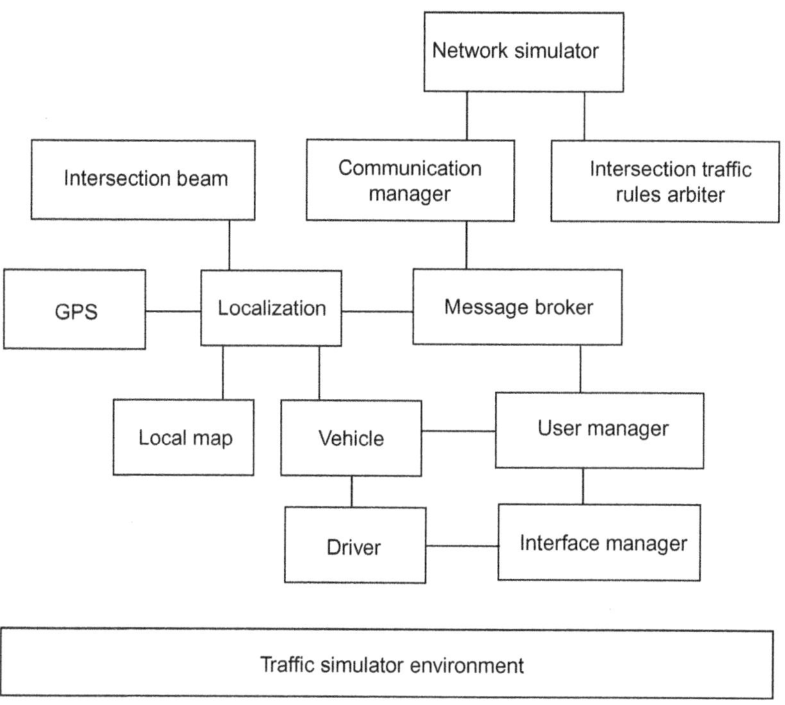

FIGURE 17.4

Multi-agent architecture [47].

agent deals with the priority of the vehicles under the ITRA supervision; the Interface Manager (IM) agent selects the most appropriate message interface to the driver; the Driver agent controls the vehicle; the Vehicle agent collects information such as speed, acceleration, brakes, etc., and communicates it to the Localization agent; the Localization agent determines the localization of the vehicle using the Global Positioning System (GPS); the Message Broker (MB) manages all internal messages based on their priority. The multi-agent architecture [47] is shown in Figure 17.4.

6 CLASSIFICATION OF SOME DISCRETE EVENT SIMULATION TECHNIQUES

In Table 17.1, we present a classification of the main techniques mentioned in this chapter to simulate discrete event systems such as wired/wireless computer networks.

Table 17.1 Classification of Some Discrete Event Simulation Techniques

Simulation Technique				Principle	Advantages	Disadvantages
	Sequential			Events in the queue are executed in a non-decreasing time-stamp order.	Easy to implement	Require large amount of memory; Inappropriate for scale-free networks.
		Replication		Multiple independent sequential simulation.	Easy to implement; Do not require LP synchronization.	Limited to stochastic and independent simulations; Make use of costly hardware (CPU).
CPU-based	Parallel	Model decomposition	Space Space Parallel	The simulation is split along the spatial dimension and is distributed among the LPs.	Simulate large networks;	Hard to implement; Require LP synchronization; Deadlock problem; Make use of costly hardware (CPU).
			Time Parallel	The simulation is split along the time dimension and is distributed among the LPs.	Simulate large networks.	Hard to implement; State-matching problem; Make use of costly hardware (CPU).
GPU-based				Aim to replace CPU hardware with GPU hardware in simulations.	Simulate large networks; Make use of cheap and ubiquitous hardware (GPU).	Hard to implement; GPU are SIMD and inappropriate for asynchronous systems; Require time modification for synchronization purposes.
Agent-based				The nodes are intelligent and are modeled as agent.	Appropriate for dynamic and intelligent networks; Appropriate for complex wireless ad hoc networks.	Hard to implement; Is platform-based.

7 CONCLUSION

The main simulation purpose is to mimic real and complex systems such as wired and wireless networks in order to have a deeper insight into their behavior. Given the large amount of memory required to simulate these networks, especially when the number of nodes is important, simulation becomes a challenging task. There are several techniques proposed in the literature. The easiest one simply consists of executing events in a unique queue in a non-decreasing temporal order. However, due to the large amount of memory required, this technique might be inefficient for large networks. To speed up the simulation, parallelism can be used. That is, the simulation is distributed among several LPs. Depending on the correlation between the simulation runs on the different LPs, one can choose either replication techniques in which the runs have to be firmly independent in order to avoid synchronization issues, or model decomposition-based simulation techniques in which the simulation runs can be correlated. The model decomposition approach divides the model into several parts along either the spatial dimension or the time dimension. This decomposition of the model allows simulating networks with large numbers of nodes. However, it introduces synchronization issues. In addition, many techniques proposed are CPU-based. Given the ubiquitousness (inexpensive) and the high parallelism of GPUs, several techniques in which GPUs are used as replacements for a CPU are proposed. However, GPUs are SIMD and are inappropriate for asynchronous systems, and GPU-based discrete event simulations require time management for synchronization purposes. In the case of complex heterogeneous networks, in which nodes are intelligent and can interact with their environment, multi-agent–based simulations are appropriate.

REFERENCES

[1] Bianchi G. Performance analysis of the IEEE 802.11 distributed coordination function. IEEE J Select Areas Commun 2000;18(3):535–47.

[2] Wagner DB. Conservative parallel discrete-event simulation: principles and practice. Ph.D. dissertation, Department of Computer Science and Engineering, University of Washington; 1989.

[3] Fishwick PA. Simulation model design and execution: building digital worlds. New York: McGraw-Hill; 1994.

[4] Banks J, Carson JS, Nelson BL. Discrete-event system simulation. Upper Saddle River, NJ: Prentice Hall; 1996.

[5] Alonso JM. An empirical evaluation of techniques for parallel simulation of message passing networks. Servicio Editorial de la Universidad del País Vasco Argitarapen Zerbitzua, Euskal Herriko Unibertsitatea; 1996.

[6] Bononi L, Luciano M, Bracuto G, D'Angelo, Donatiello L. Concurrent replication of parallel and distributed simulations. Principles of advanced and distributed simulation, (PADS 2005). Workshop on IEEE; 2005.

[7] Lin YB. Parallel independent replicated simultion on a network of workstations. ACM SIGSIM Simul Digest 1994;24(1):73–80.
[8] German R, et al. TimeNET: a toolkit for evaluating non-Markovian stochastic Petri nets. Perform Eval 1995;24(1):69–87.
[9] Heidelberger P. Statistical analysis of parallel simulations. In: Proceedings of the 1986 winter simulation conference; 1986. p. 290–5.
[10] Kiesling T, Krieger T. Efficient distributed queuing system simulation. Fak. für Informatik, Univ. der Bundeswehr München, 2006.
[11] Chandy KM, Sherman R. Space, time, and simulation. In: Proceedings of the SCS multiconference on distributed simulation: SCS simulation series; 1989.
[12] Wu H, Fujimoto RM, Ammar M. Time-parallel trace-driven simulation of CSMA/CD. In: Proceedings of the 17th workshop on Parallel and distributed simulation. IEEE Computer Society; 2003.
[13] Perumalla KS. Parallel and distributed simulation: traditional techniques and recent advances. In: Proceedings of the 38th conference on winter simulation. Winter simulation conference; 2006.
[14] Wagner DB, Lazowska ED. Parallel simulation of queueing networks: limitations and potentials. In: SIGMETRICS; 1989. p. 146–55.
[15] Konas P, Yew P-C. Parallel discrete event simulation on shared memory multiprocessors. In: Proceedings of the 24th annual simulation symposium. New Orleans: Louisiana; 1991. p. 134–48.
[16] Bruner JD, Cheong H, Veidenbaum A, Yew P-C. Chief: a parallel simulation environment for parallel systems. In: Proceedings of the fifth international parallel processing symposium; 1991. p. 568–75.
[17] Konas P, Yew P-C. Synchronous parallel discrete event simulation on shared-memory multiprocessors. In: 6th workshop on parallel and distributed simulation; 1992. p. 12–21.
[18] Prakash S, Deelman E, Bagrodia R. Asynchronous parallel simulation of parallel programs. IEEE Trans Softw Eng 2000;26(5):385–400.
[19] Schneiders J. Area virtual time. PhD. Thesis. Institute of Computing Systems Architecture, School of Informatics, University of Edinburgh; 2003.
[20] Chandy KM, Misra J. Asynchronous distributed simulation via a sequence of parallel computations. Commun ACM 1981;24(4):198–205.
[21] Fujimoto RM. Parallel discrete event simulation. Commun ACM 1990;33(10):30–53.
[22] Misra J. Distributed discrete-event simulation. ACM Comput Surv 1986;18(1):39–65.
[23] De Munck S, Vanmechelen K, Broeckhove J. Revisiting conservative time synchronization protocols in parallel and distributed simulation. Concurr Comput Pract Exper 2014;26(2):468–90.
[24] Naroska E, Schwiegelshohn U. Conservative parallel simulation of a large number of processes. Trans Soc Model Simul Int 1999;72(3):150–62.
[25] Jefferson DR. Virtual time. ACM Trans Program Lang Syst 1985;7(3):405–25.
[26] Jefferson DR. Virtual time II: storage management in distributed simulation. In: Proceedings of the 9th annual ACM symposium on principles of distributed computing. Québec, Canada; 1990. p. 75–89.
[27] Preiss BR, Loucks WM. Memory management techniques for Time Warp on a distributed memory machine. In: Proceedings of the 9th international workshop on parallel and distributed simulation. Lake Placid, NY; 1995. p. 30–9.

[28] Jafer S, Liu Q, Wainer G. Synchronization methods in parallel and distributed discrete-event simulation. In: Simulation modelling practice and theory, vol. 30. Elsevier; 2013. p. 54–73.

[29] Steinman JS. Breathing Time Warp. ACM SIGSIM Simul Digest 1993;23 (1):109–18.

[30] Suppi R, Cores F, Luque E. An efficient method for improving large optimistic PDES. In: Proceedings of the 8th international symposium on modeling, analysis and simulation of computer and telecommunication systems. San Francisco, CA; 2000. p. 351–7.

[31] Soliman HM. Throttled lazy cancellation in Time Warp parallel simulation. Simulation 2008;84(2–3):149–60.

[32] Noronha R, Abu-Ghazaleh NB. Early cancellation: an active NIC optimization for Time Warp. In: Proceedings of the 16th international workshop on parallel and distributed simulation. Washington, DC; 2002. p. 43–50.

[33] Vee VY, Hsu WJ. Pal: a new fossil collector for Time Warp. In: Proceedings of the 16th international workshop on parallel and distributed simulation. Washington, DC; 2002. p. 35–42.

[34] Chetlur M, Wilsey PA. Causality information and fossil collection in Time Warp simulations. In: Proceedings of the 2006 winter simulation conference. Monterey, CA; 2006. p. 987–94.

[35] Tay SC, Teo YM. Probabilistic checkpointing in Time Warp parallel simulation. In: Proceedings of the 8th IEEE international symposium on modeling, analysis and simulation of computer and telecommunications systems. San Francisco, CA; 2000. p. 366–73.

[36] Lin Y, Lazowska E. A time-division algorithm for parallel simulation. ACM Trans Model Comput Simul 1991;1:73–83.

[37] Jones KG, Das SR. Time-parallel algorithms for simulation of multiple access protocols. In: Proceedings of the 9th international symposium on modeling, analysis and simulation of computer and telecommunication systems; 2001. p. 49–58.

[38] Wang G, Bölöni L, Turgut D, Marinescu DC. Time-parallel simulation of wireless ad hoc networks with compressed history. J Parallel Distrib Comput 2009;69(2):168–79.

[39] Wang JJ, Abrams M. Approximate time-parallel simulation of queueing systems with losses, In: Proceedings of 24th conference on winter simulation, ACM Press; 1992. p. 700–8.

[40] Heidelberger P, Stone HS. Parallel trace-driven cache simulation by time partitioning. In: Proceedings of the 1990 winter simulation conference; 1990. p. 734–7.

[41] Wang G, Turgut D, Bölöni L, Marinescu DC. Time-parallel simulation of wireless ad hoc networks. ACM/Springer J Wirel Netw (WINET) 2009;15(4):463–80.

[42] Perumalla KS. Discrete-event execution alternatives on general purpose graphical processing units (GPGPUS). In: PADS'06: Proceedings of the 20th workshop on principles of advanced and distributed simulation; 2006. p. 74–81.

[43] Hyungwook P, Fishwick PA. A fast hybrid time-synchronous/event approach to parallel discrete event simulation of queuing networks. In: Proceedings of the 40th conference on winter simulation. winter simulation conference; 2008.

[44] Davidsson P. Multi agent based simulation: beyond social simulation. Berlin Heidelberg: Springer; 2001.

[45] Telecom Italia, Java Agent DEvelopment Framework (JADE), website Available at: <http://jade.tilab.com/>; 2015.
[46] George Mason University, MASON website, Available at: <http://cs.gmu.edu/~eclab/projects/mason/>; 2015.
[47] Fernandes P, Nunes U. Multi-agent architecture for simulation of traffic with communications. ICINCO-RA. 2008;(2).
[48] Wooldridge M, Jennings NR, Kinny D. The Gaia methodology for agent-oriented analysis and design. Auton Agents Multi Agent Syst 2000;3(3):285–312.
[49] Lin P, Winikoff M. Prometheus: a methodology for developing intelligent agents. Agent-oriented software engineering III. Berlin Heidelberg: Springer; 2003.
[50] IEEE Computer Society. The foundation for intelligent physical agents (FIPA), Website Available at: <http://www.fipa.org/>; 2014.
[51] George Mason University. MASON documentation, Available from, <http://cs.gmu.edu/~eclab/projects/mason/docs/>; 2015.
[52] Standard specification for telecommunications and information exchange between roadside and vehicle systems–5 GHz Band dedicated short range communications (DSRC) medium access control (MAC) and physical layer (PHY) specifications, ASTM E2213-03, 2003.

CHAPTER 18

An integrative approach for hybrid modeling, simulation and control of data networks based on the DEVS formalism

Rodrigo Castro[1,2] and Ernesto Kofman[3,4]

[1]University of Buenos Aires, Buenos Aires, Argentina
[2]National Scientific and Technical Research Council (CONICET), Buenos Aires, Argentina
[3]National University of Rosario, Rosario, Argentina
[4]French-Argentine International Center for Information and Systems Sciences (CIFASIS-CONICET), Rosario, Argentina

1 INTRODUCTION

1.1 MOTIVATION AND APPLICATION DOMAIN

In the last two decades, packet communication networks have become a central component in almost every branch of engineering. Either as constitutive parts of technological solutions or as services and subjects of study on their own, data networks have experienced an exponential growth in both scale of adoption and technical complexity. In many areas, to a good extent network engineering has been evolving outpacing the science.

Currently there is a lack of theoretical and practical tools that can fully explain several of the phenomena affecting the Quality of Service (QoS) of networking infrastructures, and consequently of the applications that rely on them.

In this context, the automatic control of the QoS of data networks is a discipline that has been gaining increasing interest. It involves the formal design of optimal adaptive control strategies to assign efficiently finite network resources to the service of all packet flows competing for them.

Usually, the demand imposed by packet flows is unpredictable, raising the challenge of designing flexible networks. They should accommodate an acceptable average traffic load along with sporadic peaks of bandwidth consumption, at a viable cost, without incurring over-sized designs, and guaranteeing quality requisites in terms of maximum delays, uptime, effective bandwidth, etc.

As networks grew in size and complexity, the design of efficient controllers required the support of modeling techniques and computer simulation. However, networks have certain features that impose theoretical and practical challenges to the known Modelling and Services (M&S) methodologies.

Due to the diversity of networks and its applications, the concept of QoS can change dramatically and, with it, the associated control objectives and techniques to be adopted. The network M&S community oscillates from the development of generalized models that help by describing most problems with a high level of abstraction to the development of highly specialized models that focus on solving specific issues in narrowly bounded domains. Accordingly, various and disparate modeling paradigms are adopted, more or less distant from the underlying simulation algorithms selected to execute them.

Therefore, it is natural to face integration challenges that hamper the combination of research results across different approaches. This hinders progress in the field of M&S-based network research.

This chapter describes an attempt to develop an integrative formal framework for modeling, simulation and implementation of data network controllers, suitable for a vast range of approaches and types of models, under a single formalism, fostering interdisciplinary cooperative research.

1.2 DIFFERENT TYPES OF NETWORK MODELS AND SIMULATION STRATEGIES

Models of networks and their controllers span problem domains at different hierarchical levels, involving dissimilar temporal dynamics and heterogeneous specification languages.

In upper control hierarchies, we find global policy controllers for traffic prioritization, dealing with different types of data flows, and involving control actions that take place a few times per minute, per hour or even per day.

At a lower intermediate level, the traffic shaping controllers tailor resource allocation strategies trying to enforce global policies, resorting to dynamics in the order of seconds.

Finally, at the lowest level, control algorithms make real-time decisions with a packet-by-packet granularity (dynamics of milliseconds or microseconds).

These heterogeneous goals and time scales pose a challenging scenario. Since changes in a control strategy at any level affect the entire performance, the design and testing of the different controllers should simultaneously involve the overall system.

However, neither network control design nor M&S techniques offer appropriate tools for coping with this problem. The main issue is that each level requires a different type of model representation.

Very detailed representations (at the *individual packet* level) consist of discrete event models able to capture very accurate behaviors. These models have the disadvantage of posing very high computational demands for simulating large topologies and/or considerable traffic intensities [1]. Besides, said models are not useful for the analytical study of global network properties. A possible strategy to ameliorate computational demands while retaining good degrees of detail about the traffic consists of aggregating group packets into larger discrete entities [2,3].

Alternatively, less detailed representations (at the *packet flow* level) consist of continuous time models that approximate network dynamics using averaged traffic intensity values [4–6]. Continuous models allow simulating efficiently (small execution times) systems that exhibit high complexity in their original discrete formulation. When their mathematical representations are expressed in the form of differential equations analytical treatment becomes possible, opening the door for applying automatic control theoretical approaches. Consequently, powerful and sound techniques can be used to design the network controllers and to study the stability, robustness and efficiency of the resulting systems.

Unfortunately, continuous approximations have the limitation of capturing only slow or steady-state dynamics, since all variables are described by means of time averages.

Finally, hybrid representations aim at mixing the advantages of discrete and continuous models. Typically, discrete flows are retained to offer detailed information at the per packet level, while the continuous part is used to set the system at an averaged, possibly changing, point of operation (background traffic) that would be otherwise too expensive to reach with a purely discrete technique [7–12]. Hybrid models must deal with the interactions between the foreground discrete flows and the background fluid traffic.

Other relevant dimensions worth considering are a) the stochastic characteristic of network traffic (and sometimes of topologies), and b) the need to implement the controllers to operate in real time and embedded into specific-purpose hardware. It is essential to ensure a solid formal representation of random processes governing demand patterns and other stochastic phenomena in networks, and to keep such representations consistent throughout all design stages. At the final implementation stage, it is also very relevant to retain the validity of the models obtained across all previous stages, being able to reuse them directly as the actual final pieces of software running in real time, interacting with specialized hardware for network control.

1.3 AN INTEGRATIVE SOLUTION FOR MODELING, SIMULATION AND CONTROL OF DATA NETWORKS

As we mentioned earlier, this chapter presents an integrative framework for applying hybrid M&S techniques under a unifying formalism, avoiding the need for changing the modeling languages or tools to study the systems of interest. Specifically, we resort to DEVS as the formal and practical integrative framework for M&S of data networks and control algorithms.

DEVS is able to represent exactly any discrete time or discrete event system, and to approximate continuous systems as accurately as desired. DEVS allows us to model exactly any network protocol, and by means of techniques for approximating continuous systems, to describe also continuous approximations of network flows, thus making discrete and continuous models coexist without requiring a change of formalism. Moreover, there exist tools for transforming DEVS models into pieces of software running in real time avoiding the need to tailor (or even rewrite) the logic of the models.

We analyze the feasibility of applying DEVS as an integrative solution. In so doing we identify limitations of theoretical and practical nature, and propose new tools to overcome them.

The DEVS framework was recently extended by adding stochastic elements in a formal way to compensate for the fact that the DEVS mathematical structure is essentially defined by means of deterministic functions. Also, existing numerical integration methods (required to solve numerically fluid models expressed by means of differential equations) were extended to account explicitly for the presence of nonnegligible delays—a pervasive characteristic in data network dynamics.

We will describe briefly the theoretical tool that extends the DEVS formalism, equipping it with the ability to express stochastic systems in a mathematically formal and sound way, resorting to the theory of probability spaces. Such formalism is called STochastic DEVS (STDEVS), and allows DEVS to represent stochastic phenomena such as those dominating most of the traffic patterns and control algorithms used in data networks.

Also, we will touch on a novel theoretical tool that extends the Quantized State Systems (QSS), a family of numerical integration methods that allows DEVS to approximate solutions of continuous systems expressed with ordinary differential equations (ODE). This extension permits solving delay differential equations (DDEs), a class of continuous models that appear often in distributed control of quality of service in data networks. The extension is called Delay QSS (DQSS).

These new tools transcend the data networking application domain, constituting generalized solutions for the M&S field in a broad sense.

Besides these theoretical tools, the chapter provides practical contributions, corresponding the software implementation of the mentioned results and the study of their performance on different problems.

In particular, we treat two case studies: an admission control system and a congestion control system.

The first case consists of a simple queue-server model, where a control mechanism decides which of the arriving packets are admitted in order to keep a bounded length of the queue. The system and control are first represented as continuous time models following an example of the literature. Then, the control is discretized, and, finally, the continuous flow model is replaced by a precise discrete event representation at the packet level.

The second case follows a similar approach but over a far more realistic and complex TCP protocol, proposing also a hybrid approach that combines the features of the continuous and the packet level representations.

While the first case is aimed to illustrate the proposed methodology and tools, the second case is aimed to demonstrate the possibilities they offer in complex and realistic systems.

In both examples, the novel modeling tools allow representation of all the systems with the different approaches under the same formalism reusing the submodels in a seamless fashion.

The chapter is organized as follows. Section 2 introduces the concepts and tools used in the rest of the chapter, and then Sections 3 and 4 illustrate their usage in problems of admission and congestion control. Finally, Section 5 presents conclusions about the methodologies, tools and results obtained.

2 BACKGROUND AND TOOLS

In this section, we describe the concepts and tools which are then used through the rest of the chapter.

2.1 DEVS FORMALISM

The DEVS formalism can describe most discrete systems, including discrete time systems, and through numerical approximations, continuous and hybrid systems. This generality is achieved by the ability of DEVS to represent general *discrete event systems*, i.e., any system whose input/output behavior can be described by sequences of events.

More specifically, a DEVS model [13] processes an input event trajectory and—according to that trajectory and to its own initial state—provokes an output event trajectory. Formally, a DEVS *atomic* model is defined by the following structure:

$$M = (X, Y, S, \delta_{int}, \delta_{ext}, \lambda, ta)$$

where

- X is the set of input event values, i.e., the set of all the values that an input event can take;
- Y is the set of output event values;
- S is the set of state values;
- δ_{int}, δ_{ext}, λ and ta are functions which define the system dynamics.

Each possible state s ($s \in S$) has an associated *time advance* calculated by the *time advance function* $ta(s)$ ($ta(s){:}S \rightarrow \Re_0^+$). The *time advance* is a nonnegative real number saying how long the system remains in a given state in absence of input events.

Thus, if the state adopts the value s_1 at time t_1, after $ta(s_1)$ units of time (i.e., at time $ta(s_1) + t_1$) the system performs an *internal transition*, going to a new state s_2. The new state is calculated as $s_2 = \delta_{int}(s_1)$, where δ_{int} ($\delta_{int}{:}S \rightarrow S$) is called *internal transition function*.

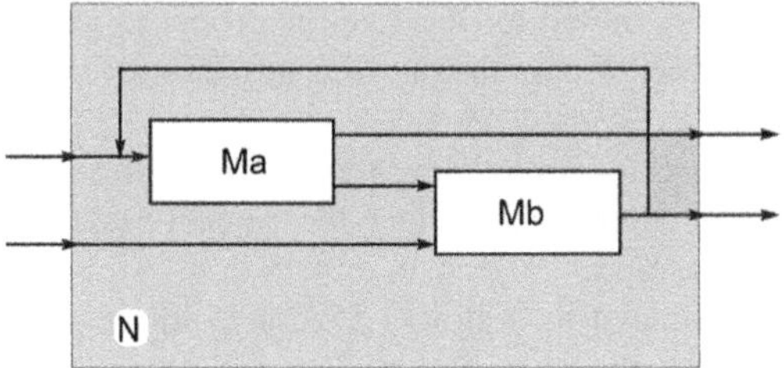

FIGURE 18.1

Coupled DEVS model.

When the state goes from s_1 to s_2 an output event is produced with value $y_1 = \lambda(s_1)$, where λ ($\lambda: S \rightarrow Y$) is called *output function*. Functions ta, δ_{int}, and λ define the autonomous behavior of a DEVS model.

When an input event arrives, the state changes instantaneously. The new state value depends not only on the input event value but also on the previous state value and the elapsed time since the last transition. If the system goes to the state s_3 at time t_3 and then an input event arrives at time $t_3 + e$ with value x_1, the new state is calculated as $s_4 = \delta_{ext}(s_3, e, x_1)$ (note that $ta(s_3) \geq e$). In this case, we say that the system performs an *external transition*. Function δ_{ext} ($\delta_{ext}: S \times \Re_0^+ \times X \rightarrow S$) is called the *external transition function*. No output event is produced during an external transition.

Atomic DEVS models can be coupled. DEVS theory guarantees that the coupling of atomic DEVS models defines new DEVS models (i.e., DEVS is closed under coupling) and then complex systems can be represented by coupling DEVS models in a hierarchical way [13].

Coupling in DEVS is usually represented through the use of input and output ports. With these ports, the coupling of DEVS models becomes a simple block-diagram construction. Figure 18.1 shows a coupled DEVS model N which is the result of coupling the models M_a and M_b.

According to the closure property, the model N can be used itself as an atomic DEVS and it can be coupled with other atomic or coupled models.

2.1.1 Simulation of DEVS models

DEVS coupled models can be easily simulated. The simplest way of doing it is writing a program with a hierarchical structure equivalent to the hierarchical structure of the model to be simulated [13].

Also, there are several modeling and simulation tools that allow users to build and simulate DEVS models, offering various features including GUIs, real-time and parallel executions, and implementing different DEVS extensions.

Among them, we can mention

- ADEVS [14], a C++ library for the simulation of DEVS models. ADEVS also supports the dynamic structure DEVS extension.

- DEVSJAVA [13], an M&S environment for the DEVS formalism written in Java. It supports parallel execution.
- DEVS-SUITE [15], an extension of DEVSJAVA with improved visualization capabilities.
- CD++ [16], written in C++ and based on the cellular extension of DEVS called Cell–DEVS. Different versions include Real-Time, Parallel and centralized simulators.
- VLE (Virtual Laboratory Environment), a multimodeling platform written in C++ and based on several DEVS extensions [17].
- PowerDEVS [18], written in C++ and oriented to DEVS simulation of continuous and hybrid systems. It shall be described in more detail in Section 2.

2.1.2 Stochastic DEVS extension

The DEVS formalism permits representation of any discrete deterministic system. However, networking models usually have stochastic features that cannot be captured by the classic DEVS language.

To overcome this limitation, a stochastic DEVS (STDEVS) extension was formalized in [19]. Taking into account that DEVS models deal with general sets, STDEVS makes use of *probability spaces theory* to define stochastic processes taking place on those general sets.

Although the formal definition of STDEVS and its theoretical basis are beyond the scope of this chapter, we shall mention that STDEVS theory ensures the well-posedness of using random number generators inside transition functions of regular DEVS atomic models.

2.2 QUANTIZED STATE SYSTEMS

As we mentioned above, continuous time and hybrid systems can be represented by DEVS models through the use of numerical approximations. While most numerical integration methods produce discrete time approximations—which can eventually be represented by DEVS—there are some algorithms that exploit the asynchronous nature of the DEVS formalism.

These algorithms are called *Quantized State Systems* (QSS) methods [20,21] and they replace the time discretization by the *state quantization.*

A continuous time system can be written as a set of ordinary differential equations (ODEs):

$$\dot{x}(t) = f(x(t),\ u(t)) \tag{18.1}$$

where $x \in \Re^n$ is the state vector and $u(t)$ is a vector of known input trajectories. The different QSS methods approximate the ODE by

$$\dot{x}(t) = f(q(t), u(t)) \tag{18.2}$$

where $q \in \Re^n$ is the quantized state vector. The state and the quantized state vectors in Equation (18.2) are component-wise related by *quantization functions.*

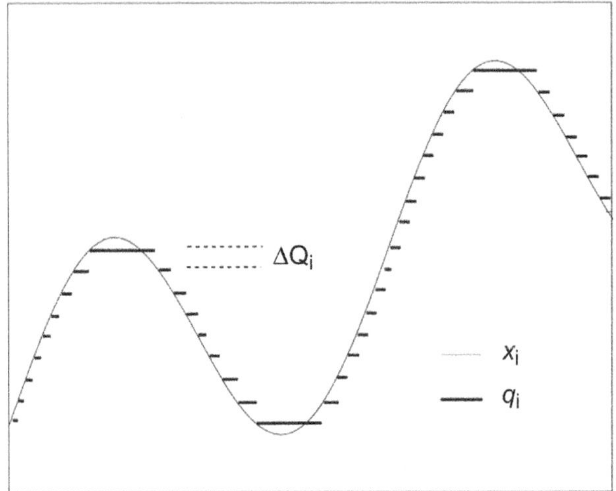

FIGURE 18.2

QSS1 quantization.

Figure 18.2 shows the quantization function corresponding to the first-order accurate QSS1 method [20], where the quantized states follow piecewise constant trajectories.

Since the quantized states $q_i(t)$ follow piecewise constant trajectories, and assuming that the input variables $u_i(t)$ are piecewise constant, the state derivatives $\dot{x}_i(t)$ also result piecewise constant. Consequently, the states $x_i(t)$ follow piecewise linear trajectories. These particular trajectory forms imply that the approximation given by Equation (18.2) can be easily solved.

Higher order QSS algorithms can be obtained using higher order quantization functions, which produce piecewise linear and piecewise parabolic quantized state trajectories, leading to QSS2 [22] and QSS3 [23] methods, respectively. The family of QSS methods is completed by linearly implicit QSS (LIQSS) methods of order 1 to 3 [24], and by backward QSS (BQSS) and centered QSS (CQSS) algorithms [25], which are efficient for simulating stiff systems.

QSS methods have nice stability and error bound properties [21], and they are particularly efficient to simulate continuous systems that exhibit frequent discontinuities [26]. In these cases, QSS methods usually simulate sensibly faster than classic discrete time numerical approximations.

2.2.1 Implementation of QSS methods

Each component of the QSS1 approximation given by Equation (18.2) can be thought of as the coupling of two elementary subsystems: a static one,

$$\dot{x}_i(t) = f_i(q_1, \ldots, q_n, u(t)) \tag{18.3}$$

and a dynamical one

$$\dot{q}_i(t) = Q_i(x_i(\cdot)) = Q_i(\int \dot{x}_i(\tau)d\tau) \qquad (18.4)$$

where Q_i is the quantization function that relates the state x_i and the quantized state q_i as depicted in Figure 18.2 (notice that it is not a function of the instantaneous value $x_i(t)$, but a functional of the trajectory $x_i(\cdot)$).

Taking into account that the quantized variables $q_i(t)$ and the input variables $u_j(t)$ follow piecewise constant trajectories, it results that both subsystems, Equation (18.3) and Equation (18.4), receive piecewise constant input trajectories and compute piecewise constant output trajectories. These piecewise constant trajectories can be represented by sequences of events in a straightforward manner.

The relation between the input and output sequences of events of these subsystems can be expressed by simple DEVS models. The DEVS representations of Equation (18.3) are called *static functions* and the DEVS representations of Equation (18.4) are called *quantized integrators* [21].

Then, the QSS approximation Equation (18.2) can be simulated by a DEVS model consisting in the coupling of n quantized integrators, n static functions (with the eventual addition of signal sources). The resulting coupled DEVS model looks identical to the block diagram representation of the original system of Equation (18.1).

Higher order QSS methods are implemented in the same way. In this case, the events represent the changes in piecewise linear or piecewise parabolic trajectories and the static functions and quantized integrators take into account not only the values but also the slopes and second derivatives of the trajectories they receive and send.

Based on these ideas, the whole family of QSS methods was implemented in PowerDEVS [18], a DEVS-based simulation platform specially designed for and adapted to simulating hybrid systems based on QSS methods. In addition, the explicit QSS methods of orders 1 to 3 were also implemented in a DEVS library of Modelica [27] and implementations of the first-order QSS1 method can also be found in CD++ [28] and VLE [17].

2.2.2 QSS and delay differential equations

The use of QSS methods allows the DEVS formalism to simulate continuous time and hybrid systems represented by ordinary differential equations. However, continuous time representations of networking systems usually contain delays, leading to *Delay Differential Equations* (DDEs) of the form

$$\dot{x}(t) = f(x(t), x(t-\tau_1(x, t)), \ldots, x(t-\tau_m(x, t)), u(t)) \qquad (18.5)$$

where $\tau_j(\cdot) \in \Re^n$ are delay functions.

For these cases, an extension of the QSS algorithms was developed in [29]. This extension also improves the simulation performance of classic DDE solvers.

2.3 PowerDEVS

PowerDEVS [18] is a general-purpose tool for DEVS simulation. It consists of two main modules: The main graphic user interface (GUI) that allows the user to

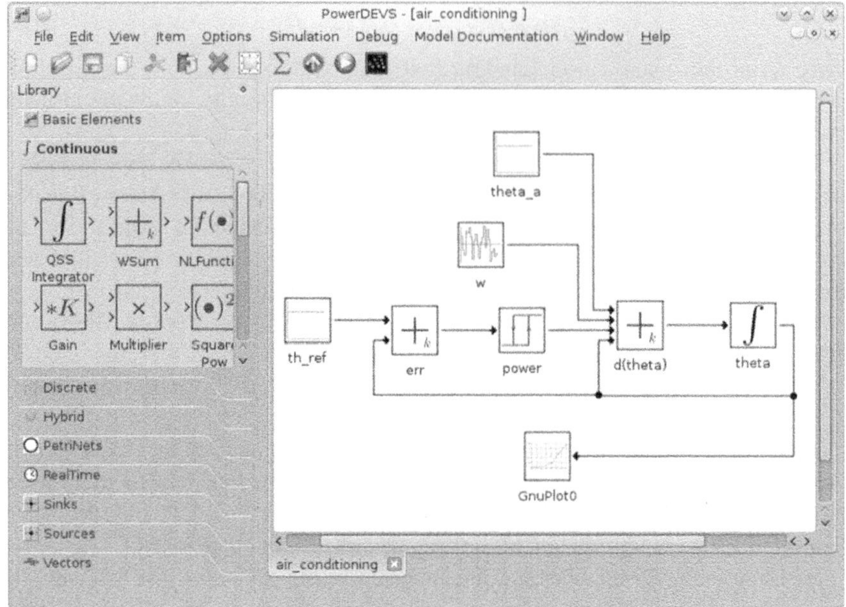

FIGURE 18.3

PowerDEVS Main Window.

describe DEVS models in a block diagram fashion as shown in Figure 18.3, and the simulation engine, which is in charge of the simulation.

An auxiliary tool (PowerDEVS preprocessor) translates the models into C++ descriptions that are compiled together with the simulation engine producing an executable simulation code.

PowerDEVS can also simulate continuous (and hybrid) models through the use of different QSS-based methods. For that goal, the PowerDEVS distribution contains a complete library for hybrid system simulation based on these numerical algorithms.

End-users can take the blocks from the libraries, build the block diagrams using the main GUI, and invoke the simulation in a straightforward way (without any knowledge about the DEVS formalism).

DEVS-aware users can easily build new blocks by defining the dynamics of the corresponding atomic DEVS model in C++ language using a simple GUI (called *atomic model editor*).

3 APPLICATION TO A PROBLEM OF ADMISSION CONTROL

Embedded networking systems include protocols and algorithms embedded in specific-purpose devices with limited resources. Control of real-time computing systems is a new area where traditional continuous control system techniques are applied to computing and/or networking systems, which are considered as the subjects to be controlled [30]. In these systems, the control objective is to keep performance metrics (delay, jitter, throughput, resource consumption, deadline miss ratio, etc.) within certain required bounds, usually specified by Quality of Service (QoS) requirements. The control actions trigger discrete events involving the allocation of available resources (pools, buffers, queues, slots, etc.) among different competing resource consumers (tasks, packets, jobs, requests, etc.) which are also discrete in nature.

Although these control systems are typically designed using ad-hoc techniques, classic control theory has been recognized as a way to exploit the theoretical and methodological background of the discipline [31]. Based on control theory, stability and transient response can be easily and systematically analyzed, and controllers can be designed following different optimal and/or robust criteria. In the last decade, control theory has been applied to many computing and networking systems, including active queue management schemes [32], network routers [33] and high-performance web servers [34], among many others.

One of the difficulties of applying control theory to these systems is related to the heterogeneous nature of the models used to analyze them. Real-time computing and networking systems are usually described by discrete event dynamic systems [35], modeled with languages such as timed Petri Nets, timed Automata or timed finite state machines. Instead, classic control theory is based on continuous-time models (differential equations) or discrete-time models (difference equations). Although there are many techniques for dealing with control of discrete-event systems [35], their goal (i.e., safety, deadlocks) is not performance control (where stability, transient response and robustness play a key role).

In order to bridge the analytical gap between event-based and time-based control techniques, systems are usually approximated by either continuous or discrete-time models. Then controllers are designed and analyzed in these domains. However, this usually leads to oversimplifications and errors, and controllers must be checked (and eventually redesigned) with more accurate models (i.e., using discrete-event M&S). Although there are very advanced tools for this phase (e.g., ns-2 [36], OMNET++ [37]) their integration with the control theoretical models is very complex. This usually forces ad-hoc customizations, without a supporting formalism guaranteeing the correctness of these procedures.

As we can see, the resulting design process involves working with models of different nature, described with different languages and obtained by specialists of different fields. This usually leads to switching between M&S tools and techniques, which yields many practical difficulties. Moreover, the implementation of the control algorithms in embedded platforms poses additional challenges as the algorithms usually must be translated and adapted to the target architecture.

In this section we propose an integrative methodology to deal with these issues, exploiting the ability of the Discrete Event System Specification framework to describe and simulate all kinds of hybrid systems.

We show the definition of a case study on Admission Control strategies to network traffic based on nonlinear control theory. The Admission Control algorithm is designed as a sequence of M&S tasks. We first model the network with a continuous-time approximation (designing a continuous-time control law). We then find a discrete-time approximation of the control law (verifying the discrete-time controller applied to the original discrete-event system).

Although not described here, the controller was finally implemented on real-time embedded network processor, reusing on it the simulation C++ code [38].

This end-to-end methodology is completely based on a unified DEVS M&S framework.

3.1 CONTROL THEORY FOR ADMISSION CONTROL

We present a case study on the use of our methodology in a control theory-based solution for designing an Admission Control mechanism. We will adopt the example shown in Figure 18.4 for a **Service Control Node**, originally presented in [39].

In our case, the block **Server System** is the plant to be controlled. It is composed by one **Queue** and one **Server**. In this example, the QoS requirement is to keep the queue length at a given reference desired value (i.e., the *set point* x_{ref} for the plant), this being the control objective.

The input signal to the plant is the *admittance rate* $\overline{u}$ packets into the queue, which is assumed to follow a stochastic Markovian process. The queue is assumed to have infinite capacity, and the instantaneous number of packets in the queue is the *output signal* x of the plant.

The servicing times at the server follow some probability distribution with a mean value of $1/\mu$ seconds. The stochastic nature of both the queue admittance rate and the servicing times represent noise that will have to be compensated by the **Admission Control** block. The **Controller** block is designed to keep the

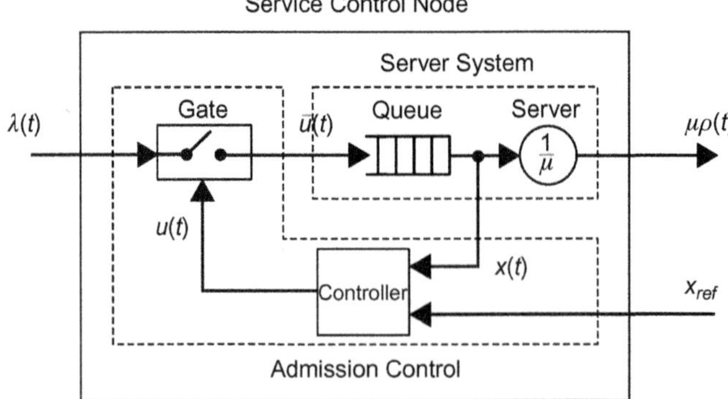

FIGURE 18.4

Service Control Node system under study.

average number of enqueued packets x at the given desired value x_{ref}. The control action consists of opening and closing a gate at the entrance of the Server System to limit the incoming traffic (*arrival rate* λ). This is achieved by the **Gate** block which is the actuator for the controller. According to the *control signal* u, the gate selectively rejects packets from λ to obtain a controlled admittance rate $\overline{u}$.

This case study generalizes a problem of admission of packets of any sort operating at an arbitrary node in an arbitrary topology. It is then abstract enough not to be tied to any particular type of network technology or topology.

In order to apply control theory techniques to the Server System, a fluid model in the form of differential equations can be derived to describe the behavior of the system. Following a *fluid flow approximation* method [40], the system can be modeled by the expression:

$$\frac{dx}{dt} = \overline{u}(t) - \mu \cdot G(x(t)) \tag{18.6}$$

where $\overline{u}(t)$ is the packet admittance rate, $1/\mu$ is the mean service time and $G(x)$ is a nonlinear function:

$$G(x(t)) = \frac{x(t) + 1 - \sqrt{x^2(t) + 2C^2x(t) + 1}}{1 - C^2} \tag{18.7}$$

C is the coefficient of variance of the service times. This approximation reproduces accurately the steady-state condition for the average number of packets in the system and the server utilization. In terms of dynamic behavior, the approximation captures some stochastic properties of the system under nonstationary traffic conditions.

The gate action can be described as follows:

$$\overline{u}(t) = \begin{cases} \lambda(t) & \textit{if } u(t) > \lambda(t) \\ \max(0,\ u(t)) & \textit{otherwise} \end{cases} \tag{18.8}$$

The authors proposed a Proportional-Integral (PI) type of continuous controller for the Control Node (see Figure 18.5) with the following law:

$$u(t) = K \cdot e(t) + \frac{K}{T_i}\int e(t)dt \tag{18.9}$$

where $e(t) = x_{ref} - x(t)$ is the error signal (difference between the actual and the desired value) that is monitored by the controller to update its control action.

The system parameters for this example are set accordingly:

$$\begin{array}{lll} \lambda = 20s^{-1} & \mu = 5s^{-1} & \mu_1 = 2s^{-1} \\ x_{ref} = 10 & C^2 = 3.7 & \mu_2 = 60s^{-1} \\ & & \alpha_1 = 0.38 \end{array} \tag{18.10}$$

The service processing times at the server are hyperexponentially distributed, combining two exponential distributions with mean service rates μ_1 and μ_2 respectively, and a bias of $\alpha_1 = 38\%$ in favor of μ_1. This yields a mean service rate $\mu = 5$

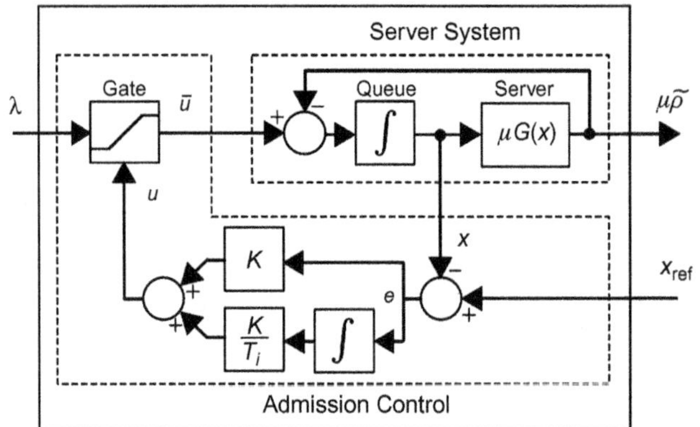

FIGURE 18.5

Continuous-time nonlinear model.

with a squared coefficient of variance $C^2 = 3.7$. The controller design consists of finding appropriate values for K and T_i according to performance requirements. In this case a linearization and pole placement procedure is followed, and the values are set to $K = T_i = 2.4$. We will refer to the system parameterized according to Equation (18.10) as the Controlled Packet Processing System (CPPS).

3.2 DEVS-BASED M&S METHODOLOGY FOR THE CONTROLLED PACKET PROCESSING SYSTEM

This section shows the use of our DEVS-based M&S methodology to assist the processes of analysis, design, verification, implementation, and validation of the PI controller for the CPPS specified before. The main results in [39] will be reproduced for comparison and validation purposes. The analysis, design and verification phases are supported by the DEVS-based tool *PowerDEVS* [21].

3.3 CONTINUOUS CONTROL, CONTINUOUS PLANT

The first goal of the process is to verify via simulations the appropriateness of the controller parameters designed for the CPPS. At this point we have a continuous-time nonlinear approximation of the Server System and a continuous-time specification of the controller.

The CPPS continuous-time model built in PowerDEVS is shown in Figure 18.6. Each block in the model represents either a Coupled DEVS model or an Atomic DEVS model. The inner details for the coupled models **Server System** and **PI Controller** are also shown in Figure 18.6. The **Weighted Sumator** models implement the required multiplication factors to implement the control law (i.e., the K and T_i parameters and proper summation/subtraction signs). A **Nonlinear Function** atomic model (a component of the **Server System** coupled model) implements the

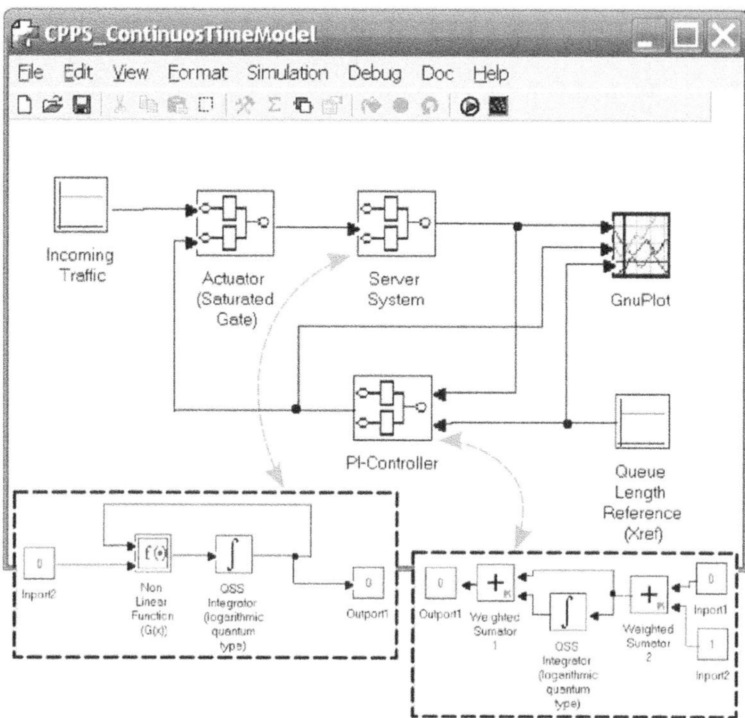

FIGURE 18.6

Continuous time CPPS model in PowerDEVS.

nonlinear expression at the right-hand side of Equation (18.6). The **QSS Integrator** models locally implement the QSS methods. In this case, we selected the third-order accurate QSS3 algorithm setting a quantum of (the global error tolerance is linearly bounded by the quantum). All the mentioned models are part of the standard libraries of PowerDEVS.

We ran the simulation of this deterministic system for 30 seconds of virtual time. Results are shown in Figure 18.7. We can see that the queue length response is slightly underdamped, and stabilizes quickly to the reference value with a settling time of about 7 seconds (with a small overshoot at the beginning). These results (obtained in a full discrete-event framework) show a qualitative close match with the simulation results shown in [39] (obtained with a discrete-time numerical integration approach) for the same parameters and signal ranges.

3.4 DISCRETE CONTROL, CONTINUOUS PLANT

The second goal of the design process is to obtain a controller that can be implemented with an algorithm in a digital computer. For this aim we translate the continuous-time controller into a discrete-time controller, while keeping (for now)

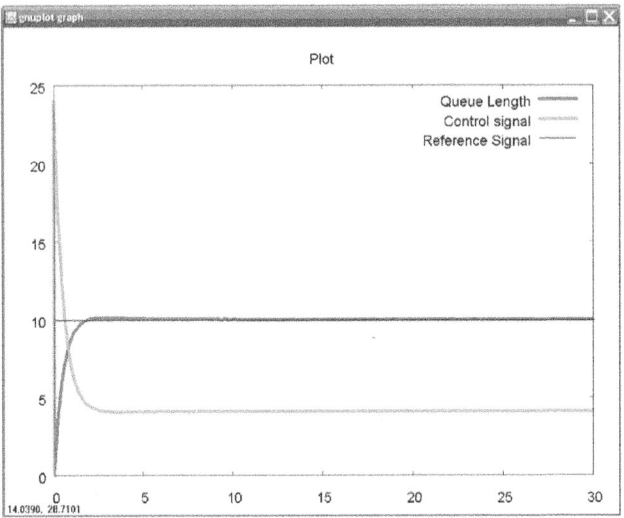

FIGURE 18.7

Continuous time CPPS simulation results. Time units correspond to seconds.

the continuous-time version of the plant. By applying the Forward Euler method [21] to the control law (11) we obtain the following discrete-time expression for the PI-controller:

$$Z_{k+1} = Z_k + e_k \cdot h;\ u_k = K\left(e_k + \frac{z_k}{T_i}\right) \tag{18.11}$$

where $k \in Z$ is the discrete-time index and $h \in \Re^+$ is the time step size such as $z_k = z(t = kh)$. Now we proceed to replace the previous **PI-controller** DEVS coupled model by its discretized counterpart, which will implement Equation (18.11). The block diagram of the discretized controller is shown in Figure 18.8, where the discrete versions of the input variables x_k and x_{ref_k} are obtained by means of **Sample and Hold** blocks, and the delay operation on the controller variable z is performed by a **Discrete Delay** block. These blocks are DEVS atomic models parameterized with a common sampling period of $h = 0.5$ seconds.

We ran the simulation of the new system with the discretized control again for 30 seconds. The results are shown in Figure 18.9. The analysis of the results shows that the response reproduces closely the properties of the continuous-time simulation results of Figure 18.7, noting that in the discrete-time case the queue length response is slightly more underdamped (with a more pronounced overshoot at the beginning). This is due to the control delay introduced by the discretized controller. The system stabilizes to the reference value in a time qualitatively similar to the continuous controller case, with a settling time of about 7 seconds. With these results we can adopt the discrete-controller design as a satisfactory solution.

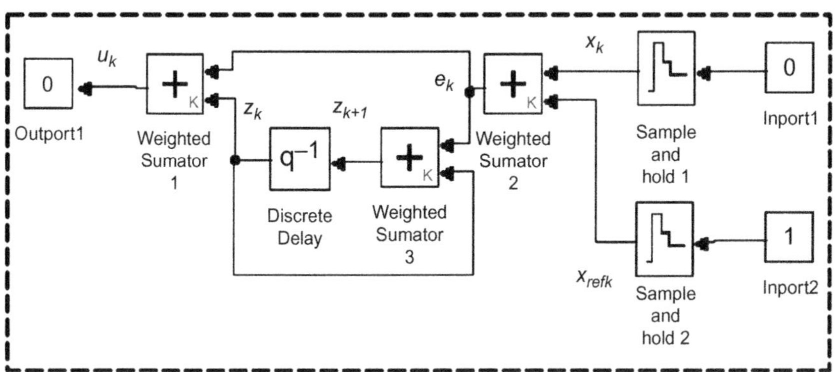

FIGURE 18.8

Discretized PI controller in PowerDEVS.

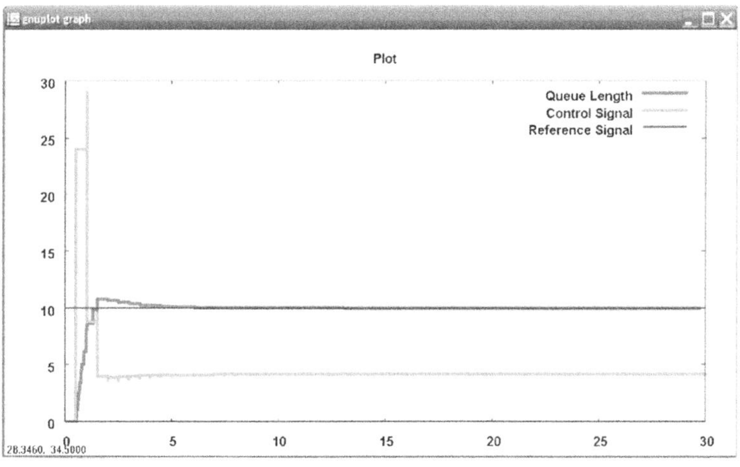

FIGURE 18.9

Discrete time CPPS simulation results. Time units correspond to seconds.

3.5 DISCRETE CONTROL, DISCRETE EVENT PLANT

Now, following a standard M&S design cycle, we should verify our controller design against a more accurate representation of the real system, getting rid of all kinds of approximations as much as possible. This constitutes the third goal of our design process.

The real M/G/1 Server System is in fact a stochastic Discrete Event Dynamic System (DEDS), which can be straightforwardly represented and executed under the DEVS framework. Thus, one of the advantages of our full DEVS-based methodology arises at this stage of the M&S process, where the verification activity against the real system involves a simple model replacement task (without switching formalisms or design tools).

Then, the continuous-time continuous-variable components of our original model are replaced by their discrete-event counterparts. By doing so, we obtain the Hybrid CPPS model shown in Figure 18.10.

Traditional M&S methodologies, on the contrary, would usually involve some kind of special-purpose language-specific implementation of the Hybrid CPPS, i.e. the M/G/1 system and the controller algorithm. In the case of [39] it consisted of a custom code implemented by the authors in the C language.

The Hybrid CPPS model in Figure 18.10 implements the following DEVS atomic models for the networking system: a **Packet Generator**, an **Admission Gate**, a **Queue** and a **Server**. These models handle discrete entities individually (which represent network packets) on a continuous-time basis. The stochastic behavior of the network traffic is considered explicitly at the packet inter-generation times and the servicing times, so no approximation is involved. The queuing of the packets is achieved by a *First In-First Out* (FIFO) policy at the

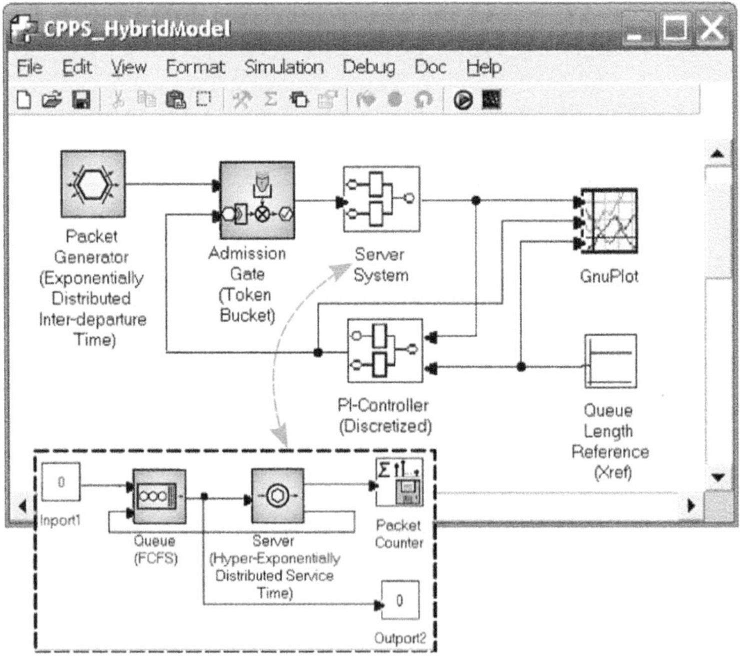

FIGURE 18.10

Hybrid CPPS model in PowerDEVS.

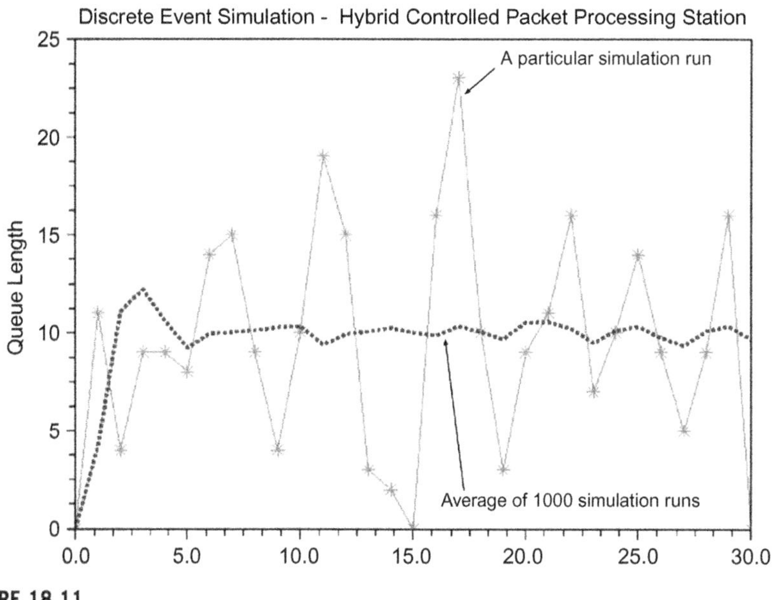

FIGURE 18.11

Hybrid discrete event CPPS simulation results. Time units correspond to seconds.

Queue block. A token bucket algorithm is implemented at the **Admission Gate** block, which is exactly like the one that would be implemented on a real-world target embedded piece of code. It generates internal tickets at a rate commanded by the control signal u coming from the Controller. Individual incoming packets will be accepted if available tickets exist for them. Otherwise, the packets are rejected. The control signal will be updated according to the time base imposed by the discrete-time controller (the control period) which in our case is set to $h = 0.5$ from previous design phases.

Given the stochastic nature of the system we performed a set of 1000 simulations (of 30 virtual seconds each) and derived relevant statistics. This task was automated with the PowerDEVS–Scilab [41] integrated environment for simulation and numerical computation. The results are shown in Figure 18.11, where the Queue Length measure is provided for the average of the set of simulations and also for a single representative realization of the set. Both curves were obtained by sampling the queue length values every 1 second.

The dotted curve shows a satisfactory regulation of the average Queue Length around the desired reference value of 10 packets (as specified by the system parameters in (12)). A qualitative comparison of our stochastic simulation results with those obtained in [39] for the discrete-event experiments (under the same conditions) showed a close match in terms of time response and excursion values for both the average and the individual realization curves.

4 APPLICATION TO A PROBLEM OF CONGESTION CONTROL

In this section we show how to use PowerDEVS to model a fluid representation of the Congestion Control used in the Transport Control Protocol (TCP), taking advantage of the novel Delay QSS methods.

Although this representation is confined to express averaged behavior of variables, it permits understanding relevant characteristics of the system (stability, convergence, robustness, etc.) in conveniently small simulation times.

Nevertheless, the price to pay is the loss of details at per packet level, such as the behavior of outlier end-to-end delays that often affect the overall performance of a communication channel.

We use PowerDEVS to develop discrete event models that simulate exactly the behavior of the network driven by the TCP/AQM type of flow control (TCP in collaboration with Active Queue Management). This involves modeling the first principles of protocols and the physical medium.

The results obtained in this second step allow verifying the fluid results of the first step, and studying in more detail fine-grained protocol behavior.

Finally, we implement a hybrid flow modeling strategy. It permits combining the advantages of enhanced computing efficiency (fluid flow case, for bulk background traffic) along with fine-grained analysis capability (discrete event case, for single selected probe flows).

4.1 ANTECEDENTS

During the last 15 years several congestion control strategies have been proposed for TCP/AQM [42] by modeling the problem as a decentralized, distributed control problem subject to variable delay conditions. The importance of counting on precise models relies in the extreme difficulty of tuning simultaneously and efficiently the numerous protocol parameters involved in TCP/AQM.

The first generation of techniques proposed soon started to perform poorly as technology evolved, higher bandwidth became affordable, and networks with very big Bandwidth-Delay-Product (BDP) [43] (a.k.a. LFN, Long Fat Networks) appeared as a pervasive scenario.

The flow control mechanism of TCP implements a *sliding window* technique [44] to limit the injection rate of packets into the network.

Only a window W worth of packets is allowed to be sent per cycle of transmission until acknowledgements are received from the destination. The goal of this regulation is to keep the channel as busy as possible (not to *waste* available resources) but without incurring into saturations that can lead to counterproductive congestion conditions (not to *overload* the available resources).

In the ideal condition, $W = BDP$ holds. I.e., the window W matches (and limits) in size the number of *in-flight and unacknowledged* packets that can be "held" within the channel. The channel is seen as a buffering element with a limited storing capacity determined by its characteristic transmission rate and end-to-end delay.

As the BDP of a channel is in principle unknown and varying, $W = W(t)$ must be adapted dynamically. When a state of congestion is inferred, W must be reduced. Usually the detection of the loss of a single packet triggers the halving of W, which is allowed to regrow seeking again for a new optimum value. On average, this mechanism wastes the available bandwidth, a condition that was initially to be avoided by TCP.

A series of design enhancements have been proposed to solve this and other problems, resorting to techniques both analytical and empirical. All of them face the challenge of credibility, provided they are meant to replace the transport layer of Internet routers currently servicing billions of active users.

In this context, an integrative process of modeling, simulation, verification and analysis of hybrid systems can play a central role for designing the protocols of the future in high capacity networks [45].

In order to increase the credibility of such processes, it is vital to rely on a robust underlying mathematical formalism, supported by an integrated modeling and simulation framework to avoid the need of switching between tools and formalisms when using different representation paradigms in a given hybrid system.

Evidently, DEVS is a natural candidate to pursue these goals. We will see in the following how DEVS provides a unified solution for the hybrid modeling and simulation of TCP/AQM.

An event-driven paradigm is the obvious and natural representation for a network and its control algorithms. But difficulties arise when trying to apply pure discrete-event approaches to high capacity networks. Independently of the efficiency of the simulation engine of choice or the level of optimization of the models, there is an intrinsic limitation in scalability when simulating thousands of high throughput packet flows.

At this point the modeling of continuous systems becomes relevant. It uses fluid (averaged) approximations of packet flows and of the protocols implementing congestion control algorithms. The continuous-time paradigm allows studying problems at any scale with a fairly low computational demand. The advantage is the increased agility in the research process, while the price to pay is a decreased representation granularity.

Several continuous models have been proposed to study the dynamics of TCP, mainly based on nonlinear differential equations [46–49]. The approximated variables are usually rates, loss probabilities, round-trip times, etc. and also some stochastic properties of the variables such as nonstationarity, burstiness, self-similarity, etc. [50]. Relying on these models, advanced algorithms are developed [32] to efficiently control other variables such as queue lengths, sending rates, maximum delays, etc.

A salient property of fluid approximations applied to distributed network control is they produce models with delays (Delay Differential Equations). In a nutshell, the laws governing the current state of the network at time t depend on delayed information of the system, measured in the near past at some $t-\tau$. In turn, these delays are dynamic, and depend not only on time but on the system state itself. Delays play a central role in the stability and performance of network controllers. Nevertheless, the usual practice is to simplify their representation using a fixed, average delay value (or even to consider them as negligible).

Few tools exist that can solve DDE [51]. Moreover, they usually provide a tough interface for expressing the model, and none of them allow dealing with hybrid systems (combining fluid-driven with event-driven systems—in general— nor with network systems (in particular).

We will present a methodology that eliminates this limitation, using an intuitive graphical user interface that resorts to the DQSS numerical methods natively provided in the PowerDEVS tool.

4.2 CASE STUDY

We shall first describe the underlying mechanisms of TCP/AQM with the main goal of presenting a hybrid M&S exercise.

The mechanisms consist mainly of two types of cooperative controllers: AIMD (for flow and congestion control at end nodes running the TCP protocol) and RED (for the admission control at intermediate routers).

We will make a series of simplifications at a *logical level* for the sophisticated algorithms that perform flow and congestion control in TCP. We shall retain only a basic set of features that are enough for exhibiting the dynamics of TCP/AQM we are interested in. For instance, we will adopt a *packet* as the minimum unit of manageable data when, in fact, TCP works at the *bit or byte* level of granularity. Also, we will model only the *Congestion Avoidance* mechanism of TCP, and will set aside the *Slow Start* mechanism that complements the first.

On the other hand, at the *physical level*, the delays caused by limited bandwidth and limited processing speed at the routers will retain their strict correlation with the size and number of packets.

This case study can be analyzed abstracting away the specificities of the possible network technologies and topologies upon which TCP operates. On the one hand, the transport layer is by definition meant to be independent of its underlying network (TCP runs only at the two end nodes of any communication channel). On the other hand, it is true that different conditions in the network, link and physical layers will impact differently on the performance of TCP. Accordingly, in the study of TCP/AQM settings, all these possible effects are subsumed mainly into two model parameters: the average end-to-end delay and the packet loss probability. Both will be explicitly dealt with in the upcoming studied example.

4.2.1 Sliding window for flow and congestion control: AIMD mechanism

The standardized TCP versions most massively deployed in the Internet worldwide are TCP-Reno [52] and TCP-NewReno [53]. These are not the newest flavors, but remain standing as versions of reference for several studies.

TCP implements a type of congestion control characterized by the so-called *Congestion Avoidance* phase. The latter is in turn dominated by the AIMD algorithm [54] by means of which the size of the sender's congestion window $W = W(t)$ is adjusted dynamically. In essence, at the sender size, the tuning is as follows:

- the reception of a correct (expected sequence number) acknowledge packet (ACK) issued by the receiver indicates the successful reception of a packet sent before. This situation permits increasing the size of W with $W_{k+1} = W_k + 1/W_k$, termed Additive Increase.
- the reception of an unexpected (noncontiguous increasing sequence number) acknowledge packet (ACK) from the receiver indicates that problems happened in the underlying network, which are interpreted as a potential congestion condition, withdrawing the size of W down to $W_{k+1} = W_k/2$, termed Multiplicative Decrease.
- when no ACK is received for a certain maximum time RTO (RTT Time Out), W is forced to some minimum values (e.g., $W = 1$). Nevertheless in our implementation we will treat an RTO event the same way we treat unexpected ACKs, thus halving the value of W.

This scheme makes TCP maintain a maximum of W unacknowledged packets injected into the network (a.k.a. in-flight packets), increasing the size of W by 1 per each RTT in the ideal condition that all expected ACKs have arrived correctly within that RTT lapse.

This condition considers that the current network state is such that it can be "filled up" with W packets without reaching saturation limits, and that a subsequent cycle can be probed safely using a window of $W_k = W_k + 1$ to try luck. Eventually, congestion occurs.

Upon detecting a condition linked to a packet loss, the sender infers congestion and limits W to half its size.

Thus, a TCP connection limits its throughput to $\lambda_{max}^{TCP}(t) = \frac{W(t)}{RTT(t)}$. It represents a fairly conservative control scheme, updated at a pace dominated by the (constantly varying) RTT.

In Figure 18.12 we show a hypothetical packet exchange sequence between a Sender *A* and a Receiver *B*. The system has so far sent successfully the full sequence of packets until SEQ:36. It is assumed that for whatever reason the window starts with value $W = 1$.

At the left margin of the figure, the stripes of numbers denote the buffer state at the Sender side. Sent IDs that have been already *acknowledged successfully* are

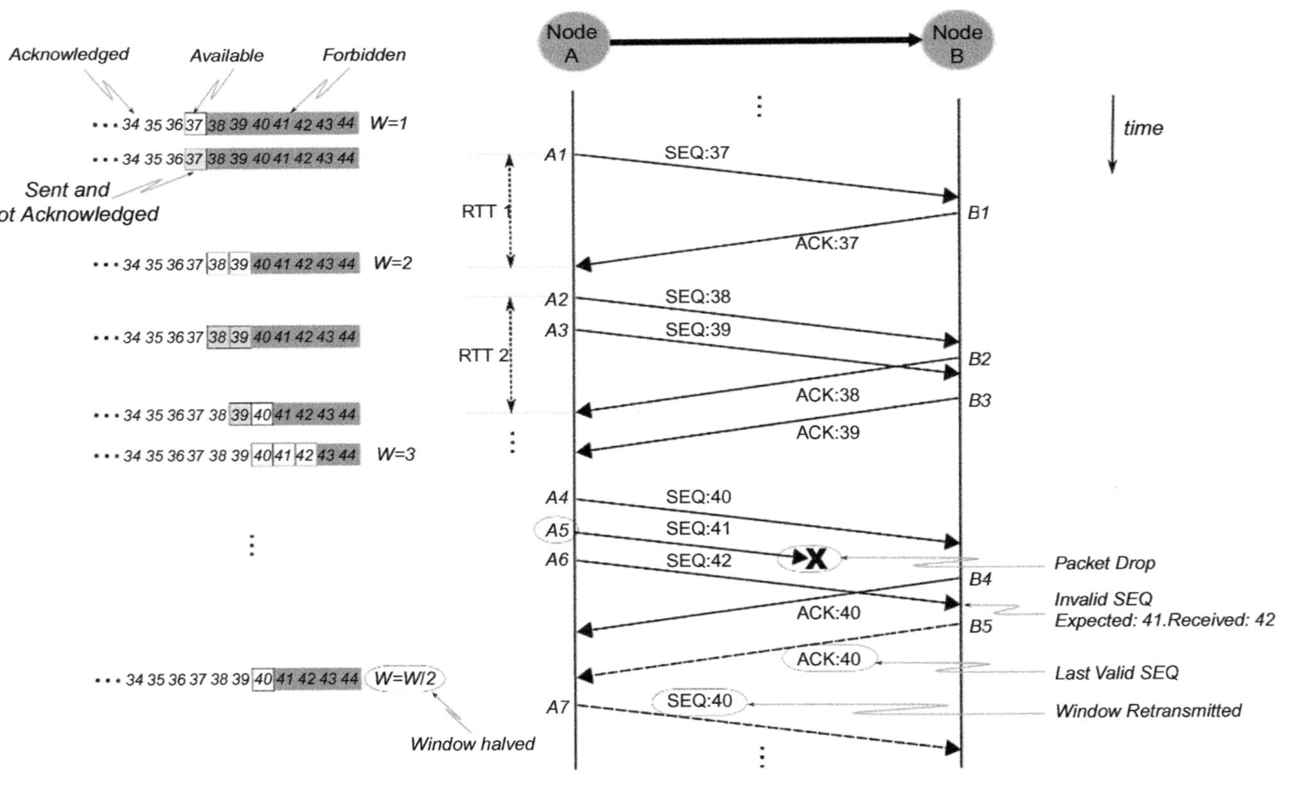

FIGURE 18.12

Sliding Window-based congestion control.

denoted in borderless slots with white background. With borderless dark greyed slots, *Forbidden* IDs are denoted which cannot be sent even if there is information waiting to be sent (the transmission is stuck waiting for the window W to move forward). By using thin borders we denote the IDs contained within W. In turn, light greyed slots contain the *Sent but Not yet Acknowledged* IDs, while the white slots contain *Usable* IDs. As can be seen, the window slides toward the higher IDs as correct ACK packets progressively arrive from B.

In a normal situation the Receiver expects to see SEQs arriving in an ordered compact sequence from the Sender. As correct cumulative IDs are received, corresponding ACK messages are sent back to the Sender. For the purpose of this analysis we will assume that an ACK number bears the SEQ number of the packet being acknowledged. This dynamic can be seen in the figure for the sequence $A1$ to $A4$, where ACK numbers are, as expected, also increasing. When all ACKs are received correctly for an entire window, the Sender increases W by 1. This can be seen upon reception of packets $B1$ and $B3$.

Nevertheless, when the Receiver gets a kth packet with sequence SEQ_k different from the next immediate expected value $SEQ_{k-1}+1$, it considers it as an invalid (out of order) packet. This is the case of A6, when SEQ:42 is received after SEQ:40, as a consequence of having lost the packet with SEQ:41.

Then the receiver returns the *last valid* ACK number, implying that it repeats (or duplicates) the sending of the last correctly received SEQ number. This is the case of $B5$. The Sender receives then a *Duplicated ACK* (DACK) which is interpreted as a "signaling" of an abnormal condition. The consequence is that upon reception of $B5$ the Sender sets its window to $W = W/2$, and falls back to retransmit the whole current window since its first packet (in this case, since SEQ:40)[1].

Several consecutive packet losses produce multiple DACKs. Depending on the version of the protocol, the congestion condition can be assumed after getting M successive DACKs. For our models in this work we will adopt the standard practice of $M = 3$, commonly referred to as the triple DACK (3DACK).

In Figure 18.13 we show a typical evolution for $W(t)$ where three packet losses are detected. The growth slope during the phase of (linear) additive increase is determined by the value of RTT.

In the next section we describe the admission control mechanism that takes place at intermediate routers, which are responsible for producing the main events leading to packet losses.

4.2.2 Active queue management and random early detection

A widely adopted active queue management (AQM) technique to cooperate with TCP/AQM is *random early detection* (RED) [55]. Each router monitors its local

---

[1]We purposively leave aside efficient complementary techniques such as Selective Acknowledge (SACK) and Fast Retransmit, which are not required for the goals pursued by the models in this work. Also, we implicitly assume that we use the Fast Recovery technique (when a duplicated ACK is detected and the window is halved) instead of setting it back to value 1.

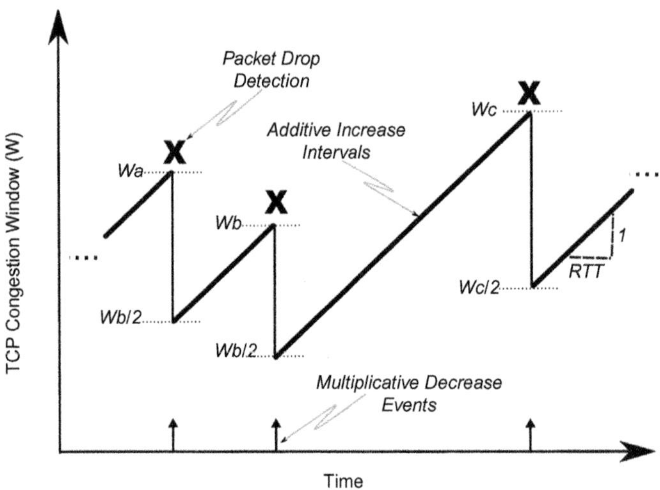

FIGURE 18.13

Temporal evolution of the Congestion Window W (during the *Congestion Avoidance* phase).

queue. When it detects that a congestion situation is imminent, but before the queue becomes full, it starts dropping packets purposively. This action ends up in packets not arriving at the receiver and hence with Duplicate ACKs arriving back to the sender, which leads at some point to the window W being reduced. I.e., the router *prevents* the congestion by forcing (indirectly) the senders to lower their (average) sending rates. RED calculates an average queue length x applying a low pass filter (moving weighted average): $x = (1 - \alpha) \cdot x + \alpha \cdot x_{inst}$, where x_{inst} is the instantaneously measured queue length, and $0 \leq \alpha \leq 1$ is the time constant of the filter.

The use of x instead of x_{inst} captures more efficiently the concept of congestion. Due to the inherent bursty AIMD takes 1 RTT, it doesn't make any sense to send congestion notifications in too reactive a fashion, as they would typically obey to bursts instead to a representative averaged queue lengths. The low pass filter tries to detect *persistent* congestion conditions.

RED features two queue length thresholds, a lower t^{min} and an upper t^{max}. They are mapped to packet discard probabilities $p = 0$ y and $p = p^{max}$ respectively. The full function $p(x)$ is shown in Figure 18.14.

Upon each change in the instantaneous queue length x_{inst} the following algorithm is applied:

- **Update** $x = (1 - \alpha).x + \alpha\ x_{inst}$
- If $x \leq t^{min}$ **Enqueue/Dequeue** a packet. I.e., no signaling action is taken.

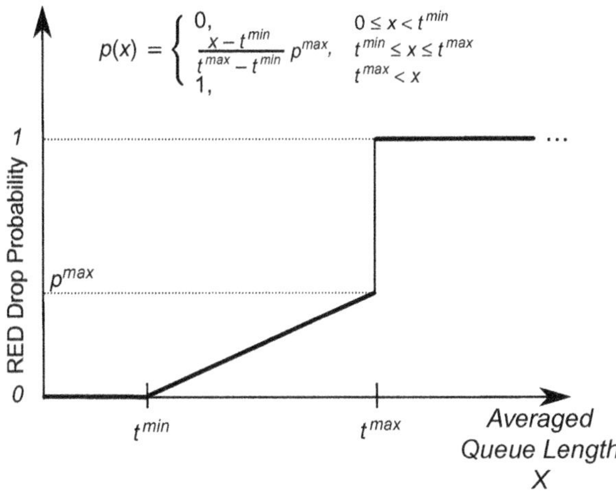

FIGURE 18.14

Probability profile of proactive packet discards. RED mechanism.

- If $t^{max} < x < t^{min}$ **Update** $p(x) = p^{max} \cdot \frac{x - t^{min}}{t^{max} - t^{min}}$ **Discard** an incoming packet with probability p. I.e., a potential congestion situation starts being signaled, proportional to the length x.
- If $x > t^{max}$ **Discard** the incoming packet. I.e., a more drastic measure is applied.

If everything goes well, RED eliminates only a small percentage of packets when x exceeds slightly t^{min}, and after 1 RTT the sender reacts reducing its rate, thus making x decrease. Therefore, congestion is prevented, average queue size is kept away from saturation, and effective throughput is kept high so as to utilize efficiently the available BDP. It is worth noting that x_{inst} can be much higher than x and t^{max}, thus allowing for instantaneous bursts to be absorbed without altering the average behavior of the control scheme. Common design guides suggest adopting $t^{max} = 2.t^{min}$ and a maximum buffer size of $3.t^{min}$.

4.3 DEVS BASED METHODOLOGY FOR ADMISSION CONTROL M&S

In this section we will follow the methodology introduced in Section 3, modeling and simulating the system under study with discrete and continuous representations, using DEVS as the underlying and unifying formalism.

We will implement a fluid approximation of TCP/AQM that permits simulating its main characteristics in very convenient simulation times. We will make use of the novel Delay QSS technique to model the dependency of the TCP congestion window W on its own historical values.

Afterwards, we present new discrete event models to simulate exactly the TCP/AQM behavior, including RED, resorting to the first principles of protocols and the physical medium.

4.3.1 Continuous controller with delays, continuous plant

As a case study, we adopt a continuous nonlinear model for TCP/AQM presented in [56]. This model approximates the dynamics for AIMD and RED in an integrated fashion, by means of a system of differential equations as follows:

$$\dot{W}(t) = \overbrace{\frac{1}{RTT(q(t))}}^{AI} - \overbrace{\frac{W(t)}{2} \times \underbrace{\frac{W(t-\tau)}{RTT(q(t-\tau))} p(x(t-\tau))}_{RED}}^{MD} \tag{18.12}$$

$$\dot{q}(t) = -\mathbf{1}_{q(t)} C + N \times \frac{W(t)}{RTT(q(t))} \tag{18.13}$$

where W is the size of the TCP congestion window, RTT is the connection round trip time, q is the instantaneous queue length, and x is the filtered version of q.

This simplified system models a single router with maximum bandwidth C (capacity), shared by N concurrent and identical flows (users), all of them subjected to a common RTT.

The first term at the right-hand side of Equation (18.12) models the Additive Increase (AI) part of AIMD, while the second term models the Multiplicative Decrease (MD).

The factor $\frac{W \cdot p}{RTT}(t-\tau)$ modulates $\frac{W(t)}{2}$. This implies that a) the multiplicative reduction of $\dot{W}$ depends on the state of the system τ seconds in the past, b) said delay will be a function of the effective TCP throughput $\frac{W}{RTT}$ also τ seconds in the past, and c) the probability that said throughput caused packet losses τ seconds in the past is given by p evaluated at that instant.

Moreover, the delay τ is variable and coincides with RTT itself, i.e., $\tau = \tau(t) = RTT(t)$. This transforms the system (18.12)–(18.13) in a *Delay* Differential Equation with *variable* and *state-dependent* delays (for RTT is in turn a function of the state variable $q(t)$).

Equation (18.13) shows, in its first term at the right of the identity, the rate C at which the router's queue is emptied. The operator $\mathbf{1}_{q(t)}$ takes the value of 1 when the queue is not empty ($q(t) > 0$), and 0 in any other case. This switch operator prevents $q(t)$ from assuming infeasible negative values. The addition of the second term models the filling rate of the queue, accepting aggregated traffic coming from N identical users.

In Figure 18.15 we show this system modeled with PowerDEVS. As can be seen, for representing the delayed MD dynamics we use a variable delay block labelled $t - tau$, which implements the DQSS methods presented in the introduction.

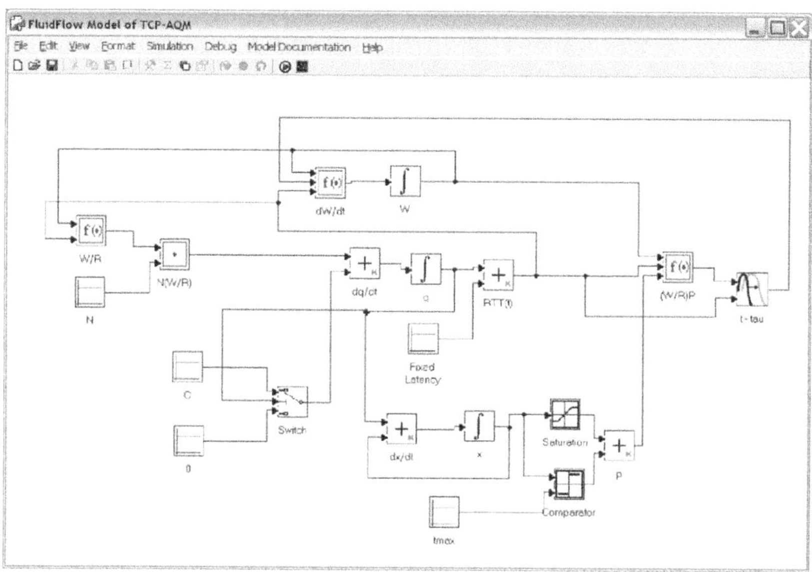

FIGURE 18.15

PowerDEVS continuous time model of TCP/AQM.

Table 18.1 Configuration Parameters for TCP/AQM

Parameter	Value	Unit	Description
BW	5.10^6	bits/s	Available bandwidth
pkt	500	Bytes	Packet size
C = BW/pkt	1250	packets/s	Maximum packet rate
t^{min}	50	packets	Minimum threshold for queue length (RED)
t^{max}	100	packets	Maximum threshold for queue length (RED)
p^{max}	0.1		Packet discard probability at t^{max} (RED)
α	10^{-3}	s	Time constant for the low pass filter (RED)
$\delta = 1/C$	8.10^{-4}	s	Time constant for the low pass filter (RED)
RTT	0.02	s	Minimum round trip time

To simulate this system we will adopt the set of parameters shown in Table 18.1, inspired partially in examples presented in [55]. For the time being we take $N = 1$, i.e., there is a single TCP flow controlled by a single AIMD scheme.

The simulation was run with final time $T_f = 100$ s, and results are shown in Figure 18.16. The oscillatory character of the system becomes evident, confirming what was expected from the original publication of reference.

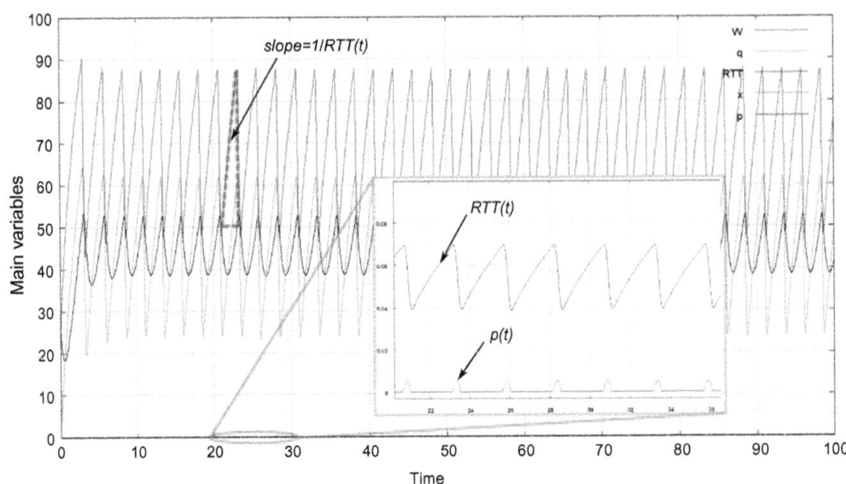

FIGURE 18.16

Simulation results. Fluid approximation of TCP/AQM parameterized according to Table 18.1. Time units correspond to seconds.

In a first place we verify the expected behavior for the AI phase, with congestion window W growing with slope $1/RTT(t)$, and is consistent with the evolution of $RTT(t)$ shown in the zoomed-in frame of Figure 18.16. The smoothing effect of queue length $q(t)$ is noticed, caused by the signal $x(t)$ that presents a smaller excursion of values while retaining the low frequency component.

Observing the time instants when $p(t)$ is greater than 0, they can be easily correlated with the brief MD phases in $W(t)$, which cause a reduction in the sending rate, a corresponding reduction of $q(t)$, and consequently the vanishing of $p(t)$, thus closing one regulation loop.

In Figure 18.17 we show in detail the effect of the block implementing the delay $(t - \tau(t))$ which, as mentioned before, is equivalent to $(t - RTT(t))$. The influence of the varying nature of $RTT(t)$ can be observed on the signal that *modulates time*.

The results permit investigating quickly and easily the effects of parameter changes on the dynamics of the system. However, once a satisfactory scenario is obtained and more detailed knowledge is required, the fluid model needs to be verified against a more realistic representation of the system providing information richer than average values.

4.3.2 Discrete controller and plant

We now present a fully discrete model of the TCP/AQM system, obeying the underlying first principles described in Sections 4.2.2 and 4.2.1. In Figure 18.18 we show the model implemented in PowerDEVS, which is a hybrid model mixing discrete event and discrete time.

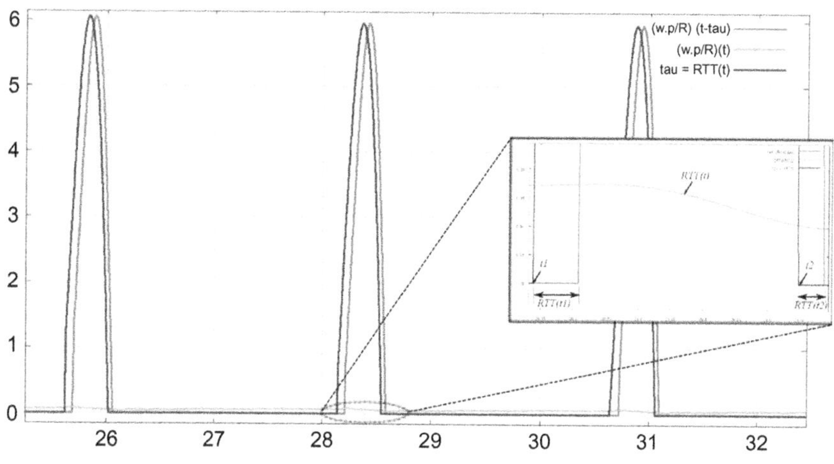

FIGURE 18.17

Simulation results. Detail of the effect of variable delays calculated with the DQSS method. Time units correspond to seconds.

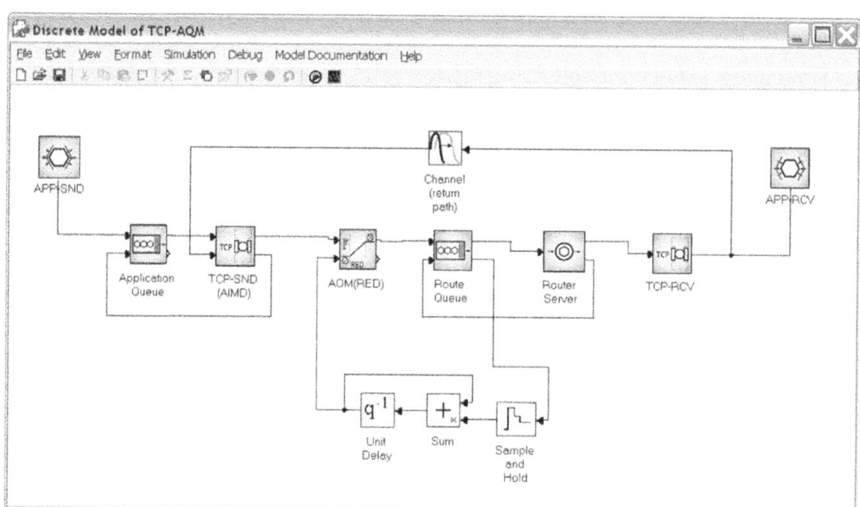

FIGURE 18.18

Discrete model of TCP/AQM implemented in PowerDEVS.

The traffic generator **APPSND** represents the "user application" that uses the underlying TCP services. APPSND uses an exponential probability distribution for the inter–generation times between packets. All blocks are capable of accepting, processing and emitting *packets*, which are hierarchical data structures nesting *header–payload–tail* substructures, imitating the way in which real-world protocols are encapsulated one into another according to the OSI model.

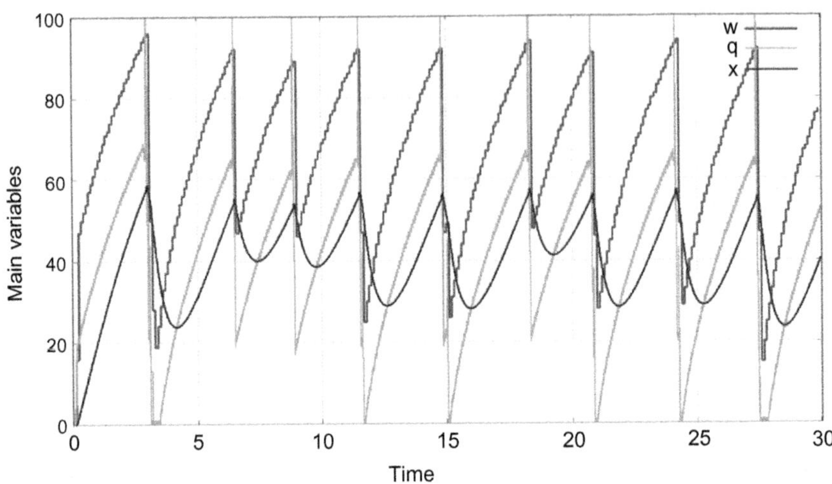

FIGURE 18.19

Simulation Results. TCP/AQM discrete approximation with parameters of Table 18.1. Time units correspond to seconds.

Simulation results are shown in Figure 18.19. The simulation time was 127 s (averaging several equivalent simulations). This is approximately 28 times slower than the fully continuous model presented in the previous section.

Again, an oscillatory pattern is observed. Results are qualitatively comparable, offering a reasonable confidence about the coherence across models. In Figure 18.20 variables W, q and x are compared for models continuous and discrete during the first 30 seconds.

Qualitatively, the excursion ranges of the variables are comparable, taking into account that the continuous system removes all dispersions around averages, which are present in the discrete system; e.g., the discrete system shows phases in which the instantaneous queue lengths drop down to zero, which is consistent with what happens in real systems.

4.3.3 Experiments with multiple identical flows

The most outstanding feature of TCP/AQM is its ability to adapt individual control actions at each TCP user according to the congestion state of the common shared network, the congestion being a result of the aggregated contribution of all individual flows, and without explicit communication across the competing users.

The continuous model of Equations (18.12–18.13) assumes that N identical users coexist and therefore calculates a single solution $W(t)$ valid for each one of the competing flows. Intuitively, W at each node must decrease inversely proportional to the number of new flows that join the network, aiming at sharing the available bandwidth in a fair way and without causing congestion.

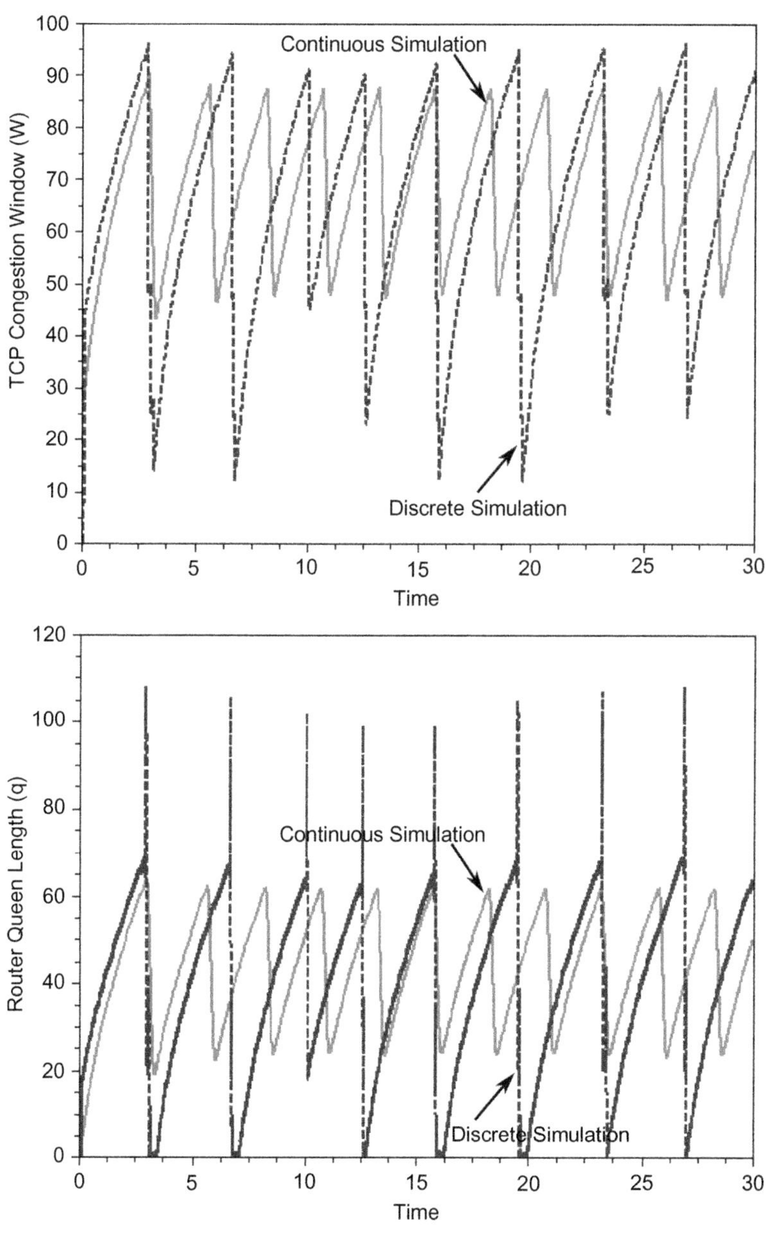

FIGURE 18.20

Simulation results. Comparison of the discrete and continuous models. Time units correspond to seconds.

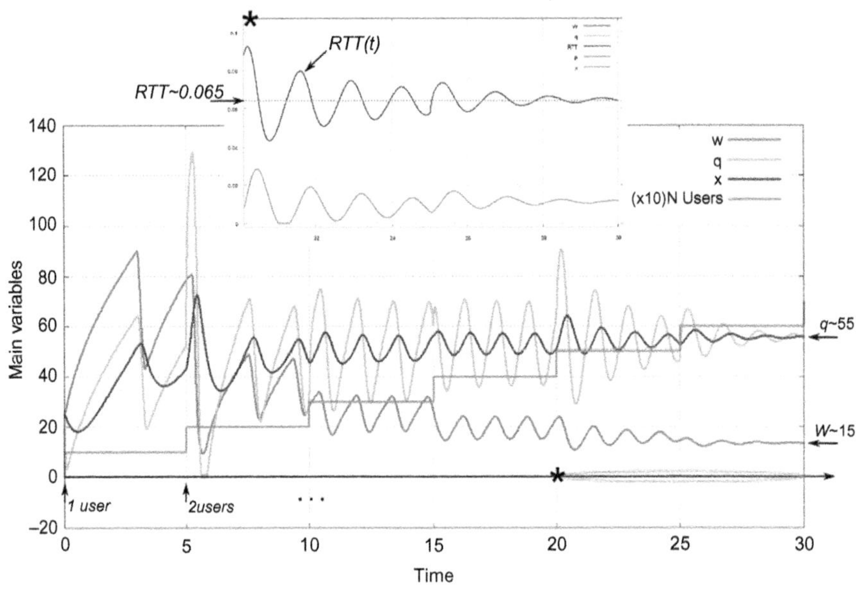

FIGURE 18.21

Fluid approximation of TCP/AQM with new users joining the system (from $N = 1$ to 6). Time units correspond to seconds.

To better illustrate this idea, we repeat the simulation of the continuous model but now assume that one extra user joins the network every 5 seconds. The result can be observed in Figure 18.21. It can be seen that the behavior for $W(t)$ is as expected, and also that the queue length $q(t)$ at the router is regulated satisfactorily around an interval (in this case between 40 and 60 packets, approximately.)

As this system converges to an equilibrium in its queue length q and round trip time *RTT*, we can try an analytical verification resorting to the classical *Little's Formula* [57]. For a queue-server system *at equilibrium* and any probability distribution of the traversing traffic, $\overline{q} = \lambda_q \cdot \overline{R_q}$ always holds, where $\overline{q}$ is the number of packets in the queue,λ_q is the incoming rate of ingress into the queue, and $\overline{R_q}$ is the average delay experienced by each packet since it enters the queue until it leaves. In our study case, the equilibrium is reached at $\overline{q} \sim 55$ packets and $\lambda_q = 1250$ packets per second, for which $\overline{R_q} \sim 0.044$ seconds must be verified. Recalling that the minimum *RTT* set for the system is $RTT = 0.02$ seconds, it must be verified that $RTT(t) = RTT + R_q(t)$ at all instants, for the queuing delay is the variable component of *RTT*. Then, at equilibrium $\overline{RTT} = RTT + \overline{R_q}$ must hold. By merging the values obtained via calculation and simulation, we verify satisfactorily that $\overline{RTT} \sim 0.065 \sim 0.02 + 0.044$.

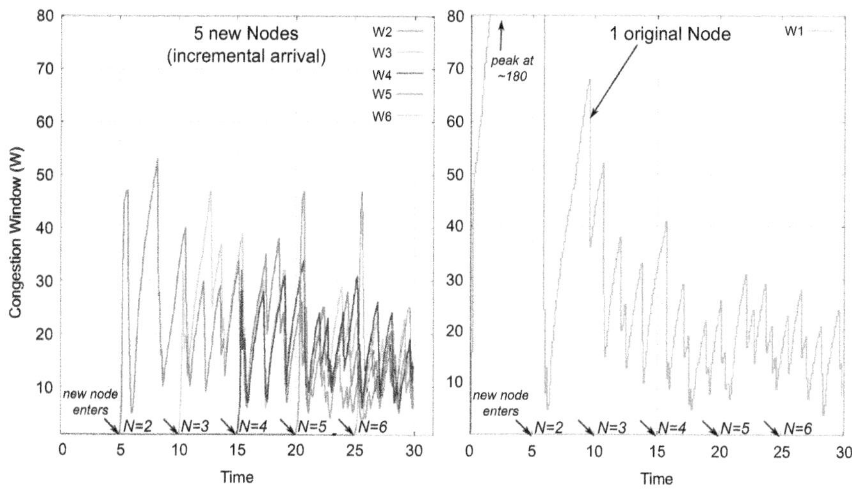

FIGURE 18.22

Discrete approximation of TCP/AQM with new users joining the system incrementally (from $N = 1$ to $N = 6$). Time units correspond to seconds.

4.3.4 Detailed analysis of W(t) with the discrete model

The experiment with the continuous model shows $W(t)$ as a valid representation identical for all N users in the system. This is a simplification that omits several interesting details about the discrete and stochastic behavior of the flows.

For a more realistic view of the concurrent dynamics of $W(t)$ at each node, we go back to the discrete model, this time around modeling $N = 6$ TCP connections sharing a common router. New TCP connections will be allowed to join the system every 5 seconds, repeating the scenario tested with the continuous model. In Figure 18.22 we show the result of the simulation for 30 seconds.

On the right-hand side we show the evolution of W for the first user that joins the system. At the left, we show W for all extra users joining every 5 seconds. The curves W_2 to W_6 show in detail the adaptation of each node to the varying overall congestion condition.

4.4 HYBRID CONTINUOUS/DISCRETE MODEL

As was mentioned before, approaches exist that try combining the computational efficiency of fluid approximations with the advantage of fine-grained details of discrete simulations (see [9,58,59] and references therein). These techniques are known as hybrid modeling and simulation, where the concept of hybridicity refers to the simultaneous and heterogeneous representation of the *same phenomena*, and not simply the coexistence of separate heterogeneous submodels that build up a more complex systems.

Experience on hybrid flow simulation presents several challenges that need to be faced regarding synchronization issues. The latter show up when trying to coordinate the time steps of the numerical integration method (required to approximate the fluid flows with differential equations) and the time advances for the discrete-event driven flows. Several techniques aim at modifying existing special-purpose discrete-event simulation engines of (e.g., ns-2, OMNET++, Opnet) either by encapsulating numerical integration methods or by building interfaces from the engines to external differential equation solvers. While these approaches exhibit acceptable empirical results, they rise from pragmatic endeavors whose validity is bound to each specific tool, lacking a formal mathematical framework that can guarantee correctness and generality.

In the next section we will show how it is possible to implement a simple strategy for hybrid flow simulation in an almost straightforward yet robust fashion, relying on the DEVS formalism. It provides formal properties such as legitimacy and closure under coupling (which are extensible to stochastic processes STDEVS), together with properties of convergence, stability and efficiency offered by the DEVS-based QSS and DQSS methods that approximate continuous systems. These tools allow for dealing with hybrid systems under a unified mathematical formalism, and a single practical tool.

4.4.1 Combination of continuous and discrete flows with DEVS

The intuitive idea is to obtain a hybrid model of the aggregated flow of packets trafficked by a router, which simultaneously represent a continuous part and a discrete part. The continuous component models the effect of N_{cont} users, and the discrete component models N_{disc} users, all of them concurrently using up the available bandwidth. Thus, both parts must correctly influence each other to accurately represent the aggregated dynamics of the total number of users N_{tot}.

We follow a simple "background traffic" strategy, through which the totality of users are assigned to the continuous part ($N_{tot} = N_{cont}$) setting the system in its operational regime. Simultaneously, we designate a single discrete user ($N_{disc} = 1$) serving the purpose of a *probe flow* permitting maximum level of detail.

We accept the assumption that $N_{cont} \gg N_{disc}$, which a priori allows neglecting the influence of the discrete flow on the continuous background bundle of flows. We still need to determine the way in which the probe discrete flow is influenced by the state of the fluid system.

We will choose the queue-server system as the portion of the system that acquires the hybrid characteristics required to model the influence of the continuous part over the discrete part, as will be described below.

Informally, a hybrid queue must represent the occupancy of the queue in terms of the continuous flow $q_c(t)$ while keeping the consistency with the state of the discrete part $q_d(t)$ that models the arrival, placement, moving and leave of individual packets.

A descriptive sketch of this idea can be found in Figure 18.23.

Upon arrival of a new discrete packet to the hybrid queue, it must "see" in front of it a number of packets equivalent to the total present in the system, i.e., it must find the $q_c(t)$ packets imposed by the continuous representation.

The idea is to build an artificial *hybrid packet* for each new discrete packet that joins the queue. It is done in such a way that its equivalent length renders valid the identity $q_c(t) = q_d(t) + q_h(t)$ at all instants. We define $q_h(t)$ as the artificial hybrid length that represents the number of fluid packets that should be interspersed in between adjacent discrete packets, thereby amounting to the total length $q_c(t)$.

For this mechanism to work, the server shall unload hybrid packets from the queue, and for each of them, simulate a service time corresponding to one discrete packet (T_d^S) plus the equivalent artificial hybrid service time (T_h^S), corresponding to the number of extra packets that was attached to it when it entered the queue.

In order to formalize this idea, we analyze the pure continuous queue-server system, then another one that is pure discrete, and then we calculate the proportion of artificial packets to be added to compose each hybrid packet. To support the calculation, in Figure 18.23 we show the Queuing Time T^Q, Service Time T^S and Response Time $T^R = T^Q + T^S$ that define the main delays that take place at the queue-server system working at a capacity of C packets per second.

For the continuous case, and considering the full system at equilibrium, we resort again to the Little's Formula:

$$C \cdot T_c^R = q_c \tag{18.14}$$

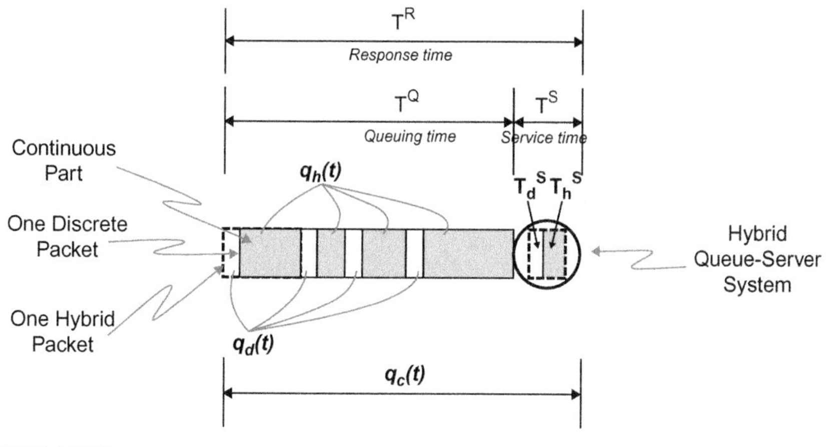

FIGURE 18.23

Hybrid queue-server system. Discrete packets (in white), equivalent continuous packets (in grey) and hybrid packets (with dotted border).

where T_c^R is the response time of the queue-server system and is the number of elements in said system, both for the continuous case. If we assume that all packets are equally sized, and that there are always packets to be processed, the service time will be $T_c^S = \frac{1}{C}$. After replacing in the previous equation, we verify:

$$T_c^R = q_c \cdot T_c^S \tag{18.15}$$

that relates the total response time, the queue occupancy and the service time (all averaged values) for the continuous system.

On the other hand, in the discrete system, retaining the assumption that all packets have the same length, the service time T_d^S will be constant and same for all. Then, the following shall hold:

$$T_d^R = T_d^S \cdot (1 + q_d) \tag{18.16}$$

where T_d^R is the response time and $q_d + 1$ is the number of elements in the whole system (which includes the one that is being processed at the Server), both for the discrete case.

According to the proposed characteristics for the hybrid system, the following shall hold:

$$T_h^R = T_c^R \tag{18.17}$$

i.e., the hybrid system and the continuous system must be equivalent from an external point of view, imposing the same delay (total response time) to the traversing packets.

Another characteristic that shall hold is:

$$T_h^R = T_h^S \cdot (1 + q_d) \tag{18.18}$$

i.e., the hybrid system and the discrete system must be equivalent from the perspective of the number of hybrid and discrete packets that are present in the queue, which explains the use of q_d in the previous equation. The hybrid service time T_h^S represents artificially part of the continuous queue.

Finally, by replacing Equation (18.18) and (18.15) in Equation (18.17) and reordering, we obtain the hybrid service time that shall calculate the new system state upon arrival of each new discrete packet:

$$T_h^S = T_c^S \frac{q_c}{1 + q_d} \tag{18.19}$$

The assumption made that all packets in the continuous part have the same length must be consistent with the discrete part, in order to merge them together into the hybrid representation.

I.e., the service time of an individual packet shall be the same in both representations, verifying $T_c^S = T_d^S$. With this consideration, and defining $Y = \frac{q_c}{1+q_d}$ as a proportion factor, we finally obtain:

$$T_h^S = T_d^S \frac{q_c(t)}{1+q_d(t)} = T_d^S \cdot Y(t) \tag{18.20}$$

4.4.2 Implementation of the hybrid flow combiner block

From the perspective of the implementation, each discrete packet brings with itself the information of its length, and therefore its service time T_d^S can be immediately determined using the factor Y. Then, the novel requisite for building a hybrid queue-server system is that the system knows about $Y(t)$ at all times. Such information is enough for converting each discrete packet into a coherent new hybrid packet.

For implementing a hybrid queue-server system that is backward compatible with other DEVS models implemented before for blocks *Queue* and *Server*, we developed the model *HybridFlow* that combines the discrete and continuous flows, avoiding to modify the preexisting libraries.

In Figure 18.24 we show this idea implemented in PowerDEVS, where we prepend the new block *HybridFlow* to the standard queue-server tandem (those used in all previous discrete examples).

The upper input port accepts discrete packets and the lower port accepts a continuous signal that represents $Y(t)$. At the output port, hybrid packets are issued with a length modified so that when it arrives to the Server the service time for the packet can get defined by Equation (18.20). When the packet comes out of the server and continues advancing through the discrete system, it will preserve its original discrete length. That is, a packet will acquire a hybrid nature only since it enters a HybridFlow block until it egresses from a Server block.

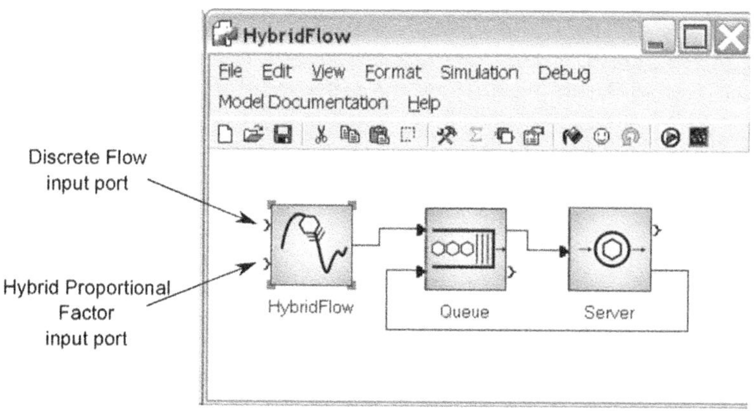

FIGURE 18.24

New HybridFlow Block for combining discrete and continuous flows.

4.4.3 Hybrid TCP/AQM model

We will apply the hybrid technique to the study case TCP/AQM presented in 4.3.1 and 4.3.2, aiming at evaluating its efficacy to reduce the simulation time of the discrete system yet providing detailed information for a discrete probe flow.

An additional advantage is the practicality of visual modeling offered by PowerDEVS. When the number of individual users grows, the explicit visual modeling of each single connection soon becomes cumbersome, as can be observed in Figure 18.25 already for the case of $N = 6$ users (cf. Figure 18.18 for $N = 1$).

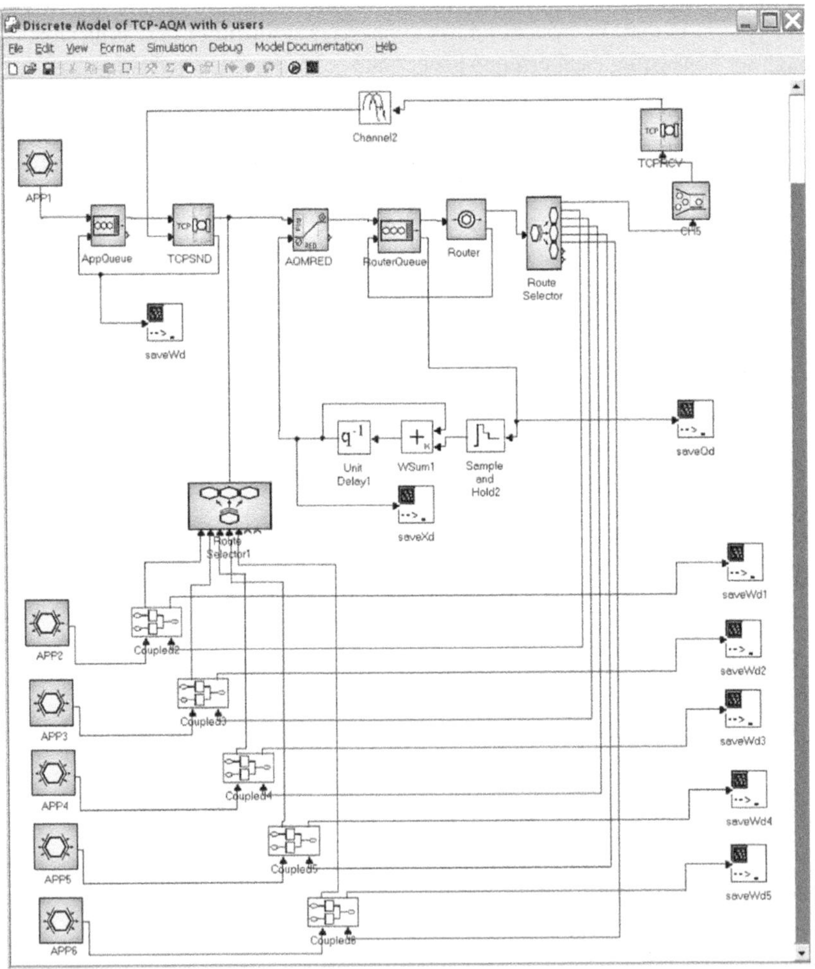

FIGURE 18.25

TCP/AQM discrete model for $N = 6$ users.

The opposite way, using the hybrid technique it is required to model explicitly only a single discrete user (the probe flow), using the continuous part to set the system at the average level of occupancy corresponding to $N=6$. The implementation of this approach is shown in Figure 18.26.

We show a comparison of selected results for the pure continuous, pure discrete and hybrid types of models, for $N=2$, 4 and 6, focusing on the state variable $W(t)$ (equivalent for all users). The results are presented in Figure 18.27, showing an acceptable qualitative match, evidencing a coherent similarity in terms of amplitudes and frequencies.

The resulting simulation times are shown in Figure 18.28, confirming our expectations. On the one hand, the execution times of the continuous model

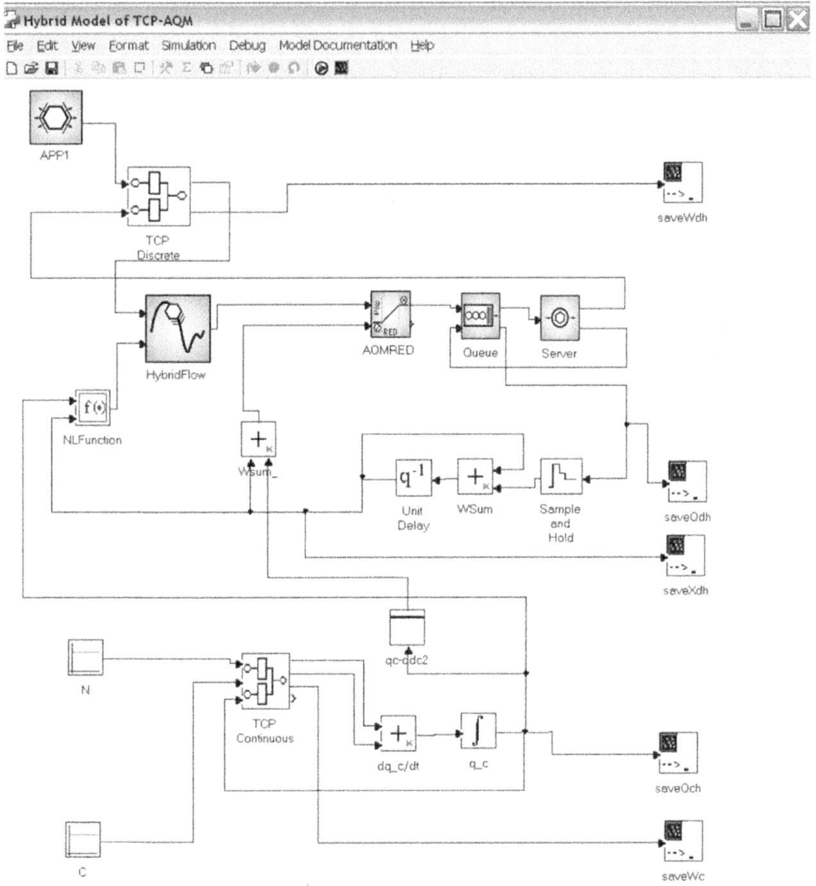

FIGURE 18.26

TCP/AQM hybrid (continuous/discrete) model.

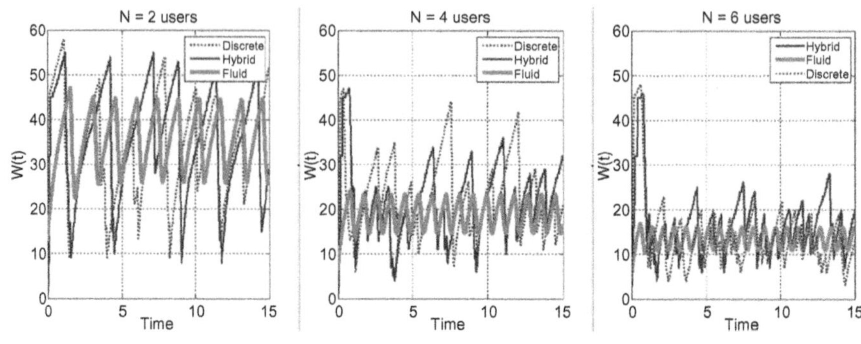

FIGURE 18.27

Simulation results for pure discrete, pure continuous and hybrid models. Scenarios with $N = 2$, 4 and 6 concurrent users. Time units correspond to seconds.

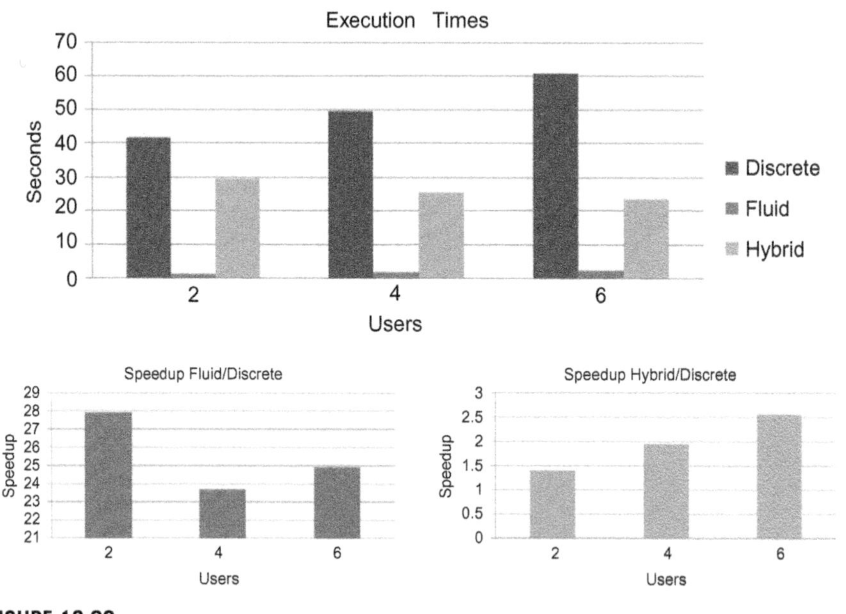

FIGURE 18.28

Simulation speed-ups for scenarios $N = 2$, 4 and 6 concurrent users (15 seconds of simulated time).

present speed-ups on the order of 20 to 30 times over the exact discrete simulation. On the other hand, when using the hybrid model the speed-ups are reduced to 1.5 to 2.5, approximately. Nevertheless, as N grows the speed-up increases. This is reasonable, since the computational load of the pure discrete model grows linearly with N_{disc}, while a) the computational load of the pure continuous model can be considered constant and independent of N_{cont} and b) the load imposed by the discrete part of the hybrid model can also be considered constant with $N_{disc} = 1$ (the probe discrete flow).

The results obtained so far show the efficacy of the proposed strategy and implementation at keeping computing times under control while retaining the granularity of details provided by discrete flows. A side effect benefit is the simplicity of visual modeling that a hybrid system offers as compared to its pure discrete counterpart as N grows.

5 CONCLUSIONS

This chapter described new theoretical and practical tools for the modeling and simulation of hybrid systems with application to data networks. In particular, the focus was oriented to the study and design of automatic controllers for meeting quality of service requirements.

We presented an integrative methodology based on the Discrete EVent Systems specification (DEVS), which permits modeling and simulating of complex systems by choosing the most convenient representation paradigm for each subsystem, thus allowing discrete time, discrete event and continuous models to coexist.

Guided by two practical examples, we showcased the need of expressing the network controllers under study in discrete and/or continuous approximations and with different levels of abstractions, according to the requirements imposed by different design phases.

The resulting design process often calls for working with models of different natures, described with different languages by specialists in different fields. This usually forces switching between tools and specification languages, or performing ad-hoc customizations to existing tools, without a coherent supporting formalism that guarantees the correctness of these procedures. As a consequence many practical difficulties arise, like increased error proneness, but most critically, hard incompatibilities between the numerical solvers and the discrete event dynamics.

Our approach removes such barriers completely in a flexible yet mathematically robust way, thanks to the generality and soundness of the DEVS formalism.

On the one hand, DEVS is able to model exactly any sort of discrete formalism, including stochastic ones. On the other hand, the modern DEVS-based QSS numerical integration solvers amalgamate smoothly with discrete event systems, thanks to its state-quantization technique that operates on a continuous time base.

In the first case study, we described an end-to-end methodology for applying control theory into a network admission control problem, which yields a stochastic hybrid system. We modeled a fully continuous approximation of the network and the controller using differential equations. Then we discretized the controller in time, keeping the network model in its fluid approximation, a typical step foreseeing the final implementation of the controller in a clock-driven digital platform. Finally, as a last verification step before the validation in a real set-up, the fluid approximation of the network is replaced by an exact discrete event model of the network, much more representative of the real system.

In the second case study, we modeled the TCP/AQM collaborative congestion control, which is the main type responsible for the dynamics of the Internet today. In this case we focused on performance issues. For a system with many flows, a fully discrete event set-up offers a maximum level of granularity. The price to pay is that the relation between the computational demand and the network size (number of flows) is at least linear. To circumvent this situation, a possible strategy is that of foreground and background flows. A single "probe" (foreground) flow is simulated in discrete form, thus accessing to a per-packet level of granularity that is representative of each single flow in the system. Simultaneously, a single aggregate background flow is simulated in fluid form, which is representative of the sum of all of the flows in the system. We implemented a hybrid representation (both continuous and discrete, concurrently) of the flows of packets, using a unified mathematical formalism and under the same tool. The computational demand of the hybrid model becomes practically insensitive to the size of the system, being dominated mostly by the demand of single discrete probe flow.

Wrapping up, we provided strong evidence that the DEVS technology offers several advantages for hybrid modeling and simulation of data networks, particularly when applied in the domain of QoS automatic newtork control.

REFERENCES

[1] Fujimoto RM, Perumalla K, Park A, Wu H, Ammar MH, Riley GF. Large-scale network simulation: how big? how fast? Modeling, analysis and simulation of computer telecommunications systems, 2003. MASCOTS 2003. 11th IEEE/ACM international symposium on. IEEE; 2003. p. 116–23.

[2] Ahn JS, Danzig PB. Packet network simulation: speedup and accuracy versus timing granularity. IEEE/ACM Trans Netw (TON) 1996;4(5):743–57.

[3] Guo Y, Gong W, Towsley D. Time-stepped hybrid simulation (TSHS) for large scale networks. INFOCOM 2000. In: Nineteenth annual joint conference of the IEEE computer and communications societies. Proceedings. IEEE, vol. 2. 2000. p. 441–50.

[4] Liu Y, Lo Presti F, Misra V, Towsley D, Gu Y. Fluid models and solutions for large-scale IP networks. ACM SIGMETRICS performance evaluation review, vol. 31. ACM; 2003. p. 91–101.

[5] Nicol DM, Yan G. Discrete event fluid modeling of background TCP traffic. ACM Trans Model Comput Simul (TOMACS) 2004;14(3):211–50.

[6] Liu B, Figueiredo DR, Guo Y, Kurose J, Towsley D. A study of networks simulation efficiency: Fluid simulation vs. packet-level simulation. INFOCOM 2001. In: Twentieth annual joint conference of the IEEE computer and communications societies. Proceedings. IEEE, vol. 3. IEEE; 2001. p. 1244–53.

[7] Riley GF, Jaafar TM, Fujimoto RM. Integrated fluid and packet network simulations. Modeling, analysis and simulation of computer and telecommunications systems, MASCOTS 2002. Proceedings. Tenth IEEE international symposium on. IEEE; 2002. p. 511–18.

[8] Kiddle C, Simmonds R, Williamson C, Unger B. Hybrid packet/fluid flow network simulation. Parallel and distributed simulation, (PADS 2003). Proceedings. Seventeenth workshop on. IEEE; 2003. p. 143–52.

[9] Gu Y, Liu Y, Towsley D. On integrating fluid models with packet simulation. Proceedings of IEEE INFOCOM. IEEE; 2004.

[10] Liu J. Packet-level integration of fluid TCP models in real-time network simulation. Simulation conference, WSC 06. Proceedings of the winter, IEEE; 2006. p. 2162–69.

[11] Chen Y, Dong Y, Xiang Z, Lu D. A hybrid simulating framework of TCP traffic at aggregated level. Communications and networking in China. ChinaCom 2008. Third international conference on. IEEE; 2008. p. 327–32.

[12] Li T, Van Vorst N, Liu J. A rate-based TCP traffic model to accelerate network simulation. Simulation 2013;89(4):466–80.

[13] Zeigler B, Kim TG, Praehofer H. Theory of modeling and simulation. 2nd ed. New York: Academic Press; 2000.

[14] Nutaro JJ. Building software for simulation: Theory and algorithms, with applications in C++. Wiley Publishing; 2010.

[15] Kim S, Sarjoughian HS, Elamvazhuthi V. DEVS-Suite: a simulator supporting visual experimentation design and behavior monitoring. Proceedings of the 2009 spring simulation multiconference. society for computer simulation international; 2009. p. 161.

[16] Wainer G. CD++: a toolkit to define discrete-event models. Softw Pract Exp 2002; 32(13):1261–306.

[17] Quesnel G, Duboz R, Ramat E, Traoré M. VLE: a multimodeling and simulation environment. In: Proceedings of the 2007 summer computer simulation conference, San Diego, California; 2007. p. 367–74.

[18] Bergero F, Kofman E. PowerDEVS. A tool for hybrid system modeling and real time simulation. Simulation 2011;87(1–2):113–32.

[19] Castro R, Kofman E, Wainer G. A formal framework for stochastic discrete event system specification modeling and simulation. Simulation 2010;86(10):587–611.

[20] Kofman E, Junco S. Quantized state systems: a DEVS approach for continuous system simulation. Trans SCS 2001;18(3):123–32.

[21] Cellier FE, Kofman E. Continuous system simulation. New York: Springer; 2006.

[22] Kofman E. A second order approximation for DEVS simulation of continuous systems. Simulation 2002;78(2):76–89.

[23] Kofman E. A third order discrete event simulation method for continuous system simulation. Latin Am Appl Res 2006;36(2):101–8.

[24] Migoni G, Bortolotto M, Kofman E, Cellier F. Linearly implicit quantization-based integration methods for stiff ordinary differential equations. Simul Model Pract Theory 2013;35:118–36.

[25] Migoni G, Kofman E, Cellier F. Quantization-based new integration methods for stiff ODEs. Simulation 2012;88(4):387–407.

[26] Kofman E. Discrete event simulation of hybrid systems. SIAM J Sci Comput 2004;25(5):1771–97.

[27] Beltrame T, Cellier FE. Quantised state system simulation in Dymola/Modelica using the DEVS formalism. In: Proceedings of the fifth international modelica conference, vol. 1, Vienna, Austria; 2006. p. 73–82.

[28] D'Abreu M, Wainer G. M/CD++: Modeling continuous systems using Modelica and DEVS. In: Proceedings of MASCOTS 2005, Atlanta, GA; 2005. p. 229–36.

[29] Castro R, Kofman E, Cellier F. Quantization based integration methods for delay differential equations. Simul Model Pract Theory 2010;19(1):314–36.

[30] Hellerstein J. Feedback control of computing systems. Wiley-IEEE Press; 2004.

[31] Arzen K-E, Robertsson A, Henriksson D, Johansson M, Hjalmarsson H, Johansson KH. Conclusions of the ARTIST2 roadmap on control of computing systems. SIGBED Rev 2006;3(3):11–20.

[32] Hollot CV, Misra V, Towsley D, Gong WB. On designing improved controllers for AQM routers supporting TCP flows. In: IEEE INFOCOM, vol. 3; 2001.

[33] Christin N, Liebeherr J, Abdelzaher TF. A quantitative assured forwarding service. In: IEEE INFOCOM, vol. 2; 2002.

[34] Robertsson A, Wittenmark B, Kihl M. Analysis and design of admission control in web-server systems. In: Proceedings of ACC'03, Denver, Colorado; 2003. p. 254–59.

[35] Cassandras CG, Lafortune S. Introduction to discrete event systems. Kluwer Academic Publishers; 2004.

[36] Issariyakul T, Hossain E. Introduction to network simulator NS2. Springer; 2008.

[37] OMNeT++ Community. OMNeT++ Discrete event simulation system. Available from: <www.omnetpp.org>; 2004.

[38] Castro R, Kofman E, Wainer G. A DEVS-based end-to-end methodology for hybrid control of embedded networking systems. In: Proceedings of ADHS'09: third IFAC conference on analysis and design of hybrid systems, vol. 3. Zaragoza, Spain; 2009. p. 74–79.

[39] Kihl M, Robertsson A, Wittenmark B. Analysis of admission control mechanisms using non-linear control theory. In: Proceedings of ISCC 2003, vol. 2. Kiris-Kemer, Turkey; 2003. p. 1306–11.

[40] Tipper D, Sundareshan MK. Numerical methods for modeling computer networks under nonstationary conditions. IEEE J Sel Areas Commun 1990;8(9):1682–95.

[41] Scilab Enterprises. Scilab: The open source platform for numerical computation. Available from: <http://www.scilab.org>; 2015.

[42] Barakat C. TCP/IP modeling and validation. Netw IEEE 2002;15(3):38–47.

[43] Liu S, Basar T, Srikant R. TCP-Illinois: A loss-and delay-based congestion control algorithm for high-speed networks. Perform Eval 2008;65(6–7):417–40.

[44] Chkliaev D, Hooman J, de Vink E. Verification and improvement of the sliding window protocol. TACAS'03: Proceedings of the ninth international conference on Tools and algorithms for the construction and analysis of systems. Berlin, Heidelberg: Springer-Verlag; 2003. p. 113–27.

[45] Floyd S, Kohler E. Internet research needs better models. ACM SIGCOMM Comput Commun Rev 2003;33(1):29–34.

[46] Padhye J, Firoiu V, Towsley D, Kurose J. Modeling TCP throughput: A simple model and its empirical validation. Amherst, MA, USA: Technical report, University of Massachusetts; 1998.

[47] Kelly F. Mathematical modeling of the internet. In: Engquist B, Schmid W, editors. Mathematics unlimited—2001 and beyond. Berlin: Springer-Verlag; 2001. p. 685–702.

[48] Grieco LA, Mascolo S. TCP Westwood and Easy RED to improve fairness in high-speed networks p. 130–146. PIHSN'02: Proceedings of the seventh IFIP/IEEE international workshop on protocols for high speed networks. London, UK: Springer-Verlag; 2002.

[49] Mascolo S. Modeling and designing the internet congestion control. In: Tarbouriech S, Abdallah CT, Chiasson J, editors. Advances in communication control networks. Springer; 2005. p. 33–6 [chapter 7].

[50] Park K, Willinger W. Self-similar network traffic and performance evaluation. New York, NY, USA: John Wiley & Sons, Inc.; 2000.

[51] Balachandran B, Kalmár-Nagy T, Gilsinn D, editors. Delay Differential Equations - recent advances and new directions. Springer Science + Business Media; 2009.

[52] Padhye J, Firoiu V, Towsley DF, Kurose JF. Modeling TCP Reno performance: a simple model and its empirical validation. IEEE/ACM Trans Netw (ToN) 2000;8(2): 133–45.

[53] Floyd S, Henderson T. RFC2582: The NewReno Modification to TCP's fast recovery algorithm. RFC Editor United States; 1999.

[54] Yang YR, Lam SS. General AIMD congestion control. ICNP. Published by the IEEE Computer Society; 2000. p. 187.

[55] Floyd S, Jacobson V. Random early detection gateways for congestion avoidance. Netw IEEE/ACM Trans 2002;1(4):397–413.

[56] Misra V, Gong W, Towsley D. A fluid-based analysis of a network of AQM routers supporting TCP flows with an application to RED. Proceedings of the SIGCOMM 2000. ACM; 2000. p. 151–60.

[57] Little JDC. A proof for the queuing formula: $L = \lambda W$. Oper Res 1961;9(3):383–7.

[58] Schormans J, Liu E, Cuthbert L, Pitts J. A hybrid technique for accelerated simulation of ATM networks and network elements. ACM Trans Model Comput Simul (TOMACS) 2001;11(2):182–205.

[59] Yi Y, Shakkottai S. Flunet: A hybrid internet simulator for fast queue regimes. Comput Netw 2007;51(18):4919–37.

PART

Next generation wireless networks evaluations

5

CHAPTER

An Ns-3 based simulative and emulative platform

19

Igor Bisio, Stefano Delucchi, Fabio Lavagetto, Mario Marchese, Giancarlo Portomauro, and Sandro Zappatore

University of Genova, Genova, Italy

1 INTRODUCTION

The deployment of a tool able to integrate simulated wireless networks and a set of real hosts and links is presented. This tool is called **Hybrid Simulative-Emulative Platform (HySEP)** in order to highlight the integration of simulated and emulated networks which are connected and work together transmitting/receiving traffic flows from/to each other.

HySEP enables simultaneous testing of wireless access technologies without the use of real network implementation, and the transmission of real traffic flows from/to real hosts that communicate with the simulated part of the network. Such a scenario assures major advantages in the study and analysis of different network configurations using different types of real traffic flows.

HySEP is used to model a heterogeneous network and is composed of a *Long Term Evolution* (LTE) based access network, simulated using Network Simulator 3 (ns-3), and of an emulated backhaul network that implements the Differentiated Service (DiffServ) QoS solution. These two components are connected assuring communication between simulated LTE terminals, called User Equipments (UEs), and a real remote host.

LTE and DiffServ have been selected because the former represents an important innovative access network, which assures excellent performance while supporting user mobility, and the latter is a consolidated and commonly used approach to assure QoS inside a backhaul network. For the sake of completeness, a brief overview of both of them is also included in this chapter.

It is very important to clarify the scope and the purpose of this chapter: it is aimed at describing the developed platform, including the configuration and the characteristics of each employed personal computer (PC), the configuration of the DiffServ protocol and the implementation of the simulated network. As a matter of fact, the goal of this chapter is to explain how HySEP has been designed and built and how to reproduce it. For these reasons, some practical configuration scripts are also included in the next sections. HySEP utilization as a test-bed for QoS validation is not presented in this work, not only because it is a subject of ongoing research but mainly because it is beyond the scope of this chapter, which is mainly focused on the platform description.

Only open source software and simple PCs, not equipped with any particular hardware, have been used to implement HySEP. This choice facilitates the distribution and the diffusion of this platform as a tool easy to be configured and used but, at the same time, with great simulation capabilities. Actually, using both simulated and emulated components, it is possible to create realistic scenarios and implement details to analyze the functionalities of expensive technologies, typically not available within an academic research laboratory.

It is necessary to evidence the important role played by the ns-3 simulator. As detailed in the next section, this software assures not only the simulation of different wireless access networks, including obviously LTE, which absorbed most of the design efforts, but also the capability of interconnecting the simulated network with real tools and links.

The rest of this chapter is organized as follows: in Section 2 an overview of the LTE architecture is reported. Section 3 describes the DiffServ solution largely used to support the QoS. Section 4 contains the description of ns-3, adopted in the proposed platform, and, in particular, of the models used to simulate the LTE network. Section 5 contains a detailed description of the HySEP structure and all the configuration steps. Conclusions and future developments are reported in Section 6.

2 THE EVOLVED PACKET SYSTEM

The Evolved Packet System (EPS) is composed of:

- **Long Term Evolution** (**LTE**), which is the radio access network [1,2].
- **Evolved Packet Core** (**EPC**), representing the core network of the system,

The EPS supports user mobility and is connected through Internet Protocol (IP) to a generic Packet Data Network (PDN) assuring the use of applications and services running on mobile terminals called User Equipment (UE), such as Voice over IP (VoIP), data transfer, web browsing and video conferences.

2.1 THE EPS NETWORK ARCHITECTURE

The EPS network architecture is shown in Figure 19.1, where dotted and continuous lines represent, respectively, control and data plane links.

2.1.1 Long Term Evolution (LTE)

The LTE access network is composed of:

- **User Equipment** (**UE**): which is a mobile terminal such as a smart-phone, portable PC, tablet.

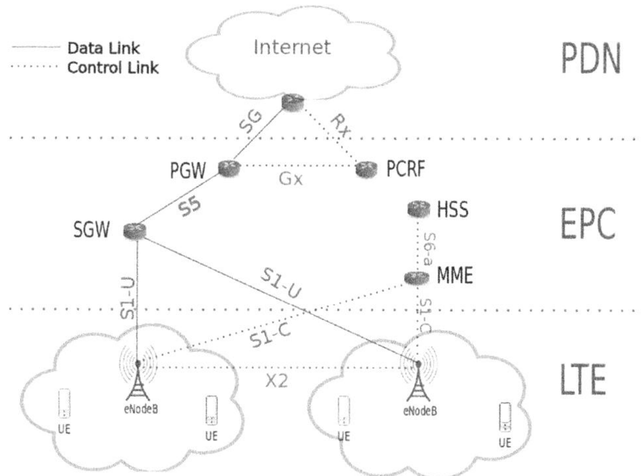

FIGURE 19.1

EPS architecture.

- **Evolved Node B** (**eNodeB** or **eNB**): which is the radio base station responsible for the control functions within the radio link, of the radio connection management for all UEs associated to it, and of assuring UE connection to EPC. The eNodeB performs network control techniques such as radio admission control, scheduling and radio channel resource allocation in order to guarantee QoS in the access network. eNodeB links to other EPS architecture components are shown in Figure 19.1.

2.1.2 The Evolved Packet Core

According to [2], the Evolved Packet Core (EPC) core network is composed of:

- **Serving Gateway (SGW):** which is the node that connects each UE to the EPC by using a tunneling protocol called GPRS Tunnel Protocol (GTP). It is connected to eNBs and Packet Data Network Gateway (PGW) as shown in Figure 19.1.
- **Packet Data Network Gateway (PGW):** which is the node that connects the EPS network with external networks. It assures a connection to a remote destination to each UE, through the assigning of an IP address.
- **Policy Control and Charging Rules Function (PCRF):** which is the control plane node that manages the QoS inside the EPS network and performs pricing actions.
- **Home Subscription Server** (**HSS**): which contains user profile information and authenticates users inside the EPS network.
- **Mobility Management Entity** (**MME**): which manages user mobility and handover execution.

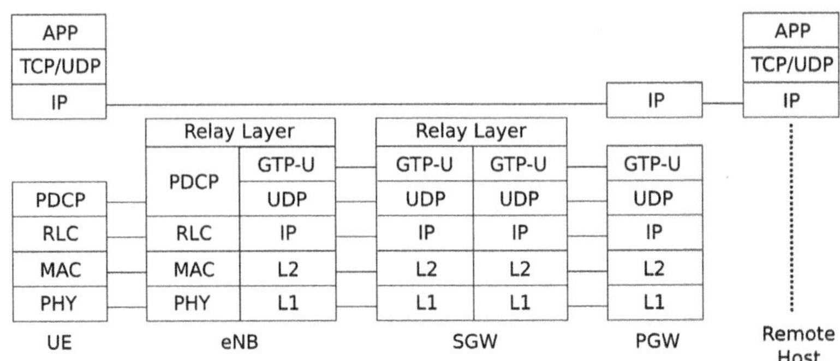

FIGURE 19.2

Data plane architecture.

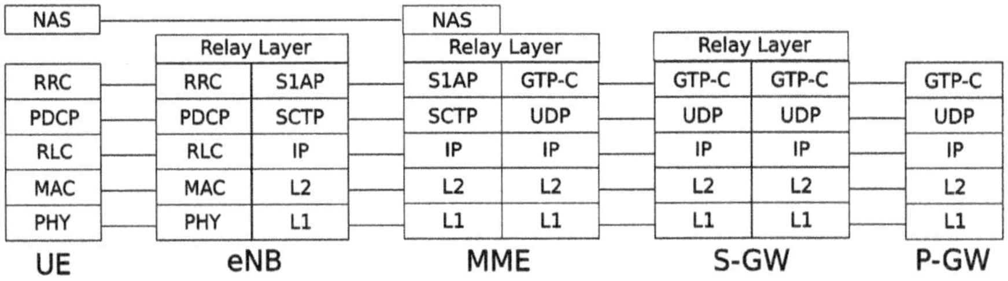

NAS
NAS
Relay Layer
Relay Layer
Relay Layer
RRC | RRC | S1AP | S1AP | GTP-C | GTP-C | GTP-C | GTP-C
PDCP | PDCP | SCTP | SCTP | UDP | UDP | UDP | UDP
RLC | RLC | IP | IP | IP | IP | IP | IP
MAC | MAC | L2 | L2 | L2 | L2 | L2 | L2
PHY | PHY | L1 | L1 | L1 | L1 | L1 | L1
UE
eNB
MME
S-GW
P-GW

FIGURE 19.3

Control plane architecture.

2.2 EPS PROTOCOL STACK

EPS data and control plane architectures are shown respectively in Figure 19.2 and Figure 19.3.

Four layers are the same in both architectures:

- **Packet Data Convergence Protocol (PDCP)** layer: which executes IP header compression and packet encryption [3].
- **Radio Link Control (RLC)** layer: which executes packets segmentation and reassembly and Automatic Repeat and reQuest (ARQ) error control procedures.
- **Medium Access Control (MAC)** layer: which executes channel access procedures and implements Hybrid-ARQ (HARQ) error control mechanisms [4].
- **Physical (PHY)** layer: which implements Multi Input Multi Output (MIMO) transmission technology.

Two other layers are defined only for the control plane:

- **Radio Resource Control (RRC)** layer: which manages the UE connection with the LTE handover operations and initializes the UE authentication.
- **Non Access Stratum (NAS)** layer: which identifies and authenticates the UE, working together with the MME.

Tunnel protocols, based on UDP/IP encapsulation, are adopted between the eNB and the SGW-PGW within both data and control plane. They are called, respectively, GPRS Tunnel Protocol User (GTP-U) and GPRS Tunnel Protocol Control (GTP-C) [5].

2.3 QUALITY OF SERVICE OVER EVOLVED PACKET SYSTEM

The "bearer" is the minimum aggregation of traffic flows within the EPS network. It identifies a single QoS-class. Its scope covers the whole EPS network, defining how the traffic flows have to be treated. Two different bearers are defined:

- **Default Bearer**: assigned to each UE when it is connected to the EPS network. Default bearers support only best effort traffic but they can be used also by the UE to require other bearers for specific traffic flows.
- **Dedicated Bearer**:which guarantees different level of QoS to the traffic flows, defining a sort of tunnel between UEs and PGWs. They belongs to two different groups: *(i)* Guaranteed Bit Rate (GBR); *(ii)* non-Guaranteed Bit Rate (non-GBR).

The QoS Class Identifier (QCI) is a number ranging between 1 and 9 that identifies the traffic class of each bearer and defines how each EPS node has to handle the traffic flows carried by the bearers within the data plane architecture. Each QCI is associated with four other parameters:

- Type of bearer: GBR or non-GBR.
- Priority level: a number in the range 1–9 (where 1 is highest priority).
- Packet Delay Budget (PDB).
- Packet Error and Loss Rate (PELR).

Similar functionalities are played by the Allocation Retention Priority (ARP) parameter within the control plane. ARP identifies the network access priority for each UE and regulates the activation and the termination of each bearer. It is composed of three values:

- **Priority**: a number in the range 1–9 (where 1 is highest priority).
- **Preemption Capability flag**: which acts as follows: if it is set to 1 the bearer is allowed to ask the termination of other bearers characterized by lower priority, in case of congestion.
- **Vulnerability Preemption flag:** which acts as follows: if it is set to 1 the bearer could be terminated in case of congestion.

3 DIFFERENTIATED SERVICES DOMAIN

The Differentiated Service (DiffServ) is a network solution aimed at classifying the IP traffic flow [6] into traffic classes. It uses six bits, called DiffServ Code Point (DSCP), part of the eight-bit field called Type of Service (TOS) inside the

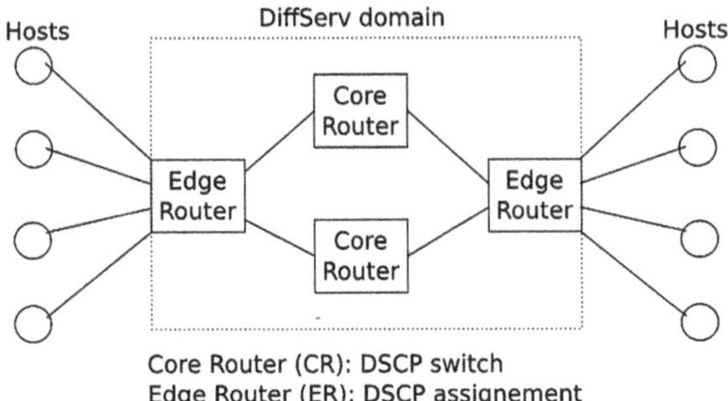

FIGURE 19.4

A DiffServ Domain.

IP header [7]. Its goal is to determine the Per Hop Behavior (PHB) that defines the packet forwarding procedure of each node. The scope of this protocol is identified by the DiffServ Domain [8], shown in Figure 19.4.

A generic DiffServ Domain is composed of two types of nodes [9]:

- **Edge Routers** (**ERs**): located at the borders of the domain. They assign the DSCP value to each packet according to the QoS information received through the incoming network. They also execute a sort of horizontal mapping between the DiffServ and the QoS protocol adopted inside the incoming domain. During this operation the ERs aggregate the traffic flows with similar QoS requirements by assigning the same DSCP values.
- **Core Routers** (**CRs**): located inside the DiffServ Domain. They forward the packets with different policies according to their DSCP values. Each traffic flow with the same DSCP value receives the same treatment inside the domain.

Traffic classes defined by the DiffServ protocol may be grouped as follows:

- **Expedited Forwarding** (**EF**): which identifies traffic with specific QoS requirements such as telephony [10].
- **Assured Forwarding** (**AF**): composed of traffic classes AF1, AF2, AF3 and AF4 [11].
- **Best Effort** (**BE**): does not assure any level of QoS.

4 THE NS-3 SIMULATOR

Network Simulator 3 (ns-3) is a free, open source software project that implements a discrete-event network simulator for research and education [12]. The project started in 2006 with the first implementation of the simulator. Its current version, ns-3.19, was developed in December 2013. Ns-3 is not an extension of Network Simulation 2 (ns-2), but it is a new simulator written completely in C++ programming language,

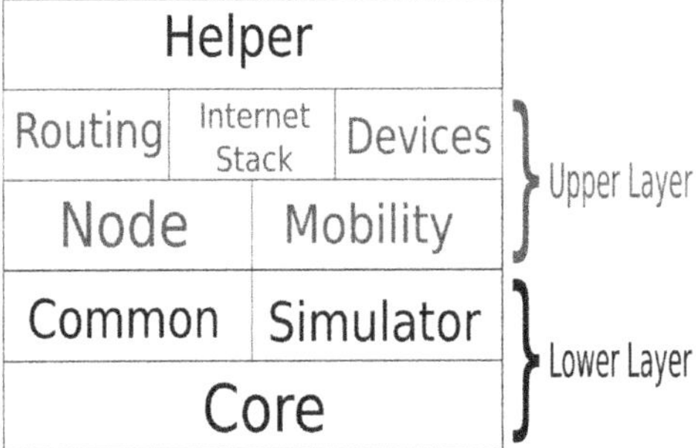

FIGURE 19.5

The ns-3 architecture.

using some optional building mechanisms in the Python programming language. Ns-3 does not use the Tool Command Language (Tcl), largely adopted inside ns-2.

Ns-3 has a modular structure, as reported in Figure 19.5. The lower layers contain the common basic model of the whole simulator, while the upper layers are aimed at adding details in the simulated scenario such as network protocols, devices and technologies. Finally, at the top of the model, the "helper" defines useful interfaces for the user to implement the characteristics of the scenario. Each element represented in Figure 19.5 is briefly described in the following:

- **Core module**: includes all the basic components used to create the simulation (random variables generator, events tracing and logging mechanisms).
- **Common module**: contains all the elements that define each transmitted packet within the simulator.
- **Simulator module**: schedules the events managing the simulation timers.
- **Node module**: is aimed at creating all the network nodes inside the simulation. It enables also the definition of protocol stacks, network interfaces and applications for each node.
- **Mobility module**: contains different mobility patterns which can be used in a mobile scenario to simulate the device movements.
- **Routing module**: enables the utilization of different routing protocols inside the simulated network.
- **Internet stack module**: enables the installation of the TCP/IPv4 and of the IPv6 protocols in each simulated node.
- **Devices module**: defines all the available technology-dependent network interfaces of a node.

4.1 THE EPS MODEL INSIDE NS-3

Ns-3 supports a very detailed model for the simulation of a LTE-EPC network. The first effort to create an open source product-oriented LTE network simulator was done

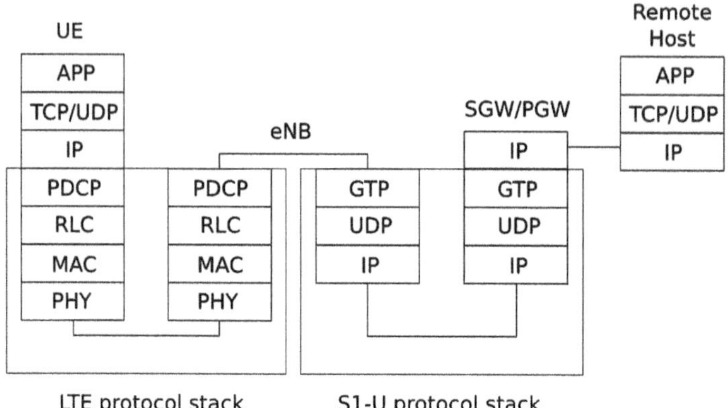

FIGURE 19.6

LTE-EPC data plane protocol stack in the ns-3 model.

by Ubiquisys, a leader organization in the development of intelligent multimode LTE/3G/WiFi small cells, and by the Centre Tecnològic de Telecomunicacions de Catalunya (CTTC), a well-known research organization in the telecommunications field. They made public the first release of the LENA project, an LTE-EPC simulator based on the ns-3 network simulator, in June 2011 [13]. The LENA project has been included inside the ns-3 version 3.19 since December 2013. In this project there are two distinct models:

- **LTE model**: which includes the radio access architecture and the UE and eNB protocol stack.
- **EPC model**: which enables the simulation of the SGW, of the PGW and of the MME inside the core network.

Ns-3 supports the simulation of a connection between UEs and remote hosts over LTE-EPC as shown in Figure 19.6. Two IP layers are used and respectively aimed at: i) end-to-end IP connection and ii) S1-U interface IP connection used to implement the GTP tunnel.

4.1.1 LTE model

This model enables the simulation of the LTE radio access network. Part of its data plane features are supported such as Resource Block (RB) Allocation, Packet Scheduling and Spectrum Management [13]. The LTE radio access network is composed of UE and eNB radio interface. A large number of UE and eNB can be simulated. Other details can be added to the simulated scenario: bandwidth and carriers used, channel interference and UE movements, including configurable speed and direction. Also the handover between LTE cells is supported.

An important LTE functionality supported by ns-3 is the allocation and the termination of the bearer inside the EPS network. In particular, the wireless segment of the bearer, called radio bearer, may include different details. For this reason, in the project, all the layers in the UE and eNB stack are implemented: Radio Resource Control (RRC), Packet Data Control Protocol (PDCP), Radio Link Control (RLC), Medium Access Control (MAC), Physical (PHY).

Another strong point of the ns-3 LTE model is the presence of a detailed log mechanism, which assures a huge amount of information and statistics about the wireless channel. More details about these functions can be found in [12].

4.1.2 EPC model

This ns-3 model enables the EPC core network simulation. It is composed of the following elements:

- eNB wired interface.
- GTP implementation over the S1-U link.
- SGW and PGW nodes, whose functionalities are implemented in a single node, as also is typically done in industrial implementations. It is worth noting that only a single SGW/PGW node can be simulated in each scenario.

The EPC network is able to transport different traffic flows through the encapsulation inside different bearers, according to their QoS requirements. Moreover, ns-3 supports packet filtering that is an important functionality implemented by EPS: the uplink traffic flows are conveyed in the appropriate bearer according to the traffic flow template of each UE. The same actions are executed by the PGW for the downlink traffic flows.

A slight weakness of the ns-3 EPC model is the implementation of the control plane model, which includes only S1-AP, X2-AP and S11 interfaces. Nevertheless, the large availability of helper modules assures the capability of implementing many control plane functionalities to the user.

The EPC model supports a very useful output model: both eNB and SGW/PGW generate pcap files for each interface. Pcap, which stands for **p**acket **cap**ture, is a programming library to capture network frames of any traffic flow transmitted or received by a network interface.

5 THE DEVELOPED TOOL

The Hybrid Simulated-Emulated Platform (HySEP) structure is composed of two different segments:

- The **Simulated Network Segment** (**SNS**) simulates different wireless access networks through the ns-3 simulator. The LTE-EPC network is simulated within this segment that is composed of a single PC (identified as PC1). Even if our action is focused on LTE, other access network architectures can be adopted inside this segment.

- The **Emulated Network Segment (ENS)** implements an emulated backhaul network and a remote host that communicates with the simulated mobile nodes inside the SNS. Mobile nodes represent the UEs in our scenario. ENS is composed of two PCs, (namely PC2 and PC3), each of them playing a different role.

A fundamental task performed by the SNS is the interconnection with the ENS, where the network is emulated and real traffic flows are handled by nodes. It is necessary to synchronize the simulated network inside PC1 with the real network within the ENS so to transmit simulated traffic flows as real inside the emulated network. This requirement justifies the ns-3 utilization: this software cnables the utilization of a "real-time" scheduler that assures the synchronization between the ns-3 time event and the clock of the PC where the simulation is executed. Otherwise the simulated network works as a normal Discrete Event Simulator (DES) and the interconnection with real networks is not possible.

The backhaul network is implemented within PC2 by using Virtual PCs (VPCs) representing the network nodes. Similarly, virtual devices such as tap and bridge are used to emulate the link. The Differentiated Service (DiffServ) is the protocol implemented inside the backhaul network to support the QoS and therefore it is installed in each VPC.

PC3 is the real remote host that implements the same applications running in the simulated UEs, such as User Datagram Protocol (UDP) client-server and Iperf. In this way it is possible to create traffic flows from the UEs to this real node and vice versa.

Figure 19.7 summarizes the HySEP structure. All PCs are connected with each other by using Ethernet cables and are also connected to the Internet so that each HSEP component can be upgraded, configured and used not only locally but also remotely.

5.1 THE SIMULATED NETWORK SEGMENT

The Simulated Network Segment (SNS) structure is shown in Figure 19.8: it includes the EPS network, which is simulated by using the ns-3 tool and the virtual network interface (tap1) used to connect the SNS and the ENS, as explained in Section 3. The ns-3 scenario is composed of the following elements:

- **User Equipment** (**UE**) represents the mobile terminals that assure network connectivity to users. Each UE is equipped with an LTE interface.
- UEs are connected with one of the available base stations, called **evolved Node B** (**eNB**), which control and manage the radio link and receive/forward the traffic to/from the associated UEs.
- As previously stated in Section 4.1.2, SGW and PGW functionalities are implemented in a single node, called, in this chapter, **sgwPgw**. sgwPgw represents the key node of the EPC network and communicates with the eNBs and with the Ghost Node.

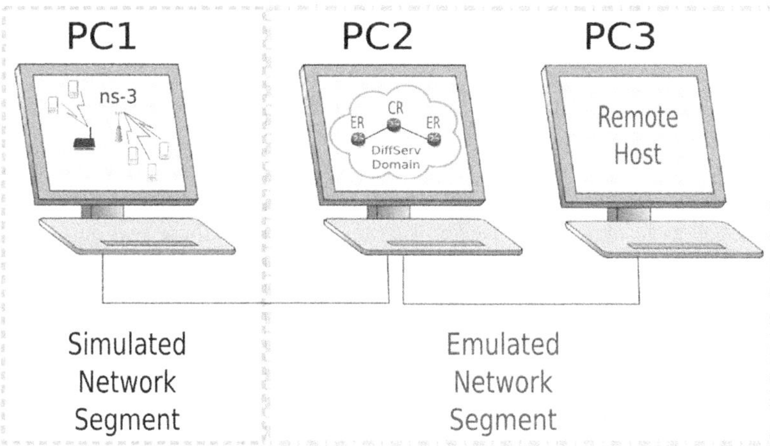

FIGURE 19.7

HySEP architecture.

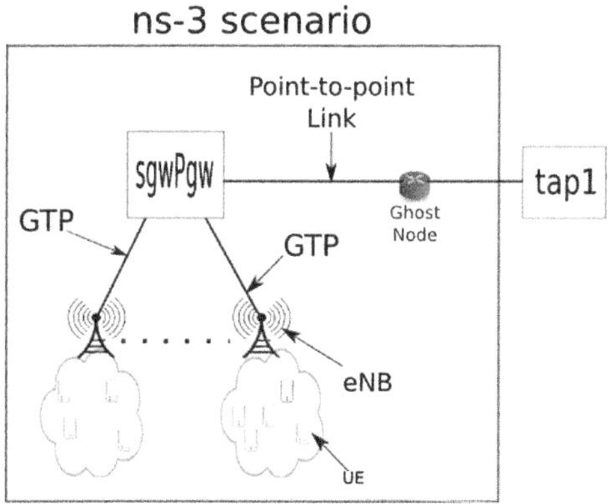

FIGURE 19.8

The Simulated Network Segment (SNS) structure.

- The **Ghost Node** is a particular type of node whose goal is to assure the interconnection between the simulated scenario and the virtual interface **tap1** that forwards the traffic to the ENS. More details about this interconnection procedure are included in Section 5.3.
- Two different types of wired links are used inside the proposed scenario: *(i)* the **GPRS Tunnel Protocol (GTP)**, which connects each eNB and the sgwPgw; and *(ii)* the **Point-to-Point** link, which is a simple link characterized only by data rate and delay, and used to connect sgwPgw node and Ghost Node.

The aforementioned simulated scenario is created by a script written in the C++ programming language.

The most important parts of the ns-3 script are reported and described in the rest of this subsection. The script includes all the header files necessary to use the methods of the classes acting as application programming interface (API). A fundamental setting that must be adopted is the *Real-time mode*, which is necessary to enable the interconnection between the ns-3 simulated scenario and the real network as well as to manage the packet forwarding between these two domains. This configuration can be fulfilled by adding the following strings to the ns-3 script:

```
GlobalValue::Bind ("SimulatorImplementationType", StringValue ("ns3::
RealtimeSimulatorImpl"));
GlobalValue::Bind ("ChecksumEnabled", BooleanValue (true));
```

Two important instances are created in order to implement LTE and EPC networks. This is done by using two helper modules, called respectively *LteHelper* and *EpcHelper*, which assure the capability of creating a minimal but complete instance of many different network utilities.

```
Ptr<LteHelper>lteHelper = CreateObject<LteHelper> ();
Ptr<EpcHelper>epcHelper = CreateObject<EpcHelper> ();
```

The successive step is to create the nodes that compose the network: UEs, eNBs and *sgwPgw*, representing jointly SGW and PGW, by using the aforementioned helpers as indicated in the following lines of code:

```
NodeContainer ueNodes;
ueNodes.Create(numberOfUeNodes);
NodeContainer enbNodes;
enbNodes.Create(numberOfEnbNodes);
Ptr<Node>sgwPgw = epcHelper->GetPgwNode ();
```

To complete the network topology it is necessary to create the "Ghost node" that is to connect the simulated node with the virtual interface tap1. This node is associated with the sgwPgw through a particular ns-3 variable, the *NodeContainer*, called in this example *gatewayContainer*.

```
NodeContainer gatewayContainer;
gatewayContainer.Create (1);
gatewayContainer.Add(sgwPgw);
```

When all the network nodes are created and properly configured, it is necessary to define the protocols used by each component of the simulated scenario. In particular, the Carrier Sense Multiple Access (CSMA) protocol can be created simply by using the appropriate helper: CSMA with supported data rate equal to 100 [Mb/s] and 1 [ms] latency is configured in the following example. When the

CSMA instance is created, the *gatewayContainer* variable is assigned to sgwPgw and Ghost Node. In this way it is possible to properly configure the link between these two nodes.

```
CsmaHelper csma;
csma.SetChannelAttribute ("DataRate", DataRateValue (DataRate ("100 Mb/s")));
csma.SetChannelAttribute ("Delay", TimeValue (MilliSeconds (1)));
NetDeviceContainer netDeviceContainer = csma.Install(gatewayContainer);
```

Similarly, the TCP/IP protocol stack is installed on the *sgwPgw* by using the available helper as reported below. It is worth noting that:

- the "Ghost node" does not require the TCP/IP functionalities, which are not installed on it for this reason;
- thanks to LteHelper it is not necessary to explicitly install the TCP/IP suite inside the UEs, because this operation is executed by this module.

```
InternetStackHelper internet;
internet.Install(sgwPgw);
```

The ns-3 configuration adopted to connect the simulated scenario with real hosts is shown below. A more detailed description of this code is reported in the next subsection.

```
TapBridgeHelper tapBridgeName;
tapBridgeName.SetAttribute ("Mode", StringValue (UseBridge));
tapBridgeName.SetAttribute ("DeviceName", StringValue (tapName));
tapBridgeName.Install (gatewayContainer.Get(0), netDeviceContainer.Get(0));
```

A successive step consists in assigning the IP address to *sgwPgw* and Ghost node, as reported below:

```
Ipv4AddressHelper ipv4h;
ipv4h.SetBase ("<net_address>", "<net_mask>");
Ipv4InterfaceContainer remoteIpIfaces2 = ipv4h.Assign(netDeviceContainer);
```

IP addresses are automatically assigned to each UE and eNB by the *LteHelper*, previously defined. When IP addresses are assigned to each network node it is necessary to set static routes used by nodes to correctly forward packets. Two different utilities, provided by ns-3, can be used to reach this goal: *(i)* Ipv4StaticRoutingHelper, and *(ii)* Ipv4StaticRouting.

```
Ipv4StaticRoutingHelper ipv4RoutingHelper;
Ptr<Ipv4StaticRouting> remoteHostStaticRouting =
      ipv4RoutingHelper.GetStaticRouting (gateway->GetObject<Ipv4> ());
remoteHostStaticRouting->AddNetworkRouteTo(Ipv4Address network,
   Ipv4Mask networkMask,
      Ipv4Address nextHop, uint32_t interface);
```

The LTE radio channel is the last part of the network to be configured. In particular, the mobility model and the interconnection between UE and eNB are defined by using, respectively, MobilityHelper and EpcHelper as shown here:

```
MobilityHelper mobility;
mobility.SetMobilityModel ("ns3::BuildingsMobilityModel");
mobility.Install (enbNodes);
mobility.Install (ueNodes);
internet.Install (ueNodes);
Ipv4InterfaceContainer ueIpIface;
ueIpIface = epcHelper- > AssignUeIpv4Address (NetDeviceContainer (ueLteDevs));

for (uint32_t u = 0; u < ueNodes.GetN (); ++u){
      Ptr< Node > ueNode = ueNodes.Get (u);
      Ptr< Ipv4StaticRouting > ueStaticRouting =
            ipv4RoutingHelper.GetStaticRouting (ueNode- > GetObject< Ipv4 > ());
      ueStaticRouting->SetDefaultRoute (epcHelper->GetUeDefaultGatewayAddress (), 1);
}
```

At the end of the script the log procedures and the overall duration of the simulation are set through the reported code:

```
p2ph.EnablePcapAll("lte-epc-wifi-remotehost", false);
Simulator::Stop (Seconds (30.0));
NS_LOG_INFO ("Simulationisrunning...");
Simulator::Run();

Simulator::Destroy();
NS_LOG_INFO ("Done.");
```

In particular, the first line reported above is necessary to display the pcap file, which is useful to analyze transmitted and received packets by each node, as stated in Section 4.1.2.

5.2 THE EMULATED NETWORK SEGMENT

The Emulated Network Segment (ENS) is composed of two machines: PC2, which implements the DiffServ Domain inside the backhaul network; and PC3, which acts as a Remote Host.

5.2.1 The backhaul network

The backhaul network can contain many nodes (i.e., routers) interconnected with each other by links. Therefore it could be useful to build a simplification of this domain that can limit costs, time and complexity with respect to a real implementation. At the same time it assures the availability of the same functions of a real network.

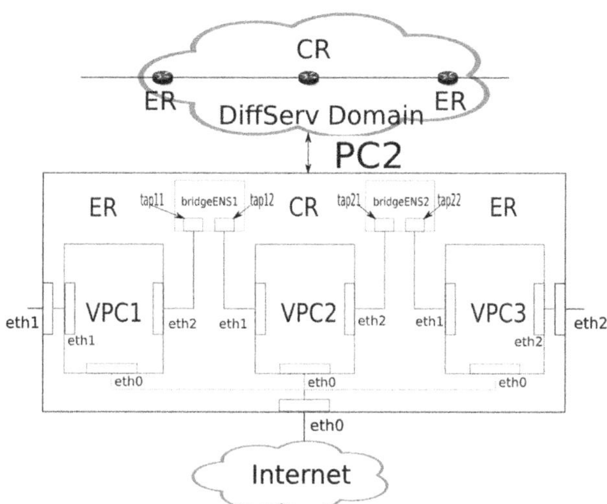

FIGURE 19.9

Backhaul network implementation by using Virtual Machines on PC2.

Table 19.1 The Hardware and Software Characteristics of PC2

Hardware Characteristics	Software Characteristics
CPU: Dual Core	OS: Linux with kernel 2.6 (32 bit)
RAM: 4 GB at least	Linux DiffServ configuration tool (Linux Traffic Control – tc)
Hard Disk: 10 GB at least	Linux Network bridging configuration tool (brctl)
Motherboard with 2 Ethernet interfaces	Virtual network device (TUN/TAP) configuration tool (tunctl)
	Oracle VirtualBox or similar virtualization software

The chosen solution is to build a network composed of Virtual PCs (VPCs) within a host PC that does not require special hardware quality and performance, but it is able to implement a complex emulated network. Obviously, the larger the emulated network size, the higher the performance required by the PC. Advantages assured by this solution are the creation of such a network inside a unique PC and the facility to export/import virtual nodes to/in other host PCs, so increasing the dimension of the domain.

The configuration adopted inside PC2 to emulate the backhaul network is shown in Figure 19.9: it is composed of three virtual machines: two Edge Routers (ERs, VPC1 and VPC3) and a Core Router (CR, VPC 2).

The host PC (i.e., PC2) used to implement such a network has the hardware and software characteristics listed in Table 19.1.

Table 19.2 VPC Hardware Characteristics

CPU	The same as the host PC
RAM	128 MB
Hard Disk	2 GB
Network interface	2 × Ethernet 10/100/1000

Three VPCs are created and managed by Oracle VirtualBox software [14]. Each of them has the hardware characteristics reported in Table 19.2.

Each VPC is equipped with a GNU/Linux operative system with a Debian 6 distribution, the packet analyzer software tcpdump, and the network statistics utility IPTraf. Both these tools are under open source license and are developed for Linux operative systems. Furthermore, the VPCs have to support the DiffServ and the procedure to receive and forward the packets, acting as a router [15].

VPCs are interconnected as shown in Figure 19.9 by using two types of virtual network devices: bridge and tap. Two different connections are created, according to the role of the VPCs. In more detail, the CR is connected to the other VPCs by using tap and bridge; ERs use tap and bridge to communicate with the CR and the real network interfaces of PC2 to communicate with the nodes outside the backhaul DiffServ domain. Observing Figure 19.9, it is possible to see that VPC1 and VPC3 are connected to PC1 and PC3, respectively, by using *eth1* and *eth2*; on the other hand, VPC1 and VPC2 are connected through *bridgeENS1*. This bridge has two interfaces: *tap11* and *tap12*, where the first index identifies the bridge and the second one identifies the interface. In practice *tap12* is the second interface of the first bridge. The following steps are necessary to create and to configure bridges and taps:

- Creation of a tap and its activation:
```
$ tunctl -t <tapName>
$ ifconfig <tapName> 0.0.0.0 up
```
- Creation of a bridge, association of the interfaces, and its activation:
```
$ brctl addbr <bridgeName>
$ brctl addif <bridgeName> <tapName>
$ ifconfig <bridgeName> up
```

The previous steps connect VPCs and enable the communications among them. The Linux Traffic Control (tc) utility can be used to add details about the links; tc is software, developed for Linux operating systems, which is used to manage traffic control settings by acting on the queuing discipline (qdisc), i.e., on the rules that define how packets have to be handled inside the network interface [16]. Using this software, it is possible to set data rate, latency, and delay of a link, for example, by using the following commands:

```
$ tc qdisc replace dev <deviceName> root handle 1: tbf rate <data rate>
    latency <latency>
$ tc qdisc add dev <deviceName> parent 1:1 handle 10: netem delay 50 ms
```

Finally, each VPC is connected directly to the Internet by using three different bridges created by the virtualization software Oracle VirtualBox. These bridges connect the network interfaces *eth0* of each VPC to the real interface *eth0* of PC2 that is connected to Internet. In this way it is possible to access each VPC from a remote host to modify and update the configuration.

5.2.2 Differentiated service domain configuration

The backhaul network described in the previous section implements the DiffServ protocol to differentiate the traffic flows in transit.

Linux operative systems offer a wide variety of network traffic control functions partially located in the kernel-space and partially in the user-space. In particular, some of them implement the mechanisms required to support the DiffServ architecture. The kernel processes the IP packets received from the network and forwards them: incoming packets are examined and, then, either forwarded directly to the output queue, as typically happens in routers, or forwarded internally to the upper layers of the protocol stack (e.g., to the transport UDP or TCP protocols). The transport layer may also convey packets generated by higher levels to the lower levels deputy to encapsulation and routing. The forwarding procedures include also the selection of the output interface as well as packets queuing at the interfaces.

Outgoing queues management is within the aim of the traffic control provided by Linux operating system. Linux Traffic Control tool can decide whether a packet must be queued or discarded (for example, when the outgoing queue reaches a given limit, or when the flow exceeds a certain rate). It can set the packet scheduling policy and the packet transmission delay.

The tc software, which implements the traffic control policy, has the following conceptual components: i) queuing disciplines (qdisc), ii) packet classes (class), iii) packet filters (filters) and iv) management policing. Each network interface has a queuing discipline that controls the enqueued packets. The simplest case is represented by a single queue served according the packet arrival order.

It is possible to define more elaborate disciplines by using the filter to identify different packet classes and to process them specifically, assuring, for instance, different priority levels to different classes as shown in Figure 19.10, for two queues.

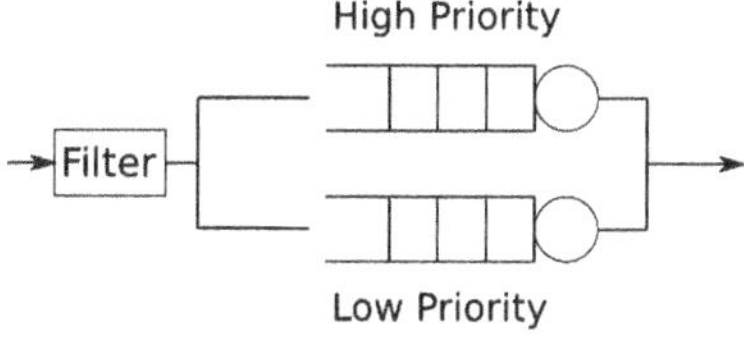

FIGURE 19.10

Example of queuing discipline with two different priority levels.

5.3 INTERCONNECTION BETWEEN SNS AND ENS

Taps and bridges used to connect VPCs inside the PC2 are also used to connect the ns-3 scenario inside the SNS with the ENS. The configuration of this connection is shown in Figure 19.11: differently from the VPCs interconnection between Edge and Core routers, in this case only one bridge interface is virtual (*tap1*), while the other is a physical PC1 interface (*eth1*). Consequently, *bridgeSNS* belongs both to the user and kernel space and communicates with the ns-3 scenario, inside the user space, by using *tap1* and with PC2 by using *eth1*, located in the kernel space. It is worth noting that the procedures necessary to create and configure *bridgeSNS* and *tap1* are the same as shown in Section 5.2.1. Furthermore, in this case, it is also necessary to disable the Ethernet filters inside the kernel space with the following code; when those filters are active no packet from ns-3 nodes can pass through *bridgeSNS*.

```
for f in /proc/sys/net/bridge/bridge-nf-*; do echo 0 > $f; done
```

The ns-3 scenario must be properly configured to enable the connection between simulated and emulated networks: a particular type of node, called a "Ghost node," is created inside the simulated scenario. The *InternetStackHelper* is installed in this node but it does not have an assigned IP address. Its function is to enable a dialogue between the ns-3 simulation and the real network acting as an alias of *tap1* inside the ns-3 framework. In other words, this node represents the interface of the *bridgeSNS* within the simulated network and it is connected to the *sgwPgw* node through a NetDevice (i.e., a simulated network interface) and a point-to-point link, as explained in Section 5.1. In this way, the traffic generated by the simulated UEs is received by the "Ghost node" and subsequently forwarded on *tap1*. Then it is transmitted to *PC2* by using *bridgeSNS* and *eth1*.

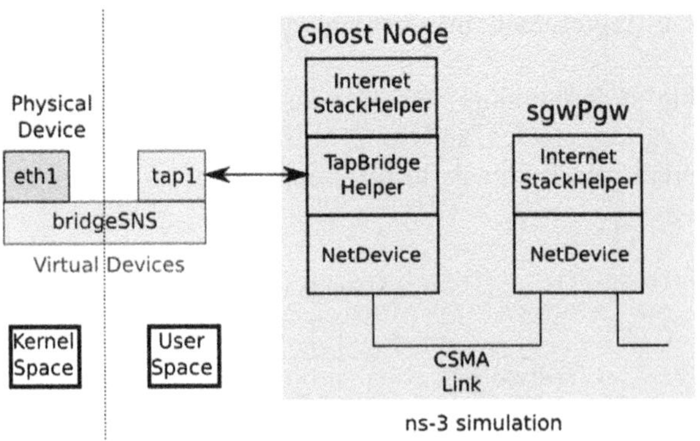

FIGURE 19.11

Interconnection between SNS and ENS by using a bridge inside PC1.

In more detail, the tap configuration is performed in the ns-3 scenario by using the class TapBridgeHelper as shown here:

```
TapBridgeHelper tapBridgeName;
tapBridgeName.SetAttribute ("Mode", StringValue ("UseBridge"));
tapBridgeName.SetAttribute ("DeviceName", StringValue ("tap1"));
tapBridgeName.Install (gatewayContainer.Get (0), netDeviceContainer.
  Get(0));
```

The first *TapBridgeHelper* parameter is *Mode*. It acts by following one of the following three modalities:

1. In the TapBridge ConfigureLocal mode the tap device is automatically created by ns-3 and the external PC is directly connected with the simulated network.
2. The UseLocal mode is similar to the previous one (i.e., ConfigureLocal) but, in this case, an existing tap device created and configured manually by the user in the host PC operative system is used.
3. UseBridge identifies a mode where the tap is configured by the user, as in UseLocal, but, in this case, the tap is used as the ns-3 interface of a virtual bridge. This mode is adopted in HySEP, as shown in Figure 19.11.

The name of the tap is another *TapBridgeHelper* parameter. It has to be set equal to the tap name created by the user in the PC host (i.e., tap1). Finally the *TapHelperBridge* is installed on the Ghost node as shown.

6 CONCLUSIONS AND FUTURE DEVELOPMENT

This chapter describes the developed tool called Hybrid Simulated-Emulated Platform (HySEP), which is composed of two different segments: *(i)* the Simulated Network Segment (SNS), which correspond to a PC where a ns-3 script is executed to simulate an LTE-EPC network and the *(ii)* the Emulated Network Segment (ENS), which is composed of two PCs: one implementing the functionalities of a DiffServ domain, by using Virtual PC as the node of the domain, and one acting as a remote host connected to the simulated UEs within the SNS.

The main contribution of this paper is the detailed description of all the elements that compose the platform. This description is divided into two parts: the theoretical presentation of the adopted technologies according to the standards defining them (i.e., LTE-EPC network and DiffServ protocol); and the implementation description of the platform. Detailed instructions and some adopted configuration scripts are reported to better explain how HySEP has been created. The chapter also describes how the two network segments are interconnected with each other and how the traffic flows pass from the emulated segment to the simulated one and vice versa. As a matter of fact, synchronization between the ns-3 simulation and the PC where the simulation is executed is fundamental to assure this communication.

HySEP is currently used for ongoing research activities.

Considering that not only the LTE-EPC network can be simulated by using ns-3, a possible development is represented by the integration of other wireless access technologies inside the simulated scenario such as Wi-Fi and Wi-MAX. Moreover, another possible goal is to integrate all these technologies in a unique simulative scenario and to develop a mobile terminal equipped with different network interfaces, one for each considered technology. In this way it is possible to test and validate different algorithms for the network selection problem, not only evaluating the performance locally, but also focusing on end-to-end results.

Other test activities are aimed at defining a set of scenarios that can be simulated through HySEP, at defining the maximum number of nodes as well as the maximum link data rate that can be supported by the platform in the SNS still maintaining real time execution and synchronization with the ENS.

REFERENCES

[1] Olsson M. SAE and the Evolved Packet Core: driving the mobile broadband revolution. Academic Press; 2009.

[2] Korowajczuk L. LTE, WIMAX, and WLAN network design, optimization and performance analysis. Chichester, West Sussex, UK: Wiley; 2011.

[3] Evolved Universal Terrestrial Radio Access (E-UTRA); packet data convergence protocol (PDCP) specification. 3GPP TS 36 323, ETSI; 2013.

[4] Evolved Universal Terrestrial Radio Access (E-UTRA); medium access control (MAC) protocol specification. 3GPP TS 36.321, ETSI, 2014.

[5] Universal Mobile Telecommunications System (UMTS); LTE; 3GPP evolved packet system (EPS); Evolved general packet radio service (GPRS) Tunneling protocol for control plane (GTPv2-C). 3GPP TS 29.274, ETSI, 2014.

[6] Blake S, Black D, Carlson M, Davies E, Wang Z, Weiss W. An architecture for differentiated services. RFC 2475, The internet engineering task force (IETF). [Online]. Available from: <http://www.rfc-editor.org/rfc/rfc2475.txt>; December 1998.

[7] Nichols K, Blake S, Baker F, Black D. Definition of the differentiated services field (DS Field) in the ipv4 and ipv6 headers. RFC 2474, The internet engineering task force (IETF). [Online]. Available from: <http://www.rfc-editor.org/rfc/rfc2474.txt>; December 1998.

[8] Nichols K, Carpenter B. Definition of differentiated services per domain behaviors and rules for their specification. RFC 3086, The Internet Engineering Task Force (IETF). [Online]. Available from: <http://www.rfc-editor.org/rfc/rfc3086.txt>; April 2001.

[9] Marchese M. QoS over heterogeneous networks. Chichester, UK: Wiley & Sons; 2007.

[10] Davie B, Charny A, Bennet J, Benson K, Boudec JL, Courtney W, et al. An expedited forwarding PHB (per-Hop Behavior). RFC 3246, The internet engineering task force (IETF). [Online]. Available from: <http://www.rfceditor.org/rfc/rfc3246.txt>; March 2002.

[11] Heinanen J, Baker F, Weiss W, Wroclawski J. Assured forwarding PHB group. RFC 2597, The internet engineering task force (IETF). [Online]. Available from: <http://www.rfceditor.org/info/rfc2597.txt>; June 1999.

[12] ns-3 Project. Network simulator 3 (ns-3). [Online]. Available from: <http://www.nsnam.org/>; 2015.

[13] Centre Tecnològic de Telecomunicacions de Catalunya (CTTC). LENA: LTE-EPC Network simulAtor. [Online]. Available from: <http://networks.cttc.es/mobile-networks/software-tools/lena/>; 2015.

[14] Oracle. VirtualBox [Online]. Available from: <https://www.virtualbox.org/>; 2015.

[15] Almesberger W, Salim JH, Kuznetsov A. Differentiated services on Linux. In: GLOBECOM 1999, GLOBECOM, vol. 1. 1999. p. 831–36.

[16] Hubert B. Linux advanced routing & traffic control [Online]. Available from: <http://www.lartc.org/> [accessed 30.01.15].

CHAPTER

A random access model for M2M communications in LTE-advanced mobile networks

20

Meriam Bouzouita[1,2], Yassine Hadjadj-Aoul[1], Nawel Zangar[2], Sami Tabbane[2], and César Viho[1]

[1]University of Rennes, Rennes, France

[2]Higher School of Communication of Tunis (SUPCOM), Ariana, Tunisia

1 INTRODUCTION

Machine type communication (MTC), also known as machine-to-machine (M2M) communication, is an ongoing standardized form of communication in mobile networks [1,2]. In opposition to the former modes of communications, M2M enables devices or machines and also terminals to communicate directly without involving the human factor. Indeed, M2M allows an automated communication between remote machines and central management applications, providing real-time control and monitoring. It constitutes an emerging market with predictions of more than 500 billion M2M devices embedded in the near future [3].

Cellular mobile networks were formally designed to support classical human-to-human (H2H) applications, which are very different from M2M applications with their particular characteristics [1,4]. Thus, mobile networks should be accommodated to support such devices by efficiently managing the different supported applications, while maintaining, at the same time, the same guarantees for the former communications.

The support of M2M communications in mobile networks, and particularly in Long Term Evolution-Advanced (LTE-A) networks [5], comes with a set of challenges and open issues for network operators (NO). In fact, the increasing number of MTC devices, with their stringent constraints in terms of energy and traffic pattern (i.e., typically represented by a small portion of data and a huge amount of signaling traffic), may reduce the efficiency of such networks and may cause severe congestion in the different levels of the cellular networks: radio access network (RAN), core network (CN) and signaling network. As a result, this will impact the user data and control planes. RAN congestion happens when a high number of MTC devices try to connect to the network at almost the same time. The CN congestion can be caused by simultaneous transmissions from a large number of MTC devices. Whenever congestion occurs, this may degrade performance of the networks by causing intolerable delays, packet loss and even service unavailability.

Different solutions have been proposed in the literature to deal with the network overload of LTE networks [6,7]. The European Telecommunications Standards Institute (ETSI) [8] and the Third Generation Partnership Project (3GPP) [9], introduced a list of solutions to overcome such problems. The Access Class Barring (ACB) scheme is a key technique used by many to alleviate the congestion of MTC devices [2]. The ACB procedure is based on an access probability called the ACB factor. If the value of this factor is set properly, this may help to insure both small contention on radio resources and low access delays for MTC devices. In the present chapter, we propose a novel dynamic model to help in computing such a factor for M2M applications in LTE-A networks.

The rest of this chapter is organized as follows. In the second section, we give an overview of MTC applications and their generic architecture. In the third section, we list some congestion and overload mechanisms. We devote the fourth section to the random access process and Physical Random Access Channel (PRACH) resources allocation. The fifth section is dedicated to a brief description of RAN overload mechanisms and specifically the ACB scheme. Then, the dynamic model and the proposed method to compute the ACB factor are presented in the sixth section. The performance evaluation of the classical random access process and the proposed solution are described in the seventh section, and finally we conclude the chapter.

2 M2M APPLICATIONS OVERVIEW

As mentioned previously, MTC communications involve many devices such as computers, embedded systems, sensors, meters, mobile devices, etc. These devices are generally the origin of data packets, which are transferred, through a network, to an application-called server or an application server (AS). As mentioned in [2], a simple architecture of M2M applications is mainly composed of three essential domains:

- MTC Device Domain: includes the MTC devices.
- Network Domain: transports messages or events between MTC devices and MTC servers.
- MTC Application Domain: where the MTC servers are localized. Theses servers are under the control of the NO.

Within the newest 3GPP agreement, two new entities were introduced (see Figure 20.1). The MTC Interworking Function (MTC-IWF) [2], which helps in

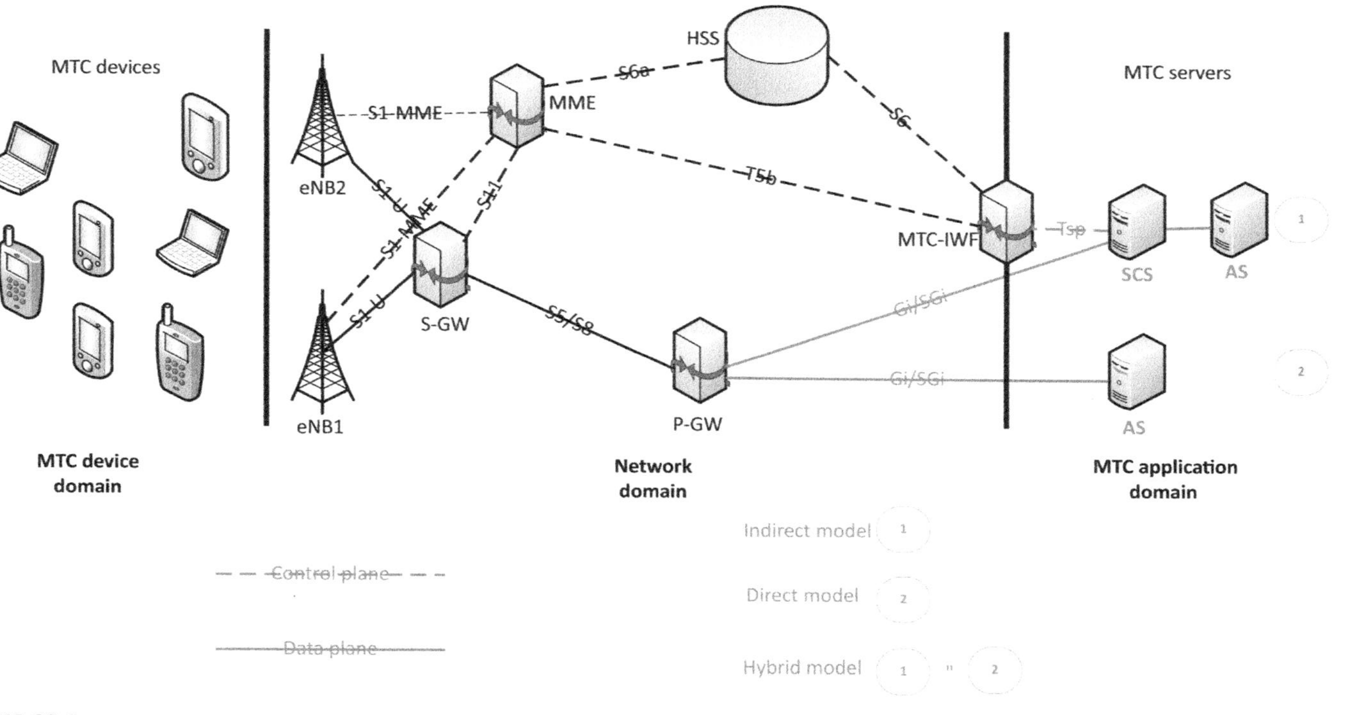

FIGURE 20.1

MTC architecture and communications scenarios.

facilitating the interworking between the MTC servers and the mobile network by authenticating MTC servers and by supporting the control plane messaging from and to these servers. The second entity is the Services Capability Server (SCS) [10] that provides optional services between the MTC devices and the MTC applications server in the external network.

Based on the SCS's provider, different architectural scenarios were proposed for M2M applications [10]. In the first model, called the "direct model," the application server connects directly to the mobile network without passing through the SCS. In the second model, called the "indirect model," the server connects to the network through the SCS if it requires additional services for MTC devices. A third model, denoted the "hybrid model," combines the direct and indirect models.

3 CONGESTION CONTROL FOR M2M APPLICATIONS OVER LTE NETWORKS

Congestion is a fundamental problem in mobile networks; it usually occurs when the aggregated demand for bandwidth exceeds the available link capacity [11]. However, when supporting M2M applications, new forms of congestion arise. Indeed, the congestion for M2M applications may concern the data plan, but also the control plan, which represents the main source of congestion in mobile networks. In order to combat such congestion, LTE network nodes should be able to intelligently reject attach and connection requests without impacting either non-MTC traffic or traffic from other MTC applications that are not causing any problem. To manage the overload and congestion network caused by MTC applications, many solutions have been developed, as detailed in [2] and [11]. These solutions can be classified as follows:

- Access Control by the RAN (i.e., eNB)
- RRC connection and channel requests' rejections by the RAN
- Connection requests' rejection by the CN (e.g., MME)

When the RAN handles overload situations, an internal congestion alarm can trigger the broadcasting of an access control message to the MTC devices to avoid further access to the network. The CN may also initiate the process by sending a notification message called OVERLOAD START to the RAN node. The OVERLOAD START message indicates the barring information (barring factor, barring time, M2M group to be blocked, etc.). Thus, time tolerant MTC devices can be treated as low priority devices, which may lead to access rejection with an extended waiting time. Indeed, in the case of massive simultaneous connection requests, it is of benefit to reject connection requests as early as possible during the access procedure to optimize resources' utilization. This is the only solution preventing signaling messages sending from MTC devices.

When the MME handles the rejection of connection/attach requests, the targeted devices can belong to a particular Access Private Network (APN), an MTC group or an MTC device access priority. Indeed, MME can perform rejection by providing a back-off time to the MTC devices in order not to re-initiate a connection/attach request immediately after a first rejection. Also, when the number of MTC devices becomes large, the back-off time should be randomized to avoid synchronous re-initiating access requests.

Access control by RAN is considered to be the most beneficial, due to the fact that there is no wasting of resources. Thus, in this chapter we focus on this mechanism by proposing a solution allowing the calculation of the barring factor.

4 RANDOM ACCESS PROCEDURE AND PRACH RESOURCES

4.1 RANDOM ACCESS PROCEDURE

The rapid growth of the number of MTC devices complicates considerably their support in LTE-Advanced Networks and may significantly increase the probability of congestion and access failure when performing the random access (RA) procedure. In fact, to attach or connect to the network, every MTC device should first accomplish the random access procedure. The RA is a process initiated by terminals in the idle state to request uplink radio resources required to send data. In LTE networks, this procedure can be achieved in two ways: contention-based and contention free-based access.

M2M devices usually perform such access in a contention manner. This contention-based RA procedure consists of four essential steps (or exchanged messages) and uses an uplink channel called the Physical Random Access Channel (PRACH) [12]. These different steps are listed below (see Figure 20.2):

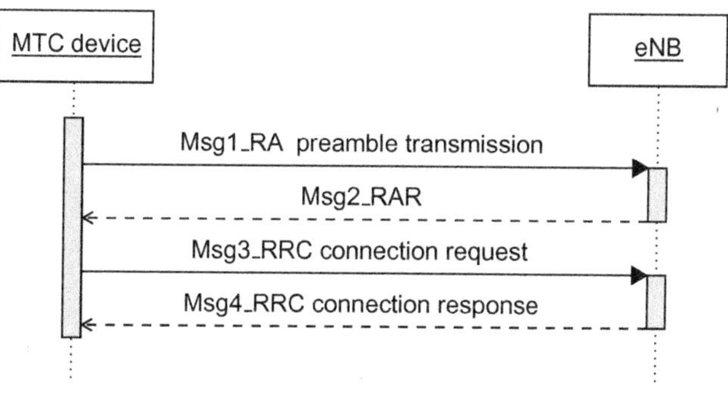

FIGURE 20.2

Contention-based RA procedure.

- **Step 1:** Random access preamble transmission: in the first message, the MTC device chooses randomly a sequence code called a preamble, among the set of available preambles. This preamble is then transmitted using the PRACH to the eNodeB (eNB). In each PRACH, if two or more MTC devices select the same preamble, the eNB will be unable to identify the initiator of the RA since the devices don't indicate their own identities in this request. Thus, eNB cannot decode any of these preambles and a collision will happen.
- **Step 2:** Random access response (RAR): when the preambles are detected, eNB assigns uplink resources to the concerned MTC devices and sends a RAR message using the Physical Downlink Shared Channel (PDSCH). If the message is not received within a time window (i.e., RAR window), a collision event is generated by the nodes. These will retransmit their preambles after a random back-off time.
- **Step 3:** RRC (Radio Resource Control) connection request: during this step, the MTC device transmits its unique identity to the network using the Physical Uplink Shared Channel (PUSCH). This is done only if msg2 (i.e., RAR message) contains the RA preamble that corresponds to the preamble transmitted in the first message.
- **Step 4:** RRC connection Response: using the PDSCH, the network sends a contention resolution message to the device. In fact, after receiving msg3 and using msg4, eNB confirms that the connection is established successfully and thus ends the RA procedure.

Thus, whenever two or more MTC devices choose the same preamble, a collision will happen. Consequently, when the number of terminals trying to access the network at the same time is larger, this leads to an excessive level of PRACH congestion and a low RA success probability.

4.2 PRACH RESOURCES

In LTE systems, every cell contains 64 preambles. These preambles are available for random access but only 54 preambles are reserved for contention based RA [13]. As mentioned previously, the RA process takes place within the PRACH. The PRACH or Physical Random Access Channel is a time frequency resource block (RB) reserved by the eNB. In LTE networks, and for each radio frame whose duration is equal to 10 ms, a limited number of PRACH resources, called also Random Access Opportunities (RAO), are available. This number depends on a parameter called the *PRACH configuration index*.

For example, and as you can see in Figure 20.3, if this parameter is equal to 6, then the number of available PRACH occasions within a radio frame is 2. As the number of preambles reserved for the contention mode is equal to 54 per one PRACH opportunity (i.e., during 1 ms). Then the total number of preambles (or RAOs) is equal to 54*2 = 108 [14].

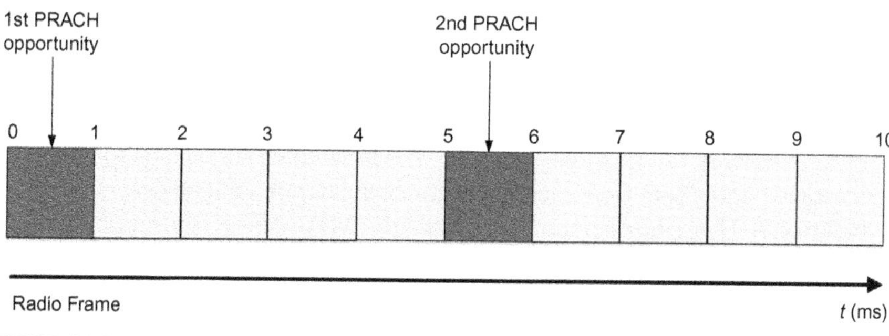

FIGURE 20.3

PRACH opportunities (case PRACH configuration index = 6) here.

5 RAN OVERLOAD CONTROL

5.1 RAN OVERLOAD MECHANISMS

To reduce the PRACH overloads, the 3GPP specified a list of different PRACH overload resolution methods to improve MTC device support in LTE-Advanced Networks. These solutions may help such networks to meet performance requirements even under excessive MTC loads [14,15]. It also helps in avoiding resources' wastage since every failed attempt consumes radio resources. In the following paragraphs, we will briefly describe such solutions.

- ***Separation of RACH resources:*** consists of affecting separate resources for MTC devices and human to human (H2H) devices. The separation can be achieved by either separating the preambles or by allocating different time-frequency resources or resource blocks (RBs). Two different approaches considering such solution are proposed in [16]. In the first scheme, the preambles are split into two groups, one for H2H devices and another for M2M devices. In the second scheme, resources are also divided into two groups: one dedicated to H2H devices and the other shared between MTC and H2H devices.
- ***Dynamic allocation of RACH resources:*** eNB may allocate additional resources for MTC devices in case of a huge load. In [17], an algorithm was proposed to dynamically change the number of RA slots according to the channel load. This solution might be effective but it depends on limited availability of additional resources.
- ***MTC-specific back-off scheme:*** the network can set the back-off time to a large value in order to delay the RA reattempts of some MTC devices, for example after a first failed access. It is expected that an extended back-off time can alleviate congestion between devices and, thus, facilitating overload resolution.
- ***Slotted access:*** the network distributes the access of MTC devices in dedicated access slots. A device is allowed to access the network during a

specified interval called AGTI (Access Grant Time Interval) and will be blocked during a FTI (Forbidden Time Interval).
- ***Pull based schemes:*** in this method, the RACH procedure is initiated by the eNB. This is feasible only when the eNB is aware of the future transmissions of MTC devices. In fact, only devices whose identities are included in the message paging sent from the base station can attempt random access procedure. Then, the operation of a particular MTC device can start only after receiving a paging from the network. The transmission can start immediately or after a back-off time depending on the paging message.
- ***Grouping of MTC devices:*** according to [2], MTC groups can be formed to help the radio resources allocation and decrease the redundant signaling to avoid congestion. In [18], a massive access management scheme, based on group optimization, was proposed to efficiently manage massive accesses on the air interface. In fact, a massive number of MTC devices request to attach to the eNB all at once. To deal with this problem, it is of benefit to group M2M devices according to their QoS characteristics and requirements.
- ***Access Class Barring (ACB) scheme:*** it is used for barring, or not, the access of an MTC specific class. To do it, an ACB factor and a barring time are defined for each MTC access class. These parameters determine if a device is blocked for a certain time from accessing the cell or not. This may effectively reduce the collision probability of transmitting a bulk of preambles over the same radio resources.

In this chapter, we proposed a RAN overload control method based on the ACB scheme. Thus, in the following subsection, we explain in more detail the principles of such an approach.

5.2 ACB PROCEDURE

As mentioned previously, the ACB concept is a solution adopted to deal with the RACH overload and to control the access attempts of mobiles over the radio interface. Originally, the ACB scheme defined 16 access classes (AC). AC in the interval 0–9 represent normal MTC terminals, AC equal to 10 represents the emergency case, and finally AC in the interval 19 specified some services with high priority requirements.

Initially, eNB broadcasts an AC barring information, that is, *ac_BarringInfo* via the System Information Block (SIB2). The AC barring information contains the following three parameters: [19]

- ACB factor p labeled *ac_BarringFactor;*
- Access barring time labeled *ac_BarringTime*;
- *Bit-string.*

The *ac_BarringFactor* $p \in [0, 1]$ determines the barring probability of a device, whereas the *ac_BarringTime* is used to determine the duration (i.e., *Tbarring*) before retrying the RA in case the device wasn't allowed to access the PRACH during the first trial. These two parameters are applied to all AC in 0–9 in the same way. SIB2 also contains the information about the allowed and barred classes.

Before performing the RA, the device should pass the ACB check. It initially draws a random number q. If this number is less than the ACB factor p, the device starts the random access procedure; otherwise the device will be blocked during a barring time. Consequently, the ACB method can help to alleviate the level of congestion by adjusting the value of p according to the load's level. However, a low value of p leads to high access delays and radio resources' under-utilization; whereas a high value of p leads to heavy traffic load. Thus, it will be interesting to design a new solution allowing computing of the appropriate and adaptive value of p.

Using the *ac_BarringTime*, the value of *Tbarring* is calculated as follows:

$$Tbarring = (0.7 + 0.6^{*} rand)^{*} ac\_BarringTime$$

where *rand* is a random number generated by the MTC device after passing a first failed ACB check and before a second attempt. The values of *ac_BarringTime* can range from 4 s to 512 s.

AC in the interval 11–15 use a *bit-string* configuration in which every bit indicates if the corresponding AC is barred or not. For AC equal to 10, SIB2 contains information related to that access class [19]. Figure 20.4 illustrates the ACB mechanism followed by a random access attempt.

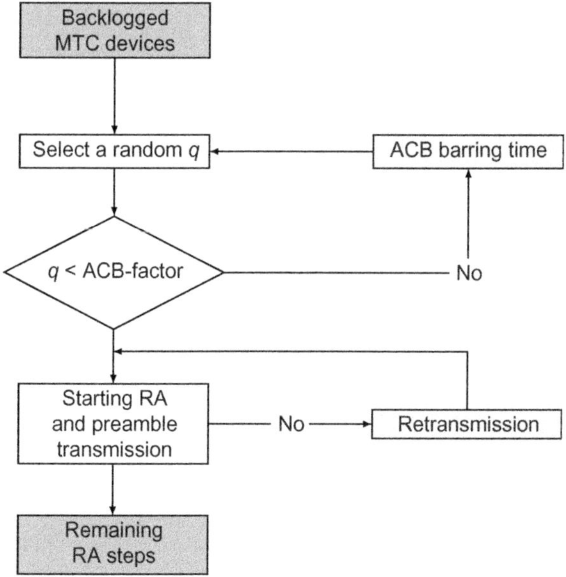

FIGURE 20.4

ACB procedure and RA diagram.

Many works have dealt with the ACB procedure. In [20], a cooperative ACB scheme was proposed. All ACB parameters in each Base Station (BS) shall be jointly decided by all BSs rather than individually decided by one BS, in order to achieve more stabilization. In [21], a prioritized random access scheme was proposed. This PRA architecture is composed of a virtual resources allocation with class dependent back-off procedures and dynamic access barring.

In [22], the author proposed an algorithm to adaptively change the ACB factor if the number of backlogged MTC devices is not known. On the contrary, if the eNB has information of backlogged devices, a method to determine the optimal ACB factor was proposed. In [23], the authors proposed to adopt an adaptive scheme. In other words, eNBs can dynamically adjust ACB factors based on network loads rather than using fixed ACB probabilities, which are not optimal. In fact, high values of barring factor may increase access latency of devices, whereas low values may lead to an excessive resource contention. The simulation results of such work showed that adaptive ACB scheme could gain better performances than those obtained with a static ACB scheme.

In this chapter, we have proposed a dynamic model to compute the value of the ACB factor.

6 RANDOM ACCESS MODEL FOR M2M APPLICATIONS

6.1 MODEL

Our model for M2M device random access, in LTE-Advanced networks, is influenced by the fluid model proposed in [24]. The proposed system model uses the following parameters and quantities:

$x_3(t)$: average number of MTC devices that succeed the RA procedure at time t.
λ: the arrival rate of MTC devices in terms of number of devices per second.
θ_1: the rate (percentage) of ACB failure.
θ_2: the rate (percentage) of RA failure (collision and retransmission).
μ: the rate (percentage) of MTC departure after performing RA successfully.
N: total number of radio resources (preambles) available during one RA time slot.
p: ACB factor (percentage).

Our system model is illustrated in Figure 20.5.

Based on this model, the evolutions over the time of $x_1(t)$, $x_2(t)$ and $x_3(t)$ are given by:

$$\frac{dx_1}{dt} = \lambda - F_1(x_1) - \theta_1 x_1 \tag{20.1}$$

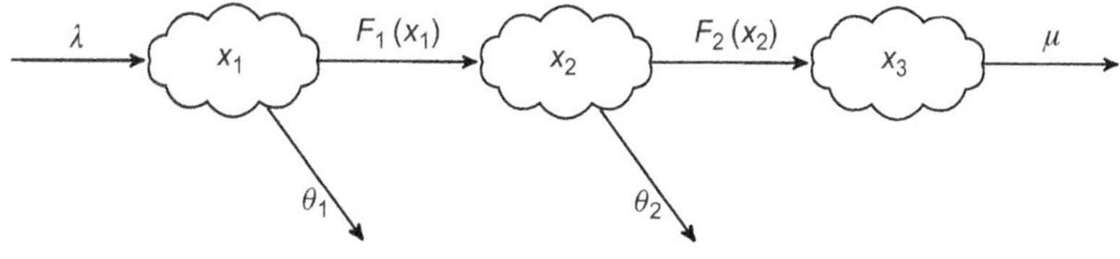

FIGURE 20.5

System model.

$$\frac{dx_2}{dt} = F_1(x_1) - F_2(x_2) - \theta_2 x_2 \tag{20.2}$$

$$\frac{dx_3}{dt} = F_2(x_2) - \mu x_3 \tag{20.3}$$

along with the constraint that $x_1(t)$, $x_2(t)$ and $x_3(t)$ should be nonnegative. $F_1(x_1)$ denotes the average number of MTC devices that passed the ACB procedure successfully. $F_2(x_2)$ denotes the average number of MTC devices that transmitted their preambles successfully. This is achieved if and only if every MTC device chooses one preamble that isn't chosen by another MTC device.

Let's now compute the values of $F_1(x_1)$ and $F_2(x_2)$. Denote X as the number of backlogged users and, among them, only Z users pass the ACB process successfully. Then, the probability to have Z users who access successfully is given by the following formula:

$$Prob(Z = i|X = n) = \binom{n}{i} p^i (1-p)^{n-i}$$

By applying Newton's binomial theorem [25], the average number is equal to: $E(Z = i|X = n) = np$. Thus, we obtain the following result: $F_1(x_1) = px_1$

In this case, x_2 devices contend for N preambles. It is similar to placing x_2 balls into N bins. Thus, the number $F_2(x_2)$ of MTC devices which transmitted their preambles successfully is equal to having only $F_2(x_2)$ bins and each one of them contains just one ball.

Denote:

$$I_j = \begin{cases} 1 & \text{if bin } j \text{ contains one ball} \\ 0 & \text{otherwise} \end{cases}$$

Y, the expected number of bins that contain just one ball, can be expressed as follows: $Y = \sum_{j=1}^{N} E(I_j)$, where, $E(I_j)$ corresponds to the probability that bin j contains exactly one ball.

$$E(I_j) = C_{x_2}^{1} \frac{(N-1)^{x_2-1}}{N^{x_2}}$$

Thus, $Y = x_2 \left(1 - \frac{1}{N}\right)^{x_2 - 1}$

Consequently, $F_2(x_2)$ is equal to:

$$F_2(x_2) = x_2\left(1-\frac{1}{N}\right)^{x_2-1} = x_2 e^{(x_2-1)Ln\left(1-\frac{1}{N}\right)}$$
$$\simeq x_2 e^{-(x_2-1)/N}$$

Finally, after approximation we obtain the following expressions for $F_2(x_2)$:

$$F_2(x_2) = x_2 e^{-(x_2-1)/N}$$

Consequently, Equations (20.1), (20.2) and (20.3) become:

$$\frac{dx_1}{dt} = \lambda - px_1 - \theta_1 x_1 = f_1(x_1, x_2, x_3, p) \tag{20.4}$$

$$\frac{dx_2}{dt} = px_1 - x_2 e^{-(x_2-1)/N} - \theta_2 x_2 = f_2(x_1, x_2, x_3, p) \tag{20.5}$$

$$\frac{dx_3}{dt} = x_2 e^{-(x_2-1)/N} - \mu x_3 = f_3(x_1, x_2, x_3, p) \tag{20.6}$$

6.2 STEADY-STATE ANALYSIS

During the rest of this section, we will study the stability of our system. To analyze the stability, we first linearize the system. In fact, linearization allows reducing the system's complexity and assessing the local stability of an equilibrium point of a system.

In the case of our system, that linearization is done around the equilibrium point $(\overline{x}_1, \overline{x}_2, \overline{x}_3, \overline{p})$, where, $\overline{x}_1, \overline{x}_2, \overline{x}_3$, and $\overline{p}$ are the values of x_1, x_2, x_3, and p, respectively, in the steady state.

To study the steady-state performance of our system, we let:

$$\frac{dx_1}{dt} = \frac{dx_2}{dt} = \frac{dx_3}{dt} = 0$$

Then, we obtain

$$\lambda - p\overline{x}_1 - \theta_1 \overline{x}_1 = 0$$
$$p\overline{x}_1 - \overline{x}_2 e^{-(\overline{x}_2-1)/N} - \theta_2 \overline{x}_2 = 0$$
$$\overline{x}_2 e^{-(\overline{x}_2-1)/N} - \mu \overline{x}_3 = 0$$

where $\overline{x}_1$, $\overline{x}_2$ and $\overline{x}_3$ are the equilibrium values of x_1, x_2, and x_3 respectively.

Table 20.1 Simulation Parameters

Parameters	Designations	Values
N_SEED	Number of seeds used for Monte Carlo simulation	8 values from the following set: {10,100,1234,6753,10000, 16000,50000,100000}
N_M2M_MAX	Maximum number of devices considered	54
N_M2M_MIN	Minimum number of devices considered	10
N_EXPERIMENTS	Number of realized experiments for the Monte Carlo simulation	10000
N_PREAMBLES	Number of available preambles	54

To simplify the resolution of such a system of equations, we suppose that $\theta_2 = 0$, which represents one of the objectives of such a system. Then, we easily obtain:

$$\overline{x}_1 = \frac{\lambda}{p + \theta_1} \tag{20.7}$$

$$\overline{x}_3 = \frac{p\lambda}{\mu(p + \theta_1)} \tag{20.8}$$

Getting $\overline{x}_1$ and $\overline{x}_3$, it will not be evident to get $\overline{x}_2$. However, it can be obtained using simulation. Thus, to find the optimal number of devices, x_2, we used the Monte Carlo simulation method. Such experiments are a broad class of computational algorithms that rely on repeated random sampling to obtain numerical results; typically one runs simulations many times over in order to obtain the distribution of an unknown probabilistic entity.

To do that simulation, we adopt a number of simulation parameters under a C-based discrete event simulator. The different parameters are specified in Table 20.1.

To find the optimal x_2, we focus on the evaluation of two metrics: the mean and the variance of the number of devices. Varying the number of M2M devices between N_M2M_MAX and N_M2M_MIN, we evaluate the mean and the variance obtained. Many seeds were tested and the results were similar. Figure 20.6 shows the obtained results.

As shown in this figure, optimal values of the mean and the variance are obtained for a number of M2M devices equal to 53 (i.e., $N - 1$). This value will be considered in the rest of the chapter. The value of $\overline{x}_2$ is the optimal value we should have using the appropriate ACB factor.

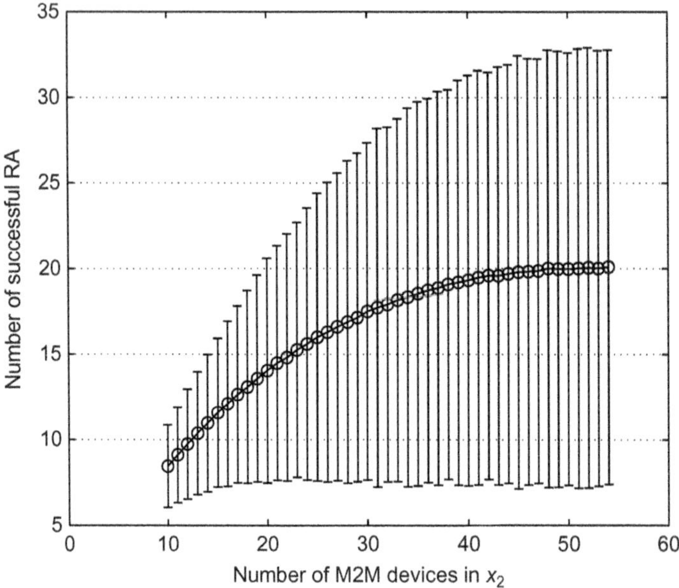

FIGURE 20.6

Successful RA.

6.3 LOCAL STABILITY

We first start, in this subsection, by linearizing our model, expressed in the system (Equations 20.4–20.6), around the equilibrium point $\left\{\bar{x}_1 = \frac{\lambda}{p+\theta_1}, \bar{x}_2 = N-1, \bar{x}_3 = \frac{p\lambda}{\mu(p+\theta_1)}\right\}$ obtained in the previous section. We, then, analyze its stability around this point.

Note $X = \begin{pmatrix} x_1 \\ x_2 \\ x_3 \end{pmatrix}$ the state vector of the system and p the entry variable or the controller output of the system. To analyze the stability of such system, we first should rewrite the previous system in the state space form:

$$\begin{cases} \dot{X} = AX + B \\ Y = CX \end{cases} \tag{20.9}$$

The matrices A, B and C are respectively the system matrix, the control matrix and output matrix [26]. Matrix A relates how the current state affects the state change $\dot{X}$. It is also called the Jacobian Matrix, which helps to determine if the system is stable or not. To analyze the system stability, we compute the eigenvalues of this Jacobian matrix. If the eigenvalues are all negatives, then the system is said to be stable.

The matrix B determines how the system input affects the state change and finally the matrix C determines the relationship between the system state and the system output.

Matrices A, B and C are given as follows:

$$A = \begin{pmatrix} \frac{\partial f_1}{\partial x_1} & \frac{\partial f_1}{\partial x_2} & \frac{\partial f_1}{\partial x_3} \\ \frac{\partial f_2}{\partial x_1} & \frac{\partial f_2}{\partial x_2} & \frac{\partial f_2}{\partial x_3} \\ \frac{\partial f_3}{\partial x_1} & \frac{\partial f_3}{\partial x_2} & \frac{\partial f_3}{\partial x_3} \end{pmatrix}, \quad B = \begin{pmatrix} \frac{\partial f_1}{\partial p} \\ \frac{\partial f_2}{\partial p} \\ \frac{\partial f_3}{\partial p} \end{pmatrix}, \quad C = \begin{pmatrix} 1 & 0 & 0 \\ 0 & 1 & 0 \\ 0 & 0 & 1 \end{pmatrix}, \text{ and } \dot{X} = \begin{pmatrix} \dot{x}_1 \\ \dot{x}_2 \\ \dot{x}_3 \end{pmatrix}$$

Then,

$$A = \begin{pmatrix} -\overline{p} - \theta_1 & 0 & 0 \\ \overline{p} & -e^{\frac{1-\overline{x}_2}{N}}\left(1 - \frac{\overline{x}_2}{N}\right) - \theta_2 & 0 \\ 0 & e^{\frac{1-\overline{x}_2}{N}}\left(1 - \frac{\overline{x}_2}{N}\right) & -\mu \end{pmatrix} \text{ and } B = \begin{pmatrix} -\overline{x}_1 \\ \overline{x}_1 \\ 0 \end{pmatrix}$$

As some eigenvalues of A are negative, the system represented in (20.9) is unstable. This means that a controller is necessary to allow the convergence towards such desired equilibrium points. Before doing that, let analyze now the controllability and the observability of our system model. The controllability and the observability represent, indeed, two major concepts of modern control system theory [27]. R. Kalman introduced these two concepts in 1960 [28]. A system is said to be controllable if we will be able to do whatever we want with the given dynamic system under the control input (i.e., we will be able to change the system states by changing the system input). On the other hand, a system is considered observable if the system states can be deduced from the system observation.

Formally, a system with a number n of states is said to be observable if $rank[C \quad CA \quad CA^2 \cdots CA^{n-1}]^T = n$, where A and C are the matrix defined previously.

Concerning the controllability, a system with n states is considered controllable if rank $[B \quad AB \quad A^2B \cdots A^{n-1}B] = n$.

It can be easily verified that the system described in (Equations 20.4–20.6) is controllable and observable. This means that a feedback controller can be designed for this system.

6.4 CONTROLLER DESIGN

In the following, the regulation of the appropriate ACB factor guaranteeing a number of devices around the desired value ($\overline{x}_2 = N - 1$) is achieved using a discrete PID (Proportional Integral Derivative) controller [29]. The reduced complexity of the PID controller and its efficient design, even in some classes of nonlinear systems, allow this controller to be one of the most common controllers used.

The discrete PID controller can be written as follows [29]:

$$u(n) = K_p e(n) + K_i \sum_{k=0}^{n} e(k) + K_d(e(n) - e(n-1)) \tag{20.10}$$

where n, u, e, K_p, K_i and K_d represent respectively the discrete step at time t, the controller output, the difference between the measured value and the reference value (the objective or the set point value), the proportional gain, the integral gain and the derivative gain.

There are several methods used for tuning the PID parameters in order to get the ideal response of the system. We considered the Ziegler-Nichols method, which has proven to be efficient for many problems [30]. First of all, derivative and integrative terms are set to zero and proportional gain is increased until a stable oscillation around the set point ($\overline{x}_2 = N - 1$) is obtained on the output system. Once the maximum gain "K_c" is achieved and the oscillation period "T_c" is obtained, we can easily calculate the proportional, integrative and derivative gains using the following equations:

$$\begin{cases} K_p = 0.6K_c \\ K_i = 2K_p/T_c \\ K_d = K_p T_c/8 \end{cases} \tag{20.11}$$

7 PERFORMANCE EVALUATION

7.1 SIMULATION PARAMETERS

Having described the details of our proposed ACB scheme and to verify the effectiveness and the accuracy of the proposed model, we now direct our focus on evaluating its performance using computer simulations using ns-3 [31].

In order to evaluate the system performance, we assume that MTC devices access the network following a Beta-based traffic. That type of traffic models a sudden M2M traffic surge where a large number of MTC devices access the network in a synchronized manner [32].

We assume that a total number (N_{MAX}) of MTC devices activate between $t = 0$ and T with the time limited beta distribution described in Equation (20.12):

$$f(t) = \frac{t^{\alpha-1}(T-t)^{\beta-1}}{T^{\alpha+\beta-1}\ Beta(\alpha,\beta)}; \tag{20.12}$$

where $\alpha > 0$, $\beta > 0$, *and* $Beta(\alpha, \beta)$ is the beta function with parameters α and β.

Then, the number of M2M arrivals (i.e., access intensity) in the ith access opportunity is given by:

$$access\ intensity(i) = N_{MAX} \int_{t_i}^{t_{i+1}} f(t)dt; \tag{20.13}$$

Table 20.2 Simulation Parameters

Parameters	Settings
Simulation Time / Distribution period (T)	10 s [32]
Number of eNBs	1
Cell bandwidth	5 MHz
N: Total number of preambles	54
Maximum number of preamble retransmissions	10
Arrival distribution	Beta distribution over T
Beta function parameters	$\alpha = 3$; $\beta = 4$ [32]
Total number of MTC devices	5000; 10000
Interarrival interval	1 frame (10 ms)
ac_BarringTime	4 s
Processing_Time	1 ms

where t_i is the time of the ith access opportunity.

The general parameter settings are summarized in Table 20.2.

To validate our model while evaluating the performance of ACB scheme, we have shown the results in terms of collision probability, success probability, number of simultaneous RACH attempts, number of successful ACB tests per second and number of devices abandoning the system after reaching maximum preamble retransmissions. Two different scenarios are considered: a first one where no RAN overload control is applied (e.g., ACB control) and a second one where devices should pass the ACB test before performing random access.

Let's now present methods for computing analytically collision and success probability. Suppose that there are N available preambles in each RA opportunity and x MTC devices present during this opportunity (x includes new arrivals and preambles retransmissions). The success probability for MTC devices (i.e., probability that a given preamble is chosen by only one user) is, as described in Section 6, equal to $F_2(x)$. Then,

$$P_{success}^{M2M} = xe^{-(x-1)/N} \tag{20.14}$$

The idle probability means that no user chooses a given preamble. It is written as:

$$P_{idle}^{M2M} = \binom{x}{0} e^{-x/N} = e^{-x/N} \tag{20.15}$$

From (20.14) and (20.15), we can easily derive the collision probability as following:

$$P_{collision}^{M2M} = 1 - P_{success}^{M2M} - P_{idle}^{M2M} = 1 - xe^{-(x-1)/N} - e^{-x/N} \tag{20.16}$$

To validate the proposed model, we present in the next section a comparison of these values against simulation results.

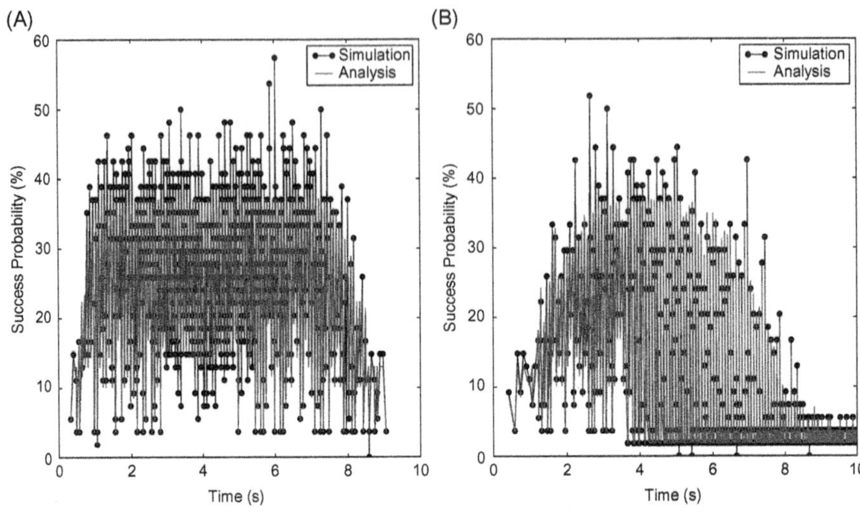

FIGURE 20.7

Success probability ($N_{MAX} = 10000$): (A) without ACB, (B) with ACB.

7.2 NUMERICAL RESULTS

In this section, we present the simulation results obtained for the number of simultaneous RACH attempts, the number of successful ACB tests per second, the number of abandons, the success probability and collision probability.

Figures 20.7 and 20.8 show the simulation results obtained respectively for success probability and collision probability against those of theoretical models. It can be seen easily that simulation results match the theoretical ones whenever an ACB control is applied or not and thus validate our proposed model.

From Figure 20.8, we can also see that collision probability, for 10000 MTC devices, exceeds 80% when no RAN control is applied, whereas this probability remains close to 20% when applying an ACB control. This demonstrates the efficiency when using an ACB control process.

Figure 20.9 illustrates the number of simultaneous RACH attempts which includes new arrivals due to beta traffic and preamble retransmissions. This number increases considerably when the total number of devices increases. In fact, when $N_{MAX} = 10000$, the number of RACH attempts reaches 180 when no access control is applied. This is almost four times the total number of available preambles (i.e., 54), which results in a huge number of collisions (see Figure 20.8). At the same time, it can be clearly seen that when applying ACB control, this number remains close to the optimal value (i.e., 53) and doesn't exceed 60. Indeed, the M2M RACH attempts are

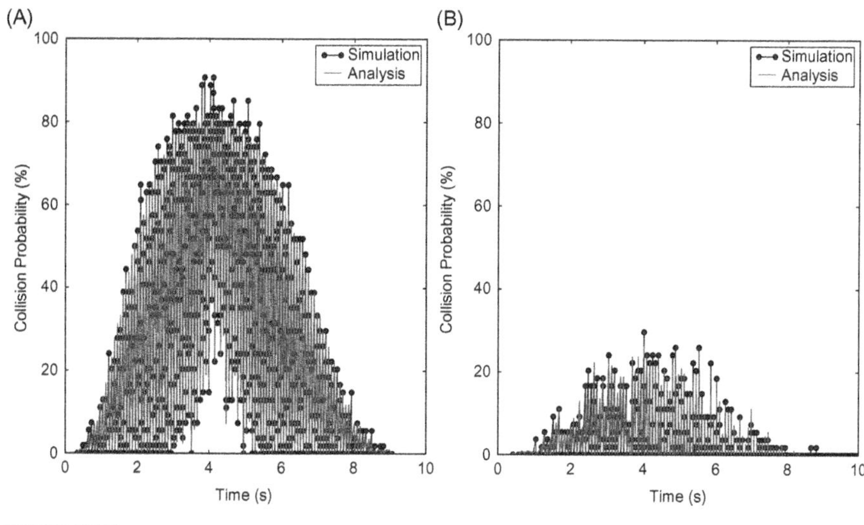

FIGURE 20.8

Collision probability ($N_{MAX} = 10000$): (A) without ACB, (B) with ACB.

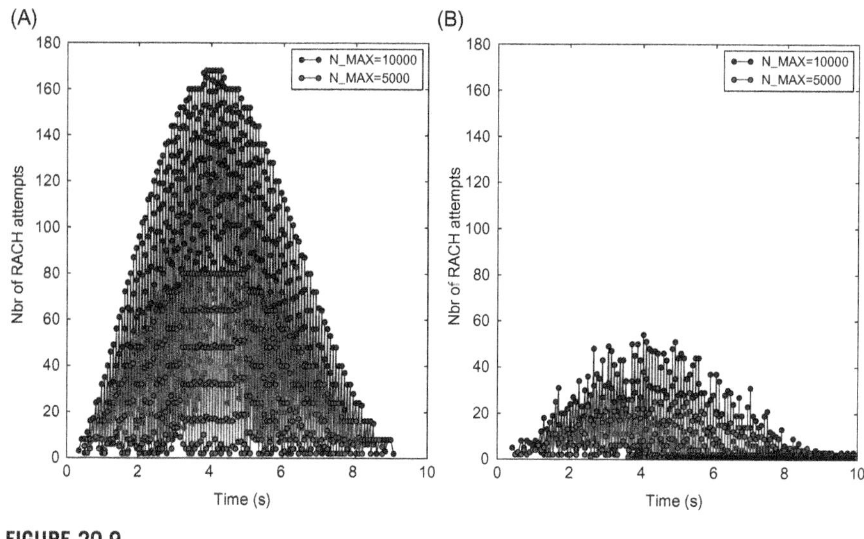

FIGURE 20.9

Number of simultaneous RACH attempts: (A) without ACB, (B) with ACB.

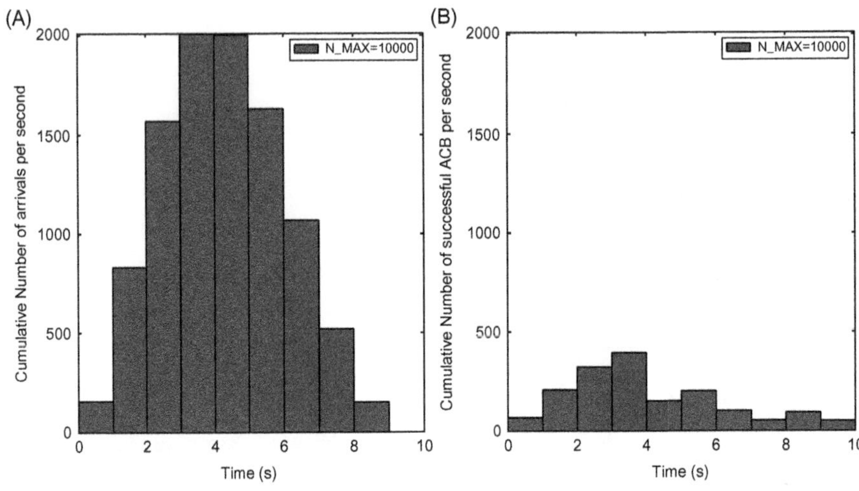

FIGURE 20.10

Number of beta arrivals per second vs. number of successful ACB tests per second.

spread in bigger intervals, which results in a reduced collision probability (Figure 20.8). This demonstrates the effectiveness of the proposed controller as it helps in regulating the ACB factor according to the overload level.

Figure 20.10 illustrates the number of successful ACB tests per second compared with the number of arrivals per second. We can clearly note that the number of MTC devices that pass the ACB test is largely lower than the one of the beta arrivals during one second. It results in reducing the number of RA trials and consequently low collision probability is observed (see Figure 20.8).

Figure 20.11 shows the number of MTC devices that abandon the system after reaching a maximum number of preamble retransmissions. This number is equal to zero when $N_{MAX} = 5000$; however, it increases considerably (reaches 60) when $N_{MAX} = 10000$. This is because of the huge number of simultaneous RACH attempts as described in Figure 20.9. In case of applying an ACB control, there is no abandon, which is one of the objectives of our proposed model.

8 CONCLUSION

In this chapter, we propose a solution to access network congestion, which is considered to be one of the most critical issues faced by the M2M communications. Such congestion is mainly caused by random accesses performed simultaneously. By using only the ACB principles, we proposed a novel fluid-based random access model for MTC devices in order to facilitate devices escaping from continuous congestions.

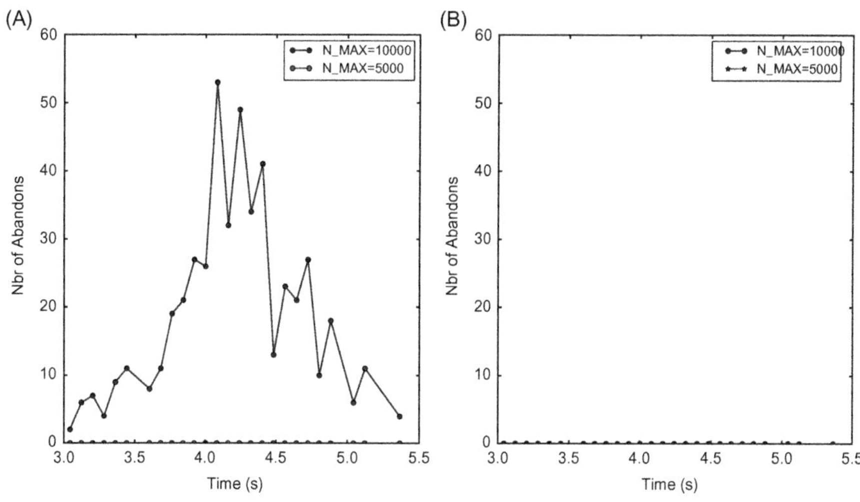

FIGURE 20.11

Number of abandons: (A) without ACB, (B) with ACB) here.

Using the Monte Carlo simulation, we find the optimal number of MTC devices that successfully passed the ACB procedure and wait for an RA attempt. After that, in order to regulate the adaptive ACB factor guaranteeing a number of devices around that reference value, we use a discrete PID controller.

The proposed model is validated using computer simulations under ns-3. Simulation results show that the proposed model can accurately predict the different considered performance metrics (i.e., collision/success probability, number of RACH attempts and number of abandons). Therefore, even when the arrival rate increases significantly, the number of RACH attempts remains close to the optimal value and the system reaches a zero number of abandons.

Finally, it should be stressed out that even if the tuning of the PID controller considers the real system model, exploiting the structure of the proposed model should allow enhancing further the performance of the M2M device RA process. This forms the focus of the authors' future research work.

REFERENCES

[1] European Telecommunications Standards Institute (ETSI). ETSI TS 102 689 V1.1.1 (2010-08), Machine-to-Machine communications; M2M service requirements.

[2] 3GPP TR 23.888 V1.6.0. Third generation partnership project; Technical specification group services and system aspects; System improvements for machine-type communications. (Release 11); 2011-11.

[3] Yankee Group. Mobile broadband connected future: from billions of people to billions of things. Commissioned by 4G Americas White Paper; 2011.
[4] 3GPP TS 22.368 V10.2.0. Third generation partnership project; Technical specification group services and system aspects; Service requirements for Machine-Type Communications (MTC); Stage 1. (Release 10); 2010-09.
[5] 3GPP TR 36.912 V11.0.0. Third generation partnership project; Technical specification group radio access network; Feasibility study for further advancements for E-UTRA (LTE-Advanced). (Release 11); 2012-09.
[6] 3GPP TS 36.413 V12.0.0. Third generation partnership project; Technical Specification group radio access network; Evolved universal terrestrial radio access network (E-UTRAN); S1 Application Protocol (S1AP). (Release 12); 2013-12.
[7] Alcatel-Lucent. The LTE network architecture: a comprehensive tutorial. Strategic White Paper. Available from: <www3.alcatel-lucent.com>; 2009.
[8] European Telecommunications Standards Institute (ETSI). Official web site: available from: <http://www.etsi.org/>; 2015.
[9] Third Generation Partnership Project (3GPP). Official web site: Available from: <http://www.3gpp.org/>; 2015.
[10] 3GPP TS 23.682 V12.0.0. Third generation partnership project; Technical specification group services and system aspects; Architecture enhancements to facilitate communications with packet data networks and applications, (Release 12); 2013-12.
[11] Ksentini A, Hadjadj-Aoul Y, Taleb T. Cellular-based machine type communication: overload control. IEEE Network 2012;26(6):54–60.
[12] 3GPP TS 36.321 V10.2.0. Medium Access Control (MAC) protocol specification; 2011-06.
[13] Sesia S, Toufik I, Baker M. LTE: the UMTS long term evolution: from theory to practice. 2nd ed. Wiley; 2011.
[14] 3GPP TR 37.868 V11.2.0. Study on RAN improvements for machine-type communications; 2011-09.
[15] Shao-Yu L, Kwang-Cheng C, Yonghua L. , Toward ubiquitous massive accesses in 3GPP machine-to-machine communications. IEEE Commun Mag; April 2011.
[16] Lee K-D, Kim S, Yi B. Throughput comparison of random access methods for M2M service over LTE Networks. In: Proceedings of the 2011 IEEE GLOBECOM Workshop; December 2011. p. 373–77.
[17] Lo A, Law YW, Jacobsson M. Enhanced random-access mechanism for massive machine-to-machine (M2M) communications. In: Proceedings of the 27th meeting of wireless world research forum; October 2011.
[18] Shao-Yu L, Kwang-Cheng C. Massive access management for QoS guarantees in 3GPP machine-to-machine communications. IEEE Commun Lett 2011;15(3).
[19] 3GPP TS 36.331 V10.2.0. Evolved universal terrestrial radio access (E-UTRA); Radio Resource Control (RRC); Protocol specification. (Release 10); 2011-06.
[20] Lien SY, Liau TH, Kao CY, Chen KC. Cooperative access class barring for machine-to-machine communications. IEEE Trans Wireless Commun 2012;11 (1):27–32.
[21] Cheng J-P, Han Lee C, Lin T-M. Prioritized random access with dynamic access barring for RAN overload in 3gpp lte-a networks p. 368–72 GLOBECOM workshops, 2011. IEEE; December 2011.

[22] Duan S, Shah-Mansouri V, Wong VWS. Dynamic access class barring for M2M communications in LTE Networks, submitted to IEEE global communications conference (GLOBECOM), Atlanta, GA; December 2013.

[23] CMCC TSG R2-113197. Performance Comparison of Access Class Barring and MTC Specific Backoff Schemes for MTC. 3GPP Meeting; 2010-08.

[24] Qiu D, Srikant R. Modeling and performance analysis of BitTorrent-like peer-to-peer networks. In: Proceedings of the 2004 conference on applications, technologies, architectures, and protocols for computer communications (SIGCOMM '04); 2004. p. 367.

[25] Benaoum HB. h-Analogue of Newton's binomial formula. J Phys A Math Gen 1998;31:L751–5.

[26] Goodwin GC, Graebe SF, Salgado ME. Control system design, vol. 240. New Jersey: Prentice Hall; 2001.

[27] Ogata K. Modern control engineering. 3rd ed. Upper Saddle River, NJ: Prentice-Hall; 1997.

[28] Kalman RE. A new approach to linear filtering and prediction problems, transactions of the ASME. J Basic Eng 1960;82(Series D):35–45.

[29] Astrom KJ, Hagglund T. ISBN 1556179421 Advanced PID control. ISA-The Instrumentation, Systems, and Automation Society; 2006.

[30] Hang CC, Astrom KJ, Ho WK. Refinements of the Ziegler Nichols tuning formula. IEEE Proc–D 1991;138(2).

[31] NS-3 Consortium. ns-3. Official website. Available at: <http://www.nsnam.org/>; 2015.

[32] 3GPP TR 37.868 V11.0.0. Third generation partnership project (3GPP). Technical specification group radio access network; Study on RAN Improvements for Machine-type Communications. (Release 11); 2011-09.

CHAPTER

Analysis and performance evaluation of the next generation wireless networks

21

Arash Maskooki, Gabriele Sabatino, and Nathalie Mitton
Inria Lille-Nord Europe, Villeneuve d'Ascq, France

1 INTRODUCTION

Data traffic consumption is exponentially growing in wireless networks; this is due to the new applications demanding high quality of services, high data rate and new advanced user terminals which have been evolving rapidly over the past decade.

Mobile data traffic is forecasted to grow by more than 24 times between 2010 and 2015, and 500 to 1000 times between 2010 and 2020 [1]. The forecasts estimate that about 50 billion devices will be connected by 2020.

Long-Term Evolution/Long-Term Evolution-Advanced (LTE/LTE-A) wireless systems, as well as the IEEE 802.11 family of technology, are expected to dominate the wireless-communication arena for the next decade. Particularly, LTE/LTE-A systems are emerging as the global choice to which all mobile-broadband operators are expected to migrate [2,3].

Other radio access technologies are available or under development such as Wireless Personal Area Network (WPAN, including Bluetooth and ZigBee) and IEEE 802.11p, for very-short-range communication between devices and for communication between vehicles.

The vision of new generation mobile systems is towards unification of various mobile and wireless networks which can be qualified as heterogeneous networks (HetNets). HetNets consist of different access networks and provide a wide range of services, including enhanced and extended mobility and accessibility.

These networks will not only help in improving existing services [4,5] but integrate intelligent algorithms for mobility management, resource management, access control, routing, etc. This chapter will discuss modeling and analysis of the performance of the next generation wireless networks. Section 2 will discuss the characteristics and development of different generations of cellular wireless network. Section 3 will present analytical methods to evaluate the performance of the next generation HetNets. Subsequently, Section 4 will discuss the simulation techniques and analysis for HetNets. Finally, Section 5 will conclude this chapter.

2 THE EVOLUTION OF CELLULAR WIRELESS SYSTEMS

The first generation (1G) mobile cellular systems were deployed in the 1980s. The first commercial cellular network was the Nordic Mobile Telephone (NMT) deployed in the Scandinavian countries. Another system developed was the advanced mobile phone service (AMPS) cellular system in the United States and the most popular Total Access Communication Systems (TACS). These technologies were based on analog systems, circuit switching and used frequency division multiple access (FDMA) radio systems.

Second generation (2G) mobile systems were introduced at the beginning of the 1990s in Europe with the Global System for Mobile communication (GSM) systems. The GSM system, developed by the European Telecommunications Standard Institute (ETSI), is a time division multiple access (TDMA) radio system with data transmission rates up to 9.6 kbps and is the pioneer of the packet switching system.

In the United States there were other lines of development such as IS-54 (North America TDMA Digital Cellular), IS-136 also known as D-AMPS, and IS-95 based on code division multiple access (CDMA), known as CDMA-One.

To provide better support for data services, ETSI developed the General Packet Radio Service (GPRS), a packet transmission system that overlays GSM and interworks with external packet data networks such as the Internet. GPRS is a 2.5 generation (2.5G) wireless communication system. The main feature is that each mobile terminal is assigned an IP address, which enables the devices in such networks to integrate with Internet easily.

Enhanced Data rates for GSM Evolution (EDGE) or Enhanced GPRS (EGPRS), considered as 2.75 generation (2.7G), is an extended version of GSM and an evolution of GPRS that uses the same radio channels and time-slots as GSM, so it does not require additional spectral resources. EDGE increases the capacity of the network and changes the data modulation, which provides a data rate three times higher than GSM (473.6 kbps uplink/downlink and even higher in Evolved EDGE).

The evolution towards third generation cellular systems (3G) was driven by the International Telecommunications Union (ITU) and referred to as International Mobile Telecommunications 2000 (IMT-2000).

The Third Generation Partnership Project (3GPP), established in 1998, defined a mobile system called universal mobile telecommunications systems (UMTS). The main feature was to evolve GSM core networks and the Radio Access

Network (RAN) technologies by adding components like base stations (BS) or Node-B and the RNC (Radio Network Controller). 3G technology was IP-based and provided multimedia services, email, web access and video conferencing. UMTS air interface technology is the WCDMA (wideband code division multiple access) that brings some advantages in terms of data transfer rate, increasing system capacity and communication quality. The 3G systems can deliver bit rates up to 2 Mbps and support of quality-of-service (QoS). The approach adopted to evolve the core network is called High Speed Downlink Packet Access (HSDPA) and is widely considered as 3.5G. The HSDPA offered a peak rate of 14.4 Mbps in a 5 MHz channel at the downlink. High Speed Uplink Packet Access (HSUPA) brings a faster upload channel with uplink speed up to 5.76 Mbps. HSPA is the combination of both HSDPA and HSUPA and its evolution resulted in the Evolved HSPA, also known as HSPA+ or HSPA Evolution, which is the enhanced version and provides some techniques developed for LTE. Peak data rates reach 28 Mbps in downlink and 11.5 Mbps in uplink (42.2 Mbps in downlink and 22 Mbps in uplink with downlink/uplink dual carrier operation and the combination of MIMO (multiple input/multiple output), antennas and 64QAM modulation). HSPA+ with HSUPA and the evolution of the CDMA 2000, known as CDMA 2000 1x EV-DO, are considered to be 3.75G.

The 3GPP started to work on the Fourth Generation (4G) cellular systems and standardized Long-Term Evolution (LTE) [6]. LTE in effect was labeled 3.9G and the motivation behind the design of this network was to provide low latency and very high data rates starting from 100 Mbps and reaching more than 1 Gbps at the downlink.

LTE is expected to substantially improve the end-user throughputs, sector capacity and reduce user plane latency, bringing a significantly improved user experience with full mobility (see Table 21.1). LTE is scheduled to provide support for IP-based traffic with end-to-end QoS and, in particular, also LTE supports a flexible bandwidth deployment and, thanks to Orthogonal Frequency Division Multiplexing (OFDM) and MIMO systems, allows high data rates.

Table 21.1 Performance LTE and LTE-Advanced

		LTE (3.9G) Release 8	**LTE-Advanced (4G) Release 10**
Peak data rate	DL	300 Mbps	1 Gbps
	UL	75 Mbps	500 Mbps
Peak spectral efficiency (bps/Hz)	DL	15	30 (up to 8 × 8 MIMO)
	UL	3.75	15 (up to 4 × 4 MIMO)
Access methodology	DL	OFDMA	OFDMA
	UL	SC-FDMA	SC-FDMA
Transmission Bandwidth		20 MHz	100 MHz

Enhancements introduced in LTE are [6]:

- Downlink peak data rate up to 100 Mbps;
- Uplink peak data rate up to 50 Mbps;
- Increased data rate for cell-edge users with respect to HSPA;
- High spectral efficiency (bit/s/Hz), increased by a factor of 3–4 with respect to HSPA;
- Round Trip Time (RTT) equal to 10 ms (it was 70 ms in HSPA and 200 ms in UMTS);
- Low latency, lower than 100 ms for idle-to-active mode switch;
- Mobility support for very high mobility speed (up to 350 km/s);
- Backward compatibility (GSM/GPRS and UMTS/HSPA);
- Voice and real-time services offered over IP network.

These enhancements are supported by several features [6]:

- OFDM/OFDMA downlink modulation;
- SC-FDMA uplink modulation;
- High channel bandwidth flexibility (1.4, 3, 5, 10, 15, 20 MHz) both in downlink and uplink directions;
- Very large section of available spectrum can be assigned to a LTE base station (including GSM or UMTS frequencies or band on 2.6 GHz);
- Transmitting and receiving diversity, spatial multiplexing and diversity with MIMO techniques;
- QPSK, 16-QAM and 64-QAM modulation schemes.

The main goals of LTE are focused on minimizing system and user equipment (UE) complexities, allowing flexible spectrum deployment in existing or new frequency spectrum management and enabling coexistence with other 3GPP Radio Access Technologies (RATs).

The definition of 4G wireless, known as the International Mobile Telecommunications Advanced (IMT-Advanced) project, was finally published by the International Telecommunications Union Radio communication Sector (ITU-R) in March 2008. In 2009 LTE-Advanced (LTE-A) (in Release 10), the 4G mobile evolution system, was defined.

LTE-A research is mainly focused on user *signal to interference noise ratio* (SINR) improvement and spectrum flexibility. The LTE-A proposes a distributed network architecture, heterogeneous network which consists of a mix of macro-, pico-, femto-cells (HeNBs: Home eNodeB, also called femto-cells), low-cost eNodeBs for indoor coverage improvement), and relay base stations. They can be directly connected to the Evolved Packet Core (EPC) or via a gateway (HeNB-GW: Home eNodeB Gateway), providing additional support for a large number of HeNBs. Self-Organizing Network (SON) concepts have been introduced in LTE standardization in order to increase the network performance and reduce the operational expenditure for operators. With the smaller transmission powers from small cells and the relative proximity of users to a small cell, more users can now be covered within the same area in a HetNet; consequently, HetNets provide "cell-splitting" gains relative to macro-only networks.

The new technology targets include better coverage, higher data rates, better QoS performance and fairness among users. 3GPP has been working on various aspects to improve LTE performance in the framework of LTE-Advanced, which includes higher order MIMO, carrier aggregation and Cooperative Multipoint Transmission (CoMP). Carrier aggregation consists of grouping several LTE Component Carriers (CC), so that devices are able to use bandwidth up to 100 MHz. Carrier aggregation can be implemented by different approaches. The first one consists of contiguous bandwidth, where five contiguous 20-MHz channels are aggregated to obtain the required bandwidth. The other approach is noncontiguous carrier aggregation. In this case, CC can be noncontiguous on the same spectrum band or noncontiguous on different spectrum bands. In an LTE-Advanced design, support of multiple antenna systems is necessary to achieve data rates of 1 Gbps in downlink and 500 Mbps in uplink within a bandwidth of 20 MHz. The key requirement for an LTE mobile station is to use two antennas for uplink and two antennas for downlink. The concept of multiple antennas became popular to increase throughput along with adaptive modulation and coding schemes. MIMO is used to increase the overall bit-rate through transmission of two (or more) different data streams on two (or more) different antennas using the same resources in both frequency and time, separated only through use of different reference signals to be received by two or more antennas. The enhanced-MIMO techniques take the name of Cooperative Multipoint Transmission (CoMP). It is a strategy to improve users' performance with the intercell interference reduction. CoMP is based on cooperation between different base stations.

There is a growing interest to interconnect and amalgamate different technologies to support future connectivity and data rates. To make our everyday life more efficient, comfortable and safer is the main objective of the Internet of Things. Services such as smart mobility, smart environment, and e-health will continue to proliferate and become more mobile.

The evolving of cellular wireless networks, fifth generation (5G), is envisioned to overcome the fundamental challenges of existing cellular networks, e.g., higher data rates, user coverage and crowded areas, energy consumption and cost efficiency. The 5G can be a complete wireless-based web application without limitation that can include full multimedia capability beyond 4G speeds, called World Wide Wireless Web (WWWW). 5G wireless networks are expected to be a mixture of network tiers of different sizes, a multitier architecture consisting of macro-cells, different types of licensed small-cells, device-to-device (D2D) networks to serve users with different quality-of-service (QoS) and different radio access technologies (RATs) that are accessed by unprecedented numbers of smart and heterogeneous wireless devices [7].

The deployment of small-cells brings an improvement in order of network coverage and, due to the changes to the functional architecture of the access network, allows data and control signals to tunnel through the Internet, enabling small-cells to be deployed anywhere with Internet connectivity [8]. Small-cells are also a very promising candidate for the backhauling of WSNs (wireless sensor networks), because terminals can use less power in comparison with other wireless systems [9]. Recent developments are moving the WSN communication towards Internet protocol (IP)-based systems, according to the IoT paradigm.

The vision of the future, 5G seems to be an enhanced heterogeneous network composed of a mixture of different radio access technologies that include WLAN technologies which can offer seamless handovers to and from the cellular infrastructure, and device-to-device communications.

Another new and important aspect of the focus of the next-generation networks is "big data." The future M2M or IoT applications will generate a vast amount of data and obviously will be a technical challenge for RANs. New network architectures may emerge from the necessity of running big data applications, making informed decisions and extracting intelligence from big data. The smart grid can be seen as a huge sensor network, with immense amounts of grid sensor data from various sensors, meters, appliances and electrical vehicles. Data mining and machine learning techniques are essential for efficient and optimized operation of the grid [8].

See Table 21.2 for a summary of this technology evolution.

3 MODELING AND ANALYSIS OF INTERFERENCE IN THE HETEROGENEOUS WIRELESS NETWORKS

In parallel with cellular system development, the wireless technologies sector provides different access systems: Wireless Local Area Networks (WLAN) 802.11 IEEE standard (Institute of Electrical & Electronics Engineers) and Wireless Metropolitan Area Networks (WMAN) 802.16 (commercialized under the name WiMAX from Worldwide Interoperability for Microwave Access). These technologies can be incorporated in the next generation of wireless systems to balance the load of the macro base station and provide better coverage. In the future wireless networks, small-cells that use these technologies are considered to be a promising solution to the increasing number of users of the next generation of wireless networks. Small-cells are equipped with low power and low cost access points which are set up based on the coverage or capacity demand in specific parts of the network and can offload parts of the demand of the macro base station. Coexistence of these technologies in a multitier heterogeneous network triggers the need for more efficient spectrum management schemes to better utilize the spectrum and limit the level of interference from different tiers of the network.

Interference is a major limiting factor in multitier heterogeneous networks due to the coexistence of several tiers in the networks. Thus, modeling and analysis of the interference provide insight into the limitations of the network and lead to

Table 21.2 Generations of Mobile Technologies [10]

Generations	1G	2G	2.5G	3G	3.5G	4G	5G
Start/ Deployment	1970–1980	1990–2004	2001–2004	2004–2010	2006–2010	2010–Now	Soon
Data Bandwidth	2 Kbps	9.6–43.2 Kbps	144–384 Kbps	2 Mbps	More than 2 Mbps	1 Gbps	Higher than 1 Gbps
Technology	Analog Cellular Technology	Digital Cellular Technology	GPRS, EDGE, CDMA	CDMA 2000 (1xRTT) UMTS, Evolved EDGE	HSDPA, HSUPA, HSPA + , CDMA 2000 (1x EV-DO), Wi-Fi, Mobile Wi-Max	Wi-Max, LTE, LTE-A, Wi-Fi	WWWW
Service	Mobile Telephony (Voice)	Digital voice, SMS, Higher capacity packetized data	SMS, MMS	Integrated high quality audio, video and data	Integrated high quality audio, video and data	Dynamic Information access, Wearable devices	Dynamic Information access, Wearable devices with AI Capabilities
Multiplexing	FDMA	TDMA, CDMA	CDMA	CDMA	CDMA	OFDMA, SCFDMA	OFDMA, SCFDMA
Switching	Circuit	Circuit, Packet	Packet	Packet	All Packet	All Packet	All Packet
Core Network	PSTN	PSTN	PSTN	Packet N/W	Internet	Internet	Internet

more practical solutions to respond to the high demand for capacity. Interference in wireless networks is the unwanted signal energy received from peer transmitters on the same network or other nearby networks. The transmitted power P_t decays exponentially with respect to the distance d from the transmitter,

$$P_r(d) = KP_t h_{rt} d^{-n} \quad (21.1)$$

where, P_r, K and h_{rt} are the received power, path loss constant, and the random channel gain variation between the transmitter and receiver respectively. Signal-to-noise and interference ratio (SINR) is defined as,

$$SINR = \frac{P_r(d)}{N + \sum_{i \in \prod} P_r(d_i)'} \quad (21.2)$$

where, N is the noise power. d_i and $P_r(d_i)$ are the distance and received power between the receiver and the ith transmitter in the set of nearby transmitters ($\prod$) respectively.

3.1 GENERAL ANALYTICAL MODELS

Legacy analytical methods for cellular wireless networks assumed hexagonal grids to model the base station (BS) locations and the coverage in the network. However, in practice, placement of the BSs is not regular. This is because the placement is performed based on the demand in specific areas and also the condition of the location and other barriers. In addition, placing small-cells would further randomize the network grid in the next generation wireless networks. Stochastic geometry is a mathematical tool that can provide the expected values of the desired parameters in a network of randomly distributed points. Stochastic geometry has been used to model and analyze ad hoc wireless networks for nearly three decades [11–18]. It has been shown that stochastic geometry can be used for modeling and analysis of the next generation multitier wireless networks as well [19–23].

In stochastic geometry models, a network is modeled by a point process that best fits the network characteristics: e.g., large networks in vast areas of coverage and randomly distributed nodes such as cellular networks can be modeled by the Poisson point process (PPP). A point process is a Poisson point process if and only if the number of points in any compact set defined on the region is a Poisson random variable. Other point processes have also been used to model wireless networks. Binomial point process (BPP) is used to model a randomly distributed wireless sensor network where the number of nodes is known and finite. Poisson cluster process (PCP) is applied to the network where nodes are clustered around specific points due to some physical constraint or MAC protocol specification such as Wi-Fi networks. Hardcore point process (HCPP) is another point process used in the networks where two nodes cannot coexist within a hardcore vicinity r_h.

Performance metrics of the network can be obtained using the network model. As an example, by assuming Rayleigh fading for point-to-point channel in the network, the cumulative distribution function (CDF) of the signal-to-noise and interference ratio (SINR) for the test receiver in the network can be calculated as follows [11],

$$F_{SINR}(\theta) = P\{SINR \le \theta\} = P\left\{\frac{KP_t h_{rt} d^{-\eta}}{N+I} \le \theta\right\} = P\left\{h_{rt} \le \frac{(N+I)\theta d^{\eta}}{KP_t}\right\} \\ = \int_u F_{h_{rt}}\left(\frac{(N+I)\theta d^{\eta}}{KP_t}\right) f_I(I)dI \tag{21.3}$$

substituting Rayleigh CDF yields,

$$F_{SINR} = 1 - exp\left(-\frac{N\mu\theta d^{\eta}}{KP_t}\right)\mathcal{L}_I(s)\Big|_{s=\frac{\mu\theta d^{\eta}}{KP_t}} \tag{21.4}$$

where,

$$SINR = \frac{KP_t h_{rt} d^{-n}}{N+1} \tag{21.5}$$

$F_{h_{rt}}$ and f_I are the cumulative distribution function and probability distribution function of the channel and interference respectively. $\mathcal{L}_I(s)$ is the Laplace transform of the probability distribution function (PDF) of the interference and is obtained based on the statistics of the point process that fits the network such as Poisson point process [24]. This method yields the CDF of SINR and hence the statistics of the other parameters of the systems such as the outage probability and maximum achievable rate.

In the same way, the average transmission rate in the network could be calculated. In [24], the average transmission rate for the downlink and in Rayleigh fading is calculated. BSs and users are modeled by homogenous PPPs. It is shown that the average transmission rate can be obtained as,

$$E(\ln(1 + SINR)) = \int_0^{\infty} e^{-\frac{N\mu d^{\eta}}{KP_t}(e^t - 1)}\mathcal{L}_I\left(\frac{\mu d^{\eta}}{KP_t}(e^t - 1)\right)dt \tag{21.6}$$

However, this method only applies for Rayleigh fading while in general the PDF of SINR may not be obtained in closed form. Another method used in literature is to obtain a lower bound using the major interference contributors. In high path loss environments ($\eta = 4$) the nearest n interferers can be included into the interference calculations where n is the parameter. For example, to obtain a lower bound for the outage probability, the vulnerability region is defined such that existence of a transmitter within this region would drive the SINR below a desired threshold θ. The probability that at least a transmitter is present in the vulnerability region is then calculated [21].

3.2 MULTITIER HETEROGENEOUS NETWORKS MODELS

The frequency reuse technique is widely used in cellular networks to mitigate the interference experienced by the users at the expense of lower spatial bandwidth efficiency. In a generic frequency reuse scheme, the total available bandwidth is divided into Δ sub-bands and used by different cells in a way that no two neighboring cells would use the same sub-band. Fractional frequency reuse (FFR) can improve the spatial frequency bandwidth efficiency by reusing a large portion of the available bandwidth for their inner cell users while assigning the rest to the borderline users. This is performed by assigning transmission power P_1 to the inner cell frequencies and P_2 to the boundary frequencies where $P_2 > P_1$. Obviously, the neighboring cells would not use the same boundary frequencies.

In general, incorporating frequency reuse schemes in stochastic geometry network models is a challenging task, as frequency reuse will violate the fundamental spatial independency in PPP. This is because the base stations with the same set of sub-bands cannot be neighbors and hence their locations are correlated. A solution is to assume that the BSs randomly choose their sub-bands. Hence, cells using the same sub-band will form a thinned version of the original PPP which is also a PPP with $\frac{\lambda}{\Delta}$ where λ is the intensity of the nodes in the original PPP and Δ is the number of sub-bands. Incorporating FFR into the network model is tricky due to the correlation it introduces to the placement of the BSs using similar boundary frequencies. This has been overcome in [25] by introducing a Threshold T_{ffr} where users with SINR lower than threshold are considered boundary users.

Using small-cells in multitier next generation cellular networks improves the network coverage and capacity. Spectrum allocation to the small-cells can be universal frequency reuse or by spectrum partitioning. In universal frequency reuse, the whole spectrum is used by macro-cells and small-cells. This improves the spatial spectrum efficiency at the cost of multitier interference: that is, small-cell users will receive interference from macro-cell and vice versa. On the other hand, spectrum partitioning will eliminate the multitier interference at the expense of lower spectrum efficiency. This is because only parts of the spectrum are used at every tier of the network. In [26], a hexagonal grid model is adopted for macro-cells while small-cells are modeled by PPP and the optimum spectrum partitioning is investigated. The trade-off between spectrum sharing and partitioning is investigated in [19]. All network tiers are modeled by PPP and Rayleigh fading is assumed for channel fading. Results show that universal spectrum sharing is optimal in sparse network deployments while spectrum partitioning is more desired in dense networks in terms of transmission capacity and outage probability.

Using centralized scheduling for spectrum access in future multitier wireless networks is inefficient in terms of delay and complexity. Hence, cognitive radio and opportunistic medium access are considered viable solutions for this type of network. Modeling and analyzing interference is crucial in this type of network, as interference is considered a major limiting factor of the efficiency of this type.

3.3 COGNITIVE RADIO NETWORK MODELS

Electromagnetic spectrum is a limited natural resource and most of the usable bands are already licensed for specific applications. In addition, a small portion of the spectrum has been released by authorities for industrial, scientific and medical purposes. However, these license free bands are increasingly getting to be overpopulated and hence they have a high level of interference. Nevertheless, studies show that the licensed spectrum is underutilized by the licensed operators. In other words, there are spatial or temporal white spaces where the spectrum is not used by the primary licensed user. Cognitive radio is a term assigned to the set of schemes where the wireless device can sense and adjust its parameters for a more efficient use of the spectrum. In future, cognitive radio devices will be allowed to access the licensed spectrum as a secondary user, conditioned on limited interference to the primary user. However, before these dynamic sharing schemes can be used in practice, their spectrum efficiency and impact on the primary users must be carefully analyzed.

This technique can be utilized to mitigate interference in the next generation heterogeneous networks where different tiers of the network use the spectrum white spaces to transmit and avoid collisions [27,11]. The interference on the primary network is defined based on a parameter pair, the maximum interference power level η and maximum probability of interference ζ [28,29]. η is in fact the maximum received power from the secondary network transmitter that can be tolerated by the primary network receiver while ζ defines how often the secondary network can exceed η before it degrades the performance of the primary network.

Stochastic geometry models can be used to evaluate the performance of cognitive radio networks. In [30], the aggregate interference by secondary user received at the primary user is obtained. The primary user network consists of one transmitter and one receiver while secondary user network is modeled by a PPP. Candidate distributions are tested against the interference model and shifted log-normal distribution is suggested as the best distribution describing the statistics of the interference. Similarly, in [31], the interference from a secondary user on a one link primary user is investigated. The probability distribution function (pdf) of interference is obtained through its characteristic function. Results show that a truncated alpha-stable function best represents the pdf of aggregate interference at the primary network receiver. The outage probability of the primary user is obtained by assuming Rayleigh fading for the links and PPP for the distribution of the node in secondary user network. In addition, lower bounds on the interference temperature are obtained using the strongest interferers. The abovementioned works assume a simple one-link network for the primary user. A more complex network model for the primary user network is investigated in [32]. Two independent PPPs are considered for primary and secondary users. Two approaches are considered to obtain the statistics of the outage and interference. In the first approach, bounds for the interference and outage are obtained by assuming a bipolar Poisson cognitive network model. In the second approach, however, the secondary user is assumed to transmit only outside the exclusion region where the received power from the primary transmitters is below a certain threshold. It is shown that the network of the secondary user outside the exclusion region can be accurately modeled by a Poisson hole process [32]. However, due to the complexity of the Poisson hole process, the secondary user network model is approximated and analyzed by the Poisson cluster process.

3.4 COGNITIVE MULTITIER HETEROGENEOUS NETWORKS MODELS

The multitier structure of the next generation wireless networks leads to a better coverage and more spectral efficiency. However, a centralized coordination of spectrum access for all tiers of the network is not an efficient solution for spectrum access management of the total network. Femto-cells are envisioned to be installed by the user or operators to respond to the demand for coverage in specific areas and hence their distribution tends to be random. Cognitive radio techniques have been investigated as a promising solution for a decentralized spectrum access management in such multitier cellular networks. In [20], the aggregate interference for a two-tier cellular network consisting of a single MBS and a network of femto-cells is obtained through its characteristic function. In their proposed scheme, MBS will transmit a busy tone to reserve the channel and hence femto-cells will defer their transmission if the received power from MBS is above a certain threshold. The outage probability and the average transmission rate are obtained through the aggregate interference statistics and the model is validated through simulations.

In [21], a more complex network model is investigated. The network is composed of two tiers where users are being served by multiple MBSs and femto-cells. Base stations, access points and users are modeled by independent PPPs. It is shown that cognitive techniques can decrease the probability of outage for femto-cells users by 60 percent. In [22], fully cognitive and semi-cognitive schemes are investigated for femto-cells in a cellular network. In the semi-cognitive technique, the femto-cell access point avoids using the channel occupied by MBSs and aggressively uses the second-tier frequencies, while in a fully cognitive scheme the access points sense and avoid using the frequencies used by other femto-cells as well. It is shown that a semi-cognitive scheme outperforms the fully cognitive scheme in terms of outage probability. This is because, although cognition with respect to all network tiers reduces the interference level, it limits the transmission opportunity severely in dense networks.

4 SIMULATION TECHNIQUES FOR THE NEXT GENERATION WIRELESS HETEROGENEOUS NETWORKS

Due to the data traffic demand in cellular networks, improvements in system spectral efficiency are necessary. One possible solution is increasing the base station deployment density. In a relatively sparse deployment of macro base stations, adding another base station does not severely affect inter-cell interference, and solid cell splitting gains are easy to achieve. However, site acquisition in a capacity limited dense urban area can get prohibitively expensive [33].

Challenges associated with the deployment of traditional macro base stations can be overcome by the utilization of base stations with lower transmit power, which are classified as pico-cells, femto-cells and relay nodes. The transmit power of the low power access points intended for outdoor deployments ranges from 250 mW to approximately 2W. They do not require an air conditioning unit for the power amplifier and are much lower in cost than traditional macro base stations, with their transmit power typically varying between 5W and 40W. Pico-cells are regular eNBs with the only difference of having lower transmit power than traditional macro-cells. They are, typically, equipped with omnidirectional antennas and are deployed indoors or outdoors often in a planned (hot-spot) manner. Femto base stations are meant for indoor use, and their transmit power is typically 100mW or less. Unlike pico, femto base stations may be configured with a restricted association, allowing access only to its closed subscriber group (CSG) members. Such femto base stations are commonly referred to as closed femtos. Relay nodes are used to extend the macro-cell coverage or fill a coverage hole. A network that consists of a mix of macro-cells and low-power nodes, where some may be configured with restricted access and some may even lack wired backhaul, is referred to as a heterogeneous network. An illustration of such networks is shown in Figure 21.1.

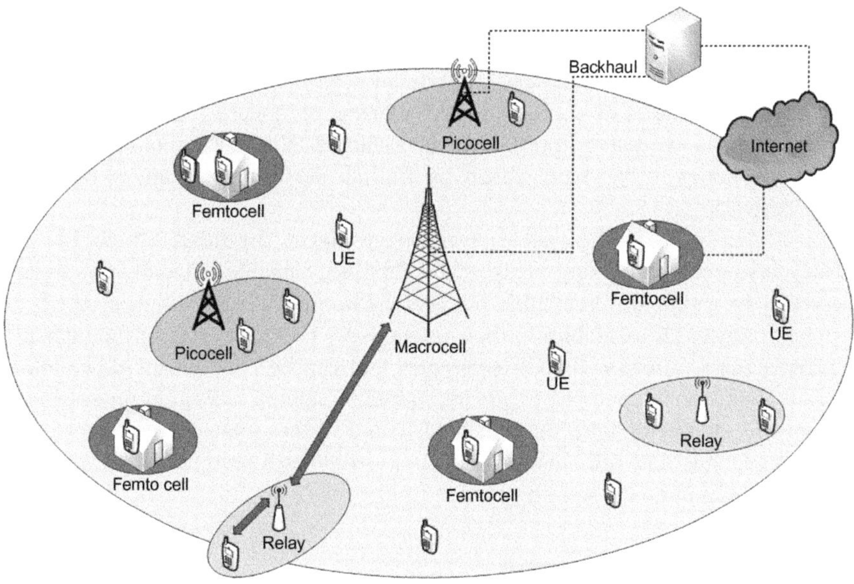

FIGURE 21.1

An illustration of a heterogeneous network.

Femto-cells have recently attracted significant attention. The use of femto-cells will benefit both users and operators. Users will enjoy better signal quality due to the proximity between transmitter and receiver and hence communicate with larger reliabilities and throughputs. Furthermore, this will also provide power savings, reduce electromagnetic interference and energy consumption. This way, more users will have access to the same pool of radio resources or use larger modulation and coding schemes, while operators will benefit from greater network capacity and spectral efficiency. There are a number of technical studies associated with various aspects of femto-cell deployments based on cellular technology. These studies consider operations, administration, and management (OAM) and self-organizing network (SON) protocols, network architecture, local IP access (LIPA), access (open, closed, and hybrid), and interference management [34–36].

Heterogeneous networks are based on different wireless technologies, whereas macro networks are based on a cellular technology while low power access points are based on WLAN [37,38]. Reduced cost is one of the main motives for the adoption of femto-cells. It is shown in [39] that in urban areas a combination of publicly accessible home base stations or femto-cells (randomly deployed by the end user), and macro-cells deployed by an operator for area coverage in a planned manner, can result in significant reductions (up to 70 percent in the investigated scenario) of the total annual network costs compared to a pure macro-cellular network deployment.

The introduction of low-power nodes in a macro network creates imbalance between uplink and downlink coverage. Due to larger transmit power of the macro base station, the handover boundary is shifted closer to the low-power node, which can lead to severe uplink interference problems as UE units served by macro base stations create strong interference on the low-power nodes.

The performance of a mixed deployment of macro-, pico-, and open femto-cells was evaluated in [40,41], which showed that the limited coverage of low-power nodes is the main reason for limited performance gain in heterogeneous networks.

In 4G networks, new physical layer design allows for flexible time and frequency resource partitioning. This added flexibility enables macro-, femto- and pico-cells to assign different time-frequency resource blocks within a carrier or different carriers (if available) to their respective UE. This is one of the inter-cell interference coordination (ICIC) techniques that can be used on the downlink to mitigate data interference [42,43]. With additional complexity, joint processing of serving and interfering base station signals could further improve the performance of heterogeneous networks [44,45], but these techniques require further study for the scenarios commonly seen in practice.

4.1 NETWORK SIMULATORS

In the network research area, the simulation is one of the most important parts as implementation of a whole network in the real world is costly and difficult.

Simulation can be done by using network simulators. Network simulators can implement the network on computer and are valuable tools to develop, test and diagnose network protocols. The simulator can model hypothetical and real-life objects on a computer and design different network topologies using various types of nodes (hosts, hubs, bridges, routers and mobile units, etc.). In a real network, a single test bed takes a large amount of time and funding while on a simulator it is possible to verify and evaluate the performance of networks by changing parameters of the network. There are many different network simulators available with different features which have gained more and more attention during the last years, e.g.: ns-2, ns-3, OMNeT++, OPNET, NetSim, GloMoSiM, JiST, etc.

Network Simulator-2 (ns-2) [46] is an open source, discrete event network simulator that is capable of simulating wired and wireless networks. Ns-2 has been developed under the VINT (Virtual Inter Network Testbed) project in 1995. It is a joint effort by the University of California at Berkeley, University of Southern California's Information Sciences Institute, Lawrence Berkeley National Laboratory and Xerox Palo Alto Research Center. Ns-2 is written in C++ and provides the simulation interface through OTcl, an object-oriented extension of Tcl. The network topology is created by OTcl scripts to simulate different network protocols with different network topologies. In ns-2, network animator (NAM) is used for the graphical view of the network. Ns-2 is the most common and widely used network simulator for research work.

The LTE/SAE model is an implementation for ns-2 to test the performance of LTE networks. The module is a project paper [47] investigating Long Term Evolution network performance from which a simulation model was made. The LTE/SAE model includes network and traffic models. The network model includes the air interfaces and S1 interface. The LTE/SAE model has different elements like the UEs (User Equipments), an eNodeB (base station which provides flow control information), an aGW (access Gateway provides flow control and Hypertext Transfer Protocol (HTTP) caching) and a main server (provide signaling services, File Transfer Protocol (FTP) and HTTP). Networks configurations, traffic models, flow control, number of elements, and bandwidth can be configured. The only limitation to the model is the eNodeB and aGateway which are permanent and fixed and cannot be moved or multiplied.

OMNeT++ (Objective Modular Network Testbed in C++) [48,49] is an open source simulator tool written in C++ that has been available to the public since September 1997 and currently has a large number of users. The most common use of OMNeT++ is for simulation of computer networks, but it is also used for modeling of multiprocessors, distributed hardware systems and performance evaluation of complex software systems. OMNeT++ is a component-based discrete event simulator. This approach promotes structured and reusable models. Model packages such as the OMNeT++ Mobility Framework and Castalia facilitate the simulation of mobile ad hoc networks or wireless sensor networks. A high-level language called Network Description (NED) is used to assemble individual components into larger components and models. In addition,

OMNeT++ has extensive graphical user interface Graphical Network Editor (GNED), and intelligence support that allows building of network topologies graphically.

SimuLTE is a simulation framework developed for OMNeT++. It is an open-source system-level simulator for LTE and LTE-A networks [50]. The SimuLTE simulates the data plane of the LTE/LTE-A Radio Access Network (RAN) and Evolved Packet Core (EPC) and the LTE user plane is compatible with the INET framework. The simulator is composed of several modules (PHY, MAC, RLC and PDCP-RRC) and allows simulation with heterogeneous eNBs (macro, micro, pico). However, SimuLTE presents some limitations where modules are not implemented (e.g., control plane model, radio bearers and handovers).

OPNET (Optimized Network Engineering Tools) [51] is simulation software with a large variety of possibilities for simulating entire heterogeneous networks with various protocols. OPNET was developed by OPNET Technologies, Inc. It had been originally developed at the Massachusetts Institute of Technology (MIT) and since 1987 has become commercial software. Recently OPNET was acquired by Riverbed. Behavior and performance of communication networks and distributed systems models can be analyzed by performing discrete event simulations. The main programming language in OPNET is C (recent releases support C++ development). The initial configuration, as topology setup and parameter setting, is achieved using a graphical user interface; simulation scenarios require writing C or C++ code.

The OPNET Modeler (Riverbed Modeler) [52] provides modules to simulate communication devices, protocols, and architectures performances. The OPNET Modeler module enables evaluation of wireless protocols such as LTE, UMTS, WiMAX, 802.11, IPv6 and power management schemes for sensor networks study.

4.1.1 Ns-3 simulator

Ns-3 (network simulator 3) is a discrete-time event-based network simulator. It can implement a large number of different protocols and it is completely free software, licensed under GNU GPLv2 license. Ns-3 is a standalone new simulator and it is not an extension of ns-2. The simulator is written in C++ language and Python and it does not support the ns-2 APIs. Some models from ns-2 have already been adapted for use in ns-3. Ns-3 is a discrete event network simulator. This means that the simulation consists of a series of independent events that change their state. Events are actions such as a sending a packet, a new node being added to the network, or a timer expiring.

Each scheduled event runs until completion without advancing the simulation time, and then the simulation time is increased to the start time for the next scheduled event.

Ns-3 is developed mostly by the same group of people that work on or have worked on ns-2 [53]. A team made up of several partners and institutions including University of Holland, Georgia Institute of Technology, University of

Washington and collaborations with INRIA at Sophia Antipolis, made a new open source project, initiated in 2006, joined by worldwide developers. The ns-3 project started because of some problems associated with the ns-2. In ns-2, C++ and OTtcl are used to build simulations and the combination of both languages is difficult to debug and it can be considered as a barrier for new developers. Different tests show that ns-2 does not scale well to large simulations, making it unsuitable for some research scenarios. Many models in ns-2 are not validated against the real world, which makes users doubt whether simulation results are the same as the results that would be measured in real-world implementations.

The nodes in the simulation with ns-3 are more realistic than ns-2. Every node is constructed out of devices and there is an Internet protocol stack that closely resembles the stack on real systems.

Network traffic generated by ns-3 can be traced and written to a file in the pcap (packet capture) format, which makes it possible to analyze it with tools like Wireshark.

Due to its ability to simulate realistic scenarios, ns-3 can be used as a virtual system or be used in test-beds. Real applications can run on top of a protocol stack implemented by ns-3 and it is also possible to run applications on real networking stacks or to run an instance of ns-3 in a network, where it interacts with "real" systems.

The ns-3 architecture consists of a core simulator part and a number of layers that add the networking-specific elements. It provides an Internet stack with implementations of protocols like TCP and UDP, as well as lower-level protocols such as various versions of 802.11. Different components and applications can be added to the nodes, after which nodes can be connected to each other. To help with building up nodes and creating a network topology, helper scripts are provided. Compared to ns-2, there are other differences as well, such as the build system.

The simulation core is implemented in src/core. Packets are fundamental objects in a network simulator and are implemented in src/network. These two simulation modules by themselves are intended to comprise a generic simulation core that can be used by different kinds of networks, not just Internet-based networks.

The first open source product-oriented LTE network simulator has been developed by Ubiquisys, the developer of intelligent cells, and the Centre Tecnologic de Telecomunicacions de Catalunya (CTTC) [54]. The development of the LTE module for the ns-3 was carried out during the Google Summer of Code 2010. This module provides a basic implementation of the LTE device, including propagation models, PHY and MAC layers. The simulator will provide a common platform for LTE femto- and macro-cells. In WCDMA networks, femto-cells and macro-cells work independently, but in LTE all cells work together as a single self-organizing network. This means that the adaptive behavior of femto-cells and macro-cells is interdependent. Simulations are important because they can evaluate product performance in densely deployed and heavily used networks, while real deployments are still in their infancy. The development of the LTE simulator is open to the community in order to foster early adoption and contributions by industrial and academic partners. The most important features provided by this module are: a basic implementation of both the user equipment (UE) and enhanced NodeB (eNB) devices, Radio Resource Control (RRC) entities for both the UE and the eNB, Adapting Modulation and Coding (AMC) scheme for the downlink, the management of the data radio bearers, Channel Quality Indicator (CQI) management, etc.

In LTE networks, a module consists of two main components:

- *The LTE Model.* This model includes the LTE Radio Protocol stack (RRC, PDCP, RLC, MAC, PHY). These entities reside entirely within the UE and the eNB nodes. This has been designed to support the evaluation of the Radio Resource Management, QoS-aware Packet Scheduling, Inter-cell Interference Coordination, and Dynamic Spectrum Access;
- *The EPC (Evolved Packet Core) Model.* This model includes core network interfaces, protocols and entities. These entities and protocols reside within the SGW, PGW and MME nodes, and partially within the eNB nodes. The EPC model provides means for the simulation of end-to-end IP connectivity over the LTE model. In particular, it supports the interconnection of multiple UEs to the Internet, via a radio access network of multiple eNBs connected to a single SGW/PGW node.

4.2 SIMULATIONS

In this section some example scenarios are simulated using ns-3 network simulator and the results will be presented and discussed. In the following simulations, the scenarios are implemented in ns-3 as shown in Figure 21.2.

The scenario includes a single macro-cell, a number of femto-cells and users of both macro-cells and femto-cells. Femto-cell coverage is CSG (Closed Subscriber Group) where only the authorized users can access the base station. Figure 21.2 shows a HetNet composed of a macro-cell (eNB), different femto-cells (HeNB) and users where MUE (macro-user equipment) and HUE (home-user equipment) indicate macro- and femto-cell users, respectively. Also the core network or EPC has been implemented, following procedures explained in [34]. Moreover, different applications are considered to generate user traffic in downlink and uplink. LTE provides users traffic with different QoS (Quality of Serivce). Each information flow is associated with a specific QoS class which constitutes a bearer. "Guaranteed Bit Rate Conversational Voice" bearer is used in the following simulations. The built-in radio channel model "Multi Model Spectrum Channel" is used for the links, while the Path Loss and Fading are implemented according to the class "ns-3: Hybrid Buildings Propagation Loss Model." Finally, the UE mobility is incorporated using the "Mobility Model" class.

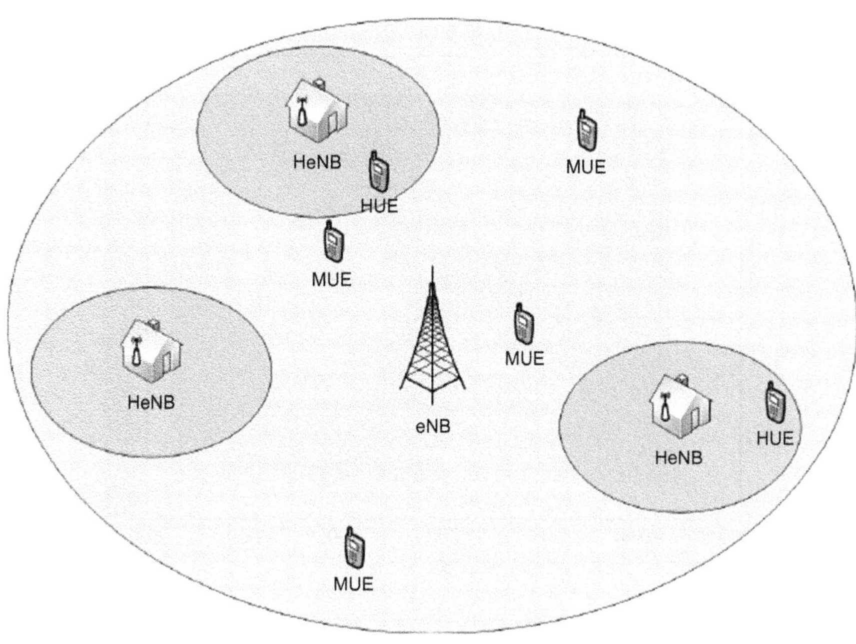

FIGURE 21.2

Simulation scenario.

The EPC network has the following parameters:

- Data Rate: 100 Gb/s;
- Delay: 10 ms;
- Maximum Transmission Unit (MTU): 1500 bytes.

The LTE network is formed by eNB, HeNB and their respective UE that has these parameters:

- Bandwidth in downlink, measured in number of Resource Blocks (RB): 25;
- E-UTRAN Absolute Radio Frequency Channel Number (EARFCN) in downlink: 100;
- Transmission Power: 10 dBm in the UE, 46 dBm in eNB and 20dB in HeNB.

The simulation time was set at 15 seconds and the packet size was 1000 bytes (1030 with the PDCP header).

The Radio Environment Map (REM) as shown in Figure 21.3 is a uniform 2D grid of values that represent the signal-to-noise ratio in the downlink with respect to the eNB that has the strongest signal at each point. Through the use of this map, we can better examine the tested scenarios, with respect to the possible interference between devices.

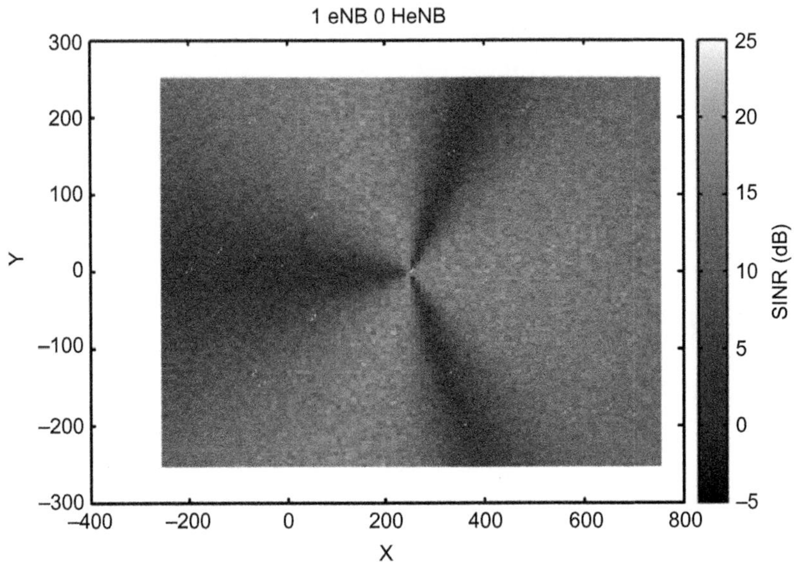

FIGURE 21.3

REM map with only one eNB without HeNB.

Simulation results presented below are referred to the metrics identified at the PDCP layer. For this set of simulations a throughput metric is taken into account, defined as the number of successful received packets during a specific time. Furthermore, two different kind of throughput are defined: the total throughput and the average throughput.

The total throughput represents the amount of total received information per unit of time and is calculated as,

$$\text{total throughput} = \frac{R_x}{T_{sim}} \tag{21.7}$$

where R_x represents the total received number of bits and T_{sim} is the time of transmission of the bits.

The average throughput is instead expressed by Equation (21.8):

$$\text{average throughput} = \frac{R_x}{T_{sim} \cdot N_{Users}} \tag{21.8}$$

which represents the amount of information received by the users per unit of time. N_{Users} is the total number of users.

In [55,56] a similar analysis is conducted. The same performance evaluation parameters were used to study the effect of femto-cells in a HetNet. Starting from

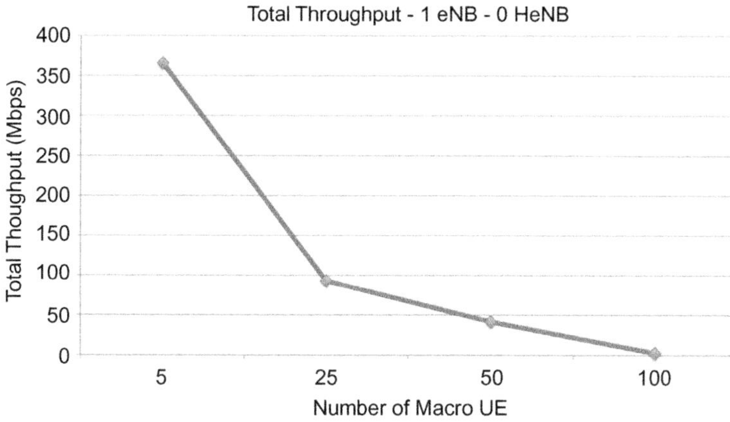

FIGURE 21.4

The total throughput for the same scenario.

different initial conditions and using a different network simulator (OMNeT++ and MATLAB based LTE simulators) the same trend is observed.

Figure 21.4 shows the total throughput vs. the number of Macro-cell users when there is no femto-cell. This configuration is chosen to evaluate the throughput vs. the number of UEs. As shown in the figure, increasing the number of users decreases the throughput. This is due to the increase in the interference received from other users. Further simulations are conducted considering a single femto-cell, HeNB as shown in Figure 21.4.

As shown in Figure 21.5 the total throughput when 25 UEs are allocated in the macro-cell and the remaining are assigned to one femto-cell, has a similar trend to that observed in the previous simulations.

On the other hand, when the users decide to connect the femto-cell and macro-cell base stations based on the power reception (load balancing), the network throughput is increased as shown in Figure 21.6.

The final scenario includes a HetNet with the presence of multiple femto-cells. The number of macro-cell users was fixed at 20. Figure 21.7 highlights how the throughput improves by increasing the number of femto-cells inside the same macro-cell. As the figure shows, with two femto-cells, increasing the number of femto UE (greater than 10 HUE) causes a drastic decrease in the throughput while increasing the number of femto-cells to five improves the network throughput as more users can be covered. However, the presence of more than five femto-cells causes interference in the HetNet system and hence degrades the SINR value and consequently the throughput.

This is shown in Figure 21.8. As shown in the figure, increasing the number of femto-cells (more than five) does not introduce further improvement in terms of throughput due to the increased level of intra-tier interference.

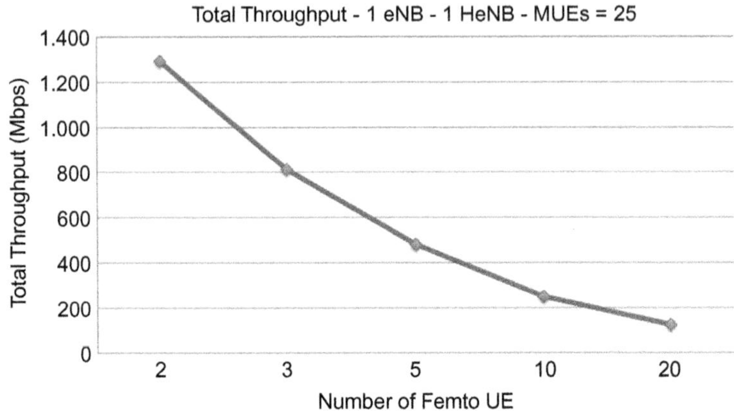

FIGURE 21.5

Total throughput in a network with an eNB, HeNB and the number of macro users set to 25. (The number of users of the femto-cell has been varied.)

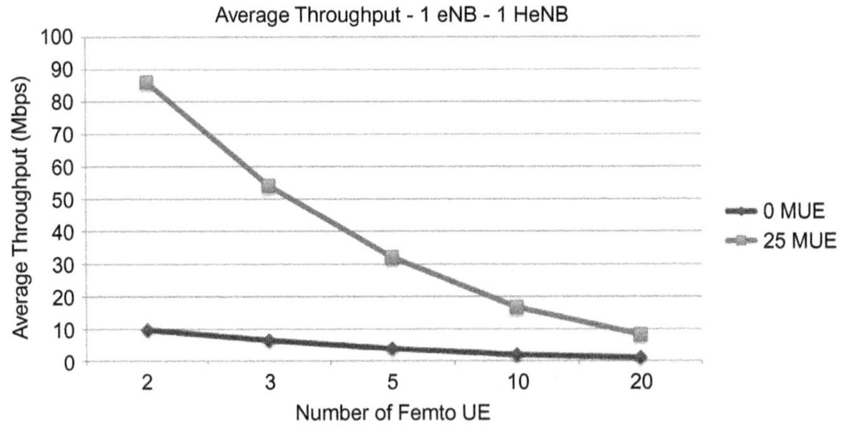

FIGURE 21.6

Comparison between the two scenarios with macro UE equal to 0 and 25, respectively.

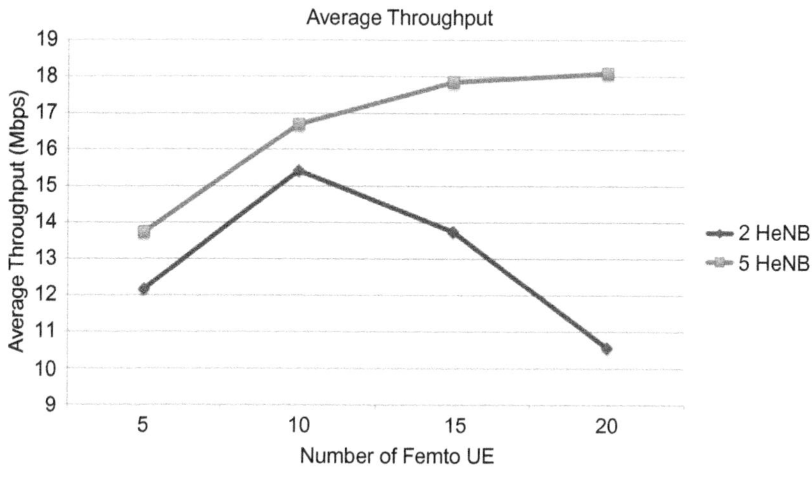

FIGURE 21.7

Comparison of average throughput in a scenario with 2 and 5 HeNB by varying the number of HUEs.

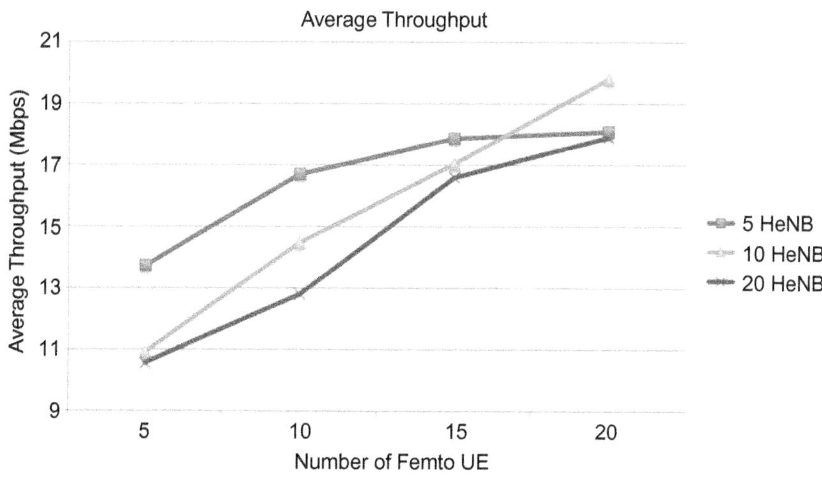

FIGURE 21.8

Comparison of the average throughput in a scenario with 5, 10 and 20 HeNB by varying the number of HUEs.

5 CONCLUSION

Spectrum is a limited natural resource and hence has to be utilized efficiently to support the increasing spectrum demand. Studies show that the spectrum is underutilized by the licensed users. This has triggered tremendous interest in using white spaces where the primary user is not accessing the spectrum in spatial, temporal or frequency domains. Heterogeneous networks (HetNet) are promising solutions to the challenge of providing coverage for a large number of users within the limited available spectrum. Such networks consist of multiple tiers where each tier can use different communication technology, transmission power and coverage range. Coexistence of these tiers brings some challenges, including the interference level, which is a major limiting factor in the performance of the next generation multitier networks. In this chapter, mathematical models to analyze the interference in such networks and subsequently the throughput are presented. Following the mathematical models, simulation techniques for these networks are discussed. HetNet performance is investigated to testify to the improvements introduced by Long Term Evolution-Advanced (LTE-A) definition. The simulation campaign has been conducted with the aim to demonstrate the pros and cons introduced by the use of the femto-cells in a HetNet scenario. To evaluate the throughput performance, a scenario where macro-cells and femto-cells coexist, is implemented in ns-3 simulator. Obtained results show that, in general, by increasing the number of femto-cells the system throughput is increased.

ACKNOWLEDGEMENT

This work is partially supported by CPER Nord-Pas-de-Calais/FEDER CIA Campus Intelligence Ambiante and by the FUI Traçaverre project. The authors would like to thank Giuseppe Araniti of University "Mediterranea" of Reggio Calabria and Prof John Cosmas of Brunel University London.

REFERENCES

[1] Nakamura T, Nagata S, Benjebbour A, Kishiyama Y, Tang H, Shen X, et al. Trends in small cell enhancements in LTE advanced. IEEE Commun Mag 2013;51(2):98–105.

[2] Dahlman E, Parkvall S, Sköld J. 4G LTE/LTE-Advanced for mobile broadband. Academic Press, Elsevier; 2011.

[3] Osseiran A, Monserrat J, Mohr W. Mobile and wireless communications for IMT-advanced and beyond. John Wiley & Sons; 2011.

[4] Boudriga N, Hassairi O, Obaidat MS. Intelligent services integration in mobile ATM networks. ACM symposium on applied computing (SAC). San Antonio, TX: ACM; 1998. p. 91–7.

[5] Third Generation Partnership Project (3GPP). 3GPP-2 X.S0004-700-E, Version 1.0.0, Wireless intelligent networks; March 2004.

[6] Third Generation Partnership Project (3GPP). Overview of 3GPP release 8 v.0.1.1. Technical report.

[7] Hossain E, Rasti M, Tabassum H, Abdelnasser A. Evolution towards 5g multi-tier cellular wireless networks: an interference management perspective. IEEE Wirel Commun; in press.

[8] Chin WH, Fan Z, Haines RJ. Emerging technologies and research challenges for 5g wireless networks. IEEE Wirel Commun 2014.

[9] Cimmino A, Pecorella T, Fantacci R, Granelli F, Rahman TF, Sacchi C, et al. The role of small cell technology in future smart city applications. Trans Emerg Telecommun Technol 2014;25(1):11–20.

[10] Sharma P. Evolution of mobile wireless communication networks: 1G to 5G as well as future prospective of next generation communication network. IJCSMC 2013;2 (8):47–53.

[11] ElSawy H, Hossain E, Haenggi M. Stochastic geometry for modeling, analysis, and design of multi-tier and cognitive cellular wireless networks: a survey. IEEE Commun Surv Tutorials 2013;15.

[12] Haenggi M, Andrews JG, Baccelli F, Dousse O, Franceschetti M. Stochastic geometry and random graphs for the analysis and design of wireless networks. IEEE J Select Areas Commun 2009;27:1029–46.

[13] Cardieri P. Modeling interference in wireless Ad Hoc networks. IEEE Commun Surv Tutorials 2010;12(4):551–72.

[14] Haenggi M, Ganti R. Interference in large wireless networks. Foundations and trends in networking, vol. 3. NOW Publishers; 2008, no. 2, p. 127–248.

[15] Weber S, Andrews JG. Transmission capacity of wireless networks. Foundations and trends in networking. NOW Publishers; 2012.

[16] Baccelli F, Blaszczyszyn B. Stochastic geometry and wireless networks. Foundations and trends in networking, vol. 1. NOW Publishers; 2009.

[17] Baccelli F, Blaszczyszyn B. Stochastic geometry and wireless networks. Foundations and trends in networking, vol. 2. NOW Publishers; 2009.

[18] Kleinrock L, Silvester JA. Optimum transmission radii for packet radio networks or why six is a magic number In: Conference record: national telecommunication conference, December 1978, p. 4.3.1–4.3.5.

[19] Cheung W, Quek T, Kountouris M. Throughput optimization, spectrum allocation, and access control in two-tier femtocell networks. IEEE J Sel Areas Commun 2012;30(3):561–74.

[20] Lima C, Bennis M, Latva-aho M. Coordination mechanisms for self-organizing femtocells in two-tier coexistence scenarios. IEEE Trans Wirel Commun 2012;11(6):2212–23.

[21] ElSawy H, Hossain E. Two-tier hetnets with cognitive femtocells: downlink performance modeling and analysis in a multi-channel environment IEEE Trans Mobile Comput Accepted.

[22] ElSawy H, Hossain E. On cognitive small cells in two-tier heterogeneous networks, In: Proceedings 9th Workshop on Spatial Stochastic Models for Wireless Networks (SpaSWiN 2013), Tsukuba Science City, Japan, May 13–17, 2013.

[23] ElSawy H, Hossain E, Kim DI. HetNets with cognitive small cells: user offloading and distributed channel allocation techniques IEEE Commun Mag Special Issue on "Heterogeneous and Small Cell Networks (HetSNets), May 2013.

[24] Andrews J, Baccelli F, Ganti R. A tractable approach to coverage and rate in cellular networks. IEEE Trans Commun 2011;59(11):3122–34.

[25] Dhillon H, Ganti R, Baccelli F, Andrews J. Modeling and analysis of K-Tier downlink heterogeneous cellular networks. IEEE J Sel Areas Commun 2012;30(3):550–60.
[26] Chandrasekhar V, Andrews J. Spectrum allocation in tiered cellular networks. IEEE Trans Commun 2009;57(10):3059–68.
[27] Asadi A, Mancuso V. A survey on opportunistic scheduling in wireless communications. IEEE Commun Surv Tutorials 2013;15(4):1671–88.
[28] Zhao Q. Spectrum opportunity and interference constraint in opportunistic spectrum access. In: ICASSP; 2007. p. 605–8.
[29] Zhao Q, Sadler BM. A survey of dynamic spectrum access. IEEE Signal Process Mag 2007;24(3):79–89.
[30] Ghasemi A, Sousa E. Interference aggregation in spectrum sensing cognitive wireless networks. IEEE J Sel Topics Signal Process 2008;2(1):41–56.
[31] Rabbachin A, Quek TQS, Shin H, Win MZ. Cognitive network interference. IEEE J Sel Areas Commun 2011;29(2):480–93.
[32] Lee C-H, Haenggi M. Interference and outage in poisson cognitive networks. IEEE Trans Wirel Commun 2012;11:1392–401.
[33] Damnjanovic A, Montojo J, Wei Y, Ji T, Luo T, Vajapeyam M, et al. A Survey on 3GPP heterogeneous networks. IEEE Wirel Commun 2011.
[34] Yavuz M, et al. Interference management and performance analysis of UMTS/HSPA + femto-cells. IEEE Commun 2009.
[35] Chandrasekhar V, Andrews JG, Gatherer A. Femto-cell networks: a survey. IEEE Commun 2008.
[36] Patel C, Yavuz M, Nanda S. Femto-cells [Industry Perspectives]. IEEE Wirel Commun 2010.
[37] Coupechoux M, Kelif J-M, Godlewski P. Network controlled joint radio resource management for heterogeneous networks. IEEE VTC 2008.
[38] Song W, Jiang H, Zhuang W. Performance analysis of the WLAN-first scheme in cellular/WLAN interworking. IEEE Trans Wirel Commun 2007.
[39] Claussen H, Ho LTW, Samuel LG. Financial analysis of a pico-cellular home network deployment. IEEE ICC 2007.
[40] Nihtila T, Haikola V. HSDPA Performance with dual stream MIMO in a combined macro-femto cell network. IEEE VTC 2010.
[41] Karimi HR, et al. Evolution towards dynamic spectrum sharing in mobile communications. IEEE PIMRC 2006.
[42] Boudreau G, Panicker J, Guo N, Chang R, Wang N, Vrzic S. Interference Coordination and Cancellation for 4G Networks. IEEE Commun Mag 2009;47(4):74–81.
[43] Khandekar A. et al. LTE-Advanced: heterogeneous networks, European wireless conference. 12–15 April 2010. Lucca. IEEE.
[44] Simeone O, Erkip E, Shamai S. Robust transmission and interference management for femto-cells with unreliable network access. IEEE JSAC 2010.
[45] Annapureddy S, Barbieri A, Geirhofer S, Mallik S, Gorokhov A. Coordinated joint transmission in WWAN, IEEE Commun. theory workshop., Cancun, Mexico, Available from: <http://www.ieee-ctw.org/2010/mon/Gorokhov.pdf>; 2010.
[46] The Network Simulator – ns-2. Website: Available from: <http://www.isi.edu/nsnam/ns/>; 2015.

[47] Qiu QL, Chen J, Zhang QF, Pan XZ. LTE/SAE model and its implementation in NS-2. In: Mobile Ad-hoc and sensor networks, MSN'09. 5th International conference on IEEE; December 2009. p. 299–303.

[48] OMNeT++ Community. OMNeT++ website. Available from: <http://www.omnetpp.org/>; 2015.

[49] Varga A. The OMNeT++ discrete event simulation system, European simulation multiconference, 2003.

[50] Virdis A, Stea G, Nardini G. SimuLTE: a modular system-level simulator for LTE/LTE-A networks based on OMNeT++. In: Proceedings of SimulTech; 2014. p. 28–30.

[51] Riverbed Technology. OPNET is now part of Riverbed. Available from: <http://www.opnet.com/>; 2015.

[52] OPNET Modeler. <http://www.riverbed.com/products/performance-management-control/network-performance-management/network-simulation.html>; 2014.

[53] ns-3 project description. Available from: <http://www.nsnam.org/docs/proposal/project.pdf>. [accessed 30.01.15].

[54] CTCC. LENA v8 documentation. Manual LENA release M8. Available from: <http://lena.cttc.es/manual>. [accessed 30.01.15].

[55] Afridi AK. Macro and femto network aspects for realistic LTE usage scenarios. PhD dis., Master of Science, Royal Institute of Technology (KTH). Stockholm: Sweden; 2011.

[56] Afridi AK. April 12 Macro and femto network aspects for realistic LTE usage scenarios: impact of increasing indoor traffic on macro network and macro effective offloading using LTE system level simulator. LAP LAMBERT Academic Publishing; 2012.

CHAPTER

Evolutionary algorithms for wireless network resource allocation

22

Nitin Sharma[1], Alagan Anpalagan[2], and Mohammad S. Obaidat[3]

[1]BITS, Pilani, RJ, India

[2]Ryerson University, Toronto, ON, Canada

[3]Monmouth University, West Long Branch, NJ, USA

1 INTRODUCTION

In the last decades, wireless communications systems have been one of the fastest growing industries in the world. Future wireless communication systems are expected to reliably provide data services with rate requirements ranging from a few Mbit/s up to some Gbits/s. The challenges to ensure the fulfillment of these requirements arise from the limited availability of frequency spectrum, the total transmit power and the nature of the wireless channel. Furthermore, in broadband applications, wireless channel encounters frequency selective-multipath fading, which leads to severe intersymbol interference (ISI) both in time and frequency, impacting the service quality and data rates. In order to meet the need of high data rates and reliable data services, orthogonal frequency division multiple access (OFDMA) has been selected as the multiple access scheme for 4th generation wireless networks, e.g., IEEE 802.16 worldwide interoperability for microwave access (WiMAX) and long term evolution (LTE).

Orthogonal frequency division multiplexing (OFDM) is a promising modulation technique which mitigates the effect of frequency selective fading inherent in high data rate environments. It is basically a type of multicarrier modulation scheme, based on the idea of dividing a given high-bit-rate data stream into several parallel lower bit-rate streams and modulating each stream on separate carriers, often called subcarriers.

Multicarrier modulation schemes eliminate or minimize ISI by making the symbol time large enough so that the channel-induced delays are an insignificant fraction of the symbol duration. Therefore, in high-data-rate systems in which the symbol duration is small, being inversely proportional to the data rate, splitting the data stream into many parallel streams increases the symbol duration of each stream such that the delay spread is only a small fraction of the symbol duration.

These individual substreams can then be sent over parallel subcarriers, while maintaining the total desired data rate. The number of subcarriers should be selected in such a way that each subcarrier has a bandwidth less than the coherence bandwidth of the channel; thus each subcarrier experiences relatively flat fading.

How to effectively allocate resources in an OFDMA [1,2] system continues to draw increasing research interest. Resource allocation in OFDMA [3–13] includes subcarrier allocation, power allocation, and bit loading. The resource allocation problem can be modeled as an optimization problem subject to various quality of service (QoS) constraints. Solutions to the resource allocation problem in OFDMA have been broadly divided into two categories: margin adaptive (MA) and rate adaptive (RA) [9]. The main objective of MA problems is to minimize the total transmission power subject to rate requirements of the users. On the other hand, the objective of RA problems is to maximize throughput in a system subject to a total transmission power constraint.

Resource allocation was tackled in [11] using an MA scheme, wherein an iterative subcarrier and power allocation algorithm was proposed to minimize the total transmit power given a set of fixed user data rates and the bit error rate (BER) requirements. In [4] the RA method was used, wherein the objective was to maximize the total data rates over all users subject to power and BER constraints. It was shown in [4] that in order to maximize the total capacity each subcarrier should be assigned to the user with best gain on it. However, there was no consideration given for the fairness of allocation among the users, which can leave some users, with low subcarrier gains, without any subcarrier being allocated to them. In [8,9,12,13] proportional fairness was incorporated by imposing a set of nonlinear constraints into the optimization problem.

Traditional mathematical programming techniques used to solve constrained optimization problems (COPs) have several limitations when dealing with the general nonlinear programming problem. Evolutionary algorithms (EAs) have been found to be successful in solving a wide variety of optimization problems. EAs refer to what is probably the most exotic group of optimization methods, generally inspired from natural phenomena or behaviors. They are said to be population-based techniques, as they work simultaneously with an entire group of solutions (called a population) and evolve iteratively towards the optimum. EAs are especially suitable for complex optimization problems where the number of parameters is large and the analytical solutions are difficult to obtain. EAs can help to determine the optimal solution globally over a domain.

2 RELATED WORK AND ALGORITHMS

Genetic algorithms (GAs) are a class of EAs based on the mechanics of natural selection and natural genetics. Single objective GAs were used to solve the resource allocation problem under consideration in [6,10]. Therein, subcarrier allocation was used as the only objective for GA, and power allocation to the users was then carried out based on the water-filling algorithm. GAs were shown to outperform other conventional iterative methods in these results. There are a few reports in the literature where EAs other than GAs were used for resource allocation [3,14–16]. The authors in [16] proposed a modified ant colony optimization (ACO) and used the same for solving the subcarrier-and-bit allocation problem for OFDMA systems. The particle swarm optimization (PSO) algorithm was used to solve the resource allocation problem in [3] through margin adaptive allocation. However, no fairness among the users was considered. In the case of large path loss differences among users, it is possible that the users with higher average subcarrier gains will be allocated most of the resources, i.e., subcarriers and power, for a significant portion of time. Because of this, the users with lower average subcarrier gains may not be able to transmit any data due to nonallocation of subcarriers to them. In [8,9,12–14] proportional fairness was incorporated by imposing a set of nonlinear constraints into the optimization problem.

There are few instances in literature where EAs were used for resource allocation in more complex OFDMA-based systems. For instance, authors in references [17,18] proposed ACO-based algorithms to solve the resource allocation problem in multicell OFDMA systems. In [17], ACO was used for subcarrier allocation followed by water filling based power allocation. Herein, the visibility was defined such that ACO selects subcarriers and power with the aim to increase the sum rate of each cell. The trail intensity was defined such that the inter-cell interference was reduced. On the other hand, in [18], the ACO was used for joint allocation of subcarriers, modulation and coding scheme (MCS) and transmit power in OFDMA cellular networks. Different combinations of user indices, MCS indices and subcarrier indices were used to represent the nodes in the graph. Specifically the subcarriers were defined as a vertex, while the combination of user and modulation type was considered as edges of the graph. Each path in the graph represented a particular solution corresponding to the allocation of these resources. In both these proposals the ACO was shown to outperform other conventional schemes used for resource allocation in OFDMA systems. Further, the authors in [19] proposed two EA-based approaches for subcarrier allocation in distributed OFDM-based cognitive radio access networks. The piecewise convex transformations were used to convert the original nonlinear integer programming problem (ILP) into an integer linear programming. In the first proposal herein, a genetic algorithm (GA) was combined with the invisible walls technique used in PSO in order to maintain diversity in the offspring population. Similar to the method in [18], the second proposal used ACO with vertices of the graph were used to represent the subcarriers and each edge represented a possible chosen modulation index of a specific radio. The performance of these algorithms was compared with the optimal solution.

However, in order to achieve optimal bound, subcarriers and power should be jointly optimized. Multi-objective GAs were used in [7,20,21] in order to solve this problem of multiuser subcarrier allocation and the optimal bit allocation jointly. In [7,20], authors used nondominated sorting GA-II (NSGA-II) with binary coding for the chromosomes; therefore, these algorithms are restricted to be used only for the cases when the number of users, bits and subcarriers are in a power of 2. Binary coded chromosomes are easy to handle as far as mutation operator is concerned, as mutation in this case simply flips a bit 1 to 0 or a bit 0 to 1. However, binary coding, apart from lengthy representation of chromosomes, poses extra burdens as it requires decoding of chromosome back to integers for fitness calculation and other integer calculations involved in each generation. This can adversely affect the convergence time.

Integer coding of chromosomes was used in our Pareto archived evolution strategy (PAES) [7] based solution for the resource allocation problem under consideration. However, since PAES does not use crossover operators, it is suitable only for problems requiring local search. Strength Pareto evolutionary algorithm-2 (SPEA-2) with integer coded chromosomes was used for joint subcarrier and power allocation in [21]. Rate maximization and total transmit power minimization are two conflicting objectives. In order to solve such an optimization problem, one may combine the two objective functions into a single function. However, the results obtained will be inferior to those obtained while considering all possible trade-offs. If the two objective functions are considered as two separate objectives, then the solutions will cater to most of the trade-offs.

Although the performance of multi-objective GAs was found to be suitable, the complexity of these algorithms can be on the higher side. In view of this low complex, fast single objective algorithm such as artificial bee colony (ABC) optimization and hybrid algorithms which combine the best traits of two or more different algorithms are preferred even for joint allocation problems.

3 SYSTEM MODEL AND RESOURCE ALLOCATION PROBLEM

An OFDMA system is considered with K users and N subchannels as shown in Figure 22.1. Serial data from all the users was fed into the resource allocation block at the transmitter, which then allocated bits from different users to different subchannels. It was assumed that each subchannel would have a bandwidth that is much smaller than the coherence bandwidth of the channel and that the instantaneous channel gains on all the subchannels of all the users would be known to the transmitter. Using this channel information, the transmitter would apply the subchannel, bit and power allocation algorithm to assign different subchannels to different users and the number of bits/OFDM symbol to be transmitted on each subchannel. Depending on the number of bits assigned to a subchannel, the adaptive modulator would use a corresponding modulation scheme, and the transmit power level would be adjusted according to the subchannel, bit, and power allocation algorithm. The principle behind adaptive modulation is simple: transmit at as high a data rate as possible when the channel is good, and at a lower rate when the channel is poor, thus limiting the number of dropped packets. Each user's data would be distributed across the set of subchannels assigned to the user. The assumption was that each subchannel would be uniquely assigned to a single user and two or more users would never share the same subchannel.

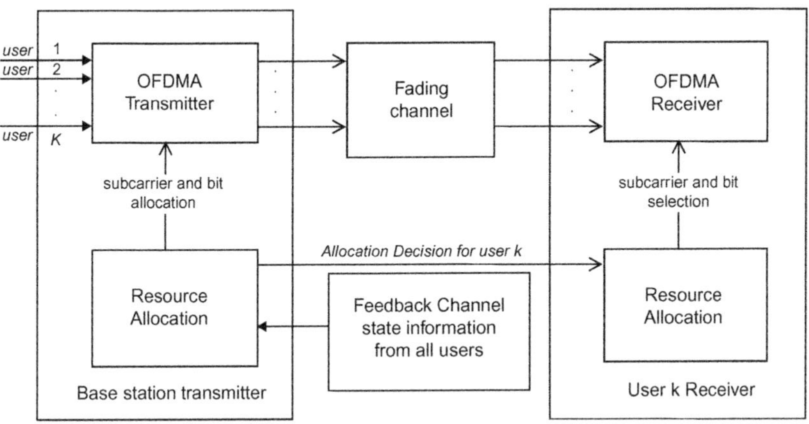

FIGURE 22.1

OFDMA system Model.

The optimization problem was formulated on the same lines as in [9]. The subchannels and power would be allocated in such a way that the total error free capacity would be maximized while satisfying the total power constraint (P_{tot}).

Defining the capacity for user k, denoted as R_k, as

$$R_k = \sum_{n=1}^{N} \frac{\rho_{k,n}}{N} \log_2 \left[1 + \frac{p_{k,n} h_{k,n}^2}{\mathrm{N}_o \frac{B}{N}} \right] \tag{22.1}$$

The optimization problem could hence be postulated as follows:

$$\max_{\rho_{k,n}, p_{k,n}} \sum_{k=1}^{K} \sum_{n=1}^{N} \frac{\rho_{k,n}}{N} \log_2 \left[1 + \frac{p_{k,n} h_{k,n}^2}{N_o \frac{B}{N}} \right] \tag{22.2}$$

subject to the constraints:

$$C_1: \quad \sum_{k=1}^{K} \sum_{n=1}^{N} p_{k,n} \le P_{tot}$$

$$C_2: \quad p_{k,n} \ge 0 \quad \forall\, k,n$$

$$C_3: \quad \rho_{k,n} \in \{0, 1\} \forall\, k, n$$

$$C_4: \quad \sum_{k=1}^{K} \rho_{k,n} = 1 \;\forall\, n$$

$$C_5: \quad R_1{:}R_2{:}R_3{:}\ldots\ldots{:}R_k = \gamma_1{:}\gamma_2{:}\gamma_3{:}\ldots\ldots{:}\gamma_1, \forall i, j \in \{1, \ldots\ldots, K\}; i \ne j$$

In (22.1), N_o is the power spectral density of AWGN, B is the total available bandwidth and $h_{k,n}$ is the channel gain for user k in subchannel n. In C_1, P_{tot} is the total available power and $p_{k,n}$ is the power allocated for user k in the subchannel n. According to C_3, $\rho_{k,n}$ can only be either 1 or 0, indicating whether subchannel n is allocated to the user k or not. C_4 restricts allocation of one subchannel to one user only. In constraint C_5, $\{\gamma_i\}_{i=1}^{K}$ is a set of predetermined values that are used to ensure proportional fairness among users. With $\alpha_k = \gamma_k / R_k$, the fairness index can be defined as:

$$F = \frac{\left(\sum_{k=1}^{K} \alpha_k\right)^2}{K \sum_{k=1}^{K} \alpha_k^2} \tag{22.3}$$

with the maximum value of 1 to be the greatest fairness case in which all users would achieve the same data rate. Since the problem formulation in (22.2) is to allocate resources to satisfy the rate constraints strictly for each channel realization, we define a quantity to measure how well the rate constraints are satisfied. The imbalance coefficient is defined as:

$$f = \left| \frac{1}{K} \sum_{k=1}^{K} \left[\alpha_k - \frac{\sum_{i=1}^{K} \alpha_i}{K} \right] \right| \tag{22.4}$$

4 EVOLUTIONARY ALGORITHMS FOR RESOURCE ALLOCATION

This section presents an introduction to the basic evolutionary algorithms (EAs) and outlines the procedures for solving problems using the simple EAs. EAs are becoming very popular with the wireless research community. The reason for the sudden popularity of EAs is simple—the gradient optimization methods that were most popular in engineering disciplines have not performed consistently across the variety of wireless communication optimization problems. The global search conducted by EAs is proving much more capable in this field of optimization. EAs are designed to search a much wider area of the design space, and could potentially provide a set of optimal solutions to a given problem. The EA approach was selected for this research because many different solutions can be expected to be found for resource allocation in OFDMA systems, and it is important for the designer to explore as much of the potential decision space as possible, before selecting a single design or control solution. Often with complicated designs, an exhaustive search of the entire design space is not feasible due to the high computational burden. EAs can help find good solutions in a much shorter time.

The EA is a stochastic global search method that mimics Charles Darwin's evolutionary theories of natural selection (survival of the fittest) [22]. EAs operate on a population of potential solutions, applying the principle of survival of the fittest to produce increasingly better approximations to a solution. In theory, it is possible for them to find true globally optimum solutions provided they exist within the decision search space and if certain optional genetic operators are included, for example "mutation."

In each generation (iteration), a new set of individuals (chromosome) is created by the process of selecting individuals (solutions) in accordance to their level of fitness (success) in the problem domain and "breeding" them together using operators similar to natural genetics. This process leads to the evolution of populations of individuals that are better suited to their environment than the individuals that they were created from, in other words, better solutions to a problem.

4.1 ENCODING

In EAs, each possible solution of the optimization problem under consideration should be encoded as a finite-length string over some alphabet. The coding techniques used can be classified into the following two categories: a binary coding and a permutation coding. The binary representation is mostly preferred for the coding of the solutions. For example, a problem with two variables, x_1 and x_2, may be mapped onto the chromosome structure as:

$$\left\{ \underbrace{1, 0, 1, 1, 1,}_{x_1} \underbrace{0, 1, 1, 1, 0, 1, 0, 0}_{x_2} \right\}$$

In the above example, a binary chromosome has been used, where x_1 contains 5 bits and x_2 uses 8 bits. The number of bits used affects the level of accuracy or the range of individual decision values that are required. A consequence of increasing the number of bits used to represent a parameter is the expansion of the decision search space size. In such a case, EA is required to search a much larger solution space, thus slowing down convergence rate. The increased chromosome length forces a problem-specific trade-off between the probability of finding a globally optimal solution and the overall algorithm run time.

The EA search process is capable of operating on coded decision variables rather than the decision variable itself. This encoding is the main strength of an EA—the genotype need not contain numerical optimization values directly and can instead contain complex encoding of systems, processes or methods, the result of which represents a solution to a problem.

The second coding scheme, the permutation coding, is used for sequencing problems such as scheduling problems and traveling salesman problems. This type of encoding is useful when individual fitness depends on positions of genes in the chromosome. For such type of problems, permutation strings of a set of integers are more natural representations than binary strings. Figure 22.2 shows examples of strings generated by the binary coding and by the permutation coding. The string generated by the binary coding consists of binary "0s" and "1s".

The binary string after being treated by EAs is often decoded to the parameter value in integer, real number, and so on. The permutation string, on the other hand, consists of numerals "1" to "n". Each numeral in the string corresponds to a job in scheduling problems or to a city in traveling salesman problems, while n is the total number of jobs or cities. Then jobs are processed according to their order in the permutation when scheduling problems are considered, or cities are visited according to their order in the permutation in traveling salesman problems.

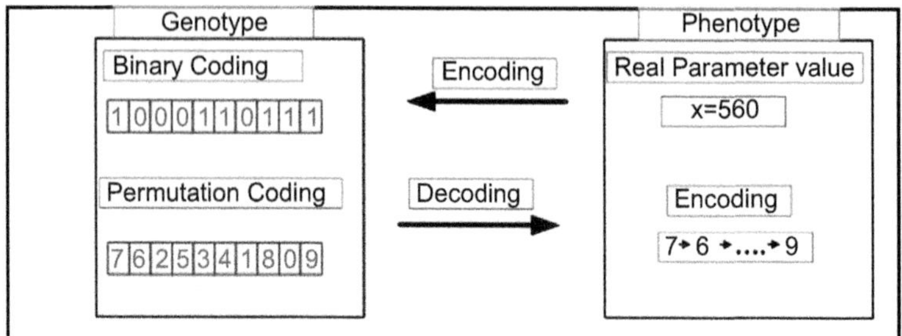

FIGURE 22.2

Genotype and phenotype.

As shown in Figure 22.2, strings that consist of binary or numeral elements are called genotype, and solutions which are decoded from strings are called phenotype. EAs acts over the strings in genotype domain and return the solution strings, which are then decoded to phenotype domain. That is, the users of EAs get final solutions of their optimization problems after the strings obtained by GAs are decoded into the solutions in the phenotype domain.

4.2 INITIAL POPULATION

EA starts with a set of randomly generated solutions of the problem called population. In GA, these randomly generated solutions are called *chromosomes*. Each chromosome's worth is assessed by the cost function. So at this point, the chromosomes are passed to the cost function for evaluation. The size of the initial population is a user-defined parameter and should be decided upon with reference to the number of variables to be optimized and the total number of solutions in the decision space. If the initial population is too small it may not reach an efficient solution. On the other hand, if the population is too large, the algorithm may not converge or may take a long time to find a solution. The chromosomes in later generations will largely be formed using the genes contained in the initial population and so the diversity of the initial "building blocks" can influence the exploration of the search space [23].

Population sizes of 30, 60 or 100 are common, but some researchers use population sizes of several hundred or more. The final choice is often decided by time taken to evaluate a single solution. In real-time applications such as in resource allocation in OFDMA systems where the wireless channels change within very short duration, the time taken by the algorithm to converge becomes a very important criterion. This becomes even more crucial when the channel is assumed to be constant during the period of allocation.

For the problem of resource allocation in OFDMA systems, the set of chromosomes leading to initial population are shown in Figure 22.3. Fig. 22.3(a) depicts population for the problem only subcarrier allocation (OSA) assuming

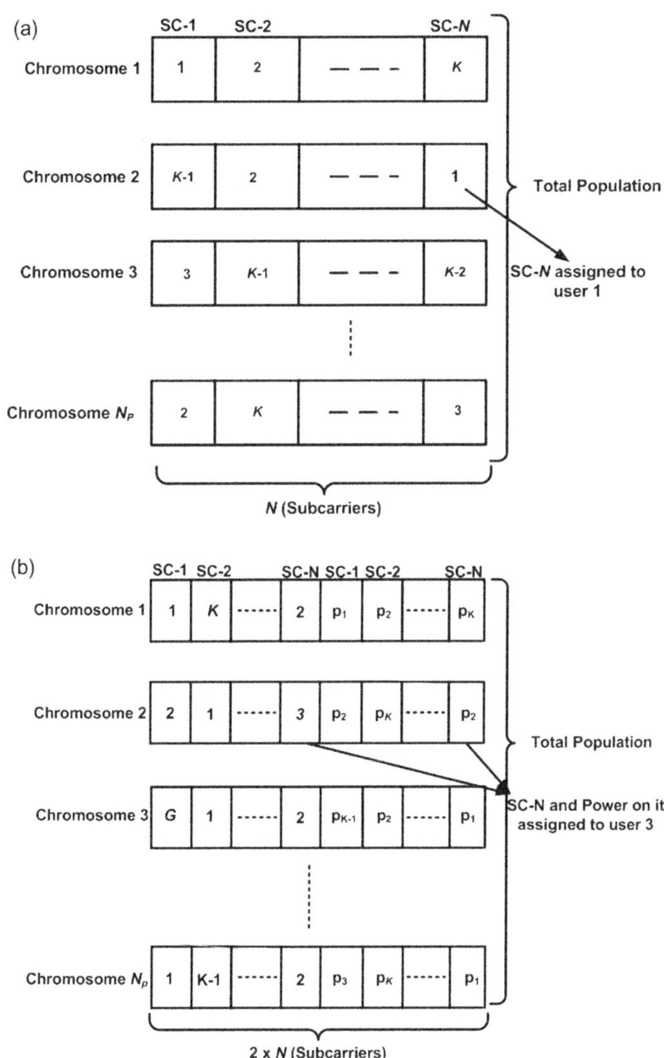

FIGURE 22.3

The set of chromosomes leading to initial population: (a) population for OSA; (b) population for JSPA.

equal power allocation in the case of separate subcarrier and power allocation, while Fig. 22.3(b) depicts the population for the problem of joint subcarrier and power allocation (JSPA).

As shown in Figure 22.3, the initial population in both cases consists of N_P chromosomes; each chromosome has a dimension equal to the number of subcarriers (N) or twice the number of subcarriers ($2N$) for the case of subcarrier allocation or joint subcarrier and power allocation, respectively. In the case of only subcarrier allocation, each element in the chromosome represents the user to which that subcarrier is allocated. In the case of joint subcarrier and power allocation, the first N elements correspond to the user to which that subcarrier is allocated and the next N elements correspond to the power allocated to the user on that subcarrier.

There exist some variants of EAs known as micro-GAs that use a very small population size of around 10 individuals in order to speed up convergence and are suitable to operate in real-time applications [24].

4.3 OBJECTIVE FUNCTIONS AND FITNESS ASSESSMENT

Once the chromosome(s) belonging to each individual in the population have been decoded into the phenotypic domain, it is possible to assess the performance, or fitness, of each of them. This assessment is done through a fitness function which characterizes an individual's performance in the problem domain under consideration. In the natural world, the performance would be an individual's ability to survive in its present environment. Depending on its environment, there can be several traits which an individual is desired to possess.

GA actually searches for a chromosome representing one of the possible solutions, with a better fitness value in the genotype domain. For example, in the case of a function optimization problem, the objective function value $f(\boldsymbol{x})$ is calculated by substituting the solution $\boldsymbol{x}$ decoded from the corresponding binary string obtained by GAs. When the function value $f(\boldsymbol{x})$ is better, the string in the genotype domain which corresponds to the solution $\boldsymbol{x}$ is assigned a better fitness value. For example, in resource allocation in OFDMA, a fitter individual will have a higher sum capacity value than a weaker solution. The assignment of fitness values establishes the basis for selection of pairs of chromosomes that will undergo crossover during reproduction. In order to keep the analogy with the process of natural selection, the fitter solutions must be selected for reproduction more often than the weaker solutions.

4.4 SELECTION OPERATOR

This operator selects individuals in the current generation to be used for constructing the next generation. This operator in GAs is analogous to the process of natural selection in biology. Fitter individuals are more capable of survival and breeding. In GAs, selection allows the search to move towards better solutions as long as the fitness is measured in terms of the objective function of the problem at hand. Therefore, the first step in selection is evaluation of fitness. The evaluation of fitness is performed in such a way that it can be decided to which extent different individuals should survive. In this "survival of the fittest" approach some individuals with very good fitness would in general be preferred to individuals with low fitness. At the same time, low fitness individuals could be lucky to survive. The set of selected individuals is usually referred to as the mating pool. In general, the purpose of the selection operation is to emphasize fit individuals in the population by giving them more chance to breed than less fit individuals.

In cases when a few individuals with low fitness survive, they allow for a continued exploration of the search space for more fruitful regions.

Selection on its own will not allow the full potential of evolution to occur. If only selection took place, the result would be that the best individual of the initial population, which may be far from optimal, would quickly dominate the entire population. As such, selection corresponds to an exploitation of the existing individuals. However, in order to evolve there is a need for discovery or exploration as well. The exploration will allow for new individuals with different features to emerge, which may or may not prove better than those already present in the population. This makes it a necessity to perform some alterations on the population in order to explore for new and better features for the individuals.

Selection preserves characteristics of fit individuals to be used to construct new offspring, and also removes bad individuals so that the overall population fitness improves over successive generations. Some of the most commonly used selection schemes are: **Elitist selection** [25], and **Roulette wheel selection** [26].

4.5 GENETIC OPERATORS

Once two individuals have been selected as parents, a number of genetic operators are applied to their chromosomes to form new offspring. The basic genetic operator is known as crossover or recombination. Like its counterpart in nature, crossover produces new individuals that have some parts of both parents' genetic material. Recombination through crossover can be severely disruptive to the candidate solutions and for that reason many researchers avoid the use of crossover altogether for some problems.

In nature, the features of an individual are contained in the genome. Since the genes interact with each other to create the different features, the genes themselves can be considered as building blocks [27]. These blocks can be recombined to form new and perhaps better features. Based on the fact that superior individuals must have had features that were superior, the building blocks of those individuals must have been better than average. Recombining these better-than-average building blocks should thus, on average, yield increasingly better features in the resulting individuals. Most commonly used crossover operators are **Single/Multi-point crossover** [28,29] and **Uniform crossover** [30].

After selection and crossover, mutations are also permitted in order to explore regions of the design space that may have already become extinct or have never been explored. In GAs, mutation is randomly applied with low probability, typically in the range 0.001 and 0.05, and modifies elements in the chromosomes. Binary mutation is quite simple; it is done by flipping a bit from 0 to 1, or the other way around, according to a specific probability. Mutation is often seen as a mechanism for ensuring the probability of searching any given string will never be zero. It also acts as a safety net to recover good genetic material that may be lost through the action of selection and crossover [31]. Each new child solution is a candidate for mutation.

Mutation is to be used cautiously as it can prevent population convergence if it is applied too often. It is important to mention that low mutation rate results in less exploration, while high mutation rate could be disruptive. With real-valued encoding, a mathematical operation is performed on values within the chromosome. The operation may be a simple multiplication or division and, again, there are many different schemes available. For example, one scheme may take a single value and change it to the maximum (or minimum) possible according to the range bounds for the variable. Others are subtler and may apply only a small change to the variable value.

It is often suggested that mutation has a somewhat secondary function, that of helping to preserve a reasonable level of population diversity—an insurance policy that enables the process to escape from suboptimal regions of the solution space. Proponents of EP, for example, consider crossover to be irrelevant, and mutation plays the major role. The advantage and the drawback of mutations is the unpredictability. Even though mutations can create new or improve on existing good features, they can also result in destruction of useful features or create unwanted features.

4.6 REINSERTION AND ELITISM

It is common to see fixed population sizes in EAs as they are easy to implement. Typically, the original population is completely replaced by the new offspring solutions and the EA is described as being steady-state [32]. If fewer individuals are produced by recombination than the size of the original population, then the fractional difference between the new and old population sizes is termed a generation gap [33]. If one or more of the highest fitness individuals are deterministically allowed to propagate through successive generations, the GA is said to use an elitist strategy.

Elitist strategies compare the fitness values of a new generation with those of the previous generation. If the highest solution from the previous generation is higher than the best solution from the new generation, an individual from the new generation is removed and the previous best is inserted to replace it. This ensures the survival of the fittest rule applies between generations as well as within them. Haupt [34] ensured elitism to occur by keeping the top 50% of each population during each generation, but this approach does not make an efficient search as there will be fewer new chromosomes contributing to the search in each generation. Elitism is a useful function, particularly when search spaces are large and good solutions prove difficult to find and maintain in the population.

4.7 TERMINATION

There are no set rules for termination of an EA. Termination criteria may be set at some given number of generations, or after some measure of convergence has been reached. Eventually, the population will tend to converge to a common point. The choice depends on the problem at hand. The basic termination criterion is when a user-specified computational budget is consumed. This budget can be measured in terms of the number of iterations or CPU time. This criterion of termination does not guarantee that a global optimum is found; it only returns the best solution found for the given budget.

It is common to simply stop a GA once a certain number of generations have been completed. Stopping the GA raises the question "Has the GA found the best possible solution to my problem?" and, unfortunately, the only way to answer the question is to perform an exhaustive search. A good practice to follow when deciding upon termination of the algorithm is to ask the question "Has the GA found a good solution?"

4.8 CONSTRAINT HANDLING

At this point, it is important to note that EAs basically are unconstrained optimization (COP) methods and require additional mechanisms to deal with constraints while solving COPs [35]. As a result, a variety of constraint-handling techniques targeted at EAs have been developed [36]. One of the most commonly used constraint handling techniques is the *penalty function method* [37]. In the penalty function method, based on the amount of constraint violation, an infeasible solution is penalized so that its chance of survival into the next generation is much reduced as compared to a feasible solution. The main drawback of the penalty function method is the requirement of fine tuning of the penalty weights. In order to address this drawback, methods based on the preference of feasible solutions over infeasible solutions have been proposed [38].

In [14], superiority of feasible solution (SFS) criterion proposed by Deb [38] was used. According to SFS, when two solutions X_i and X_j are compared, X_i is regarded superior to X_j if one of the following conditions is satisfied, tested in order: X_i is feasible and X_j is not; X_i and X_j are both feasible, and X_i has a smaller objective value (in a minimization problem) than X_j; Xi and X_j are both infeasible, but X_i has a smaller overall constraint violation than X_j. Therefore, in SFS, feasible solutions are always considered better than infeasible ones. Two infeasible solutions are compared based on their overall constraint violations only, while two feasible solutions are compared based on their objective function values only. Comparison of infeasible solutions based on the overall constraint violation aims to push the infeasible solutions to the feasible region, while comparison of two feasible solutions on the objective value improves the overall solution.

5 SELECT RESULTS AND COMPARISON

In simulations, the wireless channel was modeled as a frequency selective channel consisting of six independent Rayleigh multipaths. Furthermore, each multipath component was modeled as a Clarkes flat fading model [9]. The power delay profile was assumed to be exponentially decaying with e^{-2l}, where l is the multipath index. Therefore relative power of the six multipath components were [0, −8.69, −17.37, −26.06, −34.74, −43.43] dB. The total available bandwidth and transmit power were 1 MHz and 1 W, respectively. The power spectral density of additive white Gaussian noise was −80 dBW/Hz, and the total bandwidth of 1 MHz is divided into 64 subcarriers. The maximum path loss difference was 40 dB, and the user locations were assumed to be uniformly distributed.

The effectiveness of any EA depends on the choice of its parameters. Selection of best parameters is required in order to avoid premature convergence, to ensure diversity in the search space, to intensify the search around best solution regions, etc. Inappropriate choice of parameters may lead to premature convergence or stagnation. For instance, authors in [14] investigated the effect of the control parameters on the performance of the artificial bee colony (ABC) algorithm by manually trying some different values before each run. Based on the manual analysis, in their simulations the value of modification rate (*MR*) equal to 0.8, colony size S_n equal to 30 and the maximum cycle number (*MCN*) equal to 30 were used. Further, the value of *limit* (ν) equal to $0.5 \times S_n \times D$, where D is the dimension of the problem and S_n is the number of solutions in the population, and scout production period (*SPP*) also equal to $0.5 \times S_n \times D$ were used. Experiments were repeated 30 times, each starting from a random population with different seeds. Similar procedures were used in [20,21] in order to find out the best parameters for multiobjective genetic algorithms used therein.

In the following subsections the comparison of results obtained by some of the EAs and their comparison with other traditional optimization algorithms are presented. For all these simulations and for each of these algorithms, a total of 100 different channel realizations were used corresponding to each value of individual parameter and the average values were then used for plotting the graphs.

5.1 SUM CAPACITY VERSUS NUMBER OF USERS

Figure 22.4 shows the variation of sum capacity with the number of users, for a fixed number of subcarriers ($N = 64$). The colony (population) size S_n was fixed to 30, maximum cycle number (number of iterations) (*MCN*) and SNR were fixed to 30 and 10 dB, respectively. The number of users was varied from 2 to 16 in increments of 2.

Figure 22.4 shows the comparisons of the sum capacity achieved by the ABC based JSPA and OSA approaches with that of no fairness method in [4], linear [12], modified linear [13] and immune clonal optimization [37]. It is evident from Figure 22.4 that the use of ABC for resource allocation for OFDMA systems

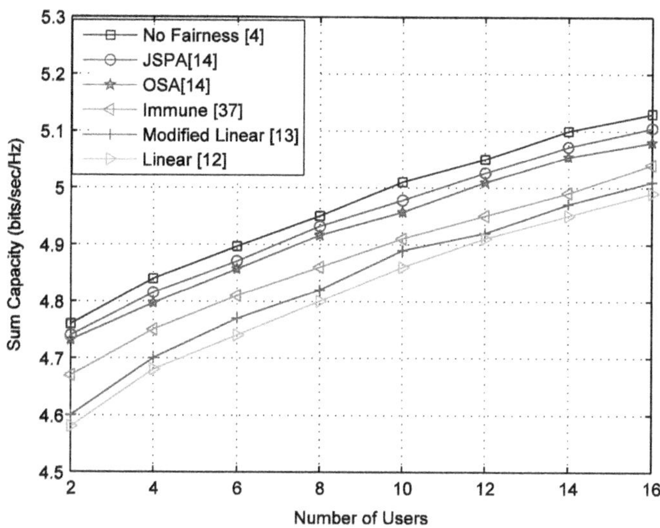

FIGURE 22.4

Sum capacity versus number of users.

provides consistently higher sum capacity than the linear method [12], modified linear [13] as well as immune clonal method [37]. It is also evident from the figure that the JSPA can provide better sum capacities as compared to OSA or separate subcarrier and power allocation [37].

Since no fairness was considered in [4], it achieves maximum capacity. Moreover, as the number of users increases, the sum capacity also increases; this is because of added multiuser diversity gain. Multiuser diversity is obtained by opportunistic user scheduling at either the transmitter or the receiver. The effect of multiuser diversity is predominant in systems with large numbers of users, as with the increasing number of users in the system, the probability that a given subcarrier is in a deep fade for all users decreases. The main advantage of using the ABC algorithm over the immune clonal method [37] is better sum capacity without the need of any fine tuning of penalty factors. Furthermore, in the JSPA approach using the ABC algorithm the subcarriers and power were allocated simultaneously in a single step as compared to a two-step method used in [12,13,37]. Note that in all the simulations, the power allocation proposed in [12] was also used for its modified version [13].

5.2 MINIMUM USER CAPACITY VERSUS NUMBER OF USERS

In this subsection, we compare the minimum user capacity obtained by multi-objective evolutionary algorithms [20,21,39] and other conventional methods used in [4,9]. The same simulation parameters were used as in the previous subsection.

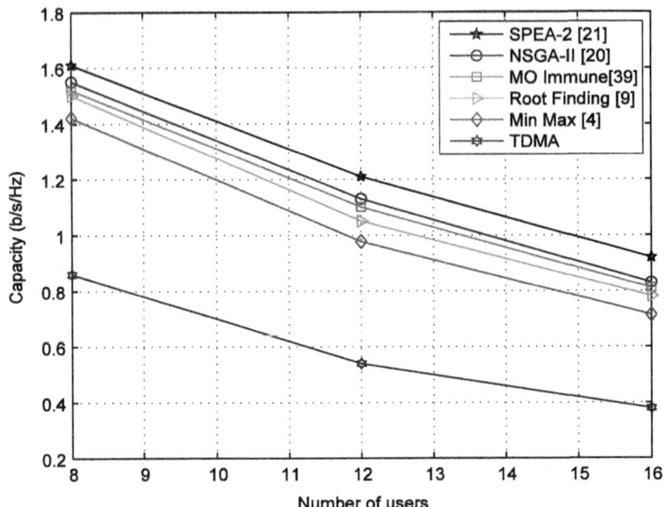

FIGURE 22.5

Minimum user data rate against number of users.

Figure 22.5 shows the comparisons of minimum user data rate obtained versus number of users for the OFDMA system. From Figure 22.5, it is evident that adaptive resource allocation achieves significant capacity gain over the nonadaptive TDMA scheme. It can also be observed that the EAs outperform conventional methods in [4,9] by consistently achieving higher minimum user data rates for all sets of users. Moreover, because the EAs in [20,21,39] maximize the minimum user data rates, fairness among the users is innately guaranteed. Note that, in order to have fair comparison with methods in [4,9], the proportional rate constraint variables γ_i in C_5 were all set to 1. Setting γ_i to 1 in the proportional rate optimization problem in (2) reduces the problem effectively to a minimum rate maximization problem considered in this paper.

Further, it can be observed from Figure 22.5 that the capacity gain over TDMA scheme increases as the number of users increase; this is because of added multi-user diversity gain. Multi-user diversity is obtained by opportunistic user scheduling at either the transmitter or the receiver. The effect of multi-user diversity is predominant in systems with large numbers of users. As with the increasing number of users in the system, the probability that a given subcarrier is in a deep fade for all users decreases. The main advantage of these EAs over the traditional scheme in [4,9] is that they provide better sum capacity together with joint allocation of subcarriers and power.

5.3 SUM CAPACITY VERSUS SNR

Figure 22.6 shows the comparison of sum capacity achieved by ABC algorithm [14] with no fairness method in [4], modified linear [13], linear [12] and immune

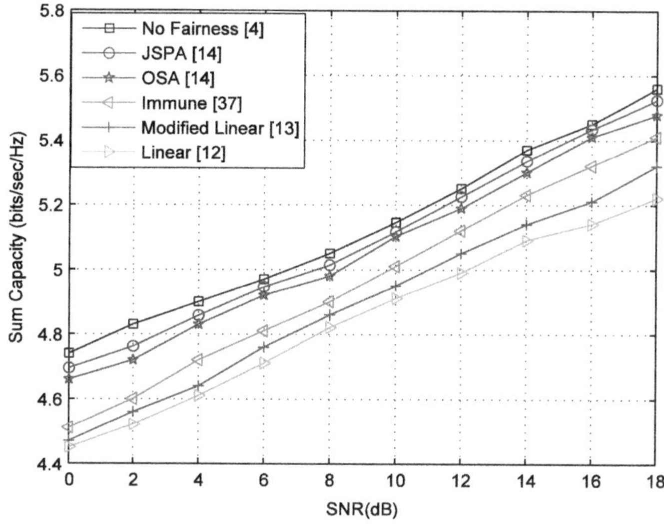

FIGURE 22.6

Sum capacity versus SNR.

clonal optimization [37] for different values of SNR. The number of users was fixed at 16 and both the maximum number of cycles and colony size were fixed to 30. As expected, it can be observed from Figure 22.6 that the sum capacity increases with the average SNR and the ABC algorithm consistently outperforms the linear [12], modified linear [13] and immune clonal optimization algorithm [37].

5.4 PROPORTIONAL FAIRNESS

Figure 22.7 shows the achieved sum capacities in a multiuser OFDM system with four users plotted against the change in proportional rate constraint defined in Table 22.1. The simulation parameters are the same as those in the previous sections. The average subcarrier power of user 1 to user 3 is the same, while the average subcarrier power of user 4 is 10 dB higher than the other three users. As the fairness index increases—that is, the imbalance coefficient (defined in (4)) becomes large—higher sum capacity is achieved. This is reasonable because user 4 has higher average SNR and can utilize the resources more efficiently. The ABC algorithm [14] achieved almost similar fairness as achieved by the schemes in linear [12], modified linear [13] and immune [37], although in the OSA algorithm only subcarrier allocation was performed as compared to both power and subcarrier allocation in [12,13,37]. This result reaffirms the fact that subcarrier allocation provides much higher capacity at lower computational cost than separate subcarrier and power allocation [12,13,37], at the cost of only slight degradation in the fairness of resource allocation. However, JSPA, where both subcarriers

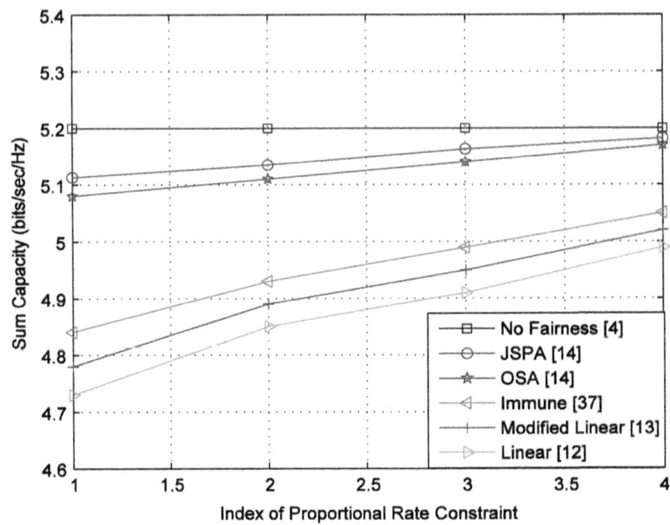

FIGURE 22.7

Sum capacity versus index of proportional rate.

Table 22.1 Proportional Fairness Comparison

Index	Rate Constraint	f(JSPA) [14]	f(OSA) [14]	f(Immune [37])	f(Modified linear [13])	f(Linear [12])
1	γ_1:γ_2:γ_3:γ_4:1:1:1:1	0	0	0	0	0
2	γ_1:γ_2:γ_3:γ_4:1:1:1:4	1.4743	1.7010	1.6875	1.6933	1.6910
3	γ_1:γ_2:γ_3:γ_4:1:1:1:8	9.232	9.8523	9.7625	9.7710	9.7648
4	γ_1:γ_2:γ_3:γ_4:1:1:1:16	32.878	33.9879	33.75	33.85	33.78

and power allocation is executed in a single step, provides much higher capacity with comparable complexity of subcarrier allocation followed by power allocation in [12,13,37].

As expected, the ABC algorithm achieved slightly lower sum capacity as compared to the algorithm proposed in [4]. Since the algorithm in [4] is not constrained by fairness requirements it allocates all the resources to the users with the best subcarrier gains on them. However, it may leave some users without any subcarrier allocated to them and hence is not fair.

Figure 22.8 and Figure 22.9 show the normalized proportions of the capacities for each user for the case of 4 users averaged over 100 channel samples. The normalized capacities are given by $R_k/\sum_{k=1}^{4} R_k$. This is compared to the normalized proportionality constraints $\{\gamma_k\}_{k=1}^{4}$. The proportional rate constraints of γ_1:γ_2:γ_3:γ_4:1:1:1:1 and γ_1:γ_2:γ_3:γ_4:1:1:1:4 were used for simulation results in

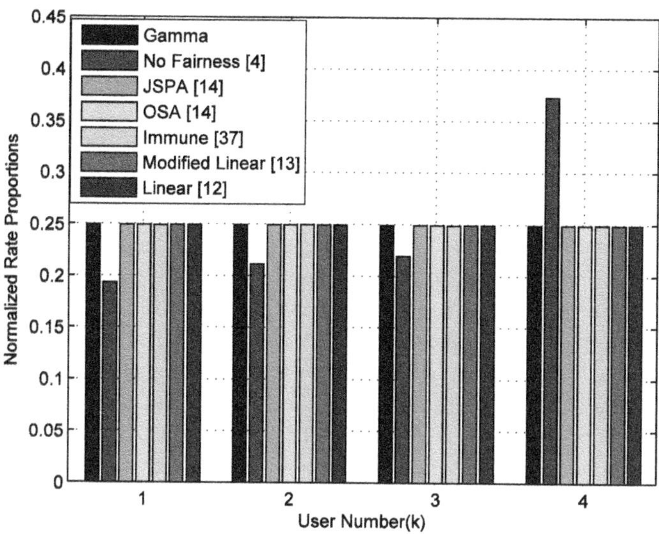

FIGURE 22.8

Normalized capacity ratios per user for 4 users averaged over 100 channels, with the required normalized proportions c shown as the leftmost bar for each user.

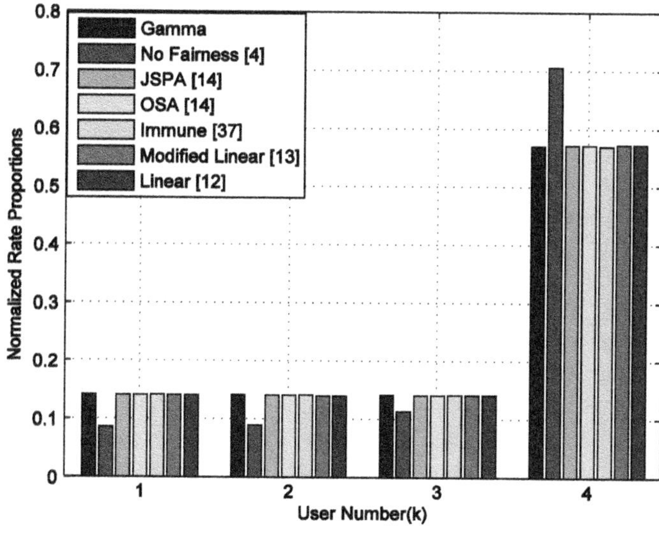

FIGURE 22.9

Normalized capacity ratios per user for 4 users averaged over 100 channels, with the required normalized proportions c shown as the leftmost bar for each user.

Figure 22.8 and Figure 22.9, respectively. The first column denotes the ideal distributions, i.e., $\gamma_k/\sum_{k=1}^{4}\gamma_k$. It can be observed that the capacity obtained by the ABC algorithm closely follows the proportional rate constraints. It can also be observed that the proportionality obtained by OSA algorithm is fairly close to the method in [12,13,37]. However, the JSPA algorithm provides slightly better proportional fairness as compared to all the other methods. This result reaffirms the fact that subcarrier allocation with equal power allocation can provide similar fairness as can be obtained by JSPA and separate power and subcarrier allocation. However, the complexity of OSA with equal power allocation is much less than that of JSPA and separate subcarrier and power allocation [12,13,37].

The algorithm in [4] allocates resources to users with the best gain and does not consider the fairness; hence when user 4 has better subcarrier conditions, it obtains almost all the resources and other users get smaller resources.

Comparison of the imbalance coefficient obtained by ABC-based approaches with that obtained by other methods considered in this paper is shown in Table 22.1. It can be observed from Table 22.1 that even the separate subcarrier and power allocation [12,13,37] achieves a slightly lower imbalance coefficient as compared to the OSA algorithm with equal power allocation. Therefore, it can be concluded that the JSPA and separate subcarriers and power allocation [12,13,37] are only slightly better in proportional fair allocation of resources as compared to the OSA with equal power allocation. The cost of better proportional fairness is higher computational cost. In wireless communication where the channels are highly dynamic, algorithms with lower computational complexity are preferred. Moreover, the assumption of perfect CSI at transmitter and receiver also restricts practical systems to use algorithms with lower computational cost.

5.5 COMPARISON WITH OPTIMAL SOLUTION

This subsection presents the comparison of the performance of ABC-based approaches with respect to optimal solutions. For this comparison the effect of variation in the fairness coefficient ratio (defined in C_5) γ_1/γ_2 on the sum capacity is achieved by ABC-based approaches in comparison to optimal solution. The fairness index defined in (22.3) can be calculated for different values of fairness coefficient ratios (shown on x-axis of Figure 22.10) used for this comparison. With the aim to reduce the time for calculating the optimal solution using exhaustive search, the number of users and subchannels were fixed to 2 and 10, respectively. A total of 200 channel realizations were simulated and the average of the sum capacities obtained were then plotted in Figure 22.10. It can be observed from the figure that, for the case of no path loss difference between the two users, the sum capacity was almost constant and hence not very sensitive to the fairness constraint ratio γ_1/γ_2. On the other hand, for the case of path loss difference between two users, the sum capacity varies significantly with the fairness constraint ratio. For instance, in the case when the user 1 mean channel power ($P_{avg}1$) was 10 dB higher than the user 2 mean channel power ($P_{avg}2$), the sum

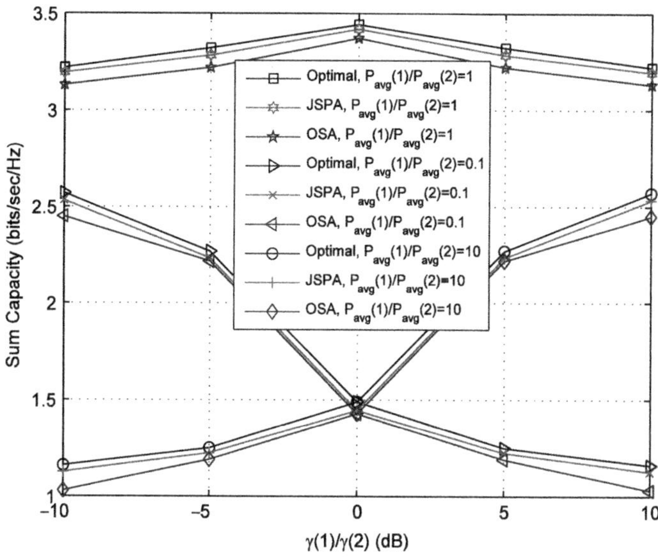

FIGURE 22.10

Performance comparison of proposed approaches and optimal algorithms for the case of two users and ten subchannels.

capacity reduced with the reduction in γ_1/γ_2. This result is inconsistent with the expected result, as γ_1/γ_2 decreases user 2 gets more priority. Therefore, user 2 is assigned a significant portion of the available resources, which, consequently, reduces the achieved capacity, because the mean channel power of user 1 is 10 dB higher than user 2. It can also be observed from Figure 22.10 that the JSPA and OSA approaches achieve about 98% and 96% of the optimal sum capacity. Although in a real wireless communication system the number of users and subchannels will be much larger, we expect the ABC-based approaches to perform even better and further close to optimum, since the EAs are expected to perform better for the case of large parameters. Further, it can be observed from Figure 22.10 that the JSPA approach provides only slight improvement in sum capacity over the OSA approach and hence it can be concluded that equal power allocation can also provide almost equivalent performance as that of joint allocation with lower computational cost.

6 CONCLUSIONS

This chapter presented the use of evolutionary algorithms to solve the problem of subcarrier allocation as well as joint subcarrier and power allocation in the downlink of OFDMA systems. The results produced by the simulations indicate that both the JSPA and OSA approaches perform better in terms of sum capacities as compared to linear [12], modified linear [13] and immune [37]. The sum capacity increases with the increased number of users. The sum capacity also increases initially with the increase in number of iterations and population size but rapidly saturates to a near optimal value. This result suggests that ABC-aided resource allocation can provide significant gain in capacity even with a small number of iterations and population size. Moreover, in ABC-aided resource allocation the search and resource allocation is performed simultaneously as compared to traditional methods [12,13] where the subcarriers are first sorted in accordance with their gains and then allocation is performed. This significantly reduces the complexity of ABC-aided allocation. It can be said that ABC-aided resource allocation is a suitable choice for practical wireless systems like WiMAX (802.16e) where the convergence rate plays a very important role as the wireless channel changes rapidly. The fact that the channel is assumed to be constant during allocation makes the convergence rate a very important parameter for wireless systems. The future scope of this work could be to use multiple antennas on both transmitter and receiver sites, which can provide further gain in capacity because of spatial multiplexing.

REFERENCES

[1] Lawrey E. Multiuser OFDM. In: Proceedings of international symposium on signal processing and its applications. 1999. p. 761–64.

[2] Tse D, Viswanath P. Fundamentals of wireless communication. Cambridge: Cambridge University Press; 2005.

[3] Gheitanchi S, Ali FH, Stipidis E. Particle swarm optimization for resource allocation in OFDMA. In: Proceedings of IEEE DSP07 conference; 2007. p. 383–86.

[4] Jang J, Lee KB. Transmit power adaptation for multiuser OFDM system. IEEE J Sel Areas Commun 2003;21:171–8.

[5] Kim I, Park IS, Lee YH. Use of linear programming for dynamic subchannel and bit allocation in multiuser OFDM. IEEE Trans Veh Technol 2006;55:1195–207.

[6] Reddy YB. Genetic algorithm approach for adaptive subchannel, bit, and power allocation. In: Proceedings of the 2007 IEEE international conference in networking, sensing and control; 2007. p. 15–7.

[7] Sharma N, Wagh S, Anupama KR. Multi-objective resource allocation in multiuser OFDM using PAES. Int J Recent Trends Eng Technol 2010;3(3):121–5.

[8] Rhee W, Cioffi JM. Increasing in capacity of multiuser OFDM system using dynamic subcarriers allocation. In: Proceedings of the IEEE international conference of vehicular technology; 2000. p. 1085–89.

[9] Shen Z, Andrews JG, Evan BL. Adaptive resource allocation in multiuser OFDM systems with proportional rate constraints. IEEE Trans Wirel Commun 2005;46:2726–37.

[10] Tang Z, Zhu Y, Wei G, Zhu J. Cross-layer resource allocation for multiuser OFDM systems based on ESGA. In: Proceedings of the 66th IEEE vehicular technology conference; 2007. p. 1573–77.

[11] Wong CY, Cheng RS, Lataief KB, Murch RD. Multiuser OFDM system with adaptive subchannel, bit and power allocation. IEEE J Sel Areas Commun 1999;17:1747–58.

[12] Wong IC, Shen Z, Evans BL, Andrews JG. A low complexity algorithm for proportional resource allocation in OFDMA systems. In: Proceedings of IEEE international workshop on signal processing systems, 2004. p. 1–6.
[13] Ashourian A, Homayoun RS, Nasab M. A low complexity resource allocation method for OFDMA system based on channel gain. Wirel Pers Commun 2013;71(1):519–29.
[14] Sharma N, Anpalagan A. Bee colony optimization aided adaptive resource allocation in OFDMA systems with proportional rate constraints. Wirel Netw 2013. Available from: http://dx.doi.org/10.1007/s11276-014-0697-y.
[15] Ahmad I, Majumder SP. Adaptive resource allocation based on modified genetic algorithm and particle swarm optimization for multiuser OFDM systems. In: Proceedings of IEEE international conference on electronics and communication engineering, 2008. p. 211–16.
[16] Ahmadi H, Chew YH. Subcarrier-and-bit allocation in multiclass multiuser single-cell OFDMA systems using an ant colony optimization based evolutionary algorithm. In: Proceedings of IEEE wireless communications and networking conference; 2010. p. 1–5.
[17] Ahmadi H, Chew YH, Chai CC. Multicell multiuser OFDMA dynamic resource allocation using ant colony optimization. In: Proceedings of IEEE 73rd vehicular technology conference: VTC2011-Spring, 2011.
[18] Malla S, Ghimire B, Reed MC, Haas H. Energy efficient resource allocation in OFDMA networks using ant-colony optimization. In: International symposium on communications and information technologies (ISCIT); 2012. p. 889–94.
[19] Ahmadi H, Chew YH. Evolutionary algorithms for orthogonal frequency division muntiplexing based dynamic spectrum access systems. Comput Netw 2012;56(14): 3206–18.
[20] Sharma N, Anpalagan A. Joint subcarrier and power allocation in downlink OFDMA systems: a multi-objective approach. Trans Emerg Tel Tech 2013. Available from: http://dx.doi.org/10.1002/ett.2736.
[21] Sharma N, Anpalagan A. Multi-objective resource allocation in multiuser orthogonal frequency division multiplexing system. IET Commun 2013;7(18):2074–83.
[22] Darwin C. On the origin of species by means of natural selection. London: J. Murray; 1859.
[23] Goldberg DE, Sastry K, Latoza T. On the supply of building blocks. In: Proceedings of the genetic and evolutionary computation conference; 2001. p. 336–42.
[24] Krishnakumar K. Micro-genetic algorithms for stationary and non-stationary function optimization. In: Proceedings of SPIE's intelligent control and adaptive systems conference; vol. 1196-32; 1989. p. 289–96.
[25] Deb K. Multi-objective optimization using evolutionary algorithms. West Sussex, England: John Wiley & Sons, Ltd; 2001.
[26] Back T. Evolutionary computation: comments on the history and current state. IEEE Trans Evol Comput 1997;1(1):3–17.
[27] Goldberg DE. The design of innovation: lessons from and for competent genetic algorithms, genetic algorithms and evolutionary computation. Norwell, MA: Kluwer Academic Publishers; 2002.
[28] Spears WM, De Jong KA. An analysis of multi-point crossover. In: Rawlins JW, editor. Foundations of genetic algorithms. 1991. p. 301–15.

[29] Neubauer A. The circular schema theorem for genetic algorithms and two-point crossover. In: Proceedings of second international conference on genetic algorithms in engineering systems; 1997. p. 209–14.

[30] Syswerda G. Uniform crossover in genetic algorithms. In: Proceedings of international conference on genetic algorithms, vol. 3; 1989. p. 2–9.

[31] Goldberg D. Genetic algorithms. Addison-Wesley; 1989.

[32] Syswerda G. A study of reproduction in generational steady-state genetic algorithms. In: Rawlings GJE, editor. Foundations of genetic algorithms. San Mateo, CA: Morgan Kaufmann; 1991.

[33] De Jong KA, Sarma J. Generation gaps revisited. In: Whitley LD, editor. Foundations of genetic algorithms 2. Morgan Kaufmann Publishers; 1993.

[34] Haupt RL, Menozzi JJ, McCormack CJ. Thinned arrays using genetic algorithms. In: Proceedings of IEEE antennas and propagation society, international symposium, vol. 2; 1993, p. 712–15.

[35] Mezura-Montes E, Coello CAC. A simple multi membered evolution strategy to solve constrained optimization problems. IEEE Trans Evol Comput 2005;9(1):1–17.

[36] Venkatraman S, Yen GG. A generic framework for constrained optimization using genetic algorithms. IEEE Trans Evol Comput 2005;9(4):424–35.

[37] Chai Z, Liu F, Qi Y, Zhu S. On the use of immune clonal optimization for joint subcarrier and power allocation in OFDMA with proportional fairness rate. Int J Commun Syst 2012. Available from: http://dx.doi.org/10.1002/dac.1395.

[38] Deb K. An efficient constraint handling method for genetic algorithms. Comput Methods Appl Mech Eng 2000;186(2–4):311–38.

[39] Zheng YC, Si-Feng Z, Lian-Feng S. Rate adaptive resource allocation in orthogonal frequency division multiple access system using multi-objective immune algorithm. Int J Commun Syst 2013. Available from: http://dx.doi.org/10.1002/dac.2539.

CHAPTER

Modeling tools to evaluate the performance of wireless multi-hop networks

23

Hakim Badis and Abderrezak Rachedi
University Paris-Est (UPEM), Cité Descartes, Marne-la-Vallée, France

1 INTRODUCTION

A wireless multi-hop network can be viewed as a set of nodes able to communicate with each other directly or beyond their transmission range by using nodes as relay points acting as routers. Multi-hop communication has several advantages such as: interference reduction, spectrum reuse increase, radio coverage extension, traffic load balancing, and energy consumption reduction. These advantages make multi-hop communication very popular, and several kinds of networks are based on it, such as Mobile Ad hoc Networks (MANETs) [1], Vehicular Ad hoc Networks (VANETs) [2,3], Wireless Sensor Networks (WSNs) [4], Wireless Mesh Networks (WMNs) [5], and so on. Their application range varies from civilian use to disaster recovery and military use. Recently, this technology has become a promising solution for next generation wireless communication systems. It is considered in the standardization process of next-generation mobile broadband communication systems such as 3GPP LTE-Advanced [6], IEEE 802.16j (mobile WiMax) [7], and IEEE 802.16m [8]. For example, Mobile Multi-hop Relaying mechanism (MMR) is integrated into intermediate nodes to forward the traffic from mobile stations (MSs) to the base station (BS). In doing this, a wireless multi-hop network can effectively extend the service coverage and improve the overall capacity performance of a wireless communication system. Moreover, if each node is equipped with multiple antennas, and using point-to-point MIMO techniques to increase the rate of every individual link, then the overall capacity of the network increases linearly with the number of antennas per node [9–11]. For this reason, both single-antenna and multi-antenna communications in multi-hop networks are considered in this chapter.

The performance of relay transmissions is greatly affected by the interference problem. Indeed, each node in a wireless multi-hop network can operate as transmitter, receiver or relay, which produces a tricky situation when several nodes in the same interference area transmit simultaneously. This situation, particularly the interference issue, has a negative impact on the performance of wireless communication. The interference depends on several factors like the locations and the number of interfering transmitters. For instance, the geometry of the locations of the nodes is an important factor because it determines the signal to interference and noise ratio (SINR) at each receiver node. The interference at a receiver node is the sum of the signal powers received from all transmitting nodes, except its own transmitter. This is why the Medium Access Control (MAC) layer protocol plays a key role since it is in charge of sharing and managing the access to the wireless link. Another important point where the interference has an indirect negative impact is the network capacity. The network capacity depends on several factors like network topology, nodes density, connectivity, mobility, etc.

Design of new MAC or routing protocols for wireless multi-hop networks is not an easy task, particularly the performance evaluation step, which is essential to know if the designed protocols work best. That is why modeling and simulation are used extensively to examine the network's behavior under different scenarios, and to optimize its performance before implementing these new protocols in the physical world. In the case of computer network design and optimization, software discrete-event-driven simulators are very important tools, able to simulate complex networks, their architectures and their protocols.

In this chapter, we study different modeling tools mainly used in capacity evaluation for wireless multi-hop networks. We focus on a stochastic modeling based on the Markov chain to evaluate the performance of MAC layer protocols. In addition, we present the relevant models based on the Conflict Graph (CG), particularly graph coloring and cliques in order to evaluate the network capacity. Finally, for large-scale networks, we discuss and analyze models based on an asymptotic approach. The main contribution consists in adapting each modeling tool from SISO to MIMO communication systems.

This chapter is organized as follows: Section 2 provides background on SISO and MIMO systems and gives a brief overview of wireless multi-hop networks: applications and challenges. Section 3 presents different modeling tools to evaluate the capacity performance of wireless multi-hop networks. The final section concludes this chapter.

2 BACKGROUND

2.1 WIRELESS MULTI-HOP NETWORKS

A wireless multi-hop network is a collection of fixed and/or mobile stations that communicate over a shared channel without requiring a fixed wireless infrastructure. In contrast to conventional cellular systems, there is no master–slave relationship between nodes such as base station to mobile stations. According to the communication range, communication between stations is performed by direct connection or through multiple hop relays. Figure 23.1 shows an example of one hop versus

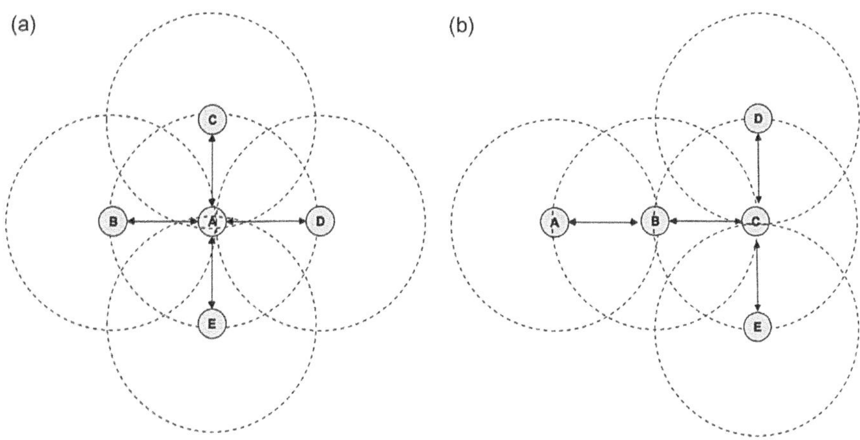

FIGURE 23.1

An example of one hop communication versus multi-hop communication: a) Case of one hop communication; b) Case of multi-hop communication.

multi-hop communication. In Figure 23.1a, node A can communicate directly with any other node (B, C, D, and E) without any assistance of an intermediate or relayed node. However, in the case of multi-hop networks (e.g., Figure 23.1b), node A needs at least one relayed node to reach nodes C, D, and E. In this network, a dynamic routing protocol is needed to ensure communication between nodes.

Several advantages of wireless multi-hop communication can be summarized as follows:

- *Interference reduction*: it is due to the reduction of transmission power (P_{TX}) where nodes use close neighboring nodes to relay packets instead of increasing P_{TX}. This allows reduction of the number of competitor nodes sharing the same channel (link).
- *Spectrum reuse increase*: it is due to the short communication range where the spectrum can be reused more frequently. The spectral efficiency increases when the coverage area decreases [12]. Thus, the availability of frequency channels per unit area increases the system capacity.
- *Radio coverage extension:* it is guaranteed by relayed nodes through multi-hop communication, and it allows reduction of the nodes' isolation.
- *Traffic load balancing:* it is due to the different potential paths to reach the destination. This allows avoidance of the congested nodes/links and selection of noncongested nodes in order to ensure load balancing between them.
- *Power consumption reduction*: it is due to short-range communication where nodes reduce their power transmission and select relayed neighboring nodes to forward packets to their destination.

However, wireless multi-hop networks have some drawbacks such as: system complexity and security. The complexity is related to the design of an efficient routing protocol able to support a large number of nodes on the one hand, and a distributed MAC protocol able to face the hidden nodes problem on the other hand. The security issue is mainly related to the link vulnerability and end-to-end security services (authentication, confidentiality, integrity, and nonrepudiation) guarantee.

Despite extensive research in networking, many challenges remain in the study of wireless multi-hop network including the development of MAC protocols that exploit the capabilities of advanced physical layer technologies like Multiple-Input Multiple-Output (MIMO) and Orthogonal Frequency-Division multiplexing (OFDM) [13]. That is why in this chapter we focus on the modeling tools able to design efficient performance models.

2.2 EXAMPLES OF EMERGING WIRELESS MULTI-HOP NETWORKS

A transmission over multi-hop networks consists of multiple low-power transmissions of data over short distances. This approach enhances the network coverage, increases the network throughput through frequency reuse and reduces the total energy consumption of all participating nodes. This is why multi-hop relaying is being currently considered in the next generation wireless communication systems. Three emerging multihop networks have attracted growing attention: Vehicular Ad Hoc Network (VANETs), Wireless Sensor Networks (WSNs) and Multihop Cellular Networks.

2.2.1 Vehicular Ad Hoc Networks

VANET is a particular case of wireless multihop network, which has the constraint of fast topology changes due to the high node mobility [2,3]. With the increasing number of vehicles equipped with computing technologies and wireless communication devices, intervehicle communication is becoming a promising field of research, standardization, and development. VANETs enable a wide range of applications, such as prevention of collisions, safety, blind crossing, dynamic route scheduling, real-time traffic condition monitoring, etc. Another important application for VANETs is providing Internet connectivity to vehicular nodes. Figure 23.2 shows an example of a VANET.

2.2.2 Wireless Sensor Networks

WSNs are composed of low-power sensor nodes equipped with sensing board, processing, and wireless communication capabilities [4]. Sensor nodes collaborate to collect and to relay sensed information to a sink node using multi-hop communication. These networks can be applied in different applications such as healthcare, military, industrial, monitoring, tracking based on multimedia sensor and many other fields [14,15]. Recently, IP-based sensor networks are attracting more attention, and are enabling the development of the Internet of Things (IoT) [16]. However, energy consumption continues to remain a barrier challenge in many sensor network applications that require long lifetimes. Figure 23.3 shows an example of a WSN.

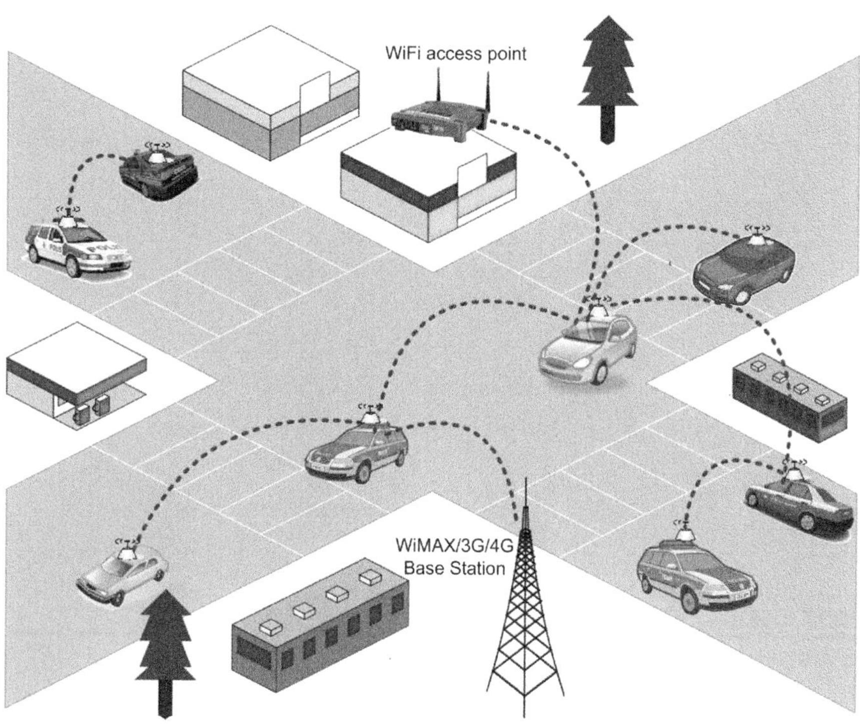

FIGURE 23.2

An example of a VANET.

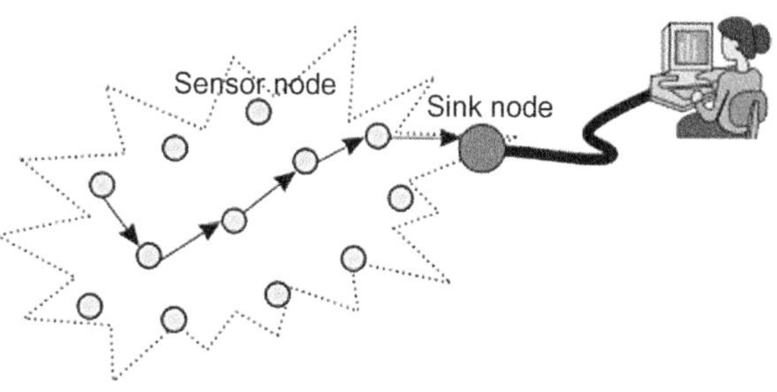

FIGURE 23.3

An example of a WSN.

2.2.3 Multihop Cellular Networks

Multihop Cellular Networks (MCNs) refers to the use of multihop relay nodes (cell phones and/or fixer relay stations) as intermediate nodes between a cell phone and its associated base station (BS) in the radio access network (RAN). This technology has the potential to offer extended cell coverage and improve the

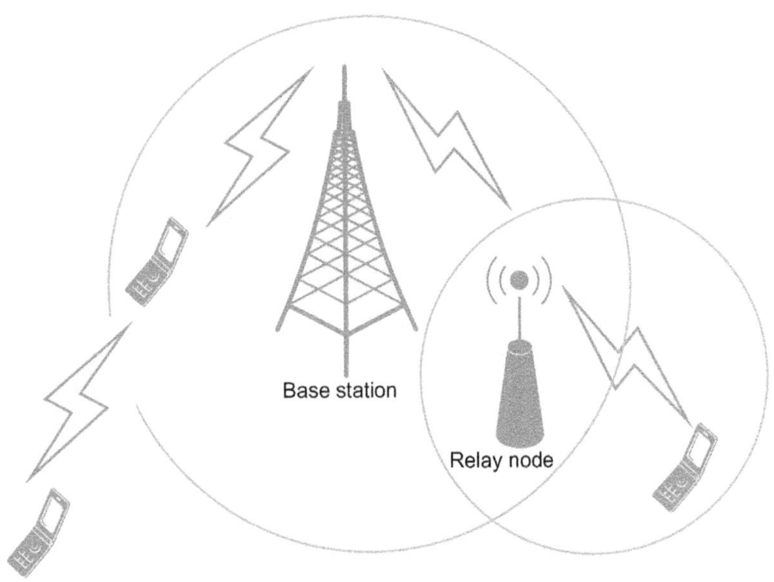

FIGURE 23.4

An example of a MCN.

capacity of the cellular networks. Recently, relay technologies have become a promising solution for the next generation wireless communication systems. It is considered in the standardization process of next-generation mobile broadband communication systems such as 3GPP LTE-Advanced [6], IEEE 802.16j (mobile WiMax) [7], and IEEE 802.16m [8]. Figure 23.4 shows an example of an MCN.

2.3 SISO VERSUS MIMO SYSTEMS

In traditional single-antenna systems, called Single-Input Single-Output (SISO) systems, both transmitter and receiver are equipped with only one antenna each, as illustrated in Figure 23.5. SISO systems are advantageous in terms of simplicity and they are relatively easy to design and implement. They are used in several radio communication technologies like radio and TV broadcast, mobile phone networks (2G and 3G), local and personal wireless technologies (e.g., Wi-Fi and Bluetooth), sensor networks, etc. The channel capacity of such systems is given by Shannon's well-known formula [17]:

$$C = B^{*}\log_2\left(1 + \frac{S}{N}\right)$$

where C is the capacity in bits per second, B is the bandwidth of the channel in hertz, and S/N is the signal-to-noise ratio. Despite the significant progress made in improving

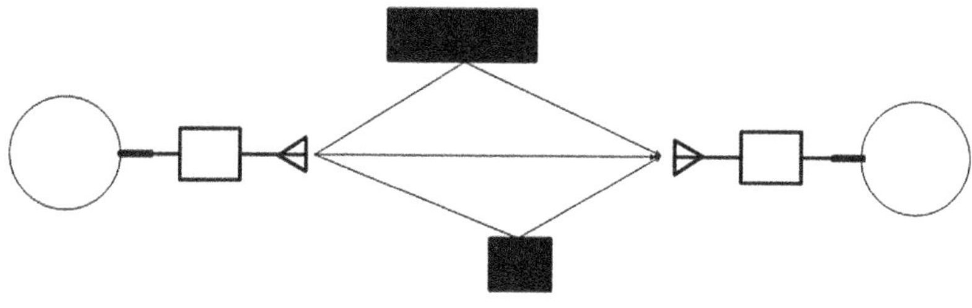

FIGURE 23.5

SISO system.

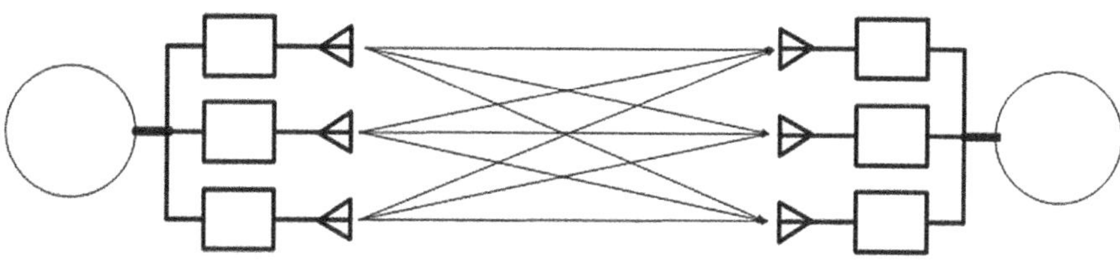

FIGURE 23.6

MIMO system.

the performance of SISO communication systems using OFDM and advanced coding and modulation schemes, they remain insufficient to meet the rapidly growing demands for high bandwidth and robustness in the next-generation networks.

Recently, the use of multiple antennas at both transmitter and receiver has attracted much interest. This system is called Multiple-Input Multiple-Output (MIMO). Figure 23.6 shows an example of a MIMO system having three antennas at the transmitter and at the receiver. MIMO system is introduced in the new communication technologies standards like the Long Term Evolution (LTE) of the third Generation Partnership Project (3GPP) [18] the second generation of the Worldwide Interoperability for Microwave Access (WiMAX 2) [8], WiFi (IEEE 202.11n) [19], etc. This growing interest in MIMO systems is attributed to their various physical layer capabilities that allow:

- an increased channel capacity at higher signal-to-noise ratios by means of spatial multiplexing techniques. The new capacity of MIMO channel is given by [20,21]

  $$C = B^{*}\log_2\left(1 + N \times M \times \frac{S}{N}\right)$$

  where N and M are, respectively, the number of transmitting and receiving antennas;
- a decreased error rate by means of spatial diversity techniques;
- an improved signal-to-noise ratios using beamforming techniques.

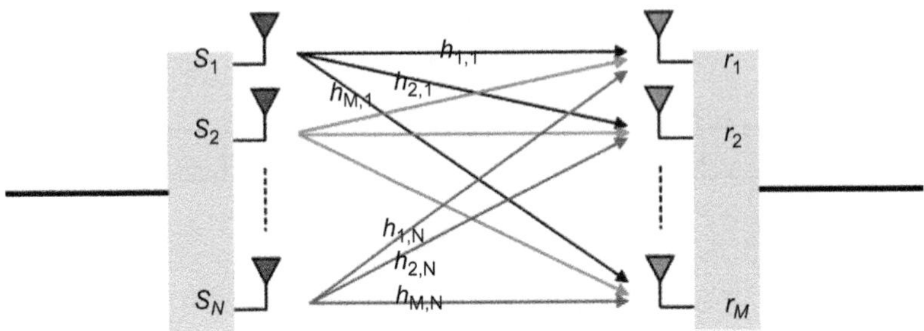

FIGURE 23.7

Spatial multiplexing.

However, these capabilities cannot be fully leveraged at the same time. The optimal strategy can be taken by the upper layers (MAC and routing) based on adapted cross-layer design approaches [22–26].

2.3.1 MIMO capabilities

By employing multiple transmitting antennas and multiple receiving antennas in conjunction with appropriately designed signal processing algorithms, MIMO has offered great benefits to wireless communications compared to conventional SISO systems. Indeed, a significant enhancement of communication quality at the physical layer has been observed in terms of link capacity, link reliability and communication range. In the following paragraphs, we briefly describe the physical layer capabilities.

Spatial multiplexing [20,27,28]. At the transmitter, the data sequence is split into N subsequences that are transmitted simultaneously using the same frequency band (see Figure 23.7). At the receiver, the subsequences are separated by means of interference cancellation algorithms, e.g., linear Zero-Forcing (ZF) [29,30], Minimum Mean-Squared-Error (MMSE) detector [31], Maximum-Likelihood (ML) detector [20,32], Successive Interference Cancellation (SIC) detector [33], etc. For a good error performance, $M \geq N$ is required. Under the spatial multiplexing technique, the capacity of MIMO systems scales linearly with min{N,M} [20]:

$$C = \min\{N, M\} \times B \times \log_2\left(1 + \frac{S}{N}\right)$$

Beamforming [34–36]. Due to antenna array geometry, radio frequency (RF) signals reach antenna elements at different times. By adjusting the initial phase of the RF signals on each antenna element, constructive superposition at the receiver can be achieved. From Figure 23.8, beamformers can reject interference while omnidirectional antennas cannot improve SNR and system capacity.

Spatial Diversity [37–39]. By sending/receiving multiple redundant versions of the same data stream and performing appropriate combining, the error rates decrease. When using diversity transmission, an appropriate pre-processing is needed to enable

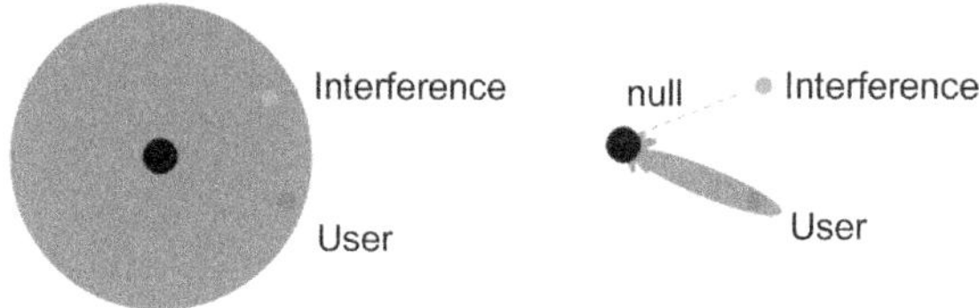

FIGURE 23.8

Beamforming.

a coherent combining at the receiver. The well-known techniques are: Alamouti's scheme [37], space-time trellis codes [38], and orthogonal space-time block codes [40]. When using diversity reception, an appropriate combining is needed. Various combining strategies are proposed [39]: Equal-Gain Combining (EGC), Selection Combining (SC), Maximum-Ratio Combining (MRC), etc.

2.3.2 Modeling of SISO and MIMO channels

The following notation will be used throughout the chapter. Scalars are given by normal letters, vectors by boldface lower case letters, and matrices by boldface upper case letters.

Under SISO systems and using OFDM, the channel can be expressed in the frequency domain as:

$$r(k) = h(k) \times s(k) + w(k)$$

where $r(k)$ is the received symbol, $s(k)$ is the transmitted symbol, $w(k)$ is the Additive White Gaussian Noise (AWGN), and $h(k)$ is a scalar value representing the gain and phase of the channel for subcarrier, k.

For MIMO systems, the received and transmitted symbols become vectors. Let us also assume that a transmitting station is equipped with N antennas and the receiving station is equipped with M antennas. The (M × N) channel matrix $\mathbf{H}(k)$ is written as:

$$\mathbf{H}(k) = \begin{bmatrix} h_{1,1}(k) & h_{1,2}(k) & \cdots & h_{1,N}(k) \\ h_{2,1}(k) & h_{2,2}(k) & \cdots & h_{2,N}(k) \\ \vdots & \vdots & \vdots & \vdots \\ h_{M,1}(k) & h_{M,2}(k) & \cdots & h_{M,N}(k) \end{bmatrix}$$

where the elements, $h_{i,j}(k)$, are each complex scalars representing the channel gain and phase from transmitting antenna j to receiving antenna i, for subcarrier k. The nth column of $\mathbf{H}$ is often referred to as the spatial signature of the nth transmitting antenna across the receiving antenna array. The MIMO channel is given after the fast Fourier transform (FFT) as in [31].

$$\mathbf{r}(k) = \mathbf{H}(k) \times \mathbf{s}(k) + \mathbf{w}(k)$$

where $\mathbf{r}(k) = [\mathbf{r}_1(k)\ \mathbf{r}_2(\text{k})\ldots \mathbf{r}_M(k)]^T$ and $\mathbf{s}(k) = [\mathbf{s}_1(k)\ \mathbf{s}_2(k)\ldots \mathbf{s}_N(k)]^T$.

In order to estimate the SISO channel in OFDM systems, several methods [31] can be employed using time or frequency domain samples. These methods are extended to MIMO channel estimation like Least-Squares (LS) method or Minimum Mean-Squared-Error linear detectors.

3 PERFORMANCE MODELS

A performance model is a model that is used to assess and to evaluate the performance of a network. The development of any performance model follows the following modeling steps: 1) understanding the network properties, (2) model construction, and (3) verification and validation. In this section, we focus on stochastic modeling based on Markov chain to evaluate the network performance in terms of throughput. The conflict graph models are presented to evaluate the network capacity. Finally, we present asymptotic capacity modeling to assess the upper and lower band network capacity. All these performance models are presented for SISO and MIMO systems.

3.1 STOCHASTIC MODELING BASED ON MARKOV CHAIN

3.1.1 Preliminary and definitions

A Markov chain is a very powerful tool used in various fields including physics, economics, engineering, genetics, and more. It is widely used because of its simplicity and flexibility. It's easy to model different systems with an arbitrary number of states and their transition matrix. It is used to model a dynamic system like wireless network that changes its states over time. Markov chains are classified into two kinds: continuous time Markov chains (CTMC) and discrete time Markov chains (DTMC). In DTMC, the state is allowed to change only at the discrete instants; that is not the case for CTMC, where the state can change at any time.

In the computer networks area, the stochastic models are selected to represent uncertainty phenomena of the network (because it depends on many unknown factors) where its behavior varies as time advances.

In this subsection, we focus on two communication systems: Single-Input Single-Output (SISO) and Multiple-Input Multiple-Output (MIMO), both in wireless multi-hop networks. We present stochastic models based on Markov chain to evaluate the network throughput and MAC protocols performance based on IEEE 802.11.

3.1.2 Case of SISO system

In carrier sense multiple access with collision avoidance (CSMA/CA) protocol, before node transmitting, it checks the state of the channel to see if it is busy or idle. If the channel is busy, the node waits until the channel becomes idle. If a collision happens, each node waits for a random time named "backoff". The backoff mechanism has a key role to avoid the interference between competitor nodes.

The well-known performance model based on DTMC to estimate the throughput at the MAC layer with IEEE802.11 is proposed by Bianchi [41]. This model assumes a single collision domain to reduce the complexity of the interference relationship. On the other hand, only one node among N competitor nodes can successfully transmit a packet at any time. It takes into account only the case of N active nodes called the "saturated case" where these nodes always have a packet to transmit. Bianchi's model consists of modeling the behavior of the backoff algorithm of a CSMA/CA in the saturated case as illustrated in Figure 23.9.

The idea behind this model is to get the throughput expression (S) according to different probabilities such as the probability that a transmission is successful (P_s), and the probability that there is at least one transmission in a given slot time (P_{tr}) where these probabilities depend on backoff parameters. Equation 23.1 shows the throughput expression [41]:

$$S_{SISO} = \frac{P_S P_{tr} E[P]}{(1 - P_{tr})\sigma + P_{tr} P_S T_S + P_{tr}(1 - P_S) T_C}$$

where $E[P]$ is the average packet payload size, T_S is the average time the channel is sensed busy, T_C is the average time the channel is sensed busy by each station during collision, and σ is the duration of one empty time slot.

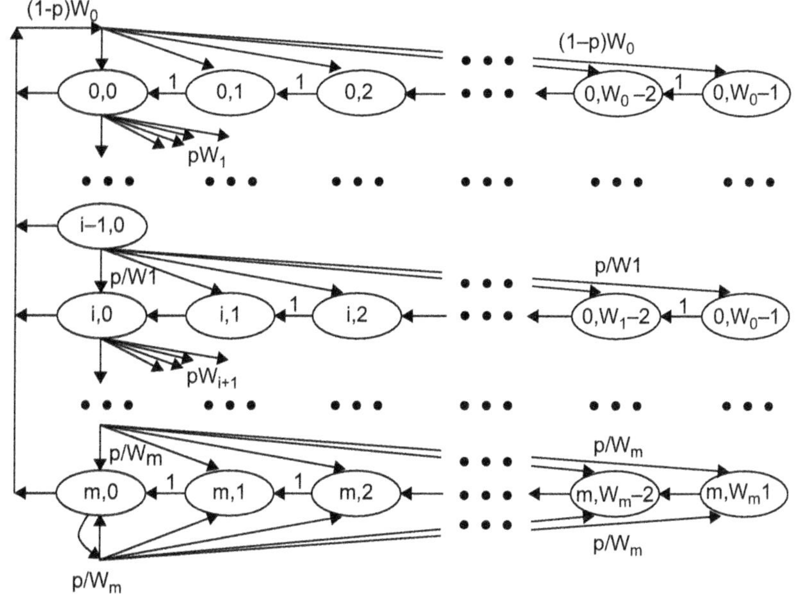

FIGURE 23.9

Markov Chain model for backoff algorithm (case of SISO) [41].

The limits of this analytical model are based on two strong assumptions: saturated traffic, and a single type of traffic assumptions. The typical traffic network is not only nonsaturated, but also is heterogeneous. Many proposed models are based on the nonsaturated case, but without taking into account the heterogeneity of the traffic source [42,43]. The main idea of these models consists of adding a new state representing the case where the buffer is empty without a significant change of the Bianchi's model.

Other proposed models taking into account the heterogeneity of traffic sources with distinct arrival rates are proposed in the literature [44,45]. The added value of these models consists of the integration of the queuing model in order to evaluate the network performance with heterogeneous traffic sources. In the network, it's impossible to predict human behavior and their communication requirements; that's why probability and stochastic processes, particularly queuing theory, are good candidates to predict the network behavior and to understand how traffic arrives. In these models the traffic arriving at the transmission queues is assumed as the Poisson process.

However, all these models are proposed for legacy IEEE 802.11. Recently, other models are proposed for the IEEE 802.11 Enhanced Distributed Channel Access (EDCA) function [46,47]. These models are applied only to the single collision domain where each link interferes with all other links as illustrated in Figure 23.10a. Figure 23.10b illustrates the case of multiple collision areas, which is a typical case of the multi-hop network.

In other words, these models can be used in the case of one hop communication (e.g. WLAN). Other models are proposed in the case of wireless multi-hop networks [48–52]. The adaptation of Bianchi's model to the multi-hop network is mainly based on the integration of neighboring node environments

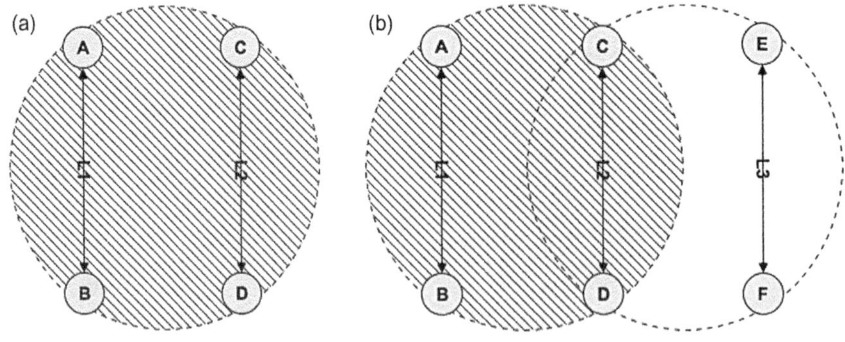

FIGURE 23.10

Collision area in both cases: one-hop and multi-hop network: a) Case of one collision area (one hop communication); b) Case of multiple collision areas (multi-hop communication).

in the throughput expression. In this kind of network the hidden-terminal problem may significantly impact the performance of MAC protocol [53]. That's why in different proposed models each node has to take into account the parameters of its neighboring nodes, such as: the probability to access a channel, the probability of failure, etc. For instance, a three-dimensional Markov chain is used to model the rate adaptation scheme in mobile ad hoc networks (MANETs) [50]. This model evaluates the performance of dynamic rate adaptation scheme able to guarantee a trade-off between throughput and relative fairness. The third dimension represents the rate that varies according to different parameters like link quality.

DTMC is not only used to evaluate the performance of the MAC layer protocol, but also to evaluate the impact of misbehaving nodes that are cheating on Binary Exponential Backoff (BEB) [54,55]. The misbehaving nodes, particularly the selfish nodes, may maliciously manipulate BEB parameters in order to create unfairness situations, or to disrupt the network services.

3.1.3 Case of MIMO

Performance models based on DTMC for MIMO systems do not require any change from those proposed for SISO systems if the MAC protocol used is independent from the underlying physical layer. For example, the MAC protocol CSMA/CA can be applied to both SISO and MIMO systems, and thus all the models previously developed for SISO systems (see Section 3.1.2) remain valid. In literature, several works [56–58] maintain the same performance models based on the original or enhanced Bianchi's two-dimensional Markov chain.

The traditional MAC protocols do not fully exploit the capabilities of the MIMO physical layer like multiple simultaneous transmissions from multiple nodes in the same collision area. Thus, some recent works [22–26] on cross-layer MAC protocols design have been proposed to offer multiple functionalities like parallel transmissions without interference. The main idea is to distribute the spatial degrees of freedom (DoFs) between spatial multiplexing, beamforming and spatial nulling to schedule multiple concurrent transmissions simultaneously. The performance evaluations of these MAC protocols based on DTMC models remain unexplored.

In the following section we show how to integrate the MIMO physical layer capabilities into the performance model. Without loss of generality, we consider only the SPACE-MAC [24] for the study to illustrate the new performance model design. The SPACE-MAC protocol is a MAC layer asynchronous protocol design. It combines beamforming with spatial multiplexing and nulling as a multiuser technology. The basic idea is that all transmitter nodes that want to initiate a new transmission and all potential receivers must handle interference to or from already ongoing transmissions. To extend the Bianchi's

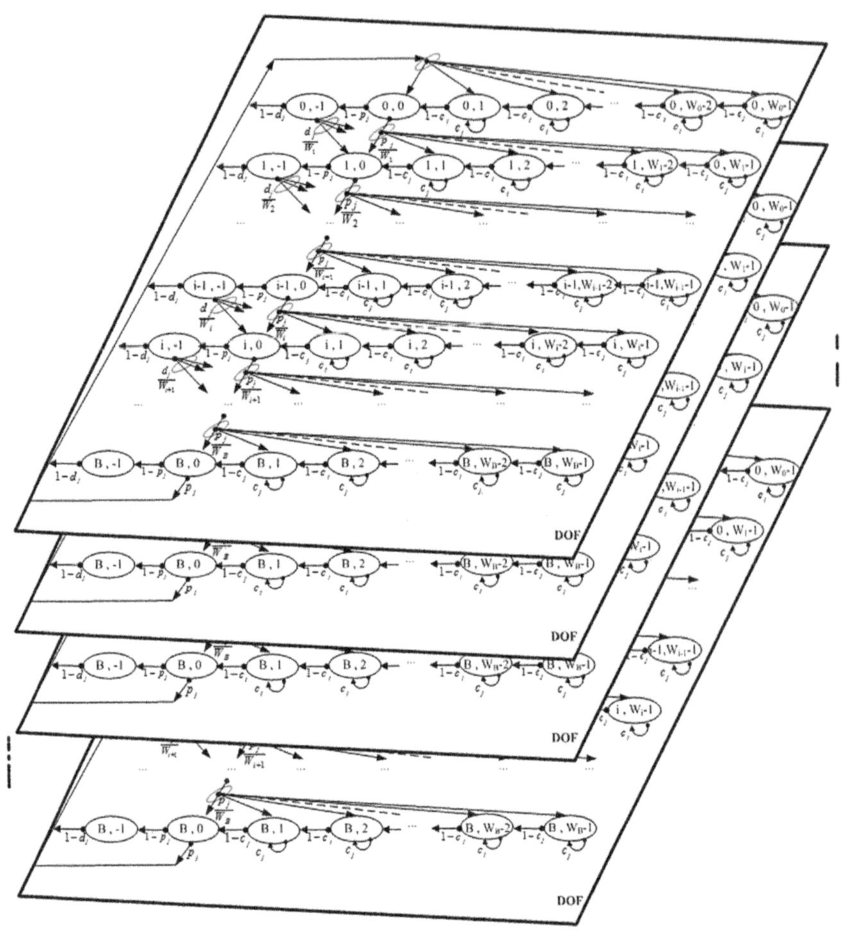

FIGURE 23.11

Three-dimensional Markov chain for backoff algorithm (case MIMO).

model to the SPACE-MAC protocol, a new dimension should be added to express the spatial dimension of MIMO communication link. In this regard, two options are available:

- Adding to each state of the Markov chain a new stochastic process, $\text{DoF}_j(t)$, representing the used DoFs for a node j at time t, where $\text{DoF}_j(t) \in [1,M]$, and M is the j's number of antennas;
- Markov chain for each DoF. Figure 23.11 shows the incremental Bianchi's model.

When a link is active, it can allocate all its DoFs for spatial multiplexing. Therefore, M data streams can be simultaneously transmitted on that link. Consequently, the normalized per-node throughput for MIMO networks S_{MIMO}

can be scaled at most by a factor of the maximum number of multiplexed data streams (M). We obtain

$$S_{MIMO} \leq M \times \frac{P_S P_{tr} E[P]}{(1 - P_{tr})\sigma + P_{tr} P_S T_S + P_{tr}(1 - P_S) T_C}$$

3.2 PERFORMANCE MODELING BASED ON GRAPH COLORING AND CLIQUE CONSTRAINTS

In graph theory, graph coloring is a special case of graph labeling; it is an assignment of labels traditionally called "colors" to the vertices or edges, or both, subject to certain conditions. It has found a number of applications in computer science such as data mining, image segmentation, clustering, image capturing, networking, etc. Since a wireless multi-hop network can be modeled as a graph, graph coloring has found its natural place to address some issues related to connectivity, scheduling, resource allocation, frequency assignment, interference reduction, capacity estimation, etc. In this section we show how to use graph coloring to find a link scheduling policy to satisfy the desired rates. Then, we compute the available link capacity to control the amount of data that could be inserted into the network.

The capacity of any link in wireless multi-hop networks is closely related to the interference relationships between all the active links on the same channel. Conflict graphs offer a way to represent and model such relationships, in which wireless links are represented by nodes, conflicts between links are represented by edges, timeslots are represented by the colors, and allocating time-slots to links can be represented by assigning colors to nodes with no adjacent vertices having the same color. The smallest number of colors needed to color the conflict graph is known as the chromatic number of the graph. If we can color our conflict graph with the fewest number of colors possible, we can schedule all the links having the same color at the same time.

All the mutual conflict links belong to the same complete subgraph (clique). At each slot time, at most one link in each clique can be active. This rule is converted to a constraint and we will show when this constraint becomes necessary and/or sufficient.

3.2.1 Preliminary and definitions

The link capacity in a single-hop network can be defined as the physical transmission bit rate of the source, determined by: the Shannon limit, the fixed modulation scheme and the bit error rate. In a wireless multi-hop network context, several links share the same transmission medium, and so the link capacity decreases when more simultaneous transmissions occur. The sum of all active data stream throughputs in the same interference area gives the consumed capacity.

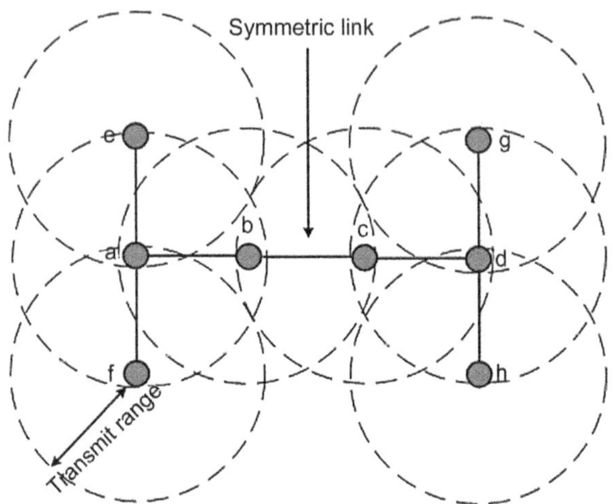

FIGURE 23.12

An example of a connectivity graph.

Consequently, the available link capacity is the difference between the link capacity (in a single-hop network) and the consumed capacity:

$$\text{Available Link Capacity} = \text{Link Capacity} - \text{Consumed Capacity}$$

A wireless multi-hop network can be seen at each time instant as an *undirected graph* in which the nodes represent wireless devices, and there is an edge between two nodes if the nodes are within transmission range of each other. The resulting graph G, called a connectivity graph, is undirected because our channel model only considers bidirectional communication links and ignores unidirectional links. Figure 23.12 shows an example of a connectivity graph. Note that graph G is not Euclidean due to the radio propagation/interference characteristics.

We use the conflict graph to model the interference relationships between links and called it the Links Conflict Graph *LCG*. Every link in connectivity graph G is represented by a node in conflict graph *LCG*. Two nodes in G are connected by an edge if the nodes corresponding to links in G cannot have simultaneous transmissions according to the protocol's interference model. For this purpose and as explained in [59], we use the following interference model: ***any link within distance H from (i,j) is a potential interfering link***. This rule is called the *distance-H interference model*.

Figures 23.13a and 23.13b show *distance-2* and *distance-4 LCGs* for the network topology presented in Figure 23.12. In the mathematical area of graph theory, a clique is a complete subgraph. A clique is maximal if it is not contained in another larger clique. In other words, a maximal clique cannot be extended

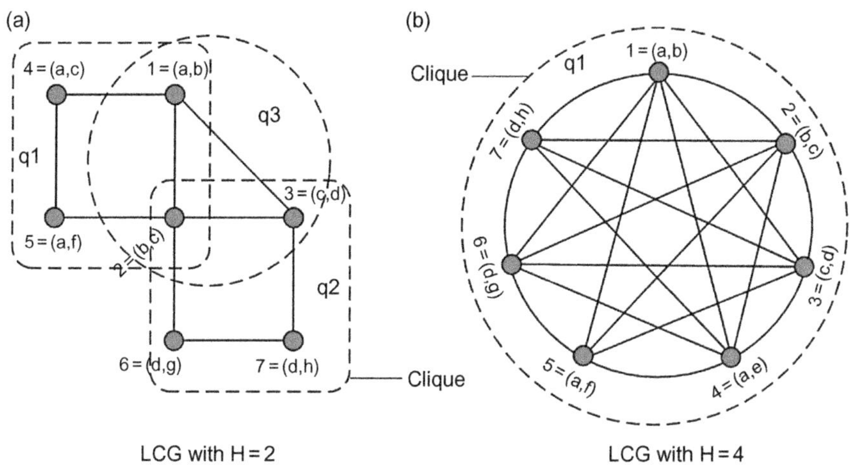

FIGURE 23.13

LCGs for network topology.

Table 23.1 Notation

Symbol	Definition
G	the connectivity graph
G-node	node in the connectivity graph
G-link	link in the connectivity graph
LCG	the link conflict graph
LCG-node	node in the link conflict graph
LCG-link	link in the link conflict graph
C_i	the capacity of a link *i* within a SISO single-hop network
$F_i\ (t)$	the instantaneous flow rate utilization on link *i* at time *t*
β	the scaling factor
Q_i	the incidence matrix of link *i*
Γ_i	the available capacity on link *i*

by including one or more adjacent nodes. Determining maximal cliques in the clique graph will allow one to determine the set of links that mutually conflict with each other in the corresponding connectivity graph. The conflict graphs shown in Figures 23.13a and 23.13b have three ({1,2,4,5}, {1,2,3} and {2,3,6,7}) and one ({1,2,3,4,5,6,7}) maximal cliques, respectively. We note that, as *distance-H* of the interference model increases, the number of maximal cliques decreases.

Table 23.1 summarizes the notation used for the link capacity estimation.

3.2.2 Link capacity estimation with SISO

The *LCG*-nodes in a maximal clique represent the maximal set of mutually contending wireless links, along which only one flow may transit at any given time on the channel. Accordingly, the sum of the rates of *LCG*-nodes in each maximal clique cannot exceed the capacity of the channel; these conditions define which we call the ***clique constraints***. Since the network must satisfy the capacity constraints for all cliques, we can write the clique constraints in a matrix form. We represent a set of flow rates as the column vector F of size n, where n is the number of links in the network G and F_i is the average flow rate assigned to link i. Let C_i be a column vector of size n with all entries equal to the channel capacity C_i. Hence we have,

$$\forall i, Q_i \times F \leq C_i \tag{23.1}$$

where Q_i is an incidence matrix, which is of order q^*n. Here, q is the number of maximal cliques that this link i belongs to, and n is the total number of links. The union of the clique matrices across all the links gives the global clique matrix Q. For example, consider the conflict graph shown in Figure 23.13a, the corresponding global clique matrix Q is given by:

$$Q = \begin{matrix} q_1 \\ q_2 \\ q_3 \end{matrix} \overbrace{\begin{pmatrix} 1 & 1 & 0 & 1 & 1 & 0 & 0 \\ 0 & 1 & 1 & 0 & 0 & 1 & 1 \\ 1 & 1 & 1 & 0 & 0 & 0 & 0 \end{pmatrix}}^{\textit{Links}}$$

Note that the flow rate F_i assigned to link i in an interval of time $[t-\tau,\ t]$, can be written as:

$$\dot{F}_i = \frac{1}{\tau}\int_{t-\tau}^{t} F_i(r)dr,$$

where $F_i(r)$ is the instantaneous flow rate utilization on link i at time r. Consequently, Equation (23.1) can be written as

$$\forall i, Q_i \times \dot{F} \leq C_i \tag{23.2}$$

It is clear that the clique constraints represent a *necessary condition* for a realizable scheduling transmission to exist, since there cannot be a feasible schedule over links that form a violated clique constraint. The challenge is to show that these conditions are also sufficient. Unfortunately, as shown in [59,60], the sufficiency is acquired only when the conflict graph is:

- A perfect graph, i.e., for every induced subgraph, the clique number equals the chromatic number;
- A unit disk graph, i.e., a planar graph in which an edge exists between two vertices if and only if their Euclidean distance is lower than a constant threshold. In this case, the clique constraints must be scaled by a factor of 0.46.

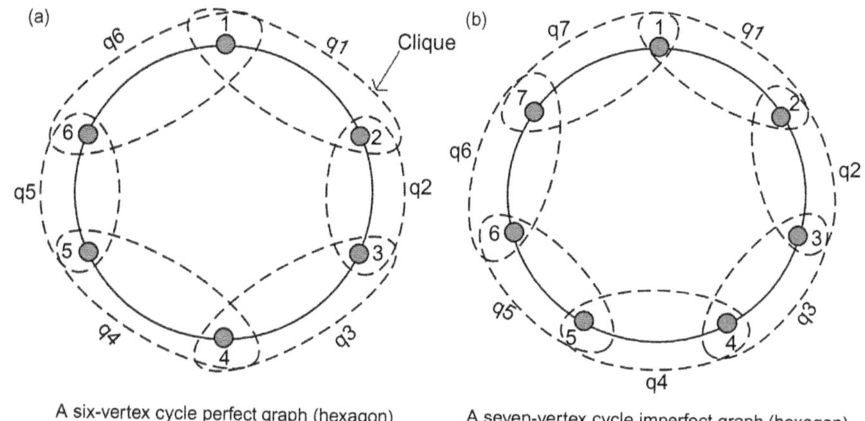

A six-vertex cycle perfect graph (hexagon)

A seven-vertex cycle imperfect graph (hexagon)

FIGURE 23.14

Perfect and imperfect LCGs.

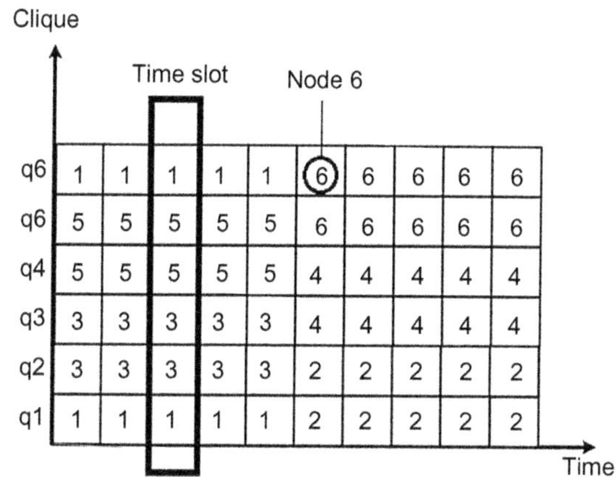

FIGURE 23.15

Sufficient conditions based on a perfect graph clique constraints.

Figures 23.14a and 23.14b illustrate an example of perfect and imperfect graphs, respectively. Let the capacity of the shared channel, C, be 10 time slots. At most 2 *LCG*-nodes may be active simultaneously. Using the clique constraints, the proposed solution for both graphs is 5 time slots for each link (0.5*C*). Figure 23.15 shows that the proposed solution for perfect graphs (see Figure 23.14a) is sufficient. However, Figure 23.16a illustrates the insufficiency of the clique constraints of the imperfect graph shown in Figure 23.14b. Indeed, a scheduling is infeasible when each active *LCG*-node takes 5 time slots. In reality only 4 time slots (0.4*C*) on each link is achievable (see Figure 23.16b).

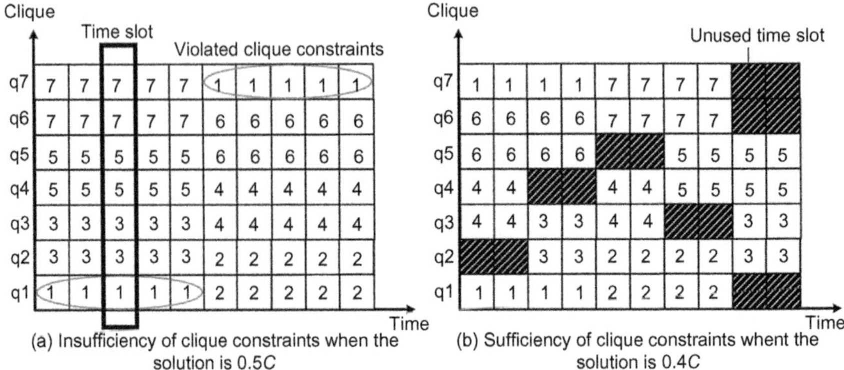

FIGURE 23.16

Insufficient conditions based on imperfect graphs clique constraints.

In a previous work [59], authors have generalized the notion of scaling factor, β, introduced to the clique constraints independently from the conflict graph type. The value of β is closely related to the interference model used:

- If distance-1 interference model is used, then the average scaling factor value is 0.29;
- If distance-2 interference model is used, then the average scaling factor value is 0.44;
- If distance-3 interference model is used, then the average scaling factor value is 0.71.

In general, and based on the notion of scaling factor, Equation (23.2) becomes:

$$\forall i, Q_i \times \dot{F} \leq \beta \times C_i \tag{23.3}$$

As the *LCG*-node can be a part of multiple cliques, it considers all the cliques that it belongs to, and takes the worst case available capacity over all the cliques. The available capacity on an *LCG*-node, i, is

$$\Gamma_i = \min\{(C_i \times \beta) - Q_i \times \dot{F}\} \tag{23.4}$$

where Γ_i is the available capacity on link i, taking into account active flows on i, as well as interference from neighboring links. For example, in Figure 23.13a, let the allocated flow on each *LCG*-node $\{1=(a,b), 2=(b,c), 3=(c,d), 4=(a,e), 5=(a,f), 6=(d,g), 7=(d,h)\}$ be denoted by $\{\dot{F}_1, \dot{F}_2, \dot{F}_3, \dot{F}_4, \dot{F}_5, \dot{F}_6, \dot{F}_7\}$. Then, the available capacity on the link between the monitor-forwarding nodes, $2=(b,c)$, is:

$$\Gamma_2 = \min\{(C_2 \times \beta) - (\dot{F}_1 + \dot{F}_2 + \dot{F}_3), (C_2 \times \beta) - (\dot{F}_1 + \dot{F}_2 + \dot{F}_4 + \dot{F}_5), (C_2 \times \beta) - (\dot{F}_2 + \dot{F}_3 + \dot{F}_6 + \dot{F}_7)\}$$

In the same way, from Figure 23.13b we can write Γ_2 as:

$$\Gamma_2 = C_2 \times \beta - (\dot{F}_1 + \dot{F}_2 + \dot{F}_3 + \dot{F}_4 + \dot{F}_5 + \dot{F}_6 + \dot{F}_7)$$

3.2.3 Link capacity estimation with MIMO

Beamforming and spatial multiplexing techniques allow data stream concentration exclusively on one active link at each time; while spatial nulling allows traffic distribution between different concurrent links. We assume that the number of antennas at each node in the wireless multi-hop network is M. Let M_i ($M_i \leq M$) be the DoF of the MIMO G-link i. At each time, the total number of DoFs allocated for beamforming, spatial multiplexing and spatial nulling cannot exceed M_i. Consequently, a G-link i can support at most M_i active data streams. Each DATA stream on the shared channel consumes one DoF. A DoF is defined by a pair of Transmit/Receive filters (precoding and receiving vectors). Two DoFs used simultaneously on the shared channel must be orthogonal to each other or have a low cross-correlation. Destructive interference can occur when two or more concurrent G-links are simultaneously using nonorthogonal or high cross-correlation DoFs due to:

- imperfect (and possibly very bad) estimation of the channel of active links in the collision area;
- mobility;
- the number of data streams in the collision area exceeds the number of available DoFs.

The first and second reasons can be avoided, respectively, by perfect channel estimation and static network assumptions. However, the third reason can be expressed in the clique constraints.

Denote $Tx(i)$ and $Rx(i)$ the transmitter and receiver of G-link i, respectively. The average flow rate, $\dot{F}_i$,a ssigned to a G-link i is the sum of the multiplexed data streams (S_i) among all available DoFs transmitted by $Tx(i)$ to $Rx(i)$:

$$\dot{F}_i = \sum_{j=1}^{M_i} S_i^j$$

The set of all flow rates assigned to each G-link in the connectivity graph G is given by

$$\dot{F} = (\dot{F}_1 \quad \cdots \quad \dot{F}_i \quad \cdots \quad \dot{F}_n)^T = \left(\sum_{j=1}^{M_1} S_1^j \quad \cdots \quad \sum_{j=1}^{M_i} S_i^j \quad \cdots \quad \sum_{j=1}^{M_k} S_n^j \right)^T$$

where n is the number of G-links in the connectivity graph G. The available capacity on an LCG-node, i, is

$$\Gamma_i = \min\{(M_i \times C_i \times \beta) - Q_i \times \dot{F}\}$$

3.3 ASYMPTOTIC CAPACITY MODELING

Asymptotic capacity analysis helps to determine how the achievable throughput of each node and/or the overall network scales as the number of nodes, n, increases. Such investigation is essential to understand and predict the behavior of large-scale networks. Transition phases from good capacity to poor capacity, or vice versa, may occur when the number of nodes increases. In this case, a threshold determination becomes a necessity.

Usually, upper and lower bounds do not match. When this happens (a gap between the upper and lower bounds), constructive methods try to lower the upper bounds and/or raise the lower bounds until they match. Consequently, algorithms are used to compute one of the bounds (the most simple) and the other is obtained by construction for a maximal matching.

This section investigates the capacity bounds modeling of large random SISO and MIMO wireless multi-hop networks. For the upper bound, critical constraints are applied. Each of them is used to obtain a related bound on the network capacity, and the minimum value between the obtained bounds determines the upper bound. For the lower bound, a constructive scheme is used to match the upper bound. It is based on several steps varying from network partitioning to scheduling and routing. After the model networks have been trained, we describe the critical constraints and the steps of the constructive scheme.

3.3.1 Random wireless multi-hop network modeling

We assume that n nodes are randomly located on the surface of sphere of unit area (S^2) or torus of unit area (T^2). These geometric topologies without borders are used to avoid edge effects, which otherwise complicates the analysis. Each node selects a destination randomly and so a node may be the destination of multiple data streams. Each single data stream can support a fixed data rate of W bits/sec. The per-node throughput $\lambda(n)$ bits/sec is defined as the minimum data rate that can be sent from each source to its destination via multi-hop routing. We use slotted time for transmissions. All transmissions employ the same nominal range or power. For the interference model, two models are proposed:

- Protocol model: Let X_i denote the location of a node i, and $r(n)$ the common range. A transmission from node i to node j is successful if for any other node k that is transmitting simultaneously,

$$\begin{cases} |X_i - X_j| \leq r(n) \\ |X_k - X_j| \geq (1+\Delta) \times r(n) \end{cases}$$

- Physical model: Let T be a subset of nodes simultaneously transmitting, and P be the common power level. A transmission from node i to node j is successful if

$$\frac{\dfrac{P}{|X_i - X_j|^{\alpha}}}{W + \sum_{\substack{k \neq i \\ k \in T}} \dfrac{P}{|X_k - X_j|^{\alpha}}} \geq \beta$$

where β is the minimum signal-to-interference ratio (SIR), W is the ambient noise power level and $\boldsymbol{a}$ ($\alpha > 2$) is the attenuation factor.

3.3.2 Critical constraints and constructive scheme

The upper bound on random wireless multi-hop networks is generally limited by three constraints: 1) connectivity constraint to ensure that the network is connected, so that every source destination pair can successfully communicate; 2) interference constraint to compute the maximum number of simultaneous transmissions on the channel according to the interference model; and 3) destination bottleneck constraint to determine the amount of what can be received by a destination node.

The most common method [10,61–68] used in literature to determine the lower bounds on the capacity of large-scale random wireless multi-hop networks is based on a constructive scheme. It consists of several steps and uses mathematical modeling.

The first step is called network partitioning. It consists of dividing the unit sphere area into small polygons, called cells. Each cell can be a square, hexagon, regular polygon or irregular polygon as in the Voronoi tessellation. The shape of cells is important to control and capture the spatial node distribution. The size of each small cell should be cleverly designed so that the maximum data rate that can be received by the nodes inside the small cell can be computed exactly. Indeed, if the size of each small cell is set too large, then the maximum number of data streams that can be received by the nodes inside the square cannot be computed exactly. On the other hand, if the size of each cell is set too small, then the maximum number of data streams that can be received by the nodes inside the square is likely to be overestimated, leading to a loose upper bound.

The second step is called interference management. It consists of bounding the number of interfering neighbors of each cell. This step allows traffic measurement on each cell and interference-free scheduling. Two cells are interfering neighbors if there is a point in one cell which is within the interference distance of some point in the other cells.

The third step is to define the scheduling method to avoid potential interference among active links. Most research work in this area considers the TDMA scheme due to its simplicity. Each cell becomes active, i.e., the nodes in the given cell can transmit successfully to nodes in the cell or in neighboring cells, at regularly scheduled cell time slots.

The fourth step is to define the routes of a packet on the polygon tessellation. Generally, packets are routed through the cells that lie along the straight line joining the source and the destination node.

The last step is to calculate the expected routes that pass through a cell and infer the expected traffic of each node.

3.3.3 Throughput capacity with SISO

In their seminal paper [61], Gupta and Kumar introduced a fixed random network model to study the throughput capacity of ad hoc networks. The capacity upper

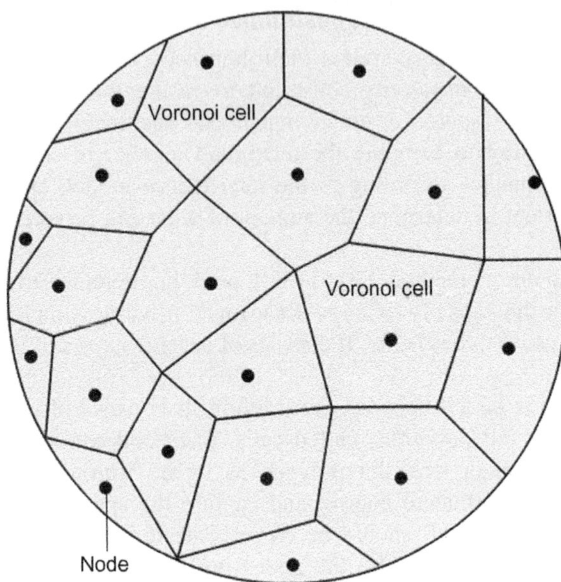

FIGURE 23.17

A Voronoi tessellation of S^2.

bound is determined using the three critical constraints (connectivity, interference and destination bottleneck). The capacity lower bound is determined by the following constructive scheme:

- Define the Voronoi tessellation (see Figure 23.17);
- Bound the number of interfering neighbors of a Voronoi cell. Two cells are interfering neighbors if there is a point in one cell which is within a distance of $(2+\Delta).r(n)$ from some point in the other cell;
- Bound the length of an all-cell transmission schedule. In the protocol model, there is a schedule for transmitting packets such that in every $(1+c_1)$ slots, each cell in V_n gets one slot in which to transmit;
- Define the routes of a packet on the Voronoi tessellation;
- Prove that each cell contains at least one node;
- Bound the mean number of routes that pass through a cell and infer the expected traffic of each node.

Gupta and Kumar have shown two main results:

- Under the Protocol Model, the order of the throughput capacity is

$$\lambda(n) = \Theta\left(\frac{W}{\sqrt{n}\log n}\right) \quad \text{bit/sec}$$

- Under the Physical Model,

$$\begin{cases} \lambda(n) = \Theta\left(\frac{W}{\sqrt{n \log n}}\right) & \text{bit/sec if feasible} \\ \lambda(n) = \Theta\left(\frac{W}{\sqrt{n}}\right) & \text{bit/sec} \quad \text{otherwise} \end{cases}$$

Considerable attention has been devoted to improve Gupta-Kumar results by adopting the same modeling method. Grossglauser and Tse [63] showed that under mobility constraints (uniform stationary distribution) a constant throughput scaling ($\Theta(1)$) per source-destination pair is feasible. In [67] the authors study the throughput and delay trade-off. They use a unit torus area and square cells instead of a unit sphere area and Voronoi tessellation, respectively. Other studies have examined the impact of multiple channels [64], sender-receiver cooperation [62], K-MPRs [68], directional antennas [66], etc.

3.3.4 Throughput capacity with MIMO

To determine the throughput capacity scaling laws for MIMO wireless multi-hop networks, the same modeling method developed for SISO systems can be extended by adding the MIMO physical layer capabilities (see Section 2.2.2). Indeed, the amount of traffic in the network increases under the impact of spatial multiplexing and interference cancellation. A transmitter can send multiple independent data streams simultaneously on a link and multiple conflicting links can be cancelled out. Thus, the critical constraints and the corresponding constructive scheme need to be enriched, which is not a simple task. Due to this difficulty, only a few papers [9,10,69] in the literature analyze the asymptotic capacity lower and upper bounds of MIMO wireless multi-hop networks. In [69], the authors give a first study on how the capacity scales from a source node to a destination node over a sequence of intermediate relay nodes. In [9], Jiang et al. extended the study to random multi-hop networks.

In [9] authors have given an interesting study. To compute the lower bound on the capacity of MIMO wireless multi-hop networks, they assume that all degrees of freedom resource at a transmitter node is allocated for spatial multiplexing. This means that the transmitting and receiving end of the conflicting links cannot cancel interference. For the upper bound, both spatial multiplexing and interference cancellation are considered to increase the overall capacity of the network. The main result obtained in [9] is that MIMO systems can have a constant improvement M on asymptotic capacity compared to the results of Gupta and Kumar:

$$\lambda(n) = \Theta\left(M \times \frac{W}{\sqrt{n \log n}}\right) \quad \text{bit/sec}$$

where M is the number of antennas at each node. To validate the obtained results, Figure 23.18 shows the capacity upper bound behavior. The capacity upper bound

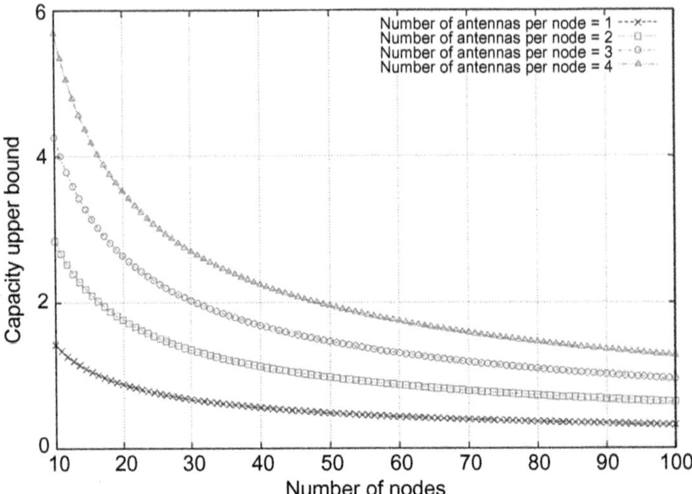

FIGURE 23.18

Capacity upper bound of random wireless multi-hop networks.

increases when the number of antennas increases and decreases when the number of nodes increases.

There are still issues that remain unaddressed in capacity scaling laws for MIMO wireless multi-hop. Indeed, the impact of beamforming, spatial diversity, mobility, etc., remains unexplored. Critical constraints and constructive schemes should consider these new constraints.

4 CONCLUSION

In this chapter, performance models designed for wireless multi-hop networks are considered. We highlight and discuss the different aspects of multi-hop communication, and its introduction in next generation wireless networks. Two physical communication systems are studied: Single-Input Signle-Output (SISO) and Multi-Input Multi-Output (MIMO). In order to evaluate the performance of these systems, we have presented three modeling tools: (i) stochastic modeling based on Markov chain; (ii) Conflict Graph, particularly graph coloring and cliques; (iii) Asymptotic approaches for large-scale networks. Stochastic modeling based on Markov chain is mainly used to evaluate the performance of MAC protocol in terms of throughput and delay. Conflict Graph is used to assess the network capacity in the case of multiple interference domains. Finally, asymptotic study for large-scale networks is presented and discussed in order to assess the upper and lower bands network capacity.

REFERENCES

[1] The Internet Society. RFC2501: mobile Ad hoc networking (MANET): routing protocol performance issues and evaluation considerations. Available from: <http://www.ietf.org/rfc/rfc2501.txt>; 1999.

[2] Aoki M. Inter-vehicle communication: technical issues on vehicle control application. IEEE Commun Mag Oct 1996;90–3.

[3] Al-Sultan S, Al-Doori MM, Al-Bayatti AH, Zedan H. A comprehensive survey on Vehicular Ad Hoc Network. J Netw Comput Appl 2014;37:380–92.

[4] Akyildiz IF, Su W, Sankarasubramaniam Y, Cayirci E. Wireless Sensor Networks: a survey. Comput Netw 2002;38(4):393–422.

[5] Akyildiz IF, Wang X, Wang W. Wireless mesh networks: a survey. Comput Netw 2005;47(4):445–87.

[6] 3GPP TR 36.912 V10.0.0, Feasibility Study for Further Advancements for E-UTRA (LTE-Advanced), 3GPP; March 2011, p. 6–61.

[7] IEEE 802.16j: baseline document for draft standard for local and metropolitan area networks part 16: air interface for fixed and mobile broadband wireless access systems; July 2009.

[8] 802.16m-2011. IEEE standard for local and metropolitan area networks, Part 16: air interface for broadband wireless access systems amendment 3: advanced air interface; May 2011.

[9] Chen B, Gans MJ. MIMO communications in Ad hoc networks. IEEE Trans Signal Process 2006;54(7):2773–83.

[10] Jiang C, Shi Y, Hou Y, Kompella S. On the asymptotic capacity of multi-hop MIMO Ad hoc networks. IEEE Trans Wireless Commun 2011;10(4):1032–7.

[11] Jindal N, Andrews JG, Weber S. Multi-antenna communication in Ad hoc networks: achieving MIMO gains with SIMO transmission. IEEE Trans Commun 2011;59 (2):529–40.

[12] Rappaport SS, Hu L-R. Microcellular communication systems with hierarchical macrocell overlays: traffic performance models and analysis. Proc IEEE 1994;82 (9):1383–97.

[13] Prasad R. OFDM for wireless communications systems. Artech House; 2004.

[14] Boulanouar I, Rached A, Lohier S, Roussel G. Energy-aware object tracking algorithm using heterogeneous Wireless Sensor Networks. In: Fourth IFIP/IEEE wireless days 2011 (IEEE WD'2011), Niagara Falls, Ontario, Canada, October 10–12, 2011.

[15] Prasanna S, Rao S. An overview of Wireless Sensor Networks applications and security. Int J Soft Comput Eng (IJSCE) 2012;2(2): ISSN: 2231-2307.

[16] Tsai C-W, Lai C-F, Vasilakos AV. Future internet of things: open issues and challenges. Wirel Netw 2014;20(8):2201–17.

[17] Shannon C. A mathematical theory of communication. Bell Syst Tech J 1948; 623–56.

[18] 3GPP, technical specification and technical reports for a UTRAN-based 3GPP system, vol. TR21.101 v0.0.8, 2009.

[19] IEEE Std 802.11-2012, Specific requirements Part 11: wireless LAN medium access control (MAC) and physical layer (PHY) specifications, 2012.

[20] Foschini GJ. Layered space-time architecture for wireless communication in a fading environment when using multi-element antennas. Bell Syst Tech J 1996; Autumn:41–59.

[21] Gesbert D, Shafi M, Shiu D, Smith PJ, Naguib A. From theory to practice: an overview of MIMO space-time coded wireless systems. IEEE J Select Areas Commun 2003;21(3):281–302.

[22] Sundaresan K, Sivakumar R, Ingram M, Chang T. A fair medium access control protocol for ad-hoc networks with mimo links. In: IEEE INFOCOM. March 2004. p. 2559–70.

[23] Tang T, Park M, Heath R, Nettles S, A joint MIMO-OFDM transceiver and MAC design for mobile Ad hoc networking. In: Workshop on wireless Ad-Hoc networks; May 2004. p. 315–9.

[24] Park J, Nandan A, Gerla M, Lee H. Space-MAC: enabling spatial reuse using MIMO channel-aware MAC. In: IEEE international conference communications; May 2005. p. 3642–6.

[25] Fakih K, Diouris JF, Andrieux G. Beamforming in Ad hoc networks: MAC design and performance modeling. EURASIP J Wirel Commun Networking 2009.

[26] Mundarath JC, Ramanathan P, Van Veen BD. A cross layer scheme for adaptive antenna array based wireless Ad hoc networks in multipath environments. Wirel Netw 2007;13(5):597–615.

[27] Telatar IE. Capacity of multi-antenna Gaussian channels. Eur Trans Telecommun 1999;10(6):585–96.

[28] Chiani M, Win MZ, Shin H. MIMO networks: the effect of interference. IEEE Trans Inf Theory 2009;56(1).

[29] Choi L-U, Murch RD. A transmit preprocessing technique for multiuser MIMO systems using a decomposition approach. IEEE Trans Wireless Commun 2004;3(1):20–4.

[30] Spencer QH, Swindlehurst AL, Haardt M. Zero-forcing methods for downlink spatial multiplexing in multiuser MIMO channels. IEEE Trans Signal Process 2004;52(2): 388–404.

[31] Barry J, Lee E, Messerschmitt D. Digital communications. 3rd ed. New York: Springer; 2004.

[32] Oussama Damen M, Gamal HE, Caire Senior Member G. IEEE on maximum-likelihood detection and the search for the closest lattice point. IEEE Trans Inf Theory 2003;49(10).

[33] Weber SP, Andrews JG, Yang X, de Veciana G. Transmission capacity of wireless Ad hoc networks with successive interference cancellation. IEEE Trans Inf Theory 2007;53(8).

[34] Lo TKY. Maximum ratio transmission. IEEE Trans Commun 1999;47:1458–61.

[35] Love DJ, Heath RW, Strohmer T. Grassmannian beamforming for multiple-input multiple-output wireless systems. IEEE Trans Inf Theory 2003;49(10).

[36] Kang M, Alouini M-S. Largest eigenvalue of complex Wishart matrices and performance analysis of MIMO MRC systems. IEEE J Select Areas Commun 2003;21: 418–26.

[37] Alamouti SM. A simple transmit diversity technique for wireless communications. IEEE J Select Areas Commun 1998;16:1451–8.

[38] Tarokh V, Seshadri N, Calderbank AR. Space-time codes for high data rate wireless communication: Performance criterion and code construction. IEEE Trans Inf Theory Mar 1998;44:744–65.

[39] Zheng L, Tse DNC. Diversity and multiplexing: A fundamental tradeoff in multiple-antenna channels. IEEE Trans Inf Theory 2003;49:1073–96.

[40] Tarokh V, Jafarkhani H, Calderbank AR. Space-time block codes from orthogonal designs. IEEE Trans Inf Theory 1999;45:1456–67.

[41] Bianchi G. Performance analysis of the IEEE 802.11 distributed coordination function. IEEE J Select Areas Commun 2000;18(3):535–47.

[42] Daneshgaran F, Laddomada M, Mesiti F, Mondin M. Unsaturated throughput analysis of IEEE 802.11 in presence of non ideal transmission channel and capture effects. IEEE Trans Wireless Commun 2008;7(4):1276–86.

[43] Lee W, Wang C, Sohraby K. On use of traditional M/G/1 model for IEEE 802.11 DCF in unsaturated traffic conditions. In: Wireless communications and networking conference, WCNC 2006. IEEE, vol. 4; 2006. p. 1933–7.

[44] Malone D, Duffy K, Leith D. Modeling the 802.11 distributed coordination function in nonsaturated heterogeneous conditions. IEEE/ACM Trans Networking 2007;15 (1):159–72.

[45] Alazemi HM, Margolis A, Choi J, Vijaykumar R, Roy S. Stochastic modelling and analysis of 802.11 DCF with heterogeneous non-saturated nodes. Comput Commun 2007;30(18):3652–61.

[46] Kosek-Szott K. A Comprehensive analysis of IEEE 802.11 DCF heterogeneous traffic sources. In: Ad hoc Networks, vol. 16; 2014. p. 165–81.

[47] Nguyen SH, Vu HL, Andrew LL. Performance analysis of IEEE 802.11 WLANs with saturated and unsaturated sources. IEEE Trans Veh Technol 2012;61(1): 333–45.

[48] Aziz A, Durvy M, Dousse O, Thiran P. Models of 802.11 multi-hop networks: theoretical insights and experimental validation. In: IEEE COMSNETS; 2011.

[49] Abreu T, Baynat B, Begin T, Guérin-Lassous I. Hierarchical modeling of IEEE 802.11 Multi-hop wireless networks. In: 16th ACM international conference on modeling, analysis and simulation of wireless and mobile systems (MSWIM); 2013.

[50] Benslimane A, Rachedi A. Rate adaptation scheme for IEEE 802.11-based MANETs. J Netw Comput Appl 2014;39:126–39.

[51] Ghaboosi K, Latva-aho M, Xiao Y, Khalaj BH. IEEE Trans Veh Technol 2009;58(7).

[52] Jiao W, Sheng M, Lui KS, Shi Y. End-to-end delay distribution analysis for stochastic admission control in multi-hop wireless networks. IEEE Trans Wireless Commun 2014;13(3):1308–20.

[53] Xu S, Saadawi T. Does the IEEE 802.11 MAC protocol work well in multihop wireless Ad hoc networks? IEEE Commun Mag 2001;39(6):130–7.

[54] Guang L, Assi CM, Benslimane A. Enhancing IEEE 802.11 random backoff in selfish environments. IEEE Trans Veh Technol 2008;57(3):1806–22.

[55] Rachedi A, Benslimane A. Toward a cross-layer monitoring process for mobile Ad hoc networks. Secur Commun Netw 2009;2(4):351–68.

[56] Zhou T, Yang Y, Eggerling SJ, Zhong Z, Sharif H. A novel distributed MIMO aware MAC protocol design with a Markovian framework for performance evaluation. In: IEEE military communications conference (MILCOM'08); 2008.

[57] Akin S, Gursoy MC. On the throughput and energy efficiency of cognitive MIMO transmissions. IEEE Trans Veh Technol 2013;62(7):3245–60.

[58] Carvalho M, Garcia-Luna-Aceves JJ. Analytical modeling of Ad hoc networks that utilize space-time coding. Fourth international symposium on modeling and optimization in mobile, Ad hoc and wireless networks; 2006. p. 1–11.

[59] Badis H. An efficient bandwidth guaranteed routing for Ad hoc networks using IEEE 802.11 with interference consideration In: Tenth ACM/IEEE MSWIM; October 2007. p. 252–60.

[60] Gupta R, Musacchio J, Walrand J. Sufficient rate constraints for QoS flows in ad-hoc networks. Ad Hoc Netw J 2007;5(4):429–43.

[61] Gupta P, Kumar PR. The capacity of wireless networks. IEEE Trans Inf Theory 2000;46(2):388–404.

[62] de Moraes RM, Sadjadpour HR, Garcia-Luna-Aceves JJ. Many-to-many communication: a new approach for collaboration in MANETs. In: IEEE INFOCOM; May 2007. p. 1829–37.

[63] Grossglauser M, Tse D. Mobility increases the capacity of Ad hoc wireless networks. IEEE/ACM Trans Networking 2002;10(4):477–86.

[64] Kyasanur P, Vaidya NH. Capacity of multi-channel wireless networks: impact of number of channels and interfaces. In: ACM MobiCom; 2005. p. 43–57.

[65] Ozgur A, Leveque O, Tse D. Hierarchical cooperation achieves linear capacity scaling in Ad hoc networks. IEEE Trans Inf Theory 2007;53(10):3549–72.

[66] Zhang J, Liew SC. Capacity improvement of wireless Ad hoc networks with directional antennas. In: ACM MobiCom, Aug. 2006. p. 17–9.

[67] Sharma G, Mazumdar R, Shroff B. Delay and capacity trade-offs in mobile Ad hoc networks: a global perspective. IEEE/ACM Trans Networking 2007;15(5).

[68] Guo M, Wang X, Wu M. On the capacity of κ-MPR wireless networks. IEEE Trans Wireless Commun 2009;8(7):3878–86.

[69] Bolcskei H, Nabar RU, Oyman O, Paulraj AJ. Capacity scaling laws in MIMO relay networks. IEEE Trans Wireless Commun 2006;5(6):1433–44.

CHAPTER

24 Modeling and performance evaluation of resource allocation for LTE femtocell networks

Ying Loong Lee[1], Jonathan Loo[2], and Teong Chee Chuah[1]
[1]*Multimedia University, Cyberjaya, Selangor, Malaysia*
[2]*Middlesex University, London, UK*

1 INTRODUCTION

Long Term Evolution (LTE), which evolved from conventional 3rd Generation Partnership Project (3GPP) systems, is expected to provide downlink and uplink peak data rates of over 100 Mbps and 50 Mbps, respectively. To support such high data rates, orthogonal frequency division multiple access (OFDMA) is chosen as the access technology. The 3GPP organization further includes the support of femtocells in the LTE standard, which has become a research area heavily pursued in both academia and industry.

An LTE femtocell is a small cell created by a low-power base station known as Home evolved NodeB (HeNB). The deployment of femtocells within a macrocell forms a two-tier LTE network, as shown in Figure 24.1. In the LTE standard, the HeNB associates with the LTE core network, i.e., the mobility management entity (MME) and the serving gateway (S-GW), via a backhaul which could be indoor broadband connections such as digital subscriber line (DSL). The plug-and-play installation feature of an HeNB makes the femtocell technology attractive and appealing from both technical and business perspectives.

The deployment of femtocells gives rise to several technical challenges in the area of resource allocation, namely interference, fairness and quality of service (QoS) [1]. In an LTE femtocell network, *cross-tier interference* occurs when the LTE macro base station, also known as evolved NodeB (eNB), interferes with an HeNB or vice versa, whereas *co-tier interference* occurs between HeNBs when they interfere with each other [2]. As the deployment of femtocells forms an additional tier to the macrocell network, the task of guaranteeing fairness for resource allocation among the users becomes more challenging. Similarly, the task of guaranteeing the

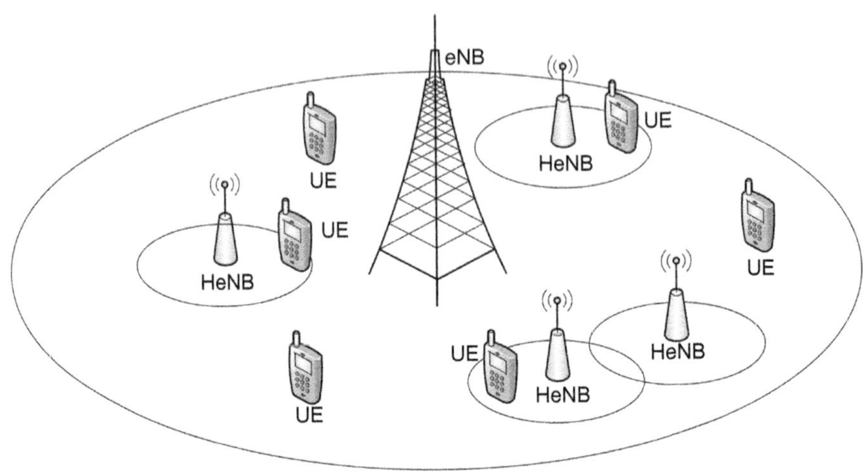

FIGURE 24.1

A two-tier LTE network consists of an eNB and several HeNBs.

QoS of each user becomes more difficult, especially when large numbers of femtocells are deployed and the available resources are limited. These issues have become the main research topics in the LTE femtocell research community.

As the femtocell resource allocation research becomes increasingly popular, performance evaluation of femtocell networks is important to study the issues of interference, fairness and QoS. Before carrying out performance evaluation of a femtocell network, one has to consider either using a real system or a model. Since cost and effort are major issues for real system implementation, models are more favored. Modeling is a process of imitating and abstracting a real system by considering the relevant aspects of the system related to the objectives of the study. Once a performance model is built, the modeled system can be studied by implementing the model in a computer simulation program.

Simulation has been widely used to investigate the performance of resource allocation schemes in cellular networks. Computer simulation is highly favored because experimental evaluations are costly and time-consuming. Simulation modeling allows researchers to quickly study certain aspects of a cellular network. As simulation models are much simpler, modifications can be made easily. Thus, simulation modeling is cost-effective, time-efficient and flexible.

In the literature, simulation modeling is the most widely used evaluation approach in the studies of resource allocation in LTE femtocell networks. A survey of these studies can be found in [1]. Many studies have developed their own simulation models for performance evaluation of LTE femtocell networks. However, these simulation models have several limitations, e.g., the radio bearers are not implemented such as those in [3–8] and several LTE QoS requirements are not considered such as those in [5–8]. Furthermore, each of the simulation models is built based on different objectives of the studies and different scenarios. Therefore, it is difficult to compare the results presented by different researchers. In [9] and [10], a simulation tool known as LTE-Sim, which can create a near-complete simulation model of LTE femtocell networks, is introduced. In LTE-Sim, various metrics can be used for performance evaluation, which are not limited to those that are used specifically for LTE network evaluation. However, dynamic resource allocation among femtocells is not implemented in LTE-Sim. Nevertheless, LTE-Sim provides users with the flexibility to modify the source code.

This chapter intends to guide the reader in carrying out performance and simulation modeling for resource allocation in LTE femtocell networks using LTE-Sim. Without loss of generality, we use the centralized dynamic frequency planning (C-DFP) [11] and distributed random access (DRA) [12] schemes as the reference femtocell resource allocation schemes for the discussion of the simulation methodology. These two schemes are chosen because one is based on a centralized model and the other is based on a distributed model. In this way, we demonstrate how these two different models can be implemented in simulation programs, and make a performance comparison in terms of signal-to-interference plus noise ratio (SINR), throughput, packet loss rate (PLR), delay, fairness and complexity.

The rest of the chapter is organized as follows. Section 2 gives an overview of LTE systems; Section 3 describes resource allocation among femtocells and introduces the C-DFP and DRA schemes. Section 4 describes resource allocation among mobile users. In Section 5, simulation modeling of LTE femtocell networks is discussed. Section 6 introduces the LTE-Sim simulator and its features for simulation modeling. Implementation of the C-DFP and DRA schemes using LTE-Sim is provided in Section 7. Section 8 explains how LTE-Sim can be used to create a simulation model for performance evaluation. Section 9 discusses and evaluates the simulation model created using different metrics. In addition, through the simulation results obtained, we discuss how the strengths and weaknesses of the two schemes under investigation can be identified. Section 10 concludes the chapter.

2 LTE SYSTEM OVERVIEW

LTE brings a major evolution to its 3GPP predecessors in terms of performance, architecture and functionality. The peak downlink and uplink rates of LTE systems are at least four times those of the 3GPP predecessors. This is the result of employing the OFDMA technology, which is robust against intracell interference and frequency selective fading. Both the LTE core and access networks are evolved from its 3GPP predecessors to support high peak rates and provide reduced latency. In LTE systems, radio resource management is decentralized to eNBs. Thus, the eNBs are responsible for performing functions such as power control, admission control, and resource allocation for their associated mobile users.

In LTE systems, radio resources are time-frequency blocks, which are assigned to user equipment (UE) every transmission time interval (TTI) of 1 ms. These time-frequency blocks are known as physical resource blocks (PRBs) [13]. In the frequency domain, each PRB is a subchannel composed of 12 OFDMA subcarriers each with a 15-kHz bandwidth. In the time domain, assuming a short cyclic prefix configuration, each PRB consists of 7 OFDM symbols, which last 0.5 ms. In the LTE downlink, PRBs can be assigned to a UE from any part of the channel bandwidth. Depending on how the resource allocation technique would be implemented, however, UEs may or may not receive PRBs, depending on the channel quality. For example, resources can be allocated to guarantee at least some PRBs for each UE even in a poor channel condition. In each TTI, the minimum number of resource units allocated to a UE is two PRBs (in the time domain). Henceforth, we refer these two PRBs as one resource tile in the rest of this chapter.

The LTE standard defines radio bearers as the logical connections or data flows established between the LTE core network and the UEs. Each radio bearer established carries a single type of data traffic, which could be best-effort, voice over Internet protocol (VoIP) and video. If a UE has three data flows, three radio bearers are established for the UE. Each radio bearer corresponds to a set of QoS parameters. These parameters specify the constraints that must be satisfied to ensure seamless communications. The constraints include delay budget, PLR and minimum guaranteed bit rate (GBR) of the radio bearer.

3 RESOURCE ALLOCATION AMONG FEMTOCELLS

The current LTE specifications [14] do not indicate any specific bandwidth allocation technique among macrocells and femtocells. As such, it is up to the operator to choose the channel allocation method. Basically, there are two channel allocation methods: orthogonal and shared channel allocation. Orthogonal channel allocation splits the channel bandwidth into two portions with one serving the macrocell and the other serving the femtocell, whereas shared channel allocation allows the macrocell and femtocell to share the whole bandwidth. Unlike shared channel allocation, split channel allocation is simpler and it maintains the original advantage of resource reuse among femtocells [15]. Thus, split channel allocation is generally preferable. For simplicity, we assume that a split channel allocation technique has assigned a portion of the bandwidth to the femtocell, and this chapter focuses only on resource allocation among femtocells.

In the literature, femtocell resource allocation schemes can be classified into centralized and distributed approaches. In the centralized approach, a central entity decides the resource allocation among femtocells. In the distributed approach, femtocells individually identify the resources available based on a mechanism that allows each femtocell to share the channel. Notably, the centralized and distributed resource allocation schemes that are commonly used for performance benchmarks are the C-DFP [11] and DRA [12] schemes, respectively, which can be found being compared with the schemes proposed in [16–18]. In the subsequent sections, we discuss network modeling and algorithms for the C-DFP and DRA schemes.

Both the C-DFP and DRA schemes adopt the following modeling assumptions:

- The bandwidth of each PRB is smaller than the coherence bandwidth of the channel, i.e., each PRB experiences flat fading;
- The time duration of each PRB is shorter than the coherence time of the channel, i.e., each PRB experiences slow fading;
- The femtocell network is perfectly synchronized.

3.1 CENTRALIZED DYNAMIC FREQUENCY PLANNING

The dynamic frequency planning (DFP) approach was originally investigated for homogeneous macrocellular networks (with only the macro tier) in [19]. Lopez-Perez et al. [11] extended it to femtocell networks and proposed the C-DFP scheme. In the C-DFP scheme, a resource broker is installed at the HeNB gateway. The resource broker is responsible for gathering resource information from HeNBs and performs resource allocation among them. The resource information contains the resource demand and interference measurements of every HeNB that associates with the resource broker.

3.1.1 Problem modeling

The C-DFP scheme models an LTE femtocell network as in Figure 24.2. Let the set of HeNBs be denoted as $F = \{1, 2, 3, \ldots, |F|\}$ with each HeNB serving a set of UEs, $U = \{1, 2, 3, \ldots, |U|\}$. Note that $|x|$ refers to the cardinality of x. In the C-DFP scheme, PRBs are allocated regularly in the frequency domain to HeNBs. In other words, when a PRB is allocated to an HeNB, other PRBs with the same subchannel frequency are also allocated to the HeNB. Let the set of PRBs in the frequency domain available to the femtocell network be denoted as $K = \{1, 2, 3, \ldots, |K|\}$.

Each HeNB has a specific resource demand in order to fulfill the throughput required by their associated UEs. The resource demand of HeNB f, denoted as D_f, can be estimated as the sum of the resource demand of all of its associated UEs, as follows:

$$D_f = \sum_{u \in U_f} D_u \tag{24.1}$$

where D_u is the resource demand of UE u and U_f is the set of UEs associated with HeNB f. In fact, each UE may have a number of data flows with each having a specific required bit rate. Therefore, the total bit rate required by the UE is the sum of the bit rates required by its data flows, as follows:

$$R_u^{\mathrm{req}} = \sum_{c \in C_u} R_c^{\mathrm{req}} \tag{24.2}$$

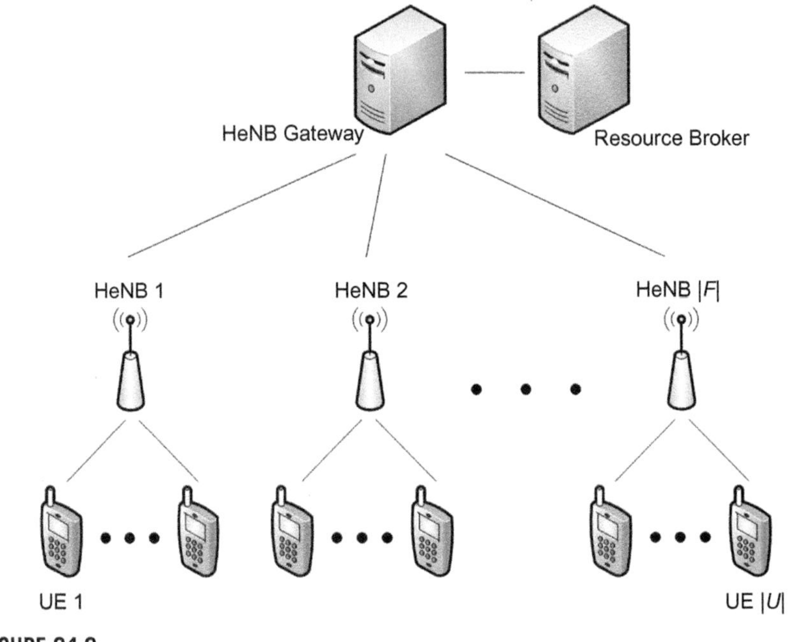

FIGURE 24.2

LTE femtocell network model.

where R_u^{req} is the bit rate required by UE u, R_c^{req} is the bit rate required by data flow c and C_u is the set of data flows associated with UE u. The number of PRBs (in the frequency domain) required by UE u can be estimated by:

$$D_u = \left\lceil \frac{R_u^{req}}{f_{PRB} S_E} \right\rceil \tag{24.3}$$

where $f_{PRB} = 180$ kHz is the bandwidth of a PRB, and S_E is the achievable spectral efficiency (bits/s/Hz) when a particular modulation and coding scheme (MCS) is selected for transmission. In LTE systems, S_E can be estimated based on the average received wideband signal-to-interference and noise ratio (SINR) from the adaptive modulation and coding (AMC) module in the medium access control (MAC) layer [20].

In the C-DFP scheme, each UE periodically sends a measurement report, which contains information on the received signal strength (RSS) of the reference signals transmitted by all HeNBs, to its serving HeNB. From this measurement report, the HeNB identifies whether an interference event has occurred. An interference event occurs between HeNB i and its neighboring HeNB j when the following inequality holds:

$$P_{ui}^{ref}(\text{dB}) < P_{uj}^{ref}(\text{dB}) + Th\,(\text{dB}) \tag{24.4}$$

where P_{ui}^{ref} and P_{uj}^{ref} are the RSSs received at UE u from the serving HeNB i and the neighboring HeNB j respectively, while Th is a protection margin that takes the aggregated interference from neighboring macrocells and fading effects into account. Subsequently, HeNB i updates the number of measurement reports N_{ij}^{MR} when it receives one that is relevant to HeNB j. If an interference event is identified, the number of interference events N_{ij}^{IE} is also updated. With the above information, the percentage of time when HeNB i is interfered with by HeNB j can be estimated as

$$w_{ij} = \frac{N_{ij}^{\text{IE}}}{N_{ij}^{\text{MR}}} \tag{24.5}$$

In this way, the interference relationships among HeNBs can be characterized by an $|F| \times |F|$ *restriction matrix* of interference restriction elements w_{ij}.

Given a number of PRBs in the frequency domain, the C-DFP scheme aims to fulfill the resource demand of each HeNB while maintaining the lowest interference level possible by mathematically solving the following optimization problem [19]:

$$\min \sum_{i\in F} \sum_{j\in F} \sum_{k\in K} \frac{w_{ij}}{D_i D_j} y_{ijk} \tag{24.6}$$

subject to:

$$\sum_{k\in K} x_{ik} = D_i \tag{24.6a}$$

$$x_{ik} + x_{jk} - 1 \le y_{ijk} \tag{24.6b}$$

$$y_{ijk} \ge 0 \tag{24.6c}$$

$$x_{ik} \in \{0, 1\} \tag{24.6d}$$

where x_{ik} is a binary variable where $x_{ik} = 1$ when HeNB i is allocated subchannel k and $x_{ik} = 0$ otherwise. Constraint (24.6a) guarantees D_i PRBs (in the frequency domain) are allocated to HeNB i. When both HeNB i and HeNB j are allocated PRB k, constraints (24.6b) and (24.6c) make $y_{ijk} = 1$ and $y_{ijk} = 0$ otherwise. The problem in Equation (24.6) is an optimization problem that aims to fulfill the resource demand of all HeNBs and avoids allocating the same subchannels to pairs of HeNBs with high interference restrictions.

3.1.2 Solution algorithm

In order to solve Equation (24.6), a greedy algorithm proposed in [19] is used. The greedy algorithm first allocates all PRBs (in the frequency domain) to all HeNBs, i.e., $x_{fk} = 1$ for all f and k. Then, the PRB that decreases the objective function value the most among all others for each HeNB is selected in each iteration. If PRB k which is allocated to HeNB f gives the largest decrement as compared to all other HeNBs, PRB k is removed from allocation to HeNB f, i.e., $x_{fk} = 0$.

The removal process continues until the resource demand of all HeNBs is satisfied. In this way, the algorithm performs dynamic resource allocation by adapting to the resource demand of each HeNB and the interference environment.

3.1.3 C-DFP execution

In order to execute the C-DFP scheme, the resource broker needs to collect from every HeNB their required demands and interference restriction. After solving Equation (24.6), the resource broker sends the resource allocation information to all HeNBs, indicating the resources that they are allowed to use. The C-DFP scheme is performed regularly with period T_{DFP}. After allocating resources among HeNBs, any scheduling method such as proportional fair scheduling can be used to allocate resources among their associated UEs.

3.2 DISTRIBUTED RANDOM ACCESS

Unlike the C-DFP scheme, the DRA scheme is decentralized to each HeNB. Specifically, it equips each HeNB with the capability to identify the best PRBs for its associated UEs, such that additional devices such as the resource broker are not needed for performing resource allocation. Nevertheless, the HeNB must have *a priori* knowledge about the set of available PRBs. In addition, each HeNB must possess a resource allocation mechanism to avoid interfering with its neighboring HeNBs. Unlike the C-DFP scheme, the DRA scheme allocates radio resources in the form of resource tiles (with each one consists of two consecutive PRBs in the time domain) to HeNBs. The PRB allocation is performed in each LTE frame. Let the set of resource tiles available in one LTE frame be denoted as $K_{\mathrm{tile}} = \{1, 2, 3, \ldots, |K_{\mathrm{tile}}|\}$, where $|K_{\mathrm{tile}}| = 10|K|$ because one LTE frame consists of ten TTIs in the time domain and $|K|$ PRBs in the frequency domain.

3.2.1 Problem modeling

The DRA scheme considers a similar resource allocation problem model as in the C-DFP scheme except that no resource broker is installed. Each HeNB is provided access to set K_{tile}. Akin to the C-DFP scheme, each UE periodically feeds back a measurement report to its serving HeNB about the RSSs received. In addition, each UE has a feedback mechanism to report collisions to its serving HeNB.

Unlike the C-DFP scheme, the DRA scheme does not model the resource allocation as an optimization problem. Instead, it employs a random hashing function whereby each HeNB receives a subset of K_{tile} from set K_{tile} in such a way that interference is avoided.

3.2.2 Solution algorithm

Figure 24.3 illustrates the pseudocode of the DRA algorithm in which resource allocation among HeNBs is carried out in a distributed manner where each HeNB obtains a number of resource tiles based on a hashing function using modulo-prime. In the first frame, each HeNB identifies its interfering HeNBs based on the

```
Map the HeNB IDs to the integer set [0, N)
if frame = 1, then
    for all f ∈ F, do
        Set x_fk = 0 for all k ∈ K_tile

        Identify the number of interfering HeNBs I
        Find a prime number p ≈ I
        Calculate r = ⌈ log2 N / ⌊log2 p⌋ ⌉
        Split the integer mapped to HeNB f's ID d into blocks of ⌊log2 p⌋ bits such that d = {d1, d2, ..., dr}, where di
        ∈ [0, p − 1]
        Generate a set of vectors H in which element is a vector h = {h1, h2, ..., hr}, where hi ∈ [0, p − 1]
        Randomly select a hash vector h
        for i = 1: ⌈ |K_tile| / p ⌉
            k_i = { (i − 1)p + ( Σ_{j=1}^{r} h_j d_j ) mod p }
            Set x_{fk_i} = 1
        end
    end
    Execute a scheduling method for PRB allocation among associated UEs
else
    for all f ∈ F, do
        Identify the set of resource tiles that experienced collisions K^col
        Calculate r = ⌈log2 N⌉
        Split the integer mapped to HeNB f's ID d into blocks of one bits such that d = {d1, d2, ..., dr}, where di ∈ [0, 1]
        Generate a set of vectors H in which element is a vector h = {h1, h2, ..., hr}, where hr ∈ [0, 1]
        Randomly select a hash vector h
        for all k ∈ K^col
            Set x_fk = ( Σ_{j=1}^{r} h_j d_j ) mod 2
        end
    end
    Execute a scheduling method for PRB allocation among associated UEs

end
```

FIGURE 24.3

Pseudocode of the DRA scheme [12].

RSSs received from the associated UEs. Then, the size of a hash table is determined based on the number of interfering HeNBs to decide which resource tiles should be used by the HeNB. After that, the HeNB allocates these resource tiles to its associated UEs using any scheduling method. In subsequent frames, each HeNB determines the resource tiles that suffer collisions. To resolve collisions, the HeNB performs hashing by using a hash table with the size of two to decide whether to use the resource tiles.

3.2.3 DRA execution

The execution of the DRA scheme is rather simple as it is performed by each HeNB individually. The RSS and collision information are directly reported to the HeNB from its associated UEs. With the information, the HeNB determines the available resources and performs packet scheduling using these resources.

4 RESOURCE ALLOCATION AMONG UEs

In the context of LTE, resource allocation among UEs is referred to as packet scheduling, which is performed at each eNB and HeNB every TTI. The objective of packet scheduling is to provide efficient and fair resource utilization for all UEs. One well-known packet scheduler is the proportional fair scheduler, which can achieve a good trade-off between the two objectives. Within each TTI t, the proportional fair scheduler allocates the available PRBs to UEs based on the following optimization objective:

$$\hat{k} = \text{argmax} \frac{R_c(t)}{\overline{R}_c(t-1)} \tag{24.7}$$

where $R_c(t)$ is the instantaneous bit rate achievable by data flow c if a PRB is allocated at TTI t and $\overline{R}_c(t-1)$ is the average bit rate achieved by data flow c in the past transmissions before TTI t. The objective of expression (24.7) is to find an optimal PRB $\hat{k}$ (in the frequency domain) that gives the maximum value of $R_c(t) / \overline{R}_c(t-1)$ to be allocated to data flow c. The $\overline{R}_c(t)$ can be determined as follows:

$$\overline{R}_c(t) = (1-\alpha)\overline{R}_c(t) + \alpha\overline{R}_c(t-1) \tag{24.8}$$

where $0 \le \alpha \le 1$. Usually, $\alpha = 0.8$ is used in the proportional fair scheduler. The process is repeated until all available PRBs are allocated.

Apparently, the data flow c with high $R_c(t)$ and low $\overline{R}_c(t-1)$ will be allocated more PRBs by the proportional fair scheduler. In this way, it maximizes the resource utilization efficiency by favoring data flows with high achievable throughput while maintaining fairness by favoring data flows with low past achieved throughput. Therefore, the resource allocation among the data flows is said to be *proportionally fair*.

5 SIMULATION MODELING OF LTE FEMTOCELL NETWORKS

A model is the abstraction or imitation of a system where only the entities and relationships relevant to the objectives of the study are abstracted. In modeling an LTE femtocell network, only entities that are relevant to packet transmission are considered and abstracted. Also, only functions that are relevant to packet transmission need to be implemented in the entities.

In simulation modeling of an LTE femtocell network, the basic relevant entities are HeNBs, UEs, channels and traffic applications. The HeNB and UE entities model the interaction between them such as packet transmission, CQI reporting, control signaling, etc. The channel entity models the fading effects and propagation losses imposed to the packets. The traffic application entity models the process of generating and receiving packets. In modeling packet transmission, other entities such as protocol stacks, MAC and physical layer are necessary. The protocol stack entity models the processes such as packet encapsulation, concatenation and segmentation. The MAC entity models the processes of scheduling packets and mapping packets to transport blocks for transmission. The physical layer entity models the packet transmission and reception over the channel.

In the next section, we introduce LTE-Sim and the tools available to create and model the relevant entities of an LTE femtocell network.

6 LTE-SIM

We employ an open-source LTE system-level simulator, namely the LTE-Sim simulator [9,10], as the simulation platform for the two schemes. Although other simulators such as those in [21,22] are available as open-source research tools, the protocol stacks implemented in these simulators are incomplete and the simulation scenarios supported are limited. Another simulator such as that in [23] needs to be purchased. On the other hand, the LTE-Sim simulator encompasses most of the aspects of an LTE system and supports various scenarios such as single-cell and multi-cell environments. LTE-Sim also has high modularity and flexibility, and thus can be easily modified for specific research goals such as evaluating resource allocation, packet scheduling, MAC or higher-level protocols, etc.

6.1 OVERVIEW OF LTE-SIM

The LTE-Sim simulator is an event-driven system-level simulator [9,10] using the object-oriented C++ programming language. The simulator encompasses near-complete sets of LTE functions and protocol stacks to create a homogeneous LTE network. These functions and protocol stacks are created using *classes*. Users of the LTE-Sim simulator can use these classes to create cells, eNBs, HeNBs, UEs and gateways. In the simulator, each eNB, HeNB or UE created contains its own physical layer, MAC layer and other upper-layer functions. Different applications which are best-effort, video, VoIP, and constant bit rate (CBR) can be created for each UE. Each application created sets up a radio bearer for the UE, in which the radio bearer corresponds to a set of QoS parameters such as the GBR, PLR and delay. The LTE-Sim simulator also provides different choices of packet scheduler in the network simulation. Besides that, a number of channel models are provided in the LTE-Sim simulator, which include indoor, outdoor, urban, suburban, rural, etc. Users can also set the channel bandwidth available to the system created from a given list to accommodate the PRBs required. Other additional features such as the selection of the UE mobility model and channel quality indicator (CQI) reporting mode are also available.

6.1.1 Simulation modeling using LTE-Sim

In LTE-Sim, an LTE network simulation model can be created by creating cells, devices (e.g., eNBs, HeNBs and UEs), physical layer, channels, protocol stacks, etc. as single entities. These entities can be created as objects using classes. LTE-Sim has defined a number of classes for modeling LTE networks. A detailed overview of classes available in LTE-Sim can be found in [9] and [10].

To create a simulation model in LTE-Sim, cells are created first which store information of their coverage (i.e., radius) and position (for multi-cell scenarios). In LTE-Sim, the *Cell* and *Femtocell* classes are defined to create macrocell and femtocell objects. For creating femtocell objects, LTE-Sim provides classes which create buildings to accommodate femtocells. Two basic building models are defined in LTE-Sim: a 5×5 apartment grid and a dual stripe. LTE-Sim provides a special function called *CreateBuildingForFemtocells()* in the *NetworkManager* class, which creates buildings and femtocells within the buildings.

After creating cells, an object of the *BandwidthManager* class needs to be created to determine the bandwidth available to the LTE network. This object will be used by the device objects such as eNB and HeNB for packet scheduling. The choice of bandwidth can be 1.4, 2, 3, 5, 10, 15 and 20 MHz which correspond to 6, 10, 15, 25, 50, 75 and 100 PRBs (in the frequency domain), respectively.

For modeling channels, LTE-Sim provides the *LteChannel* class that handles channel effects such as fading and propagation losses on both downlink and uplink transmission. Choices of propagation losses include urban, suburban and rural propagation loss models. It is noteworthy that the object of the *LteChannel* class is created for each single cell, as each cell experiences different channel conditions.

For modeling various devices, LTE-Sim defines the *NetworkNode* class which contains several common functions such as packet queuing, packet transmission and packet reception. This class is inherited to form the *Gateway*, *ENodeB*, *HeNodeB* and *UserEquipment* classes. An object of the *Gateway* class has to be created to model the traffic flows. In the *ENodeB* and *HeNodeB* classes, a record variable is defined to store the objects and information of the UEs served. These classes also call the corresponding MAC object to perform packet scheduling. In the *UserEquipment* class, variables and functions are defined to perform channel measurement and feedback, as well as position tracking. The LTE-Sim user can explicitly create any number of UEs using the *UserEquipment* class and associates them with any *ENodeB* or *HeNodeB* object. Upon creating the device objects, the objects of the *LtePhy* class and those of the *ProtocolStack* class are also created. The channel objects created need to be explicitly set up to associate with the *LtePhy* objects. In the object of the *MacEntity* class which is associated with the *ProtocolStack* object, one can explicitly set the desired packet scheduler.

After setting up the cells, devices and channels, different traffic applications can be created using the *Application* classes. These classes are *VoIP*, *TraceBased*, *InfiniteBuffer* and *CBR*, which are used for creating VoIP, video, best-effort and CBR traffic respectively. These classes also provide the options of setting the start and stop times for the applications created. Whether the traffic application is for downlink or uplink depends on the setting of the source and destination of the application.

Before running the simulation model, an object of the *Simulator* class is created to start and stop the simulation, and an object of the *FrameManager* class is created to keep count of the frame number and simulation time.

7 IMPLEMENTATION OF RESOURCE ALLOCATION SCHEMES IN LTE-SIM

We have discussed simulation modeling of femtocell networks in LTE-Sim. In this section, we demonstrate how femtocell resource allocation schemes can be implemented in LTE-Sim. In the following, we demonstrate implementation of the centralized C-DFP and decentralized DRA schemes using the LTE-Sim package (*lte-sim-r5*) [24].

7.1 IMPLEMENTATION OF C-DFP IN LTE-SIM

The major modifications or additions to the LTE-Sim source codes for implementing the C-DFP scheme are depicted by the class diagram in Figure 24.4. It is noteworthy that each box in Figure 24.4 represents a class with first row stating the class name, second row stating the variable names and third row stating the function names. In addition, the bold-lined boxes in Figure 24.4 indicate the classes that have been modified or newly added. Only functions that require modifications and definitions of new variables are shown under the classes.

7.1.1 Modifications for creating HeNBs

To create an HeNB object with C-DFP implemented, the classes, i.e., *ENodeB*, *HeNodeB*, *UeLtePhy*, *EnbLtePhy*, *EnbMacEntity* and *FemtocellUrbanArea-ChannelRealization*, require code modifications.

In the C-DFP scheme, HeNBs periodically collect measurement reports which contain the RSS of other HeNBs (including the HeNB who receives the reports) from each UE. With these measurement reports, the RSS of the HeNB is compared with those of other HeNBs to identify interference events using Equation (24.4). After that, a restriction matrix is established using Equation (24.5). For these functions, we define the *SetRestrictionMatrix()* function in the *HeNodeB* class to collect measurement reports, identify interference events and establish the restriction matrix.

Another function, namely *EstimateNumReqPRBs()*, is defined in the *HeNodeB* class for each HeNB to estimate the total resource demand of all of its associated UEs. As shown in Section 3.1.1, the resource demand of each UE can be estimated as Equation (24.3). The achievable wideband spectral efficiency can be obtained by enabling the feature of estimating the wideband SINR value in the *UeLtePhy* object. This object will incorporate the wideband SINR value in the

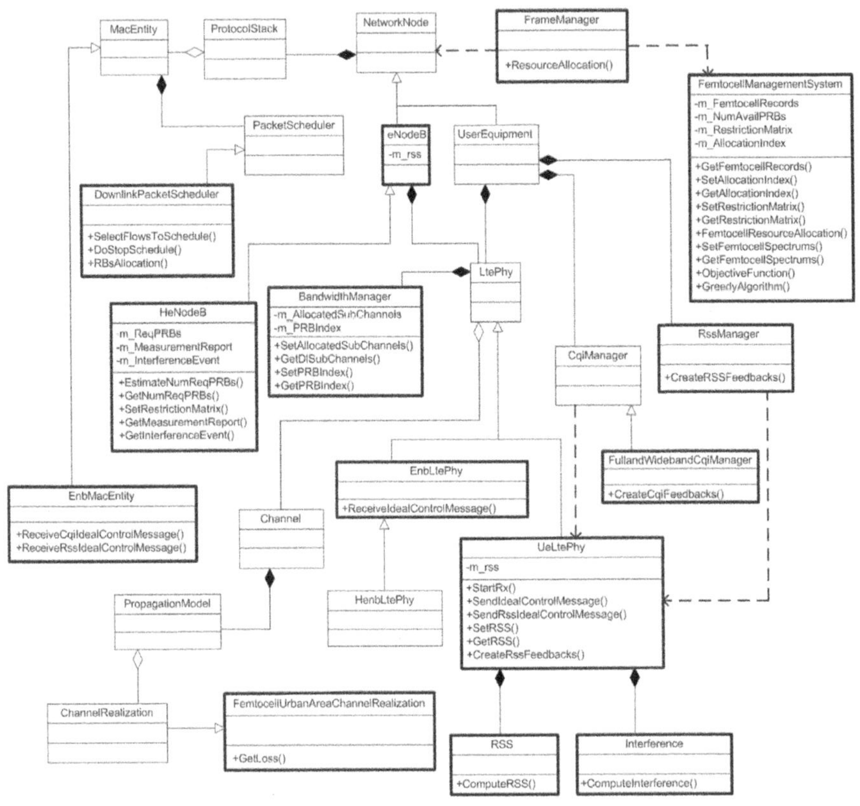

FIGURE 24.4

LTE-Sim class diagram for implementing the C-DFP scheme.

CQI feedback. The HeNB can then retrieve and use the wideband SINR value to estimate the achievable spectral efficiency using the functions in the AMC module defined in LTE-Sim.

In LTE-Sim, the *BandwidthManager* class is used to define the channel bandwidth available to the network. As LTE-Sim does not provide the feature of dynamic PRB allocation, two vector variables are defined: *m_AllocateddlSubChannels* and *m_PRBIndex*; the former stores the starting frequencies of PRBs assigned, whereas the latter stores the indices of the PRBs assigned. The *BandwidthManager::SetAllocatedSubChannels()* function is defined to assign the PRBs' frequencies into *m_AllocateddlSubChannels*. The original *BandwidthManager::GetDlSubChannels()* function is modified to access *m_AllocateddlSubChannels*. The *BandwidthManager::GetPRBIndex()* function is defined to access *m_PRBIndex*.

To simulate an LTE femtocell network, we use the femtocell channel model for urban areas, which is defined in LTE-Sim as the *FemtocellUrbanAreaChannelRealization* class. Since the *BandwidthManager* class has been modified for dynamic PRB allocation, the *FemtocellUrbanAreaChannelRealization::GetLoss()* function has to be modified accordingly. This modification is accomplished by imposing fading effects on the PRBs assigned only to the HeNB, instead of on all PRBs which were originally set up in LTE-Sim.

In the C-DFP scheme, each HeNB receives a control message containing the RSS information of its associated UEs. The *EnbLtePhy::ReceiveIdealControlMessage()* function is modified to recognize and forward the RSS control message to the MAC entity, namely the *EnbMacEntity* object. Also, the *EnbMacEntity::ReceiveRssIdealControlMessage()* function is added to store the RSS information into a vector variable in the *ENodeB* class. Furthermore, the *EnbMacEntity::ReceiveCqiIdealControlMessage()* function is modified to receive the subband and wideband CQI values.

Since the *HeNodeB* class is inherited from the *ENodeB* class as shown in Figure 24.4, several changes are made in the *ENodeB* class. These changes include the declaration of variables that store the subband and wideband CQI values, RSS values, and functions that assign values to the variables.

7.1.2 Modifications for creating UEs

For creating UE objects with implemented C-DFP, the classes such as *CqiManager*, *UeLtePhy*, *EnbLtePhy* and *Interference* are modified. Also, several new classes are added, which include *FullandWidebandCqiManager* and *RssManager*.

The UEs need to estimate the wideband SINR for the HeNBs to estimate the total resource demands required. In LTE-Sim, each UE created has a *CqiManager* object that measures the SINR experienced by each PRB used in the current transmission and convert it into a CQI value. LTE-Sim contains two *CqiManager* classes: *FullbandCqiManager* and *WidebandCqiManager* which estimate the subband and wideband CQI values, respectively. To enable estimation of both CQI values, a new class is defined. This class, namely *FullandWidebandCqiManager* is defined by inheriting the *CqiManager* class. The *FullandWidebandCqiManager::CreateCqiFeedbacks()* function is defined by merging the definitions of *FullbandCqiManager::CreateCqiFeedbacks()* and *WidebandCqiManager::CreateCqiFeedbacks()* functions. The *FullandWidebandCqiManager::CreateCqiFeedbacks()* creates a CQI value for the wideband SINR and a CQI value for each of the PRBs used by the UE. The wideband CQI can be used to estimate the total resource demands by using the functions in the *AMCModule* class for retrieving the wideband SINR value. Alternatively, one can directly use the *AMCModule::GetEfficiencyFromCQI()* to obtain the achievable spectral efficiency from the wideband CQI value for resource demand calculation.

As discussed earlier, the C-DFP scheme requires UEs to report their RSSs received from all HeNBs to their serving HeNB. For this purpose, we define a new class, namely *RssManager*. The definition of the *RssManager* class is similar to that of the *CqiManager* class and it has a function, namely *CreateRssFeedbacks()*. The *RssManager::CreateRssFeedbacks()* function allows each UE object created to report the RSS information to its serving *HeNodeB* object. However, LTE-Sim does not have a class for estimating the RSSs. As such, a new class, namely *ReceivedSignalStrength*, is created. In addition, a new control message signaling is defined for the RSSs by means of inheritance from the *IdealControlMessage* class, creating the *RssIdealControlMessage* class. The *RssIdealControlMessage* class has a similar definition with the *CqiIdealControlMessage* class. An additional function is added into the *UeLtePhy* class to send the RSS message to the *EnbLtePhy* object.

As the *BandwidthManager* class has been modified for dynamic PRB allocation, the *Interference* class has to be modified accordingly. The modification is done in such a way that interference only occurs when two HeNBs use the same PRBs.

7.1.3 Creating a resource broker object

LTE-Sim does not have a class that can create an object that models a resource broker. Therefore, we introduce a new class, namely *FemtocellManagement- System* in LTE-Sim. In the *FemtocellManagementSystem* class, a variable: *m_Femtocell Records* is declared to keep records about the associated HeNBs. Another variable *m_NumAvailPRBs* is used to store the number of PRBs (in the frequency domain) available to the network. Also, the restriction matrices obtained from all the associated HeNBs are stored in the *m_RestrictionMatrix* variable. Additionally, a variable *m_AllocationIndex* is declared to store the PRB allocation information.

A function is defined in the *FemtocellManagementSystem* class for evaluating Equation (24.6), namely *FemtocellManagementSystem::ObjectiveFunction()*. On the other hand, the greedy algorithm that is used to solve Equation (24.6) is implemented as the *FemtocellManagementSystem::GreedyAlgorithm()* function in LTE-Sim. The *FemtocellManagementSystem::GreedyAlgorithm()* function executes the greedy algorithm described in Section 3.1.2 to dynamically allocate PRBs (in the frequency domain) among the associated HeNBs while repeatedly calling the *FemtocellManagementSystem::ObjectiveFunction()* to evaluate Equation (24.6). After executing the greedy algorithm, the *FemtocellManagementSystem::GreedyAlgorithm()* function returns the PRB allocation index and stores it in the *m_AllocationIndex* variable.

In the *FemtocellManagementSystem* class, the *FemtocellManagementSystem::SetRestrictionMatrix()* function is implemented to obtain from each associated HeNBs the restriction matrix and store it in the *m_RestrictionMatrix* variable. This *m_RestrictionMatrix* variable is used to evaluate Equation (24.6) in *FemtocellManagementSystem::ObjectiveFunction()*.

After PRB allocation, the resource broker needs to set the *BandwidthManager* object of each associated HeNB such that it can only transmit on the PRBs allocated by the resource broker. To do this, the *BandwidthManager::SetFemtocellSpectrums()* function is defined in LTE-Sim. The *BandwidthManager::SetFemtocellSpectrums()* function will call from each HeNB object the *BandwidthManager* object using the *m_FemtocellRecords* variable. Then, PRBs (in the frequency domain) are allocated using the *BandwidthManager::SetAllocatedSubChannels()* function with the *m_AllocationIndex* variable.

7.1.4 Modifications in the packet scheduler class

The proportional fair packet scheduler is available in the LTE-Sim simulator. However, as the *HeNodeB* class has been modified to store subband and wideband CQI values, the downlink packet scheduler class, namely *DownlinkPacketScheduler*, must undergo a slight modification such that it uses the subband CQI values for PRB allocation among UEs.

7.1.5 Modifications in the frame manager class

An object of the *FrameManager* class is instantiated in order to keep track of the frame numbers in an LTE-Sim simulation. To simulate the C-DFP scheme, the resource broker, i.e., the *FemtocellManagementSystem* object, needs to be executed periodically with a specific interval to perform resource allocation among HeNBs. This modification is done on the *FrameManager::ResourceAllocation()* function. The size of the interval can be set accordingly by the LTE-Sim user.

7.1.6 Relationships between classes

We have discussed the implementation and modifications required to implement the C-DFP scheme in LTE-Sim. The relationships between the modified or newly created classes are depicted in Figure 24.4. In an LTE-Sim simulation, the *FrameManager* object will call the *FemtocellManagementSystem* object to execute the C-DFP scheme and the *HeNodeB* objects to execute packet scheduling. Therefore, the *FrameManager* class depends on the *FemtocellManagementSystem* class and the *NetworkNode* class (which is the class where *HeNodeB* is inherited from).

As the *HeNodeB* class inherits the *ENodeB* class, the modifications made in the *ENodeB* class also apply to the *HeNodeB* class. Similarly, the *EnbLtePhy* class inherits the *LtePhy* class, which is a part of the *ENodeB* class, so the modified *BandwidthManager* class is also a part of the *EnbLtePhy* class. The *LteChannel* class has an aggregation relationship with the *LtePhy* class; packets are sent and received by the *LtePhy* objects while the *LteChannel* object incorporates fading effects and propagation losses into the packets. The *PropagationModel* class, which is a part of the *LteChannel* class, contains an object of the *ChannelRealization* class which models the channel effects.

The *UserEquipment* object contains an object of the *CqiManager* class and an object of the *RssManager* class which are used to create CQI and RSS feedbacks. The *CqiManager* and *RssManager* objects depend on the object of the *UeLtePhy* class, which is a part of the *UserEquipment* class. This is because the *CqiManager* and

RssManager objects require the *UeLtePhy* object to send the CQI and RSS feedbacks to the *EnbLtePhy* object. Additionally, the *UeLtePhy* class contains objects of the *RSS* and *Interference* classes that estimate the RSS and interference values respectively.

The object of the *ProtocolStack* class is a part of the object of the *NetworkNode* class, which contains the object of the *MacEntity* class that performs packet scheduling. For an *ENodeB* object, the corresponding *ProtocolStack* object contains an object of the *EnbMacEntity* class which inherits the *MacEntity* class.

7.2 IMPLEMENTATION OF DRA IN LTE-SIM

The implementation of the DRA scheme in LTE-Sim is similar to that of the C-DFP scheme except that the resource broker is not needed. The newly defined or modified classes in the C-DFP scheme such as *FullandWidebandCqiManager*, *BandwidthManager*, *RssManager*, *ChannelRealization*, etc. are used in implementing the DRA scheme in LTE-Sim. However, each HeNB is equipped with functions that allocate PRBs dynamically among the associated UEs. Also, each UE is equipped with functions to report collisions back to their serving HeNB. The additions and modifications of LTE-Sim for implementing the DRA scheme are depicted by the class diagram in Figure 24.5. Similar to Figure 24.4, the boxes with bold lines in Figure 24.5 indicate the classes that have been modified or newly added.

7.2.1 Modifications for creating HeNBs

Similar to the C-DFP scheme, creating HeNB objects with DRA implemented requires code modifications on the *ENodeB*, *HeNodeB*, *UeLtePhy*, *EnbLtePhy*, *EnbMacEntity*, *DownlinkPacketScheduler* and *FemtocellUrbanAreaChannelRealization* classes. The changes are the same as that for the C-DFP scheme. However, several additional functions need to be defined to enable the DRA scheme.

As stated earlier, the *HeNodeB* class is inherited from the *ENodeB* class. As such, several changes are required on the *ENodeB* class. One of the changes is that an additional variable, namely *m_collision* is declared to store the collision information obtained from UEs. Other changes include amendments of new functions that access or assign the collision information into the *m_collision* variable. In the *EnbLtePhy* class, the *EnbLtePhy::ReceiveIdealControlMessage()* function is modified to receive control messages carrying the collision information from UEs, which will be later forwarded to the object of the *EnbMacEntity* class. Then, the *EnbMacEntity* object stores the collision information into the *m_collision* variable in the *ENodeB* class.

The DRA scheme shown in Figure 24.3 is implemented in the *HeNodeB* class as two separate functions. The first function, namely *HeNodeB::SetBandwidth-Allocation1()*, corresponds to the resource allocation algorithm in the first LTE frame. After performing random hashing, the *HeNodeB* object set its assigned PRBs using the *BandwidthManager::SetAllocatedSubChannels()* function. In the subsequent LTE frames, the *HeNodeB* object will call the second function, namely *HeNodeB::SetBandwidthAllocation2()* that corresponds to collision resolution. The *HeNodeB::SetBandwidthAllocation2()* function will access the

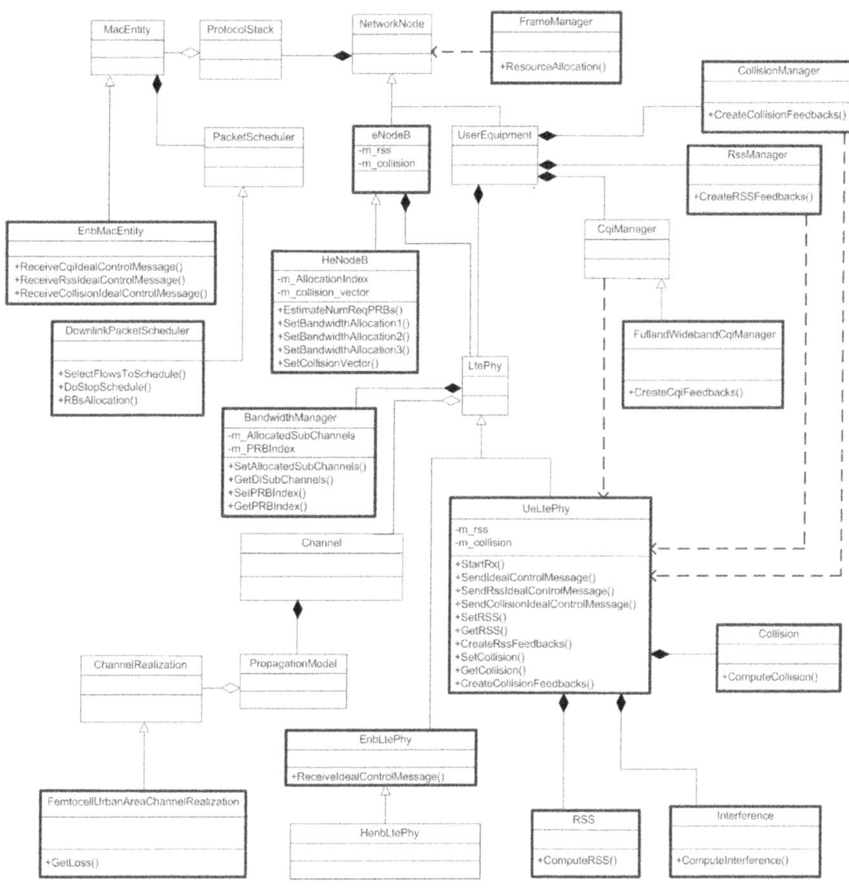

FIGURE 24.5

LTE-Sim class diagram for implementing the DRA scheme.

m_collision variable for collision resolution. The *HeNodeB::SetBandwidthAllocation3()* function is defined to set the PRBs allocated by *HeNodeB::SetBandwidthAllocation1()* or *HeNodeB::SetBandwidthAllocation2()* to be used by the HeNB. The *HeNodeB::SetCollisionVector()* function is used to store the information about the PRBs that experience collision into the vector variable, namely *m_collision_vector*, which will be used in the *HeNodeB::SetBandwidthAllocation2()* function for collision resolution.

7.2.2 Modifications for creating UEs

Creating UE objects with DRA implemented requires the same changes as in that with C-DFP implemented. In addition, the UEs are required to report collisions to their serving HeNBs. Therefore, the UEs must possess the function that can identify and report collisions. To identify collisions, the *Collision* class is created. The Collision class is similar to the *Interference* class except that the former checks if any two HeNBs are using the same PRBs and interfering with each other. To report collisions, we create the *CollisionManager* class. The functions in the *CollisionManager* class are very similar to those of the *CqiManager* class. The major difference between these two classes is that the *CollisionManager* class reports PRBs that suffer from collisions. When reporting collisions, the control message carrying the collision information needs to be defined in LTE-Sim. This control message is defined as an inheritance from the *IdealControlMessage* class, namely *CollisionIdealControlMessage*. The definition of the *CollisionIdealControlMessage* class is similar to that of the *CqiIdealControlMessage* class. An additional function is added into the *UeLtePhy* class to send the collision message to the *EnbLtePhy* object.

7.2.3 Modifications in the frame manager class

Akin to the implementation of the C-DFP scheme, the DRA scheme requires the *FrameManager::ResourceAllocation()* function to be modified in such a way that it calls the *HeNodeB::SetBandwidthAllocation1()* or *HeNodeB::SetBandwidthAllocation2()* function to execute the DRA scheme before performing packet scheduling. The DRA scheme is performed every one LTE frame, hence the *HeNodeB::SetBandwidthAllocation1()* and *HeNodeB::SetBandwidthAllocation2()* functions should only be called every one LTE frame. Also, the *HeNodeB::SetCollisionVector()* is called before the *HeNodeB::SetBandwidthAllocation2()* function.

7.2.4 Relationships between classes

The relationships between classes in the UML class diagram for implementing the DRA scheme as shown in Figure 24.5 are similar to those for implementing the C-DFP scheme, except that the former does not have an object for the *FemtocellManagementSystem* class. Additionally, the *CollisionManager* class is a part of the *UserEquipment* class and depends on the *UeLtePhy* class. Also, the object of the *UeLtePhy* class contains an object of the *Collision* class.

8 FEMTOCELL NETWORK MODELING USING LTE-SIM

We have implemented the C-DFP and DRA schemes in LTE-Sim for performance evaluation in a femtocell network. We consider an urban scenario with a 5×5 grid apartment building with each apartment having an HeNB installed. The LTE network is modeled with a channel bandwidth of 5 MHz. Each UE carries one video flow, one VoIP flow and one best-effort flow. For the C-DFP scheme, the T_{DFP} is set to 1 s. The DRA scheme runs every one LTE frame, i.e., 10 TTIs. Other simulation parameters used in LTE-Sim are shown in Table 24.1. In this

Table 24.1 Parameter Settings for the C-DFP and DRA Schemes

Parameter	Setting
Data Transmission Direction	Downlink
Channel bandwidth	5 MHz (25 PRBs in the frequency domain)
HeNB's transmission power	20 dBm (equally distributed among all PRBs)
Apartment size	100 m^2
Building type	5 × 5 apartment grid
Number of buildings	1
Number of femtocells	5, 10, 15 and 20
Number of UEs per femtocell	4
Packet scheduler	Proportional fair
Traffic	1 video, 1 VoIP and 1 best-effort
Flow duration	10 seconds
Scenario	Urban
UE speed	3 km/h

simulation, we consider four scenarios with each having different number of femtocells, as given in Table 24.1. Each scenario will be simulated with a flow duration of 10 seconds.

To create the femtocell network simulation model based on the parameter settings in Table 24.1, objects of the *Femtocell* classes can be created first using the *CreateBuildingForFemtocells()* function in the *NetworkManager* class with the building type chosen as 5 × 5 grid apartment. An object of the *Gateway* class should be created after creating the femtocells. In the C-DFP implemented simulation model, an object of the *FemtocellManagementSystem* class should be created to model the resource broker. After that, objects of the *HeNodeB* class can be created. As each HeNB could receive different PRBs after resource allocation, each *HeNodeB* object requires an object of the *BandwidthManager* class to store the information about the PRBs available. Thus, a *BandwidthManager* object is created for each *HeNodeB* object with the channel bandwidth set to 5 MHz. Also, an object of the *LteChannel* class is created for each *HeNodeB* object with the femtocell urban propagation loss model chosen and set. For each *HeNodeB* object, the proportional fair scheduler is used for packet scheduling. After that, four objects of the *UserEquipment* class can be created for each of the HeNodeB objects. For each *UserEquipment* object, an object of the *VoIP*, *TraceBased* and *InfiniteBuffer* classes can be created to model the VoIP, video and best-effort flows respectively. The source and destination of each flow is set to be the *Gateway* object and the *UserEquipment* object respectively. In order to start and stop the simulation, an object of the *Simulator* class is created with the duration set to 10 seconds. Nevertheless, an object of the *FrameManager* class is created to keep count of the frame number and simulation time.

In the LTE-Sim based femtocell network model, the following assumptions are made:

- The femtocell network is perfectly synchronized;
- The CQI feedback, RSS reporting, information transfer between the resource broker and the HeNBs and collision reporting are carried out without errors and delays.

9 PERFORMANCE EVALUATION USING LTE-SIM

In LTE-Sim, several performance evaluation tools are available. For each LTE-Sim simulation, an output trace file that contains transmission information such as packet ID, type of packet, packet size, source ID, destination ID, packet head of line delay, etc. is generated. The evaluation tools can then be used to extract this information from the trace file for evaluating certain metrics. Four main evaluation tools are available to evaluate the throughput, PLR, delay and Jain's fairness index of a simulated LTE network. As for evaluating the SINR, the LTE-Sim user is required to enable the option that allows LTE-Sim to print the downlink SINR value into the trace file.

9.1 SINR PERFORMANCE

Interference is one of the major operational issues of femtocell networks. High interference may lead to frequent occurrence of packet collisions; as a result, the achievable throughput is reduced. Since both the C-DFP and DRA schemes aim to tackle the interference generated among HeNBs, i.e., co-tier interference, it is important to measure the interference level of the LTE network. Assuming interference is the dominant factor of channel variation, we employ the downlink SINRs as the performance metric for representing the channel condition. SINR values are estimated by the UEs when they receive packets. A large value of SINR indicates that the channel quality is high.

The downlink SINR can be obtained in the LTE-Sim simulator by enabling the option that prints their values. This can be done by *uncommenting* the line, i.e., "#define TEST_DL_SINR" in the *global_config* file in the LTE-Sim folder. In this way, the trace file generated will contain the downlink SINR values measured by the UEs. In order to evaluate the downlink SINR performance of the C-DFP and DRA schemes, the SINR values can be collected as a cumulative distribution function (CDF) graph, as shown in Figure 24.6. From Figure 24.6a, it is seen that the CDF value achieved by the C-DFP scheme at an SINR value of 10 dB is 0.97. This means that 97% of the UEs that receive packets have an SINR value of less than or equal to 10 dB throughout the simulation duration. From Figure 24.6a, it is observed that the CDF reaching 15 dB is one, implying that all the UEs have SINR values of less than or equal to 15 dB.

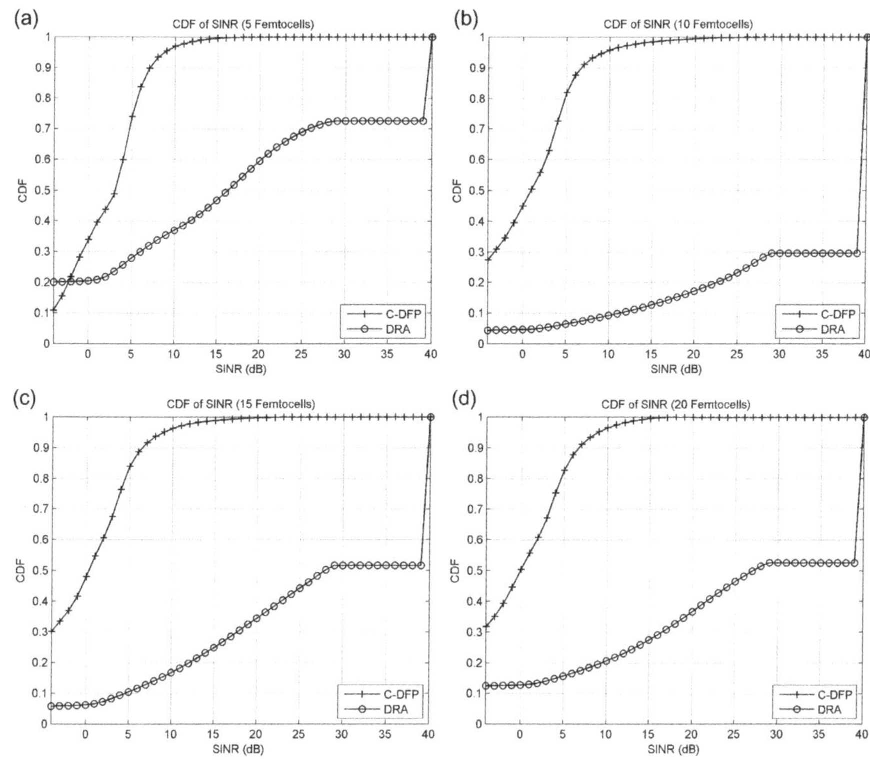

FIGURE 24.6

CDF of SINR for (a) 5 femtocells, (b) 10 femtocells, (c) 15 femtocells and (d) 20 femtocells.

Figure 24.6 shows the CDFs of SINR achieved by both the C-DFP and DRA schemes for 5, 10, 15 and 20 femtocells. In Figure 24.6a, it can be observed that the DRA scheme demonstrates a better SINR performance than the C-DFP scheme. Unlike the DRA scheme, a majority of the UEs suffer low SINR values under the C-DFP scheme. Similar trends are observed in Figures 24.6b, 24.6c and 24.6d. As the DRA scheme has a feature that stops a HeNB from transmitting on the PRB that encounters packet collisions, interference is greatly reduced. On the other hand, the main objective of the C-DFP scheme is to fulfill the resource demand of each HeNB. As a result, some PRBs which suffer low SINRs may still be allocated to the HeNB. Hence, the DRA scheme can achieve better channel conditions with less interference compared to the C-DFP scheme.

When more femtocells are deployed within the building, co-tier interference increases, which degrades the channel quality. Therefore, more UEs experience low SINR values with increased numbers of femtocells, as shown in Figure 24.6. It can be observed that the C-DFP scheme suffers greater SINR degradation compared to the DRA scheme. This is because, unlike the DRA scheme which avoids the PRBs being reused in two interfering HeNBs, the C-DFP scheme can still allocate the same PRBs to the two HeNBs if the PRBs result in an increased value of Equation (24.6).

9.2 THROUGHPUT PERFORMANCE

Throughput evaluation is one of the most important assessments for wireless networks. It gives an indication on the ability of a wireless network in delivering high data rates and achieving efficient resource utilization. Throughput evaluation for a network gives two implications. The first implication indicates the channel quality in the network; the higher the channel quality, the higher the throughput. This relationship can be justified by the following Shannon's capacity formula

$$R = B\log_2(1 + \mathrm{SINR}) \tag{24.9}$$

where R is the achievable capacity (measured in bps) and B is the channel bandwidth. From Equation (24.9), the achievable capacity is proportional to the SINR. The second implication of throughput evaluation indicates the radio resource utilization efficiency of the network. Assuming the network enjoys an interference-free environment, the radio resources given to a HeNB can be reused by other HeNBs. This increases the resource utilization efficiency and overall achievable throughput as a result of the increased bandwidth for the HeNBs due to resource reuse.

In an LTE network, throughput is evaluated as the total packet size (measured in bits) that has been successfully delivered from the transmitter to the receiver over the channel in a time interval. In downlink, the transmitter and receiver correspond to the base station and the end user, respectively. In the context of the C-DFP and DRA schemes, the base station and the end user are the HeNB and UE respectively.

As mentioned earlier, an output trace file is generated in each LTE-Sim simulation to store the transmission information. To evaluate throughput for different traffic flows such as video, the packets are first identified from the trace file. Then, the summation of all the packet sizes gives the total throughput (the summation function is provided as an evaluation tool in LTE-Sim). Thereafter, the video, VoIP and best-effort throughput performance can be plotted separately as in Figure 24.7.

Figure 24.7 shows the throughput performance of the C-DFP and DRA schemes for video, VoIP and best-effort traffic flows. It can be observed that the throughput achieved by both schemes for all types of traffic is proportional to the number of femtocells. This incremental trend can be intuitively justified by the increasing numbers of UEs transmitting data packets in the network.

The C-DFP scheme is superior to the DRA scheme for video and best-effort throughputs because the former provides better resource satisfaction to each HeNB. However, the DRA scheme is better in terms of VoIP throughput. This is due to the fact that the PRBs provided by the DRA scheme are insufficient for video flows, so most of the resources are instead used to satisfy VoIP flows that

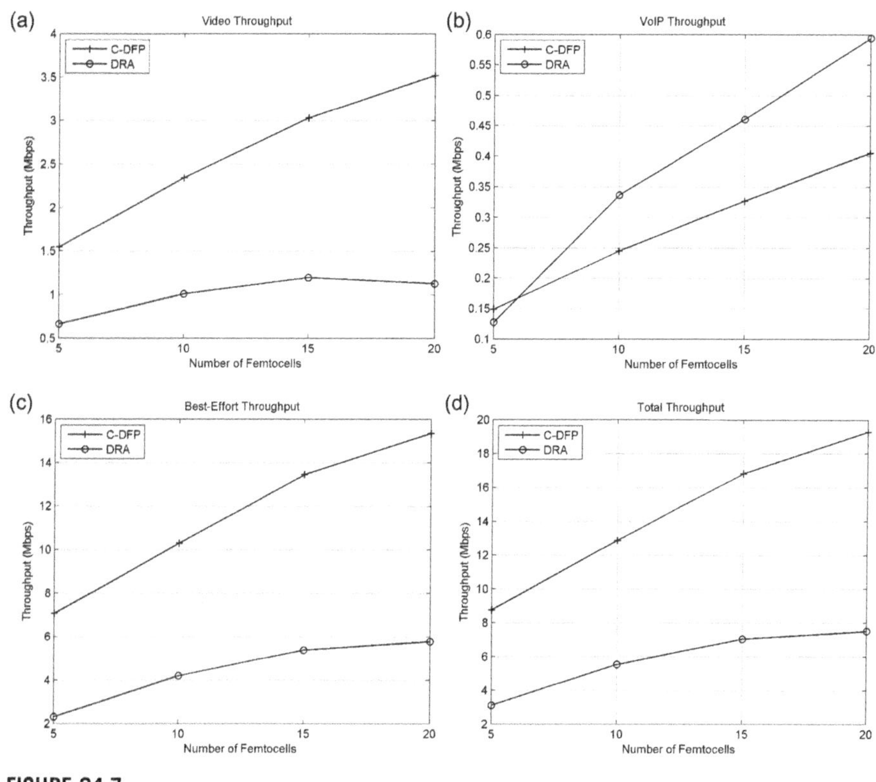

FIGURE 24.7

Throughput performance of (a) video, (b) VoIP, (c) best-effort flows and (d) the entire network.

require less resources. We notice that a dramatic increase in throughput is achieved by the C-DFP scheme as the number of femtocells increases. Although the DRA scheme obtains an increasing throughput performance trend, the increment is rather smaller compared to that of the C-DFP scheme.

Overall, the throughput of the C-DFP scheme is higher than that of the DRA scheme, as depicted in Figure 24.7d. However, this contradicts the findings found in Section 9.1 as the SINR achieved by the C-DFP scheme in the network is lower than that of the DRA scheme. This is because the C-DFP scheme constantly fulfills the resource demand of each HeNB by allocating PRBs with minimum interference. Besides that, the PRBs are not restricted to be allocated to one HeNB, but they are reused among HeNBs with the interference kept as low as possible. Thus, efficient resource reuse of the C-DFP scheme overcomes the poor channel conditions encountered in the network and achieves a higher throughput. On the other hand, the DRA scheme allocates the PRBs using a random hashing function, which does not guarantee high resource reuse and satisfaction of the resource demand of HeNBs. Moreover, the collision avoidance mechanism of the DRA scheme could remove the PRBs allocated to the HeNBs in order to avoid collision, leading to fewer PRBs available to each HeNB for transmission, which then causes some UEs to receive few or no PRBs. As a result of low PRB reuse and shortage of PRBs, the DRA scheme achieves low throughput performance despite good channel conditions.

9.3 PACKET LOSS RATE

The PLR is an important performance measure for real-time flows such as video and VoIP. Since these data flows are guaranteed certain data rates for smooth and seamless transmission, the number of packets lost or dropped during transmission must be kept low. In a transmission interval, the PLR can be calculated as follows:

$$\text{PLR} = \frac{N^{tx} - N^{rx}}{N^{tx}} \times 100\% \tag{24.10}$$

where N^{tx} and N^{rx} are the total number of transmitted and received packets, respectively. This evaluation can be easily performed by extracting all the real-time packet sizes which are both transmitted and received respectively and one of the evaluation tools in LTE-Sim can be used to calculate the PLR. The PLRs of video and VoIP flows are plotted in Figure 24.8.

Figure 24.8 illustrates the PLRs of both the C-DFP and DRA schemes for video and VoIP flows. In Figure 24.8a, the C-DFP and DRA schemes exhibit an incremental trend from 5 to 20 femtocells. This is mainly because the interference generated becomes more intense as the number of femtocells increases, resulting in more traffic flows with poor channel conditions and leads to increased packet losses. Moreover, the scheduler might have assigned more PRBs to the flows that suffer from poor channel conditions, causing inadequate PRBs for the other flows.

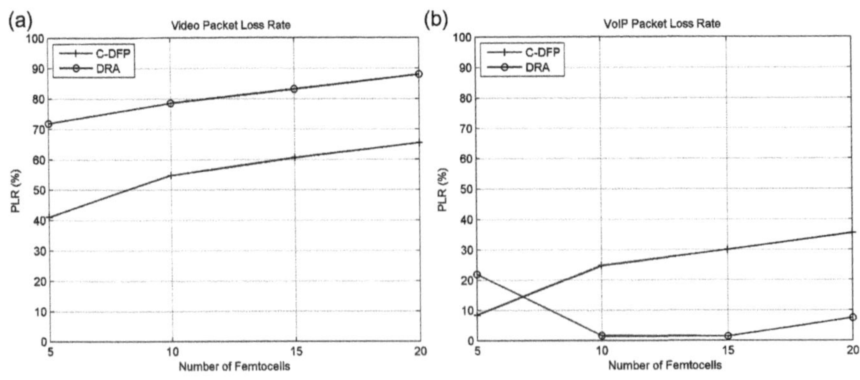

FIGURE 24.8

PLR performance of (a) video flows and (b) VoIP flows.

As a result, those flows which have not been assigned PRBs suffer longer delays in the transmission queue, which thus resulting in the increased number of packets dropped from the transmission queue. The C-DFP scheme has a better video PLR performance compared to the DRA scheme, as shown in Figure 24.8a. This is because the C-DFP scheme provides better resource satisfaction thanks to its mechanism that aims to fulfill the resource demand of the HeNBs. On the other hand, the DRA scheme has a lower PLRs compared to the C-DFP scheme for VoIP flows, as depicted in Figure 24.8b. As the PRBs provided by the DRA scheme are insufficient for the video flows, most of the PRBs are used to satisfy VoIP flows instead, thus attaining lower PLRs for VoIP flows.

In general, the C-DFP scheme outperforms the DRA scheme in terms of PLR for real-time flows. However, the PLR values achieved by the C-DFP scheme are still high, which is impractical, especially when the number of femtocells is more than five. Nonetheless, this problem can be resolved with a larger channel bandwidth. In this way, the PLR can be reduced substantially as more PRBs are available. However, this may not be of great improvement to the DRA scheme as justified in Figure 24.8 where it already suffers from high PLRs at small numbers of femtocells.

9.4 PACKET DELAY

Delay is another QoS performance measure of real-time flows. Certain real-time flows such as video and VoIP may have strict delay requirements. Therefore, it is necessary for a resource allocation scheme to be able to guarantee that packets are transmitted before their delays exceed the required threshold. Otherwise, the packets are dropped from the transmission queue.

Usually, the delay of the first packet to be transmitted is concerned. This delay is known as the head of line (HOL) delay. The HOL delay refers to the interval between the time the first packet to be transmitted pending at the packet transmission queue and the time it is received by the UE. To quantify the overall delay performance of a resource allocation scheme, all packet delays can be collected and plotted as a CDF as shown in Figure 24.9. To understand the meaning of these CDF curves, we study an example in Figure 24.9a, i.e., the C-DFP curve of VoIP delays. From this curve, it is observed that the CDF value of the packet delay of 0.01 s is 0.95, implying that 95% of the VoIP flows have packet delay values that are less than or equal to 0.01 s.

Figure 24.9 illustrates the CDFs of packet delay achieved by both the C-DFP and DRA schemes for video and VoIP flows. In Figure 24.9, it can be observed that the C-DFP scheme has a higher CDF curve compared to that of the DRA scheme for video flows. This implies that more video packets received suffer from higher delays under the DRA scheme. In terms of VoIP packet delay, the DRA scheme outperforms the C-DFP scheme. As the number of femtocells increases, the delay performance of the DRA scheme deteriorates gradually, as depicted in Figures 24.9b, 24.9c and 24.9d. The packet delay performance of the DRA scheme

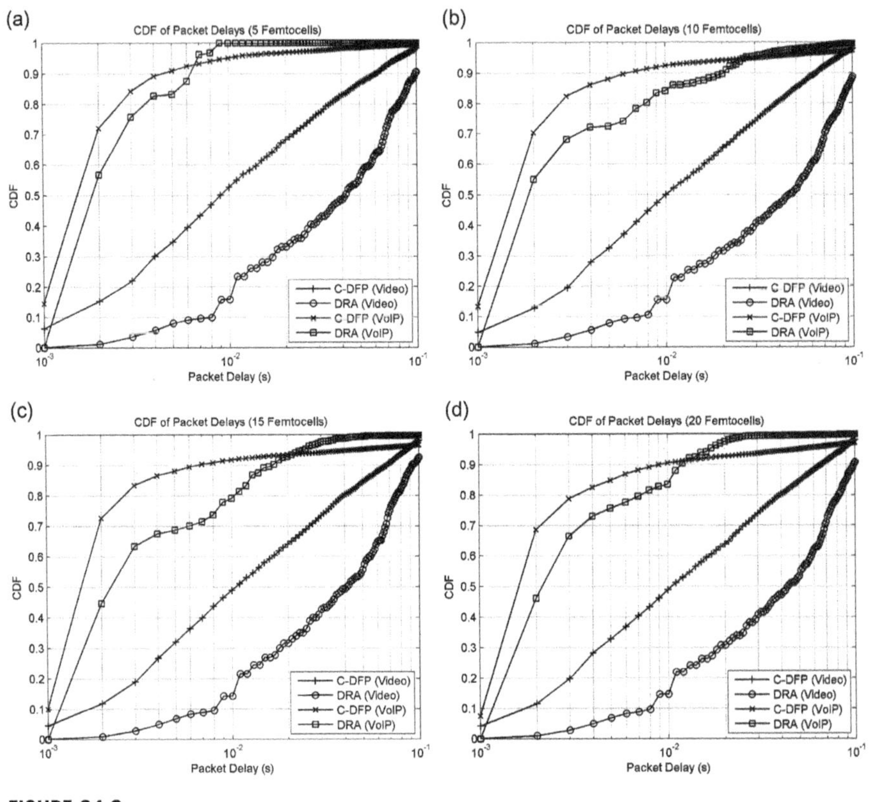

FIGURE 24.9

CDF of packet delay for (a) 5, (b) 10, (c) 15 and (d) 20 femtocells.

gradually deteriorates as the number of femtocells increases; this could be due to the fact that the contention of getting PRBs among large numbers of UEs (as a result of large numbers of femtocells) is more intense. Consequently, more packets have to be queued for longer periods before they get transmitted. It can be foreseen that the delay performance of the DRA scheme will continue to deteriorate even if the number of femtocells exceeds 20. The C-DFP scheme can maintain a relatively constant delay performance mainly due to its feature whereby each HeNB receives sufficient PRBs according to its demand.

9.5 FAIRNESS

It is desired to allocate radio resources to each HeNB fairly such that an even throughput distribution among HeNBs can be achieved. Otherwise, an unfair resource allocation will result in uneven throughput distribution whereby some HeNBs suffer from resource starvation.

When all traffic flows exhibit equivalent channel conditions, maximally fair throughput distribution and maximum network throughput can be achieved. However, in practice, the channel conditions of the traffic flows are normally different from each other. Therefore, a maximally fair throughput distribution does not necessarily lead to maximum achievable throughput and vice versa. In such situations, one may attempt to find the best trade-off between these two objectives.

The Jain's fairness index [25] is a widely known metric used for evaluating the fairness of resource allocation among traffic flows. The Jain's fairness index is defined as

$$FI = \frac{\left(\sum_{c \in C} R_c\right)^2}{|C| \sum_{c \in C} (R_c)^2} \tag{24.11}$$

where C is the set of all data flows, i.e., $C = \{1, 2, \ldots, |C|\}$. The values of FI lie in the range (0, 1] where a unity value indicates maximal fairness. In LTE-Sim, estimation of the Jain's fairness index is provided by one of the evaluation tools.

Since a wireless network may have various types of traffic flows, fairness plots can be produced separately for different traffic flows, as shown in Figure 24.10. It is noteworthy that fairness cannot be measured among different types of traffic flows, e.g., measuring fairness among the traffic flows that include video, VoIP, best-effort, etc. This is because different types of traffic flows have different transmission priorities. Usually, real-time flows such as video and VoIP have higher transmission priority compared to best-effort flows.

Using Jain's fairness index, the fairness performance of both the C-DFP and DRA schemes for video, VoIP and best-effort flows are illustrated in Figures 24.10a, 24.10b and 24.10c, respectively. It can be observed that the C-DFP scheme outperforms the DRA scheme for video flows. This is because the C-DFP scheme works by satisfying the resource demands of all HeNBs, allowing each HeNB to obtain sufficient PRBs for video packet transmission. Unlike the C-DFP scheme, the DRA scheme aims to avoid interference only. The number of PRBs assigned to each HeNB might be insufficient. Therefore, the throughput achieved by each HeNB varies significantly among each other, causing a huge degradation in video fairness. However, the DRA scheme has better fairness performance for VoIP and best-effort flows compared to the C-DFP scheme. This is because the PRBs provided by the DRA scheme are insufficient to satisfy the video flows since video packets are more resource demanding. Thus, most of the PRBs are used to satisfy VoIP flows that require less resources, and for best-effort flows which do not have specific minimum throughput demands.

It can be observed from Figure 24.10 that the fairness performance of the C-DFP scheme deteriorates as the number of femtocells increases. This is due to increased interference among the HeNBs, which leads to poor channel conditions.

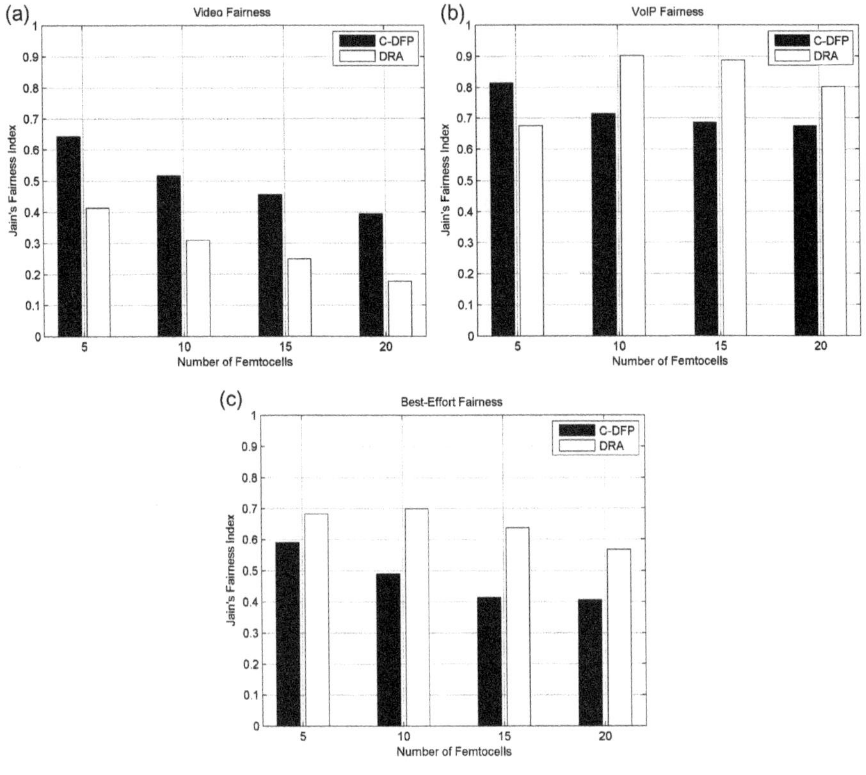

FIGURE 24.10

Fairness performance of (a) video, (b) VoIP and (c) best-effort flows.

As a result, some HeNBs receive PRBs with low channel quality, and hence achieving lower throughput. Moreover, the greedy algorithm used in the C-DFP scheme cannot guarantee to find the global optimal solution to the problem in (Equation 24.6) (see Section 3.1.2). More frequently, the algorithm finds suboptimal solutions that approximate the global optimal solution. As a result of using a suboptimal solution, the PRBs allocated to some HeNBs may experience stronger interference compared to others, thus reducing their achievable throughput. This reduction causes uneven throughput distribution among the HeNBs and degrades fairness. Similar trends are observed in the fairness performance of video and best-effort flows under the DRA scheme.

9.6 COMPLEXITY ANALYSIS

Computational complexity is an important aspect in the design of resource allocation algorithms. Since resource allocation in LTE networks is a real-time application, it is required to employ low-complexity algorithms which can be executed in a very short time. Otherwise, a computationally intensive resource allocation algorithm could delay scheduled transmissions in each base station.

In the C-DFP scheme, the resource broker employs the greedy algorithm (see Section 3.1.2) which iteratively removes a PRB from being allocated to each HeNB. Let $P = |F|$ be the number of HeNBs and $Q = |K|$ be the number of PRBs available in the frequency domain; the iterative process of the greedy algorithm requires up to P^2Q^2 iterations to complete the process of resource allocation to all HeNBs. As such, the C-DFP algorithm has a complexity of $O(P^2Q^2)$.

In the DRA scheme, each HeNB executes the DRA scheme (see Section 3.2.2) to obtain PRBs. Given I is the number of interfering HeNBs, each HeNB takes $(10Q)/I$ iterations to perform resource allocation using the DRA algorithm where $10Q$ is the number of resource tiles (where each one consists of two consecutive PRBs in the time domain) available in one LTE frame. The total number of iterations required by the DRA scheme to allocate resource tiles to all HeNBs is $(10PQ)/I$. As such, the DRA scheme has a computational complexity of $O(PQ)$.

By comparison, the DRA scheme has a significantly lower complexity than the C-DFP scheme. Therefore, execution of the DRA algorithm is faster than that of the greedy algorithm in the C-DFP scheme, making the DRA algorithm more attractive to meet the real-time requirements of LTE systems.

9.7 CONCLUDING REMARKS FOR THE C-DFP AND DRA SCHEMES

From the performance evaluation of the C-DFP and DRA schemes, as well as their complexity analyses, we can identify a number of strengths and weaknesses of these two schemes in terms of interference mitigation, spectral efficiency, QoS satisfaction, fairness and complexity, as summarized in Table 24.2.

The DRA scheme has a better interference mitigation ability compared to the C-DFP scheme because its mechanism facilitates total interference avoidance. In terms of spectral efficiency, the C-DFP scheme is better than the DRA scheme, as evidenced by their throughput performance, thanks to the fact that the former has a more efficient resource reuse feature compared to the latter. In the aspect of QoS, the C-DFP scheme is relatively better compared to the DRA scheme as

Table 24.2 Comparison Between the C-DFP and DRA Schemes

Aspect	C-DFP	DRA
Interference Mitigation	Low	High
Spectral Efficiency	High	Low
QoS Satisfaction	Medium	Low
Fairness	Medium	Medium
Complexity	High	Low

demonstrated by the superior PLR and delay performance of the C-DFP scheme over the DRA scheme. In terms of fairness, the C-DFP scheme outperforms the DRA scheme for video flows but the latter is better for VoIP and best-effort flows. From the complexity perspective, the C-DFP scheme has a substantially higher complexity compared to the DRA scheme.

In conclusion, the C-DFP scheme has an efficient resource reuse ability that compensates for its poorer interference performance in achieving high throughput. Besides that, its feature of satisfying the resource demand of HeNBs allows sufficient PRBs for the UEs to ensure QoS satisfaction. On the other hand, the DRA scheme attains channel conditions with less interference at the expense of lower resource utilization efficiency. As it does not guarantee sufficient PRBs for every UE, QoS satisfaction cannot be guaranteed. Nevertheless, the DRA scheme has significantly lower complexity compared to the C-DFP scheme.

10 CONCLUSION

The complexity of advanced wireless communication systems is a driving force behind the widespread use of simulation modeling. We have illustrated in this chapter the use of a modern simulation tool to conduct simulation modeling and performance evaluation for resource allocation schemes in LTE femtocell networks. This chapter first provided an overview of LTE systems and reviewed a centralized and a distributed femtocell resource allocation schemes, followed by a discussion of simulation methodology for femtocell networks. We introduced a powerful LTE femtocell network simulator called LTE-Sim and used it to implement and evaluate the performance of the two schemes in terms of some performance metrics such as interference, throughput, packet loss, delay and fairness. Additionally, a complexity analysis on the resource allocation schemes has been made. Finally, we showed how the strengths and weaknesses of the resource allocation schemes can be readily interpreted from the computer simulation results obtained.

GLOSSARY

3GPP	3rd Generation Partnership Project
AMC	Adaptive Modulation and Coding
CDF	Cumulative Distribution Function
C-DFP	Centralized Dynamic Frequency Planning
CBR	Constant Bit Rate
CQI	Channel Quality Indicator
DFP	Dynamic Frequency Planning
DRA	Distributed Random Access
DSL	Digital Subscriber Line
eNB	evolved NodeB

GBR	Guaranteed Bit Rate
HeNB	Home evolved NodeB
HOL	Head of Line
LTE	Long Term Evolution
MAC	Medium Access Control
MCS	Modulation and Coding Scheme
MME	Mobility Management Entity
OFDM	Orthogonal Frequency Division Multiplexing
OFDMA	Orthogonal Frequency Division Multiple Access
PLR	Packet Loss Rate
PRB	Physical Resource Block
QoS	Quality of Service
RSS	Received Signal Strength
S-GW	Serving Gateway
TTI	Transmission Time Interval
UE	User Equipment
VoIP	Voice over Internet Protocol

REFERENCES

[1] Lee YL, Chuah TC, Loo J, Vinel A. Recent advances in radio resource management for heterogenenous LTE/LTE-A networks. IEEE Commun Surv Tutorials 2014;16(4): 2142–80.

[2] Saquib N, Hossain E, Le LB, Kim DI. Interference management in OFDMA femtocell networks: issues and approaches. IEEE Wirel Commun 2012;19(3):86–95.

[3] Lien S-Y, Lin Y-Y, Chen K-C. Cognitive and game-theoretical radio resource management for autonomous femtocells with QoS guarantees. IEEE Trans Wirel Commun 2011;10(7):2196–206.

[4] Hatoum A, Langar R, Aitsaadi N, Boutaba R, Pujolle G. Cluster-based resource management in OFDMA femtocell networks with QoS guarantees. IEEE Trans Veh Technol 2013;63(5):2378–91.

[5] Domenica AD, Strinati EC. A radio resource management scheduling algorithm for self-organizing femtocells In: IEEE 21st international symposium on personal, indoor and mobile radio communications workshops (PIMRC workshops), Istanbul, Turkey; Sep. 2010. p. 191–96.

[6] Yoon S, Cho J. Interference mitigation in heterogeneous cellular networks of macro and femto cells, In: international conference on ICT convergence (ICTC), Seoul, South Korea; 2011. p. 177–81.

[7] Sadr S, Adve R. Hierarchical resource allocation in femtocell networks using graph algorithms, In: IEEE international conference on communication (ICC), Ottawa, ON, Canada; June 2012. p. 4416–20.

[8] Lopez-Perez D, Chu X, Vasilakos AV, Claussen H. Power minimization based resource allocation for interference mitigation in OFDMA femtocell networks. IEEE J Select Areas Commun 2013;32(2):333–44.

[9] Piro G, Grieco LA, Boggia G, Capozzi F, Camarda P. Simulating LTE cellular systems: an open-source framework. IEEE Trans Veh Technol 2011;60(2):498–513.

[10] Capozzi F, Piro G, Grieco LA, Boggia G, Camarda P. On accurate simulations of LTE femtocells using an open source simulator. EURASIP J Wirel Commun Networking 2012;2012(328):1–13.

[11] Lopez-Perez D, Valcarce A, de la Roche G, Zhang J. OFDMA Femtocells: a road-map on interference avoidance. IEEE Commun Mag 2009;47(9):41–8.

[12] Sundaresan K, Rangarajan S. Efficient resource management in OFDMA femtocells, In: proceedings of the ACM international symposium on mobile Ad Hoc network and computing, New Orleans, Louisiana, USA; May 2009. p. 33–42.

[13] 3GPP. Technical specification group radio access network; evolved universal terrestrial radio access (E-UTRA); Physical channels and modulation, 3rd generation partnership project (3GPP), TS 36.211, [Online]. Available from: <http://www.3gpp.org/DynaReport/36.211.htm>; Sep. 2012.

[14] 3GPP. Evolved universal terestrial radio access (E-UTRA) and evolved universal terestrial radio access network (E-UTRAN); overall description: Stage 2, 3rd Generation Partnership Project (3GPP), TS 36.300, [Online]. Available from: <http://www.3gpp.org/DynaReport/36.300.htm>; Dec. 2013.

[15] Chandrasekhar V, Andrews J. Spectrum allocation in tiered cellular networks. IEEE Trans Commun 2009;57(10):3059–68.

[16] Hatoum A, Aitsaadi N, Langar R, Boutaba R, Pujolle G. fcra: femtocell cluster-based resource allocation scheme for OFDMA networks, In: IEEE international conference on communications (ICC), Kyoto, Japan; 2011. p. 1–6.

[17] Liang Y-S, Chung W-H, Ni G-K, Chen I-Y, Zhang H, Kuo S-Y. Resource allocation with interference avoidance in OFDMA femtocell networks. IEEE Trans Veh Technol 2012;61(5):2243–55.

[18] Hatoum A, Langar R, Aitsaadi N, Boutaba R, Pujolle G. Cluster-based resource management in OFDMA femtocell networks with QoS guarantees. IEEE Trans Veh Technol 2014;63(5):2378–91.

[19] Lopez-Perez D, Juttner A, Zhang J. Optimisation methods for dynamic frequency planning in OFDMA networks, In: telecommunications network strategy and planning symposium. Budapest, Hungary; Sep.–Oct. 2008. p. 1–28.

[20] Capozzi F, Piro G, Grieco LA, Boggia G, Camarda P. Downlink packet scheduling in LTE cellular networks: key design issues and a survey. IEEE Commun Surv Tutorials 2013;15(2):678–700.

[21] Mehlführer C, Wrulich M, Ikuno JC, Bosanska D, Rupp M. Simulating the long term evolution physical layer In: proceedings of the 17th european signal processing conference (EUSIPCO), Glasgow, UK; August 2009. p. 1471–78.

[22] Ikuno JC, Wrulich M, Rupp M. System level simulation of LTE networks, In: proceedings IEEE VTC—Spring, Taipei, Taiwan; May 2010, p. 1–5.

[23] Mathworks. LTE System Toolbox™. Available: <http://www.mathworks.com/products/lte-system/>; 2015.

[24] LTE-Sim. Available from: <https://lte-sim.googlecode.com/files/lte-sim-r5.zip>; 2015.

[25] Jain R. The art of computer systems performance analysis. John Wiley & Sons; 1991.

CHAPTER

Multimedia transmission over wireless networks fundamentals and key challenges

25

Gürkan Gür[1,2]

[1]Provus-A MasterCard Company, Ayazaga, Istanbul, Turkey

[2]Bogazici University, Istanbul, Turkey

1 INTRODUCTION

Recently, there has been an unprecedented surge in the consumption of wireless multimedia content. While the demand of smart wireless devices has experienced a quantum leap and the mobile data explosion is challenging the mobile networks, the mobile network infrastructure is challenged to adapt in a sufficient and flexible manner. According to Cisco VNI [1], global mobile data traffic in 2012 was 12 times the size of the entire global Internet traffic in 2000. In 2017, it is expected that it will increase 13-fold compared to 2012, reaching 11.2 exabytes per month. The wireless network technologies ranging from body area networks to satellite Wide Area Networks (WANs) are poised to operate efficiently as seen in Figure 25.1 to address this challenge. These networks are being developed to deliver multimedia content for streaming, interactive, peer-to-peer and content distribution services with high efficiency and quality. However, Quality of Service (QoS) requirements for multimedia traffic are compounded in the case of a wireless network, where new problems arise due to the implied mobility of the users as well as due to the nature of the current IP protocols that support IP-based mobility, combined with the lossy and interrupt/outage-prone nature of the communications channel. Therefore, multimedia support in these systems needs to be improved and the traditional protocol stacks have to be re-engineered by designing more flexible and generic communication protocols [2].

Additional issues arise when considering the widely differing types of services such as streaming media, real-time communications, interactive communications, VoIP with diverse characteristics. Each of these services poses its distinct challenges such as QoS requirements including bandwidth, delay and jitter. Therefore,

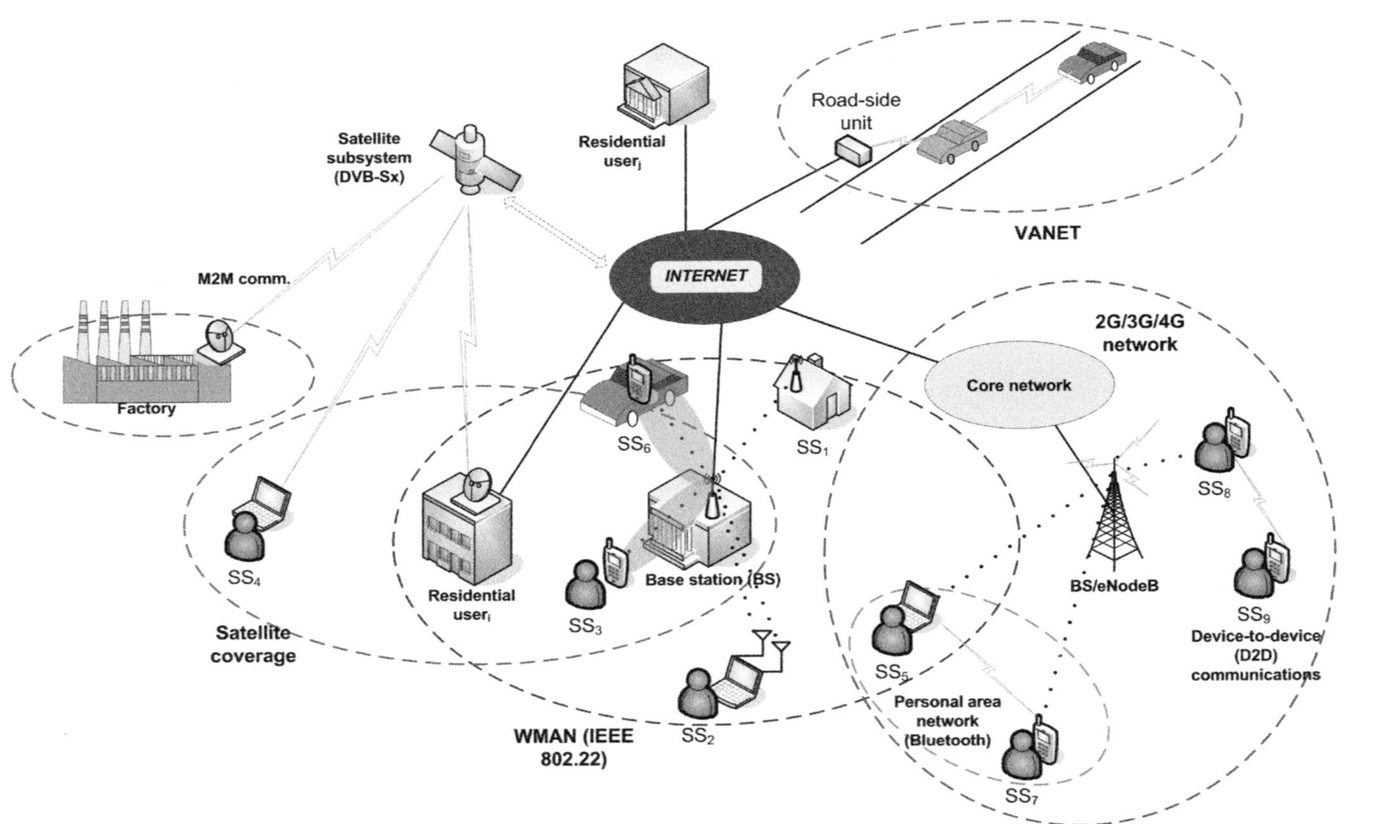

FIGURE 25.1

Heterogeneous wireless network technologies.

delivering multimedia over wireless Internet is a formidable task. The current Internet architecture is designed around a best-effort service paradigm, which does not support guaranteed delivery inherently. The advent of heterogeneous wireless networks further magnifies the volatility of network conditions and imposes greater challenges for multimedia delivery. However, future wireless all-IP networks, while offering the promise of these exciting broadband applications, are expected to consist of several, potentially incompatible, wireless access technologies that would be offered by a number of competing service providers. The diversity of access technologies, however, may drastically affect the QoS for multimedia services. To improve the perceived media quality by end users over wireless Internet, QoS requirements can be addressed in different layers ranging from application to physical layers [3]. Additionally a recent trend has been to go beyond the layered protocol architecture and adapt cross-layer solutions.

In this chapter, multimedia over wireless networks focusing on the QoS support for wireless multimedia transmission is discussed. After an overview of the multimedia transmission challenges for wireless networks, we provide a layered analysis ranging from the application to physical protocol layers. Then, we highlight a number of emerging wireless/mobile networking concepts including Cognitive Radio Networks (CRNs), ad hoc and multi-hop networks, i.e. mobile ad hoc networks (MANETs), Information-Centric Networking/Content-Centric Networking (ICN/CCN), and mobile content delivery, providing a discussion of the key challenges and multimedia networking related issues for these systems. The main emphasis is on video since it is the most complicated multimedia traffic type (for instance, it implies image transmission as a special case) and represents the most challenging case. Finally, we present and discuss emerging challenges for modeling and simulation of wireless multimedia networking in this diverse and dynamic environment.

2 MULTIMEDIA TRANSMISSION OVER WIRELESS NETWORKS

The next generation wireless networks such as 4G LTE, IEEE 802.11ac and LTE Advanced, envisioned to be the enabling technology for wireless multimedia services due to their advanced capabilities, make multimedia communication over the wireless link feasible and omnipresent. However, to achieve a high level of acceptability and proliferation of wireless multimedia, in particular wireless video, several key requirements are imperative to provide a reliable and efficient transmission: 1) easy adaptability to bandwidth variations; 2) energy efficiency; 3) robustness to data loss; and 4) support for bandwidth, power, and device scalability.

Multimedia applications in the widest sense can be grouped as *real-time* and *non-real time* multimedia. Each type of multimedia service and user in wireless networks have different QoS requirements, which brings the issue of addressing a set of problems at layers of protocol stack as seen in Figure 25.2. These problems, especially applied to multimedia and wireless networks, can be summarized as

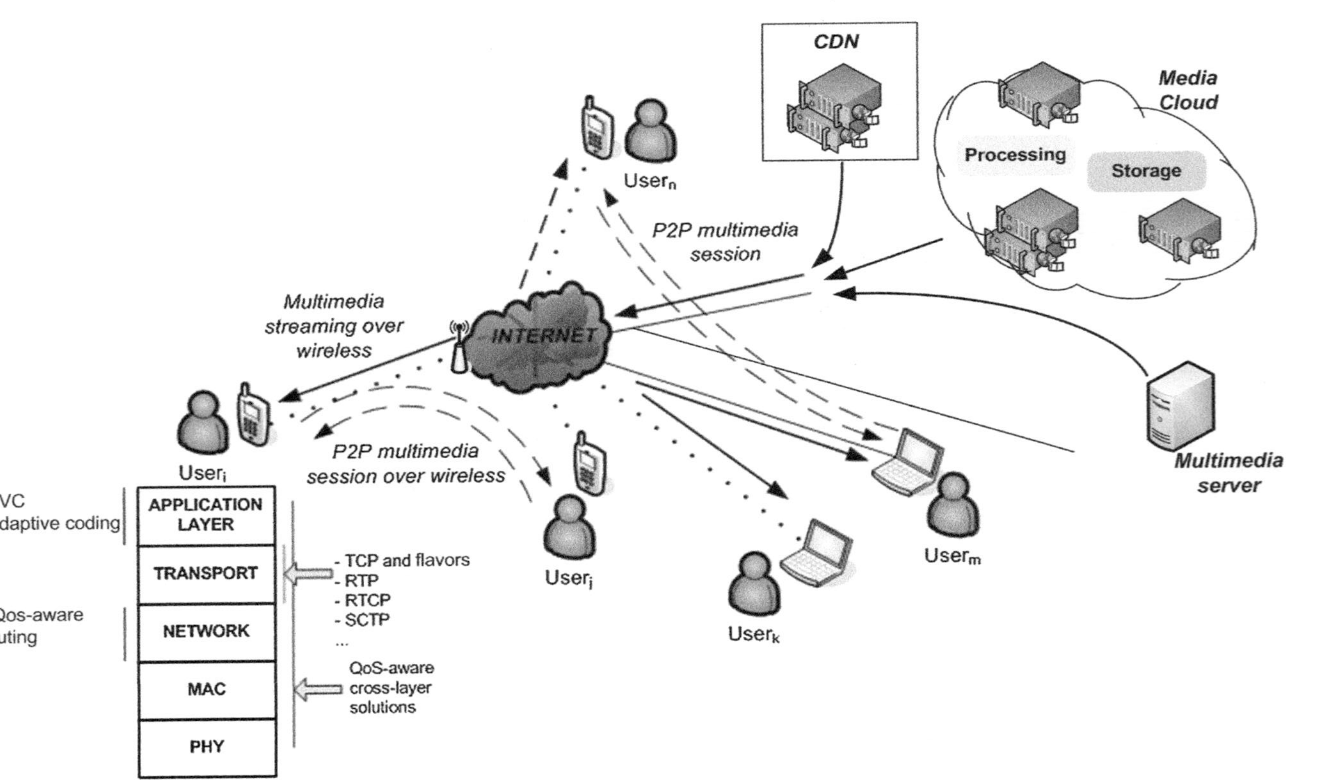

FIGURE 25.2

Various multimedia services and transmission over wireless.

Table 25.1 Traffic Classes and Their QoS Requirements [4]

Traffic Class	Content Type	QoS Characteristics	
Real-time conversational (telephony, video telephony and videoconference)	– Speech – Multimedia (Audio, Video)	– Delay and delay variation sensitive – Limited tolerance to loss and errors – Constant and variable bit rate	More challenging ↑ / Less BW-intensive ↓
Real-time streaming (e.g. audio and video broadcast, surveillance, and video-on-demand)	– Multimedia (Audio, Video)	– Delay tolerant but delay variation sensitive – Limited tolerance to loss and errors – Variable bit rate	
Near real-time interactive (e.g. web browsing)	– Data	– Delay sensitive but delay variation tolerant – Error sensitive – Variable bit rate	
Non-real-time background (e-mail, pull-based content distribution)	– Data	– Delay and delay variation tolerant – Error sensitive – Best effort	

limited bandwidth, high packet loss rate and Bit Error Rate (BER), need for mobility management, heterogeneity among users and networks (WLAN (IEEE 802.11x), PAN (IEEE 802.15, Bluetooth), cellular networks (3G/4G LTE)), security, and most importantly QoS guarantees. Providing QoS guarantees while combating these aforementioned issues is a challenging task. For instance, real-time traffic (e.g., video and voice) is highly delay-sensitive, while non-real-time traffic (e.g., TCP packets and text data transfers) can tolerate large delays. Table 25.1 enumerates various applications and their traffic classes along with some typical parameter values like bandwidth, connection time, etc.

While facing these key requirements, wireless multimedia networks are also anticipated to operate across diverse and formidable circumstances under stringent performance requirements. Consequently, multimedia communications over wireless networks pose many challenges listed as follows:

- **Different QoS requirements for different types of media:** In general, different types of media have different characteristics. Specifically, real-time media such as video and audio is delay-sensitive but more error-resilient. Non-real-time media such as Web data is less delay sensitive but requires end-to-end reliable transmission. In addition, due to scalable media encoding technologies, different parts of real-time media are of different importance for quality of user experience.

- **High packet loss rate and bit error rate:** In wired networks, packet losses are usually caused by congestion in intermediate nodes in the path. Meanwhile, wireless channels have higher BER due to interference and multipath fading. The consequential packet losses and bit errors can have detrimental effects on multimedia quality.
- **Limited energy capacity:** Comparing with fixed nodes, there is a battery lifetime constraint in mobile devices. In general, maintaining good media quality and minimizing average power consumption, including processing power and transmission power at mobile devices, are conflicting objectives. These trade-offs are especially valid for wireless ad hoc networks such as wireless multimedia sensor networks. From a multimedia coding point of view, achieving better media quality usually consumes more processing power in the source coder. From the network point of view, interference and transmission impairments such as multipath fading in wireless networks necessitate the use of high transmission power.
- **Bandwidth limitation and fluctuation:** Network conditions and characteristics in the current Internet such as bandwidth, packet loss ratio, delay, and jitter are dynamic. Meanwhile, the capacity of wireless network also fluctuates with the changing environment.
- **Shortcomings of traditional connection-oriented transport-layer protocols:** Traditional transport-layer protocol assumes congestion to be the primary cause for packet losses and unusual delay in the network. It decreases the transmission rate when packet losses occur. However, in wireless networks, the packet may also be dropped due to channel errors, thereby resulting in unnecessary reduction in end-to-end throughput when transport protocol reacts.
- **Heterogeneity among users and networks:** Receivers in multimedia delivery systems are quite different in terms of latency requirements, visual quality requirements, processing capabilities, power limitations, and bandwidth constraints. Moreover, multimedia may traverse different type of networks with different characteristics such as reliability, delay, jitter, bandwidth, and medium access control (MAC) mechanisms.

3 QoS AND QoE: THE CHALLENGE GENERATORS

The demand for enhanced QoS/QoE increases as multimedia applications/services are being widely deployed on mobile networks and devices, albeit the inherent challenges generated by these requirements themselves. Especially in time-varying wireless/mobile networks, it is becoming essential to provide adaptive multimedia transmission in order to maintain the desired QoS/QoE levels regardless of the network situation [5]. Service performance is then managed through a set of network parameters. Such metrics do not represent the entire QoS from the user perspective. Instead, they represent the network performance parameters that may affect user experience. The key network performance parameters used to manage service performance are [6]:

- *Throughput*: The effective data transfer rate measured in bits per second and a minimum rate of throughput is usually supposed to be guaranteed for specific services and applications such as video.
- *Latency*: It is defined as the delay between the time the data is sent from its origin and received at its destination.
- *Jitter*: It represents the delay variation caused by the variations in delay experienced by the packets during end-to-end transmission such as variations in queue length or in packet processing, and packets traveling through different paths.
- *Packet loss*: It represents the information losses at the network level that may occur for many reasons including network congestion, link failures and transmission errors. For wireless networks, transmission errors are more common compared to wired networks due to variations in channel conditions.
- *Availability*: It is the percentage of the time that the network is available to provide services to the end users.
- *Reliability*: According to IEEE, reliability is defined as "the probability that a system will perform its intended functions without failure, within design parameters, under specific operating conditions, and for a specific period of time."

QoE is one step further than QoS and describes the service quality as experienced by the user. Siller et al [7]. define QoE as "the user's perceived experience of what is being presented by the Application Layer, where the application layer acts as a user interface front-end that presents the overall result of the individual quality of services." In general, the relation between QoE and QoS metrics is not directly determined and is typically observed empirically because there are several QoS metrics that may impact the overall QoE. From the network perspective, there are various factors determining QoE level as shown in Figure 25.3. These factors are driven by demand and user characteristics affected by seasonality, events, holidays, and content popularity [8].

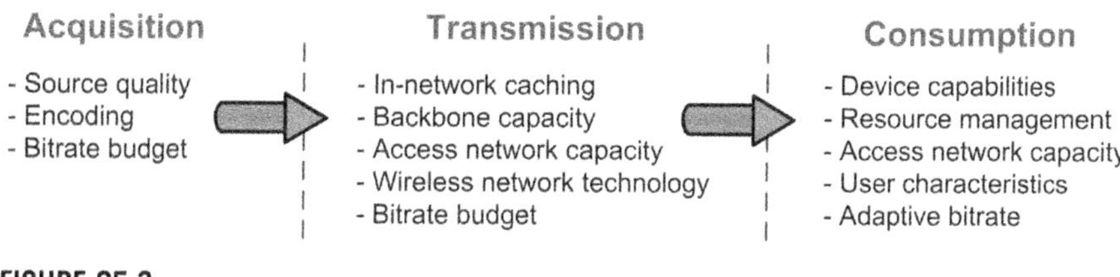

FIGURE 25.3

Factors influencing QoE [8].

There is a broad diversity of direct QoE metrics, of which some are listed below [9]:

- *Peak Signal to Noise Ratio (PSNR):* This is a basic but important metric that assesses the similarity between two different images. PSNR computes the Mean Square Error (MSE) of each pixel between the original and received images, represented in dB. Images with more similarity will result in higher PSNR values.
- *Structural Similarity (SSIM):* The main drawback of PSNR is that it does not consider how human perception works, hence in some cases it cannot detect some human perceptible video disruptions. To address this shortcoming, SSIM combines luminance, contrast, and structural similarity of the images to compare the correlation between the original image and the received one.
- *Video Quality Metric (VQM):* VQM detects human perceivable artifacts on the images, by considering blurring, global noise, and block and color distortions. This metric also uses the original video as the reference for assessment.
- *Mean Opinion Score (MOS):* MOS was originally devised for audio streams. It inherently combines various parameters such as delay, perceived jitter at the application layer, codec used for the communication and experienced packet loss and scores the picture quality of a video presentation. In the area of video QoE assessment, it can be considered as a meta-metric given that it considers values from other metrics to generate the final computed quality level perceived by the user.

For calculating PSNR, SSIM and VQM, the original reference sequence, i.e. the original video, for comparison is necessary while MOS relies on user opinion or other metrics.

QoS management refers to the mechanisms and control frameworks which are utilized for QoS/QoE provisioning. Basically, they are optimization approaches which include some model of the multimedia-based service and its QoS requirements. This optimization is carried out using mechanisms involving resource management and adaptation [10]. An example system is shown in Figure 25.4. In this proposal, resource management is used in the wider context of QoS management for multimedia.

4 LAYERED ANALYSIS

According to the traditional International Organization for Standardization/Open System Interconnection (ISO/OSI) approach, functions of communication are shared between network layers in communication networks. Thus, every layer can be implemented independently from the others. Well-defined interfaces between these layers and separation of duties are instrumental for modular design which

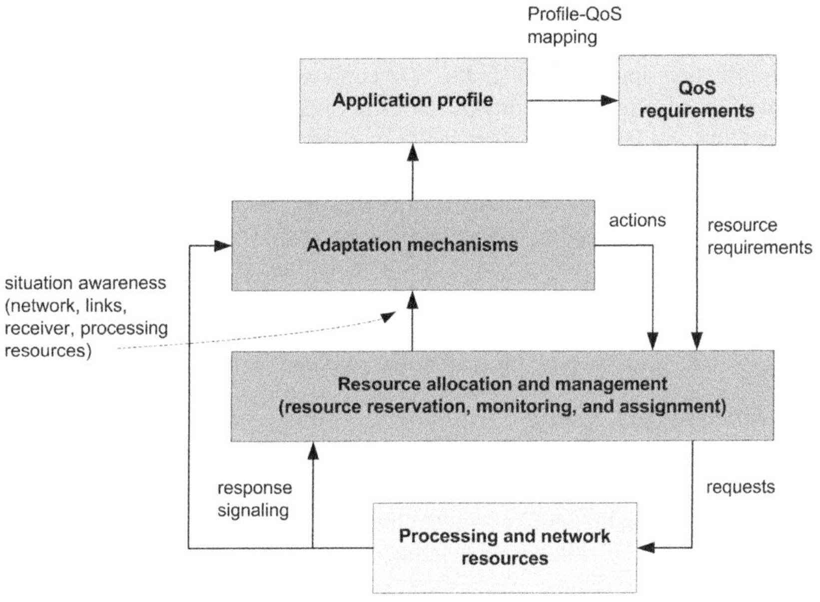

FIGURE 25.4

QoS management system for multimedia services.

lends itself to high compatibility, interconnectivity, and robustness to functionality changes. However, there is also the cross-layer paradigm of cross-layer design (CLD), which is based on the premise that the layers cannot be perfectly separated and more efficient operation is possible with an integrated cross-layer design. In this section, a layered analysis for wireless multimedia networks is provided ranging from the application to physical protocol layers extended with a cross-layer perspective in the last part.

4.1 APPLICATION LAYER

At the application layer, multimedia encoding techniques and application-based optimizations provide means for high-quality multimedia transmission in wireless networks. There are many studies in the application layer to improve wireless video streaming and multimedia communications and to develop error resilient source coding and transmission techniques such as error protection, power saving, and proxy management [11,12]. To mitigate transmission challenges such as packet loss and bit errors and maintain high-quality media delivery in wireless IP networks, error control techniques such as forward error correction (FEC) and automatic repeat request (ARQ) are instrumental. Unequal error protection (UEP) can be adopted if there is a need to take into account the varying levels of importance of different types/parts of media. To combat the power-quality dilemma, power control and joint source-channel coding (JSCC) are two effective approaches [3]. JSCC is conducted from the individual user's point of view to effectively combat the errors occurred during transmission by allocating the available bit budget between source and channel. The heterogeneous networks and diverse characteristics of receivers also require efficient proxy-caching mechanisms to address different user requirements and to exploit spatio-temporal multimedia consumption patterns. Traditional proxy servers were designed to serve web requests for noncontinuous media, such as textual and image objects in web pages. However, the varying wireless Internet conditions and diverse media characteristics of multimedia content impose challenges on how to efficiently cache both continuous and noncontinuous media.

4.1.1 Robust video coding

In general, a media encoder/decoder can be designed such that it has the ability to dynamically adjust the coding rate and other coding parameters according to the varying network conditions (bandwidth, loss, delay, etc.). For instance, scalable coding techniques are introduced to realize this type of media adaptation. The major technique to achieve scalability is layered coding technology, which divides multimedia information into several layers, and the incremental reception of layers also increases the media fidelity. Moreover, different error protection levels can be applied to different layers and path differentiation is possible for routing of transport streams.

4.1.2 Source rate control

Source rate control at the application layer aims to make the source rate match the channel condition to achieve better video quality at the receiver side [13]. Source rate control is typically adopted at the application layer to optimize the playback quality, subject to the bandwidth constraint provided by the congestion control mechanism and the QoS requirements of the multimedia application (e.g., the end-to-end delay constraint). At the receiver side, adaptive media play-out mechanism is utilized to meet the end-to-end (E2E) delay constraint by adaptively varying the play-out speed at the receiver. At the sender side, adaptive adjustment of the source rate is performed based on the channel condition and the QoS requirements of the application. For example, the proportional plus derivative (PD) controller is used in [14] to determine the source rate according to the encoder buffer state. In [15], both the encoder buffer state and the end-to-end delay constraint are considered using a virtual network buffer management algorithm for bitstream switching applications.

4.1.3 Resource allocation and scheduling

Multimedia content, and specifically, streaming video, requires per-user data rates guaranteed in order to be of useful quality [16]. Many proposed cross-layer scheduling and resource allocation methods exploit the time-varying nature of the wireless channel to maximize the throughput of the network while maintaining fairness across multiple users. These methods rely on the multi-user diversity gain achieved by selectively allocating a majority of the available resources to users with good channel quality who can support higher data rates. Many of these methods, such as the proportional fair rule, can be viewed as gradient-based scheduling policies. In these policies, during each time-slot, the transmitter maximizes the weighted sum of each user's rate, where the (time-varying) weights are given by the gradient of a specified utility function. One attractive feature of such policies is that they require only myopic decisions, and hence presume no knowledge of long-term channel or traffic distributions. However, these approaches can also be improved via introducing memory and long-term statistics. Additionally, the channel side information required for these schemes can be intricate and difficult to acquire and process in practical situations.

4.2 TRANSPORT LAYER

The main function of the transport layer is to process the data coming from the upper layer (application layer) before passing to the network layer and it has the capability of transferring data from one node to another, thus providing some level of transparency from the underlying network [2]. Generally speaking, the transport layer splits data into smaller packets, dispatches those packets and at the receiver reassembles the data in the correct sequence. Furthermore, the transport layer may provide the following services: flow control between the two ends, so-called "end-to-end" services, dividing the streams of data into chunks or packets and reassembling at the receiving end, error-checking to guarantee error-free data delivery, with losses or duplications.

Data transport protocols can be loosely grouped into two classes: *reliable transport protocols* and *unreliable transport protocols*. *Reliable protocols* such as TCP ensure that all data are delivered to the receiver using feedback mechanisms. However, these protocols do not take the delay issues into account and may deliver data sluggishly. *Unreliable protocols* such as UDP, on the other hand, do not guarantee the delivery of all packets. However, they provide less delay in the data transmission and are simpler to facilitate. Therefore they cause a smaller burden on the system resources. UDP has a low protocol overhead of 8 bytes, where the fixed part of the TCP header is 20 bytes long. Considering the nature of these two protocol types, unreliable protocols suit more to the needs of multimedia applications at the transport layer.

TCP is the most commonly used protocol at the transport layer of the network stack in IP networks, originally developed in wired networks with low bit error rate (BER) in the order of less than 10^{-8}. Generally, existing transport protocols which are initially designed for wired networks have limitations for wireless networks. Although multimedia applications need congestion control but not necessarily ordered and reliable delivery, this combination is not offered by conventional transport protocols TCP or UDP. TCP has also been designed for the early Internet with certain assumptions in mind, such as low bandwidth and low delay. For instance, when a data segment is lost, it assumes that this was most likely due to congestion (i.e., too many segments are contending for network resources). However, in wireless networks it could be due to unfavorable conditions in the transmission link [17].

Enhancing TCP to reliably handle loss, minimize errors, manage network congestion and transmit expeditiously in high throughput environments is crucial for efficient multimedia services over wireless IP networks. In that regard, the basic challenges and issues in the design of a transport protocol can be listed as following:

- ***Packet loss differentiation and estimation***: Traditional transport protocols assume that a packet loss is encountered due to congestion in the network. However, this assumption may lead to performance degradation in wireless networks where the packet losses due to transmission errors are more probable in error-prone links. Therefore, the transport mechanism must distinguish between the packet losses experienced due to network congestion and losses due to link errors.
- ***Available bandwidth estimation***: Available network condition estimation and congestion control are key issues in the transport layer for combating adverse bandwidth and delay conditions in wireless IP networks. Network information such as packet error rate, delay, and delay jitter is beneficial for high-quality media delivery. Various congestion and rate control schemes can be performed so that multimedia such as video and audio can adapt to the estimated network information in a smooth way [3]
- ***Mobility Management***: Mobility also has significant impact on perceived QoS during multimedia streaming. There is a plethora of research on how to manage the handoffs while there is an ongoing multimedia transmission.
- ***Inherent QoS Support***: The QoS issue has become vital with the increasing proliferation of wireless networks and multimedia services. For higher QoS level, QoS support mechanisms embedded in the transport layer are beneficial.

Simulation and modeling of such protocols are import components of related research efforts since they are essential for investigating and comparing transport protocol changes or proposals.

4.2.1 Transport-layer enhancements for multimedia over wireless links

To handle the severe bandwidth and delay fluctuations in wireless IP networks, available network condition estimation and congestion control are key issues needing to be addressed in the transport layer. Different congestion and rate control schemes can be performed so that multimedia, such as video and audio, can adapt to the estimated network information in a smooth way.

Considering the importance of congestion control, we first present E2E congestion control schemes, focusing on the mechanisms that are well suited to multimedia applications. Then, we discuss selected E2E mechanisms that enhance protocol efficiency and application performance.

4.2.1.1 End-to-end congestion control

At the transport layer, congestion control for streaming multimedia is adopted to make sure the users have a fair share of the network resources without flooding the links. Congestion control for the wireless scenario has to differentiate packet loss due to wireless link error from packet loss due to congestion since the transport protocol needs to decrease the sending rate only when there is congestion in the network. Moreover, it has to take into account the smoothness of the sending rate, in addition to the fairness and responsiveness of the transport protocol, to supporting the multimedia application for better playback quality [18].

The TCP congestion control mechanism decreases the congestion window when packet losses are detected in the period called *slow-start*. Audio and video, on the other hand, have "natural" rates that cannot be suddenly decreased without starving the receiver. Video could be more easily decreased simply by slowing the acquisition of frames at the sender when the transmitter's send buffer is full, with the corresponding delay. The correct congestion response for these media is to change the audio/video encoding, video frame rate, or video image size at the transmitter. These counter-actions are also related to application layer capabilities discussed in Section 4.1.

- **TCP and TCP-Friendly Congestion Control**: A number of TCP-friendly congestion control schemes for the wired channels have been proposed to provide smooth sending rates. These include the window-based schemes and the rate-based schemes, which can be further classified into probe-based and equation-based schemes. Those approaches all assume that every packet loss is an indication of congestion, which is not held for the wireless channels. As in the wireless scenario, packet losses can also be attributed to link error. Thus these mechanisms cannot be directly applied to wireless scenarios [18].
- **Rate-Based Congestion Control:** Considering TCP's limitations and the impending threat of unresponsive UDP, rate-based congestion control constitutes a plausible mechanism for media streaming applications. Rate-based protocols control the transmission rate of the connection and typically generate a smoothed flow by spreading the data transmission across a time interval. Hence, the burstiness induced by the window-based mechanisms is avoided.
- **Equation-Based Congestion Control:** Equation-based congestion control enables bandwidth estimation based on statistics of round-trip-time (RTT) and packet loss probability. In response to the bandwidth estimates, the source adjusts the transmission rate to prevent congestion. TCP-friendly rate control (TFRC) is a typical equation-based protocol which adjusts its transmission rate in response to the level of congestion, as estimated based on the calculated loss rate. Multiple packet drops in the same RTT are considered as a single loss event by TFRC leading to a more sluggish congestion control strategy.

4.2.1.2 End-to-end enhancements for wireless links

TCP is not very preferable for multimedia transmissions since reliable transmission is inappropriate for delay-sensitive data such as real-time audio and video. Due to the fact that the TCP transmission mechanism is driven by the acknowledgments from the receiver node, it takes a long time for the sender to discover the lost packet. Moreover, TCP provides the receiver with the lost packet by retransmissions. The receiver either waits for the retransmission, increasing delay and incurring an audible gap in play-out or it just discards the lost packet. These drawbacks constitute major factors on the perceived QoS.

There are numerous end-to-end enhancements for heterogeneous wired/wireless networks [19]

- *TCP Westwood:* The key innovation of TCP Westwood (TCPW) is the use of bandwidth estimate to directly drive contention window *(cwnd)* and slow start threshold (*ssthresh*). The TCP sender continuously monitors ACKs from the receiver and computes its current Eligible Rate Estimate (ERE). ERE is based on the rate of ACKs and their payload. Upon a packet loss indication, the sender sets *cwnd* and *ssthresh* based on ERE value.
- *TCP Probe:* TCP-Probe proposes a simple "passive" capacity estimation extension to TCP, to accurately estimate bottleneck link capacity of an Internet path [20]. The TCP Probe extension is packet-pair based, and it is applicable to all TCP variants. The results in [20] show that the capacity estimate provided by TCP Probe enables it to take better advantages of the capacity increase than the original TCP.
- *MULTFRC:* The applicability of TFRC protocol to multimedia transmissions has been an area of research and some modifications are proposed to improve the performance of multimedia transmission on wireless links. TFRC is originally designed for wired links; it accounts wireless losses also as congestion losses. TFRC also suffers from degraded performance due to packet reordering in the Internet. Suggestions have been made to have a rate control mechanism based on the *Explicit Congestion Notification (ECN)* marking, which works well in wireless scenarios. Similarly, Bae and Chong [21] utilize ECN marking in order to adapt the TFRC mechanism to the wireless environment. The main idea behind this scheme is that by using the ECN marking in conjunction with random early detection (RED) queue management scheme intelligently, it is possible that not only the degree of network congestion is notified to multimedia sources explicitly in the form of ECN-marked packet probability but also wireless losses are hidden from multimedia sources.
- Video Transport Protocol: Another friendly rate control protocol for wireless networks, called Video Transport Protocol (VTP), is proposed by Yang et al. [22]. The proposed scheme has a new and unique end-to-end rate control mechanism that aims to avoid drastic rate fluctuations while maintaining friendliness to legacy protocols. VTP is also equipped with an achieved rate estimation scheme and a loss discrimination algorithm, both end-to-end, to cope with random errors in wireless networks efficiently.

4.3 NETWORK LAYER

At the network layer, a fundamental tool for achieving multimedia transmission is to use multipath transport or routing where multiple paths are used to transfer data for an end-to-end session [23]. An important advantage of using multipath transport is the inherent *path diversity* (i.e., the loss processes are expected to operate independently for different paths). This feature exploits the independence of paths and thus the low probability of "all paths gone" situation. Therefore, the receiver can always receive some data during any period, unless all the paths are down simultaneously, which is much less probable compared to single path failures. In video communications, the received content is usually displayed in a continuous flow. For successful reconstruction of received video, the path used for the video session should be stable during the session. Moreover, congestion and transmission error related packet losses should be kept low and thus manageable by error control and concealment techniques. To mitigate these issues, multipath transport is highly suitable for wireless networks. In that regard, multipath transport can be combined with other mechanisms such as multiple description coding (MDC), ARQ, and FEC for multimedia delivery. Under MDC, multiple *equivalent* streams (or descriptions) are generated for a video source for transmission. At the receiver, *any* received subset of these descriptions can be combined to reconstruct the original video and the quality of the reconstructed video is commensurate with the number of received descriptions. When combined with appropriate source and/or channel coding and error control schemes, multipath transport can to take advantage of path diversity and increase robustness to packet losses and path failures, significantly improving the media quality over traditional shortest-path-routing based schemes.

Example architecture for multipath transport of video streams is depicted in Figure 25.5 [23]. Source content, i.e., raw video, is first compressed at the sender. If this compression results in multiple streams, this encoder is called *multistream encoder*. These streams can be independent (multiple MDC streams) or dependent to each other (SVC streams of base and enhancement layers). Then a *traffic scheduler* distributes these *S* streams (video packets) to *P* available paths by using an assignment algorithm. These paths are maintained by a *routing engine* which supports multipath routing. When video packets reach the receiver, they are reconstructed by a *stream constructor* via demultiplexing and resequencing them in the correct order. Finally, the video data is extracted from these reconstructed streams and decoded to be displayed to the user.

When $S > P$, multiple streams can share the same path. This is the case in Figure 25.5 where we have three streams but two paths. Stream 1 and 2 are transmitted over the same Path 1. On the contrary, when $S < P$, we have the possibility of sending multiple copies of the same stream to the receiver. This is similar to repetition coding but applied to routing rather than channel coding. In general, the number of the paths and their qualities (bandwidth, delay, packet loss, etc.) may vary over time due to network topology changes and congestion [24]. For that reason, it is important to have real-time information regarding network QoS

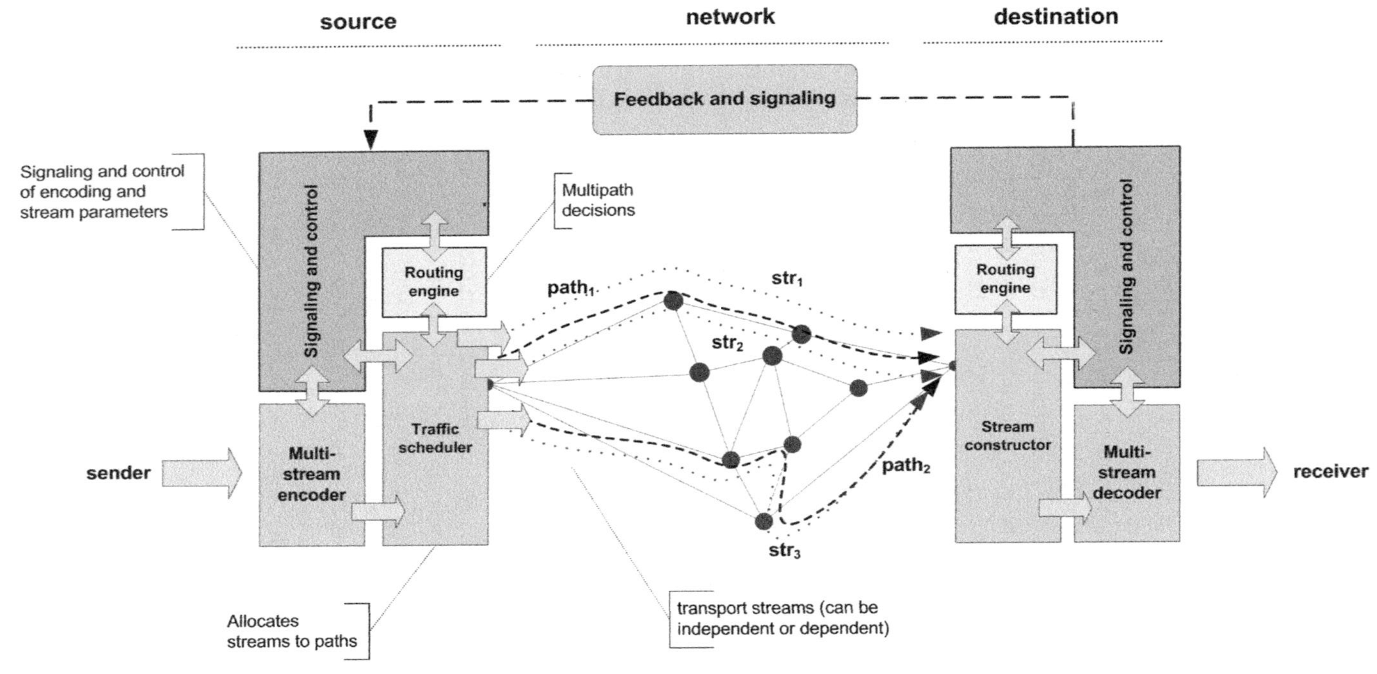

FIGURE 25.5

A general architecture for the multipath transport of real-time multimedia applications.

state. As explained in Section 4.1, such feedback can also be used to adapt the coder and transport mechanisms to network conditions (e.g., the encoder could perform rate control based on feedback information to avoid congestion in the network) in addition to facilitating multipath routing algorithms. Moreover, multipath transport can distribute traffic load in the network more evenly, resulting in low congestion and delay in the network [23].

The point-to-point architecture in Figure 25.5 can be used for two-way conversational services in addition to one-way streaming services. For the latter case, it can be extended to more general cases such as many-to-many streaming (multiple servers concurrently streaming to multiple receivers) or hierarchical streaming where multicast trees are used [23,25]. However, these capabilities and advantages of multipath routing come at the burden of higher coding redundancy, higher computation complexity, and additional delay during traffic partitioning and resequencing, and higher signaling/control traffic overhead in the network [26]. Therefore, a multistream encoder and multipath routing scheme should strive to achieve a good trade-off between coding efficiency, error resilience and throughput. In addition, these design choices are also affected by what is feasible via error control in transport layer.

4.4 PHYSICAL AND MAC LAYERS

At the physical (PHY) layer, adaptive modulation and coding (also known as link adaptation) is a very important capability of next generation networks and widely used in wireless technologies such as IEEE 802.11ac, 3GPP 3G/4G, and DVB-S2. It enables cross-layer mechanisms for configuration and monitoring of physical layer and therefore is crucial for any cross-layer design with physical layer aspects. Moreover, power control is realized from the group point of view by adjusting transmission power and other PHY parameters for a group of users so as to reduce interference.

The basic rationale behind link adaptation is to optimize throughput by selecting among the set of available rates, as given by a set of modulation and coding schemes, the one that maximizes the throughput in each "short-term" channel state [27]. In other words, the communication system utilizes higher modulation levels and higher channel coding rate when channel condition is good and use lower modulation levels and lower channel coding rate when channel condition is relatively harsh [66]. Link adaptation allows the communication system to operate around the dynamic optimal point over a time-varying channel, rather than optimally for the worst case of a static channel model, using the trade-off between link robustness and bandwidth efficiency as depicted in Figure 25.6. Therefore, this capability is crucial to cope with adverse channel conditions encountered especially in wireless communication systems.

In wireless systems, channel conditions can vary significantly due to propagation anomalies as noted in Section 1. It is therefore desirable to adapt the modulation and coding scheme to the channel conditions. In a typical adaptive modulation

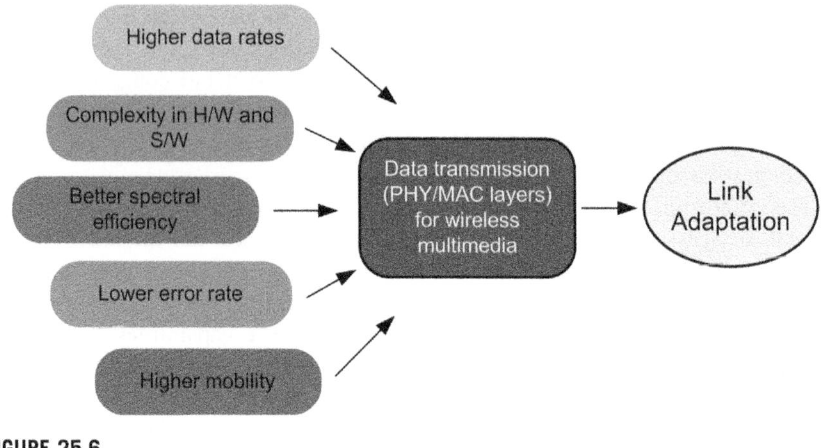

FIGURE 25.6

Design forces for link adaptation schemes.

scheme a dynamic variation in the modulation order (constellation size) and FEC code rate is possible. In practice, the receiver feeds back information on the channel, which is then used to control the adaptation. Adaptive modulation can be used in both uplinks and downlinks. The adaptation can be performed in various ways such as user-specific only, user- and time-specific, or QoS-dependent. Link adaptation results in an ultimate increase in link capacity, in that it permits the system to make the most out of a time-varying channel, instead of always operating optimally for a worst-case channel.

There are four primary mechanisms by which a system could improve link robustness at the expense of overall bandwidth efficiency [28]:

- **Adaptive modulation:** The transmitter and receiver negotiate and decide upon the most bandwidth-efficient signal constellation that can be used for the current channel condition. When the channel conditions deteriorate, bandwidth efficiency is sacrificed and a more robust, lower order modulation scheme is chosen.
- **Adaptive FEC:** The system decides to add protection bits to the transmitted data in the form of FEC codewords. This scheme again provides robustness at the expense of bandwidth efficiency. Depending on the channel conditions the strength of FEC is determined.
- **Automatic repeat request (ARQ):** Some systems adopt packet retransmissions at the link layer. In this method, unacknowledged packets are assumed to be in error and retransmitted. Retransmissions cause a reduction in overall throughput, but guarantee correct reception on subsequent trials. The system could switch ARQ on and off, depending on the channel conditions. Many systems use an optimal combination of FEC and ARQ, called Hybrid ARQ (H-ARQ), to achieve maximum bandwidth efficiency.

- **Optimal power and coding allocation:** Recently the multiple input multiple output (MIMO) technique, where both the transmitter and receiver are equipped with multiple antennas, has emerged to alleviate the multipath fading challenge [29]. MIMO technique does not try to mitigate the adverse effect of multipath fading channel. Instead, it tries to utilize the multipath richness environment in a smart way such that the received signal cannot be in deep fade. Therefore, this technique brings another dimension to link adaptation schemes, which is optimal power allocation and coding (space-time coding) among parallel sub-channels according to their respective conditions.

Typical link adaptation systems group the above parameters into "modes" or "profiles" and dynamically track a time-varying channel by switching between predefined profiles. As a general trend, more advanced wireless technologies utilize a larger number of profiles to reach higher spectral efficiencies. In addition to defining suitable operation modes, a link adaptation algorithm needs to define the operation regions for each of these modes quantitatively. This capability necessitates an appropriate metric that serves as a link quality indicator. Many practical adaptation algorithms use signal-to-noise ratio (SNR) ranges to define the operation regions for each mode. These ranges are determined to maximize objective metric(s) chosen as the quality indicator(s). Typical quality indicators are bit error rate, frame error rate, end-to-end delay or throughput. In the physical layer, additional channel knowledge could be the instantaneous channel state information (e.g., channel response) or in an average form (e.g., average SNR or statistical information of the channel coefficients).

AMC allows a more flexible PHY for multimedia transmission allowing the control of modulation and coding schemes according to physical layer conditions. This capability is also important error protection and adaptation of multimedia traffic over error-prone wireless links. However, for the system modeling perspective, this attribute is a complicating factor since models should include multiple waveforms. Additionally, more complex PHY simulation is evident due to wireless transmission involving a wide range of PHY and MAC settings.

4.5 CROSS-LAYER DESIGN

In layered architectures, networks are organized as a series of layers (or levels), each one built upon the one below it [30,31]. The main benefit of this design is modularity and abstraction. However, it has been widely recognized that a cross-layer optimization can substantially mitigate useless overhead, and thus increase the system performance. CLD allows communication to take place even between nonadjacent layers through additional entities introduced into the system's architecture. The constraints and parameter values exchanged among different layers, processes or modules in the system (packet size, delay, channel conditions, etc.) can be utilized in the context of wireless networks. This has even led to the more extreme concept of "layerless communications" [30]. As Srivastava et al. note, three main reasons to go beyond layered architectures can be listed: "the unique problems created by wireless links, the possibility of opportunistic communication on wireless links, and the new modalities of communication offered by the wireless medium" [32]. A CLD algorithm or protocol spanning across conventional protocol layers aims at solving a specific problem. In wireless mobile devices these problems include security, mobility, quality of service, and adaptation of the wireless link. For multimedia transmission, the most important vertical is the QoS plane [33,34].

The main idea behind CLD is to integrate the resources and information available in the different fractures and create a system which can be highly adaptive and QoS-efficient by sharing state information between different processes or modules in the system. However, there are also some substantial issues of cross-layer paradigm:

- Monolithic structures less open to changes and *dependency issues* [35].
- Hardware-native solutions applicable in specific domains: various requirements on signaling and control for some cross-layer designs may cause the system to be restricted to some specific hardware or use-cases.
- Backward compatibility: Since new protocols and systems based on CLD come into play with incumbent peers in operation, they may suffer from compatibility issues.
- The increased complexity of processing and communication leading to higher cost and more difficult implementations impacting system complexity and feasibility [36].

The cost-benefit analysis for CLD-related costs and potential performance gains is a critical requirement for CLD proliferation. In that regard, recent work on video rate adaptation for wireless channels has considered the use of cross-layer feedback (for example, radio link rate and quality information from the base station or access point) to further improve end-to-end performance [37,38]. Further QoS gains can be achieved with cross-layer scheduling of traffic at intermediate network nodes such as edge routers and base stations, of course, at the cost of increased complexity relative to simpler end-to-end rate adaptation [39].

5 IMPACT OF EMERGING NETWORKING PARADIGMS FROM THE PERSPECTIVE OF MULTIMEDIA

In this section, we discuss some emerging wireless network paradigms from the multimedia networking perspective. Although there are a variety of topics that could be covered regarding this aspect, we have focused on the most critical ones for the sake of brevity. We first discuss the evolution of wireless standards including OFDMA-based 4G LTE standard since it has emerged as the most dominant cellular wireless technology with a comprehensive roadmap for evolution towards

5G systems. Then, we present Cognitive Radio Networks (CRNs), followed by the Information/Content Centric Networking (ICN/CCN) concept. Finally, we briefly elaborate on mobile ad hoc networks and their challenges.

5.1 EVOLUTION OF WIRELESS NETWORK STANDARDS FROM THE PERSPECTIVE OF MULTIMEDIA-INTENSIVE SERVICES

For wireless systems, a major problem with wireless multimedia delivery is caused by the still relatively low bit-rate and large variations in the radio link bandwidth and channel error rate [40]. Although techniques for video rate adaptation developed earlier for the wired Internet to try to address this dynamism to avoid decoder starvation, the variations/fluctuations are at much shorter time scales for mobile wireless channels [41].

A second key barrier to widespread adoption of high-quality media services over wireless networks is that of limited total network capacity. While the most recent 4G cellular systems can achieve peak bit rates 10–100 Mb/s, the realistic total network throughput for a single cell is much lower and not more than 5–10 Mb/s due to the fact that peak bit rates can only be achieved for a small subset of mobile users with excellent channel propagation, while the majority of devices operate at a much lower speed hence pulling down the overall network capacity. This latter issue is typical for cell-edge users in LTE networks.

Some solutions to this capacity problem are [40]:

1. *Multicasting of media streams:* this approach offers significant capacity gains for one-to-many services but is not particularly applicable to individual media viewing as is the case for most mobile services.
2. *Small cells and HetNets:* Addition of femtocells and WiFi hot spots to offload traffic in densely populated areas takes advantage of proximity and improves spectral efficiency [42]. This approach is integrated into network standards such as LTE.
3. *Content caching at mobile devices and inside the network:* The locality principle utilized via caching mechanisms improves network capacity via shorter transmission paths and fewer nodes involved in multimedia transmission. In addition to that, P2P video delivery provides a more effective distribution of load and processing to disparate network nodes. In-network caching can operate in a flexible hierarchy ranging from large centralized caches to small but more distributed ones.
4. *Elasticity and separation of concerns via Network Function Virtualization (NFV):* The problem of maintaining a reliable QoS for media applications can also be addressed by the emerging technology of network virtualization [43]. Virtualization techniques aim to partition hardware, processing, and bandwidth resources at shared network elements such as routers and base stations into distinct segments facilitating improved isolation between traffic from distinct services such as video and data. Each virtual network can

employ its own customized resource management, scheduling and differentiation algorithms, which may take into account the specific needs of the application layer or users. These capabilities improve the network response to spatiotemporal fluctuations of demand for media services.

5. Resource management and pricing methodologies: Service-specific admission control and bandwidth pricing strategies [44] can also be used to improve overall QoS of audiovisual applications such as video streaming. They allow more efficient and adaptive usage of network resources.

For the infrastructure-based wireless networks, 3GPP LTE has been the main emerging network specification with planned releases towards 5G networks. The LTE architecture is depicted in Figure 25.7. The LTE architecture is often referred to as *two-node* architecture because, logically, only two nodes are involved—in the user and control plane paths—between the user equipment and the core network. These two nodes are (1) the base station, called *eNodeB*, and (2) the serving gateway (S-GW) in the user plane and the mobility management entity (MME) in the control plane, respectively [45]. MME and GW belong to the core network, called *evolved packet core (EPC)* in 3GPP terminology, which shows the control plane and user plane protocol stacks between the UE and the network. The services are provided to the UE in terms of evolved packet system (EPS) bearers. A finite set of possible QoS profiles—in other words, packet treatment characteristics—is defined and are identified by so-called *labels*. Each EPS bearer is associated with a particular QoS class (i.e., with a particular QoS label).

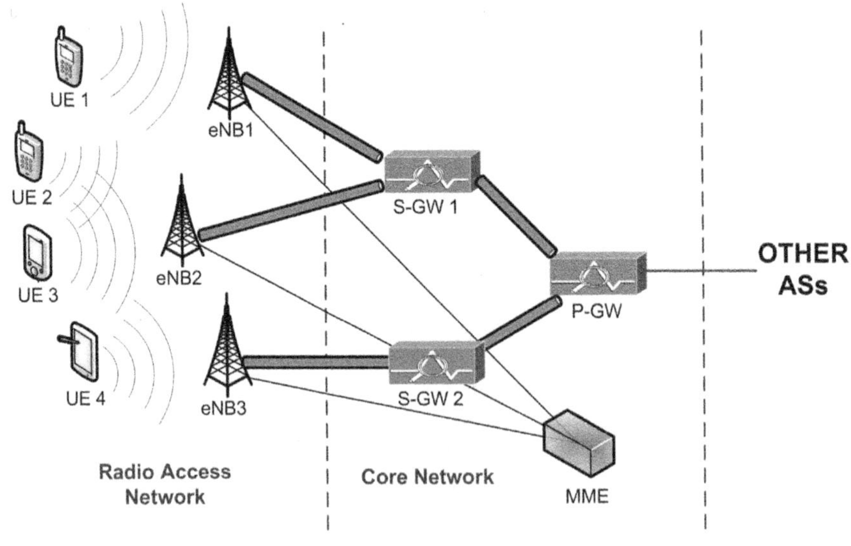

FIGURE 25.7

The 3GPP long term evolution (LTE) architecture.

The end-to-end EPS bearer can be further broken down into a radio bearer and an access bearer. The radio bearer is between the UE and the eNodeB, and the access bearer is between the eNodeB and the GW. The access bearer determines the QoS level that packets get on the transport network; the radio bearer determines the QoS treatment on the radio interface. From an RRM point of view, the radio bearer QoS is in our focus because the RRM functions should ensure that the treatment that the packets get on the corresponding radio bearer is sufficient and can meet the end-to-end EPS bearer-level QoS guarantees.

The infrastructure-based cellular networks as exemplified by 4G LTE are critical enablers (as well as potential bottlenecks) for mobile multimedia services. However, at the small scale, such as in-house or short-distance point-to-point multimedia transmission, making the distribution network completely wireless has also many advantages for the consumer relative to wired networks [40]. While some progress has been made on very high-speed short-range radios such as IEEE 802.15 and Bluetooth, this is still an emerging technology without common industry-wide standards. In addition, the technology is relatively short range (a few meters), limiting more general uses in homes and offices. The emerging 802.11ac and ad standards (the latter at 60 GHz), the next edition of IEEE 802.11x, are going to be instrumental to address local and in-home media distribution case by providing stable gigabit per second links with moderate range (10–20 m).

This discussion shows that audiovisual applications cannot treat the wireless delivery medium as a simple bit-pipe but rather incorporate rate adaptation and resilience, while further increasing compression factors to address the capacity constraints in cellular mobile systems. At the application level, developers should explore alternatives to streaming such as near-real-time file delivery, content caching, and mobile P2P, which can exploit opportunistic access and do not require continuous allocated [40].

The plethora of wireless network standards results in a highly heterogeneous environment. This situation complicates the modeling of these systems. Moreover, the validity of the models may become harder to maintain.

5.2 COGNITIVE RADIO NETWORKS

A cognitive radio is a smart wireless device that can alter its operational parameters based on interaction with the surrounding environment involving its users. This interaction can either involve passive sensing and decision making locally within the radio, or can include active interactions with other nodes in the network [46]. The cognitive radio networks (CRNs) composed of cognitive radio nodes or nodes equipped with cognitive capabilities lead to smarter networks and communication technologies [47]. In addition to intelligent end-user devices, cognitive networks with the smart core/access network nodes (e.g., cognitive femtocells) can act as an enabler for smart grids, and smart ambient environments such as homes and workplaces, and smart transportation systems).

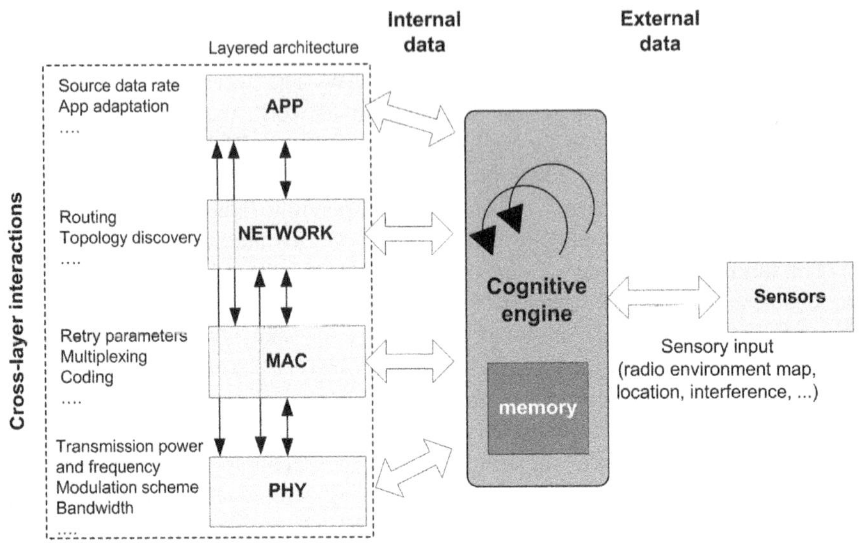

FIGURE 25.8

Cross-layer design in CRNs.

The agility and cognitive capabilities of CRs allow the optimization of various system and environmental parameters for multimedia transmission [48]. At the PHY layer, transmission power/frequency, waveform, beam profile and processing gain can be dynamically controlled. The MAC layer can consider external data such as localization or internal data such as application requirements for medium access and scheduling. Congestion control parameters and rate control parameters can be optimized at the transport layer. The cognitive engine with memory and sensory input in a cross-layer setting is shown in Figure 25.8. However, there are also fundamental challenges for CRN-based multimedia networking, the most important being the opportunistic nature of CRs. The dynamic spectrum access and secondary user requirements result in unstable and intermittent connection characteristics. Moreover, the interference from other CRs and primary users (PUs) degrade the transmission capacity. These intrinsic features complicate multimedia transmission over these networks and bring forth a challenging objective. The modeling of these systems is also challenging due to complexity stemming from diversity and degrees of freedom considering system parameters, control objectives and mechanisms. Moreover, the network environment in which these systems are embedded is highly dynamic and unpredictable due to the secondary nature of CRs.

5.3 INFORMATION/CONTENT CENTRIC NETWORKING

Information-Centric or Content-Centric Networking (ICN/CCN) has received significant attention in recent years, mainly driven by the fact that dissemination and consumption of information chunks has become the major function on the Internet. Many research groups propose and design new routing and in-network caching paradigms to create the content-centric architecture of the Future Internet [49]. Traditionally, the Internet has been a host-centric environment in which hosts generate packets that are subsequently routed to other hosts. Throughout the majority of its evolution, this function has been at the core of the Internet driven by the requirements of prominent higher-level applications such as HTTP, e-mail and FTP. However, as bandwidth availability has improved and Internet has become a ubiquitous substrate for connectivity, content-based systems like social media, video sharing and streaming, over-the-top video delivery and IPTV have become popular and emerged as the predominant traffic generators within the Internet [50].

There are many proposed designs for ICN. One of them is the Named Data Network (NDN) that is a name-based routing system for locating and delivering named data packets. Another design is PURSUIT that proposes a publish/subscribe model for data dissemination. Another one is the Data-Oriented Network Architecture (DONA). DONA is based on replacing DNS names with flat, self-certifying names, and replacing DNS name resolution with name-based anycast primitive that lives above the IP layer [51].

Internet video applications operate on top of well-defined architecture, typically with an E2E system design with prevalent protocol support such as IP for host addresses at network layer and HTTP at the application layer [52]. The multimedia applications choose protocols based on application context and protocol behavior. Video services over content-centric networks can be enabled by requesting each video packet similar to HTTP streaming. In that conventional setting, although receiver-driven stream adaptation which relies on stream quality based on end-to-end bandwidth estimation seems plausible, it is hard to estimate E2E bandwidth or other path quality metrics in CCN since content source is unknown to receiver. For live Streaming in CCN, real-time delivery of video content is possible with proactive transmission of *Interest* packets for upcoming packets. The native multicast support of CCN is another benefit. However, packet caches in routers again complicate E2E bandwidth estimation and Interest packets sent per *Data* causes overhead. System capacity can be further improved by enabling in-network caching of content. These caches assist in error recovery and cache replacement policy may be according to packet content such as frame type, which improves performance. The pros/cons of CCN for these services are summarized in Figure 25.9.

There is also the edge-caching approach where the data is cached as close as possible to the network edge such as in end-user mobile devices. This approach is different compared to server-based streaming in that streaming of content is avoided in favor of delivering video blocks/files in anticipation of user needs, utilizing low traffic periods and/or opportunistic connections to avoid use of network resources during the peak hour [40]. Caching at mobile devices is becoming increasingly feasible due to declining semiconductor memory costs,

	Advantages	Challenges
Video on demand	- Native anycast support - Inherent retransmission support via in-network packet caching	- Realization of stream concept (e.g. throughput estimation for stream adaptation)
Live streaming	- Inherent retransmission support via in-network packet caching - Packet-based differentiation via caching policies	- Link asymmetry - Stream data loss due to non-guaranteed delivery of upstream Interest packets

FIGURE 25.9

CCN and video transmission [52].

and significant effective capacity gains are possible. When the ICN/CCN paradigm is integrated with wireless networks, the network characteristics also affect the multimedia transmission drastically. For our focus in this chapter, this phenomenon is paramount since mobile networks exhibit peculiar inherent characteristics due to mobility, wireless transmission and dynamic network structure. This superposition renders the modeling and simulation of these systems highly challenging since the integration of ICN/CCN and wireless networks is yet to be explored.

5.4 MOBILE AD Hoc NETWORKS (MANETs)

According to IETF, "A mobile ad hoc network (MANET) is an autonomous system of mobile routers (and associated hosts) connected by wireless links. The routers are free to move randomly and organize themselves arbitrarily; thus, the network's wireless topology may change rapidly and unpredictably. Such a network may operate in a stand-alone fashion, or may be connected to the larger Internet." MANET nodes cooperate with each other to find routes, relay packets and facilitate communications. Such networks can be deployed instantly in situations where infrastructure is unavailable or difficult to install, and are promising to provide widespread untethered video services for users, such as first responders, search and rescue teams, and military units [23].

Mobile ad hoc networks maintain the general characteristics of wireless networks while including additional properties and restrictions. Whereas in the infrastructure wireless network there exists some fixed topology, due to the movement of the nodes in ad hoc networks and the resulting route failures and recomputations, difficulty in maintaining sessions, etc., their topology changes dynamically. Moreover, ad hoc networks have no central controller and, in most cases, each node is responsible for maintaining the information of delay, jitter, loss rate, stability, and distance for each link in order to feed routing algorithms. However, this state information is inherently imprecise due to the changes in the topology and the fact that resources such as battery, bandwidth, processing, and storage are limited.

The peculiar characteristics of a MANET that complicate QoS provision and thus multimedia communications can be summarized as follows [53]:

- *Dynamic network topology*: Due to the mobility of the nodes, the topology of mobile ad hoc networks changes continuously. This situation results in some wireless links being broken while other emerge. This characteristic is the key for the performance of routing protocols and the applications located in superior layers. Therefore, incumbent error control techniques such as FEC and ARQ should be tailored to accommodate frequent link failures and severe transmission impairments for efficient wireless video transport.
- *Asymmetry*: Due to the diversity of the MANETs, the characteristics of each node will be different.
- *Multi-hop communication*: This characteristic worsens the quality of the communication, since several wireless connections exist in an end-to-end connection. The links exhibit different characteristics, making adaptation more difficult.
- *Bandwidth*: The bandwidth available in MANETs is very small in comparison with that available in fixed networks. In addition, the wireless link suffers from very high error rates, resulting in worse quality connections.
- *Energy*: The network nodes function using the stored energy in batteries. Thus, the system is energy-limited and energy consumption plays a central role in the development of MANETs.
- *Cooperation*: Due to the lack of infrastructure, all nodes must participate in the establishment of network routes. However, this requirement implicitly assumes the cooperation of nodes, which may not always be the case. During that work, some nodes may decide not to cooperate and not to lend their resources to communications among other nodes.

These inherent characteristics of MANETs hinder the fulfillment of the multimedia services with adequate QoS levels [23]. In order to cope with these requirements, different QoS provisioning architectures and protocols can be managed at different layers. However, considering QoS requirements and solving them in a layered approach may ignore the impact of optimizations on other QoS metrics and overall system performance. For instance, an emerging networking paradigm related to MANETs is Vehicular Area (or Ad Hoc) Networks (VANETs). Scarce spectral resources, intermittent connectivity, and highly dynamic, unpredictable topology are the main challenges hindering the support of multimedia applications [54]. The network characteristics, along with the variable bit rate (VBR) nature of the traffic and the strict delay constraints, making no allowance for store-and-forward, pose a different problem from the ones previously addressed in ad hoc networks [55,56].

Few solutions are proposed for channel access and traffic forwarding to support streaming media in VANET context [57–59], and most of what does exist considers scenarios of linear network topology such as roads or do not leverage the application characteristics.

6 MAJOR CHALLENGES FOR MODELING AND SIMULATION OF WIRELESS MULTIMEDIA NETWORKING

Modeling and simulation of wireless networks are intrinsically complicated due to network node diversity, mobility, resource constraints such as energy, device capabilities, protocol diversity, application peculiarities, and rapidly evolving system specifications. Moreover, multimedia services over wireless networks further complicate this situation with the addition of multimedia service related parameters and dimensions as shown in Figure 25.10.

Major challenges and topics regarding modeling and simulation can be listed as follows (please note that some of these items are also closely linked to the more general topic of simulation and modeling of communication networks):

- *Transmission modeling for wireless and wired links:* This challenge is valid for all mobile network modeling and simulation efforts. New modulation/coding schemes and PHY techniques such as MIMO and beam-forming proposed in new network standards should be considered in the multimedia context for better link and radio models.
- *Mobility:* In the past, multimedia services could only be used in few mobile scenarios. However, the ubiquitous proliferation of wireless broadband and smart devices has dramatically changed this situation. Therefore, new models

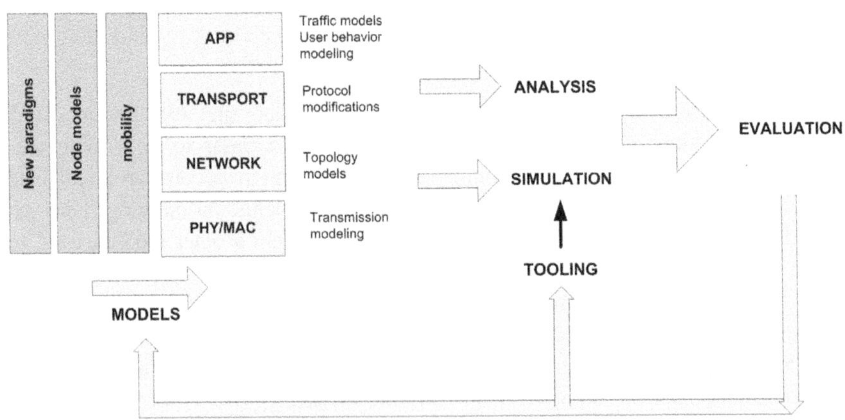

FIGURE 25.10

Modeling and simulation for wireless multimedia transmission.

and simulations should have much better support for mobility in different environments such as ultra-dense urban or high-speed scenarios.

- *Traffic modeling (applications):* Multimedia traffic has dominant characteristics such as high bit rate and jitter sensitivity. Networked multimedia applications exhibit limited tolerance to loss and errors and can have constant and variable bit rate. These features of multimedia traffic should be integrated into traffic models and generation of simulation data.
- *User behavior modeling:* For interactive applications and content-centric operation, this topic is very important. Modeling user behavior is challenging since it has the human factor and depends on sociocultural traits with a temporal dimension.
- *Node modeling:* Network nodes should have rich and realistic models for valid system analysis and simulation. This requirement is also related to how active parts of nodes such as processing and buffering are modeled and simulated. Novel queue and data processing models tailored for multimedia are beneficial to represent wireless multimedia networks in models and simulations.
- *Video coding standards:* Multimedia content is coded using various coding standards or approaches such as SVC and MDC which are very complex standards with a multitude of profiles and extensions. Control and interaction of these components are also crucial.
- *Protocol peculiarities:* The application layer is very capable in emerging wireless multimedia systems. This characteristic renders an intricate subsystem for modeling and simulation. At the network layer, QoS-aware and multipath routing schemes should be modeled and integrated to simulation frameworks. For the PHY/MAC layers, the challenges related to transmission modeling are valid and diverse mechanisms are supposed to be modeled and simulated. Moreover, there is the cross-layer design approach which results in a combination of these issues.
- *Availability of real-world traces and network topologies:* This issue is effective for all network modeling and simulation approaches/frameworks. More traces and topologies with more precise data from live systems are necessary for facilitating more realistic simulations and more valid model checking shown in the last stage of process flow in Figure 25.10.
- *New paradigms:* The wireless network infrastructure is evolving to address the emerging requirements of new services and applications as described in previous sections. New standards incorporate new paradigms such as Software-Defined Networking (SDN) and Network Function Virtualization (NFV) for improving performance, increasing flexibility, and decreasing cost of wireless networks. Moreover, cloud computing is being applied to wireless networks, e.g. to the radio segment as cloud radio access networks, where the processing is centralized in cloud centers. Heterogeneity in terms of devices and radio technologies complicate this global picture.

These challenges should be addressed in modeling frameworks and integrated into analytical and simulation tools. However, this process should have a feedback loop which will improve overall applicability and feasibility of these solutions via performance analysis according to process outcomes. Again, this last stage is also an open research topic which deserves further efforts.

7 CONCLUSIONS

The next-generation wireless networks are expected to support various applications such as voice, data, and multimedia over IP networks. Supporting surging multimedia traffic with adequate QoS levels is an important objective in the design of these systems. However, as opposed to wired networks, wireless networks are characterized by mobility, heterogeneity, substantial packet loss due to the imperfection of the radio medium, more stringent spectral constraints, and energy bottlenecks. Therefore, it is necessary to alleviate these challenges and investigate and develop countermeasures such as new transport protocols, application-layer flow control schemes (e.g., the RTP/RTCP protocol), error protection, energy-efficient design, and adaptive rate control. However, it is not a trivial problem to model and/or simulate these hybrid systems comprising diverse protocols, components, and usage patterns operating under stringent requirements.

In this chapter, some fundamental issues of multimedia transmission over wireless networks have been discussed. In the emerging wireless networks where the transmission of multimedia and IP traffic is a fundamental requirement, it is essential to provision the quality of service at each layer in the protocol stack. The traditional protocol layer functionalities and architectures must be analyzed and adapted to the wireless multimedia applications to provide satisfying multimedia services. One emerging trend is to apply cross-layer approaches in the design of these systems. Moreover, emerging wireless systems and services such as content-centric operation, CRNs, network-friendly video coding such as SVC, and MANETs pose new challenges once they are subject to provide multimedia services with high efficiency and quality. This chapter has also provided an overview for identifying and discussing these fundamental topics. Finally, we have presented and discussed major challenges for modeling and simulation of wireless multimedia networking in this diverse and dynamic environment.

REFERENCES

[1] Cisco. Cisco visual networking index (VNI): Global mobile data traffic forecast, 2013–2018 Report: Available at: <http://www.cisco.com/c/en/us/solutions/collateral/service-provider/visual-networking-index-vni/white_paper_c11-520862.html>; 2015.

[2] Gür G, Bayhan, Alagöz F. Transport protocols and QoS for wireless multimedia chapter in the book Handbook of research on wireless multimedia: quality of service and solutions. IGI Group Inc.; 2009.
[3] Zhang Q, Yang F, Zhu W. Cross-layer QoS support for multimedia delivery over wireless internet. EURASIP J Appl Sig Proc 2005;2:207–19.
[4] Marchese M. QoS over heterogeneous networks. John Wiley & Sons Ltd.; 2007.
[5] Kim A, Jeong S-H. A QoS/QoE control architecture for multimedia communications, proceedings of the 2012 international conference on information networking (ICOIN); 1–3 Feb. 2012. p. 346–49.
[6] Dutta-Roy A. The cost of quality in Internet-style networks. IEEE Spectrum 2000;37 (9):57–62.
[7] Siller M, Woods J. Improving quality experience for multimedia services by QoS arbitration on a QoE framework, Proceedings of the 13th packed video workshop. Nantes, France; 2003.
[8] Gaedtke J. Optimizing QoE in large-scale video networks (Keynote Speech), Packet Video 2013, California, USA; 2013.
[9] Serral-Gracià R, Cerqueira E, Curado M, Yannuzzi M, Monteiro E, Masip-Bruin X. An overview of quality of experience measurement challenges for video applications in IP networks, Proceedings of the 8th International conference on wired/wireless internet communications (WWIC'10), Osipov E, Kassler A, Bohnert TM, Masip-Bruin X, editors. Springer-Verlag, Berlin, Heidelberg; 2010. p. 252–63.
[10] Huang L, Kumar S, Kuo C-CJ. Adaptive resource allocation for multimedia QoS management in wireless networks. IEEE Trans Veh Technol 2004;53 (2):547–58.
[11] Politis I, Pliakas T, Tsagkaropoulos M, Dagiuklas T, Kormentzas G, Kotsopoulos S. Distortion optimized scheduling and QoS driven prioritization of video streams over WLAN, Packet Video 12–13 Nov; 2007. p. 349–55.
[12] Ghanbari M. Standard codecs: image compression to advanced video coding. London, UK: IEE Press; 2003.
[13] van der Schaar M, Turaga DS. Cross-layer packetization and retransmission strategies for delay-sensitive wireless multimedia transmission. IEEE Trans Multimedia 2007;9(1):185.
[14] Jacobs S, Eleftheriadis A. Streaming video using TCP flow control and dynamic rate shaping. J Visual Commun Image Represent 1998;9(3):211–22.
[15] Xie B, Zeng W. Rate-distortion optimized dynamic bitstream switching for scalable video streaming, IEEE international conference on multimedia and expo (ICME '04), vol. 2, Taipei, Taiwan; 2004. p. 1327–30.
[16] Pahalawatta P, Berry R, Pappas T, Katsaggelos A. Content-aware resource allocation and packet scheduling for video transmission over wireless networks. IEEE J Select Areas Commun 2007;25(4):749–59.
[17] Iren S, Amer PD, Conrad PT. The transport layer: tutorial and survey. ACM Comput Surv 1999;31(4):360–404.
[18] Zhu P, Zeng W, Li C. Cross-layer design of source rate control and congestion control for wireless video streaming. Advances in Multimedia, vol. 2007, Article ID 68502, 13 p. Available from: <http://dx.doi.org/10.1155/2007/68502>.
[19] Hu F, Sharma NK. Enhancing wireless Internet performance. IEEE Commun Surv Tutorials 2002;4(1):2–15.

[20] Marcondes CAC, Persson A, Chen L, Sanadidi MY, Gerla M. TCP Probe: A TCP with built-in path capacity estimation eighth IEEE global internet symposium, Miami, USA; 2005.

[21] Bae S, Chong S. TCP-friendly flow control of wireless multimedia using ECN marking. Sig Proc Image Commun 2004;19(5):405–19.

[22] Yang G, Sun T, Gerla M, Sanadidi MY, Chen LJ. Smooth and efficient real-time video transport in the presence of wireless errors. ACM Trans Multimedia Comput Commun Appl (TOMCCAP) 2006;2(2):109–26.

[23] Mao S, Lin S, Wang Y, Panwar SS, Li Y. Multipath video transport over wireless ad hoc networks. IEEE Wirel Commun Special Issue Adv Wirel Video 2005;12(4):42–9.

[24] Mao S, Panwar SS, Hou YT. On optimal traffic partitioning for multipath transport, Proceedings of IEEE INFOCOM 2005, vol. 4, Miami, FL; Mar. 2005. p. 2325–36.

[25] Apostolopoulos JG, Wong T, Tan W-T, Wee S. On multiple description streaming with content delivery networks, Proceedings of IEEE INFOCOM, vol. 3; 2002. p. 1736–45.

[26] Setton E, Liang Y, Girod B. Adaptive multiple description video streaming over multiple channels with active probing, Proceedings of IEEE ICME, Baltimore, MD; July 2003. p. I-509–12.

[27] Haleem M, Chandramouli R. Adaptive transmission rate assignment for fading wireless channels with pursuit learning algorithm, Conference on information sciences and systems (CISS), Princeton, USA; 2004.

[28] Ramachandran S. Link adaptation algorithm and metric for IEEE Standard 802.16, M. S. Thesis, Virginia Polytechnic Institute and State University, USA; February 2004.

[29] Nguyen HT, Andersen JB, Pedersen GF. On the performance of link adaptation techniques in MIMO systems. Wirel Pers Commun 2007;42(4):543–61.

[30] Kuran MS, Gür G, Tugcu T, Alagöz F. Applications of cross-layer paradigm for improving the performance of WiMAX. IEEE Wirel Commun Mag 2008;17(3):86–95.

[31] Chen X. Transporting compressed digital video. NJ, USA: Kluwer Academic Publishers; 2002.

[32] Motani VS. Cross-layer design: a survey and the road ahead. IEEE Commun Mag 2005;43(12):712–21.

[33] Foukalas F, Gazis V, Alonistioti N. Cross-layer design proposals for wireless mobile networks: a survey and taxonomy. IEEE Commun Surv Tutorials 2008;10(1): 1st quarter.

[34] Carneiro G, Ruela J, Ricardo M. Cross-layer design in 4G wireless terminals. IEEE Wirel Commun 2004;11(2):7–13.

[35] Kawadia V, Kumar PR. A cautionary perspective on cross-layer design. IEEE Wirel Commun 2005;12(1):3–11.

[36] Weingart T, Sicker DC, Grunwald D. Identifying opportunities for exploiting cross-layer interactions in adaptive wireless systems advances in multimedia, vol. 2007, Article ID 49604, 11 p. Available from: <http://dx.doi.org/10.1155/2007/49604.014>; 2007.

[37] van Der Schaar M. Cross-layer wireless multimedia transmission: Challenges, principles, and new paradigms S. S. N. IEEE Wirel Commun 2005;12(4):50–8.

[38] Ramos N, Dey S. A device and network-aware scaling framework for efficient delivery of scalable video over wireless networks, Proceeding IEEE 16th international symposium personal indoor mobile radio communications. Available from: <http://dx.doi.org/10.1109/PIMRC.2007.4394082>; 2007.

[39] Mukhopadhyay S, Schurgers C, Dey S. Joint computation and communication scheduling to enable rich mobile applications, In: proceedings IEEE global telecommunications conference; 2007. p. 2117–22.

[40] Raychaudhuri D, Mandayam NB. Frontiers of wireless and mobile communications. Proc of the IEEE 2012;100(4).

[41] Schulzrinne H, Casner S, Frederick R, Jacobson V. RTP: a transport protocol for real-time applications, ietf internet standard, RFC 3550, 2006.

[42] Gür G, Bayhan S, Alagöz F. Energy efficiency impact of cognitive femtocells in heterogeneous wireless networks, Proceedings of the 1st ACM workshop on Cognitive radio architectures for broadband (CRAB'13). ACM, New York, NY, USA, p. 53–60.

[43] Zhu Y, Ammar M. Algorithms for assigning substrate network resources to virtual network components. In: Proceedings 25th IEEE international conference computer communications, 2006.

[44] Hande P, Chiang M, Calderbank R, Zhang J. Pricing under constraints in access networks: Revenue maximization and congestion management. In: Proceedings IEEE INFOCOM; Mar. 2010. p. 1–9.

[45] Long Term Evolution: 3GPP LTE radio and cellular technology. In: Furht B, Ahson SA, editors. Auerbach Publications; 2009.

[46] Gür G, Alagöz F. Green wireless communications via cognitive dimension: an overview. IEEE Network 2011;25(2):50–6.

[47] Alagöz F, Gür G. Energy efficiency and satellite networking: a holistic overview. Proc IEEE (Invited Paper) 2011;99(11):1954–79.

[48] Chen S, Wyglinski AM. Efficient spectrum utilization via cross-layer optimization in distributed cognitive radio networks. Comput Commun 2009;32(18):1931–43.

[49] Wang S, Bi J, Wu J, Li Z, Zhang W, Yang X. Could in-network caching benefit information-centric networking? Proceedings of the 7th asian internet engineering conference (AINTEC'11), Bangkok, Thailand; November 09–11 2011. p. 112–15.

[50] Tyson G, Kaune S, Miles S, El- khatib Y, Mauthe A, Taweel A. A trace-driven analysis of caching in content-centric networks, proceedings of the 2012 21st International conference on computer communications and networks (ICCCN), July 2012. p. 1–7.

[51] Koponen T, Chawla M, Chun B-G, Ermolinskiy A, Kim KH, Shenker S, et al. A Data-Oriented (and Beyond) Network Architecture. In: Proccedings of SIGCOMM'07, Kyoto, Japan; August 27–31, 2007. p. 181–92.

[52] Tsilopoulos C, Xylomenos G, Polyzos GC. Are information-centric networks video-ready? 20th international packet video workshop (PV); Dec. 2013. p. 1–8.

[53] Mohapatra P, Li J, Gui C. QoS in mobile ad hoc networks. IEEE Wirel Commun 2003;10(3):44–52.

[54] Soldo F, Casetti C, Chiasserini C, Chaparro PA. Video streaming distribution in VANETs. IEEE Trans Parallel Distrib Syst 2011;22(7):1085–91.

[55] Ros FJ, Ruiz PM, Stojmenovic I. Reliable and efficient broadcasting in vehicular Ad Hoc networks, proceedings of IEEE vehicular technology conference (VTC) Spring; Apr. 2009. p. 1–5.

[56] Chen W, Guha RK, Kwon TJ, Lee J, Wirel Y-YH. A survey and challenges in routing and data dissemination in vehicular Ad Hoc networks. Wirel Commun Mob Comput 2011;11(7):787–95.

[57] Guo M, Ammar MH, Zegura EW. V3: A vehicle-to-vehicle live video streaming architecture, Proceedings IEEE International conference pervasive computing and communications (PerCom); 2005. p. 171–80.

[58] Bonuccelli M, Giunta G, Lonetti F, Martelli F. Real-time video transmission in vehicular networks. In: Proceedings IEEE mobile networking for vehicular environments (MOVE workshop), May 2007. p. 115–20.

[59] Chu Y-C, Huang N-F. Delivering of live video streaming for vehicular communication using peer-to-peer approach. In: Proceedings IEEE mobile networking for vehicular environments (MOVE workshop), 2007. p. 1–6.

CHAPTER

Software-defined wireless network (SDWN) 26

A new paradigm for next generation network management

Muge Erel, Zemre Arslan, Yusuf Ozcevik, and Berk Canberk
Istanbul Technical University, Ayazaga, Istanbul, Turkey

1 INTRODUCTION

Today's rapid proliferation of mobile data traffic as well as the high increase in user service requests stress conventional cellular wireless technology [1]. According to the Cisco Visual Networking Index (VNI) report published in 2014, by 2018 the increase of global mobile data traffic is predicted to show a nearly 11-fold increase over 2013, as seen in Figure 26.1. The amount of mobile data traffic was 1.5 exabytes per month, whereas it is expected to reach 15.9 exabytes per month by Cisco. With this explosion of mobile data traffic, various GoS demands of mobile users have also emerged, thereby stressing the macrocells to handle more users with heterogeneous GoS levels.

The stressed conventional macro cells of cellular wireless technology are not enough to serve more users and keep their service quality at an acceptable level. Moreover, macro cells could not reach indoor mobile users with better service quality, thereby increasing the number of dissatisfied users in indoor applications. To reach indoor mobile users with higher quality, different and novel management solutions such as small cells must be deployed.

The small-cell deployment solution, which was proposed to handle the increased mobile traffic demands, brings an important mobile data offloading challenge with it. This can be explained as follows. One can say that the occurrence of mobile data traffic occurs mostly in a user's home [2]. Here, because of receiving higher power from small cells and WiFi rather than macro cells inside homes, users prefer to offload their traffic to these RATs in order to have better GoS. According to the same Cisco VNI report, the level of transition from cellular technology to fixed broadband technologies such as small cells and WiFi is shown as seen in Figure 26.2. The explosion of mobile data traffic is also shown in this graph as indicated in Figure 26.1. Here, the offloaded mobile data is

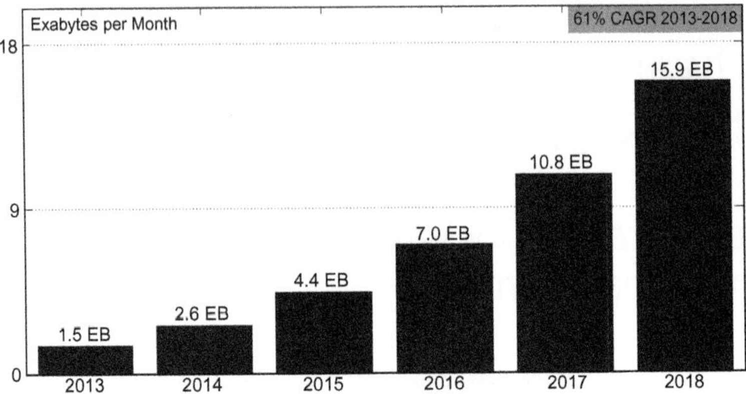

FIGURE 26.1

Growth of mobile data traffic [2].

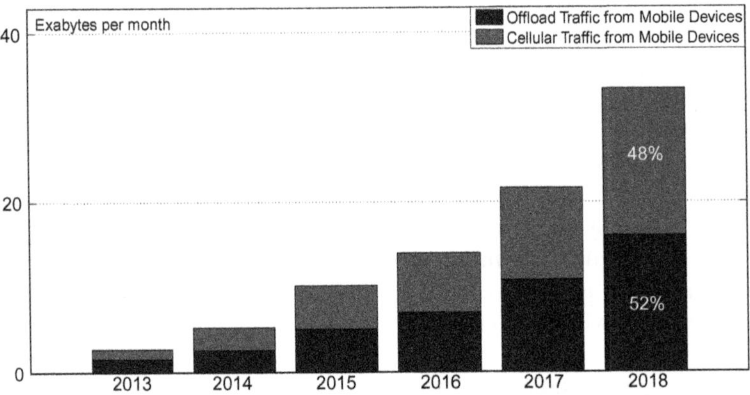

FIGURE 26.2

Offloaded mobile data traffic [2].

expected to increase from 45% (1.2 exabytes per month) to 52% (17.3 exabytes per months) by 2018. Therefore, the transitions between different RATs become more significant in next generation wireless cellular networks.

In today's wireless topologies, the total number of channels is constant and statically clustered. This static clustering limits the serving flows by conceding GoSs of each user. This clustering of channels decreases the number of flows that can be handled. To overcome this challenge, the channel constraint should be isolated using a global view of the physical topology in order to handle more flow with better GoS. Consequently, in this chapter we present the following outcomes [1,3]:

- The wireless topology in SDN approach is separated into two planes named Control Plane and Data Plane. Thanks to the virtualization capability of SDN, the complexity of the physical structure is isolated to the Controller. Therefore, the flexibility of the topology is greatly increased without any resource deployment. In other words, the channel efficiency is increased by serving more users without conceding their service qualities [3].

- The Adaptive Topology Control Algorithm in SDN Controller creates Virtual Switches by optimizing channel efficiency with respect to GoS [3].
- An Admission Control Algorithm for each Virtual Switch distributes flows fairly to available wireless channels via a global view in order to increase the scalability [3].
- A software-defined offloading framework decides the most efficient small-cell switching between different base stations [1].

The rest of the chapter is organized as follows: In Section 2 a literature survey of current and SDN solutions for mentioned challenges is given. Then, GoS optimized topology and admission control mechanisms will be studied by giving network architecture, algorithms and experimental results in Section 3. In Section 4, a similar examination will done for the software-defined offloading approach. Finally, the chapter will be summarized in the concluding section.

2 LITERATURE SURVEY

2.1 CURRENT SOLUTIONS IN RATs

In literature, in order to enhance the GoS of end-users and increase efficiency of channels, many studies emphasize the necessity of transition between different RATs. Zahir et al. indicate major challenges of heterogeneous networks by examining the interference effect and topology management in [4]. ElSawy et al. also emphasize small cell problems to be distributed randomly in locations [5]. Andrews et al. say that mobile networks require longstanding models in [6]. They have attempted to make some changes in design and implementation of heterogenous networks by summarizing changes such as topology, backhaul, interference management, etc. In [7], the authors try to achieve the required GoS in the whole network by determining typical radio network parameters of a wireless system. In order to do this, Taguchi's method in radio network optimization is used. In order to remove these challenges and enhance GoSs of end-user, transition between RATs is required. Therefore, the flexibility and scalability of the topology can be increased without making any resource deployment.

Offloading is one of the popular technologies that attempts to overcome these aforementioned challenges. Ristanovic et al. propose an approach that decreases offloading delay and enhances GoSs of mobile users. In this approach, they define a mostly visited set for WiFi users [8]. According to this user-based view, Huang et al. also introduce a new dynamic offloading algorithm that uses Lyapunov optimization to save energy on mobile devices [9]. Differing from them, Lee et al. examine base station-based power savings by increasing offloading efficiency in [10]. In [5], the small-cell deployment meets increasing demands of users and they serve as offloading spots in RATs to offload users from congested macro-cells. Singh et al enhance offloading by using a parameter called rate coverage, i.e., the fraction of users able to meet a given rate threshold to provide the GoS requirements in [11].

Cognitive Radio is another solution to increase the utilization of the spectrum by user-based dynamic channel switching. Xie et al examine energy-efficient resource allocation heterogeneous cognitive radio networks based on the Stackelberg game in [12]. Tachwali et al. also introduce an optimization for resource allocation in cognitive radio network [13]. The authors in [14] use a location-based solution to try to enhance spectrum efficiency. The authors design the Cognitive Capacity Harvesting network (CCH), which allocates suitable spectrum to secondary cognitive users.

On the other hand, Kokku et al. emphasize that the inefficient resource utilization challenge of wireless networks can be solved by the design and implementation of a Network Virtualization Substrate (NVS) for effective virtualization of wireless resources in wireless networks in [15]. The technical report in [16] indicates that Network Function Virtualization (NFV) offers rapid service, greater flexibility and improves operational efficiencies, etc.

However, these approaches are not sufficient to enhance GoS of the end-user because of being user- or base-station based approaches. Due to lack of the global view and failure to rescue from the complexity of the physical topology, the channel constraint limits their enhancement on GoSs and the number of end users. In [17], Bojic et al. prefer to control the system using a top-down approach that overcomes these topology and admission problems on wireless networks by using SDN. The aim of the authors is to enhance user quality in networks by dynamic resource allocation with the SDN approach. Therefore, the importance of SDN should be examined in detail, as in the following subsection.

2.2 BACKGROUND OF SOFTWARE DEFINED NETWORKING

A Software Defined Network (SDN) is an architecture that decouples the Data Plane, which contains dummy devices to forward the traffic, and the Control Plane, which controls how the traffic will flow through the network. The main motivation in SDN paradigm is unpredictable distributed network configurations, with all the network devices individually configured [18] and [19]. For small-scale networks, the configurations are easy to handle; however, when the scale gets larger, the network configurations get more complicated [5]. Besides the operational cost of the large-scale networks, the capital costs are the other reason to trend to SDN. Since the network logic is moved to the centralized controller the hardware may get cheaper.

In conventional networks, the Infrastructure layer (Data Plane) and Control layer (Control Plane) are combined and the Application layer does not exist. With SDN, these three layers are separated and this separation allows the network operators to control the whole network behavior from a single control software.

Therefore, this new SDN architecture moves the network intelligence to an application called Controller, as seen in Figure 26.3:

- The Infrastructure Layer: This layer holds the network devices responsible for forwarding the traffic.
- The Control Layer: It contains the software logic which communicates with the Data Plane with an OF like software interface.
- The Application Layer: This layer consists of the end-user applications for network operators to configure the control software.

SDN-based architecture has the promise to redefine network management by decoupling the Control Plane and Data Plane and SDN has also emerged as simplifying the configurations with abstraction of these two planes [1]. As indicated in [21] and [22], SDN simplifies the network management by dynamically

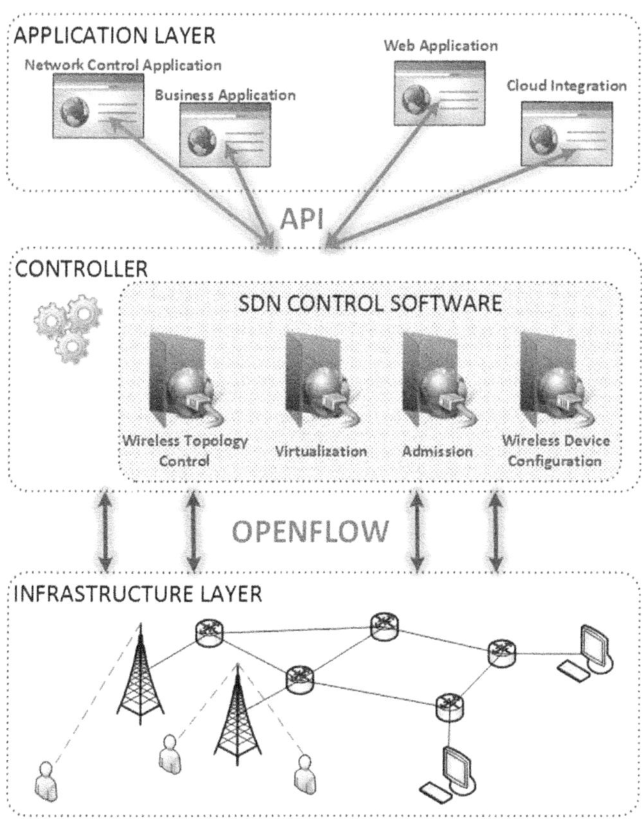

FIGURE 26.3

Software-Defined network architecture [20].

managing the traffic isolation. Moreover, SDN offers opportunities to enable network operations on a logically centralized global network view [23]. Akyildiz et al. [24] indicate that with the SDN approach network resource utilization is improved, network orchestration is simplified, and operating costs are reduced. These are also emphasized in [25] and [22] as SDN simplifies the deployment by decreasing total cost of managing enterprise and carrier networks. In [26], Kim et al. indicate the continual network state, changes in real-world deployments and it is significant to emphasize that operators require SDN-based solutions to adapt their network to frequent alteration. The OF protocol, which is a significant enabler of SDN, utilizes the abstraction of these two planes by describing the network traffic in flows on the contrary of traditional routers and switches, in which the fast packet forwarding (data path) and the high level routing decisions (control path) occur on the same device [20]. In [27], Dely et al. state that by integrating the OF to their cloud architecture, they achieved a new level of flexibility and configurability to save more energy. In [28], for the sake of low-cost implementation and deployment, the authors propose three extensions to OF protocol and flow table and also they believe deployments will be accelerated in production networks.

Briefly, the main benefits of SDN can be listed as follows:

- Centralized controllers have global network view, so the operators can program the whole network from a single point.
- Network operators can dynamically change the whole network traffic flow to meet their demands.
- SDN provides vendor-neutral control because controller software development does not depend on devices types.

OF is the first standard communication interface defined between control and forwarding layers of SDN architecture. The controller communicates to the OF switches using a secure channels (See Figure 26.4). OF switches contain flow tables which are used for forwarding and packet look-ups. The OF protocol makes able switches to update their flow table entries in order to take corresponding

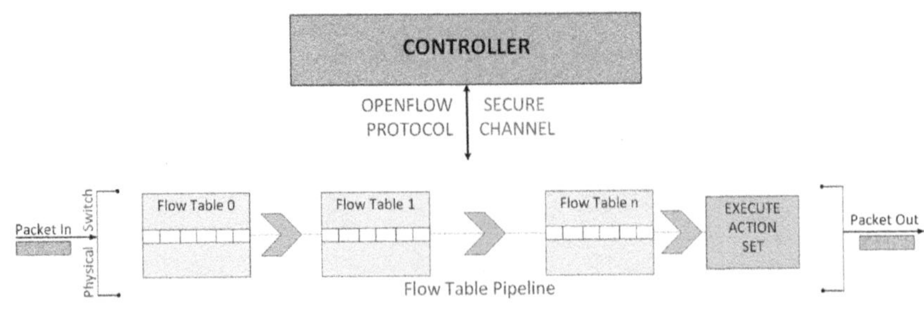

FIGURE 26.4

Idealized OF switch [29].

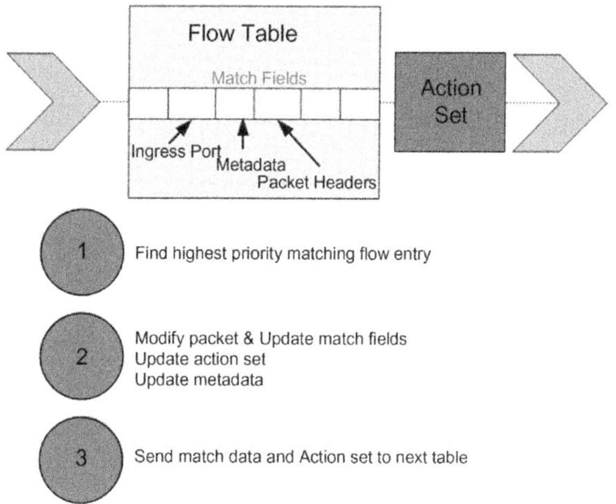

FIGURE 26.5

Steps of packet flow [30].

actions according to incoming flow and matching entries. The control channel is responsible for updating these flow tables. In addition, dummy switches forward according to their flow tables; those are installed by the controller [29].

The flow table contains flow entries, which are counters and a set of actions to apply when matching packets. The packets that are processed by the OF enabled switch are compared with the flow tables, as depicted in Figure 26.5. If a matching entry is found, the corresponding action on the packet will be performed. If it does not match, the packet will be forwarded to the controller to determine how to handle it.

There can be many flow tables in an OF switch. Therefore, packets are matched against multiple tables in the pipeline. Figure 26.4 illustrates a pipeline processing in which a packet resides in the OF switch. Tables in the pipeline can update packet header fields and add the corresponding set of actions that will be performed before the packet is sent to an output port. This set can be empty.

A flow table consists of three main fields as shown in Figure 26.6. Match fields contain packet headers and ingress ports. Counters are used to update matching packets, and instructions allow the system to modify the action set that will be applied to an incoming packet.

In SDN, the default flow management is handled as follows: All incoming flows are received by the OF switches. The switches check the destination address of each flow and if the destination address could not match any flow table entries in the switch, it is defined as newcomer flow and forwarded to the controller [3]. Here, the controller is responsible for updating the flow table entry and assigning a forwarding rule for this newcomer flow. Consequently, with the result of virtualization algorithms, the controller decides about the new forwarding enty and updates the flow tables in switches by considering fair flow distribution. These

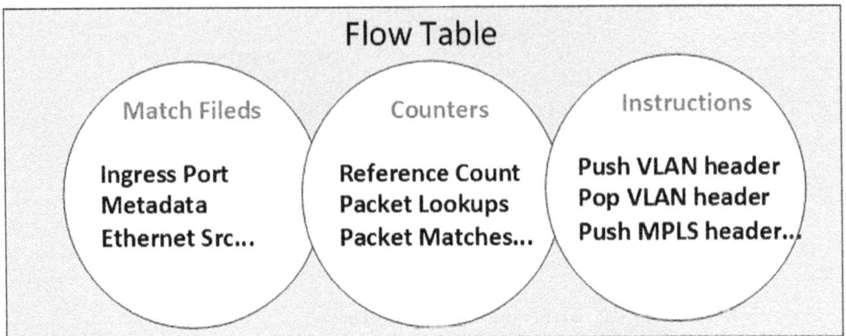

FIGURE 26.6

Main fields of a flow entry in a flow table [31].

forwarding messages are disseminated by the OF protocol to the entire OF network. However, this centralized decision characteristic causes both a flow load bottleneck and latency on overall flow traffic, bringing a communication overhead on quality of flows [32,33]. In order to remove this negative effect, some solutions exist. For example, in order to overcome the communication overhead between the data and control plane [28], proposes three extensions to OF protocol and flow tables. They claim that these approaches will accelerate the production in networks [27]. enhances the flexibility and configurability of the system to save more energy by deploying the OF to their cloud architecture.

3 GoS OPTIMIZED TOPOLOGY AND ADMISSION CONTROL MODEL FOR SOFTWARE DEFINED MOBILE WIRELESS NETWORKS [3]

3.1 NETWORK ARCHITECTURE

The two components of SDN architecture are shown in Figure 26.7 as Data and Control Planes. The Data Plane includes m OF switches, i.e., Access Points (APs), that access end-users. The flow tables of these switches are filled by the Controller; therefore, the only responsibility they have is forwarding data packets according to table entries. One of the components of the Control Plane is the Network Virtualization Layer that includes virtual switches which are virtual representers of physical switches. Each virtual switch is responsible for its physical switch cluster. Moreover, the communication between them is provided by using the OF protocol. The required data are collected periodically by virtual switches from the Data Plane and sent to the Controller that is another part of the Control Plane. In the Controller, the Topology Control Algorithm builds the Network Virtualization Layer and the Admission Control Algorithm manages flow fairly on the new created virtual topology.

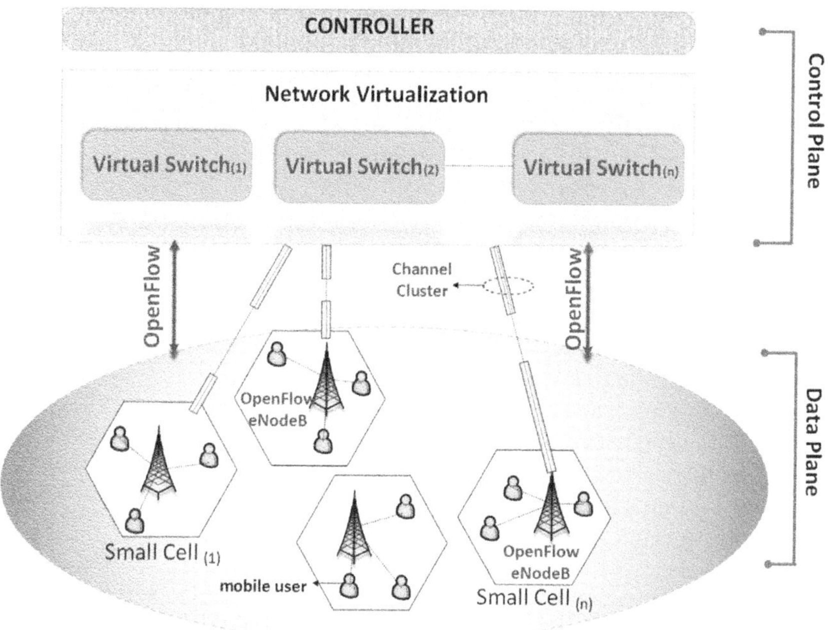

FIGURE 26.7

Network architecture for SDWN [3].

3.2 CONTROL ALGORITHMS

3.2.1 Topology Control Algorithm

The Topology Control Algorithm aims to maximize channel efficiency (ψ) with respect to the number of virtual switches (N′) and GoS of each flow. The optimization that is the major part of this algorithm is organized as follows:

$$\max \quad \psi = \frac{\sum A_j}{A_{\max}} \times 100 \quad j \in (1, 2, \ldots, N) \tag{26.1}$$

$$s.t. \quad N' < N$$

$$\frac{\dfrac{\left(\sum_{n=1}^{k_j} A_j\right)^{\left(\sum_{n=1}^{k_j} C_j\right)}}{\left(\sum_{n=1}^{k_j} C_j\right)!}}{\sum_{k=0}^{\left(\sum_{n=1}^{k_j} C_j\right)} \left(\dfrac{\sum_{n=1}^{k_j} A_j^k}{k!}\right)} \leq threshold \quad \forall_j \in N' \tag{26.2}$$

where A is the individual traffic request of a mobile user in Erlang, $A_{\max}$ is the total traffic intensity that the topology can handle where there exists only one virtual switch that has all of the channels. N is the number of physical OF

ALGORITHM 26.1 TOPOLOGY CONTROL ALGORITHM [3]

```
Require: N; Cj; Aj
  Ensure: updated N′, updated OF tables
Create and initialize an array for physical switches
for Kj ← 1 to N do
    tempN ← ceil(N = Kj)
    Create tempN Virtual Switches
    Sort physical switches according to Aj using Quick Sort
for k ← 0 to N do
        Find virtual switch that has minimum A and assign kth physical switch to that virtual switch
        Update A and C for virtual switch
        Calculate GoS for virtual switch
end for
for i ← 0 to tempN do
    if Any GoS ≥ 0.01 then
        Create virtual switches again
    end if
end for
        Update OF flow table according to new created virtual switches
end for
```

switches in Data Plane and N′ is the number of virtual switches that are responsible for many physical switches. Moreover, j is the index of the physical OF switch and C_j is the number of channels that are assigned to the jth AP, i.e., physical OF switch. On the other hand, the threshold value varies as 1%, 5%, and 10% according to [34]. If the GoS of the flow is less than 1%, the flow has perfect service quality. If it is nearly 5%, this is called an acceptable service quality level for each flow. However, if the GoS is higher than 10%, this means that flow cannot be served under these circumstances.

The pseudocode of the Topology Control Algorithm that optimizes channel efficiency as seen in Equations (26.1) and (26.2) is provided as Algorithm 26.1.

3.2.2 Admission Control Algorithm

The Admission Control Algorithm distributes flows fairly by using a created virtual switch layer using the Topology Control Algorithm. By considering the traffic requests of each flow and fairness of switches, this algorithm determines the destination fields of packets. Therefore, both the GoS of each flow can be enhanced and more users can be handled without any alteration of the physical topology. The pseudocode of this algorithm as provided as Algorithm 26.2.

ALGORITHM 26.2 ADMISSION CONTROL ALGORITHM [3]

Require: threshold, physical and virtual switch arrays

Ensure: Update OF flow table

for all physical switches that are member of tth virtual switch **do**

Calculate GoS for j^{th} physical switch

    while GoS > threshold **do**

        Find minimum $GoS_{j'}$ in virtual switch t

        min ← j'

        if min = j **then**

            return signal to Topology Control Algorithm to redesign virtual topology

        endif

        Assign maxflow according to traffic request to min physical switch

        Update GoS of each switch

    end while

end for

3.3 EXPERIMENTAL RESULTS

The performance evaluation is done for two different topologies, named Case A and Case B. In Case A, the topology has 24.96 Erlang traffic request in total, whereas in Case B it has 42 Erlang. For each scenario the total number of channels is 60 and the number of OF switches varies from 1 to 60. The results are compared with a conventional network that does not have any dynamic decision making as in the software-defined approach.

The topology control algorithm results in terms of resource efficiency is shown in Figure 26.8(a). According to this graph, both in Case A and Case B the software defined wireless network provides higher resource efficiency than conventional ones. Moreover, the admission control algorithm result is shown in Figure 26.8(b). Again, according to conventional topology, SDWN can handle more flows, whatever the threshold value is.

4 SOFTWARE DEFINED OFFLOADING MECHANISM [1]

The network architecture for SDN is represented in Figure 26.7 with two layers named Control and Data Plane. The Control Plane includes virtual representations of Data Plane components and the Controller processes data coming from the Virtualization layer that collects it from physical components. These two planes communicate by using special protocol named OpenFlow in the SDN approach.

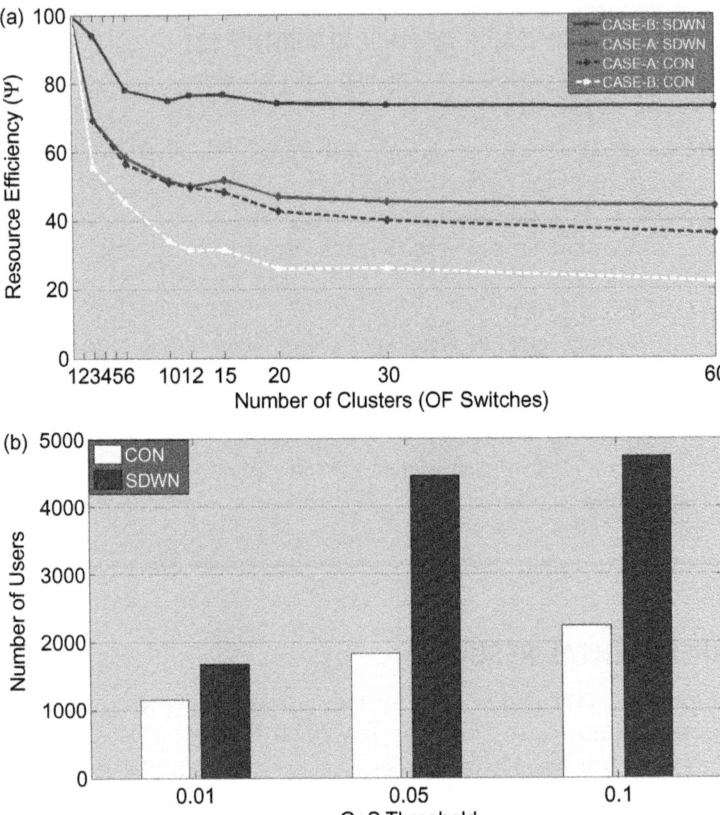

FIGURE 26.8

Comparison of SDN and conventional framework in terms of resource efficiency and maximum number of flows in the topology: (a) Resource efficiency; (b) Maximum number of flows that can be handled under different GoS thresholds.

4.1 OFFLOADING DECISION ALGORITHM

The Offloading Decision Algorithm processes the offloading strategy by using necessary data from the Entity sublayer. With this algorithm, the suitable offloading strategy for the system is defined. To do this, it uses a set of base stations M, a set of users N, and a satisfaction level of each mobile user. Satisfaction level is calculated according to user types and is rated as Gold, Silver or Bronze. The Gold user has the least tolerance on alteration of service quality, whereas Bronze has the most tolerance. Silver is between these two types. According to these user types, the dissatisfaction ratio is calculated for each end-user. With these inputs, the algorithm runs according to the pseudocode provided in Algorithm 26.3.

ALGORITHM 26.3 OFFLOADING DECISION ALGORITHM [1]

```
Require: Set of user (N), Set of Base Station (M), User Types
    Ensure: x is offloaded to y
Initialize x←1
for i←1 to n do
    index←Find most dissatisfied user by considering user types
    x←index
end for
Find most suitable Base Station j to offload that x is in the range of
according to density and pathloss of cell
j←y
Offload x to y
Update Flow Tables in topology
```

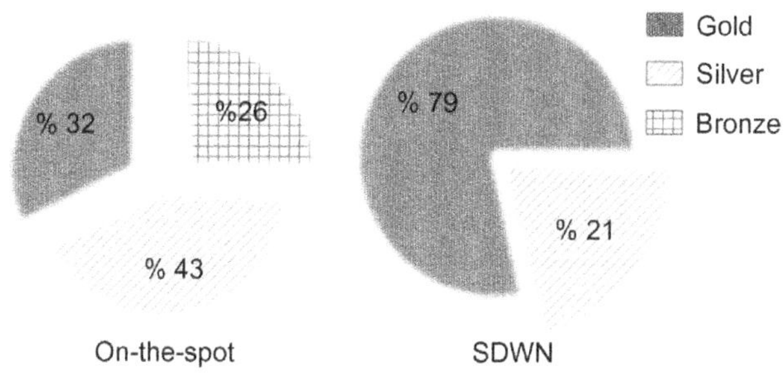

FIGURE 26.9

Comparison of on-the-spot and SDWN controlled offloading strategies.

4.2 EXPERIMENTAL RESULTS

As seen in Figure 26.9, the pie charts show the offloaded user types for on-the-spot and SDWN controlled offloading strategies. Due to having the least tolerance of change of quality of service, Gold users have higher priority than Silver users, and Silver users have higher priority than Bronze ones. Because the Offloading Decision Algorithm gives precedence to Gold users in offloading, the percentage of offloaded Gold users should be high in SDWN. However, in the on-the-spot, the percentage of offloaded user types depends on topology, due to having a user-based offloading strategy. The percentage of Gold users in on-the-spot is 32%, whereas this is increased to 79% in SDWN. This means that the service quality of Gold users is enhanced by the controlled offloading strategy with global view.

5 CONCLUSION

In this chapter, we present the implementation of the promising SDN technology into wireless network management. The virtualization capability of SDN provides an isolated, robust and flexible topology and admission control for next generation wireless deployment. Therefore, this chapter gives a deep approach for the SDN implementation into the wireless environment. After summarizing the current state-of-the-art in SDN, we present a topology and admission control mechanism, considering and optimizing the GoS of the overall network and the traffic intensity of the small cells. The presented framework offers virtualization of the data plane and isolates the resource constraint to the SDN controller. Here, the SDN-based topology control algorithm removes physical resource constraints by creating virtual switches. The admission control algorithm fairly distributes flows to the available channels, thanks to the global view provided by the SDN virtualization. Consequently, the resource efficiency and scalability of the system are increased without any change of physical wireless topology. We also present a software-defined wireless traffic offloading mechanism. It senses the wireless traffic demands in order to virtualize and realize an accurate reflection of the Data Plane in its central control plane. It then orchestrates the offloading according to the Quality of Service (QoS) based user satisfaction and types).

REFERENCES

[1] Arslan Z, Erel M, Özçevik Y, Canberk B. SDoff: A software-defined offloading controller for heterogeneous networks. IEEE wireless communications and networking conference (WCNC) Istanbul, TURKEY; April 2014.

[2] Cisco visual networking index report: global mobile data traffic forecast update, 2013–2018; 5 February 2014 [Online]. Available from: <www.cisco.com>.

[3] Erel M, Arslan Z, Özçevik Y, Canberk B. Grade of service (GoS) based adaptive flow management for software defined heterogeneous networks (SDHetN), to appear in Computer Networks (Elsevier), 01/2015;76.

[4] Zahir T, Arshad K, Nakata A, Moessner K. Interference management in femtocells. IEEE Commun Surv Tutorials 2013;15(1):293–311.

[5] ElSawy H, Hossain E, Kim DI. HetNets with cognitive small cells: user offloading and distributed channel access techniques. IEEE Commun Mag 2013;51(6):28–36.

[6] Andrews J. Seven ways that HetNets are a cellular paradigm shift. IEEE Commun Mag 2013;51(3):136–44.

[7] Awada A, Wegmann B, Viering I, Klein A. Optimizing the radio network parameters of the long term evolution system using Taguchi's method. IEEE Trans Veh Technol 2011;60(8):3825–39.

[8] Ristanovic N, Le Boudec J-Y, Chaintreau A, Erramilli V. Energy efficient offloading of 3G networks. In: IEEE eighth international conference on mobile Adhoc and sensor systems (MASS), 2011, p. 202–11.

[9] Huang D, Wang P, Niyato D. A dynamic offloading algorithm for mobile computing. IEEE Trans Wireless Commun 2012;11(6):1991–5.

[10] Lee K, Lee J, Yi Y, Rhee I, Chong S. Mobile data offloading: how much can WiFi deliver? IEEE/ACM Trans Networking 2013;21(2):536–50.

[11] Singh S, Dhillon H, Andrews J. Offloading in heterogeneous networks: modeling, analysis, and design insights. IEEE Trans Wireless Commun 2013;12(5):2484–97.

[12] Xie R, Yu F, Ji H, Li Y. Energy-ecient resource allocation for heterogeneous cognitive radio networks with femtocells. IEEE Trans Wireless Commun 2012;3910–20. Available from: http://dx.doi.org/10.1109/TWC.2012.092112.111510.

[13] Tachwali Y, Lo B, Akyildiz I, Agusti R. Multiuser resource allocation optimization using bandwidth power product in cognitive radio networks. IEEE J Sel Areas Commun 2013;31(3):451–63. Available from: http://dx.doi.org/10.1109/JSAC.2013.130311.

[14] Yue H, Pan M, Fang Y, Glisic S. Spectrum and energy efficient relay station placement in cognitive radio networks. IEEE J Sel Areas Commun 2013;883–93. Available from: http://dx.doi.org/10.1109/JSAC.2013.130507.

[15] Kokku R, Mahindra R, Zhang H, Rangarajan S. NVS: a substrate for virtualizing wireless resources in cellular networks. IEEE/ACM Trans Networking 2012;20(5):1333–46.

[16] E.G.NFV. Network functions virtualization(nfv); use cases,v1.1.1,Tech.rep. URL available at: <http://docbox.etsi.org/ISG/NFV/Open/Published/>; 2013.

[17] Bojic D, Sasaki E, Cvijetic N, Wang T, Kuno J, Lessmann J, et al. Advanced wireless and optical technologies for small-cell mobile backhaul with dynamic software-defined management. IEEE Commun Mag 2013;51(9):86–93.

[18] Goransson P, Black C. Software defined networks: a comprehensive approach. 1st ed. M. Kaufmann; June 2014.

[19] Azodolmolky S. Software defined networking with OF. Packt Publishing; October 2013.

[20] Open Networking Foundation. Software-defined networking: the new norm for networks. Palo Alto, CA, USA, White sub-chapter, Apr. 2012 [Online]. Available from: <www.opennetworking.org>.

[21] Lara A, Kolasani A, Ramamurthy B. Simplifying network management using software defined networking and OF. In: IEEE international conference on advanced networks and telecommunications systems (ANTS), 2012. p. 24–9.

[22] Shukla V. Introduction to software defined networking OF & VxLAN. CreateSpace Independent Publishing Platform; June 2013.

[23] Levin D, Wundsam A, Heller B, Handigol N, Feldmann A. Logically centralized?: state distribution trade-offs in software defined networks. In: Proceedings of the first workshop on hot topics in software defined networks, ser. HotSDN '12. New York, NY, USA: ACM; 2012. p. 1–6. [Online]. Available from: <http://doi.acm.org/10.1145/2342441.2342443>.

[24] Akyildiz IF, Lee A, Wang P, Luo M, Chou W. A roadmap for traffic engineering in SDN-OF networks. Comput Netw (Elsevier) J 2014;71:1–30.

[25] Sezer S, Scott-Hayward S, Chouhan P, Fraser B, Lake D, Finnegan J, et al. Are we ready for SDN? implementation challenges for software-defined networks. IEEE Commun Mag 2013;51(7).

[26] Kim H, Feamster N. Improving network management with software defined networking. IEEE Commun Mag 2013;51(2):114–19.

[27] Dely P, Vestin J, Kassler A, Bayer N, Einsiedler H, Peylo C. CloudMAC; an OpenFlow based architecture for 802.11 MAC layer processing in the cloud. In: IEEE globecom workshops (GC Wkshps), 2012. p. 186–91.

[28] Feng T, Bi J, Hu H. OpenRouter: OpenFlow extension and implementation based on a commercial router. In: 19th IEEE international conference on network protocols (ICNP), 2011, p. 141–42. Available from: http://dx.doi.org.10.1109/ICNP.2011.6089045.

[29] McKeown N, Anderson T, Balakrishnan H, Parulkar G, Peterson L, Rexford J, et al. OF: enabling innovation in campus networks. ACM SIGCOMM Comput Commun Rev 2008;38(2):69–74.

[30] Open Networking Foundation. OF switch specification. Version 1.3.2; April, 2013.

[31] Open Networking Foundation. OF switch specification. Version 1.1.0; February, 2011.

[32] Sünnen D. Performance evaluation of openflow switches, Semester Thesis at the Department of Information Technology and Electrical Engineering, Master's thesis, Swiss Federal Institute of Technology Zurich; 28 February 2011.

[33] Yeganeh SH, Tootoonchian A, Ganjali Y. On scalability of software-defined networking. IEEE Commun Mag 2013;51(2):136–41.

[34] ITU-T. Series E: Telephone Network and ISDN, Network Grade of Service Parameters and Target Values for Circuit-Switched Public Land Mobile Services. International Telecommunication Union, Geneva, Recommendation E.771(10/96), October. 1996.

CHAPTER

Radio resource management for heterogeneous wireless networks

27

Schemes and simulation analysis

Wahida Mansouri[1], Khitem Ben Ali[1], Faouzi Zarai[1], and Mohammad S. Obaidat[2]

[1]University of Sfax, Sfax, Tunisia

[2]Monmouth University, West Long Branch, NJ, USA

1 INTRODUCTION

The demand for accessing services while on the move, at any place and any time, has led to the current efforts towards integration of heterogeneous wireless networks. In particular, interoperability of cellular networks and Wireless Local Area Networks (WLANs), as complementary systems in providing capacity and coverage, has drawn a lot of attention. Next Generation Wireless Networks (NGWNs) are expected to be heterogeneous networks which integrate different Radio Access Technologies (RATs) such as 3GPP's Long Term Evolution (LTE) systems and WLANs where a transmission is supported if the signal-to-interference-plus-noise ratio (SINR) at the receiver is greater than some threshold. A main challenge of the heterogeneous wireless networks is the Radio Resource Management (RRM) strategy. Currently, the RRM strategies are implemented in different kinds of networks separately. None of the RRM strategies is suitable for the heterogeneous networks as each RRM strategy only considers the situation of one particular Radio Access Technology (RAT). However, the heterogeneous network requires a more efficient RRM approach to coordinate radio resources among different RATs in an optimized way. The Common RRM (CRRM) is proposed to jointly manage radio resources among different RATs in an optimized way. As described in [1], the CRRM is designed to coordinate the management of resource pools over the heterogeneous air interface in an efficient way. This efficiency depends on how to construct its functionalities. These functionalities are common admission control and common scheduling techniques.

In this chapter, we proposed a survey on resource management schemes. The remainder of this chapter is organized as follows: Section 2 presents definitions and operating principles, and Section 3 presents the call admission mechanism in wireless networks. In Section 4, we study the scheduling schemes and their characteristics. We discuss the modeling of RRM schemes in Section 5. Section 6 presents the performance evaluation of the RRM schemes using a modeling and simulation approach. Finally, we conclude the chapter in Section 7.

2 DEFINITIONS AND OPERATING PRINCIPLE

The increasing demand for advanced multimedia services combined with the resource constraints of the wireless heterogeneous networks indicate that there is a need for efficient congestion control and scheduling schemes to achieve a competent resource management combined with adequate quality of service (QoS) levels for end users. QoS provision in wireless networks is closely related to the exploitation of available network resources and the maximization of the number of users.

- Call admission control is one of the key issues in wireless mobile communications, concentrating great interest in research work on QoS. CAC algorithms are employed to ensure the admission of a new call into a resource limited network does not violate the service level agreements (SLAs) concerning ongoing calls. Simply stated, it determines the condition for accepting or rejecting a new call based on the availability of sufficient network resources to guarantee the QoS parameters without affecting the existing calls.
- Scheduling schemes: A scheduling algorithm is a set of rules that determines the task to be executed at a particular moment. It refers to the problem of determining which users will be active in a given time-slot. Although there are a number of packet scheduling algorithms that have been proposed, the design of those algorithms is challenged by supporting different levels of services, fairness, and implementation complexity, among others.

3 CAC IN WIRELESS NETWORKS

Compared to wired networks, the fluctuation in resource availability in wireless networks is much more than severe and results from inherent features such as fading and mobility. The current and next generation wireless networks are expected to provide multimedia services with different QoS requirements. A typical example is the universal mobile telecommunication system (UMTS), which is required to support a wide range of applications each with its specific QoS. There are four QoS classes defined in UMTS specifications: the conversational class, the streaming class, the interactive class, and the background class. Since multimedia services have different traffic characteristics, their QoS requirements may differ in terms of bandwidth, delay, and dropping probabilities. The radio resource management unit is responsible for the fair and efficient allocation of network resources among different users. The large demand for high capacity has led to the use of micro- and pico-sized cells. As a consequence, the handoff rate significantly increases and the handoff procedure becomes a crucial issue to ensure seamless connectivity and satisfactory QoS. Also from the user's point of view, handoff attempt failure is less desirable than blocking a new call. Due to the limited resources in wireless multimedia systems, efficient call admission control and resource reservation schemes are needed to maintain the desired QoS.

3.1 CAC SCHEMES CLASSIFICATION

CAC schemes can be classified into general categories based either on the criteria considered in the decision part of the CAC scheme or on specific design characteristics. The admission criteria considered by CAC schemes are usually related to various QoS parameters. Moreover, CAC schemes can be divided into two categories, local and collaborative schemes. The local schemes have been proposed for homogeneous networks. This can be grouped into prioritized schemes and nonprioritized schemes. In the nonprioritized admission control scheme, all types of calls are treated equally; the scheme is based on a first come, first served rule. A call is denied access only if no channel is available. In the prioritized admission control scheme, the policy is based on dropping lower bandwidth calls to serve handover or new call requests of a larger bandwidth class; thus higher bandwidth calls have a higher priority. This prioritizing policy could be useful in a heterogeneous environment where connections with a low bandwidth requirement can be reallocated to a network optimized for that specific data rate and service provisioning and thus leave high-speed connections free for users requiring high QoS. Figure 27.1 shows a classification diagram of the CAC algorithms.

3.1.1 Nonprioritized CAC schemes

In heterogeneous wireless networks, interference poses critical constraints concerning mainly the signal quality. This situation has an impact not only on network conditions but also on system capacity. Particularly in CDMA wireless networks, interference is the dominant factor affecting their performance in terms of capacity and QoS provision to end users. Thus, the SINR is an adequate metric of the signal quality. CDMA-based air interfaces are mainly influenced by interference caused by other users from the same network instead of Gaussian noise, so the effect is usually neglected, focusing mainly on SIR. Therefore, CAC schemes implemented for interference limited networks employ as admission criterion either the interference levels caused by a new incoming call or the signal quality levels achieved. Hence, interference-based CAC schemes admit new calls only if the SNR/SIR values can maintain a minimum signal quality level. The SNR/SIR levels

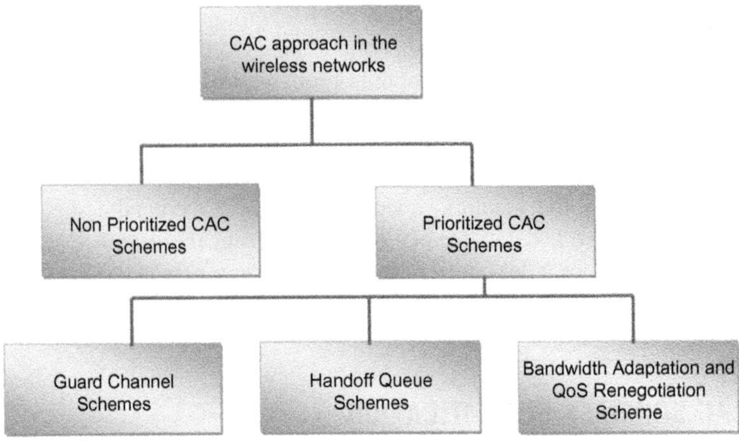

FIGURE 27.1

Classification for CAC algorithms in wireless networks.

correspond to predefined QoS levels for new and ongoing users. This simple approach offers a tool to reduce interference in wireless networks, while on the other hand constitutes an efficient admission criterion. These nonprioritized schemes (NPS) [2] are employed by typical radio technologies that have been proposed for personal communications services (PCS). In NPS, the BS handles handoff calls in exactly the same way as new calls (i.e., handoff call is blocked immediately if no resources are available). All available resources in the BS are shared by handoff and new calls. However, the drawback of this scheme is that it is difficult to guarantee the required dropping probability of handoff calls.

3.1.2 Prioritized CAC schemes

In these schemes preferential treatment is given to priority and handoff calls, to avoid unwanted call blocking and handoff dropping while maximizing channel utilization. Usually handoff calls are assigned higher priority over new calls. These strategies fall into two categories: Handoff queue [3], and Guard channel [4] schemes.

3.1.2.1 Handoff queue schemes

The principal concept of these schemes is that when resources become available, one of the calls in the handoff queue is served. If there are no available resources, call requests are queued until resources become available again. The HQ scheme needs a lot of buffers to deal with real-time multimedia traffic and sophisticated scheduling mechanisms are needed to meet the QoS requirement for delay sensitive calls in order to guarantee that the queued data will not expire before they are transmitted.

3.1.2.2 Guard channel schemes

In general, the concept of the guard channel (GC) approach was first introduced by [4]. The basic idea of GC-based admission control strategies is to reserve resources in each cell a priori to deal with handoff requests. To provide user's equipment with continuous connectivity, the system reserves backup channels referred to as *guard channels* to offer preferential treatment to priority calls and handoff calls. In such a system, call requests with lower priority are rejected if the number of available resources is less than a certain threshold. GC strategies differ in the number of guard channels to be chosen by a base station. They are called *fixed guard channel* and *dynamic guard channel*, as developed in [5], respectively. The fixed guard channel schemes reserve a fixed number of channels for handoff calls [4]. In this scheme, only one traffic class was considered. The advantage of this scheme lies in the simplicity of deployment, because there is no need to exchange control information between the base stations. However, with a small portion of handoff calls, GC schemes result not only in increased blocking probability of new calls, but also in inefficient resource utilization, because only a few handoff calls are able to use the reserved channels exclusively. On the other hand, with a large number of handoff calls, it is difficult to guarantee the service requirements of the handoff call. All these schemes proposed above are static because such GC schemes cannot adapt to quick variation of the traffic pattern. Dynamic GC schemes, reported in [6], improve the system efficiency while providing the QoS guarantees to priority calls. These schemes adaptively reserve the actual resources needed for priority calls and, therefore, accept more lower-priority calls compared to the fixed scheme. In [4], a distributed adaptive guard channel reservation scheme is proposed to give priority to handoff calls. This scheme is built upon the concept of guard channels and it uses an adaptive algorithm to search automatically the optimal number of guard channels to be reserved at each base station.

3.1.3 Bandwidth adaptation and quality of service renegotiation scheme

Wireless networks support a variety of services which can be classified into rate-adaptive applications and constant bit rates (CBR) services. In such services, e.g. voice calls, a bandwidth increase beyond the standard requirement will not improve the respective QoS. On the other hand, in rate adaptive services users specify, at their connection request, the minimum and maximum bandwidth required. Apart from specifying the bandwidth range required by every service class, rate variations may originate from the dynamic nature of the wireless environment along with the mobility of user terminals. Thus, in modern wireless networks bandwidth adaptation algorithms are employed to improve network utilization and guarantee the QoS of ongoing calls, assigning the minimum bandwidth required. When the network conditions are favorable and enough resources are available, they may be assigned to ongoing rate adaptive users according to two general strategies based on service class priorities. According to the first strategy, the available resources are fairly assigned to all ongoing users without taking into account any priorities. According to the second scheme, resources are first assigned to service class calls of high priority, until the resources are exhausted or all high priority service class calls have taken the maximum bandwidth required. If resources are still available, the scheme assigns them to service class calls of the next high priority. The procedure continues until all resources are exhausted or all calls are served with their maximum bandwidth demand. Recently, several CAC algorithms and bandwidth adaptation algorithms have been proposed for wireless networks. As described in [7], an adaptive call admission control algorithm was proposed in 3GPP LTE network. This proposed scheme encompasses the bandwidth allocation/reallocation policy and the bandwidth adaptation algorithm. In this framework we provide a CAC algorithm that takes into account the separation between incoming traffic for each class and prioritizes handoff calls (HC) over new calls (NC). Thus, we assign three classes of service for arrival calls of each class depending on their QoS profile, such as latency tolerance. Then we find the non-real-time service (NRT), real-time tolerant service and real-time intolerant service. The studied algorithm proposes a system of priority for the six classes of services in the increasing direction: NC-NRT/HC-NRT/NC-TLR/HC-TLR/NC-INTLR/HC-INTLR. These algorithms are needed to reduce the requested or already connected call bandwidth allocation. In general, when a call arrives in a certain cell, the network may either have enough resources to provide bandwidth between the minimum and the maximum demand or be congested—that is, it cannot provide the minimum bandwidth requested by the new call. In the first case the call is admitted, whereas in the second, bandwidth adaptation CAC algorithms, also known as rate-adaptive schemes, are applied to determine an optimal resource allocation aiming at serving as many users as possible while reducing the admission failure probability. This is accomplished by reducing the rate of some users as much as required, when possible, in order to accommodate the new call. Thus, for the acceptance of a new call request, a degradation procedure is triggered to degrade the amount of allocated bandwidth of some ongoing calls in the cell to attempt to allocate required resource. However, it should be mentioned that user rates cannot be reduced below the minimum rate values required to assure required QoS; thus, when all users operate at their lowest bandwidth requirement, a new call request will be rejected. Rate degradation may be enforced according to a prioritization or to a non-prioritization scheme. In the former, the rate degradation policy is first applied to service class call of the lowest priority. If the resources released are still insufficient for the admission of a new call, the calls of the next priority level are examined.

3.2 CAC IN THE NEXT GENERATION WIRELESS NETWORKS

In heterogeneous wireless networks, mobile users will be able to communicate through any of the available radio access technologies and roam from one RAT to another, using multimode terminals. In the prevailing scenario the collaborative admission control decision should be based on multiple criteria such that the optimization user satisfaction and selection of optimal RAT is achieved. As described in [8], the CAC scheme is based on the IEEE 802.21 Media Independent Handover (MIH), to seamlessly hand over mobile users between heterogeneous wireless networks for load balancing purpose. This work involves call admission control and a handoff scheme to maximize system reward for network providers and to guarantee QoS-aware seamless handoff for mobile users.

The load balancing strategies are required to efficiently utilize the available radio resources and avoid the unwanted congestion situations due to overloaded wireless networks. Such schemes should control the access of users between available networks based on current loads, predicted traffic profiles and the optimality of a network connection to a particular service request. In these heterogeneous networks, a joint call admission control (JCAC) algorithm is needed to decide whether a call will be admitted or not, and to select the most appropriate RAT for each admitted call [9]. This JCAC scheme is envisioned as user-centric in that user's preferences are considered in decision making for RAT selection.

Moreover, lots of literature has been dedicated to the signal to interference ratio (SIR)-based call admission control algorithms in the heterogeneous networks [10]. The concept of residual capacity is introduced as the additional number of calls a base station can accept such that the system wide outage probability will be guaranteed to remain below a certain level. The residual capacity is dynamically updated at each cell according to the reverse link SIR measurements at each base station.

4 SCHEDULING

The concept of heterogeneous wireless networks is based on the coexistence and interoperability of different types of radio RATs in a unified wireless heterogeneous platform. How to guarantee the QoS of the services is of great concern in the next generation wireless network, which is characterized by providing different types of services. The management of QoS from the beginning to the end implies the presence of specific scheduling mechanisms. In fact, in packet networks, link scheduling is an important mechanism to realize QoS as it directly controls packet delays. In this section, we address the problem of scheduling packets. A number of the radio resource scheduling algorithms have been proposed in the literature. But in some works, authors ignore the influence of the transmission link condition, which will make the performance result not so good when the differences of link qualities between RATs become evident.

4.1 SCHEDULING ALGORITHMS

A scheduling algorithm is a set of rules that determines the task to be executed at a particular moment. Although there are a number of packet scheduling algorithms that have been proposed in the literature, the design of those algorithms is challenged by need for supporting different levels of services, fairness, and implementation complexity and so on. Scheduling and resource allocation are essential components of wireless systems. The scheduling refers to the problem of determining which users will be active in a given time-slot. Resource allocation refers to the problem of allocating physical layer resources such as bandwidth and power among these active users.

Radio resource scheduling is a process in which resource blocks are distributed among the users. Many works have been proposed for the radio resource scheduling algorithms in the literature.

- Proportional Fairness (PF) Resource Allocation Scheme: The PF scheduling algorithm is a commonly used scheduling scheme for Orthogonal Frequency Division Multiple Access (OFDMA). It begins with the calculation of the priority for each user at each Resource Block (RB) and then the user with maximum priority is assigned the RB. The priority of kth user for j^{th} resource block in time is calculated by dividing the requested data rate for the kth user over the jth RB in time n by the low-pass filtered averaged data rate of the kth user. The algorithm continues to assign the RB to the user with next maximum priority until all RBs are assigned or all users have been served with RBs.
- Softer Frequency Reuse based Resource Scheduling Algorithm: The designed frequency scheduler runs in such a way that the cell edge users have the greater probability to use the frequency band with higher power and the cell center users have the higher probability of using the frequency band with lower power.
- Resource Scheduling Algorithm, based on Dynamic Allocation: This scheme performs efficient radio resource utilization in different types of network traffic. Conversational class traffic is transmitted on the network in small chunks which are considerably smaller than the packets of streaming class traffic. In this algorithm the equal allocation of the radio resources is ensured.

4.2 SCHEDULING ALGORITHMS CLASSIFICATION

The scheduling can be divided into two classes as described in [11]: channel-independent scheduling and channel-dependent scheduling. Figure 27.2 illustrates the scheduler classification.

4.2.1 Channel independent scheduling schemes

Channel independent strategies were first introduced in wired networks and are based on the assumption of time-invariant and error-free transmission media. The simplest examples of those scheduling algorithms are:

- First In First Out (FIFO): This has the principle of a queue processing technique or servicing conflicting demands by ordering process by

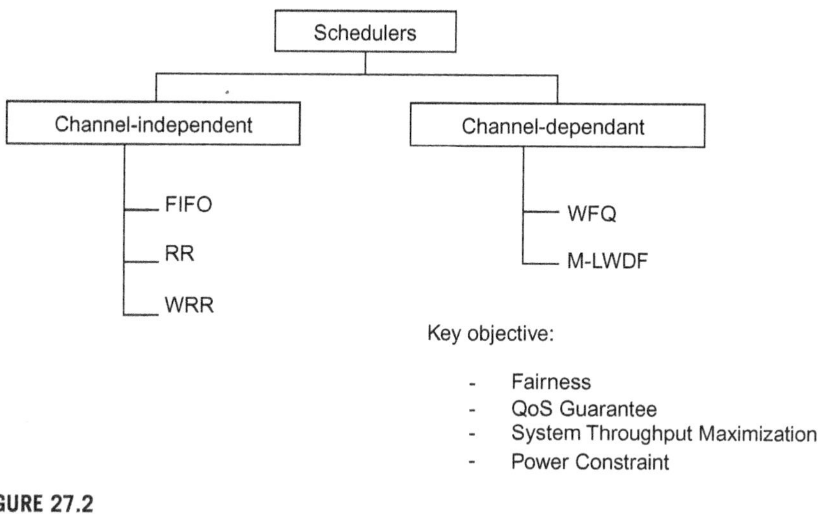

FIGURE 27.2

Scheduling schemes classification.

First-Come, First-Served: what comes in first is served first, what comes in next waits until the first is finished. In this case a single queue exists; thus the order of the arrival of packets determines the order in which they are forwarded to the output link. This mechanism is not used for networks that require a classification of the flow to guarantee the quality of services for the end users.

- Round Robin (RR): RR is one of the simplest scheduling algorithms designed especially for a time-sharing system, where the scheduler assigns time slots to each queue in equal portions without priority. The Round Robin algorithm consists of serving the queues one after the other. If the current queue contains a packet, it will be served; else the algorithm selects the next queue for service.
- Weighted Round Robin (WRR) scheduling is used to facilitate controlled sharing of the network bandwidth. WRR assigns a weight to each queue; that value is then used to determine the amount of bandwidth allocated to the queue.

4.2.2 Channel-dependent scheduling schemes

Channel-dependent scheduling may be used to take advantage of favorable channel conditions to increase the throughput and system spectral efficiency. Examples of those scheduling algorithms are:

- Weighted Fair Queuing (WFQ Scheme): Here, the packets are grouped into various queues. A weight is assigned to each queue which determines the fraction of the total bandwidth available to the queue. The weight may depend on both channel quality and the number of packets in the queue of individual users. It provides balanced resource utilization between fairness and efficiency.

- Modified Largest Weighted Delay First (M-LWDF): It combines both channel conditions and the state of the queue with respect to delay in making scheduling decisions [12]. It guarantees that the probability of delay packets does not exceed the discarded bound below the maximum allowable packet loss ratio.

Another comparative study of scheduling algorithms for real-time traffic was proposed in [13]. These scheduling algorithms are classified in two types: Fixed Priority Scheduling and Dynamic Priority Scheduling. The Rate Monotonic (RM) and Deadline Monotonic (DM) are fixed priority scheduling. The RM tasks with smaller periods get higher priorities. The DM tasks are assigned priorities according to their deadline. The task with the shortest deadline is assigned the highest priority. The Dynamic Priority Scheduling contains:

- The Earliest Deadline First algorithm (EDF), which maintains a list of waiting packets to be executed. This list is sorted by deadline with the first packet having the earliest deadline. The priority of each packet is decided based on the value of its deadline. The task with nearest deadline is given highest priority and it is selected for execution.
- The Least Laxity First (LLF) scheme is a dynamic scheduling method. For every task ready to run at the given moment the difference S between the time until deadline D and the remaining computation time C is computed. This difference, called slack or laxity, can be seen as an inverted priority value. The task with the smallest S-value is the one to be executed next. Whenever a task other than the currently running one has the smallest slack, a context switch will occur.

4.3 SCHEDULER DESIGN

An efficient scheduling scheme must take into consideration some important parameters in the scheduling decision such as signal strength, interference and channel quality. Due to the rapidly and instantaneously changing nature of radio channel quality there must be a fast enough scheduling algorithm to compensate for the changing in channel conditions. Figure 27.3 illustrates the scheduler design which consists of three main function modules.

As described in [14], the scheduling scheme is based on the different modules: classifier, channel quality and scheduler.

4.3.1 Classifier

This module is responsible for sorting the incoming packets into different services classes and forwarding them to the corresponding queues. DiffServ allows the definition of multiple traffic classes. A packet is classified according to the Differentiated Services Code Point (DSCP) field. Applications with real-time constraints such as voice does not tolerate variation of the delay between the sent and

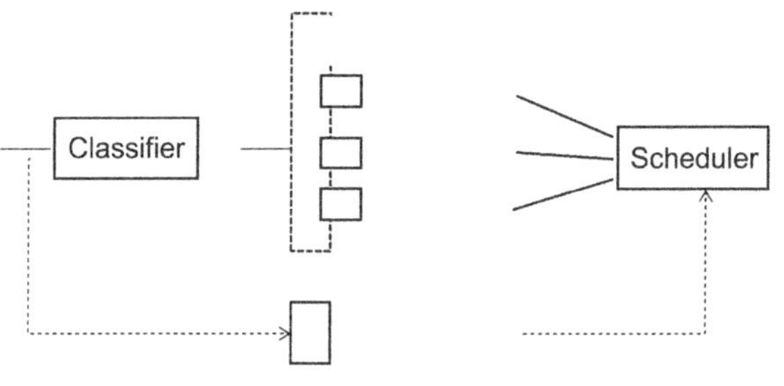

FIGURE 27.3

Scheduler design.

the received packets. The guarantee of these constraints becomes more difficult in technologies which take into account the mobility of the users. We need to differentiate the class of service for each packet. Such classes of services are delay intolerant, which means that the intolerant applications such as VoIP require a limited delay and cannot tolerate a delay higher than this limit. Another class tolerates an additional delay like video traffic and another class is not exigent in term of delay such as data traffic.

4.3.2 Channel quality

The Media Independent Handover Function (MIHF) provides different information to the channel quality module. This module uses some existing MIH services such as *Link_Up*, *Link_Down*, *Link_Going_Down* and *Link_Get_Parameters* in order to determine the link status and QoS parameters of different links like data rate, throughput and other information. The MIES defines the functions of event classification, event filtering, and event reporting to upper layers. Specifically, different events such as *Link_Up*, *Link_Down*, *MIH_Link_Up*, and *MIH_Link_Down* are defined. *Link_Up* and *Link_Down* events are generated by lower layers (layer 1 or layer 2), and these events are notified to the MIHF. Then, the MIHF reports these situations to upper layers by triggering *MIH_Link_Up* and *MIH_Link_Down* events. Access to network-specific information like network identifier is provided by the Information Elements (IE) of the MIH Information Services (MIIS). The major concern of the quality should be interference. If the estimated SINR is larger than a threshold γ, the channel is considered good. The channel quality module sends a message to the scheduler module which will use this information without any channel scanning procedure. Each flow is associated with a channel, which is in one of the three states, namely, good, medium or bad state, at any time instant.

4.3.3 Scheduler

This module makes a scheduling decision according to the delay constraint, the classes of service, the channel status and the type of connection (handoff or new call). It assigns priorities according to the packets deadlines: the shorter the deadline, the higher priority is assigned to the packet. Then, it serves the packets according to their priority index to the corresponding link. The algorithm should be able to dynamically identify the packet with the earliest deadline and to take into consideration the instantaneous channel conditions. The packet with nearest deadline and the best channel quality is given highest priority and it is selected for being served. The packet will be assigned the lowest priority if its deadline is the furthest.

4.4 SCHEDULING IN THE NEXT GENERATION WIRELESS NETWORKS

The literature provides examples of the different approaches that have been proposed for packet scheduling in the fourth generation systems. The LTE introduces enhanced data link mechanisms to support successful implementation of new data services across the network. The incorporated scheduling mechanisms can significantly contribute to this goal. The proposed scheduling scheme presented in [15], provides a fair, class based, delay jitter controlled packet scheduling scheme that manages handoff and buffer occupancy for 4G wireless access systems. This algorithm depends on the pilot signal strength measured at a mobile terminal. A connection has one of the following flow states: normal, handoff warning or handoff processing.

In [16], the authors propose a new scheduling scheme and introduce criteria for fairness regarding resource distribution. The basic principle for the downlink scheduler in LTE is to dynamically determine which terminal is supposed to receive Downlink Shared Channel transmission and on what resources, at every 1 ms interval. The overall objective of the scheduling schemes is to take advantage of the channel variations between the mobile users and to schedule transmissions to a mobile user on resources with good channel conditions. Other recent work presented in [17] proposes a new scheduling algorithm which makes use of an Assignment Model for resource allocation to all selected users during each Transmission Time Interval (TTI). In order to counteract the time-varying and frequency-selective nature of the wireless channel, the scheduler needs to allocate resources to appropriate users which have good channel conditions based on the selected metric every TTI. The metric is used as transmission priority of each user on each resource block, and is defined based on channel quality, and status of transmission queues, buffer state, QoS requirements and resource allocation history.

5 MODELING OF RRM SCHEMES

5.1 MODELING OF CAC ALGORITHMS IN WIRELESS NETWORKS

There are many call admission control algorithms proposed in the literature. Some of these works focus attention on performance modeling of wireless networks based on a hierarchical overlay infrastructure. They are categorized as

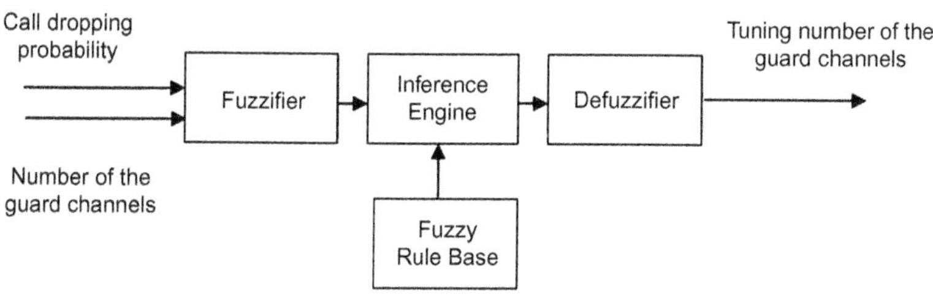

FIGURE 27.4

Block diagram of the proposed fuzzy logic model for CAC.

Utility-function based CAC and computation Intelligence CAC. In the Utility-function based CAC the incoming calls are admitted based on some utility or cost function based on multiple criteria. These algorithms are very optimal algorithms and in most of the cases are complex in nature and pose high computational overhead. The majority of the computational intelligence based CAC algorithms incorporate evolutionary approaches such as genetic algorithms (GA) [18], fuzzy logic [19] and artificial neural networks (ANNs) [20]. The concept of fuzzy logic has been extensively applied in characterizing the behavior of nonlinear systems [21]. In [19], an intelligent admission control model based on a fuzzy logic model is proposed to handle the imprecision for the concept of the guard channels as illustrated in Figure 27.4. The fuzzy call admission controller is implemented on the Mamdani inference scheme, taking relative mobility of user and number of reserved channels for handoff calls as input parameters [19]. Therefore, the input linguistic parameters of the model are set as call dropping probability and number of the guard channels. The output linguistic parameter is set as the tuning number of the guard channels.

The main advantages of the fuzzy logic approach [21] are that it is easy to understand and build a predictor for any desired accuracy with a simple set of fuzzy rules. Due to less computational demand there is no need of a mathematical model for estimation and also for fast estimation of future values. The limitations of the fuzzy logic approach are that first it works on single step prediction; and, second, the fuzzy logic does not have learning capability. A fuzzy neural approach for making the call admission control decision in multi-class traffic based next generation wireless networks is proposed in [20]. The proposed fuzzy neural call admission control (FNCAC) scheme is an integrated CAC module that combines the linguistic control capabilities of the fuzzy logic controller and the learning capabilities of the neural networks. The model is based on recurrent radial basis function networks, which have better learning and adaptability that can be used to develop intelligent systems to handle the incoming traffic in a heterogeneous network environment. Neural networks (NNs) are flexible soft computing frameworks for modeling a broad range of nonlinear problems [20]. One significant advantage of the neural network based approach over other

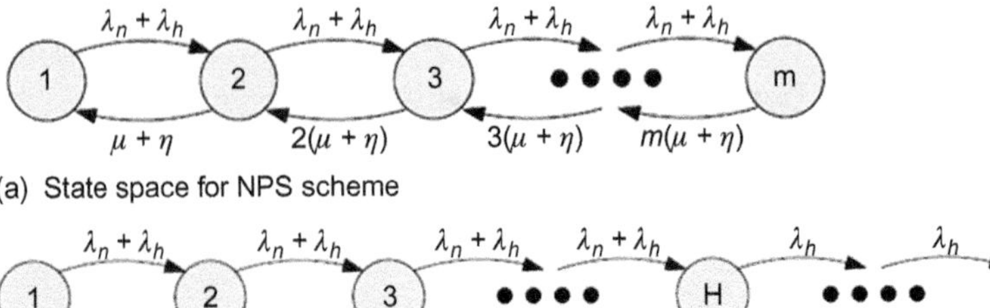

FIGURE 27.5

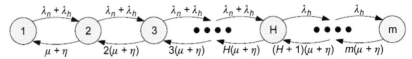

The Markov chain model of the existing PS with guard channel and NPS scheme.

classes of nonlinear models is that NNs are universal approximation tools that can approximate a large class of functions with a high degree of accuracy.

There are very few works reporting on the usage of mathematical models in CAC. In [22], Stochastic Reward Net (SRN) models are constructed to analyze the performance of some call admission control algorithms in the wireless networks. The SRN is an extension of Petri net (PN) [23], which is a high-level description language for formally specifying complex systems.

Typically, the analytical model for CAC mechanisms in heterogeneous wireless networks is modeled using a higher order Markov model. In [24], the authors focus on such classical modeling techniques based on the Markov chain analysis. They formulate the admission problem as an average cost Markov decision problem. Authors, in [25], present a comparative performance analysis based on the Markov model for different existing CAC mechanisms such as NPS scheme and guard-channel scheme (GCS) as PS. They consider a system at single cell that can be modeled again using a one-dimensional Markov process where each state is defined by the number of ongoing calls. Figure 27.5 shows the proposed comparative Markov model for the two different CAC schemes.

5.2 MODELING OF SCHEDULING ALGORITHMS IN WIRELESS NETWORKS

A variety of link scheduling algorithms under different interference models have been studied in order to achieve high performance and low complexity. In [26], the main model used in wireless networks is the SINR model or the physical interference model. The authors describe the model and analyze the complexity of scheduling wireless links with analog network coding capability. The authors study two models with different definitions of network coding. In the first model, they assume that a receiver is able to decode several signals simultaneously. In the second model, they assume that in a two-way relay channel, routers are able to forward the interfering signal of a pair of nodes that wish to communicate together, and nodes are able to decode the "collided" message. An instance of the scheduling problem is constructed in the geometric SINR model for each model. The nodes are distributed in the Euclidean plane. The authors present NP-hardness proofs for both scenarios and propose a scheduling algorithm that explores analog network coding opportunities to achieve superior throughput capacity.

The design of a distributed low-complexity scheduling algorithm becomes even more challenging when taking into account a physical interference model. The work proposed in [27] addresses the problem of scheduling in a graph-based interference models. They propose a scheduling algorithm under SINR-based interference model. The authors assume that some scheduling schemes have been developed assuming "theoretical" graph-based interference models, which only consider a binary interference relation between each pair of links. Moreover, this cannot capture accumulative nature of wireless interference from multiple transmitters. Hence, they propose to take into account the SINR among all activated links.

Such works model the scheduling algorithms by a Markov decision process, as presented in [28] where the authors propose both a mathematical model and a scheduling algorithm for emerging new services in wireless networks. The proposed scheduling scheme maximizes the expected revenue for the service provider. The model takes into account the revenue, penalty, and time sensitivity of data to be downloaded to the customer's end device. To guarantee the quality of service and enhance system's data rate, the Radio Resource Management (RRM) scheduling mechanisms play a very crucial role to guarantee the QoS performance for different services. In [29], the authors model and evaluate the performance of Round Robin, Proportional Fairness and Max Rate scheduling algorithms for downlink LTE cellular Network.

6 PERFORMANCE EVALUATION OF RRM SCHEMES

6.1 PERFORMANCE METRICS

The common performance metrics used to evaluate the CAC and scheduling schemes are: the blocking ratio, packet loss ratio and average packet waiting time.

6.1.1 Blocking ratio

This is defined as the probability that the call cannot be accepted. We can identify two types of blocking ratio:

- Handoff call dropping probability (HCDP): HCDP is defined as the fraction of handoff attempts that is denied access because of lack of resources.
- New call blocking probability (NCBP): NCBP is defined as the fraction of new calls blocked because of lack of resources.

6.1.2 Packet loss ratio

The packet loss ratio represents the ratio of the number of lost packets to the total number of sent packets. Each packet has a deadline before which it must be executed, and if this is not possible, the scheduler tries to minimize the number of lost packets due to deadline expiry. Guarantees on meeting these timing constraints and how the system handles those packets that cannot meet their deadline are the objectives of the scheduling algorithm. The fraction of packets dropped, due to deadline violation, is used to evaluate the loss performance of a scheduling scheme. The fraction of the dropped packets, due to deadline violation, is used to evaluate the loss performance of a scheduling schemc. Whenever this fraction is small, the scheduler is suitable for scheduling real-time traffic.

6.1.3 Average packet waiting time

The average packet waiting time represents the ratio of the total waiting time of packets to the total number of packets. The control of delays is often of vital importance for real-time applications such as audio and video streaming. Each packet has an expiry time beyond which the packet is of no use to the end user. The objective of the scheduler is to transmit each packet before its expiry. Expiry occurs when a packet has been waiting in the base station queue for a time greater than its deadline without being served. Such a packet is dropped by the system.

6.1.4 Resource utilization ratio

This is the ratio of the number of allocated resources for user equipment during the whole simulation time.

6.2 SIMULATION RESULTS

6.2.1 Simulation of CAC schemes

In the simulation results, we should verify whether the CAC approach can guarantee the QoS requirements. The performance evaluation metrics considered here are: the blocking ratio, the packet loss ratio and the resource utilization ratio. In this section, the performance of the CAC schemes is evaluated via simulation, using as an example a heterogeneous wireless network; two RATs (LTE and WLAN) supporting two classes of calls: namely real-time and non-real-time traffic. Figure 27.6 and Figure 27.7 illustrate the blocking probability and the packet loss ratio, respectively, for a proposed CAC scheme [8]. This work involves call admission control and a handoff scheme to maximize system reward for network providers and to guarantee QoS-aware seamless handoff for mobile users.

Figure 27.7 presents the packet loss ratio as a function of number of users. The packet loss ratio in the case of NRT applications is more important than the RT applications. This can be justified by the fact that the CAC algorithm serves the RT packets in priority. Consequently, serving RT packets with the highest priority minimizes the number of packets dropped due to deadline expiry.

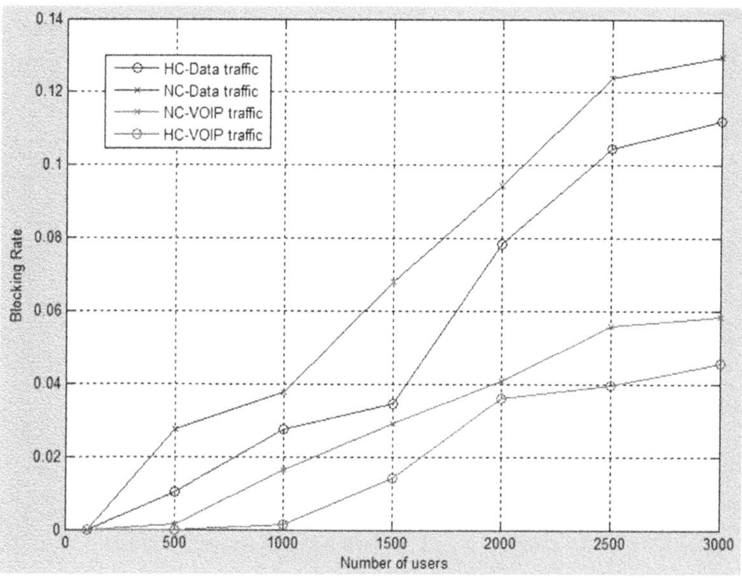

FIGURE 27.6

Handoff and new calls blocking probability vs. number of users.

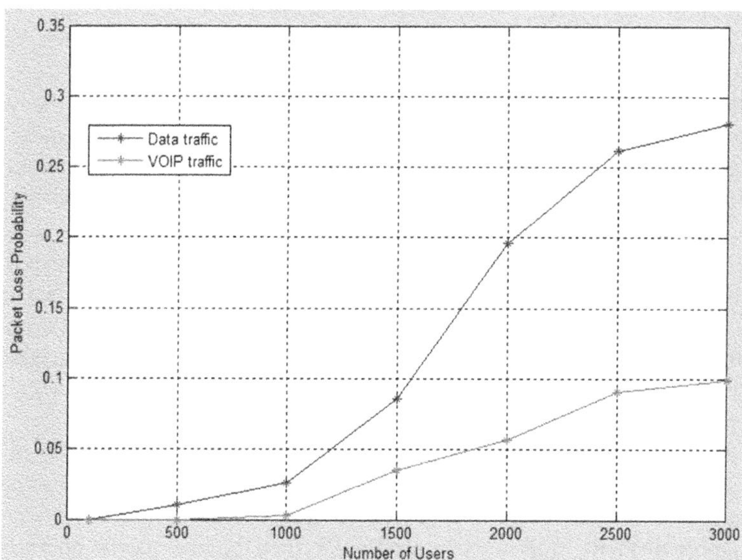

FIGURE 27.7

Packet loss ratio vs. number of users.

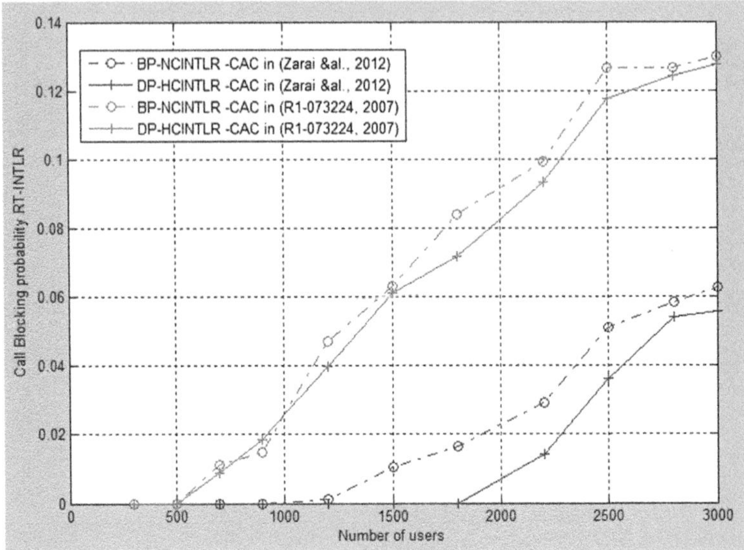

FIGURE 27.8

Blocking probability for real-time intolerant vs. number of users.

Figures 27.8, 27.9 and 27.10 represent a comparative study between two different proposed CAC algorithms: a PS CAC in [7] and a NPS CAC in [30]. The objective of these mechanisms is to improve resource utilization while keeping blocking and dropping probability as low as possible.

In the first scheme [7], the authors propose new methods to reduce the handoff blocking probability in the 3GPP LTE wireless networks. This reduction is based on an adaptive call admission control scheme that provides QoS guarantees and gives the priority of handoff calls (HC) over new calls (NC) in admission by applying a resource allocation policy that takes into account the separation between incoming traffic for each class of service and prioritizes handoff calls over new calls. They address an adaptive resource allocation mechanism that gives a threshold resource block capacity for each service's classes and accommodates an arrival call to an overloaded cell. A degradation procedure is triggered to degrade the allocated resource of some ongoing calls in the cell. The degradation rate depends on the type of arrival call (HC or NC).

In the second scheme [30], if the sum of the required number of physical resource blocks (PRBs) per time transmission interval by a new user requesting admission and existing users is less than or equal to the threshold which is the total number of PRBs in the system bandwidth, then the new user is admitted.

Figure 27.8 and Figure 27.9 show the blocking probability for the intolerant real-time service and the blocking probability of the tolerant real-time service, versus number of users, respectively. We note that the application of the first

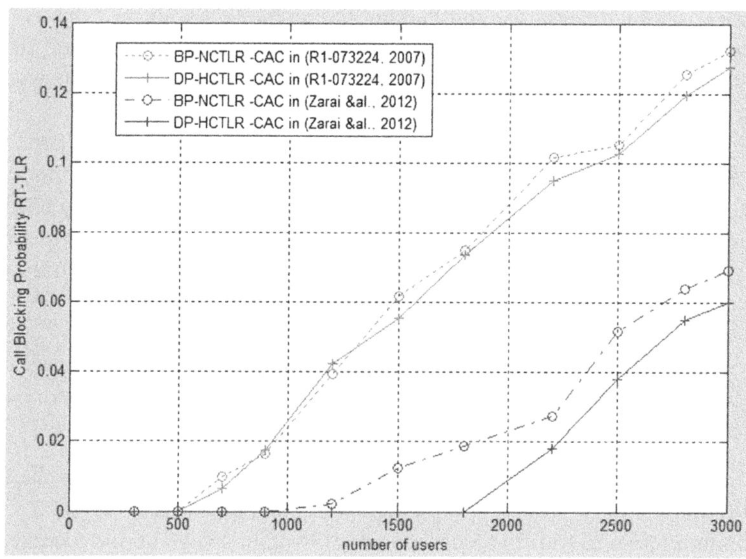

FIGURE 27.9

Blocking probability for real-time tolerant vs. number of users.

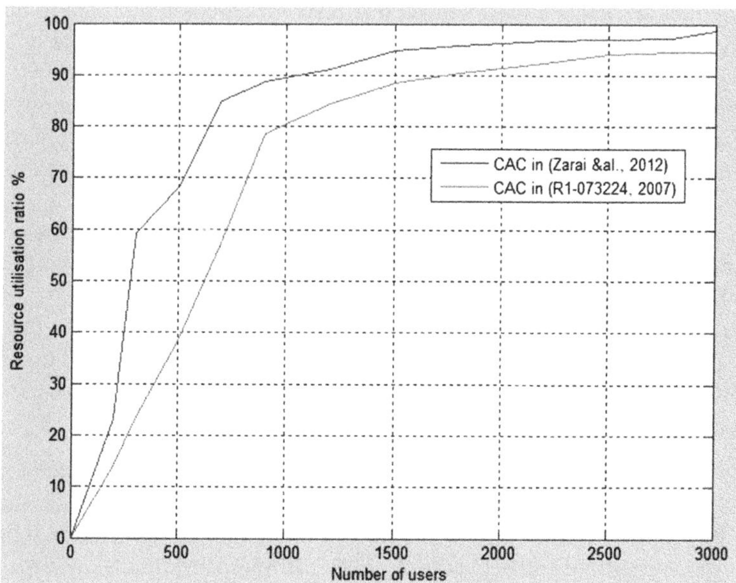

FIGURE 27.10

Resource utilization ratio vs. number of users.

CAC [7] gives a light increase in blocking rate compared with the other solution. One reason for this difference is that [7] gives the priority of handoff call over new call in admission decision and it introduces the priority notion between the different classes of service in terms of latency tolerance.

Figure 27.10 shows the result of the physical resource blocks utilization according to the number of user equipments for the two schemes. We notice that the utilization ratio of resource in [7] can reach a higher resource utilization ratio compared to [30]. The Zarai et al. [7] scheme introduces an efficient adaptive which makes it possible to offer customer services by adjusting the attribution of resources intelligently.

6.2.2 Simulation of scheduling schemes

Since the resources in wireless networks are limited, some specific factors should be considered in the scheduling policy as follows [31]:

- *QoS requirements:* An efficient scheduling algorithm could improve the QoS of the different types of service classes. The main parameters are the minimum reserved traffic, the maximum allowable delay and the tolerated jitters.
- *Fairness:* Aside from assuring the QoS requirements, the bandwidth resources should be allocated fairly. Thus, fairness represents one of the most challenging problems in the scheduling approaches.
- *Channel Utilization:* It is the fraction of time used to transmit data packets. A scheduling mechanism has to check that resources are not allocated to the users that do not have enough data to send, thus resulting in wastage of resources.
- *Complexity:* Since the BS has to handle many simultaneous connections and decisions have to be made, the scheduling algorithm must be simple, fast and should not have a prohibitive implementation complexity as it serves different service classes in various constraints.
- *Scalability*: The scheduling algorithm should efficiently operate as the number of connections increases. It represents the capacity to handle a growing number of flows.
- *Cross-layer design*: A scheduling algorithm should take into account the characteristics of different layers (e.g., the adaptive modulation and coding (AMC) scheme). It is significant to consider the burst profile in such scheduling policy in order to improve system performance.

In the simulation results, we should verify whether the scheduler can assure the QoS requirements. An important consideration is how to maximize the total system throughput. The metrics here could be the maximum number of supported MSs or the capacity of the networks. In [32], the authors estimate the capacity of the network. Figure 27.11 and Figure 27.12 show the capacity of two different service classes (Real-time: RT and Non Real-time: NRT, respectively) in heterogeneous wireless networks containing different Radio Access Technologies

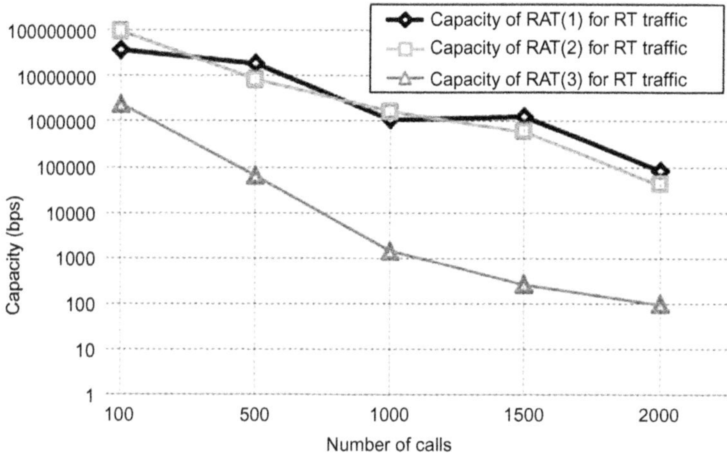

FIGURE 27.11

RATs capacity for RT calls.

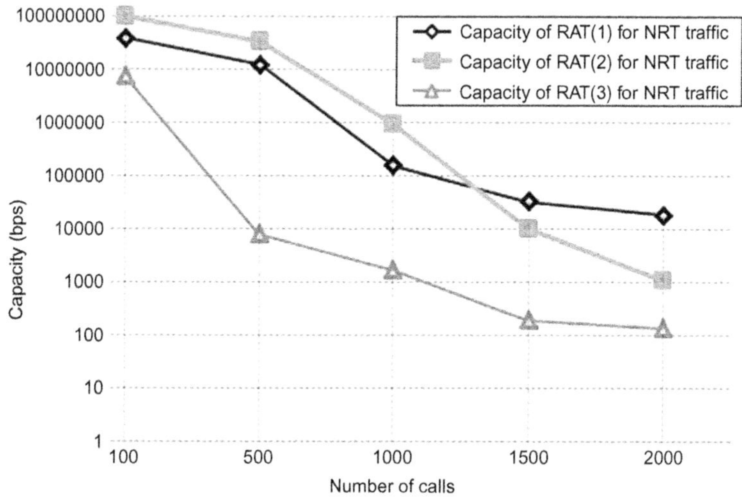

FIGURE 27.12

RATs capacity for NRT calls.

(RATs) where RAT 1, RAT 2 and RAT 3 are Long Term Evolution (LTE), Wireless Mesh Network (WMN) and Wireless Local Area Network (WLAN), respectively, versus number of calls, which varies from 100 to 2000.

The performance evaluation metrics considered here are: the blocking ratio, packet loss ratio and average packet waiting time. In this section, the performance

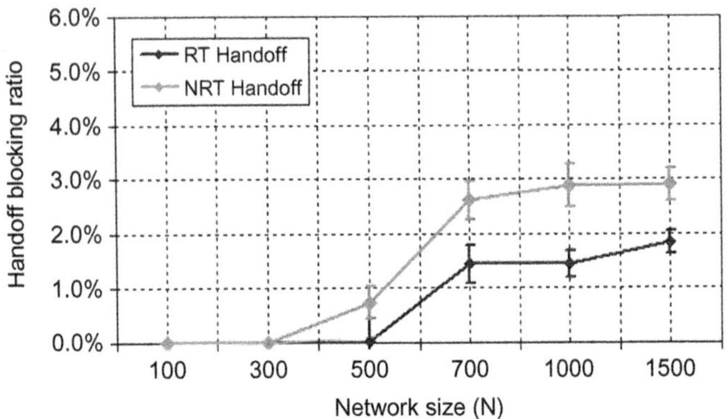

FIGURE 27.13

Handoff calls blocking probability vs. number of users.

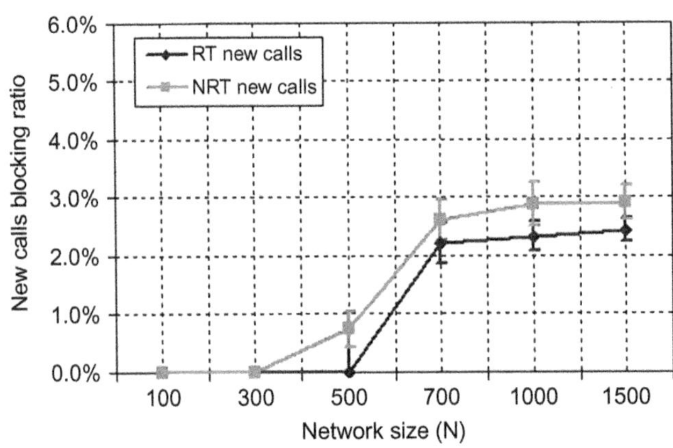

FIGURE 27.14

New calls blocking ration vs. number of users.

of the scheduling schemes is evaluated via simulation, using as an example of Heterogeneous Wireless Networks supporting two classes of calls; namely real-time and non-real-time traffic. Figure 27.13 illustrates the blocking ratio of the handoff calls for real time (RT) and non-real-time (NRT) traffic versus the size of the network.

Another example of blocking probability is presented by Figure 27.14 which shows the new calls blocking ratio with two types of traffic (RT and NRT) versus the network size (number of mobile nodes in the network).

The fraction of packets dropped, due to deadline violation, can be used as a good performance measure of a scheduling scheme. This fraction is required to

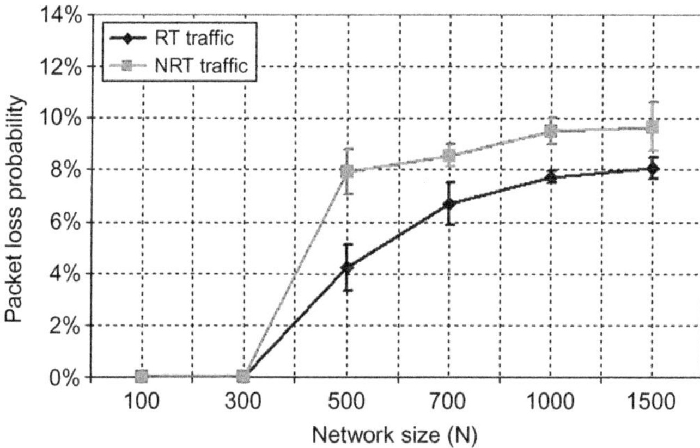

FIGURE 27.15

Packet loss probability vs. number of users.

be small. Figure 27.15 illustrates the packet loss ratio as a function of the network size (number of users in the networks).

The variation of the packet loss probability with different number of mobile nodes carrying various types of traffic is presented in Figure 27.15. As shown in the latter figure, RT traffic experiences less packet loss probability.

7 CONCLUSION

In this chapter, we have provided an extensive survey of crucial elements that provide and enhance QoS on radio resource management. All the relevant proposals such as admission control and scheduling are discussed in detail. We focus on resource management issues as crucial elements to support QoS in wireless Networks. CAC algorithms are important for wireless networks not only for providing the expected QoS requirement to mobile users, but also to maintain network consistency and prevent congestion. Due to the nonpreemptive nature of packet transmission, the problem of achieving fairness and maximizing the overall throughput as a scheduling objective has attracted the attention of many researchers in this domain.

In this chapter, the importance of CAC and scheduling schemes in wireless networks for providing QoS guarantees has been investigated. The key idea of studying the different mechanisms, apart from offering a comprehensive study of CAC and scheduling process in wireless networks, is to focus on the CAC and scheduling methods as a powerful tool to provide the desired QoS level to mobile users along with the greatest network resource exploitation. We present a classification and a description of CAC and scheduling algorithms reported in the literature. Then, we provide a deep analysis of existing models of CAC and scheduling schemes. In order to evaluate the different schemes and to verify whether they can guarantee the QoS requirements, we present some modeling and simulation results at the end of this chapter.

REFERENCES

[1] Pérez-Romero J, Sallent O, Agustí R, Karlsson P, Barbaresi A, Wang L, et al. Common radio resource management: functional models and implementation requirements. IEEE 16th international symposium on personal, indoor and mobile radio communications (PIMRC), Berlin; 2005. p. 2067–71.

[2] Lee JH, Jung TH, Yoon SU, Youm SK, Kang C. An adaptive resource allocation mechanism including fast and reliable handoff in IP-Based 3G wireless networks. IEEE Pers Commun 2002;7(6):42–7.

[3] Gaasvik P.O., Cornefjord M, Svensson V. Different methods of giving priority to handoff traffic in a mobile telephone system with directed retry. IEEE vehicular technology conference gateway to the future technology, 1991. p. 549–53.

[4] Hong D, Rapport SS. Traffic model and performance analysis for cellular mobile radio telephone systems with prioritised and non prioritised handoff procedures. IEEE Trans Veh Technol 1986;35(3):77–92.

[5] Pati HK. A distributed adaptive guard channel reservation scheme for cellular networks. Int J Commun Syst 2007;20(9):1037–58.

[6] Ojesanmi OA, Famutimi RF. Adaptive threshold based channel allocation scheme for multimedia network. Int J Comput Sci Netw Secur (IJCSNS) 2009;9(1):260–5.

[7] Zarai F, Ali KB, Obaidat MS, Kamoun L. Adaptive call admission control in 3GPP LTE networks. Int J Commun Syst 2012;25(12):1–13. Wiley.

[8] Ali KB, Zarai F, Obaidat MS, Kamoun L. QoS scheme based IEEE 802.21 MIH in heterogeneous wireless networks (Hwns). 6th international conference on information technology ICIT, 2013. p. 1–11.

[9] Falowo OE, Chan HA. Joint call admission control algorithm for fair radio resource allocation in heterogeneous wireless networks supporting heterogeneous mobile terminals. consumer communications and networking conference (CCNC), Las Vegas, NV; 2010. p. 1–5.

[10] Almugheid AT, Yousef S, Aboud SR. Performance evaluation of quality metrics for single and multi cell admission control with heterogeneous traffic in WCDMA networks. Int J Eng Technol 2014;4(1):1–11.

[11] Singh D, Singh P. Radio resource scheduling in 3GPP LTE: a review. Int J Eng Trends Technol (IJETT) 2013;4(6):2405–11.

[12] Ramli H, Basukala R, Sandrasegaran K, Patachaianand R. Performance of Well Known Packet Scheduling Algorithms in the Downlink 3GPP LTE System. IEEE Malaysia International Conference on Communications (MICC), Malaysia, Kuala Lumpur; 2009. p. 815–20.

[13] Kaladevi M, Sathiyabama S. A comparative study of scheduling algorithms for real time task. Int J Adv Sci Technol 2010;1(4):8–14.

[14] Mansouri W, Zarai F, Mnif K, Kamoun L. New scheduling algorithm for wireless mesh networks. The 2nd international conference on multimedia computing and systems (ICMCS'11), Morocco; 2011. p. 1–6.

[15] Hussain M, Gregory MA. An efficient, fair, class based, delay jitter controlled packet scheduling algorithm for 4G broadband wireless access systems. international conference on communication systems, Singapore; 2010. p. 326–30.

[16] Talevski D, Gavrilovska L. Novel scheduling algorithms for LTE downlink transmission. Telfor J 2012;4(1):20–5.

[17] Chadchan SM, Akki CB. A fair downlink scheduling algorithm for 3GPP LTE networks. Int J Comput Netw Inform Secur 2013;34–41.

[18] Sathyapriya R, Malathy S. Dynamic channel allocation and revenue based call admission control using genetic algorithm. Int J Adv Res IT Eng 2012;1(5):33–46.

[19] Asuquo DE, Williams EE, Nwachukwu EO, Inyang UG. An intelligent call admission control scheme for quality of service provisioning in a Multi-traffic CDMA network. Int J Sci Eng Res 2013;4(12). p. 1522–161.

[20] Ramesh HSB, Gowrishankar, Satyanarayana PS. A QoS provisioning recurrent neural network based call admission control for beyond 3G networks. Int J Comput Sci Issues 2010;7(5):7–15.

[21] Mahesh G, Yeshwanth S, Manikantan UV. Survey on soft computing based call admission control in wireless networks. Int J Comput Inform Technol 2014;5 (3):3176–80.

[22] Yue M, James HJ, Trivedi KS. Call admission control for reducing dropped calls in code division multiple access (CDMA) cellular systems. INFOCOM 2000. Nineteenth annual joint conference of the IEEE computer and communications societies, Tel Aviv; 2000. p. 1481–90.

[23] Ciardo G, Muppala JK, Trivedi KS. SPNP: stochastic petri net package. proceedings of 3rd international workshop on petri nets and performance models, 1989. p. 142–50.

[24] Chowdhury MZ, Jang YM, Haas ZJ. Call admission control based on adaptive bandwidth allocation for wireless networks. J Commun Netw 2013;15(1):15–24.

[25] Lee J, Kim H. Performance analysis of adaptive QoS handoff mechanism using service degradation and compensation. Int J Softw Eng Appl 2013;7(6):127–36.

[26] Goussevskaia O, Oswald YA, Wattenhofer R. Complexity in geometric SINR. 8th ACM international symposium on mobile Ad Hoc networking and computing, MobiHoc'07, Montreal, Quebec, Canada; 2007. p.100–9.

[27] Ryu J, Joo C, Kwon TT, Shroff NB, Choi Y. DSS: distributed SINR-based scheduling algorithm for multihop wireless networks. IEEE Trans Mob Comput 2012;12 (6):1120–32.

[28] Massey WA, Ramakrishnan KG, Aravamudan M, Pai G. Scheduling algorithms for downlink services in wireless networks: a markov decision process approach. IEEE global telecommunications conference, (GLOBECOM), vol. 6; 2004. p. 4038–42.

[29] Nsiri B, Nasreddine M, Ammar M, Hakimi W, Sofien M. Modeling and performance evaluation of scheduling algorithms for downlink LTE cellular network. The tenth international conference on wireless and mobile communications (ICWMC), Seville, Spain; 2014. p. 60–4.

[30] R1-073224. Way forward on power control of PUSCH. 3GPP TSGRAN WG1 49-bis.; 2007.

[31] Chaari L, Saddoud A, Maaloul R, Kamoun L. A comprehensive survey on WiMAX scheduling approaches. In: Hincapie RC, Sierra JC, editors. Quality of service and resource allocation in WiMAX. Intech; 2012. p. 25–58.

[32] Mansouri W, Mnif K, Zarai F, Obaidat MS, Kamoun L. Capacity analysis of heterogeneous wireless networks under sinr interference constraints, 10th International conference on wireless information networks and systems (WINSYS), Iceland; 2013. p. 233–9.

PART

6

Modeling and simulation for system security

CHAPTER

DoS detection in WSNs

28

Energy-efficient designs and modeling tools for choosing monitoring nodes

Quentin Monnet and Lynda Mokdad

Université Paris-Est, Créteil, France

1 INTRODUCTION

1.1 WIRELESS SENSOR NETWORKS

Smart cities or the *Internet of Things* are foreseen to deeply change people's daily lives. Such projects will interconnect a multitude of devices and bring many functions and services to the end users through an extensive use of sensors. Ambient light, temperature, air pollution degree measurement, or traffic monitoring are just a few examples of applications involving those sensors. There will be sensors everywhere, to gather amounts of data that human beings alone could not measure: sensors deployed as networks can perform constant measuring tasks over wide—and sometimes hard to access—areas.

Such networks are called *wireless sensor networks* (WSNs). The sensors (or *nodes*) are small devices able to gather data on their physical environment. They communicate with one another through radio transmission, but they have low resources at their disposal: limited computing power, limited memory, as well as a limited battery [1,2]. They are often dropped into hostile areas (by helicopter for instance), or may generally be difficult to access, so the batteries must be considered as single-use. The sensors have to self-organize themselves and to deploy low-consuming routing algorithms so as to create a functional network. All relevant data is typically forwarded to an entity called the *base station* (BS), which does not have the same limitations as the sensors, and acts as an interface between the WSN and the user (or the external world) as displayed on Figure 28.1.

Wireless sensor networks may be deployed for all kinds of applications, some of them being crucial. For instance there is a lot at stake when sensor networks are used for watching forest fires. Critical cases also involve all military uses of the sensors: they can be used to detect the presence of biological, chemical or nuclear agents, or to monitor infantry units over battlefields [3]. Such contexts

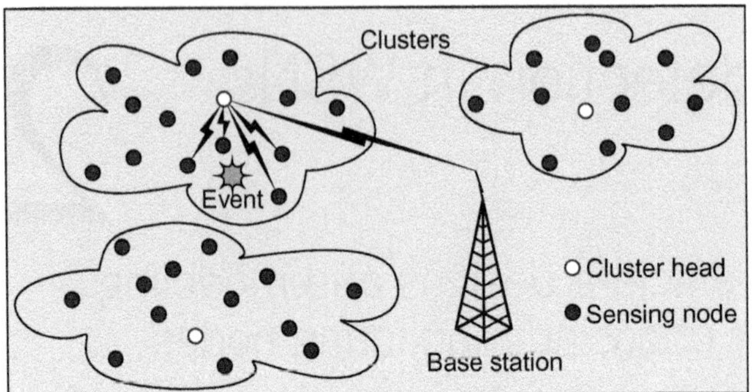

FIGURE 28.1

Clustered wireless sensor networks scheme.

bring strong requirements in terms of security guarantees to the network. Various works deal with ways of preventing unauthorized access to data, or with the necessary precautions to guarantee data authenticity and integrity inside WSNs [4,5]. But confidentiality and authentication both are of little use if the network is not even able to deliver its data correctly.

1.2 DENIAL OF SERVICE IN WSNs

Denial of Service (DoS) attacks indeed aim at reducing, or even annihilating, the network's ability to achieve its ordinary tasks, or trying to prevent a legitimate agent from using a service [6]. Because of the limited resources of their nodes, WSNs tend to be rather vulnerable to DoS attacks. Concrete attacks include jamming the communications, monopolizing the channel ("greedy" attacks) or attempting sleep deprivation on "normal" sensors, for example. They are launched from the outside as well as from the inside of the network: a compromised sensor node can be used in order to send corrupted data at a high rate, either to twist the results or to drain the nodes' energy faster. Attacks can target all layers of the network, although we mainly focus here on the Media Access Control (MAC) and routing layers. The problem we tackle is the development and analysis of detection mechanisms which are efficient both in terms of detection (i.e., they guarantee a high rate of detection of compromised nodes) and in terms of energy (i.e., they guarantee a balanced energy consumption throughout the network).

1.3 CLUSTERED WSNs

One way to save some battery power during communications may reside in the choice of the network architecture and of the protocol used to route data from a sensor to the BS. In a hierarchical WSN, the network is divided into several clusters. The partition is done according to a clustering algorithm such as LEACH [7,8], HEED [9], or one based on ultra-metric properties [10,11]. In each cluster, a single common node is designated and it becomes a cluster head (CH), responsible for directly collecting data from the other nodes in the cluster. Once enough data has been gathered, the CHs proceed to data aggregation [12]. Then they forward their data to the BS. CHs are the only nodes to communicate with the BS, either directly, through a long-range radio transmission, or by multi-hopping through other CHs (see Figure 28.1).

So as to preserve the nodes' energy as long as possible, the network reclustering is repeated periodically, with different nodes being elected as CHs. Note that clustering is not limited to a "single-level" partition. We can also subdivide a cluster into several "subclusters." The CHs from those "subclusters" would then send their aggregated data to the CHs of their parent clusters.

1.4 DoS DETECTION: FROM STATIC TO DYNAMIC GUARDING POLICIES

In a hierarchically organized WSN, a control node (cNode in the remainder of this chapter) is a node that is chosen to analyze the traffic directed to the CH of the cluster it belongs to, and potentially detect any abnormal behavior. Therefore, cNodes provide us with an efficient way to detect DoS attacks occurring in the network. Note that cNodes are only meant to detect DoS attacks, thus they do not perform any sensing, nor do they send any data (apart from attack detection alarms). cNodes-based detection was first presented in [13], but the authors do not mention any periodical (cNodes) re-election scheme. One can suppose that the renewal of the election occurs each time the clustering algorithm is repeated. In [14], we proposed a dynamic approach: cNodes are re-elected periodically (any node in a cluster may be chosen, except the CH) with the election period selected to be shorter than that between two network clusterings. Intuitively such a dynamic approach (in comparison to that of [13]), leads to more uniform energy consumption while preserving good detection ability.

1.5 OUR CONTRIBUTION

We propose a dynamic renewal of the designation process of the cNodes. The process itself can be performed by applying different algorithms: nodes can self-elect as cNodes, or can be designated by a central authority (cluster head or base station) depending on several criteria. Two ways to proceed are presented here. We will first consider a self-election model and address the problem of validating the above conjecture on energy consumption and efficiency by means of modeling techniques. More specifically, this consists in the following aspects:

1. We present a number of numerical results obtained by simulation of DoS detection on WSN models by means of the network simulator ns-2. In particular we simulate models of grid topology WSN including DoS (*static* and *dynamic*) detection policies.

2. We present a characterization of Markov chain models for representing DoS detection mechanisms and detail relevant steady-state measures analytically (i.e., we give the expression for the probability of detection of attacks in the Markov chain model).
3. We present formal models of the DoS detection mechanisms expressed in terms of Generalized Stochastic Petri Nets (GSPN). In combination with GSPN models we also present a number of performance and dependability properties formally expressed in terms of the Hybrid Automata Stochastic Logic (HASL) [15].

We then propose another designation process which is based on energy, in order to obtain an even better load balancing. We propose to designate the sensors for the cNode position according to their residual energy, but we show that several problems occur with deterministic election. Indeed, compromised nodes could see a flaw to exploit in order to take over the cNode role and decrease the odds of being detected by announcing high residual energy. We address this issue by introducing a second role of surveillance: we choose "vNodes" responsible for watching over the cNodes and for matching their announced consumption against the mathematical model. We also recommend that every node in the cluster be monitored by at least one cNode, to prevent all the cNodes from being elected inside the same spatial area of the cluster during each election iteration. Once again, simulation results indicate a better load balancing.

1.6 CHAPTER STRUCTURE

The remainder of this chapter is organized as follows: in Section 2 we give an overview of DoS attack detection for cluster-based WSNs. Section 3 presents the self-election method: it includes network topology and protocols introduction (subsection 3.1) and simulation results (subsection 3.2). In Section 4 we use modeling tools to represent our network under attack: we provide the structure of Markov chains for modeling our WSN (subsection 4.1, as well as the application of statistical model checking performance analysis to Petri net models of attacked networks (subsection 4.2). Section 5 proposes a second way to designate the nodes by using their residual energy (subsection 5.1), coming with associated results from simulated experiments (subsection 5.2). Finally Section 6 permits us to sum up our contribution and to consider future work leads.

2 RELATED WORK

This section is divided into three parts: security in wireless sensor networks, denial of service specific mechanisms, and clustering algorithms and energy preservation.

2.1 SECURITY IN WSNs

Denial of service is not the only type of attack a WSN should resist to. Security in general in sensor networks has attracted quite a lot of interest during the last few years. Hence it has been the subject of many studies in literature, as well as several state-of-the-art articles [16,17].

Confidentiality and integrity must be ensured to prevent attackers access to or tampering with sensitive data. A number of solutions have been proposed [8], many of them involving strong [4] and/or homomorphic [18] cryptography, some relying on other mechanisms such as multi-path based fragmentation of the packets [19] or game theory [20].

Authentication brings to participants the guarantee that the peer they are communicating with truly is what it pretends to be; that is another important point. It has been deeply investigated as well [21]. Many lightweight proposals for key management in WSNs have been suggested [22,23].

Apart from those, there have been a variety of proposals to secure other elements, on a basis than any information about any aspect of the network might be valuable to an attacker. Hence there are approaches, for instance, to secure the geographical location of the nodes through epidemical information dissemination [24] as well as through more conventional mechanisms [25].

2.2 DoS-SPECIFIC MECHANISMS

Denial-of-service attacks embrace many different attacks, which can target all layers of the network [26]. Jamming the radio frequencies as well as disturbing the routing protocols are just two examples of ways to harm the network. In reaction to these, a number of solutions have been proposed [27]. As stated in the introduction, we focus in this paper on inside attackers attempting to bend the MAC protocol parameters to their needs, be it to achieve better performances for themselves (greedy attacks) or to generally harm the network (jamming attacks or sleep deprivation). To detect such attackers, many solutions rely on trust models [28,29] with agents applying a set of rules [30] on traffic to attribute a trust value to each of the nodes in the network. Below are outlined some notable proposals.

Back in 2001, most work focused on making WSNs feasible and useful. But some people already involved themselves into security. For instance, SPINS (Security Protocols for Sensor Networks) was proposed in [31] to provide networks with two symmetric key-based security building blocks. The first block, called SNEP (Secure Network Encryption Protocol), provides data confidentiality, two-party data authentication and data freshness. The second block, called μTESLA ("micro" version of the Timed, Efficient, Streaming, Loss-tolerant Authentication Protocol) assumes authenticated broadcast using one-way key chains constructed with secure hash functions. No mechanism was put forward to detect DoS attacks.

The best way to detect for sure a DoS attack in a WSN is simply to run a detection mechanism on each single sensor. Of course, this solution is not feasible in a network with constraints. Instead of fitting out each sensor with such mechanism, it is proposed in [32] to resort to heuristics in order to set a few nodes equipped with detection systems at critical spots in the network topology. This optimized placement enables distributed detection of DoS attacks as well as reducing costs and processing overheads, since the number of required detectors is minimized. But those few selected nodes are likely to run out of battery power much faster than normal nodes.

Some works examine the possibility of detecting the compromising of nodes as soon as an opponent physically withdraws them from the network. In the method that is developed in [33], each node keeps a watch on the presence of its neighbors. The Sequential Probability Radio Test (SPRT) is used to determine a dynamic time threshold. When a node appears to be missing for a period longer than this threshold, it is considered to be dead or captured by an attacker. If this node is later redeployed in the network, it will immediately be considered as compromised without having a chance to be harmful. Nothing is done, however, if an attacker manages to compromise the node without extracting the sensor from its environment.

In [34], a revised version of the OLSR protocol is proposed. This routing protocol called DLSR aims at detecting distributed denial of service (DDoS) attacks and at dropping malicious requests before they can saturate a server's capacity to answer. To that end, the authors introduce two alert thresholds regarding this server's service capacity. The authors also use Learning Automata (LAs), automatic systems whose choice of next action depends on the result of its previous action. There is no indication in their work about the overhead or the energy load resulting from the use of the DLSR protocol.

A novel broadcast authentication mechanism can also be deployed so as to cope with DoS attacks in sensor networks such as in [35]. This scheme uses an asymmetric distribution of keys between sensor nodes and the BS, and uses the Bloom filter as an authenticator, which efficiently compresses multiple authentication information. In this model, the BS or sink shares symmetric keys with each sensor node, and proves its knowledge of the information through multiple MAC values in its flooding messages. When the sink floods the network with control messages it constructs a Bloom filter as an authenticator for the message. When a sensor node receives a flooded control message, it generates their Bloom filter with its keys and in the same way the sink verifies message authentication.

Much of our work relies on the work of Lai and Chen who proposed in [13] a system detection based on static election of a set of nodes called "guarding nodes" which analyze traffic in a clustered network. When detecting abnormal traffic from a given node, "guarding nodes"—we call them cNodes—identify it as a compromised node and inform the cluster head of it. On reception of reports from several distinct cNodes (to prevent false denunciation from a compromised node), the CH virtually excludes the suspicious node from the cluster. The authors show the benefit of their method by presenting numerical analysis of detection rate. Although the method is efficient for detecting rogue nodes, the authors do not give details of the election mechanism for choosing the cNodes. Also, there is no mention in their study of renewing the election in time, which causes the appointed cNodes to endorse heavier energy consumption on a long period.

2.3 CLUSTERING ALGORITHMS AND ENERGY PRESERVATION

A lot of approaches intended to bring security into a WSN are cluster-based [36]. But the main purpose of clustering a sensor network usually resides in scaling possibilities, improved nodes management and energy savings brought by partitioning. Several clustering algorithms have been proposed [37]. They generally aim at determining which nodes in the network will be the cluster heads, often basing the choice on energetic considerations. Basically, choosing a cluster head in a network is not so different than selecting cNodes in a cluster. But in the latter case we have some additional constraints on security.

One of the easiest clustering algorithms to implement, and probably one of the most used, is the LEACH algorithm [38].

2.3.1 LEACH functioning

LEACH is likely one of the easiest algorithms to apply to recluster the network. It is a dynamical clustering and routing algorithm. It splits a set of nodes into several subsets, each containing a cluster head. This CH is the only node to assume the cost-expensive transmissions to the BS.

Here is the LEACH detailed processing. Let P be the average percentage of clusters we want to get from our network at an instant t. LEACH is composed of cycles made of $\frac{1}{P}$ rounds. Each round r is organized as follows:

1. Each node i computes the threshold $T(i)$:

$$T(i) = \begin{cases} \dfrac{P}{1 - P \cdot \left(r \bmod \dfrac{1}{P}\right)}, & \text{if } i \text{ has already been CH yet} \\ 0, & \text{if } i \text{ has not been CH} \end{cases}$$

   Each node chooses a pseudo-random number $0 \leq x_i \leq 1$. If $x_i \leq T(i)$ then i designates itself as a CH for the current round. $T(i)$ is computed in such a way that every node becomes CH once in every cycle of $\frac{1}{P}$ rounds: we have $T(i) = 1$ when $r = \frac{1}{P} - 1$.
2. The self-designed CH inform the other nodes by broadcasting a message with the same transmitting power, using carrier sense multiple access (CSMA) MAC.
3. The other nodes choose to join the cluster associated to the CH whose signal they receive with most power. They message back the CH to inform it (with the CSMA MAC protocol again).

4. CHs compile a "transmission order" (time division multiple access, TDMA) for the nodes which joined their clusters. They inform each node at what time it is expected to send data to its CH.
5. CHs keep listening for the results. Normal sensors get measures from their environment and send their data. When it is not their turn to send, they stay in sleep mode to save energy. Collisions between the transmissions of the nodes from different clusters are limited thanks to the use of code division multiple access (CDMA) protocol.
6. CHs aggregate, and possibly compress the gathered data and send it to the BS in a single transmission. This transmission may be direct, or multi-hopped if relayed by other CHs.
7. Steps 5 and 6 are repeated until the round ends.

It is possible to extend LEACH by adding the remaining energy of the nodes as a supplementary parameter for the computation of the $T(i)$ threshold.

Note that each node decides whether to self-designate itself as a CH or not. Its decision does not take into account the behavior of surrounding nodes. For this reason, we can possibly have, for a given round, a number of CHs very different from the selected percentage P. Also, all the elected CHs may be located in the same region of the network, leaving "uncovered" areas. In that case, one can only hope that the spatial repartition will be better during the next round.

2.3.2 LEACH improvements

There are a number of proposals derived from LEACH, to improve either its efficiency [39,40] or its security. In [41], the authors propose to add security mechanisms via a revised version of LEACH protocol. SecLEACH uses random key predistribution as well as μTESLA (authenticated broadcast) so as to protect communications. But the authors do not mention any mechanism to fight DoS attacks.

In [42], the authors propose another way to secure the LEACH protocol against selfish behaviors, using elements from game theory. With S-LEACH, the BS uses a global Intrusion Detection System (IDS) while LEACH CHs implement local IDSs. The interactions between nodes are modeled as a Bayesian game, that is, a game in which at least one player (here, the BS) has incomplete information about the other player(s) (in this case: whether the sensors have been compromised or not). Each node has a "reputation" score. Selfish nodes can cooperate (so as to avoid detection) or drop packets. The authors show that this game has two Bayesian Nash equilibriums which provide a way to detect selfish nodes, or to force them to cooperate to avoid detection.

2.3.3 Other algorithms

Other possible clustering algorithms include HEED [9], which is designed to save more energy than standard LEACH, and could lead to a better spatial repartition of the CHs inside the network. But in our network, all the sensors have the same initial available energy, and every one of them is able to directly reach the BS if need be. Under those assumptions, LEACH may not consume more energy than HEED protocol, and remains easier to use.

Note that, aside from clustering, the importance of energy issues in WSNs has led to proposals of several mechanisms to cut down its consumption [43], based for example on packet priority [44].

3 PSEUDO-RANDOM SELF-ELECTION OF THE cNODES

3.1 DETECTION OF DoS ATTACKS

3.1.1 Wireless sensor networks

We focus on the problem of detecting denial of service (DoS) attacks in a WSN. We recall that a WSN consists of a finite set of sensors plus a fixed base station (BS). Traffic in a WSN (mainly) flows from sensor nodes towards the BS. Furthermore since WSN nodes have inherently little energy, memory and computing capabilities, energy efficiency is paramount when it comes with mechanisms/protocols for WSN management. Also communications between sensors and the BS rely on wireless protocols. In the following we assume that the nodes' mobility is limited or null.

Our goal is to set an efficient method to detect compromised nodes which may try to corrupt data, or to saturate the network's capacity, by sending more data than it should. In this case, efficiency can be measured in two respects:

- the detection rate of the compromised node(s);
- the network's lifetime, as we want to spend as little energy as possible.

In order to achieve these goals we focus on the following techniques: hierarchical network clustering, and dynamical election of control nodes responsible for monitoring the traffic.

3.1.2 Hierarchical clustering

The class of WSNs we consider is that of hierarchically cluster-based networks. The set of sensors has been partitioned into several subsets called "clusters." Those clusters are themselves split into subclusters. For more clarity, we will call 1-clusters the sets resulting from the first clustering of the global set, and k-clusters the subset issued from the splitting of any $(k-1)$-cluster. The successive clusterings are carried out with the use of any existing clustering algorithm, such as LEACH [7,8] or HEED [9] algorithms based on ultra-metric properties [12], etc. Each cluster contains a single cluster head (CH), designated among the normal nodes. The CH is responsible for collecting data from the other nodes of the subset. To follow up our naming conventions, we will call k-CHs the CHs belonging to the k-clusters. The k-CHs send the data they gathered to their $(k-1)$-CH, the "0-CH" being the base station. In that way, the k-CHs are the only nodes to send packets to the $(k-1)$-CHs. Normal nodes' transmissions do not have to

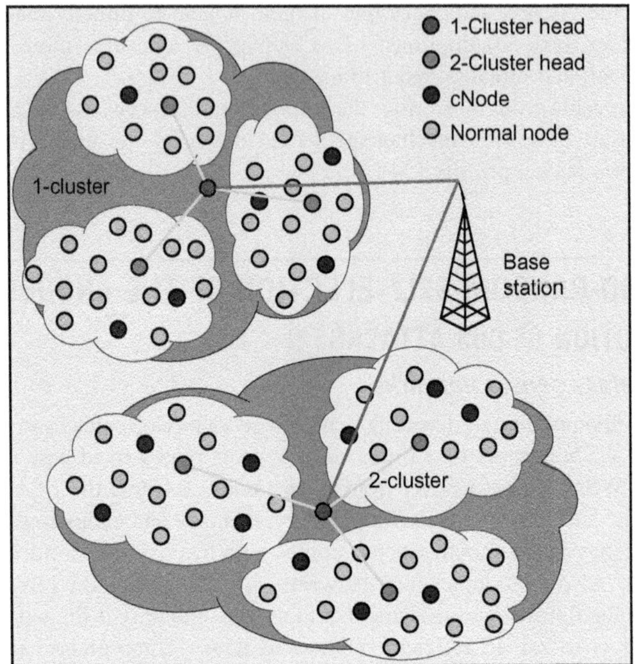

FIGURE 28.2

Scheme of a twice clustered WSN.

reach the base station directly, which would often consume much more energy than communicating with a neighbor node. An example of a 2-clustered network is displayed in Figure 28.2.

3.1.2.1 k-LEACH

Once a clustering algorithm has been applied to the network to determine a first set of clusters, nothing prevents us from applying it again on each cluster. This is the way we got our k-clusters: we applied the LEACH algorithm k times recursively. We call those recursive iterations the k-LEACH algorithm. In practice, we had k equal to 2, for the following reasons:

- so as to save more energy than what we would do with 1-LEACH;
- so as to have a finer clustering of the network, in order to elect control nodes in each of the 2-clusters, to maximize the cover area and the probability of detecting compromised nodes.

3.1.3 Attacks detection through cNodes

Along with normal nodes and cluster heads, a third type of node is present in the lower k clusters of the hierarchy (see also Figure 28.3).

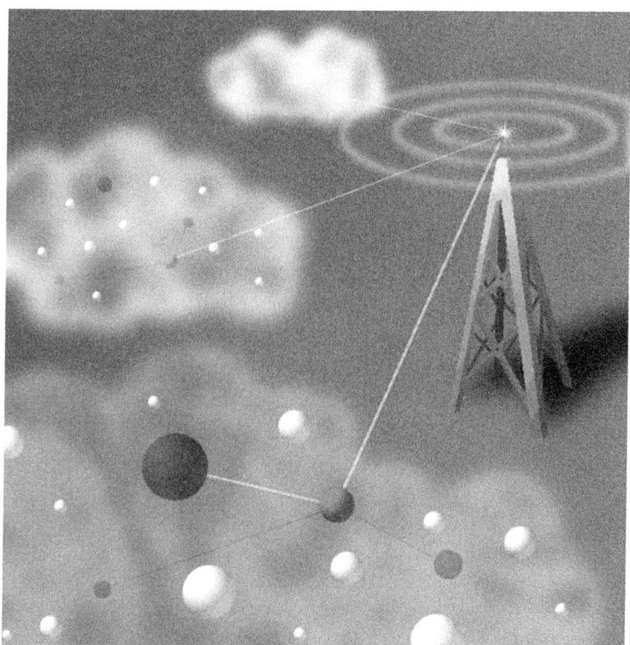

FIGURE 28.3

Cluster-based sensor network with cNodes.

The cNodes—for control nodes—were introduced in [13] to analyze the network traffic and to detect any abnormal behavior from other nodes in the cluster.

Control nodes watching over the input traffic allow the detection of various types of denial of service attacks. This is achieved with agents running on the cNodes and applying specific rules on overheard traffic [29]. Each rule is used to fight against one (sometimes a few) specific attack(s): jamming, tampering, black hole attacks, and so on. Each time a cNode notices that a rule is broken by a node, it raises a bad behavior for this node, and sends an alert to the cluster head. Following are some example rules:

- Rate rule: assuming that minimal and maximal rates for data each node sends are enforced, a bad behavior will be reported if those rates are not respected. With this rule, monitoring agents should be able to detect negligence (if minimal rate is not reached) or flooding (if maximal rate is exceeded) attacks.
- Retransmission rule: a cNode overhearing a packet supposed to be retransmitted by one of its neighbor (the neighbor node is not the final destination for this packet). If the concerned neighbor does not forward the packet, it may be undertaking a black-hole (full dismissal of packets) or a selective forwarding attack.

- Integrity rule: a bad behavior will be raised if a neighbor of the node running the monitoring agent tampers with a packet before forwarding it. Applying this rule assumes that the nodes are not expected to proceed either to data aggregation or compression before forwarding.
- Delay rule: forwarding a packet should not exceed a threshold delay.
- Replay rule: a message should be sent no more than a limited number of times.
- Jamming rule: an unusually high number of collisions (compared to average, or concerning only some nodes) may be related to the presence of a jamming node. If jamming is done with random noise, without legitimate packets containing a node identifier, it may be difficult to detect the source of it, but several cooperating agents should be able to detect it.
- Radio transmission range rule: a node sending messages with an unexpectedly high power may be trying to launch a hello flood (it tries to appear in the neighbor list of as many nodes as possible) or wormhole attack (it redirects a part of the overheard traffic to another part of the network). Hence it may be considered as a bad behavior.

In the rest of this study, we will not describe in detail each one of the mentioned attacks, nor will we detail the associated solutions to counter them. When details are needed, we will consider only one example: flooding attacks. The model of a flooding attack is the following: a malicious node sends a high amount of data to prevent legitimate nodes from communicating by saturating the medium, or by establishing too many connections with the receiver node [45]. In wireless sensor networks, it is also used to drain the energy of neighbor nodes.

So cNodes analyze the input traffic for the 2-CH of their 2-cluster, and watch out for abnormal traffic flows. Detection takes place whenever a rule is broken. In that case the cNode sends a warning message to the CH. In order to prevent a compromised cNode to declare legitimate nodes as compromised the detection protocol requires that the CH receives warnings by a minimum number of distinct cNodes before actually recognizing the signaling as an actual anomaly. Once the CH has received warnings from a sufficiently large number of distinct cNodes it starts ignoring the packets coming from the detected compromised sensor. cNodes may also monitor output traffic of the CHs and warn the BS if they come to detect a compromised CH.

cNodes are periodically elected among normal sensors. The guarding functionality of cNodes may lead to energy consumption higher than that of "normal" (i.e., sensing) nodes. In order to maximize the repartition of the energy load, we propose a scheme by which a new set of cNodes is periodically established with an election period shorter than the length of a LEACH round (that is, the period between two consecutive CH elections). We propose three possible methods for the election process: self-election as for the CHs, election processed by the CHs, and election processed by the BS.

3.1.3.1 Distributed self-election

A first possibility to elect the cNodes is to reuse the distributed self-designation algorithm defined for the election of the CHs. With this method, each non-CH node chooses a pseudo-random number between 0 and 1. If this number is lower than the average percentage of cNodes in the network that was fixed by the user, then the node designates itself as a cNode. Otherwise, it remains a normal sensor.

This method has two drawbacks. First, each node has to compute a pseudo-random number, which may not be necessary with other methods. Second, each node chooses to designate (or not) itself, without taking into account at any moment the behavior of its neighbors. As a result, the election proceeds with no consideration for the clustering that has been realized in the network. Indeed it is unlikely that the set of elected cNodes will be uniformly distributed among the 2-clusters that were formed, and it is even possible to end up with some 2-clusters containing no cNodes (thus being completely unprotected against attacks).

A possible workaround for this second drawback could be a two-step election: in a first round nodes self-designate (or not) themselves. Then they signal their state to the 2-CHs they are associated to. In the second round, the 2-CHs may decide to designate some additional cNodes if the current number of elected nodes in the cluster is below a minimal percentage.

3.1.3.2 CH-centralized election

A second possibility is to get the cNodes elected by the 2-CHs. In this way, each 2-CH elects the required number of cNodes (i.e., corresponding to user specifications). For example, if the 2-cluster contains 100 nodes and the desired percentage of cNodes in the network is 10%, the 2-CH will compute 10 pseudo-random numbers and associate them with node IDs corresponding with sensors of its 2-cluster. This solution is computationally less demanding as only the 2-CHs have to run a pseudo-random number generation algorithm. However, it has yet another drawback: if a CH gets compromised, it won't be able to elect any cNode in its cluster, thus leaving the cluster open to attacks. As with the LEACH protocol, every sensor node becomes, sooner or later, a CH; the problem may occur for any compromised node hence propagating, potentially, throughout the network. Note that nothing prevents a compromised sensor to declare itself as a CH node to the others at any round of the LEACH algorithm.

This method is the one that we have implemented in our ns-2 simulation whose outcomes will be discussed in subsection 3.2. It is also the method we consider in all of Section 4 for modeling.

3.1.3.3 BS-centralized election

A third method consists of a centralized approach where the BS performs cNodes election. With this method CHs send the list of nodes that compose their clusters to the base station and the BS returns the list of elected cNodes. Observe that, opposite to sensor nodes, the BS has no limitation in memory, computing capacity or energy. Thus the clear advantage of BS-centralized election is that all costly operations (i.e., pseudo-random numbers calculation) can be reiterated in a (virtually) unconstrained environment (i.e., the BS). This technique is explained in detail in [14].

From a robustness point of view, note that this method is not completely safe either. In fact, if a compromised node was to declare itself as a CH, its escape method to avoid detection would consist of declaring its cluster as empty (i.e., by sending an empty list instead of the actual sensors in its cluster to the BS). In this case, the BS would not elect any cNode in its cluster, hencc the compromised CH would not be detected. To avoid such a situation, the BS should react differently in case it receives an indication of empty cluster from some nodes. Specifically, in this case, the BS would have to consider that nodes not detected as or by CHs might not simply be dead, and thus still consider them as eligible cNodes. The main drawback of this method is that the distributed nature of election (together with its advantages) is completely lost.

3.1.4 Dynamical selection process

The dynamical renewing of the selection process is an essential part of our proposal. Many of the recent intrusion detection systems proposed for WSNs tend to be lightweight, to consume little energy. We believe that a dynamical renewing of the selected cNodes helps a lot to balance the load inside the cluster. Depending on the application running in the network, maybe this balancing is not worth the constraints induced by periodical re-election, but generally energy preservation is a priority in WSNs and distributing the consumption among all the nodes helps to maintain the highest possible amount of nodes in activity for as long as possible. Also, lightweight IDSs themselves may be designed to minimize the disparities in energy consumption inside a network, but we argue that with a system as simple as the cNodes, simulation results indicate that the savings are not negligible.

A second thing to consider is that the recursive clustering as well as the dynamic renewing of the monitoring nodes can be used with other detection systems than the cNodes we use here. If an IDS is good at preserving energy and balancing nodes, but needs to be run only by a subset of the nodes in the network, dynamical selection processes presented in this work can be applied so as to select the sensors which will run the system (provided the monitoring sensors do not need any specific hardware that would differentiate them from the "normal" nodes).

3.2 NUMERICAL RESULTS

In an attempt to validate the efficiency of the proposed method, we have developed an implementation of an example WSN by means of an existing simulative framework, the ns-2 Network Simulator. In this subsection, we present a selection of numerical results obtained by simulation of ns-2 models of WSN systems equipped with DoS detection mechanisms. The experiments we present are

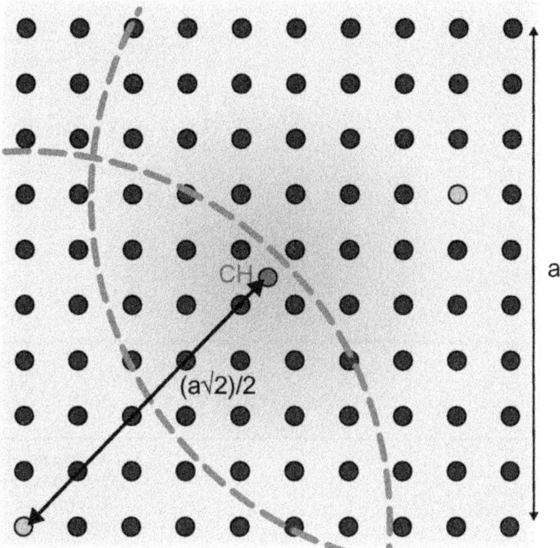

FIGURE 28.4

A 10 × 10 regular-grid cluster of size *a*.

referred to one cluster consisting of a (10 × 10) regular grid topology with the following characteristics (see Figure 28.4):

- grid is a square of size a;
- cluster head is placed at the center of the grid (i.e. red node in Figure 28.4);
- the grid contains 100 (sensing) nodes regularly displaced;
- each node can communicate directly with the cluster head (i.e., the transmission power is such that all nodes—for example: the nodes in green in Figure 28.4—can reach a circle of radius $a\sqrt{2}/2$. In this way all nodes, included corners, can reach the CH). No power adjustment is done by the nodes for transmission.

In such network cNodes (represented in green in Figure 28.4) are elected periodically either using the static approach or using the dynamic election mechanism described in previous subsections. We have designed our experiments focusing on two performance measures: the rate of detection of attacks and the overall energy consumption. Table 28.1 reports about the (range of) parameters considered in our simulation experiments.

3.2.1 Detection rate

In order to evaluate the considered performance measure, namely attack detection rate, we have considered the parameters given in Table 28.1. We have assumed that the traffic generation follows a Poisson distribution with rate λ. This rate is low (10 kbits/s) for normal nodes. Compromised nodes are trying to flood the network; hence they send numerous messages in order to saturate the medium and/or

Table 28.1 Simulation Parameters

Parameter	Value
Simulation time	100–3,600 s
Rate	10–800 kbits/s
Packet size	500–800 bytes
Nodes number	100 (+cluster head)
cNodes number	0–30
Compromised nodes number	1–10
Nodes queue size	50

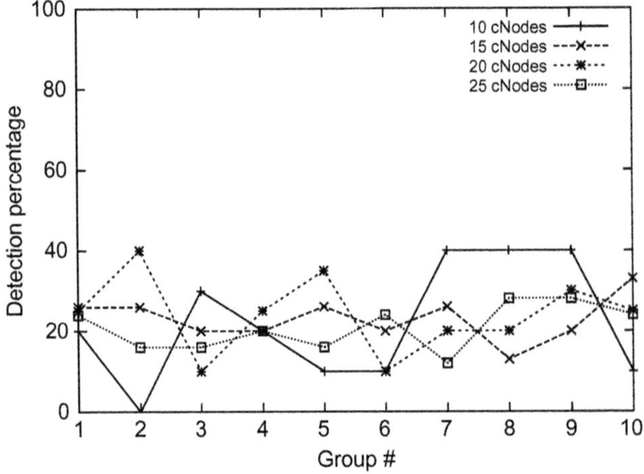

FIGURE 28.5

Detection versus group.

to exhaust the resources of the other nodes. Their transmission rate is much higher, and was set at 800 kbit/s. In the experiments we have considered a cluster with 100 nodes.

Figure 28.5 represents the detection rate for different numbers of cNode groups and for groups of different sizes. The same node is considered compromised in all the graphs. Notice that for 10 cNodes, the group 2 did not detect any attack. With 15 cNodes, in average 3 nodes detect an attack in each group. We also note that when we increase the number of cNodes (20 and 25), the behavior remains similar, which suggests that we do not need to use more nodes than 15 nodes in each group.

Above $\lambda = 4$ packets/s, the dynamic method detects more attacks than the static one. To enhance this difference, we give other results in Figure 28.6 below for an average of 10 compromised nodes.

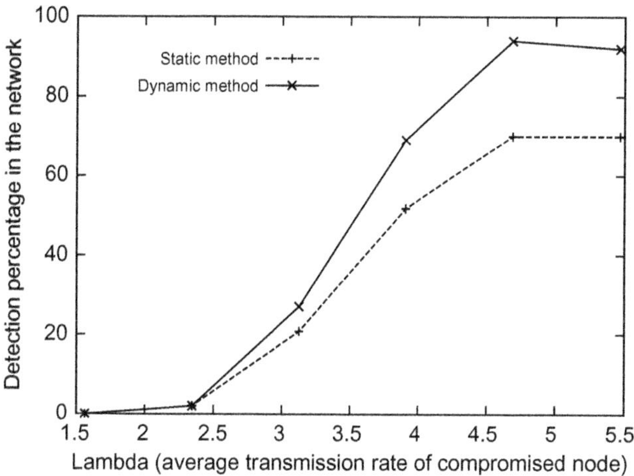

FIGURE 28.6

Detection versus lambda.

Table 28.2 Simulation Parameters

Parameter	Value
Number of sensor nodes	100
Simulation time	500 seconds
Reception consumption	0.394 W
Emission consumption	0.660 W

In Figure 28.6 we notice that, as the average transmission of attacking nodes increases, our dynamic solution detects more attacks than the static solution.

3.2.2 Consumed energy

All the simulations that were run to produce the results presented in this subsection used the parameters given in Table 28.2.

Figure 28.7 shows the average energy consumption for all nodes (except for the cluster head and the flooding compromised node, which consume much more than usual nodes, and act in the same way for both methods) at the end of the simulation, for various percentages of elected cNodes. The number of cNodes goes from 0 (no detection) to 30 % (nearly one third of the nodes).

Note that the "normal nodes" (non-cNodes sensors) do not receive messages from their neighbors, as they are "sleeping" between their sending time slots (see LEACH detailed functioning).

The average consumption is the same for the static and dynamic methods: both methods use the same quantity of normal and cNode sensors.

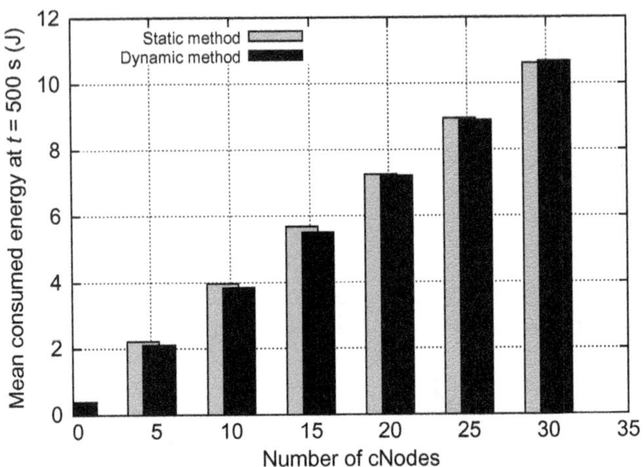

FIGURE 28.7

Average energy consumption.

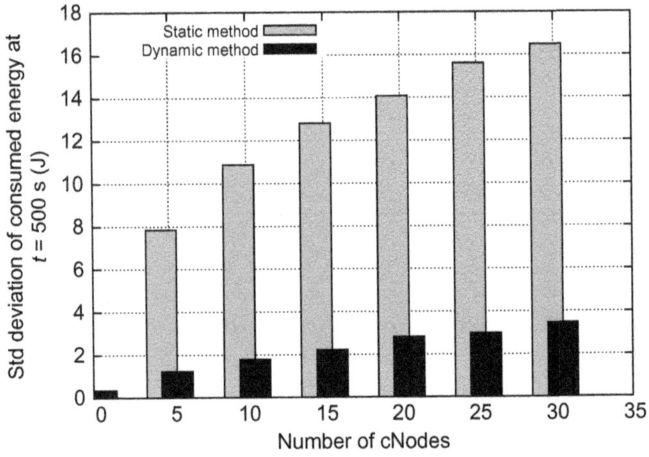

FIGURE 28.8

Energy consumption standard deviation.

Figure 28.8 depicts the standard deviation for the energy consumption at the end of the simulation Once again, the cluster head and the compromised node are not taken into account.

One can observe that the standard deviation is much higher for the static solution: only the initial (and not re-elected) cNodes have a significant consumption over the simulation time, while the consumption is distributed among all the periodically elected nodes in the dynamic solution.

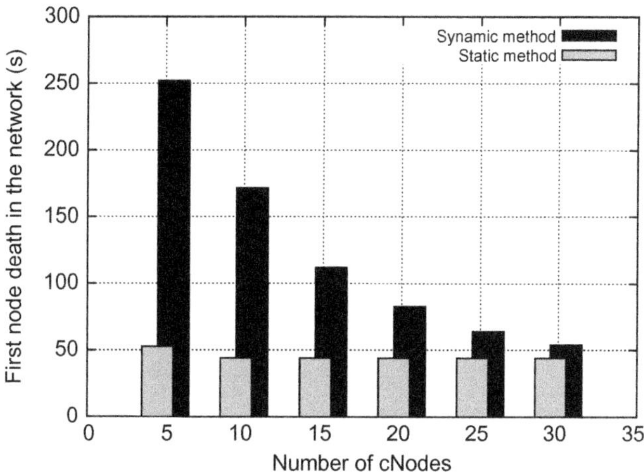

FIGURE 28.9

First death in the network.

Table 28.3 Simulation Parameters

Parameter	Value
Number of sensor nodes	100
cNodes percentage	10%
Simulation time	3,600 seconds
Reception consumption	0.394 W
Emission consumption	0.660 W
Initial energy amount	10 J

For Figure 28.9, we have supposed that the nodes have an initial energy of 4 J. This is a small value, but 500 seconds is a small duration for a sensor lifetime. A lithium battery (CR1225) can offer something like 540 J, and a LR06 battery would provide something like 15,390 J. Note that the compromised node was given an extra initial energy (we did not want it to stop flooding the network during the simulation). However, we set the initial energy to 4 J, and we notice for the first node's death for several percentages of cNodes.

As the cNodes are re-elected and the consumption is distributed for the dynamic method, the first node to run out of battery power logically dies later (up to 5 times later with few cNodes than in the static method).

3.2.3 Node death and DoS detection

The duration of this new simulation was extended to one hour (3,600 seconds). 10 % of the sensors are elected as cNodes. The initial energy power was set to 10 J. So the considered parameters are given in Table 28.3.

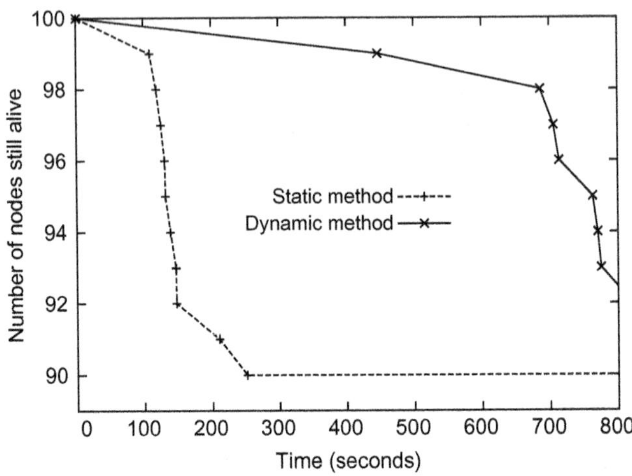

FIGURE 28.10

Nodes remained alive.

Figure 28.10 shows the evolution of the number of alive nodes in time.

As for the previous subsection, the non-cNodes sensors barely consume any energy regarding the cNodes' consumption (cNodes consume each time they analyze a message coming from one of their neighbors; other sensors don't). In the static method, elected cNodes consume their battery power, and die (at about $t = 150$ seconds). That is why the first ten sensors die quickly, whereas the other nodes last much longer (we expect them to live for 5 hours). For the static method, the cNodes are re-elected, so the first node to die lives longer than for the previous method. It is a node that was elected several times, but not necessarily *each* time. Only two nodes have run out of energy at $t = 700$ seconds for the dynamic method. But at this point, the number of alive nodes decreases quickly, and there is only one node left at the end of the first hour of simulation. Note that this was not reported on the curve above.

It is obvious that the nodes die much faster in the dynamic method, given that cNodes, the only nodes whose consumption is significant, are re-elected, whereas there are no more consuming cNodes in the network for the static method after the ten first nodes are dead. Hence, it is interesting to consider how many nodes do effectively detect the attack as the time passes by. This is what is shown in Figure 28.11. The average number of cNodes which detected the attack (out of 10 cNodes) is presented for each 60 second-long period.

After the fourth minute, every cNode is dead for the static method, and the compromised node is no longer detected. With the dynamic method, a raw average of 6.5 out of 10 cNodes detect the compromised nodes during each 10 second-long period corresponding to the dynamic election. The flooding sensor is still detected by more than one node after half an hour.

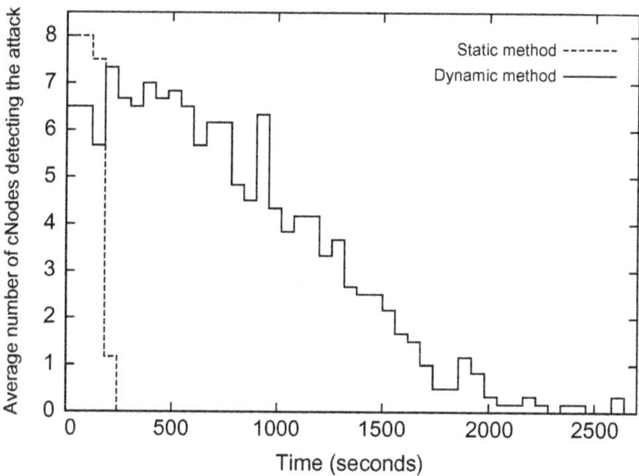

FIGURE 28.11

DoS detection.

Simulations are a fine way to obtain results from a proposed algorithm. But they do not have the rigor of mathematical models. In the next section we will attempt to model our network under attack using formal verification tools.

4 MODELING WITH MARKOVIAN PROCESSES AND GSPN MODEL

4.1 MODELING USING MARKOV CHAINS

Continuous Time Markov Chains (CTMC) are a class of discrete state stochastic processes suitable to model discrete-event systems that enjoy the so-called *memory less* property (Markov property): i.e., systems in which the future evolution depends exclusively on the current state (and not on the history that led into it). It is well known that in order to fulfill the Markov property, delay of events must be exponentially distributed.

In this subsection we describe how to structure continuous time Markov chain (CTMC) models for modeling of a WSN subject to DoS attacks and equipped with DoS detection functionalities. To illustrate the CTMC modeling approach we focus on a specific (sub)class of WSN corresponding to the following points:

- The network consists of a single cluster containing one CH, N sensing nodes and K cNodes.
- (Exactly) one among the N sensing nodes is a compromised node.
- Sensing node i $(1 \leq i \leq N)$ generates traffic according to a Poisson process with rate λ_i.

- The compromised node c generates traffic according to a Poisson process with rate $\lambda_c \gg \lambda_i$.
- Each cNode periodically performs a detection check with period distributed exponentially with rate μ. On detection of abnormal traffic a cNode reports the anomaly to the CH.
- The network topology corresponds to a connected graph: each node can reach any other node in the cluster.

The dynamics of WSN systems agreeing with the previous characterization can straightforwardly be modeled in terms of a $K\cdot(N+1)$-dimensional CTMC. States of such a CTMC consist of K-tuples $x=(x_1,x_2,\ldots,x_k)$ of macro-states $x_K=(x_{k_1},x_{k_2},\ldots,x_{k_N},x_{k_d})$ encoding the number of overheard packets by cNode k. More precisely, component $x_{k_j}(1\leq j\leq N)$ of macro-state x_k is a counter storing the total number of packets sent by node j and overheard by cNode k, whereas component x_{k_d} is a boolean-valued variable which is set to 1 on detection, by cNode k, of abnormal traffic. We also consider a *threshold function* $f{:}\mathrm{N}^N\rightarrow\{0,1\}$ which is used (by cNodes) to decide whether traffic rate has exceeded the "normal" threshold. The arguments of f are an (N)-tuples $(n_1,\ldots,n_N)$, where $n_i\in\mathrm{N}$ is the number of overheard packets originating from node i.

We illustrate the transition equations for such a CTMC. For simplicity we illustrate only equations regarding transitions for a generic macro-state x_k: the equations for transitions of a generic (global) state $x=(x_1,x_2,\ldots,x_K)$ can be straightforwardly obtained by combination of those for the macro-states. In the following x_{k_c} denotes the counter of received packets from the compromised node.

$x_k \rightarrow$ Normal transmission
$\rightarrow(x_1,\ldots,x_{k_i}+1,\ldots,x_{k_c},\ldots,x_{k_n},0)$ with rate $\lambda_i\neq\lambda_c$

$\rightarrow$ Transmission by compromised nodes
$\rightarrow(x_1,\ldots,x_{k_i},\ldots,x_{k_c}+1,\ldots,x_{k_n},0)$ with rate λ_c

$\rightarrow$ Check and Detection of abnormal traffic
$\rightarrow(0,\ldots,0,\ldots,0,\ldots,0,1)$
with rate $\mu\times 1_{f(x_k)\geq \text{threshold}}$
$\rightarrow$ Check and No-Detection of abnormal traffic
$\rightarrow(x_1,\ldots,x_{k_i}+1,\ldots,x_{k_c},\ldots,x_{k_n},0)$ with rate $\lambda_i\neq\lambda_c$
with rate $\mu\times 1_{f(x_k)<\text{threshold}}$

We assume that in the initial state all counters x_{k_i} as well as the Boolean flag x_{k_d} are set to zero. The above equations can be described as follows. When cNode k is in state x_k a "Normal transmission" from node i $(1\leq i\leq N, i\neq c)$ takes place at rate λ_i leading to a state such that the corresponding counter x_{k_i} is incremented by one, leaving all remaining counters unchanged. Similarly a "Transmission by the compromised node" c happens with rate λ_c leading to a state such that the corresponding counter x_{k_c} is incremented by one. Finally checking for abnormal traffic conditions happens at rate μ and whenever the controlling function f detects that in (macro) state x_k the number of overheard packets from any node is above the considered threshold ($f(x_k) \geq$ threshold) , the detection flag x_{k_d} is raised (i.e., alarm is sent to the CH), and counters x_{k_j} are all reset (so that at the next check they are updated with "fresh" traffic data). On the other hand, if traffic has not been abnormal over the last $Exp(\mu)$ duration ($f(x_k) <$ threshold) counters x_{k_j} are reset while the detection flag is left equal to zero.

The detection probability for cNode k (DPk) can be computed in terms of the steady-state distribution of the above described CTMC in the following manner:

$$DPk = \sum_{x_{k_1},\ldots,x_{k_N}}^{\infty} \pi(x_{k_1},\ldots,x_{k_N},x_{k_d}=1)$$

where $\pi(x_{k_1},x_{k_2},\ldots,x_{k_N},x_{k_d})$ denotes the steady-state probability at (macro)state $x_k=(x_{k_1},x_{k_2},\ldots,x_{k_N},x_{k_d})$ of the CTMC.

4.1.1 Discussion

The described CTMC modeling approach relies on the assumption that the period with which detection checking is performed is an exponentially distributed random variable. Indeed, such an assumption may introduce a rather significant approximation, as in reality detection checking happens at intervals of fixed length, or even "continuously." Therefore stochastic modeling of DoS attacks detection requires us to exit the Markovian sphere and to consider non-Markovian stochastic processes (more specifically, periodic detection checking can more accurately be modeled by means of deterministic distributions). We discuss non-Markovian modeling of DoS detection mechanisms in subsection 4.2.

4.2 NON-MARKOVIAN MODELING AND VERIFICATION OF DoS

We have pointed out that using Markov chains to model DoS detection mechanisms may inherently imply a significant approximation. To obtain more accurate models of DoS detection it is necessary to resort to a more general class of stochastic processes, namely the so-called Discrete Event Stochastic Processes (DESP, also often referred to as Generalized Semi-Markov Processes, or GSMP). The main characteristics of DESP are that these processes allow for representing generally distributed durations, rather than, as with CTMC, being limited to exponentially distributed events.

In this subsection we present a modeling approach of DoS detection in terms of Generalized Stochastic Petri nets (GSPN) [46], a class of Petri nets suitable for modeling stochastic processes. By definition, the GSPN formalism is a high-level language for representing CTMC. However, herein we refer to its straightforward extension where *timed-transitions* can model generally distributed durations. Such extended GSPN (eGSPN in the following) becomes a high-level language for representing DESP. Furthermore, eGSPN is also the formal modeling language supported by the COSMOS [47] statistical model checker, a tool that allows for verification of (sophisticated) performance measures in terms of the Hybrid Automata Stochastic Logic (HASL) [15].

In the following we provide a succinct description of both the GSPN modeling formalism and the HASL verification approach, before describing their application to the DoS attack detection case.

4.2.1 Generalized stochastic Petri nets

A GSPN model is a bipartite graph consisting of two classes of nodes, *places* and *transitions* (Figure 28.12). Places (represented by circles) may contain *tokens* (representing the state of the modeled system) while transitions (represented by bars) indicate the events the occurrence of which determine how tokens "flow" within the net (thus encoding the model dynamics). The state of a GSPN consists of a *marking* indicating the distribution of tokens throughout the places (i.e., how many tokens each place contains). Roughly speaking, a transition is enabled whenever all of its input places contains a number of tokens greater than or equal to the multiplicity of the corresponding input arc (e.g., transition T1 in the left-hand part of Figure 28.12 is enabled, while T2 is not). An enabled transition may fire consuming tokens (in a number indicated by the multiplicity of the corresponding input arcs) from all of its input places and producing tokens (in a number indicated by the multiplicity of the corresponding output arcs) in all of its output places. Such an informally described rule is known as the Petri net firing rule. GSPN transitions can be either timed (denoted by empty bars) or immediate (denoted by filled-in bars, e.g., transition T2 in left-hand side of Figure 28.12). Generally speaking, transitions are characterized by: (1) a distribution which randomly determines the delay before firing it; (2) a priority which deterministically selects among the transitions scheduled the soonest, the one to be fired; (3) a weight, which is used in the random choice between transitions scheduled the soonest with the same highest priority. With the GSPN formalism the delay of timed transitions is assumed exponentially distributed, whereas with eGSPN it can be given by any distribution with nonnegative support. Thus, whether a GSPN timed-transition is characterized simply by its weight $t \equiv w$ ($w \in \mathbb{R}^+$ indicating an $Exp(w)$ distributed delay), an eGSPN timed-transition is characterized

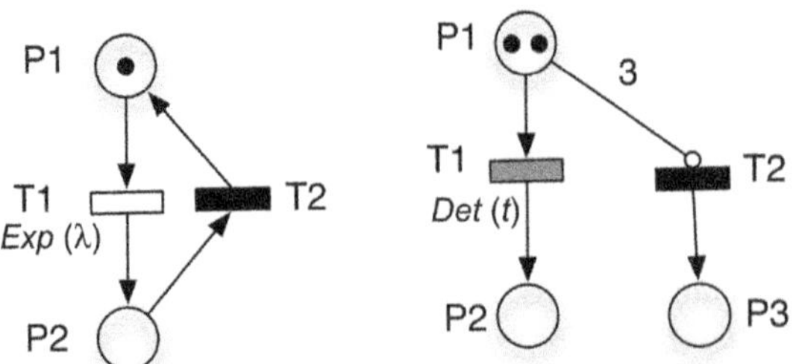

FIGURE 28.12

Simple examples of eGSPN: timed transitions, immediate transition and inhibitors arcs.

by a triple: $t \equiv (\text{Dist-t}, \text{Dist-p}, w)$, where Dist-$t$ indicates the type of distribution (e.g., Unif, Deterministic, LogNormal, etc.), Dist-p indicates the parameters of the distribution (e.g., $[\alpha, \beta]$) and $w \in \mathbb{R}^+$ is used to probabilistically choose between transitions occurring with equal delay.[1]

In the following we describe how eGSPN models can be derived for modeling WSN scenarios with DoS mechanisms. More specifically, in our eGSPN models we will use only two types of timed transitions, namely: *exponentially distributed* timed transitions (denoted by empty bars, e.g., T1 on the left-hand side of Figure 28.12) and *deterministically distributed* timed transitions (denoted by blue-filled-in bars, e.g., T1 on the right-hand side of Figure 28.12). In our Petri nets models we will also extensively exploit inhibitor arcs, an additional element of the GSPN formalism. An inhibitor arc is denoted by an edge with an empty-circle in place of an arrow at its outgoing end (e.g., the arc connecting place P1 to transition T2 in the right-hand side of Figure 28.12). In the presence of inhibitor arcs the semantics of the GSPN firing rule is slightly modified, thus: a transition is enabled whenever all of its input places contain a number of tokens greater than or equal to the multiplicity of the corresponding input arc and strictly smaller than the multiplicity of the corresponding inhibitor arcs (e.g., transition T2 in the right-hand part of Figure 28.12 is also enabled, because P1 contains less than 3 tokens).

Having summarized the basics of the syntax and semantics of the eGSPN formalism, we now describe how it can be applied to formally represent WSN systems featuring DoS mechanisms.

4.2.2 Modeling DoS Attacks with eGSPN

We describe the eGSPN models we have developed for modeling DoS attacks in a grid-like network. For simplicity we illustrate an example referred to as a 3×3 grid topology. The proposed modeling approach can easily be extended to larger networks.

In a WSN with DoS detection mechanisms the functionality of sensing nodes is different from that of cNodes. Here we describe GSPN models for representing: (i) sensing nodes, (ii) statically elected cNodes, and (iii) dynamically eligible cNodes.

4.2.2.1 GSPN model of sensing nodes

Sensing node functionality is trivially simple: they simply keep sending sensed data packets at a pace which (following subsection 4.1) we assume, being exponentially distributed with rate λ_i. This can be modeled by a simple GSPN that consists of a single exponentially distributed timed-transition (labeled TX) with no input places (i.e., always enabled) and with as many outgoing arcs leading to the input buffer of the neighboring nodes (represented by dashed places labelled "InBuff_{i_j}" in Figure 28.13). Note that transition TX in Figure 28.13 has no input places, which means (according to the Petri net firing rule) that it is always (i.e., perpetually) enabled. Note also that TX is an exponentially distributed timed transition with rate λ_i, which complies with the assumption that each sensor node

---

[1] a possible condition in case of non-continuous delay distribution

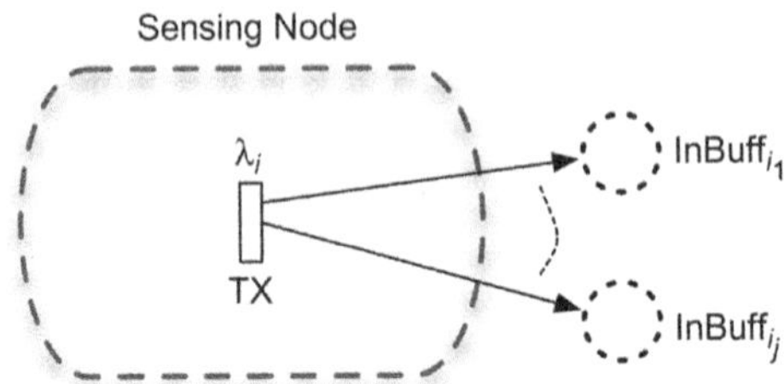

FIGURE 28.13

GSPN model of a sensing node.

performs a sensing operation every δ_s time with $\delta_s \sim \text{Exp}(\lambda_i)$. To summarize: the sensing functionality of a specific node in WSN is modeled by a single timed-transition provided with as many outgoing arcs as the number of neighbors of that node. The complete sensing functionality of a WSN can be modeled by combining several such GSPN modules.

4.2.2.2 GSPN model of cNodes

A cNode functionality, on the other hand, is entirely devoted to monitoring of traffic of the portion of the WSN it is guarding. From a modeling point of view, a distinction must be made between the case of statically elected cNodes (as in [13]) and that of dynamically eligible cNodes (as in [14]). In fact, with dynamic cNodes election each node in the network can be elected as cNode; therefore each node can switch between a sensing-only functionality and a controlling functionality. On the other hand, static cNodes will be control-only nodes.

GSPN models for both static and dynamic cNodes are depicted in Figure 28.14(a) and Figure 28.14(b), respectively. A cNode detects an attack whenever the overheard traffic throughput (i.e., number of overheard packets per observation period) exceeds a given threshold ρ_{attack}. Place "InBuff" (Figure 28.14 (a)) represents the input buffer of a node, where packets received/overheard from neighbor nodes are placed. The "InBuff" place receives tokens (corresponding to overheard packets) through input arcs originating from neighbors sensing-node modules (i.e., the input arcs of place "InBuff" are the output arcs of the timed-transition representing the corresponding sensing activity of each neighbor node).

To model the traffic monitoring functionality of cNodes we employ two mutually exclusive, deterministically distributed timed transitions labeled "checkYES" and "checkNO" in Figure 28.14(a) and Figure 28.14(b). They correspond to the periodic verification performed by the cNode to check whether the frequency of incoming traffic has been abnormal (over the last period). At the end of each (fixed) interval $[0, \Delta]$ either: transition "checkYES" is enabled, if at least k packets have been received (*i.e.* place "InBuff" contains at least k tokens); or transition "checkNO" is enabled, if less than k packets have been received (i.e., place "InBuff" contains less than k tokens); in the first case (i.e., "checkYES" enabled)

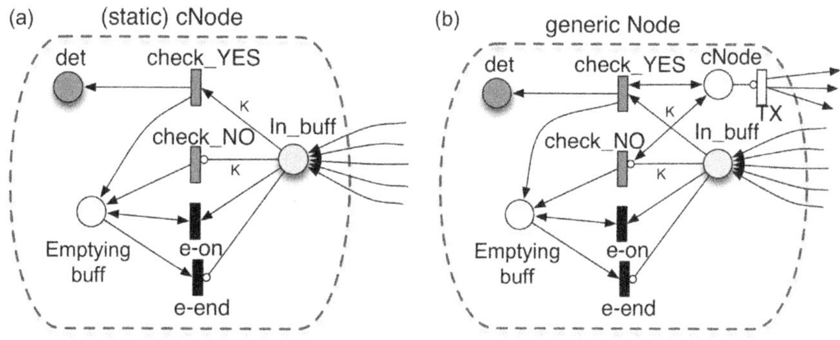

FIGURE 28.14

GSPN components representing cNodes behavior in a WSN with DoS detection mechanisms: (a) GSPN model for a statically elected cNode; (b) GSPN model for a dynamically eligible cNode.

a token is added in the output place "det" representing the occurrence of an DoS detection, otherwise (i.e., "checkNO" enabled) no tokens are added to place "det". After firing of either the "checkYES" or the "checkNO" transition, the emptying of the input buffer starts by adding a token in place "empty." This enables either immediate transition "e-on" (which iteratively fires until the input buffer is empty), or "e-end" which represents the end of the emptying cycle. Note that buffer emptying does not consume time, and it is needed in order to correctly measure the frequency of traffic at each successive sampling interval $[0, \Delta]$.

The GSPN model for the dynamic cNodes (Figure 28.14(b)) is a simple extension of that for static cNodes obtained by adding an auxiliary place "cNodes" and an auxiliary exponentially distributed timed-transition "TX". This is needed because with dynamically elected cNodes, each node in the network may periodically switch from sensing-only to controlling-only functionality, hence the corresponding GSPN model must represent both aspects. The GSPN model for the dynamic cNodes (Figure 28.14(b)) is a simple extension of that for static cNodes obtained by adding an auxiliary place "cNodes" and an auxiliary exponentially distributed timed-transition "TX". This is needed because with dynamically elected cNodes, each node in the network may periodically switch from sensing-only to controlling-only functionality; hence the corresponding GSPN model must represent both aspects. If the auxiliary place "cNode" contains a token, then the "controlling" functionality (i.e., the left part of the GSPN) is switched on, and in that case the GSPN of Figure 28.14(b) behaves exactly as that of Figure 28.14(a). Conversely, if place cNode is empty then the "sensing" functionality is switched-on (i.e. transition "TX" is enabled due to the inhibitor arc between place "cNode" and transition "TX") while the "controlling" part of the net is disabled (*i.e.* in this case the net of Figure 28.14(b) behaves exactly as that of Figure 28.13).

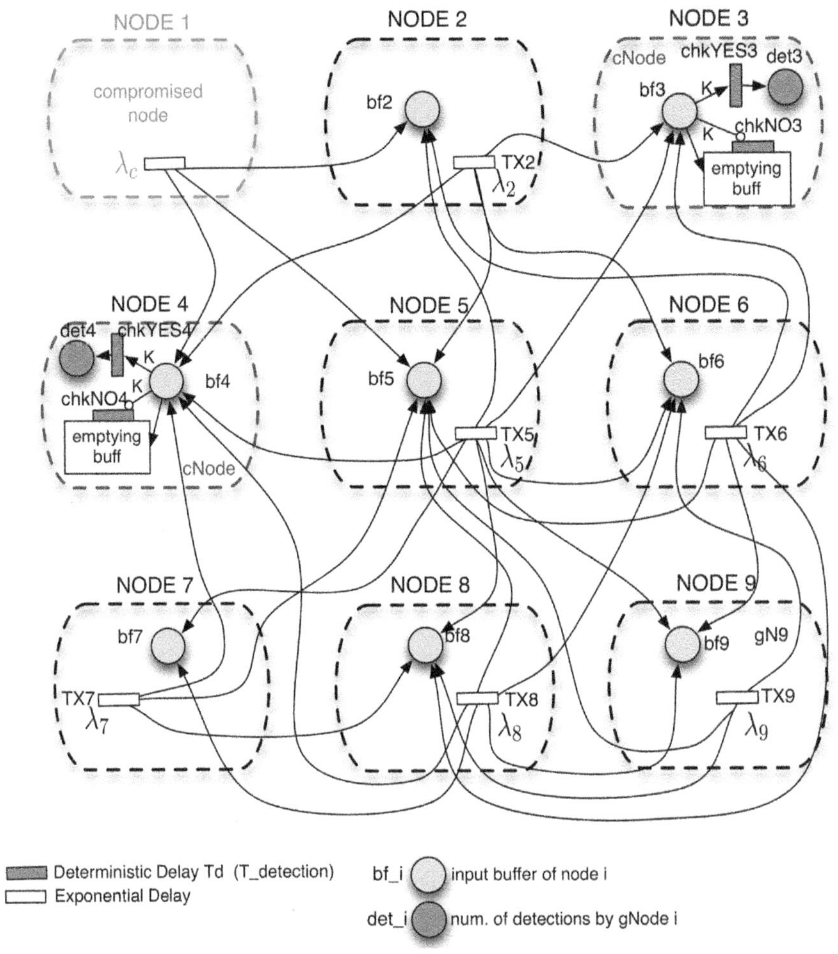

FIGURE 28.15

GSPN model of a 3 × 3 grid-topology with 1 (fixed) compromised node and 2 static cNodes.

The above described GSPN models for sensing-nodes, static cNodes and dynamic cNodes can be used as basic building blocks to compose models of specific WSN topologies. In the following we provide examples of GSPN for 3 × 3 WSN grid-topology equipped with DoS detection functionalities.

4.2.2.3 GSPN model of DoS detection with static cNodes

Figure 28.15 illustrates a complete GSPN model for a 3 × 3 grid topology representing an example of DoS detection with static election of cNodes (as in [13]). In particular in this example we consider the presence of 2 cNodes (i.e., node 3 and 4)

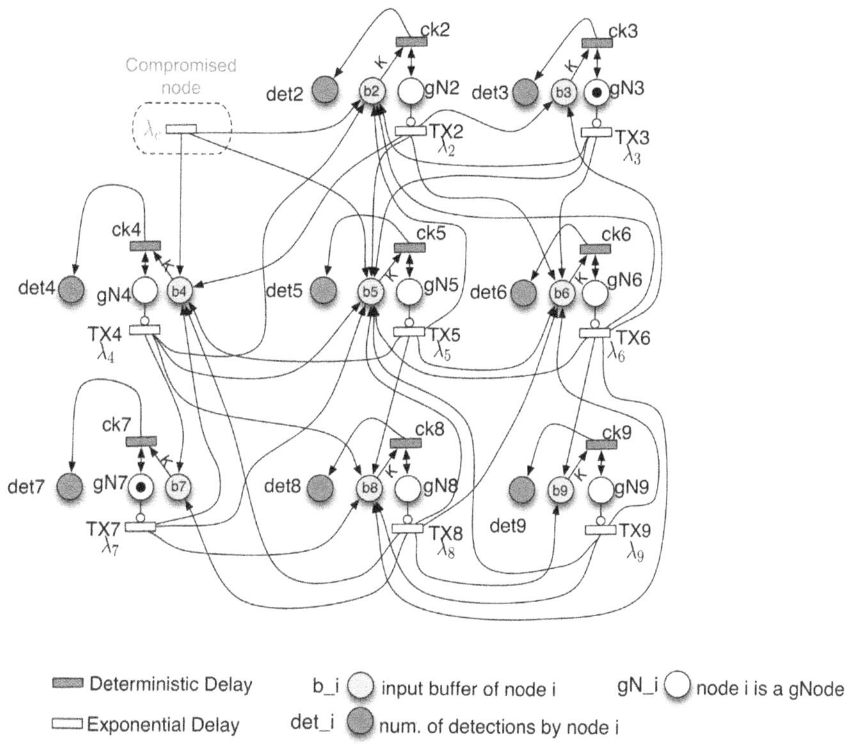

FIGURE 28.16

GSPN model: the traffic part in a 3 × 3 topology with 1 (fixed) compromised node and 2 randomly elected cNodes.

and 1 compromised node (i.e. node 1). Note that for simplicity the "emptying buffer" part in the GSPN modules of the cNodes (i.e., node 3 and 4) is depicted as a box (i.e., the content of that box corresponds to the subnet responsible for emptying the "inBuff" place as depicted in Figure 28.14(a) and Figure 28.14(b)).

This model can be used to study the performances of DoS detection with static cNodes in many respects, such as: measuring the expected number of detected attacks within a certain time bound, or also, for example, assessing the average energy consumption of cNodes. In the next subsection we describe how to build GSPN models of WSNs with DoS detection and dynamic election of cNodes. The resulting GSPN is more complex than that for statically elected cNodes, as it must include an extra module, namely a GSPN module for periodically electing the cNodes.

4.2.2.4 GSPN model of DoS detection with dynamic cNodes

Figures 28.16 and 28.17 illustrate the GSPN model of a 3 × 3 grid topology for the case of DoS detection with dynamic election of cNodes (as in [14]).

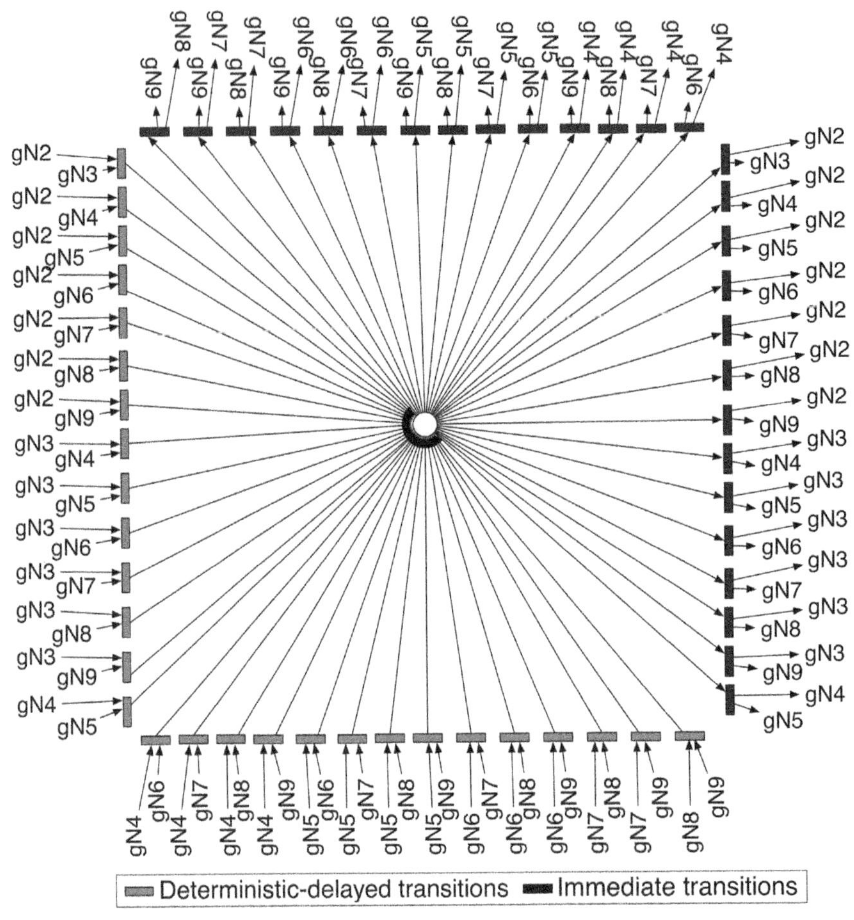

FIGURE 28.17

GSPN model: the random election policy part: 2 cNodes are elected out of 8.

For simplicity the model has been split into two parts: the actual network topology part (Figure 28.16) and the cNodes random election mechanism (Figure 28.17). The network model (Figure 28.16) is obtained by composition of node's GSPN component in the same fashion as for the model of the WSN for DoS detection with static cNodes, only now all nodes must be reconfigurable as either sensors or controllers (thus the basic GSPN components used to build the network topology are those of Figure 28.13(b)). The cNodes election component (Figure 28.17), on the other hand, consists of a single place, n mutually exclusive deterministically distributed timed-transitions (blue-filled) and n mutually exclusive immediate transitions (black-filled) (with $n = \binom{8}{2} = 28$, as we assume that, at each round, 2 cNodes are elected out of 8 possible candidates, thus, for simplicity we rule out the compromised node from the eligible ones). The deterministically distributed timed-transitions (blue-filled) of Figure 28.17 correspond to all possible different pairs of "cNode" places. At the end of each selection period only (exactly) one such timed-transition will be enabled and will fire retrieving, in this way, the tokens from the current pairs of active cNodes and inserting one token in the only (central) place of the net in Figure 28.17. At this point all 28 immediate transitions will become enabled and a random choice will take place resulting in the selection of only (exactly) one of them. The selected transition will fire and by doing so will insert one token into each "cNode" place of the corresponding pair of cNodes to which it is connected, activating, in this way, the controlling functionality of the newly elected cNodes.

4.2.3 HASL verification of DoS detection models

One of the main motivations for developing GSPN models of discrete-event systems is that a fairly large and well established family of formal methods can be applied to analyze them. Recently a new formalism called Hybrid Automaton Stochastic Logic (HASL) has been introduced which provides a unified framework both for model checking and for performance and dependability evaluation of DESP models expressed in GSPN terms. In essence, given a GSPN model, we can express sophisticated performance measures in terms of a HASL formula and apply a statistical model checking functionalities to (automatically) assess them. In the following we informally summarize the basics about the HASL verification approach, referring the reader to [15] for formal details.

4.2.3.1 HASL model checking

Model checking is a formal verification procedure by which given a (discrete-state) model M and a property formally expressed in terms of a temporal logic formula φ, an algorithm automatically decides whether φ holds in M (denoted $M \models \varphi$). In the case of stochastic models (i.e., stochastic model checking [48]) formulae are associated with a measure of probability and verifying $M \models \varphi$ corresponds to assess the probability of φ with respect to the stochastic model M. HASL model checking extends this very simple concept in the sense that an HASL formula can evaluate to any real number (thus it can represent a measure of probability as well as other performance measures). To do so HASL uses Linear Hybrid Automata (LHA) as machineries to encode the dynamics (i.e., the execution paths, or trajectories) of interest of the considered GSPN model. An LHA, simply speaking, is a generalization of Timed Automaton where clock-variables are replaced by real-valued data-variables. In practice a formula of HASL consists of two parts:

- an LHA used as a selector of relevant of timed execution of the considered DESP (path selection is achieved by synchronization of a generated DESP trajectory with the LHA).

- an expression Z built on top of data variables of the LHA according to the following syntax and which represent the measure to be assessed:

$$Z ::= E(Y)|Z + Z|Z \times Z$$
$$y ::= c|Y + Y|Y \times Y|Y/Y|\text{last}(y)|\text{min}(y)|\text{max}(y)|\text{int}(y)|\text{avg}(y)$$
$$y ::= c|x|y + y|y \times y|y/y$$

The informal meaning of an HASL expression Z is as follows: x is a data-variable of the LHA automaton associated to the expression. y is an (arithmetic) expression of data-variables. Y is a path random variable, i.e. a variable which is evaluated against a synchronization path, a path resulting by the synchronization of a trajectory of the DESP with the LHA associated to the formula. The basic operators (i.e., last(y), min(y), max(y), int(y), avg(y)) on top of which a path variable Y is built have intuitive meanings. In particular: last(y) indicates the last value of expression y along an accepted synchronized path; min(y)/max(y) indicates the minimum (maximum) of y along a path; int(y) the integral of y along a path; avg(y) the average of y along a path.

The HASL statistical model checking procedure works as follows:

- It takes a GSPN model and an HASL formula.
- It iteratively generates trajectories of GSPN model state-space and synchronizes them with the LHA.
- The trajectories that have been "accepted" by the LHA are considered in the estimation of the measure of interest, the others are dropped.

4.2.4 HASL formulae for DoS models

Having seen the nature of HASL verification, we provide here a few examples of HASL formulae (i.e. LHA + expression) which can be used to assess performance measures of the DoS (GSPN) models presented in the previous subsection. Such formulae may be readily assessed through the COSMOS model checker and the results can be used to compare different DoS detection mechanisms.

The LHA we present are based on the following data-variables:

- x_t: global time
- x_{d_i}: number of attacks detected by cNode i $(1 \leq i \leq N)$
- x_{TX_i}: number of data transmitted by node i $(1 \leq i \leq N)$
- x_{bf_i}: flow of packets in buffer of node i $(1 \leq i \leq N)$

The LHA in Figure 28.18 is a template automaton that can be used for calculating different measures of a node (either a sensing or a cNode) of a WSN model. It refers to GSPN models (Figure 28.15, Figures 28.16/28.17). It consists of 2 locations and refers to the 4 data-variables above described. In the initial-location (l1) the rate of change (*i.e.* the first derivative) of data-variables is indicated (inside the circle). The global time variable x_t is incremented with rate $\dot{x}_t = 1$ following the linear flow of time. Counter variables x_{d_i} and x_{TX_i} (used to count occurrences of events) are unchanged in location l1 (*i.e.* their rates are zero). Finally variable x_{bf_i}

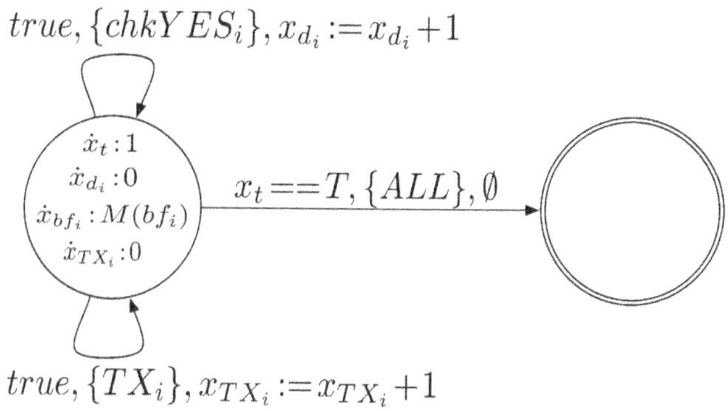

$true, \{chkYES_i\}, x_{d_i} := x_{d_i} + 1$

$\dot{x}_t : 1$
$\dot{x}_{d_i} : 0$
$\dot{x}_{bf_i} : M(bf_i)$
$\dot{x}_{TX_i} : 0$

$x_t == T, \{ALL\}, \emptyset$

$true, \{TX_i\}, x_{TX_i} := x_{TX_i} + 1$

FIGURE 28.18

An LHA for assessing relevant measures of DoS GSPN models.

is incremented with rate proportional to the number of tokens in the input buffer of cNode i (*i.e.* $\dot{x}_{bf_i} = M(bf_i)$); this data-variable can be used to measure the average length of overheard packets by cNode i, and thus to measure the average energy consumption of a cNode. The two self-loops transitions on location l1 are used to increment the counter variables x_{d_i} and x_{TX_i} on occurrence of the associated events in the GSPN model. For example transition l1 $\xrightarrow{true,\{chkYES_i\},x_{d_i} := x_{d_i}+1}$ l1 indicates an occurrence of the GSPN transition labeled $chkYES_i$ (i.e., detection of an attack by cNode i) the variable x_{d_i} is incremented by 1. Transition l1 $\xrightarrow{x_t == T,\{ALL\},\emptyset}$ l2 from l1 to the accepting location l2 indicates when the synchronization stops and the processed path is accepted. Precisely, this happens as soon as $x_t == T$, where $T \in \mathbb{R}$ denotes a time-bound, that is: as soon as the observed trajectories is such that the simulation time is T. In this case, no matter which GSPN transition is occurring (*i.e.* synchronization set is $\{ALL\}$) the transition from l1 to l2 will fire and the path generation will stop by accepting the path. In other words the LHA in Figure 28.18 trivially accepts all paths of time duration T. The value of the 4 data variables collected during synchronization of the LHA with the GSPN model will be then used for estimating relevant Z expressions. In the following we describe a few examples of Z expressions that can be used in association to the LHA in Figure 28.18 to evaluate relevant measures of the DoS GSPN models.

- $Z_1 \equiv E(\text{Last}(x_{d_i}))$: the expected number of detected attacks by cNode i after T time units
- $Z_2 \equiv E(\text{Last}(x_{d_i} + x(d'_i)))$: the sum of attacks detected by cNode i and i" after T time units
- $Z_3 \equiv E(\text{Last}(x_{TX_i}))$: the expected value of packets transmitted by node i after T time units
- $Z_4 \equiv E(\text{Avg}(x_{bf_i}))$: the expected cumulative flow of packets received by node i within T time units

5 ENERGY-BASED DESIGNATION OF THE cNODES

5.1 cNODES SELECTION MECHANISM

Electing the cNodes is not an easy task. In subsection 3.1.3 we exposed and compared three ways to elect them:

- pseudo-random election by the base station;
- pseudo-random election by the cluster head;
- pseudo-random election by the nodes themselves.

We assumed that election should be random so that compromised nodes would not be aware of which node could control the traffic. We did not consider the remaining energy during the cNodes election. But monitoring the traffic implies to keep listening for wireless transmission without interruption. Hence cNodes will have a greater energy consumption than normal nodes. Given that preserving energy is an essential issue in the network, we now prefer to ensure load balancing rather than assuring a pseudo-random election, and thus to consider the residual energy of the nodes during the election. This choice also raises new issues and makes us define a new role for the nodes in the cluster[2].

5.1.1 Using vNodes to ensure a secured deterministic election

The issue with energy measurement is that no agent in the network is able to measure the residual energy of a given node *N*, but the node itself. The neighbor nodes of *N* may record messages sent from *N* and compute a rough estimate, but as they know neither the initial amount of energy of *N* (at the network deployment) nor the energy *N* spent for listening, estimates can not be used to obtain values precise enough so as to reliably sort the nodes according to their residual energy.

So the only way to get the residual energy of a node is to ask this node. The election algorithm we propose is described as follows:

1. During the first step, each node evaluates its residual energy and sends the value to the cluster head;
2. Having received the residual energy of all nodes in the cluster, the cluster head picks the *n* nodes with the highest residual energy (where *n* is the desired number of cNodes during each cycle) and returns them a message to assign them the role of cNode.

It is a deterministic selection algorithm that eliminates any random aspect from the process. The rule is simple: nodes possessing the highest residual energy will be elected. Given that the cNode role implies consuming more energy

[2]The recursive *k*-clustering is used in this Section in the same way as in former Section. And yet for simplicity we will only mention "clusters", as the solution is not dependent of the depth of recursive clustering.

(cNodes listen to surrounding communications most of the time), rotation of the roles is theoretically assured. But the deterministic aspect is also a flaw that may be exploited by compromised nodes. This is a crucial issue: we cannot neglect compromised nodes as the whole cNodes mechanism is deployed in the sole purpose to detect them!

More precisely, the problem may be stated as follows. Compromised nodes will be interested in endorsing a cNode role, as it enables them:

- to reduce the number of legitimate cNodes able to detect them;
- to advertise the cluster head about "innocent" sensing nodes to have them revoked.

When a pseudo-random election algorithm is applied, a compromised node (or even several ones) can be elected during a cycle, but it will lose its role further in time, for later cycles. Even with a self-election process, compromised nodes can keep their cNode role as long as they want, but they cannot prevent other (legitimate) nodes to elect themselves, too. With deterministic election, however, they can monopolize most of the available cNode roles. They only have to announce the highest residual energy value at the first step of the election to get assured to win. If there are enough compromised nodes to occupy all of the n available cNode roles, then they become virtually immune to potential detection.

To prevent nodes from lying when announcing their residual energy, we propose to assign a new role to some of the neighbors of each cNode. Those nodes— we call them vNodes, as for *verification* nodes— are responsible for the surveillance of the monitoring nodes. Once the cNodes election is over, each neighbor to a cNode decides with a given probability whether it will be a vNode for this cNode or not. A given node can act as a vNode for several cNode (in other words, it can survey several neighbor cNode).

If this role consumes too much energy, it is not worth deploying vNodes: we should rather use pseudo-random election for the cNodes. So vNodes must not stay awake and listen most of the time, as cNodes do. Instead they send, from time to time, requests to the cNode they watch over, asking it for its residual energy. They wait for the answer, and keep the value in memory.

Once they have gathered enough data, vNodes try to correlate the theoretical model of consumption of the cNode they survey and its announced consumption, deduced from broadcast messages (during elections) and answers to requests from vNodes. Four distinct cases may occur:

1. The announced consumption does not correlate (at all) with the theoretical model: there is a high probability the node is compromised and seeks to take over cNode role. It is reported to the cluster head.
2. The announced consumption correlates *exactly* with the theoretical model: the node is probably a compromised node trying to get elected while escaping detection (in other words, the rogue cNode adapts its behavior regarding to the previous point). It is easy to detect the subterfuge as values received from

the rogue node and the ones computed by the vNodes are exactly the same. It is reported to the cluster head.

3. The announced consumption correlates roughly with the theoretical model, but does not evolve in the same way (with regard to the model) as the real consumption locally observed by the vNodes (local (in time) evolution of the announced consumption does not "stick" to the one of the surrounding vNodes, which should roughly rise or decrease during the same periods). The node is probably compromised, trying to escape detection by decreasing its announced energy with random values. It is reported to the CH.
4. The announced consumption correlates roughly with the theoretical model and evolves in the same way as the traffic observed by vNodes. Whether the node is compromised or not, it has normal behavior and is allowed to act as a cNode.

If a given vNode is in fact a malicious node, it could lie about the integrity of the cNode it watches. To prevent that, the cluster head must receive multiple reports (their number exceeding a predetermined threshold) from distinct vNodes before actually considering a cNode as compromised. To some extent, this also makes the scheme resilient to errors from the vNodes.

In that way, nodes are allowed to act as cNodes only if they announce plausible amounts of residual energy. Assuming that this role consumes more energy than sensing only, the nodes elected as cNodes will sooner or later see their residual energy drop below the reserve of normal sensing nodes, which implies that they will not get re-elected at the next election. Note that the cases 2 and 3 make a compromised node decrement its announced energy as the time goes by. Even if inconsistency may be noticed and the compromise detected, this simple behavior ensures that the rogue node will stop being elected at some point in time.

Thus, the interest of vNodes can be summarized as follows: a compromised node cannot ensure the takeover of the cNode role at each cycle without cheating when announcing residual energy, and hence being detected by the vNodes. Detecting rogue cNodes, or forcing them to give up their role for later cycles, are the two purposes of the vNodes. The vNode role does not prevent a node from processing to its normal sensing activity (requests to cNodes must not occur too often, or too much power will be drained from the vNodes). The state machine of the nodes is presented in Figure 28.19.

5.1.2 Cluster coverage in case of heterogeneous activity

Deterministic election of the cNodes does not only introduce a flaw that compromised nodes could try to exploit. There is a second problem, independent from the node's behavior, which could prevent the detection of compromised nodes. If a region of the network happens to produce more traffic activity than the other parts of the network, the energy of its nodes will be drawn faster. In consequence, none of the n nodes with the highest residual energy (n being the desired number of cNodes during each cycle) will be located inside this region, and some nodes

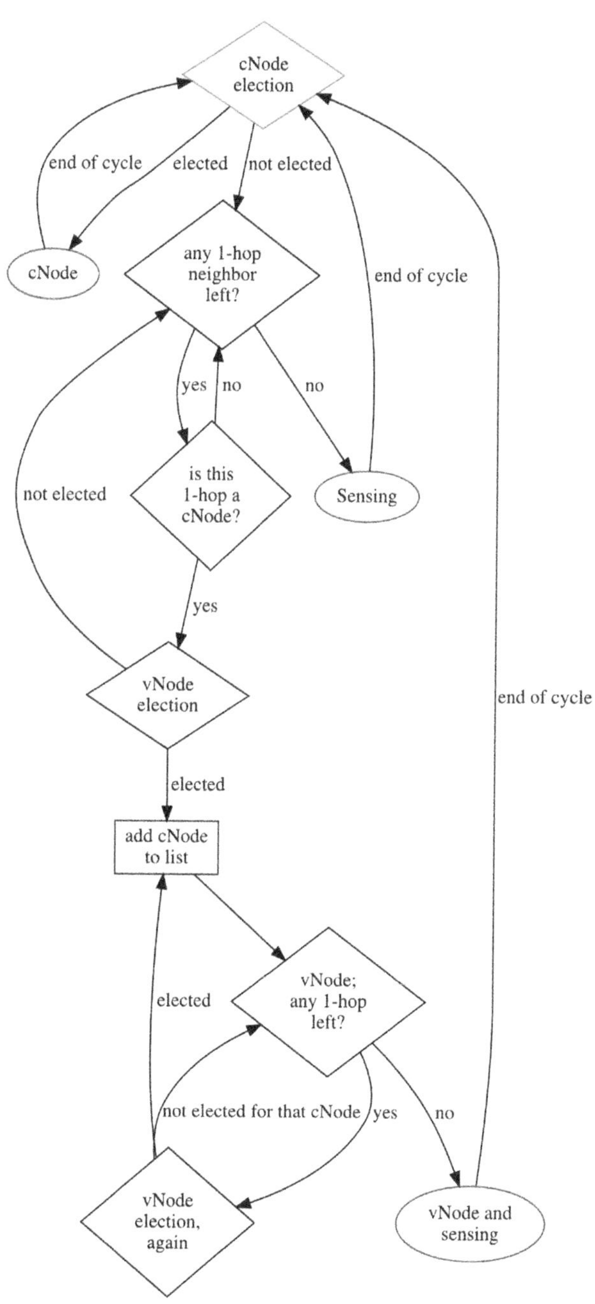

FIGURE 28.19

State machine of the (non-CH) nodes.

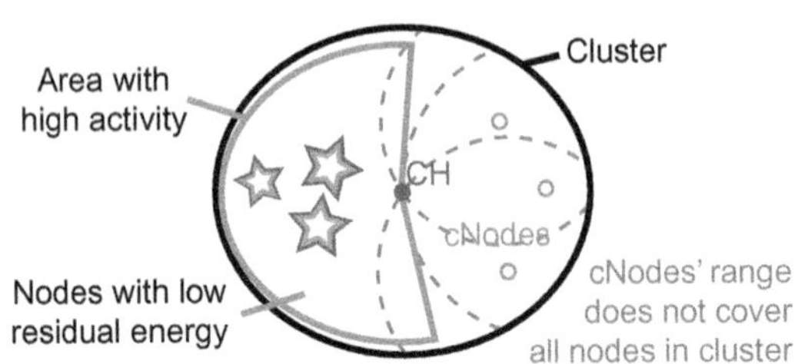

FIGURE 28.20

Illustrative scheme: cNodes are elected inside the area with less activity (thus with more residual energy) and do not cover nodes from the opposite side of the network.

may not be covered for surveillance as long as traffic does not fade, possibly for all cycles. Figure 28.20 illustrates this problem.

To address this issue we need to ensure that every node in the network is covered by at least one cNode. So the election process we presented in subsection 5.1.1 needs to be modified. The correct version is as follows:

1. During the first step, each node evaluates its residual energy and broadcasts the value;
2. The cluster head listens to all values. Other nodes also register all messages they hear into memory;
3. All nodes send to the CH the list of their 1-hop neighbors[3];
4. The CH picks the n nodes among those with the highest residual energy, such that the n nodes cover all other nodes in range[4]. If needed, it selects some additional nodes to cover all the cluster;
5. The CH returns a message to selected nodes to assign them the role of cNode.

Note that some clustering algorithms (such as HEED [9], for example) provide other election mechanisms (for cluster heads, but that can also be used for selecting cNodes) based on residual energy. We do not want to use it because energy only takes part in the process as a factor for the probability that the nodes declare themselves elected. Instead we prefer nodes to broadcast their residual energy in order to enable surveillance by the vNodes.

5.1.3 Observations

cNodes apply a very basic trust based scheme to the cluster: when a sensor node breaks a rule, for example by exceeding a given threshold for transmitted packets, it is considered as untrustworthy. There are many other trust based schemes in the literature, most of them more advanced than this one (see Section 2). The cNodes could implement several other trust mechanisms (by lowering a score

[3]We do not deal with the case of compromised nodes cheating at this step of the process. Indeed they could announce extra virtual neighbors to try to escape from coverage.

[4]The details of the algorithm executed by the cluster head at this step are not given in this study.

on bad behavior for each node, for instance). As more complex mechanisms would create additional overhead, we prefer to limit ourselves to this simple method in this study.

5.2 SELECTION IN PRACTICE: RESULTS FROM SIMULATION

We have undertaken simulation of our second proposal regarding the energy consumption in order to compare it with the previous model (using pseudo-random election for cNodes). We used the ns-3 framework to proceed[5].

In the new proposal, the vNodes are to model the theoretical consumption of the cNodes they watch over. We have chosen to use Rakhmatov and Vrudhula's diffusion model [49] to compute the consumption. This choice was driven by several reasons:

- It provides a pretty accurate approximation of real consumption, taking into account chemical processes internal to the battery such as rate capacity effect and recovery effect.
- It is one of the models already implemented in ns-3. So in our case it is an absolutely perfect theoretical model. It remains "theoretical" as vNodes use this model to compute the expected behavior of cNodes according to the few packets they sometimes hear. Meanwhile, real cNodes consumption computed by the ns-3 core takes into account every packet actually sent or received by cNodes, also including packets that vNodes cannot hear (because of distance or sleep schedule). So the values computed by vNodes and the ns-3 core will not always be the same, which allows us to use the model.

Rakhmatov and Vrudhula's diffusion model refers to the chemical reaction happening inside the battery electrolyte, and is summarized by the following equation:

$$\sigma(t) = \underbrace{\int_0^t i(\tau)\mathrm{d}\tau}_{l(t)} + \overbrace{\int_0^t i(\tau)\left(2\sum_{m=1}^{\infty} e^{-\beta^2 m^2 (t-\tau)}\right)\mathrm{d}\tau}^{u(t)}$$

where:

- $\sigma(t)$ is the apparent charge lost from the battery at t;
- $l(t)$ is the charge lost to the load ("useful" charge);
- $u(t)$ is the unavailable charge ("lost in battery" charge);

[5]To perform the simulations in this Section we switched to ns-3, third major version of the *network simulator* tool. There were two reasons for this: the first one was that we found it more practical to implement the different applications for the nodes (*e.g.* vNode application) through ns-3 architecture; the second reason was that we have realized this work later: as development and support for ns is progressively moving toward the third major version, so did we.

Table 28.4 Simulation Parameters

Parameter	Value
Number of nodes	30 (plus 1 CH)
Number of cNodes	4
Probability for vNodes selection	33%
Delay between consecutive elections	1 minute
Simulation length	30 minutes
Cluster shape	Squared box
Cluster length	Diagonal is 2 × 50 meters
Transmission range	50 meters
Location of the nodes	CH: center; others: random
Mobility of the nodes	Null
Average data sent by normal nodes	1024 bytes every 3 seconds
Data sent by vNodes (per target cNode)	1024 bytes every 5 seconds

- $i(t)$ is the current at t;
- $\beta = \frac{\pi\sqrt{D}}{w}$, where D is the diffusion constant and w the full width of the electrolyte.

In practice, computing the first ten terms of the sum provides a good approximation (this is also the default behavior of ns-3, by the way).

We launched several simulation instances and chose to focus on the energy consumption and load balancing in the cluster. When we implemented our solution, we set the parameters of the simulation as detailed in Table 28.4.

We obtained the residual energy values for each node at each minute of the simulation. From this data we draw the average residual energy of the nodes (excluding cluster head) as well as the standard deviation. Average residual energy per minute in the batteries of the nodes is displayed in Figure 28.21.

Increasing values at $t = 11$ minutes and $t = 15$ minutes with the use of the proposed solution traduce the recovery effect of the batteries. As expected, our proposal causes increased global energy consumption. This is due, of course, to the new vNode role. vNodes have to wake up periodically to send requests to neighbor cNodes and to wait for an answer: this is energy-consuming. The estimated overhead for our solution appears on Figure 28.22.

Standard deviation of residual energy value in the nodes at each minute of the simulation is presented on Figure 28.23.

During the first minutes of simulation, our solution creates a higher disproportion in load balancing due to the introduction of vNodes (there are more nodes assuming demanding functions). But after the first seven minutes or so, the standard deviation with our method falls below the standard deviation of previous method. This is the consequence of a better load repartition over the nodes with

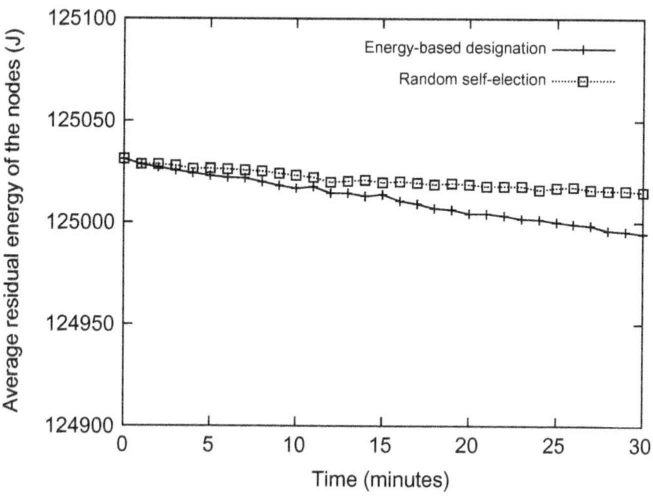

FIGURE 28.21

Average residual energy of the nodes (excluding cluster head).

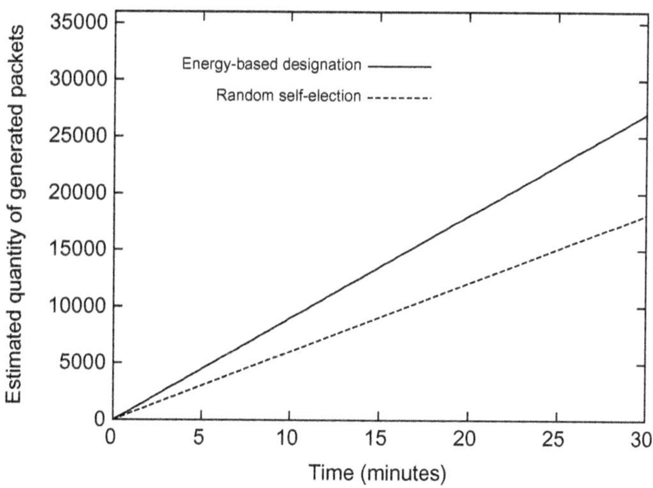

FIGURE 28.22

Estimated number of generated packets during the simulation.

our solution. The difference between standard deviation with and without our simulation may look small: this is due to the model of the simulation we implemented. Given that we have a good pseudo-random numbers generator, when the number of elections get high, all nodes will roughly assume cNode role the same number of times in simulation *not* using our solution. As sensing nodes

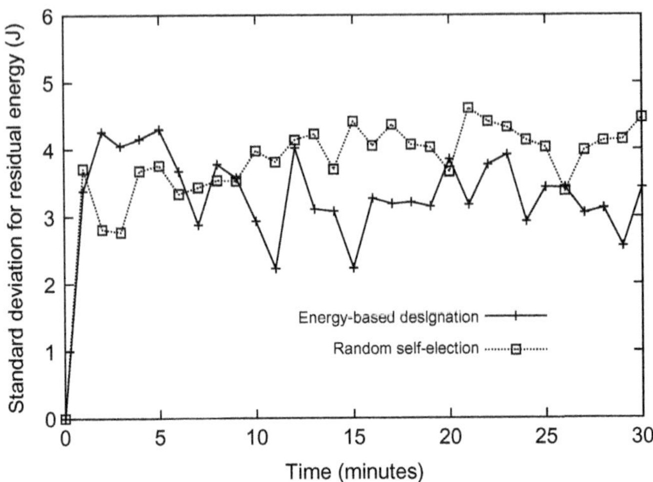

FIGURE 28.23

Standard deviation for residual energy of the nodes.

all have the same activity, a correct repartition of the cNode roles over the time leads to a good energy balance. But in a situation where sensing nodes have different activity levels—for instance, if there is an area in the cluster when measured events occur much more often than in the other parts of the cluster—the consumption would not be equilibrated between all the nodes with the previous method; whereas our solution would deal well with this case, since cNodes are elected according to residual energy. Thus simulations show that the use of vNodes leads to a higher energy consumption, but electing cNodes on residual energy provides a better load repartition in the cluster.

6 CONCLUSION

Detection of DoS attacks is a fundamental aspect of WSN management. We have considered a class of DoS detection mechanisms designed to operate on clustered wireless sensor networks: cNodes are used to monitor traffic of the nodes and to detect denial of service attacks (e.g., flooding, black hole attacks). In the literature two basic election approaches have been proposed: static versus dynamic election. In this chapter, we have proposed two distinct election algorithms related to the dynamic approach to elect those cNodes, in an attempt to provide a better load balancing in the network.

The first one is a self-election similar to the process involved in LEACH clustering algorithm. With this solution we presented different modeling approaches for obtaining models of WSNs with DoS functionalities. First we described how

Markov chains model should be structured for modeling DoS attack and detection, pointing out that, because of the nature of DoS detection, Markovian models may inherently come with some significant approximation. Hence we presented formal non-Markovian models of DoS detection in terms of Generalized Stochastic Petri Nets, a high level formalism for generic Discrete Event Stochastic Process. We have illustrated how a model of WSNs with DoS can be built "incrementally" by combination of small GSPN modules of single (sensing/controlling) nodes up to obtaining a model of the desired network. We have also stressed how the GSPN formalism is naturally well suited for modeling of the dynamic random cNodes election policy. Expressive performance measures of the DoS GSPN models can be formally written and assessed by means of the recently introduced Hybrid Automata Stochastic Logic. We have then presented numerical results obtained with virtual WSN implementation via the ns-2 simulator. They confirm the intuition that cNodes' dynamic allocation guarantees a more uniform energy consumption (throughout the network) while preserving a good detection capability.

The second designation algorithm is based on the residual energy of the sensors. We have addressed several issues related with the use of this deterministic selection. Compromised nodes trying to systematically take over the cNode role are forced to abandon it for later cycles, or be detected, by vNodes. The vNode role is a new role we introduced to survey the cNodes by matching their announced energy consumption with a theoretical model. The issue of areas of the cluster uncovered by cNodes, depending on the activity in the cluster, is addressed by enforcing covering of the whole cluster: the cluster head is to designate additional cNodes if needed. The results we have obtained through simulations show that, even though using our simulation causes a higher global consumption of energy in the cluster, it provides an even better load repartition between sensors.

Working with clusters ensures a good scalability of network management. The detection system is also flexible, as cNodes can endorse various trust-based models, and monitoring rules can be set to fight against several types of denial of service attacks. Future developments of this work should include the execution of actual verification experiments on the presented GSPN models by means of the COSMOS statistical model checker, as well as the extension of the proposed modeling approaches to consider more complex networks (different topologies and scales, and areas with different activity levels).

REFERENCES

[1] Ben-Othman J, Yahya B. Energy efficient and QoS based routing protocol for wireless sensor networks. J Parallel and Distrib Comput 2010;70(8):849–57.

[2] Bernard T, Fouchal H. A low energy consumption MAC protocol for WSN. In: Proceedings of the 2012 IEEE International Conference on Communications (ICC'12). Ottawa, ON, Canada; 2012. p. 533–37.

[3] Claycomb WR, Shin D. A novel node level security policy framework for wireless sensor networks. J Netw Comput Appl 2011;34(1):418–28.
[4] Simplicio Jr MA, de Oliveira BT, Barreto P, Margi CB, Carvalho T, Naslund M. Comparison of authenticated-encryption schemes in wireless sensor networks. In: Proceedings of the 36th annual IEEE conference on local computer networks. Bonn, Germany; 2011. p. 454–61.
[5] Ben-Othman J, Bessaoud K, Bui A, Pilard L. Self-stabilizing algorithm for efficient topology control in wireless sensor networks. J Comput Sci 2013;4(4):199–208.
[6] Hu F, Sharma NK. Security considerations in ad hoc sensor networks. Ad Hoc Netw 2005;3(1):69–89.
[7] Heinzelman WR, Chandrakasan A, Balakrishnan H. Energy-efficient communication protocol for wireless microsensor networks. In: Proceedings of the IEEE 33rd Hawaii international conference on system sciences, vol. 8, Maui, HI, USA; 2000.
[8] Ozdemir S, Xiao Y. Secure data aggregation in wireless sensor networks: a comprehensive overview. Comput Netw 2009;53(12):2022–37.
[9] Younis O, Fahmy S. HEED: a Hybrid, Energy-Efficient Distributed clustering approach for ad-hoc sensor networks. IEEE Trans Mob Comput 2004;3(4):366–79.
[10] Fouchal S, Lavallée I. Fast and flexible unsupervised clustering algorithm based on ultrametric properties. In: Proceedings of the seventh ACM symposium on QoS and security for wireless and mobile networks (Q2SWinet'11), Miami, FL, USA, 2011.
[11] Fouchal S. A new clustering algorithm for wireless sensor networks. In: Proceedings of the fourth IEEE international workshop on performance evaluation of communications in DIStributed systems and WEb based service architectures (PEDISWESA'12), Nevşehir, Turkey; 2012.
[12] Liang Y. Efficient temporal compression in wireless sensor networks. In: Proceedings of the 36th Annual IEEE conference on local computer networks. Bonn, Germany; 2011. p. 470–78.
[13] Lai GH, Chen C-M. Detecting denial of service attacks in sensor networks. J Comput 2008;4:18.
[14] Guechari M, Mokdad L, Tan S. Dynamic solution for detecting denial of service attacks in wireless sensor networks. In: Proceedings of the 2012 IEEE international conference on communications (ICC'12), Ottawa, Canada; 2012.
[15] Ballarini P, Djafri H, Duflot M, Haddad S, Pekergin N. HASL: an expressive language for statistical verification of stochastic models. In: Proceedings of the fifth international ICST conference on performance evaluation methodologies and tools (VALUETOOLS'11). Cachan, France; 2011a. p. 306–15.
[16] Dak AY, Yahya S, Kassim M. A literature survey on security challenges in VANETs. Int J Comput Theory Eng 2012;4(6):1007–10.
[17] Alam S, De D. Analysis of security threats in wireless sensor network. Int J Wirel Mobile Netw 2014;6(2):35–46.
[18] Ben Othman S, Bahattab A, Trad A, Youssef H. Confidentiality and integrity for data aggregation in WSN using homomorphic encryption. Wirel Pers Commun 2014.
[19] Monnet Q, Mokdad L, Ben Othman J. Data protection in multipaths WSNs. In: Proceedings of the eighteenth IEEE symposium on computers and communications (ISCC'13), Split, Croatia; 2013.
[20] Shi H-Y, Wang W-L, Kwok N-M, Chen S-Y. Game theory for wireless sensor networks: a survey. Sensors 2012;12(7):9055–97.

[21] Guo P, Zhu J, Cheng YP, Uk Kim J. (1). Authentication mechanism on wireless sensor networks: A survey. In: Proceedings of the second international conference on information technology and computer science (ITCS'13). Beijing, China: 2013. p. 425–31.

[22] Guo P, Wang J, Zhu J, Cheng Y, Uk Kim J. (2). Construction of trusted wireless sensor networks with lightweight bilateral authentication. Int J Secur Appl 2013;7(5): 225–36.

[23] Bawa H, Singh P, Kumar R. An efficient novel key management scheme for enhancing user authentication in a WSN. Int J Comput Netw Inf Secur 2013;5(1):56–64.

[24] Kazatzopoulos L, Delakouridis C, Anagnostopoulos C. WSN location privacy scheme enhancement through epidemical information dissemination. Int J Commun Netw Inf Secur 2014;6(2):162–7.

[25] George CM, Kumar M. Cluster based location privacy in wireless sensor networks against a universal adversary. In: Proceedings of the International conference on information communication and embedded systems (ICICES'13). Chennai, India: 2013. p. 288–93.

[26] Varshovi A, Sadeghiyan B. Ontological classification of network denial of service attacks: basis for a unified detection framework. Scientia Iranica 2010;17(2):133–48.

[27] Singh SK, Singh MP, Singh DK. A survey on network security and attack defense mechanism for wireless sensor networks. Int J Comput Trends Technol 2011; May–June issue.

[28] Momani M, Challa S. Survey of trust models in different network domains. Int J Ad Hoc Sens Ubiquitous Comput 2010;1(3):1–19.

[29] Fernández-Gago MC, Roman R, López J. A Survey on the applicability of trust management systems for wireless sensor networks. In: Proceedings of the third international workshop on Security, Privacy and trust in Pervasive and Ubiquitous computing (SECPerU'07). Istanbul, Turkey; 2007. p. 25–30.

[30] Rohbanian MR, Kharazmi MR, Keshavarz-Haddad A, Keshtgary M. Watchdog-LEACH: a new method based on LEACH protocol to secure clustered wireless sensor networks. Adv Comput Sci Int J 2013;2(3):105–17.

[31] Perrig A, Szewczyk R, Wen V, Culler D, Tygar JD. SPINS: security protocols for sensor networks. Wirel Netw 2002;8(5):521–34.

[32] Islam MH, Nadeem K, Khan SA. Optimal sensor placement for detection against distributed denial of service attacks. In: Proceedings of the 2009 international conference on advanced computer control (ICACC'09). Singapore; 2009. p. 675–9.

[33] Ho J-W. Distributed detection of node capture attacks in wireless networks. In: Chinh HD, Tan YK, editors. Smart wireless sensor networks. InTech; 2010, [chapter 20] p. 345–60.

[34] Misra S, Krishna PV, Abraham KI, Sasikumar N, Fredun S. An adaptive learning routing protocol for the prevention of distributed denial of service attacks in wireless mesh networks. Comput Math Appl 2010;60(2):294–306.

[35] Son J-H, et al. Denial of service attack-resistant flooding authentication in wireless sensor networks. Comput Commun 2010;33(13):1531–42.

[36] Ghosal A, DasBit S. A lightweight security scheme for query processing in clustered wireless sensor networks. Comput Electr Eng 2014.

[37] Abbasi AA, Younis M. A survey on clustering algorithms for wireless sensor networks. Comput Commun 2007;30(14–15):2826–41.

[38] Handy MJ, Haase M, Timmermann D. Low Energy Adaptive Clustering Hierarchy with Deterministic Cluster- Head Selection. In: Proceedings of the fourth IEEE international workshop on mobile and wireless communications networks. Stockholm, Sweden; 2002. p. 368–72.
[39] Reddy BB, Rao KK. A modified clustering for LEACH algorithm in WSN. Int J Adv Comput Sci Appl 2013;4(5):79–83.
[40] Chawla N, Jasuja A. Algorithm for optimizing first node die (FND) time in LEACH protocol. Int J Curr Eng Technol 2014;4(4):2748–50.
[41] Oliveira LB, Ferreira A, Vilaça MA, et al. Signal Process 2007;87(12):2882–95.
[42] Mohi M, Movaghar A, Zadeh PM. A Bayesian game approach for preventing DoS attacks in wireless sensor networks. In: Proceedings of the 2009 international conference on communications and mobile computing (CMC'09), Kunming, Yunnan, China; 2009.
[43] Anastasi G, Conti M, Di Francesco M, Passarella A. Energy conservation in wireless sensor networks: a survey. Ad Hoc Netw 2009;7(3):537–68.
[44] Sivakumar P, Amirthavalli K, Senthil M. Power conservation and security enhancement in wireless sensor networks: a priority based approach. Int J Distrib Sens Netw 2014.
[45] Rehana J. Security of wireless sensor network. Technical report. Helsinki University of Technology; 2009.
[46] Ajmone Marsan M. Modelling with generalized stochastic Petri nets. John Wiley & Sons; 1995.
[47] Ballarini P, Djafri H, Duflot M, Haddad S, Pekergin N. COSMOS: a statistical model checker for the hybrid automata stochastic logic. In: Proceedings of the 8th international conference on quantitative evaluation of systems (QEST'11). Aachen, Germany; 2011b. p. 143–4.
[48] Kwiatkowska M, Norman G, Parker D. Stochastic model checking. In: Bernardo M, Hillston J, editors. Formal methods for the design of computer, communication and software systems: performance evaluation (SFM'07), vol. 4486. Springer; 2007. p. 220–70. Lecture Notes in Computer Science.
[49] Rakhmatov D, Vrudhula S. An analytical high-level battery model for use in energy management of portable electronic systems. In: Proceedings of the international conference on computer aided design (ICCAD'01). San Jose, CA, USA; 2001. p. 488–93.

CHAPTER

Formal methods of attack modeling and detection

29

Tarik Guelzim and Mohammad S. Obaidat
Monmouth University, West Long Branch, NJ, USA

1 INTRODUCTION TO COMPUTER SYSTEM THREATS

In many engineering disciplines, including the automobile and aviation industries, techniques such as *failure mode and effect analysis* (FMEA) have been used to identify risks and propose ways to control and attenuate them. FMEA, for instance, is used in early system design stages to ensure that common issues are addressed by the future car or plane. By identifying the risks, systems can be created more realistically [1]. This knowledge allows identification of the potential failures and tweaking of the design in order to prevent them at best.

While engineers have thus far been effective in using failure data analysis to improve their design, system and software engineers have not [1]. One of the major reasons is that organizations are wary of communicating such information for the fear of affecting public confidence, as well as the fear that attackers will use this data in order to plan and execute attacks on their systems, which often handle private data such as personal or banking information.

The paradigm of pervasive computing has thus far made the security challenges more difficult [2,3]. With attacks being designed to be more and more sophisticated, detecting and preventing them in real time is no longer adequately possible with passive log-based monitoring tools. To cope with this task, specialized software called intrusion detection systems (IDS) are designed to protect systems by detecting and taking corrective actions in real time. In addition, algorithmic solutions, both deterministic and probabilistic, are used to increase the precision and accuracy of analyzing and detecting attacks.

Throughout this chapter, we consider two major notions [4]:

- **Threat**: Any action that leads to access, copy, modification or compromise of an organization asset in an intentional or unintentional way.
- **Vulnerability**: A weakness, a flaw or a hole in an application, product or asset that makes it infeasible to prevent an attacker from gaining privileged access to an organizational system.

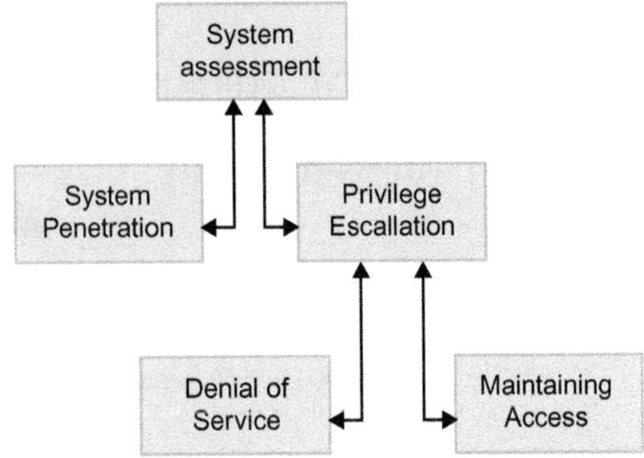

FIGURE 29.1

Classical attack methodology.

1.1 ANATOMY OF AN ATTACK

Understanding how attackers target applications and systems gives some insight on how to prevent this, to a certain degree, during the design phase.

The attacking methodology is summarized in the following paragraphs.

As illustrated in Figure 29.1, a classical attack methodology [5] revolves around these steps:

- **System assessment**: This is usually the first step an attacker takes. Understanding the characteristics of the system, open ports, applications, used technologies, operating systems, hosting services, private infrastructure, applied security and organizational procedures, business workflow. . ., etc. The attacker uses this information to plan an initial attack.
- **System penetration**: having planned and enumerated a list of vulnerabilities from step 1, the attacker uses the collected knowledge to gain entry to the system.
- **Escalate privileges**: After compromising the system, the next step would be to obtain elevated privileges, ideally an administrator account. A good preventive defense mechanism is to use "least privileged account" throughout the application. First access to a system shall always be routed to this type of account.
- **Maintain access**: After gaining access to the system for the first time, the challenge is to maintain the access to implement other attacks. It simply allows the attacker to set the basis for a later system penetration by installing back doors and back channels to simplify the access. Track-covering techniques are also used to hide any suspicious trace from logs or audit software.
- **Deny service**: This is a common technique used by attackers who cannot gain access. Denial of Service (DoS) attacks attempt to prevent other users from establishing network connections by flooding the server queue. A more sophisticated form of DoS is the distributed (DDoS), which attempts to enable a large number of machines that are geographically uncorrelated to flood the server queue. DOS and DDOS are catastrophic when they target critical infrastructure. A well-known denial of service attack is the SYN flood attack [6,7].

1.2 TOP LEVEL THREATS

Many security threats have surged as the technology and systems evolved [8]. They are summarized as follows:

- **Viruses**: These are disruptive small programs that replicate upon execution of infected software. The malicious code propagates throughout the system. Nevertheless, viruses do not rely on the network to propagate but rather on code execution.
- **Trojan-horses**: They are more advanced forms of viruses. Trojan-horses implement back doors to simplify future access to the host exploited.
- **Worms**: They are self-replicating shells that rely on networks to download their payload to the infected host. Later, they propagate through a network by exploiting system vulnerabilities.
- **Foot-printing**: along with viruses, this is one of the oldest techniques to plan attacks. It involves port scanning in order to determine operating system information, installed servers, server names, patch level, etc. This collected knowledge is then used to choose the right attack.
- **Password breaking**: Many systems disable anonymous connections given the security threats they present. Attackers are only left with establishing authenticated connections using valid accounts. Brute force attacks are still very common given the weak password tokens chosen by users.
- **Denial of Service**: Denial of service (DoS) attacks attempt to exhaust system resources by forcing the target host to handle many requests at once. They are also often aimed at many targets of the infrastructure at the same time in order to make the damage more severe.
- **Code injection & execution**: An attacker alters the sequence of executed program instructions by injecting code at a specific memory location and altering indirectly a CPU instructor pointer to that malicious code region. Although many technologies have been introduced by the Windows operating system such as Data Execution Protection (DEP) and SELinux for Linux systems, new innovative ways are still being discovered to compromise a program's execution using such attacks.
- **Unauthorized access**: Accessing system resources using valid credentials or inadequate access control (ACL) configuration often leads to elevated privileges in a system.

1.3 CLASSIFYING THREATS AND ATTACKS: STRIDE

Given the multitude of existing attacks, we need a well-defined classification model to organize them. One of the most commonly used ones is the Microsoft STRIDE classification method [5] to categorize different threats, described as follows:

- Spoofing: Any attack that allows gaining access to a system using the identity of a legitimate user. Such a method requires stealing credentials. As an entry point, the attacker starts privilege elevation to gain wide access to the compromised system.

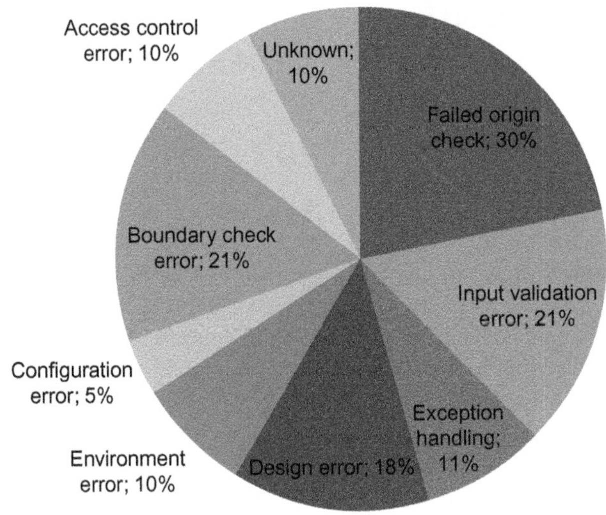

FIGURE 29.2

Vulnerabilities breakdown by type [9].

- **T**ampering: Tampering is any unauthorized modification of data so as to compromise the integrity of the information flowing intra- or inter-system.
- **R**epudiation: This relies on mathematical irrefutable evidence proving that a user or system has originated an action. This type of attack attempts to exploit weaknesses in the system to compromise its repudiation properties, but not the repudiation concept itself.
- **I**nformation disclosure: Any attack that results in unwanted and unauthorized exposure of nonaccessible data (example: a document, a database table, etc.)
- **D**enial of Service: any vulnerability making a system or resource inaccessible for normal use.
- **E**levation of privilege: Any vulnerability that allows a limited account to gain access and assume the identity of a higher privileged user in the system.

1.4 THREATS AND VULNERABILITIES BREAKDOWN

In order to explore and define techniques as well as tools to prevent attacks, it is necessary first to understand and analyze the attack landscape of systems and its breakdown structure. The data in Figure 29.2 from [9] gives historical vulnerabilities by type.

As explained in Figure 29.2, 75% of all major vulnerabilities are due to failed verifications of either input data or data boundary check (see the Heartbleed Bug case study). Poor design alone amounts for 18% of vulnerabilities in systems. A mature system development process, such as system design security reviews

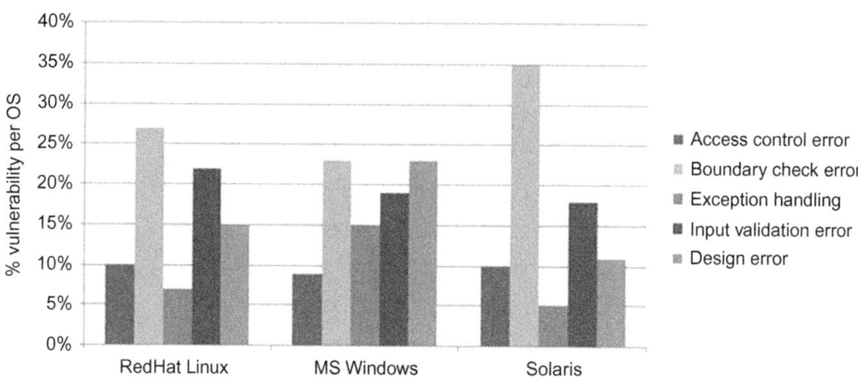

FIGURE 29.3

Vulnerability distribution by OS [9].

and white hat testing, can be effective tools to aid in detecting these weaknesses and preventing them before being used by attackers.

If we filter this data to see how effective operating systems are, we get the graph shown in Figure 29.3. In this figure, we notice that boundary checking accounts for one-third of all attacks in all three operating systems; input validating is the second largest attack vector and design errors impact all three operating systems, with a slight advantage of *NIX systems over Windows ones. These statistics demonstrate that the most basic programming and design errors today are behind some of the major attacks on systems.

2 MODELING COMPUTER SYSTEM ATTACKS

Designing effective and secure systems is best accomplished through the use of modeling. Effective modeling is not easily achievable, as it requires the following two important characteristics to be present:

- **Simplicity:** i.e., presenting complex attack data and information in "understandable" terms. This is often important in case we want to explain the potential threats to the system to people that are not well-versed in the security topic yet who are required to make decisions on what direction the system is going.
- **Relevance:** attempts to organize the information in a practical way for real-world problems. In addition, the model must take into account the characteristics of potential attackers.

Security researchers today acknowledge that various attacks are easy to start and scale without any advanced knowledge of the inner computer system's workings or advanced computer science or cryptography topics. Major security

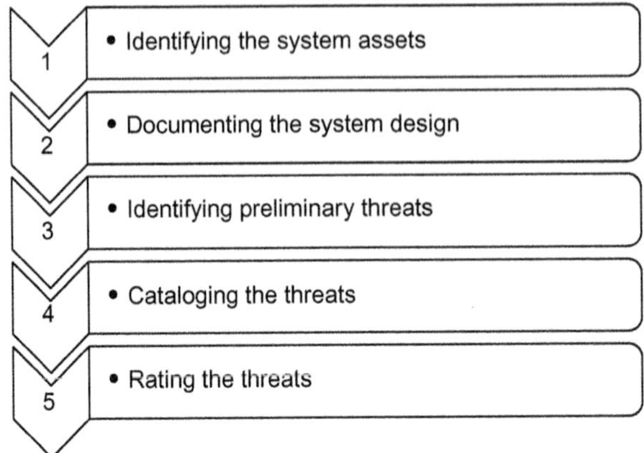

FIGURE 29.4

Iterative threat modeling process.

researchers and agencies have reported that recent and very sophisticated attacks are carried out using either Botnets or dormant low activity computers in order to disguise and not raise warnings or suspicions. All of these facts must be taken into consideration when modeling attacks.

In this chapter, we explore some practical modeling techniques.

2.1 THREAT MODELING PRINCIPLES

As the basis of a mature design process, threat modeling is an iterative process that spans the entire system or application life cycle [5]. There are two major reasons why this is important:

- During the first stages of the system design phase, it is very difficult to identify all possible threats. A few companies have throughout the years maintained threat catalogs that they refer to, as part of a capitalization process, whenever they start the design process.
- Given the evolving nature of information systems, threat severity changes with time while other threats emerge due to requirement changes. Maintaining a catalog can prove to be very valuable.

In a nutshell, threat modeling involves the activities shown in Figure 29.4. These modeling steps can be described as follows:

- Identifying system assets requires identifying parts of the system that must be protected. With limited resources, not all system components can be protected so a priority list must be established. To the reader, this idea might sound counterintuitive as we preach for integrated system security. Targeted asset protections can be illustrated, such as a military network for example. These

networks carry very sensitive information with tens of thousands of nodes distributed across the globe. It is simply not feasible to secure this entire network, and it is also not necessary. Protecting assets (hardware, data, network packets, etc.) based on their value will give greater results.
- Documenting the system architecture allows one to understand the subsystems, trust boundaries, component coupling and information flow. This information will later allow a rapid response to threats.
- Identifying the preliminary threats is considered a design tool for system architects. It allows for guiding of architecture reviews and technical risk mitigation.
- Cataloging the threat facilitates managing the system failures as they evolve but also creates a knowledge base for recurring system design.
- Rating the threats assigns a severity and probability of occurrence to each threat.

2.2 D.R.E.A.D THREAT RATING MODEL

Rating threats can be very different from one team to another, as viewpoints are mixed. In product management, as an example, how a technical team member views a particular threat is going to be different from the view of the sales or marketing team, or even executive management. Each has a set of concerns that should be addressed when rating the threat. To reconcile the point of views, the DREAD model allows these differences to be managed using a meta-data approach, described as follows:

- **D**amage potential: how big the damage is if the vulnerability is exploited.
- **R**eproducibility: how often the threat can be played.
- **E**xploitability: how complex it is to start the attack.
- **A**ffected user: the data set that is affected by the threat.
- **D**iscoverability: how easy it is to find the vulnerability.

Each of these elements is rated High (1), Medium (2) and Low (3) and their sum gives a single threat value. The higher the final sum, the more important the threat is.

2.3 MODELING ATTACKS USING ATTACK TREES

An attack tree (AT) is a visual aid tool used to model and understand attack vectors [9]; the term was coined by Schneier [10] in the late 90s. An attack can be modeled as follows using an AT:

- The root node of the tree is the ultimate goal of the attacker.
- Child nodes are different viewpoint representations of this attack or subgoals.
- Attack trees have two types of nodes: and-nodes (conjunctive) and or-nodes (disjunctive).
- Child nodes of an and-node must all be executed, while in the case of an or-node, one may not be.

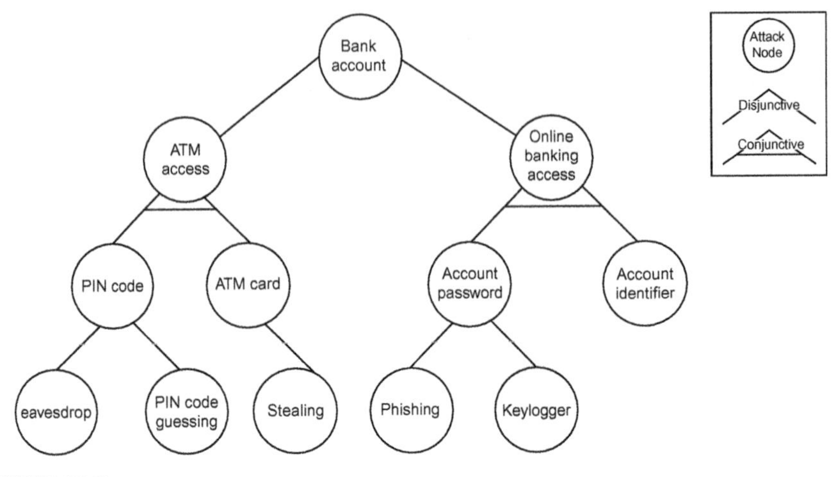

FIGURE 29.5

Attack tree modeling applied to simplified bank account attack.

Figure 29.5 shows an example of an attack tree for a simplified bank account attack. Modern banking services offer online banking access in addition to traditional ATM access. In ATM access, both the bank card and a personal identification number (ATM) are required to withdraw, deposit or access account balances. Similarly, online banking requires a user account and password to open a bank session in the client browser. Simple attacks can range from eavesdropping the PIN to extracting the password using a key logger installed in a compromised machine.

Analyzing an attack tree allows questions to be answered such as:

- What is the attack?
- What is the internal structure of the attack?
- Under what conditions can the attack can occur?
- Etc.

Attack trees can also be extended by the mathematical formalism of forests, directed acyclic graphs (DAG) or graph theory, for short. For example, by means of algebraic properties, attack trees may be transformed structurally (using the reduction rule, for instance) to represent the same attack in simplified form while preserving all the semantics of the original.

2.4 ATTACK TREE CAPABILITY ANALYSIS

Representing attacks in AT form is a first step in understanding the different dimensions of volatilities and their impact on a system [10]. As we saw in the previous section, simple algebraic properties may apply to attack trees such as association, reduction, transitivity, and so forth. Nevertheless, in order to understand the attack and its "business impact form," we are required to decorate these mathematical properties with other meaningful attributes. For example, we can assign weights, cost units, to each vertex from the bottom up. A simple product would indicate the risk/cost ratio that may further be used to prioritize the different scenarios given by the formula:

Risk = Probability of an event × Resulting damage

This formula is often considered too generic and rather simplistic for practical situations. This is why capability-based analysis is introduced. It relies on the following simple premise:

IF they want to AND they can THEN they WILL

In other words, if attackers have the motivation to exploit a system and the necessary capabilities such as money, skill and mostly the willingness to accept consequences, then they may be expected to attack the system.

Given this explanation, [10] is often rewritten as follows:

Risk = Threat × Vulnerability × Resulting damage

So we can write:

Event probability = Threat × Vulnerability

Using the attack tree concept described in the previous section, we can make a good estimate of the event probability of an attack. This is often done by associating each leaf node with the resources required to carry the attack. The resources required at any node in the attack tree are a direct reflection of the vulnerability to be exploited.

3 ANALYSIS AND DETECTION OF COMPUTER SYSTEM ATTACKS

Although attack trees are also used in attack analysis as well as in modeling, there exist other techniques that rely on machine learning algorithms to analyze data based on input parameters. Each of these has its characteristics that makes it suitable for solving a particular data analysis problem: attack trees are used for a priori analysis while machine learning and statistical tools are used for static and real-time analysis and detection.

In their real implementations, these modeling algorithms are at the core concepts of intrusion detection systems (IDS) [4]. IDS have many variations and are categorized as follows:

- *Signature-based intrusion detection systems*: This IDS category works by comparing data packets with the signatures and attributes database of known intrusions.
- *Host intrusion detection systems (HIDS):* This IDS category is installed on targeted hosts to monitor system log activity, critical system files and to detect abnormal activity.

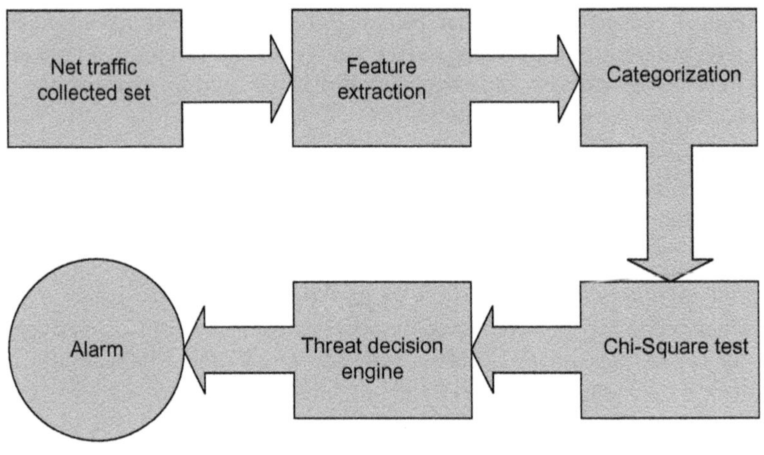

FIGURE 29.6

Chi-square detection framework.

- *Network intrusion detection systems (NIDS*): NDIS are used to detect attacks against a number of networked systems inside the same network segment. They are usually embedded in routers and gateways to prevent attacks before they reach the targeted nodes
- *Network node intrusion detection systems (NNIDS)*: NNIDS are used on critical systems such as databases or network storage systems where sensitive information is stored and accessed via network.

In this chapter, we present a survey of the statistical and machine learning tools.

3.1 STATISTICAL ANALYSIS

Top-down statistical techniques are employed when we have a first and often vague idea as to the relationship we are looking for. Statistical analysis techniques can be categorized as linear, nonlinear and decision trees, and can be calculated with respect to time frame or on a per-host basis or per service [11]. Having statistical models will lead to fully automated attack detection.

3.1.1 Chi-square test

The chi-square test is a commonly used test for the goodness of fit of an observed distribution to a theoretical one. In [12–14], the authors described using the chi-square test modeling technique on entropy values of the TCP packet headers. The chi-square test uses data training techniques to categorize and determine how far observed packet distributions are from expected packets.

In Figure 29.6, the detection model [13] works as follows:

- Network traffic is collected in a simulated normal set-up.
- Collected samples are preprocessed and data features are extracted.

- Extracted data features are categorized into expected traffic.
- Received traffic, i.e. observed traffic, is compared to the expected traffic using chi-square.
- A decision engine launches alarms based on preconfigured thresholds.

3.1.2 Multivariate correlation analysis

Multivariate correlation analysis measures how a variable can be predicated using a linear function of a set of other variables. In [15], the Distributed Denial of Service SYN flooding attack was analyzed using multivariate correlation algorithms. The authors showed that the method is highly accurate in detection of malicious network traffic. In addition, the linear complexity of this algorithm makes it practical for near real-time use.

Although there are a few issues with the statistical-based detection methods to determine with a high certainty whether the network packet distribution is normal or not, authors in [14] have suggested complementing the above methods with clustering algorithms. In [15,16] for instance, a new method relied on time-to-live (TTL) value to calculate packet hop count information. In public accessed systems, understanding the origin of the traffic and the growth path of traffic in different regions can give some good insights as to whether source packets can be accepted or simply dropped.

3.1.3 Hidden markov models

A hidden Markov model (HMM) is a statistical tool that models a system as a Markov process with unknown parameters. It attempts to determine the hidden parameters from the observed ones [17]. HMMs can be used to detect complex internet attacks that have big noise ratio due to change in action sequence at each and every execution of the "same" attack. In these attacks, a perpetrator may use a different set of actions within each step to mask and throw off traditional IDS systems that rely on attack "fingerprint" detection. Faced with continuously morphed attacks, these engines may become useless. For this reason, HMMs have been demonstrated to be very useful to detect a multi-step attack problem. In [18], authors applied HMMs to system call data and concluded that the best performance was obtained when mapping application system calls to a number of states, i.e. Markov process. They were also demonstrated to perform better than neural networks when trained to the same datasets [19].

3.1.4 IDS using static analysis

One of the many challenges in intrusion detection is organizing and categorizing attacks while keeping the false alarm ratio low [20]. Using a static analysis model of an application behavior allows creating a host-based intrusion detection system. In [20], the authors argue that formal methods are not sufficient for building and deploying secure systems. In addition to formal methods, we need to ensure that the program execution is consistent with the program's source code. For this reason, a framework is necessary to detect the cases when an application is penetrated and exploited to harm other parts of the system. This can be accomplished

BOX 29.1 PSEUDO-CODE TO OPEN AND VERIFY THE VALUE OF THE FIRST LINE.

```
Function f1: a ←h_Handle
a.readLine()? "value_1" : "value_2"; // read first line
Function f2:
  Handle = open("file")
  F1(Handle)
  Close(Handle)
Exit
```

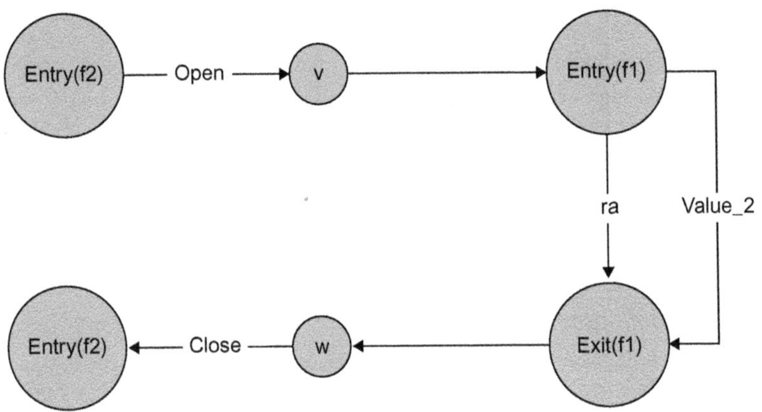

FIGURE 29.7

Control flow for simple program.

by precomputing a model of expected application behavior, which is built statically from program source code, and then have programs monitor system call traces for compliance with this model.

In order to build a model, we can use control flow analysis. A control flow graph $G = \{V, E\}$, where G is the graph composed of vertices (V) and edges (E), is associated with the source code.

Let's look at an example program.

The pseudo-code in Box 29.1 opens a file and verifies if the first line value equals "value_1" or "value_2" and then closes the file.

In Figure 29.7, we attempt to translate the pseudo-code to simplified control flow notation. A static analysis based IDS would monitor this program system call against the control flow in this figure, and then raise an alarm if the same program executes an additional instruction that might have been injected by a virus, for instance.

3.2 MACHINE LEARNING

Machine learning algorithms are designed to improve automatically through experience. This property is of great use in an intrusion detection system because

Table 29.1 Example of Selected Features and Their Potential Information Significance

Feature	Interpretation
Number of TCP FIN flags	Average duration/all services
Number of ICMP packets	Average duration/current host
Number of RST flags	Destination bytes
Number of SYN flags	Destination IP
Number of packets to services	Bytes transferred / current services

Table 29.2 Test Results of Neural Network Based Intrusion Detection

	Correct Normal Prediction	False Negatives	Correct Attack Prediction	False Positives
Mixed attacks	100%	0%	24%	76%
Single attack	100%	0%	100%	0%

attacks tend to vary greatly and become more and more sophisticated. In complex attacks, no well-known patterns are immediately available, which renders statistical analysis of no practical use. In this chapter we explore some of the most commonly used machine learning algorithms applied to IDS problems.

3.2.1 Feature selection

Feature selection is a commonly used data preprocessing technique and is used in combination with other machine learning techniques to create "knowledge." Most of the time, using all domain features may seem impractical for use or can be intractable in computation time. Table 29.1 lists some examples of selected features in a system.

Features are extracted based on rank with respect to their contribution to the system information and are weighted accordingly. Note that features may carry different information based on their context of interpretation during the machine learning phase.

3.2.2 Neural networks

In [21,22], the authors present a neural network based approach to detect malicious network traffic. The proposed IDS analyses tcpdump data to developed time framed traffic intensity patterns. The network traffic is synthesized into the following:

- The number of times a host is accessed via its services in a time boxed interval;
- The ports that are monitored from which the network frequency data is attached.

Table 29.2 shows the results of such a technique employed by [22].

As this table demonstrates, the test results of the experiment led to the following performance conclusions:

- Neural networks are suitable for detecting single types of attacks with 100% correct prediction.
- Neural networks lag behind in detecting mixed attack types simultaneously, with only 24% correct prediction.

3.2.3 Genetic algorithms

Genetic algorithms have emerged from the field of computational biology and are trained to learn various patterns which make them applicable for intrusion detection applications [23]. In [24,25], the authors trained the REGAL system to learn the first order logic on multi-model descriptions. The REGAL system relies on relational databases to understand the relation between the different tuples. In [26], the authors classified particular feature anomalies in a 2-bit system as *normal* and *dangerous*. Given the simplicity of this decision "rating" system, it proves to be of practical use in scenarios where near real-time attack detection is required.

3.2.4 Fuzzy logic

Fuzzy logic is derived from fuzzy set theory and deals with finding an approximate rather than a definite, precise pattern. In [27,28], the authors have described the use of fuzzy data mining techniques to extract patterns from network traffic data in order to detect or classify normal from malicious activity. To detect anomalous behavior, fuzzy association rules are generated from system and network data and a similarity set is computed from a "normal" system activity. Most of the work in this area is based on Borgelt's prefix trees because of their practicality in reducing the number of rules and increasing the accuracy of the IDS. Using fuzzy logic in detecting anomalies demonstrated that it allows generating abstract enough yet flexible patterns for anomaly detection.

4 CASE STUDIES

4.1 INTRUSION DETECTION SYSTEM USING NETWORK SECURITY MONITORING

Intrusion detection systems (IDS) are designed to deal with the many modern attacks; nevertheless, system administrators complain about the multiple "false positives." A few causes of this are the use of encryption, code obfuscation and also polymorphism (object-oriented paradigm), which makes it harder to detect by signature matching algorithms.

Network security monitoring (NSM) has been introduced to give an edge on the above limitations. They are applications (may also take the form of a large system) that collect, analyze and escalate warnings to detect warnings. NSM integrates IDS software at its core along with customized processes to warn and escalate the right threat, at the right risk level and at the right moment. NSM defines two major types of warnings, which we will describe in the following sections.

4.1.1 Event data

Event data is triggered and generated by IDS systems given some preconfigured rules.

As an example, using "snort," which is a lightweight IDS, we can set up rules describing a violation. Network packets are checked one by one against the rules. Any match would trigger alerts that are logged and/or sent to qualified personnel for analysis (such as system administrators). The rather basic rule shown in the pseudo-code in Box 29.2 as an illustration would log any packet that enters a network and that is destined for an application port that is less than 6000.

4.1.2 Session data

In order to be meaningful, event data needs to be complemented with session data. Session data contains information about network connections, source IP address and port, destination IP address and port, and packet payload. Event data is a very important "piece of evidence" in case of an attack analysis. It allows us to have an insight into what the attacker is searching for and what information he/she is attempting to get hold off or has already compromised.

Given the rule above, the traffic in Table 29.3, will generate an alarm because the destination port is less than 21 as well as the destination machine IP is in the IDS protected range. Once the alarm is triggered, data recording on hard disk will

BOX 29.2 PSEUDO-CODE CONFIGURATION RULE TO LOG SPECIFIC TCP PACKET.

```
log tcp any any -> 192.168.3.0/24 :6000
```

Table 29.3 Example of Session Data Log (Truncated)

Time	Source IP	Source Port	Destination IP	Destination Port	SP	SB	DP	DB
00:01:43	173.17.89.36	5678	192.168.3.2	21	5	64	7	256
00:02:09	173.17.89.36	5678	192.168.3.2	21	6	48	9	56
00:02:15	173.17.89.36	5678	192.168.3.2	21	10	10	15	64

- ***SP**: source packets. The number of packets sent by the source*
- ***SB**: Source bytes. The number of bytes sent by the source*
- ***DP**: Destination packets. The number of packets sent by the destination*
- ***DB**: Destination bytes. The number of bytes sent by the destination*

start in order to allow further analysis. A first analysis would successfully give us the origin of the attack: IP 173.17.89.36, and the data that the attacker is attempting to access. Nevertheless, given that attackers masquerade behind multiple anonymized networks and machines, e.g., using The- Onion-Router (TOR) network, the node that gets logged in our case is only the "last exit node," for example. Determining the "real" origin of the network would require more difficult analysis from specialized forensic analysis teams.

We would simply know that the attacker attempted to establish a communication to port 21. This is the standard port for FTP service. Clearly, this attack is an attempt to seize a specific file or multiple files. We need to parse all the raw packets that we would have logged in order to replay the traffic. Many tools exist to simplify this task, such as "Wireshark" network analyzer tool.

This is of course a basic example to illustrate analysis in the realm of security; however, merging these basic rules together can result in very robust and complex intrusion detection systems.

4.2 THE HEARTBLEED BUG: AN EXAMPLE OF INPUT DATA BOUNDARY VALIDATION ATTACK

The heartbleed bug [29] is a security vulnerability in the OpenSSL software. OpenSSL is a cryptographic library used ubiquitously as an implementation of SSL/TLS protocols in web servers and applications. This bug allows stealing of information that is protected by SSL/TLS, putting at risk the privacy of data that transits the internet. First impact analysis suggests that at the time this bug was discovered (April 2014), 17% of all secure web servers in the internet (around half a million) were affected. Security specialists and companies have described this bug as catastrophic, since a successful attack analysis suggested that it is possible to steal private keys and cookies as well as passwords.

4.2.1 Anatomy of the vulnerability

The heartbleed bug is caused by a buggy implementation of the SSL/TLS extension Heartbeat protocol (RFC 6520), which allows testing the secure communication channel by sending a heartbeat request message. Due to an input validation error, when the originating message is sent with a normal payload but with a larger length than the message itself (normally it should be the payload's length), the server responds with up to 64 kilobytes of memory that was previously used by the OpenSSL library. An attacker can thus replay the same attack over and over and retrieve 64KB at each run. Successful post-attack analysis could retrieve the server private master key. This could enable passive eavesdropping and man-in-the-middle attacks.

Further analysis has shown that the attack is not just a one way—i.e. from client to server—but can also be exploited in reverse, known as the *reverse heartbleed*. A server may be able to read client's memory data. Google has also

BOX 29.3 PSEUDO-CODE, RFC 6530 HEARTBEAT MESSAGE STRUCTURE.

```
Struct {
  HeartbeatMessageType type;
  uint16 payload_length;
  opaque payload[HeartbeatMessage.payload_length];
  opaque padding[padding_length];
} HeartbeatMessage;
```

BOX 29.4 PSEUDO-CODE, RFC 6530 HEARTBEAT MESSAGE SSL STRUCTURE.

```
struct ssl3_record_st {
  unsigned int length; /* Number of available bytes */
  /** code eliminated to illustrate the bug */
  unsigned char *data; /* A pointer to the data */
  /** code eliminated to illustrate the bug */
} SSL3_RECORD;
```

BOX 29.5 OPENSSL FAULTY INPUT VALIDATION CODE.

```
*bp+ + = TLS1_HB_RESPONSE;
s2n(payload, bp);
memcpy(bp, pl, payload);
```

announced that Android version 4.1.1, also known as jellybean, is vulnerable to the attack. This leaves us with over 50 million devices that remain unpatched.

4.2.2 Attack analysis

To understand the vulnerability, we are required to analyze the OpenSSL source code.

The Standard RFC 6530 heartbeat message is shown in Box 29.3.

This heartbeat message is encapsulated inside the SSL3_RECORD (Box 29.4).

In this pseudo-code, we describe the basic C structure where `SSL3_RECORD->data` points to the payload and `SSL3_RECORD->length` points to the data size.

The issue is that the length field is never used and checked against in subsequent code and only the value `payload_length` sent by the sender is used (Box 29.5).

`Payload` is never verified against the real payload size. `Memcpy C function` will just copy memory up to the given size. In a typical scenario, the following data exchange would occur between client and server agents.

The heartbeat message is sent to the server. Table 29.4 shows the structure of the HeartbeatMessage record client request.

Table 29.4 HeartbeatMessage Record Request

HeartbeatMessage record		
Type	**Length**	**Data**
TLS1_HB_REQUEST	65535 bytes	1 byte

Table 29.5 HeartbeatMessage Record Response

HeartbeatMessage record			
Type	**Length**	**Data**	**Server private memory**
TLS1_HB_REQUEST	65535 bytes	65535 bytes	Data in this section is returned too

As shown in this table, the client sends a heartbeat message record request with type TLS1_HB_REQUEST with the "real" record length of 1 byte but an "intentionally wrong" data length of 65K bytes.

The server will respond as shown in Table 29.5.

As shown here, the server parses the client request that totals the "intentionally wrong" 64K bytes and creates a server response with 65K bytes message that includes server private memory using the `memcpy` C function. The attacker can now iterate through server memory 65K at a time.

5 CONCLUSIONS

This chapter provides an up-to-date summary of the trends in modeling and analyzing computer system attacks. With attacks becoming more sophisticated at various levels, providing a categorization model simplifies understanding them as well as evaluating their risk. The STRIDE model was explained and suggested for this purpose, given its simplicity. Modeling computer attacks is often considered a system design tool nowadays. Instead of awaiting attacks and then providing corrective measures, modeling potential attacks can lead to preventive solutions that attempt to thwart them. Attack trees are widely used modeling tools to visualize the attack vector with a complete view of both conjunctive and disjunctive vulnerability scenarios that might lead to the attack. Techniques to analyze and detect system attacks revolve around both statistical and machine learning methods. Statistical analysis allows attack detection based on static properties and attributes of data such as the fingerprint of traces and logs patterns in order to match observed data to the normal. Machine learning methods, on the other hand, allow training algorithms to learn the system under normal behavior and detect anomalies as well as learn new ones.

As attacks become more sophisticated, the necessity for novel detection and analysis techniques becomes essential. Nevertheless, prevention through a mature system development practice such as reviews, continuous system audits and system improvements remains the main aspect of effective system security.

REFERENCES

[1] Obaidat MS, Guelzim T. In: Obaidat MS, Misra S, editors. Fundamentals and issues with cooperative networking, in cooperative networking. Wiley; 2011. p. 7–20.

[2] Sabahi F, Movaghar A. Intrusion detection: a survey, systems and networks communications, 2008. ICSNC '08. 3rd International Conference on; 26–31 October 2008. p. 23,26.

[3] Guelzim T, Obaidat MS. In: Obaidat MS, Denko M, editors. Security and privacy in pervasive networks, in pervasive computing and networking. Wiley; 2011. p. 161–73.

[4] Obaidat MS, Boudriga N. Security of e-systems and computer networks. Cambridge University Press; 2007. p. 12–4.

[5] Meier JD, Mackman A, Dunner M, Vasireddy S, Escamilla R, Murukan A. Threat and countermeasures. [Online]. Available from: <http://msdn.microsoft.com/en-us/library/ff648641.aspx>; June 2006.

[6] Eddy W. TCP SYN Flooding attacks and common mitigations, IETF, RFC 4987, [Online]. Available from: <http://tools.ietf.org/html/rfc4987>; August 2007.

[7] Wang H, Zhang D, Shin KG. Detecting SYN flooding attacks, proceedings of the IEEE computer and communications societies (INFOCOM 2002), vol. 3; June 2002. p. 1530–39, p. 23–7.

[8] Jin C, Wang H, Shin KG. Hop-count filtering: an effective defense against spoofed DDoS traffic. In: Proceedings of the 10th ACM conference on computer and communications security, (CCS 2003); October 2003. pp 30–41.

[9] Chen S, Xu J, Kalbarczyk Z, Iyer RK. Security Vulnerabilities: From Analysis to Detection and Masking Techniques. Proc IEEE 2006;94(2):407–18.

[10] Schneier B. Attack trees: modeling security threats. dr. dobbs' journal; 1999.

[11] Krügel C, Toth T, Kirda E. Service specific anomaly detection for network intrusion detection. proceedings of the 2002 ACM symposium on applied computing, SAC '02, Madrid, Spain; March 2002. p. 201–8.

[12] Feinstein L, Schnackenberg D, Balupari R, Kindred D. Statistical approaches to DDoS attack detection and response, DARPA information survivability conference and exposition, DISCEX; April 2003. p. 303–14.

[13] Ye N, Chen Q. An anomaly detection technique based on a chi-square statistic for detecting intrusions into information systems. Qual Reliab Engng Int 2001;17:105–12.

[14] Obaidat MS, Boudriga N. Fundamentals of Performance Evaluation of Computer and Telecommunication systems. Wiley; 2010.

[15] Xiang Y, Li Z. An analytical model for DDoS attacks and defense, computing proceedings of the 2006 International multi-conference on global information technology, ICCGI '06, Athens, Greece; August 2006. p. 66.

[16] Jin C, Wang H, Shin KG. 2003. Hop-count filtering: an effective defense against spoofed DDoS traffic, In: Proceedings of the 10th ACM conference on computer and communications security, CCS '03; October 2003. p. 30–41.

[17] Ourston D, Matzner S, Stump W, Hopkins B. Applications of hidden Markov models to detecting multi-stage network attacks, Proceedings of the 2003 IEEE annual hawaii international conference on system sciences, HICSS; January 2003. p.10, 6–9.

[18] Warrender C, Forrest S, Pearlmutter B. Detecting intrusions using system calls: alternative data models, proceedings of the 1999 IEEE symposium on security and privacy, 1999. p.133–45.

[19] Xie Y, Yu S-Z. A novel model for detecting application layer DDoS attacks, proceedings of the 2006 IEEE first international multi-symposiums on computer and computational sciences, IMSCCS'06, vol. 2; 20–24, June 2006. p. 56–63.

[20] Wagner D, Dean D. Intrusion detection via static analysis, Proceedings of the 2001 IEEE symposium on security and privacy, S&P 2001; May 2001. p. 156–68.

[21] Lee SC, Heinbuch, DV. Training a neural-network based intrusion detector to recognize novel attacks, Proceedings of the 2001 IEEE transactions on systems, man and cybernetics, part a: systems and humans, vol. 31, no.4, July 2001. p. 294–9.

[22] Shun J, Malki HA. Network intrusion detection system using neural networks, Proceedings of the 2008 IEEE fourth international conference on natural computation, ICNC'08, vol. 5, 18–20; October 2008. p. 242–6.

[23] Ektefa M, Memar S, Sidi F, Affendey LS. Intrusion detection using data mining techniques, Proceedings of the 2010 IEEE 2010 International conference on information retrieval & knowledge management, CAMP; 17–18, March 2010. p. 200–303.

[24] Neri F. Comparing local search with respect to genetic evolution to detect intrusions in computer networks, Proceedings of the 2000 IEEE congress on evolutionary computation, vol. 1; July 2000. p. 238–43.

[25] Neri F. Mining TCP/IP traffic for network intrusion detection, In: de M'antaras RL, Plaza E. editors. In: Proceedings of the 2000 european conference on machine learning, ECML 2000, vol.1810, Barcelona, Spain; May 2000. p. 313–22.

[26] Dasgupta D, Gonzalez FA. An intelligent decision support system for intrusion detection and response, Proceedings of international workshop on mathematical methods, Models and architectures for computer networks security, MMM-ACNS, St. Petersburg. Russia, Springer; May 2001. p. 21–3.

[27] Dickerson, JE, Dickerson JA. Fuzzy network profiling for intrusion detection, Proceedings of NAFIPS 19th International conference of the north american fuzzy information processing society, NAFIPS; July 2000. p. 301–6.

[28] Florez, G., Bridges, S.M., Vaughn, R.B., An improved algorithm for fuzzy data mining for intrusion detection, Proceedings of the 2002 Annual meeting of the north american fuzzy information processing society, NAFIPS; June 2002. p. 457–62.

[29] CODNOMICON. The heartbleed bug. [Online]. Available from: <http://heartbleed.com>; April 2014.

CHAPTER

Security analysis of computer networks 30

Key concepts and methodologies

Gürkan Gür[1,2], Şerif Bahtiyar[1], and Fatih Alagöz[2]
[1]*Provus-A MasterCard Company, Ayazaga, Istanbul, Turkey*
[2]*Bogazici University, Istanbul, Turkey*

1 INTRODUCTION

The prevalent diffusion of computer and information networks into the daily life and critical operations of all organizations and citizens is an ongoing phenomenon. Accordingly, the Internet has become an indispensable component of our civilization's communication and interaction substrate. However, this vast interconnection facilitated by ubiquitous communication infrastructure has generated a tremendous increase in the variety and quantity of cyber-threats. Traditional IT (information technology) security threats have expanded into new sources of threats, such as social networking, cloud computing, smart mobile devices or "bring your own device" (BYOD) policies [1]. The security requirements for a computer network are critical for its performance since it comprises the fundamental requirements of network availability, usability, data privacy and trustworthiness.

Network security is a discipline concerned with protection of computer networks and related systems against security threats while ensuring trusted and controlled access to network assets [2]. These threats may include malware, data leakage, various exploits, and Denial-of-Service (DoS) attacks. An important part of this field deals with the security modeling and analysis of computer networks since these efforts provide a common ground and understanding of such systems. Moreover, they provide analytical and empirical support for security functions such as attack detection, risk analysis, countermeasure selection, and network monitoring.

Network security is concerned with vulnerabilities and attacks exploiting these vulnerabilities. Network attackers venture to identify vulnerabilities based on services open on systems and collect relevant information for launching a successful attack [3]. An attack can take many forms, such as a Trojan attack,

DoS/Distributed DoS (DDoS) attack, or a scan attack. Some of these attacks and exploits are linked in time and space (for instance, malware such as bots can be used for DDoS). Exploitation of web vulnerabilities using SQL injection and cross-site scripting, which are injection-based attacks in which malicious codes are injected and executed into trusted web sites for exploits, are very common, with the Internet becoming the main communications medium. Password-based attacks and eavesdropping try to gain access to confidential data and compromise data integrity. On the other hand, DoS/DDoS targeting availability of systems uses a large number of compromised hosts and it is very difficult to detect the original source of such an attack [3]. Malware such as worm/Trojan/bot can be very instrumental in facilitating many attack types in a pervasive manner. Another technically simpler but yet effective attack type is social engineering. It uses human error or weakness to infiltrate any system despite the layers of defensive security mechanisms/controls for gaining access to unauthorized resources.

On the defense or countermeasure side, firewalls are the first line of defense for network-based attacks. For addressing security requirements, firewalls have evolved into more advanced solutions with the addition of intrusion detection and prevention, i.e., Intrusion Detection and Prevention Systems (IDS/IPS). Recently, Unified Threat Management (UTM) solutions have emerged, which envisage an all-inclusive security appliance with multiple security functions entailing firewall, network intrusion prevention and gateway antivirus/anti-spam with additional capabilities such as Virtual Private Network (VPN) and content filtering. For malware protection and defense, antivirus and anti-malware tools are instrumental, especially for host-based settings. Moreover, the deployment of IPv6 (Internet Protocol version 6) with IPSec (Internet Protocol Security) and the utilization of Secure Sockets Layer (SSL) in IPv4 are important approaches for strengthening the security of computer networks. Especially, SSL has enabled the explosive growth of electronic commerce and secured transactions over the Internet.

As outlined above, the major components and services of an ICT (Information and Communication Technology) system and evident threats render a rather complex environment for security requirements. In that regard, network security analysis is critical to facilitate a secure network environment. This topic is also closely related to security modeling of computer networks. This chapter discusses security analysis for computer networks focusing on security threats, network attacks, and how to analyze networks for security. We provide a comprehensive overview, which is instrumental for security analysts to investigate the security level of a network effectively. First, we present security issues in computer networks (also for ICT systems) focusing on key concepts such as security objectives, threats, attack and countermeasures. Then we elaborate on the security analysis for networks, describing analysis methodologies. Finally, we present some research challenges and open problems that are expected to drive the research on network security.

2 FUNDAMENTAL SECURITY OBJECTIVES IN COMPUTER NETWORKS

The Confidentiality, Integrity and Availability, abbreviated as CIA, constitute the core principles of information security. From the computer network point-of-view, they have to be analyzed for achieving and maintaining network security to provide robust and dependable network-based services.

These core security objectives are intertwined with a plethora of security issues which aim to realize these objectives:

- **Security-aware system design and deployment:** The design of computing systems and networks should take security into consideration at their very early stages. The security issues should also be addressed during the system deployment so that systems and networks are set up with emphasis on security requirements [86].
- **Attack detection:** Network flows and activities have to be processed for aggregate metrics and events have to be identified. Subsequently, these data have to be processed for attack or intrusion detection. This process has to be quick, reliable, repeatable and with low complexity. Moreover, it has to be applicable in different environments.
- **Vulnerability identification and patching:** Vulnerabilities in a network have to be identified and patched in a timely manner. On the software side, the vendors and programmers have to discover or evaluate in-field data to identify the vulnerabilities and prevent exploits via published patches. The availability of an appropriate patch is not sufficient. The network or system administrators need to apply them to vulnerable network hosts and devices.
- **Countermeasure selection:** Once an attack or intrusion is detected, there are potential countermeasures to be applied determined by the capabilities of the security infrastructure and available human resource. This is a complicated problem with a multidimensional structure.
- **How to enforce a countermeasure:** After a countermeasure is selected, the actual enforcement is a difficult task. The heterogeneity of countermeasure enforcement points and the expected cost of countermeasures are also challenging.
- **Lack of security testing:** Most of the vulnerabilities are nondeliberately caused by substandard system design/development practices. Additionally, when the systems are deployed and in operation, they should be tested against security threats via periodic efforts such as penetration testing. However, these actions are costly and time-consuming activities and are usually ignored.
- **Human factor:** User behavior and awareness for security is a critical concern in ensuring the security of computer networks and information assets. Since the users are part of the system, they can be instrumental in both ways: They can mitigate the security risk of a computer network by following best security practices or they can capacitate attackers for exploitations by errors or weaknesses. Since users are assumed to be the weak link in information security, the effect of the human factor cannot be overlooked.

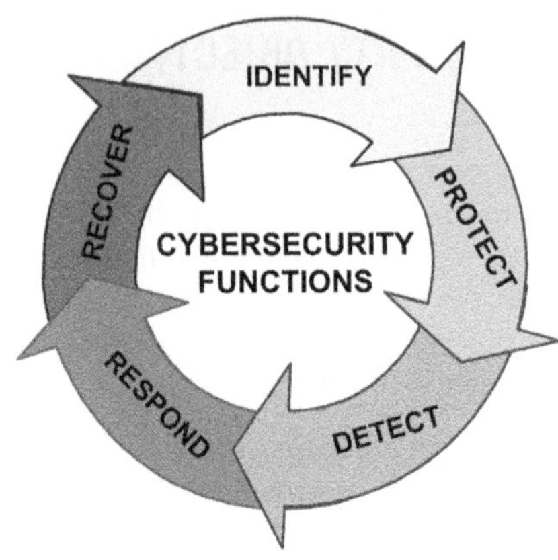

FIGURE 30.1

Cybersecurity functions for infrastructure protection.

According to the *Framework for Improving Critical Infrastructure Cybersecurity* by National Institute of Standards and Technology (NIST), at the highest level, basic cybersecurity activities involve five functions: *Identify*, *Protect*, *Detect*, *Respond*, and *Recover* as shown in Figure 30.1 [4]. These functions can be performed in parallel, merged or organized in different configurations according to the context and environment of an ICT system. They provide a common language for management of cybersecurity risks by organizing information, enabling risk management decisions, addressing threats, and improving by learning from previous activities [4]. The functions also help show the return on investment for security-related activities in cybersecurity. For example, investments in pre-attack security analysis support protection efforts while online network analysis enables proper response and recovery actions, resulting in reduced impact to the delivery of network services.

There are also some security issues that are becoming more acute with the recent advances in Internet, devices and software. The leading ones can be listed as:

- *Heterogeneous environments:* The networks are becoming more and more heterogeneous with the integration and deployment of different systems, devices and software.
- *Hyperconnectivity of users and systems:* The "network" aspect of information security has reached an unprecedented level since the proliferation of Internet and smart mobile devices have enabled "anytime, anywhere" connectivity.
- *Circulation of software from untrusted and unknown developers:* The Internet and the spread of mobile devices have enabled the widespread circulation of free software from unknown sources dramatically. The retrieval and installation of such software is usually effortless and straightforward, leading to *harder-to-control* computation and communication environments.

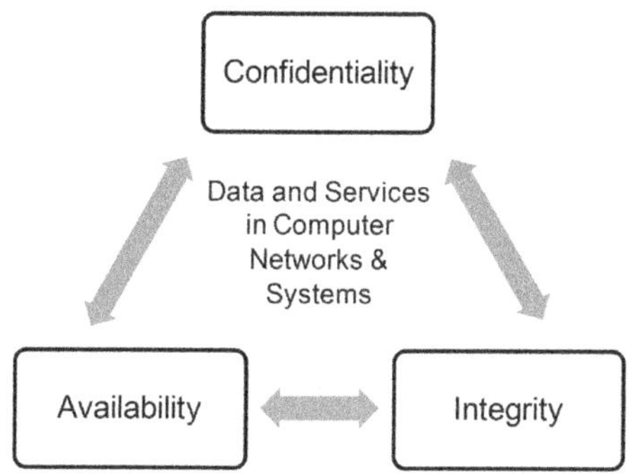

FIGURE 30.2

CIA triad.

- *Cyberphysical systems (CPSs):* According to [5], CPSs are "integrations of computation and physical processes." Embedded computers and networks monitor and control the physical processes, usually with feedback loops where physical processes affect computations and vice versa. The security analysis of such systems requires comprehending a holistic view of computers, software, hardware, networks, and physical processes. This integrated structure brings forth the challenge of "scoping," which is to determine the correct scope of network security analysis.

Due to these circumstances, the complexity of network security analysis and modeling for supporting fundamental security objectives has become extremely high.

The proliferation and the increasing complexity of computer networks and systems have made security an important issue for modern societies. Security of computer networks and systems is almost always discussed within information security that has three fundamental objectives, namely confidentiality, integrity, and availability. The objectives form a *CIA triad* that is also known as the container for both data and computer network systems security and shown in Figure 30.2. For instance, the NIST standard FIPS PUB 199 (Standards for Security Categorization of Federal Information and Information Systems) lists confidentiality, integrity, and availability as key security objectives.

2.1 CONFIDENTIALITY

Confidentiality prevents the disclosure of sensitive information to unauthorized users or systems on computer networks. Sensitive information refers to the information that should be kept confidential. Loss of confidentiality leads to the unauthorized disclosure of sensitive information. In literature, confidentiality is used to provide data confidentiality and privacy. Data confidentiality prevents unauthorized entities from accessing confidential information whereas privacy ensures entities can control or influence information related to them.

Data confidentiality assures that confidential data or information is not made available to unauthorized entities in the system. For instance, cybersystems are connected to each other and they are managed via computer networks and systems. Information about critical components of cyber physical systems and information related network components that connect the systems may be confidential information. Disclosure of such information may have a huge cost for a corporation. An example of this is the well-known attack on a nuclear facility with advanced malware *Stuxnet* [6] that is a computer worm designed to attack Siemens industrial programmable logic controllers. Stuxnet spreads via network systems that have Microsoft Windows operating systems. If the network information of the nuclear facility were kept secret, Stuxnet would not be designed to attack the facility and would not damage the facility capabilities. Another example related to data confidentiality on computer networks is about preserving personal payment-enabling data, specifically card holder data (CHD), which are sensitive information in payment networks. During payment processing, CHD are transmitted over the Internet, an open network system, between buyer, merchants, and processors. Different encryption methods are used to preserve confidentiality of CHD.

Recently, the concept of privacy has become more significant than ever. Privacy in computer networks assures entities to control information related to them, such as what information related to them may be collected and by whom and to whom that information may be disclosed. For instance, privacy of people that use social networks should be ensured since the networks connect many people who may have conflicts of interests. Different access control methods are applied to such networks to provide privacy, such as role based access control.

2.2 INTEGRITY

In computer networks and systems, the term integrity covers both data and systems. Generally, integrity assures the accuracy and consistency of data and systems, which means guarding against improper modification or destruction of data and systems in an unauthorized or undetected manner. A loss of integrity is the unauthorized change or destruction of data or systems.

Data integrity assures that data are modified only in a specified and authorized manner on computer networks and systems. For instance, assume that electronic health records (EHRs) are stored in a centralized repository and many organizations are able to access EHRs via the Internet. Hospitals and medical insurance companies are some of the organizations related to these data. In this case, unauthorized access with write permission disrupts the integrity of EHRs that may result in financial losses and health problems for patients.

System integrity assures that a system performs its intended functions in a continuous manner, free from deliberate or inadvertent unauthorized modification of the computer network or system. For example, let us consider a production system in a factory that is able to be configured remotely via a network connection. Assume also that an unauthorized entity subverts the access control mechanism of the system and changes the configuration of the systems intentionally to sabotage the production, where an improper configuration results in physical damage of some components of the production system. This case shows that system integrity is significant for some systems that are connected to computer networks and systems.

2.3 AVAILABILITY

The availability objective ensures that computer networks and systems work properly and services are accessible and are not denied for authorized users. Specifically, availability ensures timely and reliable access to information and services on computer networks and systems. A loss of availability leads to the disruption of access to the information and services on the systems.

Availability is the most important security service for some services on computer networks and systems. Highly available systems or services remain available at all times. Consider the Point-Of-Sale (POS) services of a payment network, such as the network of MasterCard. An interruption of the authentication system of POS gives rise to huge financial losses for many corporations, such as financial institutions and merchants that use debit or credit cards for payments. Moreover, unavailable POS services decrease user satisfaction with the corporation.

3 VULNERABILITY AND MALWARE IN COMPUTER NETWORKS

Security attacks are carried out by using vulnerabilities of computer networks. Most of the time, an attacker uses malicious software (malware) to reach the vulnerabilities and then to accomplish its security attack.

3.1 VULNERABILITY

Vulnerability is defined as a problem such as a programming bug or common configuration error that allows a system to be attacked or compromised. In IETF RFC 2828 (Internet Engineering Task Force Request for Comments), vulnerability is defined as "A flaw or weakness in a system's design, implementation, or operation and management that could be exploited to violate the system's security policy." Factor Analysis of Information Risk (FAIR) defines vulnerability as "The probability that an asset is unable to resist the actions of a threat agent." Actually, there are many definitions for vulnerability in the context of computer networks and systems. The definitions are directly or indirectly related to assets of computer networks and systems. We categorize assets for computer networks and systems as follows:

- **Hardware**: These include computer systems and data processing equipment, data storage and communication devices, local and wide area communication links, bridges, routers, wireless access points, and so on.
- **Software:** These include operating systems, network protocols, communication applications, system utilities, and all other applications that are used in computer networks and systems.
- **Data:** These include all logical data storages like files and databases that contain sensitive information, such as CHD of a financial institution or simply password files.

Vulnerability is a potential threat for one or more than one security objectives under some security attacks related to assets of computer networks and systems. Therefore, we categorize vulnerabilities according to their effects on security objectives and assets. In the first category, a computer network or system resource may be corrupted under threats that are capable of exploiting vulnerabilities of the resource. The main target of this category is integrity of the assets. In the next category, a resource can become leaky if it contains vulnerabilities. These kinds of vulnerabilities can expose threats to confidentiality objective. In the last category, a resource is unavailable or it fails to satisfy performance requirements if the resource contains some vulnerabilities. Vulnerabilities of this category can affect availability of resources. A detailed classification of vulnerabilities according to their effects on assets is shown in Figure 30.3.

Since there is no bug-free software, bug-related vulnerability has become a threat for computer networks and systems. The most general countermeasure against bug-related vulnerability is patching software. When vulnerability is discovered, the system developers or administrators develop and release relevant patches that can be applied to fix them. However, this is a costly and difficult process. First of all, the identification vulnerability is not straightforward and can take time. Moreover, it may become a reactive process. For instance, the Heartbleed bug is a serious vulnerability in the open source OpenSSL cryptographic software library that is widely used to secure communications over the Internet. This vulnerability can be classified as software vulnerability related to poor coding. OpenSSL has no lack of security in design or inadequate software design but the vulnerability is totally related to human factors in security. The security vulnerability of OpenSSL allows stealing the information protected by SSL/TLS encryption used to secure the Internet. It is suspected that many network devices are never patched to fix the Heartbleed bug because the systems are either hardened or infrequently updated. The vulnerability could live on for years in some networked systems like hardware, home automation systems, and home Internet routers. Thus, vulnerability may have a continuous effect on computer networks and systems. For this reason, computer networks and systems need to be

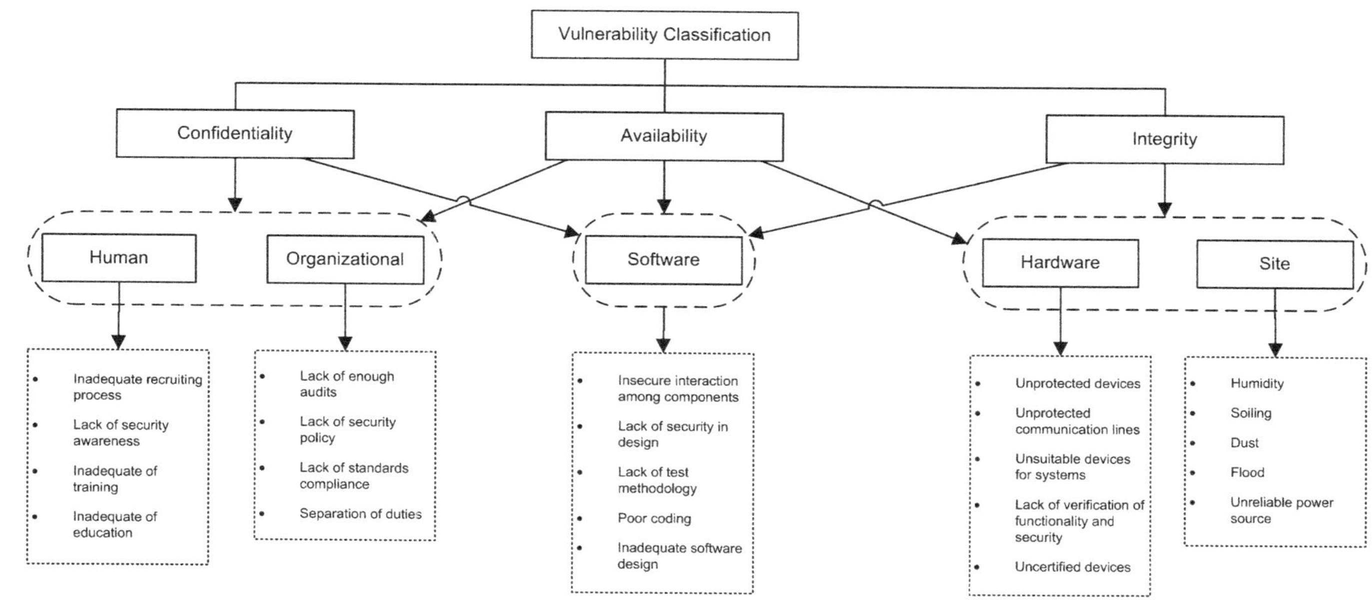

FIGURE 30.3

Classification of vulnerabilities according to their effects on assets.

analyzed and the security levels of such systems need to be assessed continuously. To accomplish these tasks, it is important to determine appropriate security analysis and assessment methods for dynamic computer networks and systems.

3.2 MALWARE

Malware is malicious software used to attack computer networks and systems to harm the systems intentionally or obtain sensitive information without permissions of owners [7–9]. Malicious software is inserted into computer and network systems with the intent of compromising the integrity, availability, and confidentiality of targeted systems' data, applications, and hardware. Therefore, malware is one of the most significant threats to computer networks and systems. For instance, viruses and worms are some instances of malware seen in the wild.

The motivation of early malware was usually to highlight some security vulnerabilities or simply to show off technical ability, but the motivation has changed over many years. Recently, there is a flourishing underground economy based on malware [10], where it is no longer a fun factor, but the expectation is that money can be made.

There are many approaches to classify malware [8,11–15] regarding computer networks and systems. Although different properties of malware may be used, one of the most accepted approaches is to classify malware into two broad categories, namely propagation and payload [16]. The first category of the classification is related to the spread or propagation method(s) of malware. In this category, malicious software fragments may attach themselves to some executable content, may use network connections to spread from system to system, or may involve social engineering methods to subvert access control of targeted systems. In the other category of the classification, the target is reached by actions that are performed with payloads of malware. Once malware is active on the target system, the next step is what actions it takes, executing payload(s). Some payloads like viruses and worms may be designed for data destruction or system corruption on the infected system after certain conditions are met. Some other payloads subvert the computer networks and system resources of the infected systems, using them to attack the system, such as payloads of bots. There are payloads that are designed to steal or gather information on the infected system to be used for malicious purposes, such as key loggers and spyware. The last category of payload that is used by malware is to hide malware presence on the infected network or system to provide covert access to that network or system. We have explained several types of malware to show the broad diversity of malware types that affect computer networks and systems, as follows:

- *Viruses* are the first instances of malware seen in the wild. A virus is a piece of software code that can attach itself to other software codes [17] by using a viral mode of infection within several propagation mechanisms. It can replicate itself and spread to other programs in the same host, including kernel programs, via many ways, such as network connections and portable devices. Actually, in a computer network, the ability to access documents, applications, and system services on other computers provides an environment for the spread of viruses. Viruses dominated malware in earlier years in computer networks and systems because of the lack of adequate authentication and access control mechanisms on such systems.

- A *worm* is parasitic malicious software that can replicate itself and propagate to computer networks and systems [18] independently from other programs. This means that worms generally use computer networks and systems to propagate from one host to another one by using vulnerabilities of operating systems. Specifically, a worm is software that seeks out more systems to infect and then infected systems serve as an automated launching pad to carry out attacks to new systems. Most of the time, worms use network connections to spread from one system to another system. The first worm spread over the Internet is the Morris [19]. Recently developed advanced worms have become a significant threat for computer networks and systems. Stuxnet and Duqu are instances of such advanced worms seen in the wild [6,20].
- A *Trojan horse* is malicious software that appears to be useful. However, the real purpose of a Trojan horse is to perform malicious actions in the background, such as stealing card holder data from payment systems [11,15]. Trojan horse does not add itself to other software like viruses. Once it infects a system, Trojan horse can download and install additional malicious code to the system via network connections.
- *Spyware* is malicious software installed on computer and network systems. It collects sensitive information without the permission of users [21]. Passwords and credit card numbers are instances of sensitive information. Spyware sends sensitive information to the attacker and it may not be easily detected during this process because it is often hidden from users. Key loggers are instances of spyware.
- *Rootkit* is malicious software that is able to hide itself from the user in a computer and network system. It has stealthy processes that cannot be easily detected with normal malicious software detection methods [22] because it may subvert malware detection tools. Each level of a computer and network system may contain rootkit [23], which circumstance complicates the detection process of rootkit. Recently, rootkit is used to make some other software stealthy. For example, a rootkit may be used with spyware to obtain card holder data from a specific payment institution.

Malware is used to accomplish distributed attacks over computer networks, such as doing attacks with botnets. A *botnet* is a set of infected computer and network systems that are controlled remotely by a third party [24]. The third party is called the *bot master* and is known as a *command-and-control* server. Malware code in each computer and network system is known as a bot that allows its bot master to control the computer system [25]. A bot in computer and network systems is created by using malicious software, such as a virus and a worm. Botnets are significant threats for computer networks because they may be developed for organized crime attacks. For instance, a botnet may be used to accomplish distributed denial of service attacks against critical systems.

Mobile systems and networks have become pervasive and the number of attacks with malware targeting mobile systems has been increasing rapidly [26]. Actually, mobile systems provide many different services with a wide variety of access technologies, such as Wi-Fi, Bluetooth, and UMTS (Universal Mobile Telecommunications System). Since services on mobile systems and networks are on the rise, they also become an ideal target for malware authors [27–29]. Reports of antivirus labs, such as Symantec [30], ESET [31], and McAfee [32] have shown the increasing numbers of malware targeting mobile systems. Malware targeting mobile systems is known as mobile malware [33] and the number of newly created mobile malware is on the rise. For instance, mobile malware targeting Android platforms has shown rapid increase in the last few years. Mobile botnets and spyware like Android/Funsbot.A botnet are also on the rise. It seems that malware authors are focusing more on developing mobile malware to target computer networks and systems [34,35].

The exponential growth of computer networks and systems has made cyberspace an attractive target for cyber attackers, with disastrous consequences. Malware is the primary arsenal to carry out malicious goals, either by exploiting vulnerabilities or using specific attack tools. Analyzing malware is the first step to counter attacks carried out with malware.

4 SECURITY THREATS AND ATTACKS IN COMPUTER NETWORKS

The security attacks aim to compromise the five major security goals for network security (extended from CIA requirements): *Confidentiality*, *Availability*, *Authentication*, *Integrity* and *Nonrepudiation*. To serve these aims, a network attack is commonly composed of five stages [3]:

1. Preparation and reconnaissance phase including information gathering to discover and identify vulnerabilities/weaknesses to be exploited and tool configuration
2. Assessing vulnerabilities and potential exploits
3. Launching the attack
4. Core attack operations such as confidential data compromises or integrity damages
5. Post-attack operations such as getting rid of traces and removing/hiding evidence

Security analysis aims to identify a network's situation with respect to these phases.

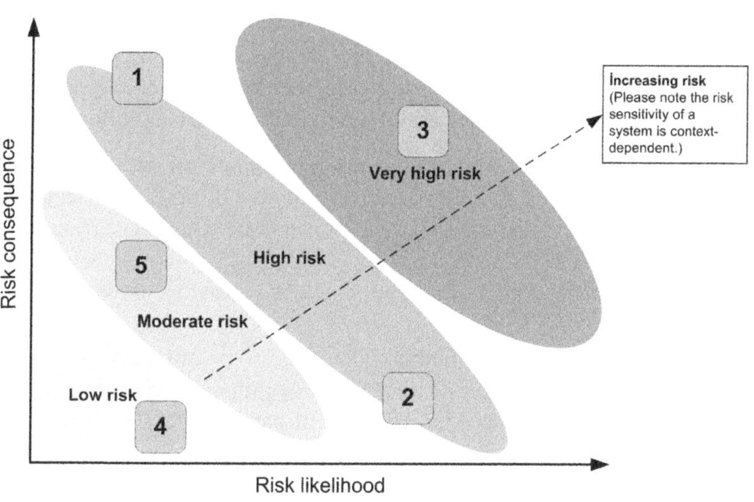

FIGURE 30.4

The different risk regimes for a computer network based on likelihood and consequence of risk(s).

Risk analysis is a fundamental constituent of security analysis of computer networks. According to ISO (International Organization for Standardization) 27005, *risk* is defined as "the potential that a given threat will exploit vulnerabilities of an asset or group of assets and thereby cause harm to the organization," quantified as a function of the likelihood of an event and its consequence [36]. Therefore, security risk management is the active process of identifying, analyzing, and responding to risk considering potential attacks, vulnerabilities and assets.

The risk regime taxonomy for a computer network is shown in Figure 30.4. The risk regime is a function of the potential consequence and the likelihood of risk(s) that the system is subject to. The sensitivity of a system to an exploit or attack utilizing vulnerabilities depends on the context. Similarly, the likelihood of such an issue may change among different contexts. For instance, *point 1* marked with numbered square in the figure can correspond to the case of an insider attack. These attacks are relatively less likely, however, with a large potential impact. These cases translate to *point 5* due to smaller but still sizable risk consequences. In contrast to *point 1*, the system may have relatively less valuable assets or services but high probability of attacks. That may render the operation at *point 2* in terms of risk. When the system subject to risk is less critical with good protection, the risk regime operates at *point 4* under favorable risk conditions. *Point 3* can be the case where a poorly protected but critical system is in question. Mission-critical systems with high rewards for attackers are also potential candidates for this regime.

In computer networks and systems, security attacks are generally classified into two groups, namely active attacks and passive attacks. Passive attacks are used to obtain information from targeted computer networks and systems without affecting the systems. Main properties of passive attacks are as follows:

- The goal is to obtain transmitted information from communicated networks.
- In comparison to active attacks, passive attacks are difficult to detect because these attacks do not affect network communications and do not change targeted information.
- Most of the time encryption of communications is used to prevent passive attacks.
- There are two types of passive attacks:
  - *Release of communication content:* If the communication contains sensitive data, such as card holder data, this type of passive attack reveals sensitive information.
  - *Traffic analysis:* This type of passive attack is applied to masked communication content, where capturing the content is not enough to extract sensitive information. Encryption is a way to mask network communications. Simply, an intruder observes the frequency of network traffic and then it guesses the nature of the traffic to extract useful information regarding sensitive data.

Active attacks may modify communication contents or may create false contents in computer networks and systems. The main properties of active attacks are as follows:

- The goal is to alter computer networks and systems resources or alter their operations.
- Active attacks are easier to detect than passive attacks.
- It is difficult to prevent active attack. Most of the time, prevention against active attacks requires physical protection of computer networks and systems.
- Detection of active attacks is more significant than prevention because mitigation of the effects of active attacks is cheaper and easier than prevention.
- There are four types of active attacks:
  - *Replay*: An intruder captures communication data and retransmits the captured data for creating unauthorized traffic.
  - *Masquerade:* An intruder pretends to be a legitimate user. This attack is accomplished with some other active attacks.
  - *Modification of communication content:* Some parts of communication content are modified, delayed, or reordered to produce an unauthorized effect.
  - *Denial of service:* This attack prevents normal behavior of computer networks and systems and services running on these systems. This attack has diverse targets and it is one of the most commonly seen attack types on contemporary computer networks and system.

5 DEFENSE MECHANISMS AGAINST SECURITY ATTACKS

A number of detection and defense mechanisms have emerged in the last decade to tackle attacks on computer networks and systems [37–39]. It is significant to classify these mechanisms to better select the appropriate defense mechanism for organizations with specific network infrastructure and security requirements. In this section, we classify defense mechanisms according to analyses of malware and responses to attacks on computer networks and systems. The taxonomy of defense mechanisms is shown in Figure 30.5.

Attackers usually use vulnerabilities of systems to attack a large number of computer networks and systems. Once an attack has been initiated by malware, the success probability of the attack is high because malware starts the attacks if the computer networks and systems meet specific requirements. Therefore, it is significant to detect and remove malware from computer networks and systems. In this chapter, the first category of defense mechanisms of computer networks and systems are those mechanisms responsible for detecting and removing malware from such networks and systems.

The best defense mechanism prevents malware from infecting computer networks and systems, which means it does not allow malware to get into such networks and systems in the first place, or it blocks the ability to affect computer networks and systems. Actually, this goal is nearly impossible to achieve since contemporary computer networks are highly dynamic and they interact with many people and cybersystems. The prevention mechanisms consist of policy, awareness, vulnerability mitigation, and threat mitigation, which can be described as an ideal defense option for computer networks and systems.

- **Policy**: A suitable security policy of computer networks and systems related to malware is probably the most significant preemptive countermeasure against attacks. For example, setting an appropriate access control mechanism reduces the number of accesses to critical components of computer networks and systems, which prevents many attacks on computer networks and systems.
- **Awareness**: Recently user awareness has become critical to secure computer networks and systems. User awareness directly reduces malware related to

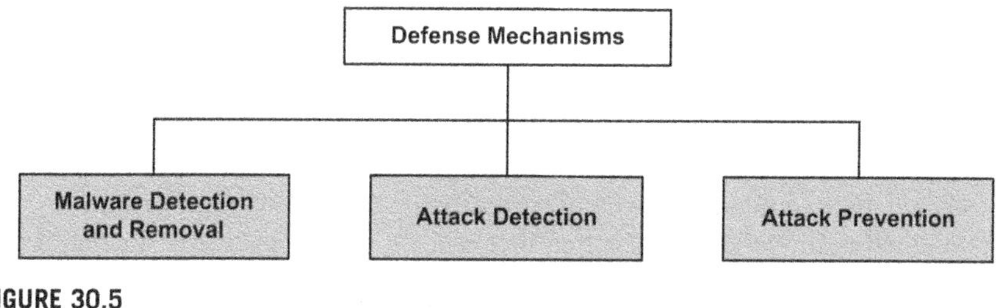

FIGURE 30.5

Taxonomy of Defense Mechanisms.

social engineering attacks. Security orientations of organizations are instances for defense mechanisms related to this category.

- **Vulnerability Mitigation**: Since it is impossible to implement vulnerability-free computer networks or systems, minimizing vulnerabilities that may be used for attacks with malware prevents many attacks to computer networks and systems. For instance, applying all recent patches reduces vulnerabilities that may be exploited on computer networks and systems.
- **Threat Mitigation**: Removing unnecessary services and applications from computer networks and systems reduces vulnerabilities and possible malware attacks.

It is impossible to prevent all malware attacks on computer networks and systems. If prevention mechanisms fail, then mitigation mechanisms regarding malware attacks are used to secure computer networks and systems. Mitigation mechanisms should detect, identify, and then remove malware from such networks and systems.

- **Detection**: Once malware infection has occurred on computer networks and systems, it is important to determine the occurrence and the location of the malware.
- **Identification**: If malware is detected, the next step is to identify the specific malware.
- **Removal**: Once the malware is identified, the last step is to remove the malware and its traces from the computer networks and systems.

Mitigation mechanisms may run on different places of computer networks and systems. For instance, some anti-virus programs may run on host systems to monitor and detect malware activity by analyzing executions and behaviors of some programs, such some critical processes on the hosts. Another example is intrusion detection systems that may take place as a part of perimeter security mechanisms running in an organization's network. A more detailed explanation of common defense mechanisms related to attack mitigations of malware is as follows:

- **Host-Based Scanners**: These scanners give software the maximum access to information, where the behavior and the activity of malware may be observed precisely. Host-based anti-virus software is an example of a host-based scanner.
- **Generic Decryption**: It enables detection of even the most complex polymorphic malware in a quick manner. Generic decryption mechanisms have CPU (central processing unit) emulation, virus signature scanner, and emulation control module. Some anti-virus software contains generic decryption mechanisms. Briefly, these mechanisms simulate the execution of a targeted code to detect malware. Depending on simulation time and the simulated code, generic decryption mechanisms decide possible damage of the simulated code on the host.
- **Host-Based Behavior Blocking Software**: This software is an integral part of the host operating system and runs with it to detect malicious behaviors by

monitoring system activities. Host-based behavior blocking can block suspicious activity, malware activity, in real time; such mechanisms are especially preferred by time critical systems. Actually, these defense mechanisms have some limitations. For instance, the code should run on targeted systems that may have a malicious effect on such systems.

- **Spyware Detection and Removal**: These mechanisms are dedicated to spyware malware detection and removal that are seen as complementary anti-virus software.
- **Rootkit Countermeasures**: Rootkits are one of the most difficult malware to detect, so defense mechanisms against such malware require integration of different security level solutions. Both network-based intrusion detection systems and host-based intrusion detection systems are used to detect and remove rootkits from computer networks and systems.
- **Worm Countermeasure**: There is an overlap between virus detection and removal mechanisms and worm detection and removal mechanism. Differently from viruses, worm propagation generates considerable network activity. Worm countermeasure mechanisms use such activity to detect work in computer networks and systems. Specifically, these defense mechanisms may use many activities of worm for detection, such as signature based worm scan filtering or rate limiting.
- **Distributed Intelligence Gathering Approaches**: These defense mechanisms collect data from a large number of both host-based and perimeter sensors to detect and correlate malware attacks on computer networks and systems. Digital immune system of IBM (International Business Machines Corporation) is an instance of these approaches. These approaches are more effective than traditional host-based approaches because they can detect malicious activity on computer networks and systems.

There are many classification approaches of defense mechanisms against attacks on computer networks and systems. Most of the time, classification approaches consider specific attack types or attacks with specific malware for classification purposes. For instance, defense mechanisms against DDoS flooding attacks are classified in [37]. Recently, DDoS attacks were launched by botnet computers that are simultaneously and continuously generating a huge amount of network traffic or service requests to the targeted system. On the other hand, DDoS defense mechanisms have put an end to some DDoS attacks, such as attacks that use script-kiddies, which could download a tool and launch an attack against any website. Simply, defense mechanisms against DDoS flooding attacks are categorized according to detection and response as shown in Figure 30.6.

Since botnets are widely used for DDoS attacks, another classification of defense mechanisms concerns botnet attacks [38]. This classifies defense mechanisms according to behavior, detection and defense against attacks. Specifically, botnet behavior helps to detect such malware and botnet behavior classification consists of five features, as shown in Figure 30.7. In addition, Botnet detection

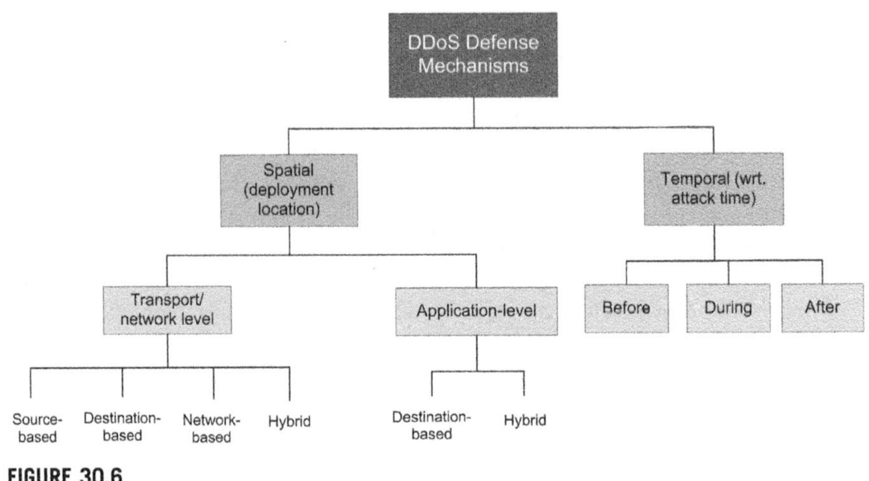

FIGURE 30.6

A taxonomy of defense mechanisms against DDoS flooding attacks [37].

mechanisms are classified according to critical components of botnet and defense mechanisms are classified according to goals of the mechanisms as in Figure 30.7, which goals are preemptive or remedial. A similar classification of DDoS attack detection mechanisms about power networks is shown in Figure 30.8, where detection mechanisms may become passive, proactive, or hybrid [40].

Computer networks and systems interact with many systems in cyberspace so defense mechanisms may be highly context dependent [41]. In this case, it may be impossible to classify all mechanisms in an appropriate manner. For this reason, we classified defense mechanisms according to their main purposes, namely detection and prevention as shown in Figure 30.5.

Attack detection mechanisms are implemented with intrusion detection systems (IDSs) in computer networks and systems. Intrusion detection is described as a service that monitors and analyzes computer system events to find and provide real-time or near real-time warning of attempts to access system resources in an authorized manner [42]. IDSs are classified as follows [16]:

- **Host-based IDS**: Monitors the characteristics of a single host and the events occurring within that host for suspicious activity. Host-based IDS use two general approaches to detect attacks, which are explained as follows.
  - **Anomaly detection:** This approach uses a collection of data to distinguish behavior of entities for detecting attacks. Then statistical tests are applied to determine legitimate users and attacker. More specifically, threshold detection and profiling is used in this approach.
  - **Signature detection:** This approach uses a set of rules or attack patterns that may be used to detect attacks of intruders.

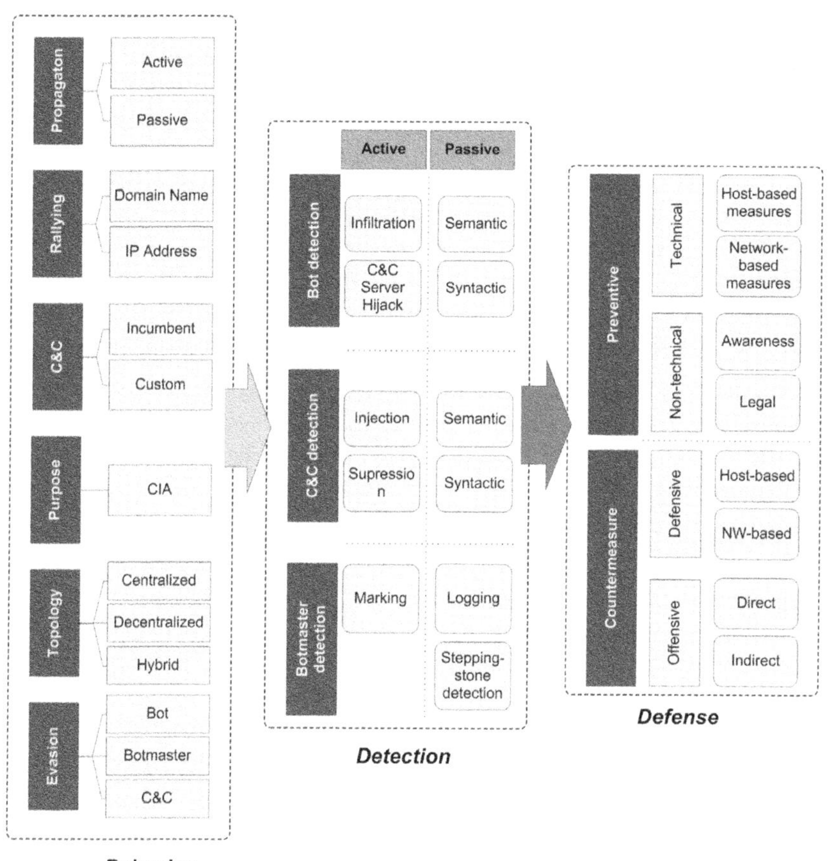

FIGURE 30.7

Attributes of botnet behavior, detection and defense [38].

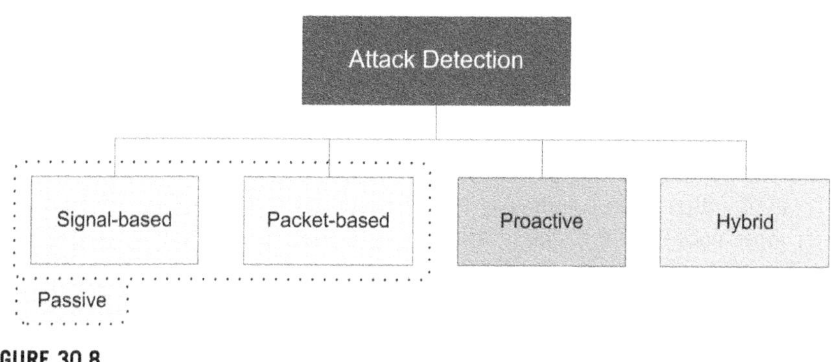

FIGURE 30.8

Classification of DoS attack detection mechanisms [40].

- **Network-based IDS**: Monitors network traffic for particular network segments or devices and analyzes network, transport, and application protocols to identify suspicious activity. A network-based IDS monitors traffic at selected points on computer networks and systems. Then, it examines the traffic in real time or close to real time attempting to detect intrusion patterns. Network-based IDSs may examine network, transport, and/or application level protocols activity to detect attacks. The main difference between host-based IDS and network-based IDS is that network-based IDS examines networks traffic directed toward vulnerable computer networks and systems whereas host-based IDS examines user and software activity on a host. Similar to host-based IDS, network-based IDS uses anomaly and signature based approaches to detect malicious activity.

Firewalls and intrusion prevention systems (IPSs) implement prevention mechanisms against attacks on computer networks and systems. Most of the time, a firewall is defined according to its capabilities. We have determined that the three main capabilities of firewalls are the following:

- **Control Point:** A firewall defines a control point to keep unauthorized users out of protected networks and systems. It simplifies security management by aggregating many defense mechanisms on a single system.
- **Monitoring:** A firewall monitors security related events of computer networks and systems, such as implementing logging, alarming, and auditing functionalities on firewalls.
- **Platform for Nonsecurity Functions:** A firewall may contain non-security related functions, such as a network address translator. This capability reduces costs of computer networks and systems implementations.

Computer networks and systems run many different protocols and software. For example, a computer network may be connected to the Internet by using TCP/IP (Transmission Control Protocol/Internet Protocol) reference model. In this case, each layer of TCP/IP reference model runs different protocols and applications; therefore the protocols can be protected with firewalls that are considered properties of each layer. For this reason, there are different types of firewalls. In this chapter, we classify firewalls into four groups as follows:

- **Packet Filtering:** It applies a set of rules to all IP packets crossing the border of the protected network. This type of firewall uses a list of rules based on matches to fields in the IP or TCP header.
- **Stateful Inspection:** This firewall tightens up the rules for TCP traffic by creating a directory of outbound TCP connections and allows incoming traffic for those packets that fit the profile of one of the entries in this directory. A packet filter firewall makes filtering decisions on an individual packet basis and does not consider any higher-layer context in contrast to a stateful inspection firewall.
- **Application Level Gateway:** It is sometimes called application proxy. It acts as a relay of application level traffic. Application layer firewall is considered

to be more secure than packet filters because it needs to scrutinize only allowable traffic. However, this firewall needs more processing power than the packet filter firewall.

- **Circuit Level Gateway:** The firewall is a standalone system for certain applications. A circuit level gateway does not permit an end-to-end TCP connection. It set sets up two TCP connections, one between itself and a TCP user on an outside host. Once the two connections are established, the firewall relays TCP segments from one connection to the other without examining the contents.

An IPS implements prevention mechanisms on computer networks and systems. It is a functional extension to a firewall that adds IDS properties to the firewall. More specifically, an IPS is an inline network based IDS that can discard packets as well as detect suspicious traffic. Similar to IDSs, IPSs can be classified as host-based IPS and network-based IPS, which have similar properties with IDSs.

6 SECURITY ANALYSES OF COMPUTER NETWORKS

6.1 FORMAL SECURITY ANALYSES

The formal modeling and verification approach is one of the most effective tools for network security analysis with maximal precision [43]. It serves the common goals of network security analysis, which are to improve the quality of the system specification and to check for the existence of security deficiencies. Moreover, it may enable a more systematic understanding of security issues and facilitate systematic testing of network-related implementations.

A security model is a formal description of security related aspects and mechanisms of a system using formal methodology. A model is formal if it is specified using a formal language, which is defined as a language with well-defined syntax and semantics such as finite state automata (FSM) and predicate logic [43]. That model includes a base system component and a security component related with a satisfaction requirement as shown in Figure 30.9. The former defines what the system does while the security component is an abstraction of security requirements. The satisfaction relation ensures that these security requirements are met

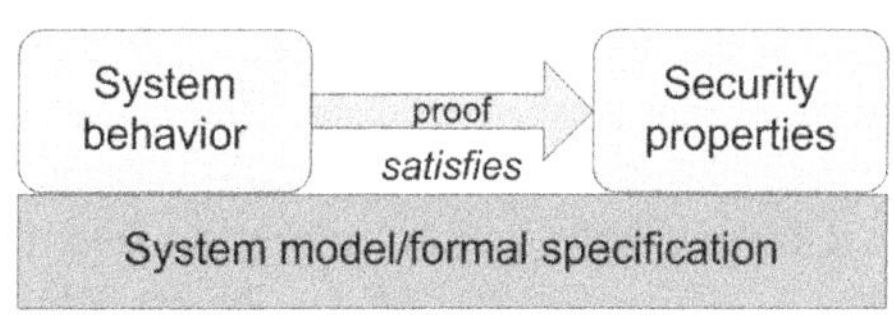

FIGURE 30.9

The structure of a formal security model [44].

and typically verified via formal methods. Therefore, security analysis is carried out based on correspondence between system description and security properties adopted in the formal modeling.

According to [43], there are four classes of practically relevant formal security models:

1. **Automata Models:** A model checking specification consists of two parts [45]. One part is the model: a state machine defined in terms of variables, initial values for the variables, and a description of the conditions under which variables may change value. The second part is temporal logic constraints defined over states and execution paths. Conceptually, a model checker visits all reachable states and verifies that the temporal logic properties are satisfied over each possible path, that is, the model checker determines if the state machine is a model for the temporal logic formula via exploration of the state space temporal logic constraints over states and execution paths. For the analysis to be performed, the constituent elements of the model such as vulnerability description, connectivity and required function need to be developed. In [46], Mao et al. describe a new approach, namely logical exploitation graphs, to represent and analyze network vulnerability. Their logical exploitation graph generation tool illustrates logical dependencies among exploitation goals and network configuration. Their approach reasons all exploitation paths using bottom-up and top-down evaluation algorithms in the Prolog logic programming engine.
2. **Access Control Models:** Classical access control models, like the traditional Bell-LaPadula model [43], relying on access control rules with security labels on objects and clearances for users, lack the modeling capabilities for current practical systems. Role-based access control (RBAC) models mapping subjects to roles in a hierarchical structure and then relating roles to access rights to subjects are proposed to address this shortcoming. In [47], a formal model of the computer network is constructed using graph-theoretic tools with packet filter functions classifying the message flow thus constraining the reachability of the entities on the network. Access control lists and routing policies are reflected into the model by means of packet filtering functions that are associated with edges of the graph.
3. **Information Flow Models:** These models describe how information may flow between which domains in a very abstract way such that they can capture also indirect and partial flow of information [48,49]. An example is the confidentiality of data output from a system. The critical issue is not whether any output contains confidential data but rather depends on it [44]. The concept of *noninterference* is a fundamental information flow property in that regard. Basically, if there is no information flow from one group of processes to another, the first group is said to be noninterfering with the other. This means that the processes in the first group cannot reveal any secret information such as passwords or encryption keys to the entities in the

second group. In return, the processes in the second group cannot be corrupted by the ones in the first group. This expressive capability provides the basis for modeling confidentiality and integrity requirements between processes.

4. **Cryptoprotocol models:** Probably the most successful class of security models are cryptoprotocol models describing the message traffic of security protocols. The formal and mathematical design of cryptographic schemes is suitable for formal verification. Virtually all formal methods have been employed for cryptoprotocol verification [50] extending to industrial size protocols. Mostly secrecy and authentication goals can be specified and then verified automatically using model-checkers tailored for this application.

Formal methods can only address certain aspects of security, those related to computer networks and system design. Some aspects of security do not lend themselves to formal methods (e.g., computer hacking, tampering, and social engineering) [51]. Since current ICT systems are very complex, it is also nontrivial to have methods scaling with network size and security flaws and vulnerabilities are hard to find using formal analysis. In Figure 30.10, a cost-benefit analysis for formal methods is shown. The benefit is a function of number of users and their importance level, i.e., as the number of users and their relative importance increase, the investment on formal analysis is more reasonable. On the difficulty aspect, the cost increases with increasing system and property complexity. The operation domain of the system renders formal analysis feasible or infeasible due to complexity. Event it is feasible, it may unreasonable due to small return-on-investment for the analysis efforts. Therefore, security analysis based on formal methods typically focuses on specific aspects of system in operation. However, these analyses are precise and can be automated.

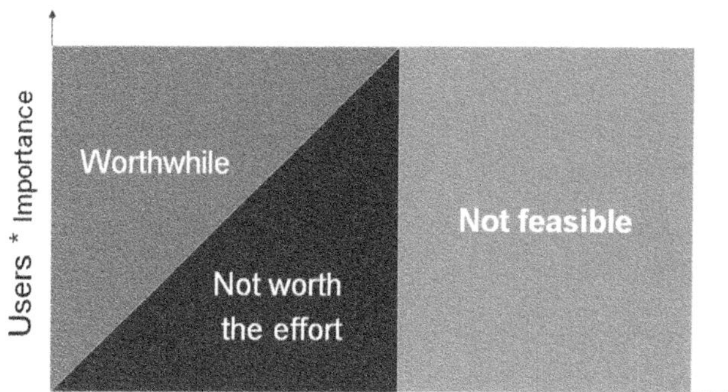

FIGURE 30.10

The cost-benefit analysis for formal methods [52].

6.2 AUTOMATED SECURITY ANALYSES

For general network monitoring and analysis, there are various tools utilizing Internet Control Message Protocol (ICMP) and the Simple Network Management Protocol (SNMP), such as HP OpenView, NetXMS, and OpenNMS. Although they support network discovery and monitoring tasks in a remote and automated setting, they are not designed and purpose-built for security analysis or evaluations.

For automated security analysis, network security tools are crucial since they allow offline and online analysis of computer networks in an automated and scalable manner. These tools are generally grouped into two classes [3] as shown in Figure 30.11.

6.2.1 Defense tools

a. *Information gathering* which is directly linked to the information support requirement for security analysis. These tools are required to construct an extensive and reliable knowledge base containing semantically rich and exhaustive information about a protected network [53]. Such tools typically provide network scanning/mapping and sniffing/traffic analysis functionalities. For instance, Vigna et al. present *NetMap*, a security tool for network modeling, discovery, and analysis in [53]. It relies on a comprehensive network model that integrates network information into a cross-layer structure. The NetMap model entails information regarding network topology, infrastructure, and deployed services. For network traffic analysis, Wireshark is a network sniffer and packet analyzer used extensively for network

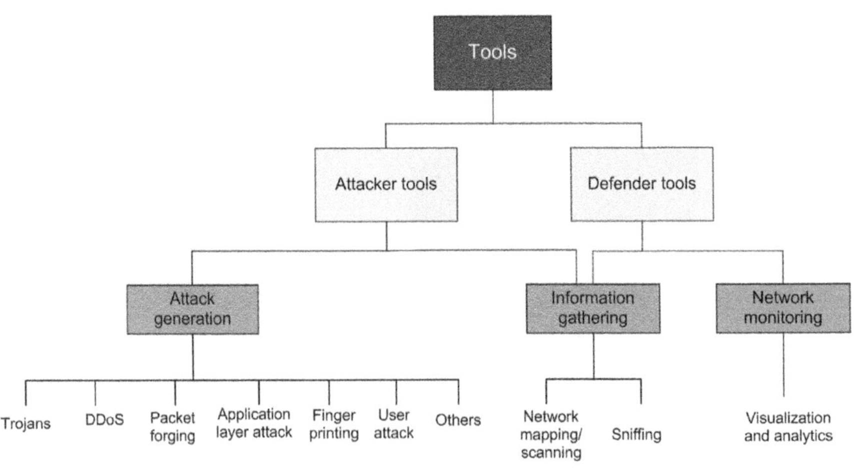

FIGURE 30.11

Taxonomy of network security tools [3].

monitoring and root cause analysis. Metasploit Framework is a tool for developing and executing exploit code against a remote target machine.

Actually, information gathering tools are used by both defenders and attackers. Before launching an attack, intruders need to know properties of the network, such as ideal nodes to launch attacks. Therefore, intruders first collect information about networks, such as IP addresses and operating systems, to find vulnerable systems in such networks using different information-gathering tools. After gathering sufficient amount of information, intruders apply their attacks to the networks.

Similar to intruders, defenders need to know vulnerabilities of their networks to be able to protect them. Therefore, information-gathering tools are used by defenders and most of the time they are classified as defender tools.

Information-gathering tools are further classified as sniffing and mapping tools. Sniffing tools are used to capture, visualize, and analyze network traffics. Tcpdump, Ethereal, Snoop, and Ngrep are some of these sniffing tools. On the other hand, mapping tools are used to identify active hosts on a network. They provide a complete status report about network hosts, ports, etc. Nmap, Vmap, and Unicornscan are instances for network mapping tools.

b. *Network monitoring:* These tools provide for visualization and analysis support for security experts. This is a very important issue since the burgeoning security related data have become much harder to interpret with emerging systems, services, and threats.

With proliferation of various malware, a diverse number of attack attempts on computer networks and systems have become an inevitable fact. Therefore, monitoring computer networks and systems has been an essential activity for defenders. To accomplish an effective monitoring, visualization tools are used for monitoring and analysis purposes. These tools assist defenders to mitigate attacks or effects of attacks. For instance, the Network Traffic Monitor tool is used to present and scan detailed traffic scenarios to analyze network traffic. Some other network monitoring tools are Rumint, EtherApe, NetViewer, etc. Briefly, there are various network monitoring tools; however, from a defender point of view, the best monitoring tool needs to perform monitoring in real time to be able to detect abnormal activities in computer networks and systems.

6.2.2 Attack tools

Both intruders and defenders use attack tools. Intruders use them to take over computer networks and systems. On the other hand, defenders use attack tools to analyze their systems. Therefore, there are various tools for attacking computer networks and systems. For instance, Nessus is a comprehensive vulnerability scanner. SAINT (System Administrator's Integrated Network Tool) is computer software used for scanning computer networks for security vulnerabilities, and exploiting found vulnerabilities. OpenVAS (Open Vulnerability Assessment

System), initially GNessUs, is a framework of several services and tools offering a vulnerability scanning and vulnerability management solution.

Attack tools may use various vulnerabilities to accomplish their attacks. Moreover, malware helps intruders to perform their attacks with high success. For this reason, there are many attack tools that perform attacks according to specific vulnerability and malware. On the other hand, new malware has increased considerably as exemplified by McAfee Labs findings in [32]. Their data show that new malware has almost quadrupled from 2011 to 2013. Increasing numbers of new malware shows potential different attack tools that may use new malware. For instance, those tools can realize application layer attacks (browser and server attacks), malware, DoS/DDoS, packet forging/malforming, and user attacks. The application layer attacks form a large set of different attacks such as SQL (Structured Query Language) injection, buffer overflow, cross-site scripting and URL (Uniform Resource Locator) misinterpretation. A classification of attack tools is presented in [3]. Recently, attack tools are available in the Internet for free so it is easier also for a novice to perform an attack to computer networks and systems. If the systems have insufficient security protection, the novice performs successful attacks. Thus, it is important to be aware of potential attacks that may be performed by attack tools.

6.3 SOFT SECURITY ANALYSIS

Traditionally, formal methods and on the shelf automated tools have been used to analyze security of computer networks and systems. With the increasing proliferation and complexity of computer networks and systems, uncertainty has become an unavoidable fact for these systems; that situation has changed the game regarding security analysis. Specifically, formal methods have been inappropriate to analyze all security properties of computer networks and systems. Moreover, automated tools have provided limited analysis options for such systems. Recently, soft security modeling and analysis methods have been widely used to complement formal analysis methods and traditional security analysis tools.

Soft security analyses and modeling methods aim to mitigate uncertainty and subjectivity. Most of the time, trust, risk, and reputation methods are used to cope with uncertainty and subjectivity so they are considered as soft security analyses and modeling methods in literature. In contemporary computer networks and systems, soft security analyses methods are used with traditional security analyses methods to have more precise analyses results. Since trust, risk, and reputation deal with uncertainty, they are interdependent as shown in Figure 30.12. Detailed explanations of soft security analysis methods are as follows:

Trust. Trust is an interdisciplinary subject studied in many fields of science. It has different meanings and many properties. For instance, trust is highly context dependent and subjective. Therefore, the definition of trust is significant to construct an analysis method of security for computer networks and systems. Moreover, some properties of trust may contradict properties of security [54].

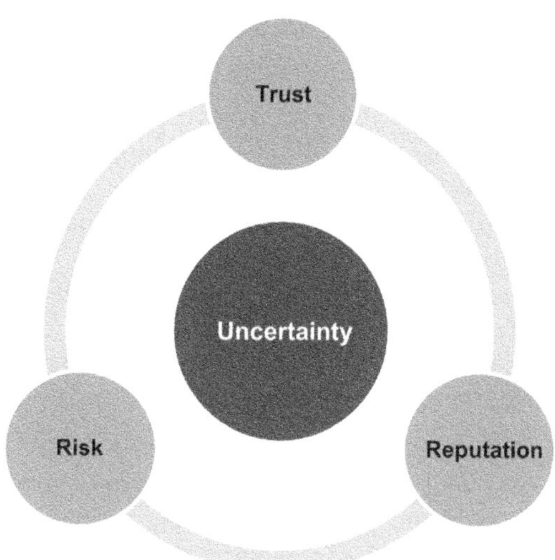

FIGURE 30.12

Soft security analyses methods and their relations.

Trust-based security analyses of computer networks and systems have two steps. First, a trust model of a security system is constructed. Following the construction, the model is analyzed with both context specific analysis approaches and conventional analysis approaches. For instance, trust information may be extracted from the security system of a service for security analysis purposes, where the service may be a network service [55].

Trust based modeling and analysis approaches have been widely used in computer networks and systems. These trust approaches may be designed for different contexts and purposes, but they may be applied to analyze security of computer networks and systems. For instance, the Eigen-Trust algorithm is used to decrease the number of downloads of inauthentic files in a peer-to-peer file-sharing network [56]. PeerTrust is another research related to peer-to-peer networks [57]. Actually, there are many other trust models related to trust modeling and analyses for computer networks and systems [58,59]. These models can be applied for security analysis of computer networks and systems with conventional analysis models to have better accuracy of analyses.

Risk. Risk is probably the best known and most commonly used soft security analysis method of computer networks and systems. Moreover, risk is used to construct trust and reputation models for analyzing security of systems [60].

The goal of risk-based analyses is to provide information necessary for security management to make reasonable decisions related to resources of computer networks and systems. Actually, there are many security risk methods defined in various standards, such as ISO/IEC 31010:2009 [61], and different risk classification methods. In this chapter, we follow the classification of risk approaches in [16] for analysis purposes of computer networks and systems as follows:

- ***Baseline:*** Implement a basic level of security control to analyze and control security of systems, such as industry best practice.
- ***Informal:*** Conduct semiformal risk analysis, such as individuals performing quick analyses of computer networks and systems.
- ***Detailed:*** Use formal structures to analyze systems and it provides the greatest degree of assurance.
- ***Combined:*** Combine baseline, informal, and detailed approaches to provide reasonable levels of protection and analysis as quickly as possible.

In the literature, there are many risk models and analysis approaches. For instance, trust and risk analysis is formalized for security architecture of systems to ensure that the systems are protected according to their stated requirements and identified risk threshold [62]. In another research work, trust and risk are used in role-based access control policies [63]. Trust mechanisms based on risk are used for evaluation of peer-to-peer networks [64]. Some other researches related to trust and risk for computer networks and systems are included in [65–69].

Reputation. Most of the time reputation is used to analyze security computer networks and systems. Reputation is defined according to the context it is applied to, and perceptions of people. Since perceptions of people vary, there are different definitions of reputation related to computer networks and systems. For example, a reputation definition for online systems is "*Reputation is what is generally said or believed about a person's or thing's character or standing.*" [70]. There are also many similar definitions for network systems in literature [71]. Although reputation has many different definitions in literature, it is about general belief of entities in computer networks and systems.

Reputation systems make some types of abstractions to decrease the amount of data to make analyses easier, which has some advantages. For instance, it decreases the need for space to store data. However, abstract representations cause loss of information. Therefore, it is impossible to verify properties of past behavior given only the reputation information [72]. Specifically, these analysis methods generally aggregate ratings via computer networks to have information about social networks [73]. Google's Page Rank algorithm uses a reputation mechanism to analyze web page popularity on the Internet [74].

Soft analysis methods are complementary solutions to formal analysis methods and automated security analysis tools. Recently, these solutions have become primary analysis methods for computer networks and systems. With the proliferation of complex and dynamic computer networks and systems, it seems that soft analysis methods are essential to properly analyze security of computer networks and systems.

7 EMERGING TOPICS AND RESEARCH CHALLENGES FOR SECURITY ANALYSIS

According to Trustwave 2014 Security Pressures Report [26], IT professionals are most pressured to use the cloud and mobile applications, which are also identified as posing the greatest security risk to their organizations (Figure 30.13). More specifically, when they were asked about the top three emerging technologies they feel the most pressure to use/deploy, respondents named the cloud (25%), mobile applications (21%), big data (19%), bring-your-own-device (BYOD) (18%) and social media (17%). They also ranked the emerging technologies they perceive as the greatest security risk sources as mobile applications (22%), the cloud (22%), BYOD (21%), social media (20%) and Big Data (15%). In addition to that, there are other security challenges that complicate security analysis, such as DDoS, insider attacks and governmental cybersecurity actions.

These emerging issues also hint of the potential research directions for security analysis research on computer networks. Accordingly, these can be listed as follows:

The loss of visibility and control created by the IT services approach and cloud computing. According to Alert Logic Cloud Security Report 2014 [75], the attacks on cloud computing systems are expected to increase with consolidation of ICT services on clouds, i.e., "traditional enterprise workloads are increasingly moving to the cloud." With more and more companies adopting cloud based

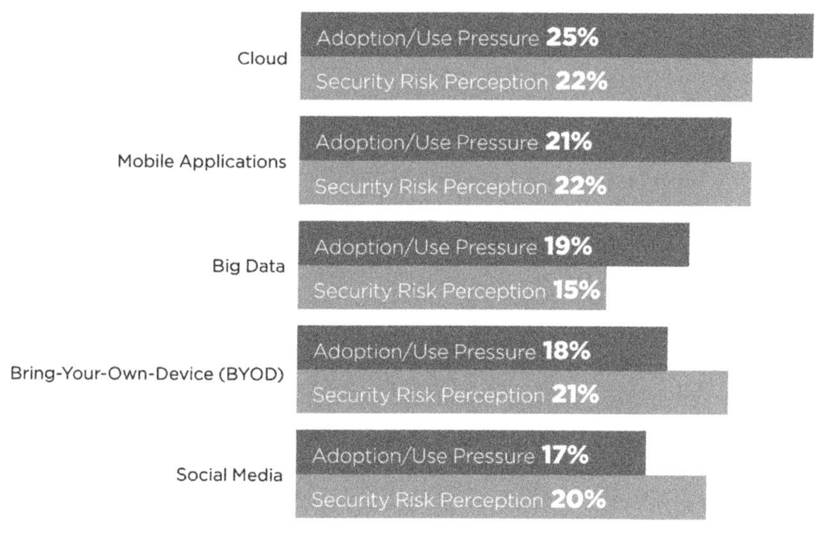

FIGURE 30.13

Emerging technology security gap.

Source: Trustwave Security Pressures Report, 2014.

systems due to modern working practices, the attacks and threat level are perpetually increasing. Moreover, this trend has been creating a loss of visibility and control magnified by the IT services externalization. Another significant security threat related to this trend is "Shadow IT," which is where employees install unapproved applications in the cloud that have not been evaluated and secured by the IT support, therefore meaning security standards are not met. The advantages of cloud based services such as workplace flexibility and service elasticity need accompanying security investments on software, hardware and employee training.

Mobile applications and devices. The widespread proliferation of networked mobile devices such as smartphones has profound effects on computer networks with more advanced and always-connected devices communicating seamlessly for richer and more immersive services. For instance, according to IDC Worldwide Quarterly Mobile Phone Tracker [76], the worldwide smartphone market has reached a new level with one billion units shipped in a single year for the first time in 2013. Specifically, vendors shipped a total of 1,004.2 million smartphones worldwide, up 38.4% from the 725.3 million units in 2012. These devices have a variety of network interfaces and host feature-rich mobile applications [13]. They are supported by centralized online application stores allowing users to discover and download applications easily. Although these application capabilities have improved the user experience and the utility of mobile devices, this open nature of mobile device computing and lack of extensive security inspection have eased the dispersion of mobile malware. Moreover, vulnerabilities in the mobile device frameworks and operating systems provide exploits for attackers in a large-scale manner. Common infection techniques of mobile malware are *Repackaging*, *Malvertizing*, *Browser Attacks*, *Update Attack*, and *Drive-by Download* [13]. According to Symantec Internet Security Threat Report 2013, 2012 saw a 58% increase in mobile malware families compared to 2011 [77]. The year's total accounted for 59% of all malware to-date. In 2014, the dominant mobile threat class was user tracking with information theft and traditional threats following that [30].

The perpetual OS (Operating System) and application changes of mobile devices with new generations of hardware deployed constantly makes security analysis a difficult task for these systems. The devices themselves are also multi-interface and multipurpose appliances with sophisticated designs. That setting exacerbates the security challenge.

Big Data. Centralization and availability of large quantities of complex data collected, stored and transmitted to internal and external systems on a global scale for processing and transformations put Big Data applications under security focus. Moreover, the usage of Big Data applications continues to outgrow security efforts and enterprises are being forced to seek new ways to address privacy and security requirements [78]. However, the big data is inherently networked and resides in an interworked environment since it are typically collected from multiple sources and processed in a distributed manner. Therefore, network-wise security analysis of such systems is paramount to enable privacy and security of Big Data environments and applications.

The increasing size and proliferation of Distributed Denial-of-Service (DDoS) attacks. The simple strategy behind a DoS attack is to deny the use of system services/resources to legitimate users and degrade the system availability. The fundamental mechanism for DoS is to send a flood of superfluous network traffic to the target so that it cannot respond to genuine requests for services or information. Botnet tools available on the Internet provide attackers with massive DDoS resources and a high level of anonymity against countermeasures. Other DoS attacks include the physical destruction of computer hardware and the use of electromagnetic interference, designed to destroy unshielded electronics via current or voltage surges [79]. The scale of these attacks has been increasing recently. According to Prolexic 2014-Q1 Global DDoS Global Attack Report [80], compared to the same quarter one year ago, total attacks increased 18% as seen by Prolexic. Although average attack duration went down 24%, from 17.38 to 22.88 hours, average peak attack bandwidth increased 114% from 4.53 Gbps to 9.70 Gbps and peak packets-per-second rate went up by 87% from 10.60 Mpps to 19.80 Mpps.

The increasing bandwidths for Internet end-points and hyperconnectivity have eased the practicality of these attack types. Moreover, the proliferation of cyber-physical systems has transformed previously uncharted territory such as Machine-to-Machine (M2M) networks into a part of the problem domain. A potent network security analysis needs to consider DDoS attacks for supporting security efforts.

The threat from inside: Insider attacks. In contrast to external attacks, an insider attacker is a system user who has been granted authorized access to the network and masks his identity, behavior, or both, for the purpose of compromising the system. Since they are supposed to access network assets for their work or obligations with the network owner, they usually have a substantial amount of knowledge about the network architecture and targeted assets. Typically, this kind of attack is relatively rare. However, the impact of such an attack is relatively major compared to external attacks. According to UK BIS 2013 Information Security Breaches Survey [81], 31% of large organizations surveyed said they have had one or more staff-related incidents of misuse of confidential information in 2012. Moreover, 49% suffered from loss or leakage of confidential information due to staff-related actions.

The main difficulty with insider attacks is the "nativeness" of the attacker for the network, which hinders incumbent security analysis and mechanisms which are usually focused on protecting the perimeter of the network. Therefore, insider attacks may be performed unnoticed for very long durations since the vast majority of attention is given to external threats. The potential areas that need to be addressed in preventing and detecting insider attacks include periodic network audit and system hardening, establishment of an enterprise wide security policy, and effective surveillance and monitoring [82]. The systems that can be deployed to assist in combating against insider attacks include application and network level monitoring, network and host-based intrusion detection systems, and anomaly-based intrusion detection systems.

Social-engineering based attacks. Social-engineering based attack refers to the process of deceiving people into giving away access or confidential information to perform actions that potentially lead to the leak or breach of proprietary or confidential information [30]. That attack can result in damage to an organization's operation, resources, or public image. Social engineers use various strategies and tools to trick users into disclosing confidential information, data or both. A very common tactic used by attackers is to pretend to be someone else to trick the victim into disclosing confidential information and data. Some such techniques are "phishing" or "baiting" to obtain private information fraudulently.

The human factor involved in social-engineering based attacks makes the task of security analysis towards them rather burdensome since the modeling of exploitations in human-decision processes is not trivial for common security analysis techniques. An effective prevention technique is to establish security protocols, policies, and procedures for handling sensitive information at various levels of an organization.

Governmental interventions and attacks by the government agencies and institutions. Cybersecurity is regarded as one of the most fundamental national security issues [83]. Governments and official agencies are using the Internet for cyberattacks on information technology, manipulation of information, or espionage, targeting critical systems including electricity, air traffic control, financial markets and government computer networks. Moreover, private enterprises may be direct victims of such attacks. According to 2007 McAfee Virtual Criminology Report, more than 100 countries already possessed or were developing capabilities to conduct offensive cyber operations [84]. Some examples of such incidents are the 2007 Estonia DDoS attack, Stuxnet malware damaging gas centrifuges in the Iranian uranium enrichment facility and 2012 DDoS attacks against the U.S. financial sector [85]. These large-scale operations on a global scale pose challenges especially for official authorities from the perspective of network security. The adequate capabilities of cyberdefense infrastructure are critical for preventing serious effects on the present and/or future of defensive or offensive effectiveness. The security analysis for preparation and recovery of such attacks are formidable due to scale, network heterogeneity and diversity of stakeholders.

7.1 CHALLENGES AND THEIR IMPLICATIONS

These emerging factors result in the following implications for network security analysis:

- **Complexity:** The security analysis may become very complex with the increasing system size and diversity. This implication is also magnified with the diversification of network functions and services.

- **Validity:** The evolution of systems accelerates and their functionalities may change very frequently. Accordingly, the analysis models may rapidly become obsolete. A typical example is the mobile cellular networks, which started as the infrastructure mainly for voice-based communications, then evolved into richer mediums, finally becoming a ubiquitous environment providing broadband IP-based services ranging from communications, infotainment, social networking to navigation.
- **Generality:** The analysis methods need to be more tailored towards specific systems rather than being widely applicable. This situation limits the reusability of models.
- **Usability:** Security analysis methods are less accessible for end-users, i.e., security experts, relying on deeper automation with more complicated usage.
- **Effectiveness:** Novel targeted and coordinated attacks weaken the strength of conventional analysis methods. Moreover, defending the network boundary is becoming almost unattainable since it is extremely obscure where the boundary is any more. Novel approaches are necessary to strengthen network analysis, thus network protection and recovery.

In Table 30.1, we list the main issues and relevant implications from the "adversity" perspective, which represents how those issues or dimensions of network security analysis are affected because of these emerging topics. For instance, when network security analysis tries to tackle Cloud Computing, its complexity is "adversely" affected at a high level, i.e., complexity substantially increases. Their validity is also significantly decreased due to still-ongoing maturization of cloud computing environments. However, they are still generally applicable to different domains. The main trait observed for these five factors is the spill-over effects of increasing complexity and generality.

Table 30.1 Effect of Emerging Trends/Issues on Network Security Modeling from the "Adversity" Perspective

	Complexity	Validity	Generality	Usability	Effectiveness
Cloud computing	High	High	Low	Low	Medium
Mobile devices	Medium	High	High	Medium	Medium
Big Data	High	High	Low	Medium	Medium
DDoS	High	Low	High	Low	High
Insider	Low	Low	High	High	Medium
Social-engineering	Low	Low	High	Medium	Low
Governmental	Low	Medium	Low	Low	Low

8 CONCLUSION

In this chapter we have presented and discussed security analysis for computer networks focusing on methodologies and approaches to analyze networks for security. We have also described security issues in computer networks (also for ICT systems) focusing on key concepts such as security objectives, threats, attacks and countermeasures. The security is a prerequisite for adequate performance and user experience of computer networks. Therefore, security analysis of computer networks and ICT systems in general, are crucial for facilitating the secure and ubiquitous operation of information systems.

The evolving functions and roles of computer networks into a pervasive substrate providing global communications have posed new security problems or rendered some incumbent ones much more vital. In that regard, such emerging research challenges and open problems entailing mobile devices/applications and cloud computing are expected to drive the research on network security.

ACKNOWLEDGEMENT

This work is supported by EUREKA ITEA2 Project ADAX with project number 10030.

REFERENCES

[1] Rizzo C, Brookson C. ETSI security white paper: security for ICT – the work of ETSI, ETSI. Available from: <http://www.etsi.org/images/files/ETSIWhitePapers/etsi_wp1_security.pdf>; Jan. 2014.

[2] Schneider D. The state of network security. Netw Secur 2012;2012(2):14–20.

[3] Hoque N, Bhuyan MH, Baishya RC, Bhattacharyya DK, Kalita JK. Network attacks: taxonomy, tools and systems. J Netw Comput Appl 2014;40:307–24.

[4] National Institute of Standards and Technology (NIST), Framework for Improving Critical Infrastructure Cybersecurity, Feb. 2014. Available from: <http://www.nist.gov/cyberframework/upload/cybersecurity-framework-021214.pdf>.

[5] Derler P, Lee EA, Vincentelli A-S. Modeling cyber-physical systems. Proc IEEE 2012;100(1):13.

[6] Langner R. Stuxnet: dissecting a cyberwarfare weapon. IEEE Secur Privacy 2011;9(3):49–51.

[7] Moser A, Kruegel C, Kirda E. Exploring Multiple Execution Paths for Malware Analysis. Security and privacy SP'07. IEEE Symposium on, 20–23 May 2007; p. 231,245.

[8] Chen Z, Roussopoulos M, Liang Z, Zhang Y, Chen Z, Delis A. Malware characteristics and threats on the Internet ecosystem. J Syst Softw 2012;85(7):1650–72.

[9] Szor P. The art of computer virus research and defense. Addison Wesley Professional; 2005.

[10] Zhuge J, Holz T, Song C, Guo J, Han X, Zou W. Studying malicious websites and the underground economy on the Chinese web. In: Proceedings of the seventh workshop on economics of information security (WEIS 2008), Dartmouth College, Hanover, NH, USA; June 25–28, 2008.

[11] Egele M, Scholte T, Kirda E, Kruegel C. A survey on automated dynamic malware-analysis techniques and tools. ACM Comput Surv 2012;44(2): Article no. 6:1–6:42 pages.

[12] Islam M, Tian R, Batten LM, Versteeg S. Classification of malware based on integrated static and dynamic features. J Netw Comput Appl 2013;36(2):646–56.

[13] Seo S-Y, Gupta A, Sallam AM, Bertino E, Yim K. Detecting mobile malware threats to homeland security through static analysis. J Netw Comput Appl 2014;38:43–53.

[14] Shabtai A, Tenenboim-Chekina L, Mimran D, Rokach L, Shapira B, Elovici Y. Mobile malware detection through analysis of deviations in application network behavior. Comput Secur 2014;43:1–18.

[15] Kim W, Jeong O-R, Kim C, So J. The dark side of the internet: attacks, costs and responses. Inf Syst 2011;36(3):675–705.

[16] Stallings W, Brown L. Computer security: principles and practices. 2nd ed. Pearson: Prentice Hall; 2012.

[17] Cohen F. Models of practical defenses against computer viruses. Comput Secur 1989;8:149–60.

[18] Mishra BK, Pandey SK. Dynamic model of worms with vertical transmission in computer network. Appl Math Comput 2011;217:8438–46.

[19] Spafford EH. The internet worm incident. In: Proceedings of the second European software engineering conference, Coventry, UK: University of Warwick; 1989. September 11–15, p. 446–68.

[20] Yan Y, Qian Y, Sharif H, Tipper D. A survey on cyber security for smart grid communications. IEEE Commun Surv Tutorials 2012;14:998–1010.

[21] Farley R, Wang X. Roving bugnet: distributed surveillance threat and mitigation. Comput Secur 2010;29:592–602.

[22] Bravo P, Garcia DF. Proactive detection of kernel-mode rootkits. August 22–26, p. 515–20. Sixth international conference on availability, reliability and security, ARES 2011. Austria: Vienna University of Technology; 2011.

[23] Baliga A, Ganapathy V, Iftode L. Detecting kernel-level rootkits using data structure invariants. IEEE Trans Dependable Secure Comput 2011;8:670–84.

[24] Li Z, Goyal A, Chen Y, Paxson V. Towards situational awareness of large-scale botnet probing events. IEEE Trans Inf Forensics Secur 2011;6:175–88.

[25] Wang P, Sparks S, Zou CC. An advanced hybrid peer-to-peer botnet. IEEE Trans Dependable Secure Comput 2010;7:113–27.

[26] Trustwave 2014 Security Pressures Report, Trustwave. Available from: <https://www.trustwave.com/Resources/Library/Documents/2014-Security-Pressures-Report/>; 2014.

[27] Miller C. Mobile attacks and defense. IEEE Secur Privacy 2011;9:68–70.

[28] Kim H, Shin KG, Pillai P. Modelz: monitoring, detection, and analysis of energy-greedy anomalies in mobile handsets. IEEE Trans Mob Comput 2011;10:968–81.

[29] Jain AK, Shanbhag D. Addressing security and privacy risks in mobile applications. IT Prof 2012;14:28–33.

[30] Symantec Internet Security Threat Report 2014, Symantec. Available from: <http://www.symantec.com/content/en/us/enterprise/other_resources/b-istr_main_report_v19_21291018.en-us.pdf>; 2014.

[31] ESET Threat Radar, Feature Article: Postcard from Hallmark hoax, ESET. Available from: <http://www.eset.com/us/resources/threat-trends/Global_Threat_Trends_March_2014.pdf>; 2014.

[32] McAfee Labs: Threats Report, Fourth Quarter 2013, McAfee 2014. Available from: <http://www.mcafee.com/us/resources/reports/rp-quarterly-threat-q4-2013.pdf>.

[33] Chandramohan M, Tan HBK. Detection of mobile malware in the wild. IEEE Comput 2012;45:65–71.

[34] Jackson JT, Creese S. Virus propagation in heterogeneous Bluetooth networks with human behaviors. IEEE Trans Dependable Secure Comput 2012;9:930–43.

[35] Baldini G, Sturman TA, Biswas AR, Leschhorn R, Godor G, Street M. Security aspects in software defined radio and cognitive radio networks: a survey and a way ahead. IEEE Commun Surv Tutorials 2012;14:355–79.

[36] ISO/IEC 27005:2007 Information Technology: Security Techniques—Information Security Risk Management, International Organization Standardization (ISO); 2007.

[37] Zargar ST, Joshi J, Tipper D. A survey of defense mechanisms against distributed denial of service (DDoS) flooding attacks. IEEE Commun Surv Tutorials 2013;15(4):2046–69.

[38] Khattak S, Ramay NR, Khan KR, Syed AA, Khayam SA. A taxonomy of botnet behavior, detection, and defense. IEEE Commun Surv Tutorials 2014;16(2):898–924.

[39] Jang-Jaccard J, Nepal S. A survey of emerging threats in cyber security. J Comput Syst Sci 2014;80:973–93.

[40] Wang W, Lu Z. Cyber security in the smart grid: survey and challenges. Comput Netw 2013;57(5).

[41] Ten CW, Manimaran G, Liu CC. Cybersecurity for critical infrastructures: attack and defense modeling. IEEE Trans Syst Man Cybern Part A: Syst Humans 2010;40:853–65.

[42] Shirey R. RFC 2828 (Internet Security Glossary), 2000, Network Working Group, 2014. Available from: <http://www.rfc-base.org/txt/rfc-2828.txt>.

[43] von Oheimb D. Formal methods in the security business: exotic flowers thriving in an expanding niche. FM 2006: formal methods, lecture notes in computer science (LNCS), vol. 4085, 2006. p. 592–97.

[44] Mantel H. A uniform framework for the formal specification and verification of information flow security. PhD diss., Saarland University; 2004.

[45] Ritchey RW, Ammann P. Using model checking to analyze network vulnerabilities. Security and privacy, S&P 2000. In: Proceedings of IEEE symposium on 2000; 2000. p. 156,165.

[46] Mao H-D, Zhang W-M, Chen F. An approach for network security analysis using logic exploitation graph. Computer and information technology, CIT 2007. Seventh IEEE international conference on 16–19 Oct. 2007. p. 761,766.

[47] Matousek P, Rab J, Rysavy O, Sveda M. A formal model for network-wide security analysis. Engineering of computer based systems, ECBS 2008. 15th annual IEEE international conference and workshop on the, p. 171,181, March 31 2008–April 4 2008.

[48] von Oheimb D. Information flow control revisited: Noninfluence = Noninterference + Nonleakage. ESORICS 2004, Lecture Notes in Computer Science, vol. 3193; 2004. p. 225–43.

[49] McLean J. Security models and information flow. Research in security and privacy. In: Proceedings of IEEE computer society symposium on 1990; 7–9 May 1990. p. 180,187.

[50] Jürjens J. Automated security verification for crypto protocol implementations: verifying the Jessie project. Electron Notes Theor Comput Sci 2009;250:1.

[51] Federal Office for Information Security BSI Study 875, Formal Methods for Safe and Secure Computer Systems, Available from: <https://www.bsi.bund.de/DE/Publikationen/Studien/formal_methods_study_875/study_875.html>; December 2013.

[52] Mitchell JC. Formal methods and computer security. USENIX'02, invited talk, 2002.

[53] Vigna G, Valeur F, Zhou J, Kemmerer RA. Composable tools for network discovery and security analysis. Computer Security Applications Conference, 2002. Proceedings. 18th Annual, 2002. p. 14,24.

[54] Gollmann D. Why trust is bad for security. Electron Notes Theor Comput Sci 2006; 157:3–9.

[55] Bahtiyar Ş, Çağlayan MU. Extracting trust information from security system of a service. J Netw Comput Appl 2012;35(1):480–90.

[56] Kamvar SD, Schlosser MT, Garcia-Molina H. The eigentrust algorithm for reputation management in P2P Networks. In: Proceedings of the 12th international conference on World Wide Web, Budapest, Hungary, May 20–24, 2003.

[57] Xiong L, Liu L. PeerTrust: supporting reputation-based trust for Peer-to-Peer electronic communities. IEEE Trans Knowl Data Eng 2004;16:843–57.

[58] Xiaonian W, Runlian Z, Shengyuan Z, Chunbo M. Behavior trust computation model based on risk evaluation in the grid environment. World Congress Softw Eng 2009;3: 392–6.

[59] Hussain OK, Chang E, Hussain FK, Dillon TS, Soh B. Risk in trusted decentralized communications. In: Proceedings of the international workshop on privacy data management in conjunction with 21st international conference on data engineering (ICDE PDM 2005); 2005. p. 1198.

[60] Bahtiyar Ş, Cihan M, Çağlayan MU. A model of security information flow on entities for trust computation. Tenth IEEE international conference on computer and information technology, CIT 2010, Bradford, West Yorkshire, UK; June 29–July 1, 2010.

[61] ISO/IEC 31010:2009 – Risk Management – Risk Assessment Techniques, Available from: <https://www.iso.org/obp/ui/#iso:std:iso-iec:31010:ed-1:v1:en>; 2014.

[62] Andert D, Wakefield R, Weise J. Trust modeling for security architecture development Tech. rep.. Sun Microsystems, Inc.; 2002.

[63] Dimmock N, Belokosztolszki A, Eyers D, Bacon J, Moody K. Using trust and risk in role-based access control policies. Proceedings of the ninth ACM symposium on access control models and technologies, SACMAT'04, New York, NY, USA; 2004. p. 156–62.

[64] Xiao-yong L, Feng Z. Developing P2P trust computing mechanism based on risk evaluation properties. Second international conference on advanced computer control (ICACC), vol. 3; 2010. p. 409–13.

[65] Lund M, Solhaug B, Stlen K. Evolution in relation to risk and trust management. IEEE Comput 2010;43:49–55.

[66] Solhaug B, Elgesem D, Stølen K. Why trust is not proportional to risk. In: Proceedings of the second international conference on availability, reliability and security (ARES'07); April 2007, p. 11–18.

[67] Zimmer JC, Arsal RE, Al-Marzouq M, Grover V. Investigating online information disclosure: effects of information relevance, trust and risk. Inf Manage 2010;47(2): 115–23.

[68] Yu W, Jie L. Trust risk of enterprise knowledge sharing within collaborative commerce. IEEE chinese control and decision conference (CCDC 2008), 2008. p. 4500–04.

[69] Asnar Y, Giorgini P, Massacci F, Zannone N. From trust to dependability through risk analysis. In: IEEE ARES'07: proceedings of the second international conference on availability, reliability and security. Washington, DC, USA; 2007. p. 19–26.

[70] Massa P. A survey of trust use and modeling in real online systems chapter. Trust in E-services: technologies, practices and challenges. Idea Group Inc.; 2007. p. 51–83.

[71] Jøsang A, Ismail R, Boyd C. A survey of trust and reputation systems for online service provision. Decis Support Syst 2007;43:618–44.

[72] Krukow K. Towards a theory of trust for the global ubiquitous computer. Ph.D. Thesis, Department of Computer Science, University of Aarhus, Denmark; 2006.

[73] Jøsang A, Hayward R, Pope S. Trust network analysis with subjective logic. In: Proceedings of the 29th Australasian computer science conference, vol. 48; 2006.

[74] Brin S, Page L. The anatomy of a large-scale hypertextual web search engine. Comput Netw ISDN Syst 1998;30:107–17.

[75] Alert Logic Spring 2014 Cloud Security Report, Alert Logic. Available from: <http://www.alertlogic.com/resources/cloud-security-report/>; 2014.

[76] IDC Report, 2013 IDC Worldwide Quarterly Mobile Phone Tracker. Available from: <http://www.idc.com/tracker/showproductinfo.jsp?prod_id = 37>; 2014.

[77] Symantec Internet Security Threat Report 2013, Symantec. Available from: <http://www.symantec.com/content/en/us/enterprise/other_resources/b-istr_main_report_v18_2012_21291018.en-us.pdf>; 2013.

[78] ISACA White Paper: Five Key Questions to Improve Big Data Governance, 2013. Available from: <http://www.isaca.org/Knowledge-Center/Research/Research Deliverables/Pages/Privacy-and-Big-Data.aspx>.

[79] Geers K. Strategic cyber security. CCD COE Publications; July 2011.

[80] Prolexic Report: 2014-Q1 Global DDoS Global Attack Report, 2014. Available from: <http://www.prolexic.com/knowledge-center/prolexic-download/prolexic-quarterly-global-ddos-attack-report-q114.html>.

[81] UK: The Department for Business, Innovation and Skills (BIS) Report, 2013 Information Security Breaches Survey. Available from: <http://www.pwc.co.uk/assets/pdf/cyber-security-2013-technical-report.pdf>; 2013.

[82] Kemp M. Barbarians inside the gates: addressing internal security threats. Netw Secur 2005;2005(6):11–13.

[83] Cavelty MD. The militarisation of cyberspace: why less may be better. In: Proceedings of 2012 fourth international conference on cyber conflict, NATO CCD COE Publications; 2012.

[84] 2007 McAfee Virtual Criminology Report, McAfee. Available from: <http://www.softmart.com/mcafee/docs/McAfee NA Virtual Criminology Report.pdf>; 2007.

[85] Iasiello E. Cyber attack: a dull tool to shape foreign policy. In: Proceedings of 2013 fifth international conference on cyber conflict, NATO CCD COE Publications; 2013.

[86] Anderson RJ. Security engineering: a guide to building dependable distributed systems. 2nd ed. John Wiley & Sons; 2008.

Index

Note: Page numbers followed by "*b*," "*f*," and "*t*" refer to boxes, figures, and tables, respectively.

D

E

F

J

K

L

O

P

S

T

X

Z

Printed and bound by CPI Group (UK) Ltd, Croydon, CR0 4YY
03/10/2024
01040324-0007